Handbook of Environmental Engineering 23

Series Editors

Lawrence K. Wang
PhD, Rutgers University, New Brunswick, New Jersey, USA
MS, University of Rhode Island, Kingston, Rhode Island, USA
MSCE, Missouri University of Science and Technology, Rolla, Missouri, USA
BSCE, National Cheng Kung University, Tainan, Taiwan, ROC

Mu-Hao Sung Wang
PhD, Rutgers University, New Brunswick, New Jersey, USA
MS, University of Rhode Island, Kingston, Rhode Island, USA
BSCE, National Cheng Kung University, Tainan, Taiwan, ROC

The past 30 years have seen the emergence of a growing desire worldwide to take positive actions to restore and protect the environment from the degrading effects of all forms of pollution: air, noise, solid waste, and water. The principal intention of the Handbook of Environmental Engineering series is to help readers formulate answers to the fundamental questions facing pollution in the modern era, mainly, how serious is pollution and is the technology needed to abate it not only available, but feasible. Cutting-edge and highly practical, HEE offers educators, students, and engineers a strong grounding in the principles of Environmental Engineering, as well as providing effective methods for developing optimal abatement technologies at costs that are fully justified by the degree of abatement achieved. With an emphasis on using the Best Available Technologies, the authors of these volumes present the necessary engineering protocols derived from the fundamental principles of chemistry, physics, and mathematics, making these volumes a must have for environmental pollution control researchers.

More information about this series at http://www.springer.com/series/7645

Lawrence K. Wang • Mu-Hao Sung Wang
Yung-Tse Hung
Editors

Hamidi Abdul Aziz
Consulting Editor

Solid Waste Engineering and Management

Volume 1

Springer

Editors
Lawrence K. Wang
Lenox Institute of Water Technology
Latham, New York, USA

Mu-Hao Sung Wang
Lenox Institute of Water Technology
Latham, New York, USA

Yung-Tse Hung
Cleveland State University
Cleveland, Ohio, USA

Consulting Editor
Hamidi Abdul Aziz
Universiti Sains Malaysia
Pulau Pinang, Malaysia

ISSN 2512-1359 ISSN 2512-1472 (electronic)
Handbook of Environmental Engineering
ISBN 978-3-030-84178-2 ISBN 978-3-030-84180-5 (eBook)
https://doi.org/10.1007/978-3-030-84180-5

This Springer imprint is published by the registered company Springer Nature Switzerland AG
The registered company address is: Gewerbestrasse 11, 6330 Cham, Switzerland

Preface

The past 75 years have seen the emergence of a growing desire worldwide that positive actions be taken to restore and protect the environment from the degrading effects of all forms of pollution – air, water, soil, thermal, radioactive, and noise. Since pollution is a direct or indirect consequence of waste, the seemingly idealistic demand for "zero discharge" can be construed as an unrealistic demand for zero waste. However, as long as waste continues to exist, we can only attempt to abate the subsequent pollution by converting it to a less noxious form. Three major questions usually arise when a particular type of pollution has been identified: (1) How serious are the environmental pollution and natural resources crisis? (2) Is the technology to abate them available? (3) Do the costs of abatement justify the degree of abatement achieved for environmental protection and resources conservation? This book is one of the volumes of the Handbook of Environmental Engineering series. The principal intention of this series is to help readers formulate answers to the above three questions.

The traditional approach of applying tried-and-true solutions to specific environmental and natural resources problems has been a major contributing factor to the success of environmental engineering, and has accounted in large measure for the establishment of a "methodology of pollution control." However, the realization of the ever-increasing complexity and interrelated nature of current environmental problems renders it imperative that intelligent planning of pollution abatement systems be undertaken. Prerequisite to such planning is an understanding of the performance, potential, and limitations of the various methods of environmental protection available for environmental scientists and engineers. In this series of handbooks, we will review at a tutorial level a broad spectrum of engineering systems (natural environment, processes, operations, and methods) currently being utilized, or of potential utility, for pollution abatement, environmental protection, and natural resources conservation . We believe that the unified interdisciplinary approach presented in these handbooks is a logical step in the evolution of environmental engineering.

Treatment of the various engineering systems presented will show how an engineering formulation of the subject flows naturally from the fundamental principles and theories of chemistry, microbiology, physics, and mathematics. This emphasis on fundamental science recognizes that engineering practice has in recent years become more firmly based on scientific principles rather than on its earlier dependency on an empirical accumulation of facts. It is not intended, though, to neglect empiricism where such data lead quickly to the most economical design. Certain engineering systems are not readily amenable to fundamental scientific analysis, and in these instances we have resorted to less science in favor of more art and empiricism.

Since an environmental solid waste engineer must understand science within the context of applications, we first present the development of the scientific basis of a particular subject, followed by exposition of the pertinent design concepts and operations, and detailed explanations of their applications to natural resources conservation or environmental protection. Throughout the series, methods of mathematical modeling, system analysis, practical design and calculation are illustrated by numerical examples. These examples clearly demonstrate how organized, analytical reasoning leads to the most direct and clear solutions. Wherever possible, pertinent cost data or models have been provided.

Our treatment of environmental natural resources engineering is offered in the belief that the trained engineer should more firmly understand fundamental principles, be more aware of the similarities and/or differences among many of the engineering systems, and exhibit greater flexibility and originality in the definition and innovative solution of environmental system problems. In short, the environmental natural resources engineers should by conviction and practice be more readily adaptable to change and progress.

Coverage of the unusually broad field of environmental science, technology, engineering, and mathematics (STEM) has demanded expertise that could only be provided through multiple authorships. Each author (or group of authors) was permitted to employ, within reasonable limits, the customary personal style in organizing and presenting a particular subject area; consequently, it has been difficult to treat all subject materials in a homogeneous manner. Moreover, owing to limitations of space, some of the authors' favored topics could not be treated in great detail, and many less important topics had to be merely mentioned or commented on briefly. All authors have provided an excellent list of references at the end of each chapter for the benefit of the interested readers. As each chapter is meant to be self-contained, some mild repetition among the various texts was unavoidable. In each case, all omissions or repetitions are the responsibility of the editors and not the individual authors. With the current trend toward metrication, the question of using a consistent system of units has been a problem. Wherever possible, the authors have used the British system (fps) along with the metric equivalent (mks, cgs, or SIU) or vice versa. The editors sincerely hope that this redundancy of units' usage will prove to be useful rather than being disruptive to the readers.

The goals of the Handbook of Environmental Engineering (HEE) series are: (1) to cover entire environmental fields, including air and noise pollution control, solid waste processing and resource recovery, physicochemical treatment processes, biological treatment processes, biotechnology, biosolids management, flotation technology, membrane technology, desalination technology, water resources, natural control processes, radioactive waste disposal, hazardous waste management, and thermal pollution control; and (2) to employ a multimedia approach to environmental conservation and protection since air, water, soil, and energy are all interrelated.

This book (*Solid Waste Engineering and Management, Volume 1*) and its two sister books (*Solid Waste Engineering and Management, Volumes 2 and 3*) of the Handbook of Environmental Engineering (HEE) series have been designed to serve as a mini-series of solid waste engineering and management textbooks as well as supplemental reference books. We hope and expect they will prove of equally high value to advanced undergraduate and graduate students, to designers of natural resources systems, and to scientists and researchers. The editors welcome comments from readers in all of these categories. It is our hope that the three solid waste engineering and management books will not only provide information on solid waste and natural resources engineering, but will also serve as a basis for advanced study or specialized investigation of the theory and analysis of various natural resources systems.

This book, *Solid Waste Engineering and Management, Volume 1,* covers the topics on: introduction to solid waste management; legislation for solid waste management; waste transportation and transfer station: characterization and measurement of solid waste; mechanical volume reduction; combustion and incineration; composting processes for disposal of agricultural and municipal solid wastes; sanitary landfill operation and management; solid waste systems planning; practices of solid waste processing and disposal; and landfilling and its environmental impacts.

This book's first sister book, *Solid Waste Engineering and Management, Volume 2,* is still being written by contributors and it may cover the topics on: sustainable solid waste management; single waste stream processing and material recovery facility (MRF); construction and demolition (C&D) waste management and disposal; recovery of plastic waste; solid waste and marine litter management; sewage sludge waste disposal; restaurant waste recycle and disposal; sanitary landfill type and design; landfill leachate collection and characterization; and landfill aftercare management plan.

This book's second sister book, *Solid Waste Engineering and Management, Volume 3,* is still being written by international solid waste experts, and it may cover the topics on: solid waste management in the tourism industry; rubber tire recycling and disposal; electronic and electrical equipment waste disposal; health-care waste management; energy recovery from waste; composting by black soldier fly; biodrying of municipal solid wastes; landfill leachate treatment; health and safety considerations in waste management; and innovative bioreactor landfill and its leachate and landfill gas management.

The editors are pleased to acknowledge the encouragement and support received from Mr. Aaron Schiller, Executive Editor of the Springer Nature Switzerland AG, and his colleagues during the conceptual stages of this endeavor. We wish to thank the contributing authors for their time and effort, and for having patiently borne our reviews and numerous queries and comments. We are very grateful to our respective families for their patience and understanding during some rather trying times.

Lenox Institute of Water Technology
Latham, NY, USA
Lawrence K. Wang

Lenox Institute of Water Technology
Latham, NY, USA
Mu-Hao Sung Wang

Cleveland State University
Cleveland, OH, USA
Yung-Tse Hung

Universiti Sains Malaysia
Pulau Pinang, Malaysia
Hamidi Abdul Aziz

Contents

About the Editors

Lawrence K. Wang has served the society as a professor, inventor, chief engineer, chief editor, and public servant (UN, USEPA, New York State) for 30+ years, with experience in the entire field of environmental science, technology, engineering, and mathematics (STEM). He is a licensed NY-MA-NJ-PA-OH Professional Engineer, a certified NY-MA-RI Laboratory Director, a licensed MA-NY Water Operator, and an OSHA Instructor. He has special passion, and expertise in developing various innovative technologies, educational programs, licensing courses, international projects, academic publications, and humanitarian organizations, all for his dream goal of promoting world peace. He is a retired acting president/professor of the Lenox Institute of Water Technology, USA; a senior advisor of the United Nations Industrial Development Organization (UNIDO), Austria; and a former professor/visiting professor of Rensselaer Polytechnic Institute, Stevens Institute of Technology, University of Illinois, National Cheng-Kung University, Zhejiang University, and Tongji University. Dr. Wang is the author of 750+ papers and 50+ books, and is credited with 29 invention patents. He holds a BSCE degree from National Cheng- Kung University, Taiwan, ROC, an MSCE degree from the University of Missouri, an MS degree from the University of Rhode Island, and a PhD degree from Rutgers University, USA. Currently, he is the book series editor of CRC Press, Springer Nature Switzerland, Lenox Institute Press, World Scientific Singapore, and John Wiley. Dr. Wang has been a delegate of the People to People International Foundation; a diplomate of the American Academy of Environmental Engineers; a member of ASCE, AIChE, ASPE, WEF, AWWA, CIE, and OCEESA; and a recipient of WEF Kenneth Research Award (NY), Five-Star Innovative Engineering Award (Inventing and installing the first DAF drinking water plant in North and South Americas) and Korean Pollution Control Association Award (Developing DAF technology in South Korea).

Mu-Hao Sung Wang has been an engineer at the New York State Department of Environmental Conservation; an editor of CRC Press, Springer Nature Switzerland, and Lenox Institute Press; and a university professor of the Stevens Institute of Technology, National Cheng-Kung University, and the Lenox Institute of Water Technology, totally being a government official, and an educator in the USA and Taiwan for over 50 years. She is a licensed professional engineer and a diplomate of the American Academy of Environmental Engineers (AAEE). Her publications have been in the areas of water quality, modeling, environmental sustainability, solid and hazardous waste management, NPDES, flotation, industrial waste treatment, and analytical methods. Dr. Wang is the author of over 50 publications and an inventor of 14 US and foreign patents. She holds a BSCE degree from National Cheng-Kung University, Taiwan, ROC; an MS degree from the University of Rhode Island, RI, USA; and a PhD degree from Rutgers University, NJ, USA. She is the co-series editor of the Handbook of Environmental Engineering series (Springer Nature Switzerland), coeditor of the *Advances in Industrial and Hazardous Wastes Treatment* series (CRC Press of Taylor & Francis Group); and the coeditor of the Environmental Science, Technology, Engineering and Mathematics series (Lenox Institute Press). She is a member of AWWA, NYWWA, NEWWA, WEF, NEWEA, CIE, and OCEESA.

Dr. Yung-Tse Hung has been Professor of Civil Engineering at Cleveland State University, Cleveland, Ohio, USA, since 1981. He is a Fellow of the American Society of Civil Engineers, a licensed Professional Engineer in Ohio and North Dakota, and a Diplomate of American Academy of Environmental Engineers. He has taught at 16 universities in 8 countries. His research interests and publications have been involved with biological treatment processes, solid wastes, hazardous waste management, and industrial waste treatment. Dr. Hung is credited with over 470 publications and presentations, 28 books, 159 book chapters, in water and wastewater treatment. He received his BSCE and MSCE degrees from National Cheng-Kung University, Taiwan, ROC, and his PhD degree from the University of Texas at Austin, USA. He is the Editor-in-Chief of *International Journal of Environmental Pollution Control and Management, International Journal of Environmental Engineering,* and *International Journal of Environmental Pollution Control, Environmental Sciences and Studies,* and Co-Editor of the *Advances in Industrial and Hazardous Wastes Treatment* series (CRC Press of Taylor & Francis Group), and the *Handbook of Environmental Engineering* series

(Springer). Dr. Hung is also the Chief Editor of the *Handbook of Environment and Waste Management series* (World Scientific Singapore), and the Permanent Executive Director and Ex-President of OCEESA (Overseas Chinese Environmental Engineers and Scientists Association).

Dr. Hamidi Abdul Aziz is a professor in environmental engineering in the School of Civil Engineering, Universiti Sains Malaysia. Professor Aziz received his PhD degree in civil engineering (environmental) from the University of Strathclyde in 1992. He is currently the Head of the Solid Waste Management Cluster (SWAM), Universiti Sains Malaysia. He has 29 years of teaching and research experiences in the field of environmental engineering, mainly related to solid waste management and landfill technology, water and wastewater treatment, leachate treatment, bioremediation, pollution control, and environmental impact assessment. To date, he has graduated over 100 PhD and MSc students and has published over 200 ISI papers, authored a few books, and has become an editor and sits in editorial boards of a few international journals. Malaysia Academy of Sciences awarded him the Top Research Scientist of Malaysia award in 2012. In 2020, he was recognized as one of the the top 2% scientists in his field in a global list compiled by the prestigious Stanford University, USA, for the year 2019.

Contributors

Salem S. Abu Amr Faculty of Engineering, Department of Environmental Engineering, Karabuk University, Karabuk, Turkey

Hamidi Abdul Aziz School of Civil Engineering and Solid Waste Management Cluster, Engineering Campus, Universiti Sains Malaysia, Pulau Pinang, Malaysia

Raul R. Cardenas Jr. Department of Civil Engineering, Polytechnic Institute of New York, Brooklyn, NY, USA

Jarir S. Dajani Department of Civil Engineering, Stanford University, Stanford, CA, USA

Fadugba O. George School of Civil Engineering, Engineering Campus, Universiti Sains Malaysia, Pulau Pinang, Malaysia

Azhar Abdul Halim School of Environmental and Natural Resource Sciences, Faculty Science and Technology, Universiti Kebangsaan Malaysia, Selangor, Malaysia

Mohd Azhar Abd Hamid School of Civil Engineering, Engineering Campus, Universiti Sains Malaysia, Pulau Pinang, Malaysia

Mohamad Haziq Mohd Hanif School of Industrial Technology, Universiti Sains Malaysia, Pulau Pinang, Malaysia

Siti Hafizan Hassan School of Civil Engineering, Universiti Teknologi MARA, Pulau Pinang, Malaysia

Yung-Tse Hung Department of Civil and Environmental Engineering, Cleveland State University, Cleveland, OH, USA

Mohamad Anuar Kamaruddin School of Industrial Technology, Universiti Sains Malaysia, Pulau Pinang, Malaysia

Omar Fawzi Suleiman Khasawneh School of Civil Engineering, Engineering Campus, Universiti Sains Malaysia, Pulau Pinang, Malaysia

Amin Mojiri Department of Civil and Environmental Engineering, Hiroshima University, Higashihiroshima, Japan

Faris Aiman Norashiddin School of Industrial Technology, Universiti Sains Malaysia, Pulau Pinang, Malaysia

Puganeshwary Palaniandy School of Civil Engineering, Engineering Campus, Universiti Sains Malaysia, Pulau Pinang, Malaysia

Harris Ramli Engineering Campus, Universiti Sains Malaysia, Pulau Pinang, Malaysia

Siti Fatihah Ramli School of Civil Engineering, Engineering Campus, Universiti Sains Malaysia, Pulau Pinang, Malaysia

Nor Habsah Md Sabiani School of Civil Engineering, Engineering Campus, Universiti Sains Malaysia, Pulau Pinang, Malaysia

Abdubaki Mohamed Hussen Shadi School of Industrial Technology, Universiti Sains Malaysia, Pulau Pinang, Malaysia

P. Aarne Vesilind Department of Civil Engineering, Duke University, Durham, NC, USA

Lawrence K. Wang Lenox Institute of Water Technology, Latham, NY, USA

Lenox Institute of Water Technology, Newton, MA, USA

Krofta Engineering Corporation, Lenox, MA, USA

Mu-Hao Sung Wang Lenox Institute of Water Technology, Latham, NY, USA

Lenox Institute of Water Technology, Newton, MA, USA

Dennis Warner Gennett, Fleming, Corddry, and Carpenter, Inc., Harrisburg, PA, USA

Eugene A. Glysson Department of Civil Engineering, University of Michigan, Ann Arbor, MI, USA

Mohd Suffian Yusoff School of Civil Engineering, Universiti Sains Malaysia, Pulau Pinang, Malaysia

Ridzuan Zakaria School of Chemical Engineering and Solid Waste Management Cluster, Engineering Campus, Universiti Sains Malaysia, Pulau Pinang, Malaysia

Sharifah Akmam Syed Zakaria School of Civil Engineering, Engineering Campus, Universiti Sains Malaysia, Pulau Pinang, Malaysia

Siti Nor Farhana Zakaria School of Civil Engineering, Engineering Campus, Universiti Sains Malaysia, Pulau Pinang, Malaysia

Chapter 1
Introduction to Solid Waste Management

Hamidi Abdul Aziz, Salem S. Abu Amr, P. Aarne Vesilind, Lawrence K. Wang, and Yung-Tse Hung

Abstract An increase in population growth, industrial development, and urbanization has led to increasing solid waste generation. Complications associated with solid waste can be dated back to ancient history. The waste produced and collected in an urban area is called municipal solid waste (MSW), mainly associated with the wastes produced from domestic, industrial, commercial, and institutional areas. The amount and composition of waste vary by country. New and effective strategies are generally needed to design urbanization models, and policies are required for effective solid waste management. All aspects of waste storage, collection, transportation, sorting, disposal, and related management are included in solid waste management. It does not stop after collection only, but what needs to be done with the wastes is part of the important aspects of the whole management protocol. Basic waste data are included in this chapter. These include their types, sources, quantity, and compositions. Next, the functional elements of the waste management system are discussed, which among others, includes the aspects of storage, collection, transportation, recovery and processing, composting, thermal treatment, and the

H. A. Aziz (✉)
School of Civil Engineering, Engineering campus, Universiti Sains Malaysia, Nibong Tebal, Pulau Pinang, Malaysia

Solid Waste Management Cluster, Science & Engineering Research Centre (SERC) Engineering Campus, Universiti Sains Malaysia, Nibong Tebal, Pulau Pinang, Malaysia
e-mail: cehamidi@usm.my

S. S. Abu Amr
Faculty of Engineering, Department of Environmental Engineering, Karabuk University, Karabuk, Turkey

P. A. Vesilind
Department of Civil Engineering, Duke University, Durham, NC, USA

L. K. Wang
Lenox Institute of Water Technology, Latham, NY, USA

Y.-T. Hung
Department of Civil and Environmental Engineering, Cleveland State University, Cleveland, OH, USA

L. K. Wang et al. (eds.), *Solid Waste Engineering and Management*, Volume 1, Handbook of Environmental Engineering 23,
https://doi.org/10.1007/978-3-030-84180-5_1

final disposal. The legislation related to waste is also discussed, followed by the descriptions of the integrated solid waste management.

Keywords Solid waste · Municipal Solid Waste (MSW) · Waste management · Recycling and recovery · Waste processing · Waste disposal

Acronyms

APCr	Air Pollution Control Residues
ASME	American Society of Mechanical Engineers
C&I	Commercial and industrial
C&D	Construction and demolition
CBA	Cost-Benefit Analysis
BFR	Brominated flame retardants
CFC	Chlorofluorocarbons
HCFC	Hydrochlorofluorocarbons,
EIA	Environmental Impact Assessment
EPA	Environmental Protection Act
EU	European Union
HFA	Humic and fulvic acids
ISWM	Integrated solid waste management
LCA	Life Cycle Assessment
MSW	Municipal solid waste
MFA	Material Flow Analysis
PWCS	Pneumatic waste conveyance system
RCRA	Resource Conservation and Recovery Act
RA	Risk Assessment
RMA	Rubber Modified Asphalt
SEA	Strategic Environmental Assessment
SoEA	Socio-economic Assessment
SA	Sustainable Assessment
S/S	Solidification/stabilization
TDA	Tyre-Derived Aggregate
UNEP	United Nations Environment Programme
US	United States
USEPA	US Environmental Protection Agency
UK	United Kingdom
VFA	Volatile fatty acids

Nomenclature

%	Percentage
$	American dollar
Per capita	head/person or individual

1.1 Introduction

An increase in population growth, industrial development, and urbanization has led to increasing solid waste generation. Solid waste is produced as results of activities from several sources such as residential areas, marketing places, restaurants and food areas, public and industrial installations, waterworks and sewage facilities, construction, and agricultural sites. Solid waste remains a critical issue in many countries. Complications associated with solid waste can be dated back to ancient history. Because of the new inventions, technologies, and services, waste has been changed quantitatively and qualitatively over time. Its generation rates and composition differ from one country to another and lifestyle. Several variables affect the characterisation of the waste based on the economic conditions, policies on waste management, industrial structure, lifestyle-changing and living standards, culture, and geography. Urbanization can be considered as one of the primary factors for increasing solid waste generation due to the urban population. New and effective strategies are generally needed to design urbanisation models, and policies are required for effective solid waste management.

Almost everything we do creates some form of waste [1]. The disposal of solid waste becomes a significant problem globally, especially in developing countries. The waste generation rate is generally proportional to the degree of economic growth and the age of the urban population. Despite this expansion, effective waste management remains a challenging task. Generally, only a small portion of the raw materials consumed will be used to make a product; the remainder will be thrown away.

All aspects of waste storage, collection, transportation, sorting, disposal, and related management are included in solid waste management. It normally involves an integrated approach covering all of the above activities. It does not stop after collection only, but what needs to be done with the wastes is part of the important aspects of the whole management protocol. The abundance of solid waste generated without proper management and handling can cause serious problems to a society, such as the spread of diseases, bad odour, and environmental pollution. Thus, a proper solid waste management system is necessary.

1.1.1 Definition of Solid Waste

Generally, solid waste is any substance in a solid form that is unwanted/unused and/or unvalued and is discarded or discharged for disposal. However, the definition varies by country. The US Environmental Protection Agency (USEPA) defined MSW as any amount of waste that contains any items thrown away after use, such as packaging products, plastic bags and papers, plastic bottles and containers, and batteries, which are generated from households, hospitals, schools, and institutions [1]. The Resource Conservation and Recovery Act (RCRA) of the United States, passed in 1976, defines 'solid waste' as garbage or refuse; sludge from a wastewater treatment plant, water supply treatment plant, or air pollution control facility; and other discarded material resulting from manufacturing, commercial, mining, and agricultural operations, as well as community activities. The Indian rules for Management and Handling of solid waste in 1999 defined municipal waste as the materials that include commercial and residential wastes which are generated from municipal, industrial, and other sources.

The waste produced and collected in an urban area is called 'municipal solid waste' (MSW), mainly associated with the wastes produced from domestic, industrial, commercial, and institutional areas. In the United States, it is referred to as waste or garbage; and in the United Kingdom, it is referred to as rubbish. It is a form of waste made up of commonplace objects discarded by the general public. In Malaysia, municipal solid waste (MSW) is outlined as any scrap materials, other unwanted surplus substances, or rejected products that occur as a result of human activity, except scheduled wastes such as sewage and radioactive wastes, as defined by the Solid Waste and Public Cleansing Management Act 2007 (Act 672). The origin of solid waste influences its characteristics.

1.1.2 Sources of Solid Waste

There are a few different forms of solid waste that come from things that people throw away. MSW, or garbage, is made up of a variety of things that people discard. Packaging, food, furniture, electronics, yard trimmings, tyres, and appliances are among these products.

While there are several different ways to classify the waste sources, the following are the most common: domestic or residential, commercial (restaurants, grocery stores, other businesses), institutional (such as offices, schools, domestic hospital wastes), non-hazardous industrial (like offices, cafeterias, packaging, but not the process waste), construction and demolition (C&D), agricultural, and municipal activities (street cleaning, garden waste, etc.). In several countries, household waste accounts for 85–90% of total MSW material for the majority of local governments. MSW includes biodegradable organic matter and is one of the most difficult fractions to deal with because it is difficult to sort when combined with the other fractions.

1.2 Waste Generation and Quantity

Understanding the sources and types of solid wastes is required for the design and operation of the functional elements associated with solid waste management. Data on the composition and waste production rate rates are equally important.

For proper waste management, the quantity and composition of produced MSW are critical. The management of these waste materials is at the heart of all solid waste management activities at the local, regional, and subregional levels, as well as at the state and federal levels. As a result, it is important to learn as much as possible about MSW. The amount and composition of waste vary by country. Some of the data are presented below.

The US Environmental Protection Agency has released a number of statistics on waste management in the US [3]. Figure 1.1 depicts the per-person municipal solid waste generation per day between 1960 and 2018. MSW generation per person increased from around 1.22 kg per day in 1960 to 2.3 kg per day in 2018. Approximately 292 million tons of MSW were produced in 2018 (Fig. 1.1). Approximately 94 million tons of waste is recycled or composted, resulting in a recycling and composting rate of 32.1% (Fig. 1.2). In addition, other food management pathways processed about 18 million tons of food (6.1%) (Fig. 1.3). With energy recovery, over 34 million tons of MSW (11.8%) were combusted. Eventually, over 146 million tons (50.0%) were dumped on the ground (Fig. 1.3).

The generation, recycling, composting, combustion with energy recovery, and landfilling of MSW have all changed dramatically over the last few decades [2]. From under 10% of produced MSW in 1980 to 35.0% in 2017, the combined recycling and composting rate have increased (Fig. 1.2). Recycling alone (without

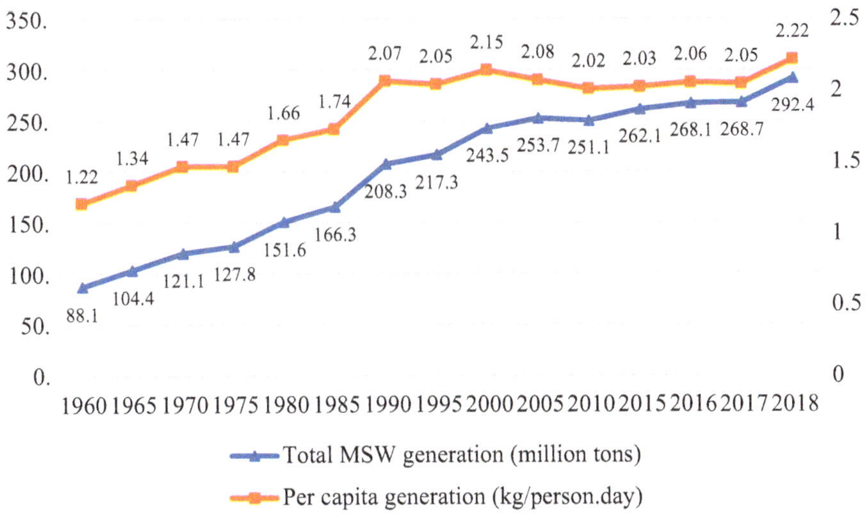

Fig. 1.1 MSW generation rates in the US from 1960 to 2018 [3]

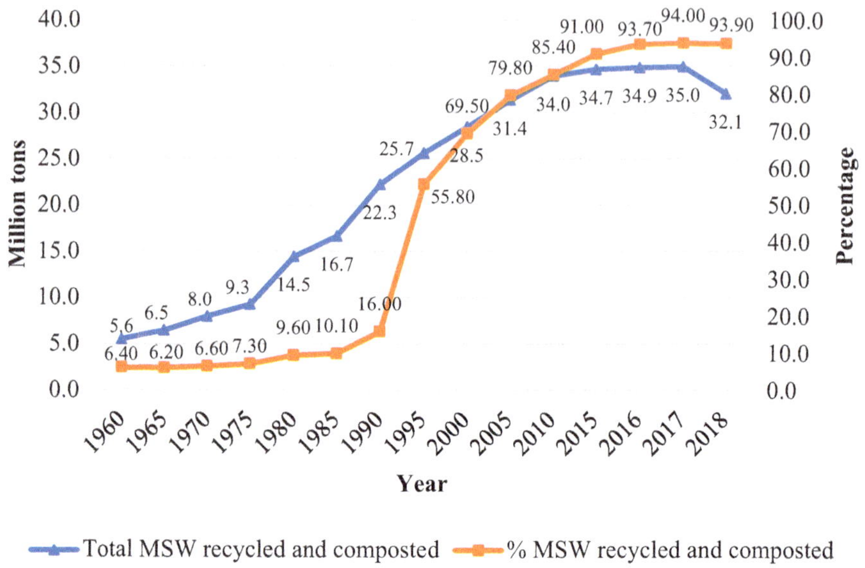

Fig. 1.2 Total municipal solid waste (MSW) recycled and composted in the US from 1960 to 2018 [3]

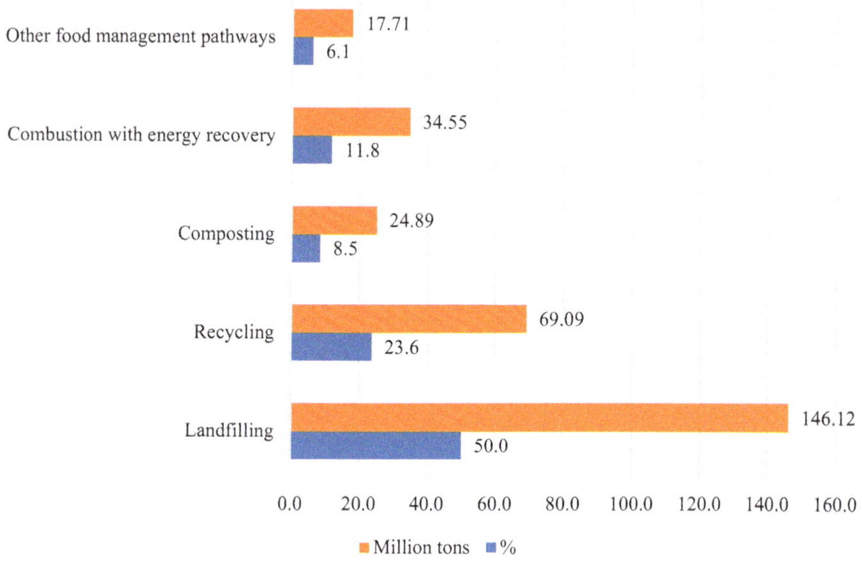

Fig. 1.3 Management of municipal solid waste (MSW) in the US in 2018 [3]

composting) increased from 14.5 million tons (9.6% of MSW) in 1980 to 69 million tons (23.6%) in 2018. Despite the fact that more tons of waste were recycled in 2018 than ever before, the recycling rate fell to its lowest level since 2006. Composting was almost non-existent in 1980, but by 2018 it had risen to 24.9 million tons (8.5%).

USEPA revised its food calculation methodology in 2018 to better capture flows of surplus food and food waste in the food system [3]. As shown in Fig. 1.3, other food management pathways accounted for 17.7 million tons (6.1%) of the total. In 1980, combustion with energy recovery accounted for less than 2% of total generation or 2.8 million tons. In 2018, 34.6 million tons of MSW were burnt with energy recovery, accounting for 118% of the total MSW produced. From 145.3 million tons in 1990 to 146.1 million tons in 2018, the overall volume of MSW sent to landfills has risen by under one million tons.

In Hong Kong, about 3700 tons of putrescible waste is dumped daily in landfills in 2019 [4]. The highest percentage of MSW generated in Hong Kong was a putrescible waste. In 2019, a total of 5.71 million tons of solid waste was disposed of at strategic landfills. The average daily quantity was 15,637 tons per day (tpd), down 2.8% from the previous year (Table 1.1). Domestic, agricultural, commercial and industrial (C&I) waste are all contained in MSW. The sum of MSW disposed of in 2019 was 11,057 tpd (4.04 million tons), down 3.2% from 2018. The shift can be attributed in part to local social unrest, which wreaked havoc on society and caused the local economy to contract in the second half of 2019. Excluding the population growth from the equation, the MSW disposal rate was 1.47 kg/person/day in 2019 versus 1.53 kg/person/day in 2018. Domestic waste makes up the bulk of MSW. In 2019, it disposed of 6554 tpd (2.39 million tons), a decrease of 2.4% from 2018. In comparison, the volume of C&I waste disposed of in 2019 was 4503 tpd (1.64 million tons), down 4.5% from 2018. In general, the amount of C&I waste produced is proportional to the rate of consumptions. The decline in C&I waste disposal in 2019 may be attributed in part to the local economy's contraction.

The changes in the per capita municipal solid waste (MSW) production between 1995 and 2018 in the EU is presented in Table 1.2 [5]. Table 1.3 presents the rate of MSW generation in Europe [6]. Also, Eurostat statistics reported that in 2019, 1.38 kg of municipal waste per capita/day were generated in the EU, and 48% of municipal waste was recycled (material recycling and composting) [7].

Table 1.1 Hong Kong's total solid waste at a landfill in 2019 [4]

Waste category		Average daily quantity (tons per day)	Year-on-year growth rate
1	Municipal solid waste (MSW)	11,057	−3.2
	(i) Domestic	6554	−2.4
	(ii) Commercial and industrial	4503	−4.5
2	Construction waste	3946	−3.3
3	Special waste[a]	635	−8.1
4	Total waste received at landfill	15,637	−2.8

[a]Does not include special waste not disposed of at landfill

Table 1.2 Waste generation rate based on region [5]

Region	MSW generation rate (kg/capita/day)	Region	MSW generation rate (kg/capita/day)
Asia		*Oceania*	
Central	0.93	Australia and New Zealand	1.64
Eastern	1.32	Melanesia	3.23
South-eastern	1.26	Polynesia	3.70
Southern	1.37		
Western	1.89	*Africa*	
		Northern	1.12
Europe		Eastern	0.79
Eastern	1.01	Middle	0.52
Northern	1.32	Southern	0.90
Southern	1.29	Western	0.49
Western	1.62		
America			
Caribbean	2.14		
Central	1.59		
South	1.18		
Northern	2.63		

According to the World Bank [8], the world generates 2.01 billion tons of urban solid waste per year, with at least 33% of it not being treated in an environmentally friendly manner. The average amount of waste produced per person per day is 0.74 kg, but it varies widely, varying from 0.11 to 4.54 kg. While having just 16% of the world's population, high-income countries produce about 34% of the world's waste or 683 million tons. By 2050, the total amount of waste generated in low-income countries is estimated to increase by more than threefold (Fig. 1.4).

As reported by the United Nations Environment Programme (UNEP), Association of Southeast Asian Nations (ASEAN) countries have a total population of 625 million people, accounting for 8.8% of the global population. By 2020, the population is predicted to reach 650 million inhabitants, with urban areas account for more than half of the total population is concentrated in this region. According to reports, Asian cities will produce the most waste in 2025, with 1.8 billion tons (up from 0.28 billion tons in 2012) [9].

In ASEAN, the MSW generation rate is 1.14 kg/capita/day (Table 1.4). The following is the order of total annual MSW generation: Indonesia produces the most urban waste, with 64 million tons per year, followed by Thailand (26.77 million tons per year), Vietnam (22 million tons per year), the Philippines (14.66 million tons per year), Malaysia (12.84 million tons per year), Singapore (7.5 million tons per year),

Table 1.3 MSW generation rate in Europe [6]

Country	Toons per year
Austria	61,225
Belgium	63,152
Czech Republic	25,381
Denmark	20,982
Estonia	24,278
Finland	122,869
France	323,474
Germany	400,072
Hungary	15,908
Iceland	1067
Italy	163,995
Korea	180,367
Latvia	2533
Lithuania	6644
Luxembourg	10,130
Netherlands	141,024
Norway	11,197
Poland	182,006
Portugal	14,739
Slovak Republic	10,607
Slovenia	5517
Spain	128,959
Sweden	141,626
Turkey	75,535
United Kingdom	277,281

Myanmar (0.84 million tons per year), and Lao Peoples' Democratic Republic (PDR; 0.07 million tons per year). Organic waste accounts for the majority of MSW in all ASEAN countries (about or more than 50%), with the exception of Singapore, where organic waste represents just 10.5% of the total MSW. Other types of waste, such as plastic, metals, and paper are popular in MSW dumps. Aside from MSW, emerging waste sources in ASEAN countries include healthcare waste, e-waste, industrial waste, and construction and demolition waste.

Looking ahead [8], global waste is projected to reach 3.40 billion tons by 2050, more than twice the rate of population increase over that time period. Overall, waste generation and profits have a good relationship. In high-income countries, daily waste generation per head is expected to rise by 19% by 2050, compared to 40% or more in low- and middle-income countries.

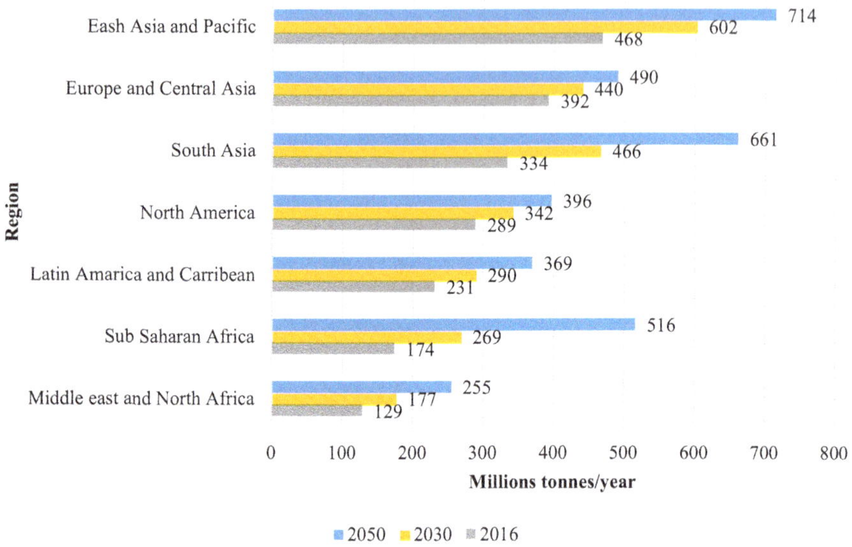

Fig. 1.4 Generation municipal solid waste by region [8]

Table 1.4 MSW generation in ASEAN countries [9]

No	Country	MSW generation kg/capita/day in 2009	Annual MSW in 2009 (metric tons)	Projected MSW in 2025 (kg/capita/day)
1	Brunei Darussalam	1.40	210,480	–
2	Cambodia	0.55	1,089,429	–
3	Indonesia	0.77	64,000,000	1.0
4	Lao PDR	0.69	77,380	0.80
5	Malaysia	1.17	12,840,000	1.40
6	Myanmar	0.47	12,840,000	0.60
7	Philippines	0.53	14,660,000	0.80
8	Singapore	1.10	7,514,500	1.10
9	Thailand	1.10	26,770,000	1.50
10	Vietnam	0.57	22,020,000	0.70

1.3 Types and Composition of Solid Waste

1.3.1 Types of Solid Wastes

In their study, Hussein et al. [10] had established that the majority of municipal solid waste produced in the developing country originates from households (55–80%), succeeded by the market or commercial areas (10–30%). The latter is made up of a variety of variable quantities produced by industries, streets, institutions, and a variety of other sources. Solid waste from such sources is usually high and heterogeneous. In general, waste characteristics differ depending on the sources of waste. It is a significant need for the classification and characterisation of these wastes for any effective treatment or disposal methods. The separating of generated solid wastes is considered as one of the most important and effective methods of solid waste management to provide useful information about the quality of the separated wastes for any potential utilisation. Table 1.5 presents the common source and types of solid waste.

1.3.2 Composition of Solid Waste

Social and economic factors, such as population growth, demand growth, changes in consumption habits, and waste management system technological advancement, have all had a significant impact on the waste aspect [11]. The composition of waste depends on locality and differs according to the income class, indicating different consumption trends (Fig. 1.5) [8]. High-income countries produce less food and green waste, accounting for 32% of total waste, and more dry waste that can be recycled, such as plastic, paper, cardboard, metal, and glass, accounting for 51% of total waste. Food and green waste are produced in 53% and 57% of middle- and low-income countries, respectively, with the proportion of organic waste rising as economic development levels decline. In low-income countries, just 20% of the materials used in building are recyclable. Aside from waste streams that are associated with wages, there is little variation in waste streams across regions. With the exception of Europe, Central Asia, and North America, which produce more dry waste, all regions produce about 50% or more organic waste on average. A standard urban solid waste in China contains 55.9% food residue, 8.5% paper, 11.2% plastics, 3.2% textiles, 2.9% wood waste, 0.8% rubber, and 18.4% non-combustibles [12].

In the US, during 2018, some 8.9% of the total MSW was generated, consisted of textiles, rubber, and leather materials (Fig. 1.6) [3]. The average daily MSW (by composition) at landfills in Hong Kong in 2019 is shown in Fig. 1.7.

MSW compositions in ASEAN countries are given in Table 1.6. It can be noted that food waste/organics constitute between 45% and 60% in all ASEAN countries, except Myanmar (73%). They are mostly disposed of in a landfill or dumpsite, except for Singapore.

Even though there are many variations, however, MSW can be broadly divided into several types such as food waste, paper, plastic, textiles, glass, wood, metal, metal cans, rubber, leather, scraps, and bulky waste.

Table 1.5 Sources of solid wastes within a community

Source	Typical activities or locations where waste is produced	Types of wastes
Residential	From various types of houses with different income groups	Food wastes/organics, paper, cardboard, plastics, textiles, leather, yard wastes, wood, glass, bottles, tin cans, drink cartons, aluminium, other metals, ashes, garden waste, special wastes (including bulky items like discarded furniture, domestic e-waste, household hazardous wastes, batteries, oil, and used tyres
Commercial	Restaurants, supermarket, mini market, grocery shops, hotels, motels, shops, service stations, automobile workshops, laundrette, etc.	Paper, plastics, cardboard, wood, food waste, glass, metals, special wastes (see above), hazardous wastes, etc.
Institutional	Schools, higher leaning institutions, prisons and government detention centres, hospitals, governmental offices, training centres, etc.	As above in commercial
Construction and demolition	New construction sites, road repair and renovation sites, broken pavement, demolition of buildings	Concrete, wood, steel, tar, glass, dirt, etc.
Municipal services (excluding treatment facilities)	Landscaping wastes, street cleaning, grass cutting, tree trimming, drain cleaning, dead animal wastes, parks and beaches, other recreational areas	Street rubbish, sidewalks, vacant lots, trees branches, debris, grass, general wastes from parks, beaches, and recreational areas, etc.
Treatment plant sites; municipal incinerators	Treatment systems for water, wastewater, and industrial waste, etc.	Wastes from treatment plants, mostly sludges, bottom and fly ashes, and slag
Municipal solid waste	All of the above	All of the above
Industrial	Construction, manufacturing – light and heavy, fabrication, chemical plants, refineries, power plants, process waste, etc.	Wastes from industrial processes, scrap materials, etc. Non-industrial wastes – rubbish, food wastes, ashes, special wastes (see commercial), hazardous wastes
Agricultural	Dairies, feedlots, farms, field and row crops, orchards, vineyards, etc.	Spoiled food wastes, agricultural wastes, rubbish, used packaging from fertiliser, etc., hazardous wastes

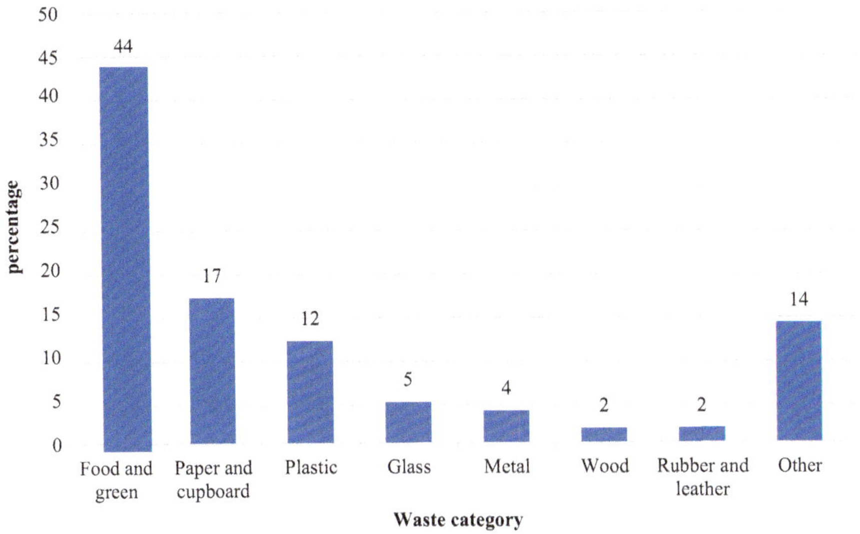

Fig. 1.5 Global waste composition [8]

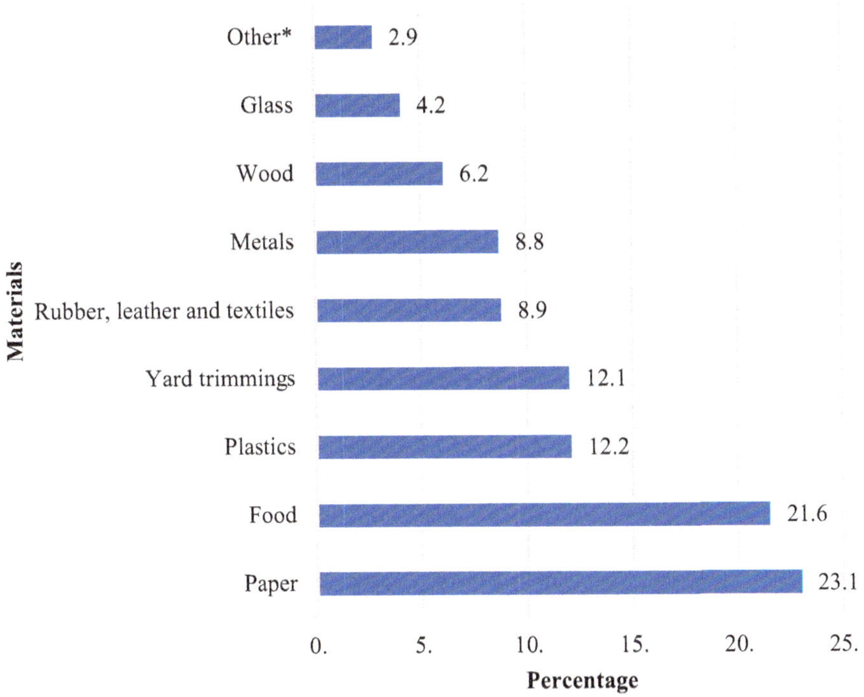

Fig. 1.6 By material distribution of MSW stream generated in the United States in 2018 [3]

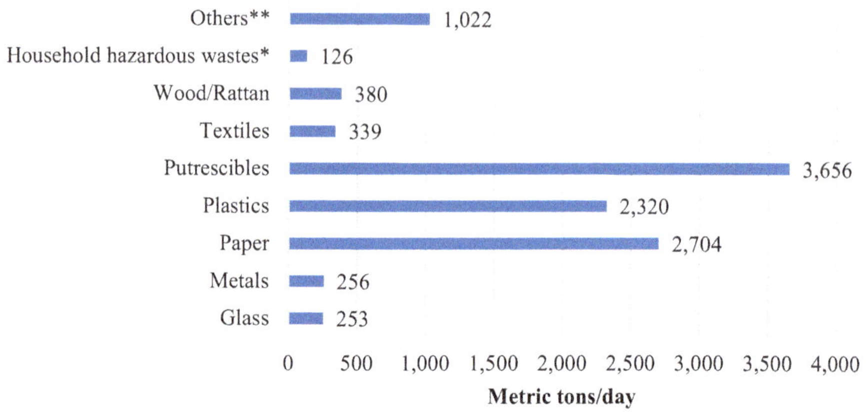

Fig. 1.7 Average daily MSW (by composition) at landfills in Hong Kong in 2019 [4]

1.4 Special Types of Solid Waste

1.4.1 Industrial Solid Waste

Residues are produced by almost every industry. Industrial solid waste is any material that is made useless during manufacturing and processing processes, as well as any waste generated by the industrial operation. Depending on the nature of the industries or businesses, solid waste may be in the form of hazardous or non-hazardous. The hazardous waste, such as waste chemicals, containers, sludge, solvents, etc., is managed as a separate waste stream that requires special handling and disposal. This is not covered in this chapter. Although the amounts can vary, the non-hazardous consists mostly of cupboard and paper waste, non-hazardous scrap metals, garbage, and related wastes and materials to residential waste.

The origin of the waste generated by industry has a substantial effect on the composition of the waste produced. Animal hide manufacturing, for example, generates a lot of biodegradable waste (animal parts), while construction generates a lot of excavated dirt, rock, and demolition waste (bricks, stones, wood, glass, etc.). As a result, industrial waste is normally treated and disposed of by the industry, which also employs advanced technologies.

There is no universally accepted classification system for industrial solid waste. However, it is more useful to divide the industry into three large groups, each with distinct types of operation and, as a result, distinct waste-generating characteristics. Extractive industries, basic industries, and manufacturing are the three groups.

Table 1.6 MSW composition in ASEAN countries [9]

No	Country	MSW composition (%)									
		Food/Organics	Paper	Plastic	Metal	Glass	Textile	Rubber	Grass/wood/etc.	C&D[a]	Others
1	Brunei Darussalam	36	18	15	4	3					
2	Cambodia	60	9	15		3	1	1			
3	Indonesia	60	9	14	4.3	1.7	3.5	5.5			2.4
4	Lao PDR	64	7	12	1	7	5	3			
5	Malaysia	45	8.2	13.2		3.3					27.3
6	Myanmar	73	2.24	17.8		0.5	1.1				5.2
7	Philippines	52	8.7	10.6	4.2	2.3	1.6				
8	Singapore	10.5	16.5	11.6	20.8	1.1	2.1		8.6	16.9	11.9
9	Thailand	64	8	17.6	2	3	1.4	1	1		
10	Vietnam	55	5	10	5	3		4			

[a]C&D – Construction and demolition

1.4.1.1 Extractive Industries

There are factories where raw materials are extracted from the earth and sold in their natural state. Mining, quarrying, agriculture and food, and logging are four extractive industries that generate significant amounts of solid waste. Solid wastes produced by such factories are simply natural components or products. These factories concentrate wastes in particular areas; the wastes are natural products of the earth and its living organisms, but they vary in nature, with some being inert materials and others being biodegradable organic matter.

(a) Mining. Waste produced during the extraction, beneficiation and processing of minerals is referred to as mining waste. The Mining Waste Exclusion under Subtitle C of the Resource Conservation and Recovery Act (RCRA) has exempted most extraction and beneficiation wastes from hard rock mining (the mining of metallic ores and phosphate rock) and 20 particular mineral processing wastes (see sidebar below) from federal hazardous waste regulations [13].

Mineral capital accounts for more than 70% of all resources used in the global economy (in terms of volume) [14]. Strong rocks, gravels, clays, pebbles, sands, limestones, chalks, siftings of fine fractions, dump tailings of ferrous and non-ferrous metal ores, sulphur ores, apatite–nepheline concentrates, coal wastes, halite flotation wastes, phosphorite screenings, phosphoric ore fines, and other mining wastes are examples. Mining and refining sectors are among the most polluting industries, with wastes accounting for more than 90% of mined mineral raw materials. Mine dumps are technogenic geological structures made of rocks or sediments that differ from background rocks in composition (chemical, particle size, bacteriological) and properties (physical and mechanical, capacity for filtration and absorption). The shapes and types of mine dumps are mainly determined by the technical processes used [14].

Per year, mining solid waste amounts to more than a billion tons, with an additional 23 billion tons of cumulative waste from the previous 30 years distributed across the United States. The extraction and processing of mineral ores are expected to produce 1.6 billion metric tons (1.8 billion tons) of mineral processing waste in the United States per year [15]. Mineral processing waste accounts for close to half of all solid waste generated in the United States each year. Mineral wastes accumulated over decades of past mining operations are estimated to be worth at least 50 billion metric tons (55 billion tons) [16]. Despite the fact that many mining sites are in remote locations, virtually every state has large amounts of mineral processing waste.

(b) *Quarrying*. Quarrying is the method of removing stone from the ground. This form of extractive industry involves open pit or strip mining, as well as the quarrying of glass sand, stone, and sand and gravel. The issue of solid waste is similar to that of mining, except that the amounts involved are much lower (approximately 0.5–5% of mining wastes).

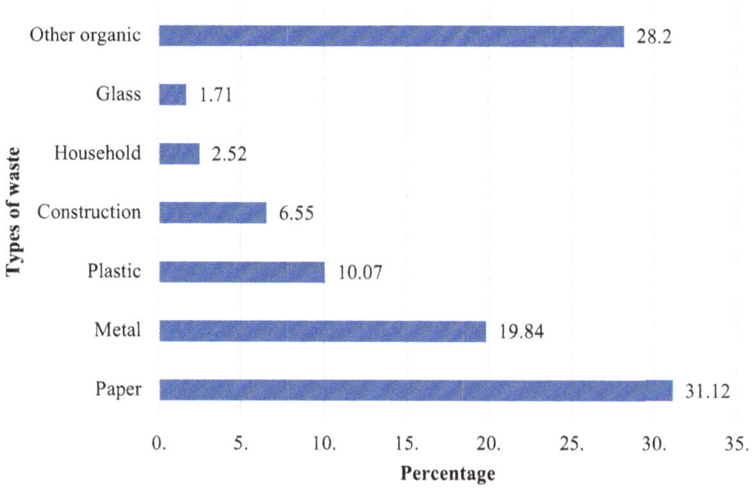

Fig. 1.8 In 2014, the following materials made up the majority of waste produced by mining, quarrying, and oil/gas extraction in the United States [17]

According to the US Chamber of Commerce, the share of waste produced by mining, quarrying, and oil/gas extraction in the United States in 2014 was broken down by content (Fig. 1.8). It was found that more than 30% of the total waste in the mining, quarrying, and oil/gas extraction industries consisted of paper.

(c) *Agriculture.* Agriculture is one of the most important sectors in the world. Agricultural waste refers to discarded or unsaleable products resulting solely from agricultural activities that are directly linked to growing crops or raising animals for the primary purpose of benefit or livelihood. Animal wastes are biomass materials because they are extracted from plants that have been eaten as food, either directly or indirectly via the food chain. Crop and orchard residues, as well as forest trash, make up plant residues. The amount of plant residue left on farms far outnumbers the number of crops brought to market. Grapefruit-bearing plants, vegetables, date palm and palm oil fronds, grass, stubble, leaves, hulls, tree limbs, and other litter are examples of farm waste. The majority of these wastes are burned to eradicate plant diseases and pests; however, a limited portion is used for mulch, ensilage, animal bedding, and other purposes.

In 2012, the biggest concentrated animal feeding operations (CAFOs) generated 369 million tons of manure, about 13 times the waste produced by the whole US population of 312 million people [18]. Between 2013 and 2016, the overall EU livestock population remained stable, accounting for 131 million livestock units [19].

Because of the large quantities generated and the concentration of animals at central processing points, livestock, and poultry wastes (mostly manure) are a major source of concern. A feedlot with 10,000 cattle produces approximately 300 tons

(270 metric tons) of solid waste/day, while a poultry operation with 270,000 hens produces approximately 40 tons (36 metric tons) of manure per day.

(d) *Food*. According to studies [20], about 30% and 40% of all food produced in the United States is thrown away. This equates to a total of 133 billion pounds ($161 billion). In the United States, overall food loss is estimated at $218 billion, while in Canada, it is $31 billion. Food waste accounts for 21–33% of agricultural water consumption in the United States. Wasted food is grown on 18–28% of our productive cropland. Each year, the USDA reports that about 4% of planted crops, or 66,500 acres, go unharvested. Across all forms of produce, manufacturing and processing (such as canning, freezing, drying, and pre-cutting) results in around 4% food loss. Despite the fact that manufacturing processes are becoming more efficient, the amount of food waste generated by expired, recalled, or unsold full packaged products are staggering. In terms of supermarket food waste, 10% of food in grocery stores, or 43 billion pounds, will never be consumed. In the case of produce, about 12% of fruit and 11.5% of vegetables are never sold, and food waste accounts for 30% of a grocery store's garbage [20].

(e) *Logging*. Logging is a form of data collection. Non-commercial sections of trees and brush, such as tops and cut-offs, are used in logging waste. Slash is a term used to characterize this sort of material. On an average, every year about 25 million tons of logging debris are left in the forest, or about 1 ton for every 1000 board feet of logs harvested. This debris is a breeding ground for insects and tree diseases as well as a serious fire hazard. Any woody material can be sold as firewood or as a source of fuel for outdoor wood boilers.

1.4.1.2 Basic Industries

Basic industries use extractive industries' products as raw materials and refine them into refined materials that other industries may use to make consumer goods. Metal sheets, tubes, cables, coke, industrial chemicals, paper, lumber and plywood, plastics, bottles, and synthetic fabrics are examples of products in this industrial group. The composition of solid waste produced by these industries is more diverse. Only a small percentage of refined products end up as solid waste, and all of the waste can be recycled right in this industrial category. The seven most popular industries that produce solid waste are listed below.

(a) Metals. Mined ores are transported to a manufacturing facility, where the metal is processed and refined. This process produces substantial solid wastes, such as slag, which accounts for 20% of steel ingot output. Aluminium and copper, on the other hand, each generate about 5 million tons (4.5 million metric tons) of inert waste each year. A subsequent step, in which ingots are shaped into shapes, produces lesser amount of solid wastes. The majority of these are product trimmings and residues from other refined materials used in the process.

(b) Chemicals. This industry creates the largest range of solid wastes, from slurries to dry solid cake, flammable organic tars to inert inorganic salts, hazardous materials like chromates to popular salt. They all come from one of three places: unreacted raw materials, pollutants in raw materials, or chemical reaction by-products. Organic solid wastes are often tars that form as unwanted by-products of chemical reactions, while inorganic solid wastes are often unreacted raw materials or contaminants in the raw material or mine.

(c) *Paper.* The processing of paper and paperboard produces two types of solid waste: residues from the materials used in the process and residues from the finished product. The first category includes tree bark, wood fibre, paper pulp, and inert filler, while the second category includes trimmings and waste. In their analysis, Simao et al. [21] reported that global pulp and paper mill production is increasing every year. As a result, the amount of waste generated is also rising. In 2013, world paper production totalled 403 million metric tons, while pulp production totalled 179 million metric tons, with the top 10 pulp producers being the United States, China, Canada, Brazil, Sweden, Finland, Japan, Russia, Indonesia, and Chile. The annual waste generation is most likely in excess of 1 million metric tons.

(d) *Plastics.* Plastics are a type of material that can be used in a number of ways. Plastic wastes are mostly trimmings or off-spec materials produced by basic factories that turn basic chemicals into plastic sheets or other forms used by fabricators. According to the USEPA, 13.7 million tons of plastics were discovered in durable goods in 2018 (Fig. 1.9) [3].

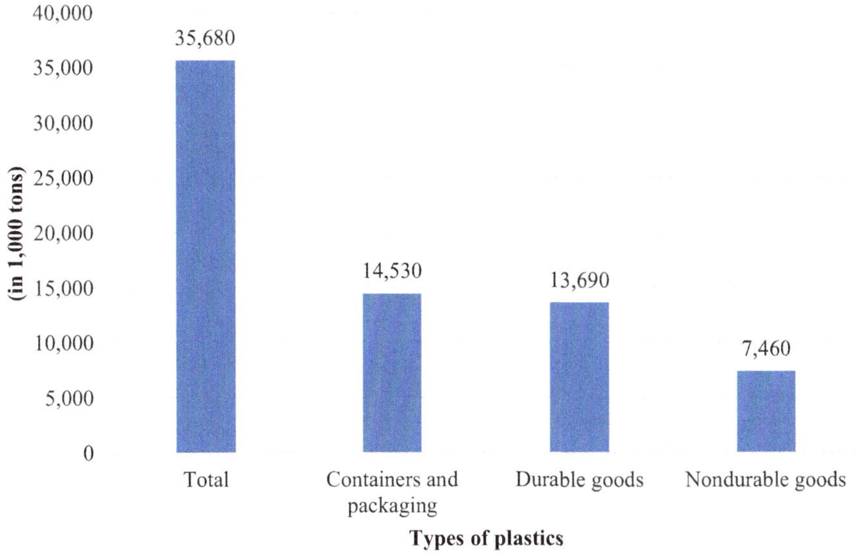

Fig. 1.9 The US municipal solid waste: plastics generation by category 2018 [3]

Table 1.7 MSW glass products in the US, 2018 [3]

Category of product	Amount generated (thousand tons)	Recycled amount (thousand tons)	Recycled amount %	Combustion (with energy recovery) (thousand tons)	Landfilled (thousand tons)
Durable goods[a]	2460	<100	<0.05%	330	2130
Packaging and containers					
Bottles (beer and soft drink)[b]	4650	1840	39.6	550	2260
Bottles (wine and liquor)	1810	720	39.8	210	880
Other bottles and jars	3330	500	15.0	550	2280
Glass containers (total)	9790	3060	31.3	1,310	5420
TOTAL GLASS	12,250	3060	25.0	1,640	7550

[a]Appliances, furniture, and consumer electronics made of glass
[b]Inclusive of carbonated and non-carbonated drinks

(e) *Glass*. In the basic glass sector, the bulk of solid waste is recycled within the industry. Cullet (glass fragments) from breakage and trimming, off-grade material, and slag from the purification of glass sand make up the majority of this waste. Table 1.7 shows the MSW glass goods in the United States in 2018 [3].

(f) Textiles. Cotton, linen, and wool are the three most important essential textile industries. Cotton textile mills produce wastes like strapping and burlap used in baling, as well as comber wastes and fibres damaged during storage and shipping, all of which are recycled. Linen textile mills produce waste that is similar to flax waste. Wastes such as fibre, twine, wool fat, and dirt are produced during the preparation of wool. Residues from spinning, weaving, and trimming operations also contribute to the overall solid waste image.

While textile waste is not considered hazardous, the waste generated by the textile and fashion industries has significant environmental implications. Textiles and garments that have been discarded may be processed to be reused and recycled. Textile waste is typically disposed of in landfills. Belgium had the largest amount of textile waste sent to landfills in 2016, with an average of 8.4 kg per capita, according to data (Fig. 1.10) on waste generation in European Union (EU) countries. Following Belgium, Czechia, Portugal, Italy, and Austria had the most landfilled textile waste per person [22].

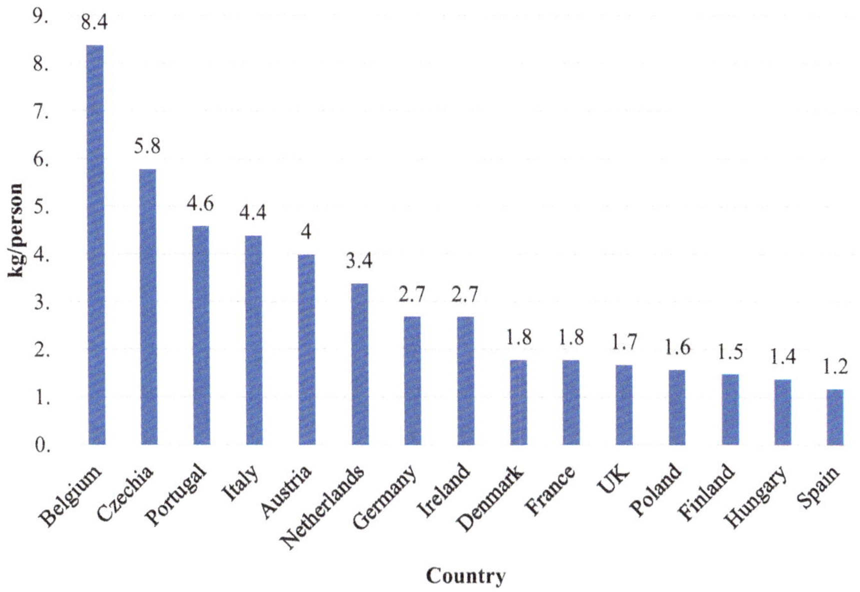

Fig. 1.10 The yearly total quantity of landfilled textile waste per person in the European Union (EU) in 2016, by country (in kg) [22]

(g) Wood Products. Furniture, other durable goods (e.g., cabinets for electronic equipment), wood packaging (crates, pallets), and other miscellaneous items are all sources of wood in municipal solid waste (MSW). In 2018, there were 2.8 million tons of wood in MSW that were combusted. In that year, 8.2% of MSW was combusted with energy recovery. Landfills received 12.2 million tons of wood in 2018. This accounted for 8.3% of all MSW landfilled in that year [23].

Every year, approximately 7.4 million tons of post-consumer wood waste are produced in the United Kingdom. Timber-based wastes from furniture production, fencing, infrastructure (such as telegraph poles and railway sleepers), agriculture and horticulture, and the do-it-yourself (DIY) sectors are all included in this figure. Although it is vital to optimizing material recovery, there are concerns about polluted wood, increased environmental impacts from recovery paths, and addressing the issue of how to better handle specific waste wood sources [24].

1.4.1.3 Manufacturing Industries

Conversion and fabrication industries make up this manufacturing market. They transform basic industry products into goods that are purchased by the general public. The solid waste generated by these industries primarily comes from the residues of the raw materials used by these industries. Furthermore, unlike simple industries, which can recycle most of their own rejects and trimmings, industrial industries can rarely use such residual wastes, and must rely on a secondary industry to reclaim them. However, the steps needed to transfer goods from one business sector to another place create a secondary solid waste burden on the recipient (in the form of packaging and containers). Since the number of manufacturing industries and the resulting variety of solid wastes produced is so large, only a few representative categories are discussed here.

(a) *Packaging*. The packaging industry uses a variety of materials, including aluminium, steel, glass, plastics, cardboard, corrugated paperboard, and plastic and paper laminates. The solid waste stream produced by this industry is dependent on the type of material used and the variety of activities performed at the facility, and it accounts for only a small portion of the total material used.

In 2018, 24% of the US packaging industry's market value was generated from the corrugated segment. The overall value of the US packaging industry was expected to be about 170 billion dollars that year, up from 144 billion dollars in 2011 [25].

(b) *Automotive*. Automotive industries generate a lot of solid waste. The discarded wastes include various components as tyres, generators, batteries, carburettors, wheels, bumpers, hub caps, and hundreds of other products that make up an automobile are by far the most significant component of automobile assembly plant waste. Painting and upholstering often add a jar to the solid waste stream, as well as material residues.

In 2020, vehicle sales were forecast to decline to just under 64 million units, down from nearly 80 million units in 2017. The sector has seen a downward trend as the global economy has slowed, and the coronavirus pandemic has spread through all major economies. The two most critical divisions of the auto industry are commercial vehicles and passenger cars. In terms of both sales and demand, China is one of the world's largest auto markets. In 2018, China's automotive sales dropped for the first time; the industry crashed in February 2020 but is now showing signs of recovery [26].

(c) *Paper Products*. Paper products, such as books, magazines, facial and toilet tissue, paper towels and napkins, and newspapers, created high-quality solid waste from paper trimmings and filled paper residues. They can be recycled in secondary industry.

In 2025, the market value of paper goods in the United States is expected to reach approximately US$109 billion, up from an approximate market value of US$81 billion in 2018 [27].

(d) Hardware. This is the metals industry, which manufactures machines, tools, utensils, and devices that are used by all forms of companies and the general public. Trimming and sizing of tubes, plates, and structural forms, boring and

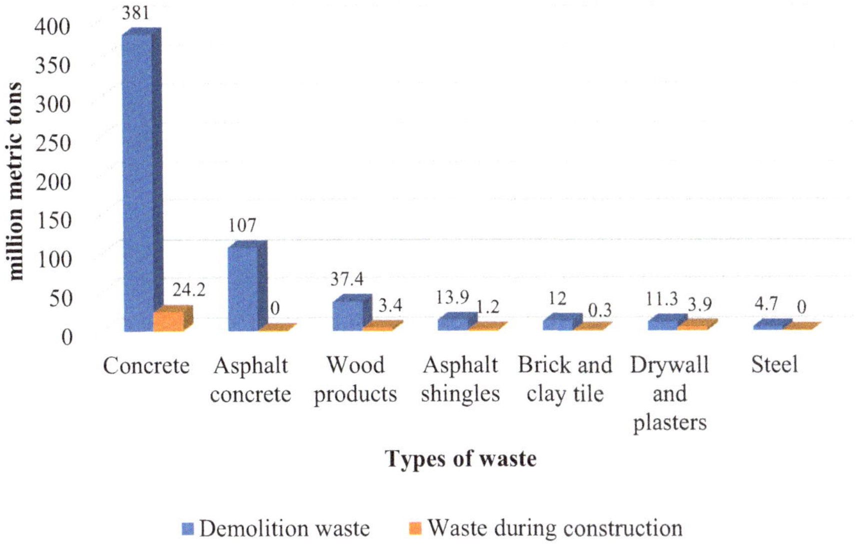

Fig. 1.11 The amount of material debris produced during construction and demolition in the United States in 2018 is depicted in Fig. 1.11 [3]

machining of metals, and miscellaneous residues from casting and forging processes make up the majority of the solid waste produced by this industry.

(e) Soft Goods. Leather, textiles, and plastics are used to make commercial goods in this industry. The residues from the processing of the materials are the most significant source of solid waste.

1.4.2 Construction and Demolition Waste

1.4.2.1 Definition

Materials that are unwanted or created during construction, demolition, or reconstruction activities of buildings, roads, and bridges are referred to as 'construction waste'. Concrete, wood, metals, bricks, glass, rocks, and asphalt are among the most popular heavy and bulky materials used in construction and demolition.

1.4.2.2 Quantity

The amount of construction and demolition waste differs from country to country based on the difference in the activities. Some of the statistics are discussed here.

In the United States, building waste accounts for one-third of all waste (2020, *Recycling magazine*). The amount of construction and demolition waste generated

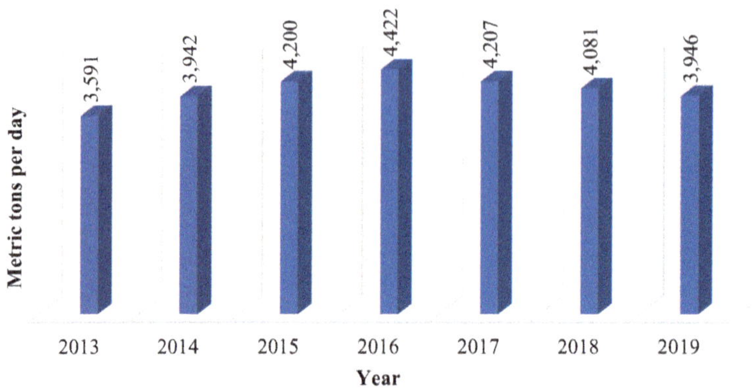

Fig. 1.12 Solid waste disposal to landfill in Hong Kong [4]

in the United States has risen significantly over the last 30 years. There were 600 million tons of construction and demolition waste in 2018, up from 135 million tons in 1990. With 143 million tons ending up in landfills, this represents a 300% increase [14].

In 2018, the amount of material debris produced during construction and demolition in the United States was broken down by material type in Fig. 1.11. Approximately 3.4 million tons of wood products were produced as waste during construction during this period.

In Hong Kong, nearly 4000 tons of total construction waste was disposed of daily in landfills in 2019 (Fig. 1.12). In Hong Kong, construction waste was generated at a lower rate than municipal solid waste.

The volume of non-hazardous solid C&D waste produced in the Abu Dhabi Emirate in 2019 was 3.7 million tons (Fig. 1.13).

1.4.2.3 Management

Many recyclable materials can be contained in construction waste. Crushed debris can be recycled in construction projects. Waste wood can also be recovered and recycled.

Waste management fees, based on the 'polluter pays' concept, will help reduce construction waste levels [29]. In 2019, a study method for optimizing the construction waste management fee in China was presented. China has a significant waste management issue, with most of its landfills located in urban areas. The study found that metal, wood, and masonry waste have separate waste management fees of $9.30, $5.92, and $4.25, respectively. Waste management cost $0.12 per m^2, or just under 11 ft^2, on average [30].

As for building design, renovation, and demolition in the European Union (EU), there is currently a heavy focus on recycling building materials and following a cradle-to-grave ethic. Depending on the political structure, their recommendations

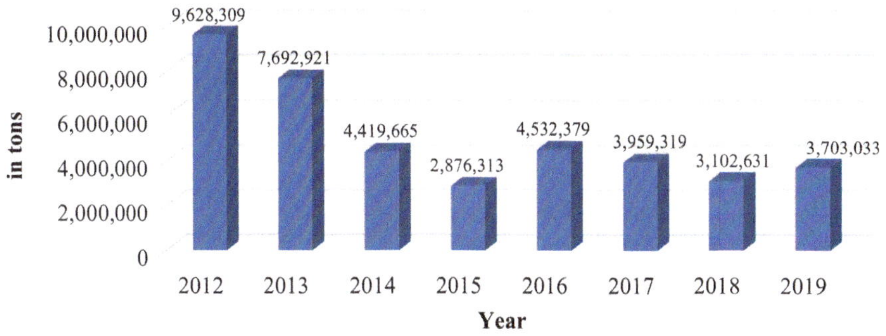

Fig. 1.13 Waste statistics in the Abu Dhabi Emirates [28]

are much easier and simpler at the regional or local level. In Austria, recent changes have been made in the processing of discarded wood products to be burned in the production of cement, reducing the carbon footprint of both products [31]. When issuing demolition and construction permits, the EU urges local governments to ensure that a high-quality waste management strategy is adopted, and they stress the significance of post-demolition follow-ups to assure that the established plans are followed. They also propose using taxes to reduce landfills' economic benefit, resulting in a situation where recycling becomes a financially viable option.

The landfill tax has had the greatest impact in Belgium, Denmark, and Austria, which have all reduced landfill disposal by more than 30% since the tax was implemented. Denmark has lowered landfill use by over 80% and has a recycling rate exceeding 60%. All staff conducting builders or construction waste clearance in the United Kingdom are required by law to be employed by a CIS registered company [32]. However, waste generation in the United Kingdom continues to rise, though at a slower pace. Although the United States does not have a national landfill tax or payment, many states and local municipalities do receive taxes and fees on solid waste disposal [33].

In 2008, the Malaysian Construction Industry Development Board (CIDB) published Guidance on Construction Waste Management [34]. The following items should be included in the building and demolition waste definition and strategy:

(a) Priorities based on the Solid Waste Management Hierarchy, which is widely recognised globally (Fig. 1.14).
(b) The cradle-to-grave approach.
(c) Device that is accommodating to the community.

The Solid Waste Management Hierarchy definition, as shown in Fig. 1.14 (right), is globally recognized and prioritizes waste prevention, reduction, reuse, recycling, waste treatment, and disposal in order of priority. The SWM Hierarchy prioritizes 'Waste Minimization' and the 3Rs: 'Reduce, Reuse, Recycle', with 'Treatment' (including Composting and Thermal Treatment) and 'Disposal', which involves landfilling, receiving the lowest priority.

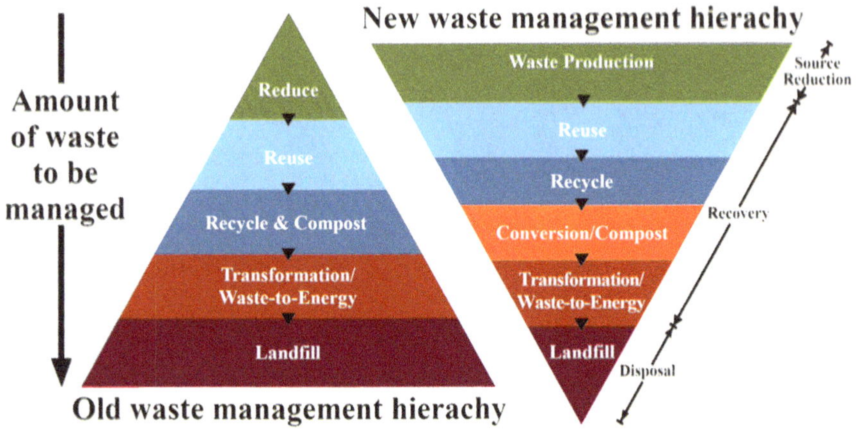

Fig. 1.14 Solid waste management hierarchy

(a) *Waste Minimization/Reduction*

Waste reduction activities can save money by lowering material prices and lowering disposal costs. A systematic approach to waste minimization/reduction is recommended, which involves the contractual, construction, and site operation stages.

(b) *Reuse*

As much as possible, materials that can be reused on-site should be identified. Some of the most reused items include door sets, faucets and plumbing, fencing, plywood and chipboard, shelving and racking, siding and shutters, gutters, roof tiles, etc.

(c) *Recycle*

Recycling building materials reduce recycling costs, which saves money. It helps to keep construction sites cleaner and safer by reducing waste going to the landfill.

Waste recycling can be done in three ways:

Separation of sources: For each form of waste, several boxes are used. It does, however, take up more room and necessitates close supervision.

Commingled recycling: Recycling that is blended together. All is sorted off-site by the hauler. Commingled recycling takes up the least amount of storage room and is the best choice for sites with limited space.

Hybrid recycling: This recycling method incorporates site isolation and mixed recycling. One box for wood, one for concrete, and one for non-recyclable waste, for example. Hybrid recycling combines the benefits of both methods. It seeks the best balance between the weight and sorting effort. By operating in stages, the total number of boxes can be decreased. It decreases the amount of work for sorting haulers, reducing hauling fees.

Table 1.8 Some of the waste identification which could be done on-site

Types of wastes	Status of recyclables
Concrete	Recyclable
Bricks	Reusable
Plasterboards	Recyclable
Paint	Reusable
Timber	Recyclable/reusable
Pipework	Recyclable
Packaging	May be used for landscaping

(d) *Waste Minimization Plan*

Rather than modern technology, waste minimization needs a shift in mindset and common sense. Sometimes, waste minimization options are free to incorporate and offer immediate benefits with little or no effort.

(i) *Identifying and Exploration of Waste Minimization Opportunities*

Table 1.8 summarizes the steps involved in finding waste reduction opportunities

(ii) *Implementation of the Waste Management Plan*

(a) To the greatest degree possible, recyclable materials are separated from waste materials by form.
(b) Have clearly labelled containers for recyclable materials by form. Non-recyclable content does not make up more than 10% of the container's volume. If the 10% cap is surpassed, include other storage options for recyclable materials before they are removed from the project site.
(c) Separation of inert and non-inert materials for site formation and reclamation.
(d) Higher-grade usage, such as an inert waste road sub-base, is also possible if the necessary standards are met.

(iii) *Disposal/Recycling Services*

Contractors must ensure that their waste is disposed of at locations that have been approved by the local government. The Local Authorities closest to the project site will be given a list of permitted construction waste disposal sites. Contractors can sell recyclable products to local recyclers.

(iv) *Disposal/Recycling Services*

Contractors must ensure that their waste is disposed of at locations that have been approved by the local government. Contractors can sell recyclable products to local recyclers.

(v) *Implementation as a Whole*

It is necessary to create an organizational chart with responsibilities. A designated on-site waste manager (also known as a Safety, Health and Environment – SHE manager or a site manager) must be named. The following are his/her responsibilities:

1. Supervising the disposal of construction waste.
2. Managing campaigns aimed at reducing waste.
3. Coordination of other employees' operations.
4. Numerous services, such as employee awareness programmes, training programmes, and safety and health events, are carried out.

Staff awareness and training programmes are needed to gain their commitment. All levels of workers should be involved in awareness and communication activities. Furthermore, all employees must be briefed on the safety and health aspects of all construction activities, especially those involving waste management. Construction waste handling, segregation, and transportation procedures must provide safety and health precautions. Clients, consultants, and contractors will also help to reduce waste during the contractual stage of a project. The form of contract should be agreed upon by both parties. They should also notice the materials used, the project's team and workforce, the building methods and techniques, and any possible waste sources. Finally, content and waste audits should be conducted in order to find areas where future projects can change.

1.4.3 Electronic Waste (E-waste)

1.4.3.1 Definition

The acronym 'E-waste' stands for 'electronic and electrical waste'. Almost any household or commercial object with circuitry or electrical components falls into this category. It describes electrical or electronic devices or broken, non-working or old/obsolete electric and electronic appliances such as PC, TV, washing machine, air conditioner, and refrigerator.

1.4.3.2 Categories

The most common E-waste categories are as follows:

1. Temperature exchange equipment: Cooling and freezing equipment is more often termed as cooling and freezing equipment. Refrigerators, air conditioners, freezers, and heat pumps are examples of typical appliances.
2. Lamps: Typical equipment includes fluorescent lamps, high-intensity discharge lamps, and LED lamps.

3. Screens and monitors: Common equipment includes laptops, televisions, monitors, notebooks, and tablets.
4. Large equipment: Typical equipment includes washing machines, dishwashing machines, clothes dryers, electric stoves, large printing machines, copying equipment, and photovoltaic panels.
5. Small equipment: Vacuum cleaners, microwaves, ventilation machines, toasters, electric kettles, electric shavers, scales, calculators, radios, video cameras, electrical and electronic toys, small electrical and electronic tools, small medical devices, and small monitoring and control instruments are examples of typical equipment.
6. Small IT and telecommunication equipment: Mobile phones, GPS systems, pocket calculators, routers, personal computers, printers, and telephones are examples of popular equipment.

1.4.3.3 Impacts

Because of the exponential growth in demand for electrical and electronic devices, as well as the disposal after use, e-waste is becoming a global concern. If not properly treated, discarded electronic waste can pose a health and environmental risk.

Every part of our lives is increasingly entwined with technology. Semiconductors and sensors are now standard in a wide range of items that previously lacked them, resulting in smart homes, wearable watches, Internet TVs, and much more. To make it worse, the product lifespans are decreasing. When the batteries run out, several items will be discarded and replaced with new ones.

In developed countries, informal e-waste production may have negative health and environmental implications. There are some of the most significant factors for properly treating and recycling household e-waste:

1. Mercury, brominated flame retardants (BFR), and chlorofluorocarbons (CFCs), or hydrochlorofluorocarbons (HCFCs) are all harmful additives or hazardous substances found in e-waste. If e-waste is not adequately handled and is merely tossed out with the trash, potentially ending up in a landfill, both human health and the environment are jeopardized. The growing amount of e-waste, low collection rates, and non-environmentally sound disposal and treatment of this waste stream pose serious environmental and human health risks. Annually, 50 tons of mercury and 71 kilotons of BFR plastics are discovered in undocumented e-waste flows around the world, which is mainly released into the atmosphere and has an effect on the health of those exposed.
2. E-waste disposal that isn't handled properly leads to global warming. CFC (Freon) gases, which have a high global warming potential and lead to ozone depletion, are used in refrigerators and air conditioners. Increased UV radiation hitting the Earth's surface and skin cancers will result from ozone depletion. If e-waste products are not recycled, they will not be able to replace primary raw materials and hence will not be able to offset greenhouse gas emissions from the

production and refinement of primary raw materials. Finally, some of the refrigerants used in temperature exchange equipment are greenhouse gases. A total of 98 Mt of CO_2 equivalents were released into the environment as a result of poorly managed discarded refrigerators and air conditioners. In 2019, this represented around 0.3% of global energy-related emissions (IEA).
3. E-waste contains a plethora of hazardous materials that damage both humans and animals, as well as the environment. If e-waste is not adequately handled and is merely tossed out with the trash, potentially ending up in a landfill, both human health and the environment are jeopardized. E-waste contains a plethora of hazardous materials that damage both humans and animals, as well as the environment.
4. Other toxic components also present in e-waste:

   (a) Lead and cadmium are present in printed circuit boards (PCBs), which are used in a number of electronic appliances (cause brain damage, cancer, etc.).
   (b) Mercury has the potential to damage the brain and nervous system.
   (c) Lead has the potential to harm the brain and impair the normal processes of water and soil systems.
   (d) Brominated flame retardants (BFR): Under some conditions, brominated dioxins and furans, which are carcinogens, will recombine with un-oxidized carbon in smelter emissions.
   (e) Beryllium: Beryllium or beryllium-containing dust, mist, or fume inhaled by susceptible people may cause a chronic lung disease called berylliosis, and beryllium is a probable human carcinogen.
   (f) Hazardous chemical additives (such as phthalates) can leach from polyvinyl chloride (PVC) parts of electronic goods when they are discarded.
   (g) Arsenic is known to cause carcinogen in the skin, lungs, bladder, liver, and kidneys, with evidence of lung cancer. Death, inhibition of growth, photosynthesis, and reproduction, as well as behavioural effects of a particular flora/fauna organism, are all potential effects of arsenic on the environment.

1.4.3.4 Quantity

The amount of e-waste generated is increasing, and the types of e-waste are becoming more diverse. Since global market demand continues to rise, e-waste is among the strongest-growth sources of waste. Global e-waste production reached 53.6 million metric tons in 2019. It is expected to keep growing over the next decade. Global e-waste generation is expected to reach nearly 80 million metric tons by 2030, according to estimates. Asia is the continent with the most e-waste, with China producing the majority of it [35] (Fig. 1.15).

Various figures on e-waste in the world have been published by Forti et al. (2020) (Fig. 1.16). They discovered that the planet produced a staggering 53.6 Mt of e-waste in 2019, averaging 7.3 kg per capita. Since 2014, global e-waste generation has risen by 9.2 Mt, and is predicted to rise to 74.7 Mt by 2030, nearly twice in just

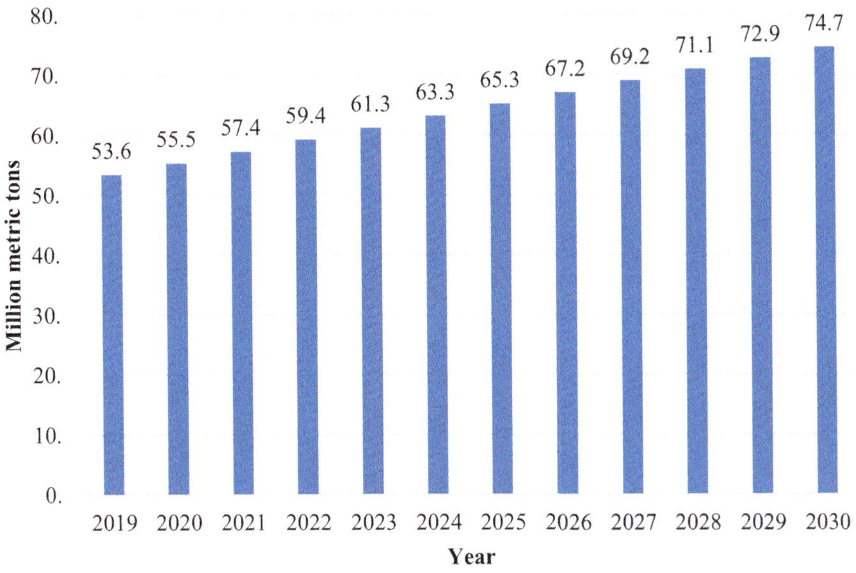

Fig. 1.15 Projected global electronic waste generation from 2019 to 2030 (in a million metric tons) [35]

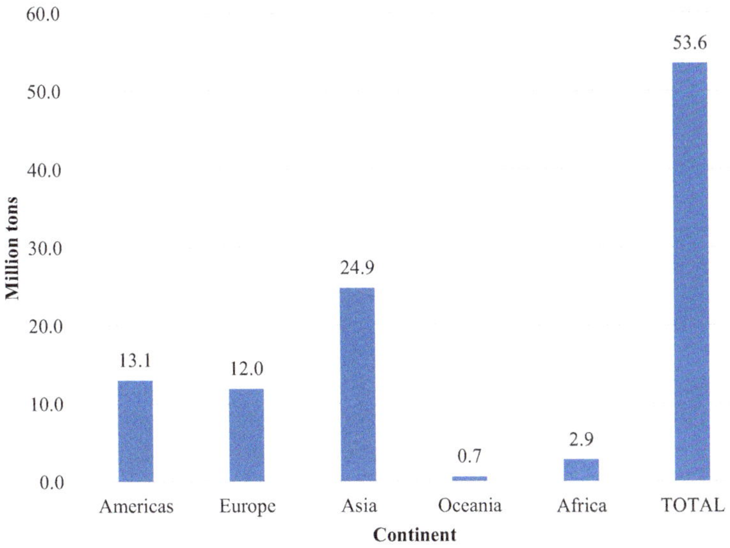

Fig. 1.16 E-waste generation worldwide in 2019 [35]

16 years. Higher electrical and electronic equipment (EEE) usage rates, short life cycles, and limited repair options are causing the increasing amount of e-waste. Asia produced the most e-waste in 2019, with 24.9 million tons, succeeded by the

Table 1.9 E-waste generation in ASEAN countries in 2014 [9]

No	Country	Kilogram/inhabitant	Annual amount (metric kilotons)
1	Brunei Darussalam	18.1	7
2	Cambodia	1	16
3	Indonesia	3	745
4	Lao PDR	1.2	8
5	Malaysia	7.6	232
6	Myanmar	0.4	29
7	Philippines	1.3	127
8	Singapore	19.6	110
9	Thailand	6.4	419
10	Vietnam	1.3	116

Americas (13.1 million tons) and Europe (12 million tons), with Africa and Oceania producing 2.9 million tons and 0.7 million tons, respectively. With 16.2 kg per capita, Europe ranked first in the world in terms of e-waste production. Oceania came in second (16.1 kg per capita), followed by the Americas (13.3 kg per capita), with Asia and Africa coming in third and fourth, respectively, with 5.6 and 2.5 kg per capita. They also reported that global e-waste generation is expected to hit nearly 80 million metric tons by 2030, according to estimates. Asia is the continent with the most e-waste, with China producing the majority of it. In 2019, formal reported collection and recycling totalled 9.3 Mt, accounting for 17.4% of total E-waste produced. Since 2014, it has risen by 1.8 Mt, or approximately 0.4 Mt per year (Table 1.9).

1.4.3.5 E-waste Recycling

E-waste composes a wide range of recyclable metals and materials. E-waste may be used as a resource. The Tokyo Organising Committee of the Olympic and Paralympic Games, for example, has launched the 'Tokyo 2020 medal initiative', which aims to create medals for the games out of recycled metals recovered from small waste electronic devices, such as discarded cell phones.

However, significant quantities of e-waste are also not being recycled properly. They end up in areas where there aren't yet any proper recycling facilities. Many countries, on the contrary, have recognized the importance of properly recycling e-waste and are working to introduce long-term solutions.

Both formally and informally, e-waste recycling can be carried out. Disassembling the electronics, sorting, and categorizing the contents by material, and cleaning them is all part of proper or structured e-waste recycling. After that, the items are mechanically shredded in preparation for further sorting using advanced separation technologies.

Based on Cho [36], developed countries shipped nearly 23% of their e-waste yearly to developing countries. This is still ongoing, although the European Union and 186 countries have ratified the Basel Convention, which seeks to reduce hazardous waste movement from developed to developing countries. The Restriction of Hazardous Substances Directive of the European Union, on the other hand, corresponds to the whole EU market and thus has the power to set better standards for all electronic goods sold in the EU. Its regulations requiring manufacturers to contribute to the cost of recycling have caused a 35% e-waste recycling rate, which is higher than the United States.

1.4.3.6 What Can We Do About E-waste

Some of the 'green options' on e-waste disposal are as follows:

(i) *Designing Better Products*

Electronics manufacturers are encouraged to produce goods that are more durable, secure, repairable, and recyclable. Above all, this leads to using lesser quantity of toxic materials. This could be implemented by applying a cleaner production system that supports a proper Environmental Management System (EMS) – ISO14000 system in the company. A political will and commitment from the top are necessary to materialize this.

(ii) *Repair*

It is also important to repair and reuse faulty devices.

(iii) *Extended Producer Responsibility*

Extended supplier liability mandates that manufacturers be accountable for the disposal and management of their goods after the end of their useful lives. The concept behind the life cycle or cradle-to-grave theory is to recycle waste materials and use them to create new items. This option requires appropriate in place policy and probably legislation by the authority.

(iv) *Facilitate Convenient Recycling*

A proper drop-off recycling centre will facilitate user to send e-waste for recycling. On top of that, a proper roadmap should be structured so that the disposed of items are properly collected and processed. In most situations, processing and transportation are the beginning steps in the e-waste recycling flow. Recycling companies set up collection bins or electronic take-back booths in strategic areas and shifted the collected e-waste to recycling facilities. Materials in the e-waste stream are processed and separated into renewable goods that can be used to produce new items after being collected and transported to recycling facilities. The basis of electronics recycling is effective content separation. After that, the e-waste is shredded into small bits for further processing. The iron and steel are then separated from the waste stream on the conveyor by a strong overhead magnet, which is then ready for sale as recycled steel. Aluminium, copper, and circuit boards are separated from the content stream by mechanical processing. Glass and plastics can be separated using water separation technology. The separated materials are then sold as functional raw materials for the manufacture of new electronics or other items after completion of the shredding, sorting, and separation processes.

1.4.4 Radioactive Waste

1.4.4.1 Definition and Sources

Radioactive waste is a substance that has been contaminated with radionuclides or contains radionuclides and is no longer usable. Radionuclides are volatile atoms of a given element that spontaneously decay or disintegrate, releasing energy in the form of radiation.

Radiation affects all on the planet. Radiation at levels higher than normal background radiation, on the other hand, can be dangerous. High levels of radiation, such as that emitted from high-level nuclear waste, can also kill you. Based on the duration of exposure, the amount of radiation, and the decay process, radiation can cause cancer, birth defects, and other anomalies. For thousands of years, high-level radioactive waste from nuclear plants may be dangerous.

Military weapons production and testing, mining, electrical power generation, medical diagnosis and treatment, consumer product development, manufacturing, and treatment, biological and chemical research, and other industrial uses have all developed nuclear waste. The front-end source is from the front end of the nuclear fuel cycle and is usually alpha-emitting waste from the extraction of uranium. It often contains radium and its decay products. Isotopes produced in nuclear reactors make up the back end of the nuclear fuel cycle. These are mainly spent fuel rods, which contain fission products that emit beta and gamma radiation. Beta particle and gamma ray emitters are common in radioactive medical waste. Many of these can be disposed of by allowing them to decompose for a brief period of time before discarding them as daily garbage. Y-90 is used to treat lymphoma (2.7 days), I-131 is used to treat thyroid cancer (8.0 days), Sr-89 is used to treat bone cancer (52 days), Ir-192 is used for brachytherapy (74 days), and Co-60 and Cs-137 are used for brachytherapy and external radiotherapy (5.3 years and 30 years respectively). Alpha, beta, neutron, and gamma emitters can be found in industrial waste. Radiography uses gamma emitters, while neutron emitting sources are used in a variety of applications.

1.4.4.2 Form and Half-Life

The physical and chemical characteristics of radioactive waste can differ greatly. It may be a solid, a liquid, a gas, or even a mixture of the three, such as sludge. Water, dirt, paper, plastic, metal, ash, glass, ceramic, or a combination of several physical types can all be contained in nuclear waste. Low-level waste (LLW), intermediate-level waste (ILW), and high-level waste (HLW) are the three types of nuclear waste, based on the level of radioactivity and the period of time it remains hazardous. Waste can be radioactive for seconds, minutes, or even millions of years, depending on the radionuclides it contains.

A half-life is a duration it takes for a given amount of radioactive material to degrade to half its original value. A radionuclide's half-life can range from fractions of a second to millions of years. Sodium-26 (half-life: 1.07 s), hydrogen-3 (half-life: 12.3 years), carbon-14 (half-life: 5730 years), and uranium-238 are several examples of radionuclides with a variety of half-lives (half-life of 4.47 billion years). A radionuclide's decay process is the mechanism by which it releases excess energy spontaneously. Alpha, beta, and gamma emission are typical pathways for radioactive decay. Alpha decay is a mechanism in which excess energy is released by the ejection of two neutrons and two protons from the nucleus of heavy atoms such as uranium-238 and thorium-234. The ejection of a beta particle, which is the same as an electron, from the nucleus of an excited atom is known as beta decay. Strontium-90 is an example of a beta-emitter commonly found in radioactive waste. The nucleus of an atom is always in an excited state after an alpha or beta decay and still has excess energy. Instead of releasing energy through alpha or beta decay, energy is lost through gamma emission, which is a pulse of electromagnetic radiation emitted from an atom's nucleus.

Figure 1.17 [37] shows the amount of radioactive waste generated in France from 2016 to 2040, broken down by waste type and measured in cubic metres. According to estimates, there will be 6.9 million cubic metres of high-level waste in 2040. Figure 1.18 [37] depicts the economic sector's share of radioactive waste volume produced in France in 2018. The nuclear industry accounted for approximately 60% of France's nuclear waste in that year.

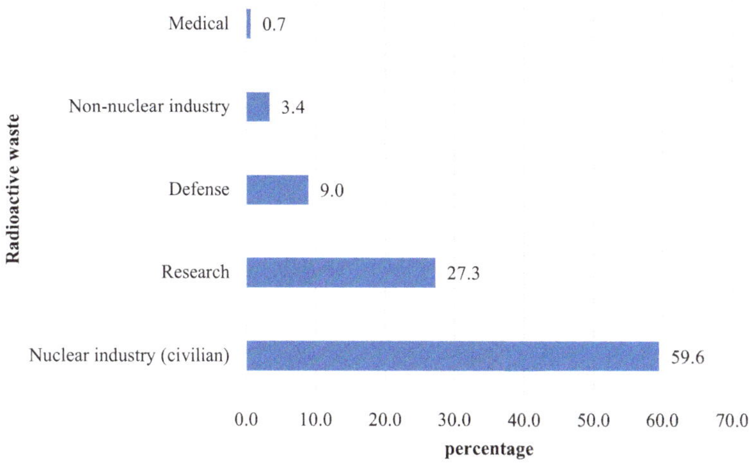

Fig. 1.17 The share of radioactive waste volume generated in France in 2018, by economic sector [37]

Fig. 1.18 The projected volume of radioactive waste produced in France from 2016 to 2040 [37]

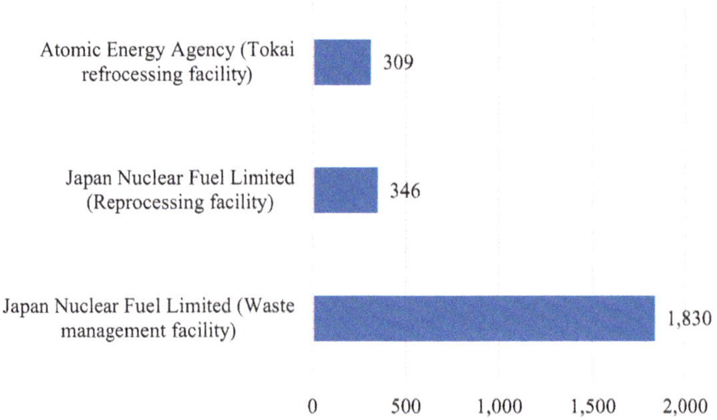

Fig. 1.19 Amount of high-level radioactive waste stored in Japan in 2019 by the company(in units) [38]

Figure 1.19 shows the volume of high-level radioactive waste deposited in Japan in 2019 by the organization (in units). It reveals that Japan Nuclear Fuel Limited's waste treatment facility has the most high-level radioactive waste [38].

1.4.4.3 Radioactive Waste Management

(i) *Introduction*

Possession, transportation, handling, storage, treatment, and final disposal of radioactive waste are all part of radioactive waste management, which aims to protect people and the environment. Human exposure to high-level radioactive waste can be dangerous, even fatal, if treated improperly.

The method of transport, the packaging required, and the labelling required to allow for the shipping of particular waste are all determined by the classification and physical size of radioactive waste. For nuclear waste, there are international transportation standards as well as more stringent legislation in individual countries. Only if the radionuclides present in the waste are directly inhaled or swallowed will certain radioactive wastes, such as some forms of transuranic waste, cause biological effects in humans. Humans can manage most low-level nuclear wastes without causing any noticeable biological effects.

(ii) *Commonly Accepted Management and Disposal Options*

The activities required for properly managing and disposing of radioactive waste can be divided into four categories, that is, reducing the amount of waste generated, conditioning and packaging to allow for safe handling and transportation, interim storage, and final disposal.

Since the radioactivity of the wastes decays over time, there is a strong incentive to store high-level waste for around 50 years before disposing of it. Low-level waste disposal is simple and can be done almost anywhere in a safe manner. Used gasoline, on the other hand, is usually stored underwater for at least 5 years and then in dry storage. The best option for the final disposal of the most radioactive waste generated is deep geological disposal, according to most experts. Options and examples of radioactive waste disposal are presented in Table 1.10 [39].

The bulk of low-level radioactive waste (LLW) is usually sent to land-based disposal for long-term management immediately after packaging. This means that a suitable disposal method has been developed and is being introduced around the world for the vast majority (90% by volume) of all waste forms generated by nuclear technologies.

The first step in storing used fuel classified as high-level radioactive waste (HLW) is to allow radioactivity and heat to decay, making handling much safer. Used fuel may be deposited in ponds or dry casks at reactor sites or centrally.

1.4.5 Litter

Litter is made up of man-made waste materials like paper cups, aluminium cans, cardboard boxes, fast food wrappers, or plastic bottles, and cigarette butts that have been improperly disposed of. It has a human impact on the environment and continues to be a significant environmental concern in many countries. Litter will stay visible for a long time before biodegrading, with certain products made of condensed glass, styrofoam, or plastic potentially staying in the atmosphere for years. Litter can travel long distances and end up in the world's oceans. Litter may have an

Table 1.10 Options and examples of radioactive waste disposal [39]

Option	Types of waste	Examples
Near-surface disposal: At ground level, or in caverns hundreds of metres below ground level	LLW or short-lived ILW	Many nations, including Finland, France, Japan, the Czech Republic, the Netherlands, Sweden, the United Kingdom, Spain, and the United States, have adopted it for LLW
		For LLW and short-lived ILW, it is used in Sweden and Finland
Deep geological disposal: Mined repositories are located between 250 and 1000 metres deep, while boreholes are located between 2000 and 5000 metres deep	Long-lived ILW and HLW (including used fuel)	The majority of countries have looked into deep geological disposal, and it is now a strategy
		It is used in the United States for transuranic waste that is related to protection
		France and Sweden were chosen as preferred locations
		Finland and the United States of America
		The collection of geological repository sites has begun in Canada and the United Kingdom

adverse effect on one's quality of life. Litter has a significant visual effect. Rainwater fills open containers such as paper cups, cardboard food packages, plastic drink bottles, and aluminium drinks cans, offering mosquito breeding grounds. Furthermore, if a spark or lightning flash hits litter, such as a paper bag or cardboard box, it may cause a fire.

Plastic/polystyrene fragments were the most typical form of litter found on British beaches (Fig. 1.20). This was significantly more than the second-most common form of litter, cigarette stubs, with 143 pieces per 100 metres of the beach [40].

Litter can trap or poison animals in their natural habitats. Cigarette butts and filters pose a risk to wildlife, having been discovered in the stomachs of fish, whales, and birds that mistakenly ate them for food. Animals may also get stuck in the garbage, causing serious pain. The plastic used to keep soda cans together, for example, can wrap around animals' necks and suffocate them as they grow. Broken glass lacerating the paws of dogs, cats, and other small mammals is another example of how litter can affect animals.

Litter is a major environmental issue faced by many countries in the world. Despite the fact that developing countries lack the resources to address the problem, developed-world market economies are capable of generating more litter per capita due to higher disposable product consumption.

It is preferable to prevent rather than cure. The most important thing is awareness. Many organizations exist with the goal of increasing awareness and implementing programmes, such as clean-up activities. World Cleanup Day is a global initiative. Specific litter movements are exemplified by TrashTag and Plogging. To fix the issue, some countries and local governments have enacted legislation.

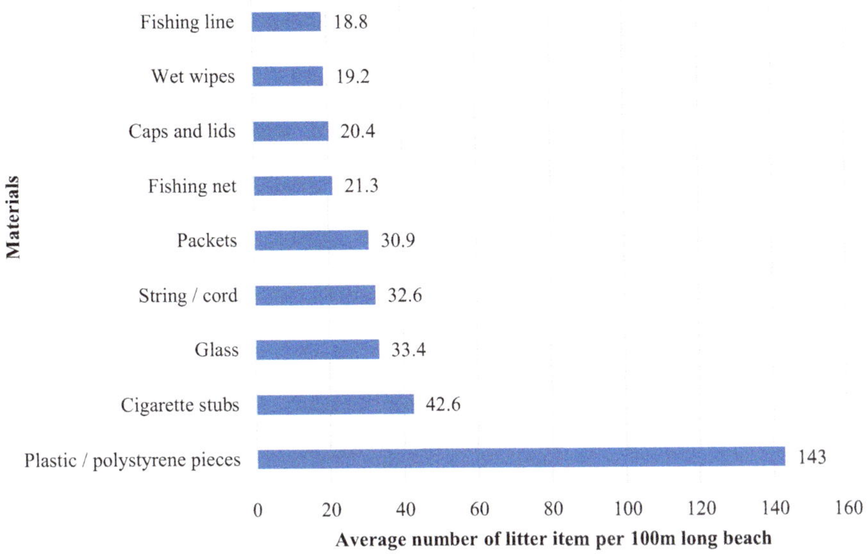

Fig. 1.20 The typical forms of litter found per 100 metres of beach in the United Kingdom (UK) in 2019 [40]

On-the-spot fines for individuals imposed by authorized officers in public or on public transportation, as well as littering from a car, are examples of actions that result in fines. Litter traps may be used to catch litter before it enters rivers from storm drains. Litter traps, on the other hand, are only suitable for large or floating litter and must be managed. Volunteers clean up trash and dispose of it, either individually or with the help of groups. Clean-up activities are often held in which participants search a region in a line to assure that no litter is missed. Litter clean-up activities can be promoted by organizations, as well as separate advertising campaigns to discourage littering. Local governments can also have appropriate municipal waste containers or street bins to be used as a safe location for litter disposal and collection.

1.4.6 *Scrap Tyre*

1.4.6.1 Introduction

Tyre recycling, also known as rubber recycling, is the method of repurposing waste tyres that have become unfit for application on vehicles due to wear or irreparable damage. Due to the large volume generated and their non-biodegradable nature, these tyres are a difficult source of waste and can take up valuable landfill space. Despite the fact that many junked tyres are burned haphazardly in open-air dumps, this activity should come to an end with the implementation of stricter air pollution regulations. Scrap tyres burn so hotly that they melt the grates of traditional incinerators, so most cities avoid incinerating them. Even when special incinerators are designed, the majority amounts of particulates, sulphur and nitrogen oxides, as well as hydrocarbons, are released, causing air pollution. Tyres do not lend themselves well to landfill disposal due to their low bulk density and resilience to biodegradation; whole tyres resist compaction and burying and rise to the surface, posing other issues such as disease vector breeding grounds and fire hazards.

Vulcanized rubbers have a strongly crosslinked structure and chemical composition, which contains toxic components such as leachable heavy metals; recycling waste tyres pose major environmental concerns [41].

1.4.6.2 Quantity

In the United States alone, around 270 million discarded tyres are produced each year. This necessitates the development of simple, energy-efficient, and cost-effective methods for recycling waste tyres [42]. In 2018, approximately 72.4 million metric tons of scrap iron and steel were processed (Table 1.11), while nearly 783,000 metric tons of plastics (only PET bottles) were handled in this country [43].

Between 1994 and 2010, the European Union increased tyre recycling from 25% to nearly 95% of annual discards, with approximately half of the end-of-life tyres being used for oil, mainly in cement manufacturing [44].

1.4.6.3 Tyre Recycling

Tyres are poorly degradable waste materials. Various solutions to reuse discarded tyres have been implemented in the world. While waste reduction initiatives will not solve the tyre issue, they will help to reduce the number of discarded tyres. According to the US Tyre Manufacturers Association [9], 16% of waste tyres are still landfilled, and these tyres are difficult to degrade and can stay in the environment for a

Table 1.11 Volume of scrap processed by type in the US in 2018 [43]

Types of scrap	Volume (thousand metric tons)
Zinc	72
Plastics (PET bottles)	782.9
Lead	1349
Copper	1783
Aluminium	5462
Electronics	5500
Paper	47,800
Iron and steel	72,400
Tyres (in thousand tyres processed – absolute number)	116,000

long period duration. Retreading, incineration for energy recovery, pyrolysis to obtain gas and carbon black, and shredding to create small particles used as fillers in a wide range of matrices such as asphalt, concrete, and polymers account for 86% of scrap tyres recycled. These recycling practices not only contribute to environmental conservation but also to the economic development of a number of industries, including artificial reefs, flood prevention, breakwaters, floatation devices, athletic tracks, playground surfaces, rubberized composites, and many others [45]. Retreading, incineration, pyrolysis, and mixing are the most common methods for recycling waste tyres (composites).

Figure 1.21 illustrates the recycling rates of different items in the urban waste stream in the United States in 2018. In that year, the tyre recycling rate was about 40%.

(i) *Tyre Retreading*

Retread is a tyre remanufacturing process that replaces worn tread with new tread. It's achieved by removing the tread and replacing it with a new one using cold or hot processes. As opposed to making a new tyre, it retains about 90% of the rubber in spent tyres and saves about 20% on material costs. Retread should not be confused with remoulding, which is a higher-quality version. It was scarcely used for passenger vehicles as of 2008, owing to road discomfort, safety concerns, and the advent of cheaper tyre brands on the market.

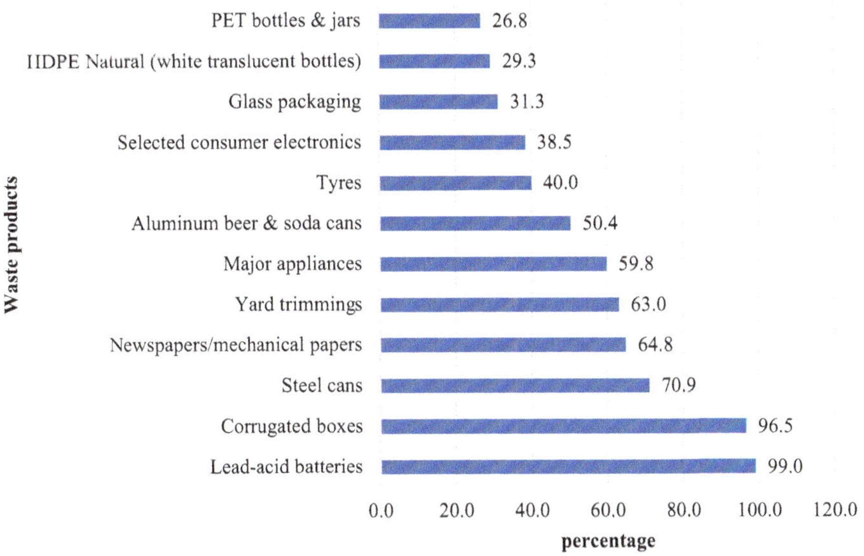

Fig. 1.21 In 2018, the United States' recycling and composting rates for a variety of items were calculated [3]

(ii) *Incineration*

Incineration of scrapped tyres may be used to generate steam. Waste tyres, which have a calorific value of 32.6 MJ/kg, are used to generate steam, electrical energy, pulp, paper, lime, and steel [46].

(iii) *Pyrolysis*

Tyre pyrolysis is a method of recycling tyres into useful materials for further manufacturing. Tyre pyrolysis, put simply, is the process of putting shredded tyre material into a big, sealed chamber and then heating the material up in the absence of oxygen. This will turn the waste tyre into useful goods like steel, carbon black, and tyre-derived fuel. The steel and carbon black can be used to make new tyres as well as a variety of other products. The tyre pyrolysis oil is used to produce energy for a variety of industries as a more environmentally sound alternative to many traditional combustion products. Cement plants often use this kind of fuel when it is available.

(iv) *Tyre-Derived Products*

Tyres may be repurposed in a variety of ways. New products made from waste tyres are beneficial than combustion or other low-multiplier output while also reducing waste and reducing pollution and emissions from recycling operations. Construction materials, artificial reefs, tyre-derived aggregate (TDA), ground and crumb rubber – also known as size-reduced rubber – can be used in both paving style projects and mouldable products are among the new waste-derived products. Rubber Modified Asphalt (RMA), Rubber Modified Concrete, and as a substitute for aggregate are examples of these forms of paving.

(v) *Alternative Fuel*

In cement production, used tyres are fed mid-kiln to a pair of long cement kilns. It is used as an alternative fuel in the production of Portland cement, a crucial component of concrete. Whole tyres are often inserted into cement kilns by rolling them into the upper end of a preheater kiln or dropping them through a slot in the middle of a long wet kiln. In any case, the high gas temperatures (1000–1200 °C) cause the tyre to burn almost immediately, completely, and smokelessly. Alternatively, tyres can be chopped into 5–10 mm chips and injected into a pre-calciner combustion chamber in this shape. Since cement production requires some iron, the iron content of steel-belted tyres is advantageous to the process [47].

1.4.7 Solid Waste from Air and Water Pollution Controls

Sulphur oxides and particulates are the most common air contaminants that can result in solid waste residues. The processing of solid waste residues is unaffected by other air contaminants. Because of the wide variety of substances released into receiving waters, determining crucial water contaminants capable of generating solid residues is more challenging.

Aside from power plants, non-ferrous smelting and processing, chemicals and allied products, paper and allied products, cement and dry products, steel, and hazardous wastes are all major industrial sectors that generate solid wastes from air and water pollution control. These industries, along with power plants, contributed roughly 90% of all solid wastes from air and water pollution control.

Large quantities of ash, as well as carbon dioxide and other gases, are emitted during coal combustion. Fly or flue ash is the fine particle ash that rises with the flue gases, while bottom ash is the heavier ash that does not rise; these two types of ash are referred to as coal ash. Also known as boiler slag, coal slag is a recycled product created by coal-burning plants. The slag is cooled by vitrification, which causes it to solidify and form sharp, angular granules of various sizes.

Mangli et al. [48] reported in their analysis that global coal ash output is expected to be about 600 million tons per year, with fly ash accounting for about 500 million tons [48]. Every day, a 500 MW thermal power plant emits around 500 metric tons of fly ash. By 2020, if fly ash is disposed of in ash ponds (usually in the form of slurry), the total land available for ash disposal would be about 82,200 ha, assuming a 0.6 ha per MW rate [49].

Despite recent developments in and increased use of renewable energy sources, coal-fired power plants still produce about 40% of the world's electricity. This figure is substantially higher in some nations, such as India, where it is about 70%, and South Africa, where it is over 90%. Large quantities of ash, as well as carbon dioxide and other gases, are produced during coal combustion. A fly or flue ash is a small particle of ash that rises with the flue gases, while bottom ash is harder ash that does not rise; these two types of ash are referred to collectively as coal ash [50].

APCr stands for air pollution control residues and is usually made up of ash, charcoal, and lime. It's hazardous waste that's actually being disposed of in a secured landfill or is being treated for non-hazardous disposals, such as washing or stabilization. According to government estimates from July 2015, about 300,000 tons of APCr are generated in the UK per year, though this figure is expected to rise dramatically in the year ahead as a major portion of domestic waste is handled in incinerator facilities. By 2020, this figure could reach 600,000 tons [51].

The average amount of sludge generated per capita per day was calculated to be 0.04 kg dry matter, corresponding to a 246 L per capita and day wastewater production rate [52]. This corresponds to a daily intake of 35 to 85 g dry solids per population equivalent (IWA, 2021; IWA, 2022) [53]. Between 2007 and 2013, China's total sludge generation rose by 13% annually, resulting in 6.25 million tons of dry solids produced in 2013. Sludge generation per capita in China is lesser than in developed countries [54].

Separation procedures, solidification/stabilization (S/S), and thermal methods are the three types of effective treatments for APC residues [55]. Fly ash was once freed into the atmosphere, but because of its potentially toxic impacts, it is now extracted from flue towers using electrostatic precipitators or other particle filtration equipment. After that, it can be thrown away or recycled into Portland cement.

Benefits of 'Bottom Ash' and 'Fly Ash' readily combine with calcium hydroxide to form required compounds in the cement manufacturing process, making it a less costly alternative to clay, sand, limestone, and gravel. Fly ash produces solid, long-lasting concrete that is chemically resistant. Bottom ash can also be utilized as a building material. As stated by the European Coal Combustion Products Association, bottom ash is used 46% of the time in manufacturing, while fly ash is used 43% of the time [50].

Electricity generated from sewage sludge digestion increased over the years in the United Kingdom, peaking at some 1.05 terawatt-hours in 2019. Between 2010 and 2019, figures increased by 352 gigawatt-hours [56]

1.5 Functional Elements of a Waste Management System

There is no overarching solution to waste management that can be extended to all waste sources; however, the implementation of a hierarchy ranking approach for solid waste management is the most adapted management strategies in many countries.

The management of this generated waste is a great challenge for the public and local authority. MSW management is an environmentally friendly method in managing waste, which consists of planning, administration, organization, generation, storage and collection, transportation, processing and recovery, and disposal methods that follow the waste management hierarchy (Fig. 1.14).

For example, open dumping and waste burning are common in the majority of ASEAN countries, according to UNEP [8]. Composting and anaerobic digestion of organic wastes, as well as the recovery of useful recyclables like paper, plastic, and metal, are common in ASEAN. The informal sector, on the other hand, is more responsible for recycling. Singapore, however, is an exception to the rest of ASEAN, as it has a sound and well-organized waste management system in place. Due to its limited land resources, Singapore chooses waste-to-energy (WTE) via incineration as its primary waste management choice.

1.5.1 Onsite Handling and Storage

Each country and area has its own approach to waste management. The unsafe handling of MSW can pollute water and soil, as well as have a significant effect on public health. In most cases, waste is deposited after it is produced at its source before being collected and transported to a disposal site.

The type of container, the location of the container, the effect on public health, and the waste disposal methods should all be taken into consideration when onsite storage of solid wastes. The capacities and types of containers used for on-site waste storage are determined by the characteristics of collected solid waste, the frequency of collection, and the container's usable capacity.

In several countries, there are two forms of waste storage: commingled waste storage and separated waste storage. Initial or mixed wastes are referred to as commingled waste. Commingled waste storage was the most common method, with 84% of households storing waste in this manner and the rest separating organic waste from other wastes. A collective bin has been set up for a group of houses or a particular neighbourhood. In general, the type of waste container chosen is determined by the area's primary operation. A waste bin or a wheelie is commonly used to collect waste directly from a residential home. Meanwhile, communal bins are given for low and medium-sized buildings, and residents may either pass their waste into the communal bins themselves or use the building maintenance personnel's services. In Malaysia, for example, spiral waste bins (SWB) are found in apartments and condominiums where large amounts of waste are produced. Due to its ability to compact waste, SWB allows for a higher storage volume [57]. This will result in a more hygienic and effective environment, but it will be costly to introduce. In Malaysia, these systems are used at Kuala Lumpur International Airport (KLIA), Complex, Kastam Kelana Jaya, and Kompleks Maju Junction [58]. Local governments or private companies typically have a rolled-on/rolled-off (RORO) bin, a 12 m^3 bin that can be rolled on and off the trailer in the industrial and wet market.

In Malaysia, three types of collection systems are used for waste segregated at source: kerbside collection using traditional and specially built vehicles, incidental kerbside collection by a voluntary agency, and residents transporting the separated waste to drop-off and recycling centres. As a result, people must use their own bins and separate their trash from the source. Containers of any kind, such as boxes and drums, may be used. However, recycling bins can be found in some suburban areas as well as many other public places to raise recycling awareness among the general public.

Some of the common bins being used are shown in Fig. 1.22.

1.5.2 Waste Collection

Waste collection is a step in the waste management process. For the preservation of public health, environmental quality, and safety, proper solid-waste collection is critical. It is the transportation of solid waste from its point of origin to a materials processing plant, transfer station, or landfill disposal site. As part of a municipal landfill reduction scheme, waste disposal also involves the kerbside collection of recyclable items that are legally not waste. The collection operation consumes roughly 50–70% of the total amount needed for solid waste management (collection, transport, processing, recycling, and disposal) [59]. In Malaysia, for example, solid waste management accounts for almost half of the local authority's (municipal) operating budget, with waste collection accounting for the other half.

Fig. 1.22 Some of the typical storage bins

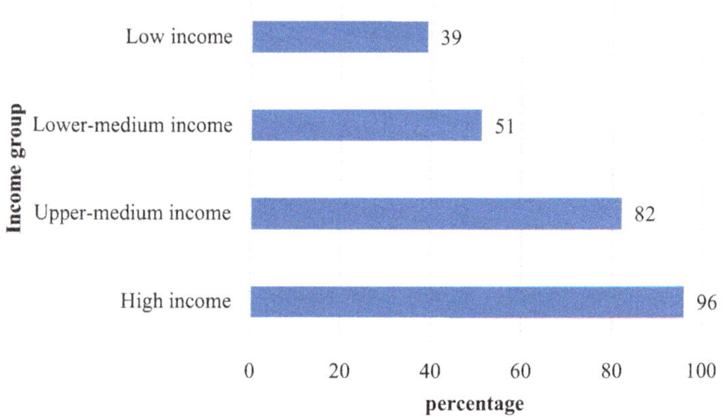

Fig. 1.23 Waste collection rates by income level (%) [8]

Waste collection is a significant part of waste management, but rates differ a lot depending on income levels, with upper middle- and high-income countries having an almost universal waste collection. In cities, low-income countries collect about

Fig. 1.24 Example of the house-to-house collection

48% of waste, but outside of cities, this percentage drops to just 26%. Sub-Saharan Africa collects about 44% of waste, while Europe and Central Asia, as well as North America, collect at least 90% of waste (Fig. 1.23) [8].

The collection system is influenced by the type and source of waste. For household waste, a door-to-door collection system is mostly adopted where household waste is removed from the home (Fig. 1.24). For areas not fully accessible to collection vehicles, a common bin is normally provided shared by many residents or streets. For a high rise building or flats, a communal bin is used, and the collection is normally done daily.

Commercial and non-hazardous waste collection is normally done primarily using dumpsters from a commercial bin. The recyclable material collection is adopted for the collection of recyclable materials separated at the source of generation.

The collection system also depends on various factors, which include but not limited to

- The served area
- Types and tonnage/volume of waste generated
- Presence or absence of waste recycling facility
- Types of waste treatment system – landfill, composting, anaerobic digester, incineration, etc.
- Economic constraints
- Types of the collection vehicle

Commercial and non-hazardous industrial waste is typically handled by a hauled container system or a stationary container system.

(i) A hauled container system is one in which a waste storage container is transported to a disposal site, drained, and then returned. Alternatively, the empty and loaded containers are usually swapped on-site by the truck.
(ii) A stationary container system is one in which waste storage containers remain at the point of generation after being emptied.

Waste disposal is a time-consuming task that accounts for almost three-quarters of the overall cost of solid waste management. While the job is often done by public employees, it is often more cost-effective for the municipality to contract out the collection services to private companies. Each collection vehicle is usually served by a driver and two to three collection staff. These are typically sealed, compacting trucks with capacities ranging from 10 to 30 cubic metres. The truck will come to a halt at each residence where the bin is stored (front or back of the street). This collection system's routing should be optimized to save time and fuel consumption. Compaction in the truck would be reduced to less than half of its total volume.

Choosing the best collection route is a difficult task, particularly in populated and dense cities. An optimal route is one that allows the most efficient use of labour and equipment, and selecting one necessitates computer calculations that account for all of the numerous design variables in a large and complicated network. Frequency of collection, haulage distance, service type, and environment are all variables. Collection of waste in rural areas can be particularly challenging due to low population densities and high unit costs.

Since food waste decomposes easily, refuse collection is usually performed at least once a week. In a hot climate, however, the collection is usually done three or four times per week. Commercial assets such as hypermarkets and wet markets have regular collections.

Many cities now have source separation and recycling systems, in which households and businesses separate recyclables from garbage and deposit them in separate bins for collection. Residents may also carry recyclables to drop-off centres in certain cities. The municipality usually assigns a dedication collection system to do the collection.

Fig. 1.25 Example of compactor truck

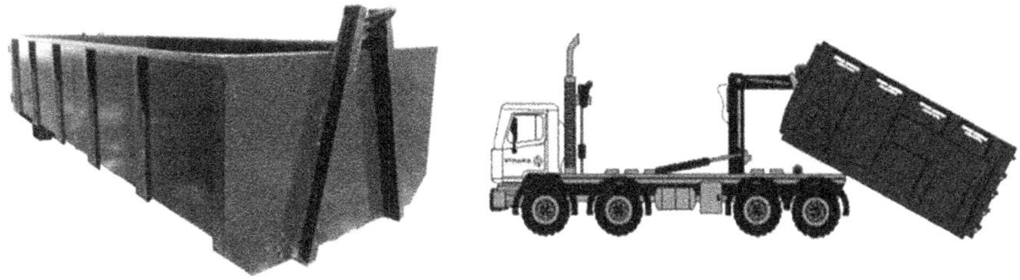

Fig. 1.26 Ro–ro bin

Fig. 1.27 Open truck

There are a few types of collection vehicles. The most common are as follows:

(i) Compactor truck (Fig. 1.25). It receives wastes from small (household) bins and medium (communal) bins.

(ii) Roll-on–roll-off (Ro–Ro) (Fig. 1.26). The trucks haul large bins (Ro–Ro bins) to disposal sites.

(iii) Open trucks (Fig. 1.27). Open trucks used to cart landscaping and grass, cutting wastes to landfill. A net is often fastened on top of the load to prevent waste from dropping. However, some private contractors still use open trucks to collect and transfer household wastes.

The waste collection could also be categories in terms of the primary and secondary system. These are detailed in Table 1.12 [60].

Table 1.12 Options for primary and secondary waste collection [60]

Vehicle	Comments
Primary	
Wheelbarrow	Recommended for waste collection from households located in narrow streets to a communal collection point. Required maintained street surface
Handcart	Stable for waste transfer in long distance, especially on the road with bad surfaces. It is recommended for the door-to-door waste collections in crowded areas
Cycle cart	Can move up to 3 m^3 of waste to a communal bin or a transfer station
Tractor	Higher costs than all other options; however, it recommended transferring a large volume of waste for long distances
Secondary	
Truck bin lifter	Suitable for collecting and transferring communal bins from residential and commercial areas
Enclosed light truck	Suitable for waste collection from narrower streets
Flatbed crane truck	It is recommended for waste collection from transfer stations, markets, and industrial areas
Compactor	Expensive method for waste collection and transfer. Not suitable for high-density wastes. It required high skills for maintenance. It is recommended for low-density waste with large volumes

1.5.3 Pneumatic Waste Conveyance System (PWCS)

The PWCS is an automated waste collection system that collects household waste via a vacuum-type underground pipe network and transports it to a sealed container through underground pipes. The waste is then collected on a daily basis by trucks for disposal. The storage of the entire waste is automated, which decreases manpower requirements while increasing efficiency. The PWCS (Fig. 1.28) decreases the environmental and sanitary problems that open refuse collection methods cause [61].

The PWCS scheme has many advantages. Some of them are as follows:

- The entire refuse collection process is automated.
- Manpower is limited, as is the need for manual labour.
- Improved working standards.
- Removes noxious odours produced by refuse chutes.
- Reduces spills during garbage collection, resulting in a more sanitary and safer climate.
- Pleasant to the atmosphere.
- Reduces the need for chute cleaning.
- Pest infestation is reduced.
- Waste decomposition is minimized.
- Encourages waste separation for recycling at the collection point.

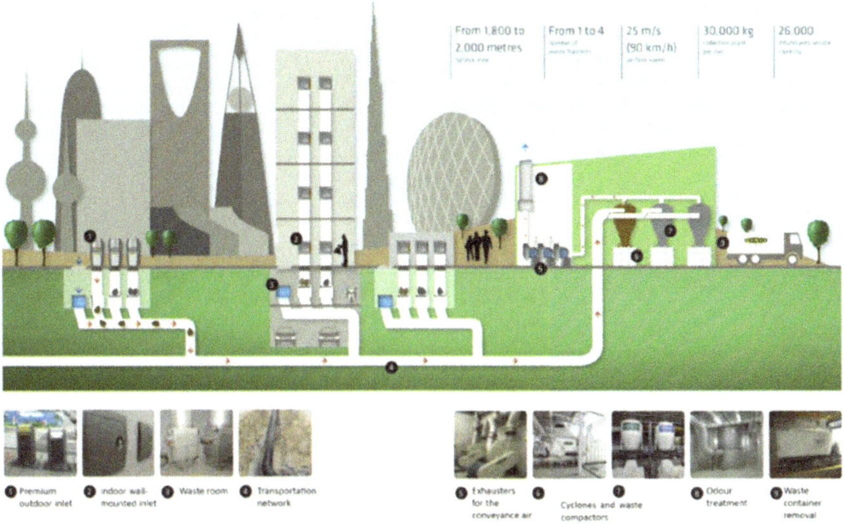

Fig. 1.28 Basic layout of a waste pneumatic collection system [62]. Reprinted from Chafer, M., Sole-Mauri, F., Sole, A., Boer, D., Cabeza, L.F. Life cycle assessment (LCA) of a pneumatic municipal waste collection system compared to traditional truck collection. Sensitivity study of the influence of the energy source, *Journal of Cleaner Production*, 231 (2019), 1122–1135, with permission from Elsevier

1.5.4 Transfer Station

If the waste's final destination is not close to where it was created, at least one transfer stations may be needed. A waste transfer is a method of decreasing the cost of managing recyclable waste while increasing the volume of recyclable waste. A transfer station is a central location where refuse from multiple collection vehicles is consolidated into a bigger vehicle, like a tractor trailer. The waste is then transported to a recycling or disposal plant, usually over long distances. Open-top trailers normally can transport up to 76 cubic metres (100 cubic yards) of non-compacted waste to a centralized processing or disposal facility [63]. Enclosed compactor trailers with ejector mechanisms are sometimes used. Multiple collection trucks dump directly into the transport vehicle at a direct discharge station. In a storage discharge station, trash is first drained into a storage pit or onto a platform, and then the solid waste is hoisted or pushed into the transport vehicle using machinery. Large transfer stations normally have the capacity to handle over 500 tons of garbage every day. If the waste's final destination is not close to where it was created, at least one transfer stations may be needed. A waste transfer is a method of decreasing the cost of managing recyclable waste while increasing the volume of recyclable waste.

There are a few options in the transferring activities of waste. The common transfer activities that take place at transfer stations are reloading (transferring from smaller truck to larger truck), compaction, or separation of the waste (Fig. 1.29) [63].

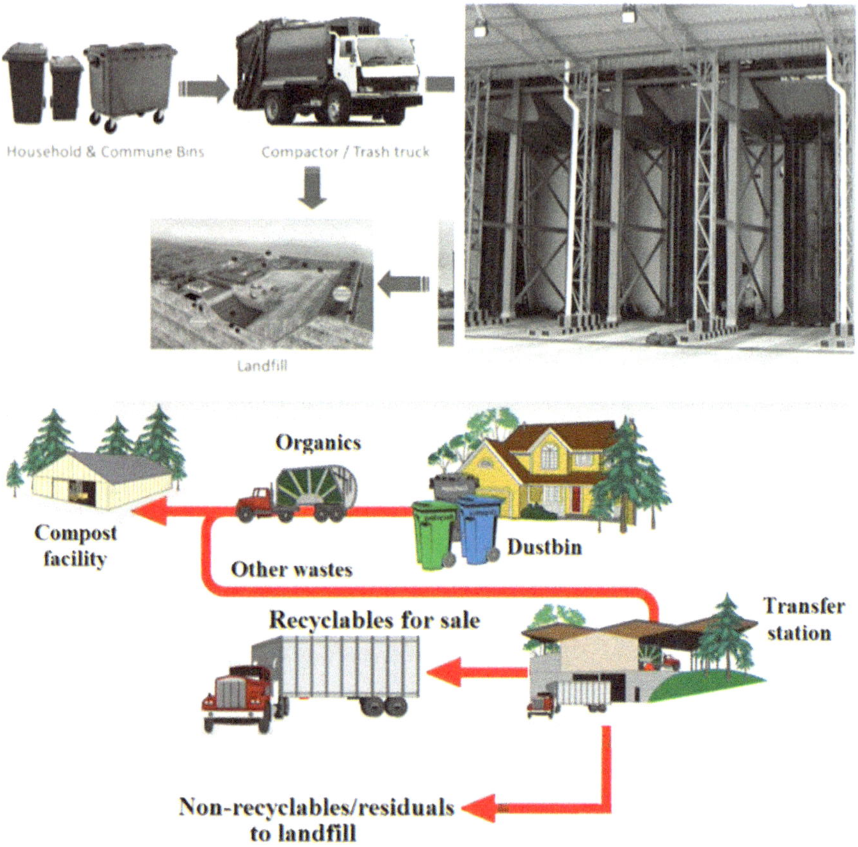

a) Example of a transfer station in Malaysia (top)

b) Another example (bottom)

Fig. 1.29 A typical transfer station [63]. (**a**) Example of a transfer station in Malaysia (top). (**b**) Another example (bottom)

1.5.5 Waste Processing and Recovery

Solid waste management employs a variety of technologies. Recycling is the method of extracting the economic value of products and resources from waste that would otherwise be discarded. It basically refers to the gathering and processing of recycled materials into new types that can be used as raw materials for new goods. Although the word is most often associated with municipal waste, it may also refer to industrial or other types of waste. All of the processes, tools, technology, and facilities used to increase the performance of the other functional components, as well as the reuse of recycled materials and energy conversion from solid waste, are included in the functional aspect of recycling.

In order to recover recyclable materials from MSW, there are three specific methods that can be used:

- Separation at the source by the owner of the building, company, or organization. This is the most straightforward and efficient approach. Additional processing may or may not be needed.
- Collecting mixed recyclables and processing them at consolidated materials recovery facilities (MRFs).
- Mixed MSW collection with processing at mixed-waste processing or front-end processing facilities to extract recyclable products from the waste stream.

Recycling has two main advantages: It conserves natural resources and landfill space, thereby extending the life of landfills. As the process of recycling involves the collection and transport of materials requiring substantial amounts of energy and labour; it, therefore, creates more job opportunities. Recycling discourages the disposal of material that can either be reused or recycled into something useful, and it helps to provide a source of raw material that has monetary value. For example, the sale of aluminium, newspapers, cardboard, glass, plastic, and other recycled materials can potentially reduce a community's waste management costs. Inevitably, recycling practice, in a way, contribute to the conservation of natural resources. Recycling 1 ton of paper, for example, saves 17 mature trees, 7000 gallons of water, 3 cubic yards of landfill space, and 2 barrels of oil [58]. A proper materials recovery system will reduce the waste management operating cost by means of recyclables sell, energy-saving, and lower volume of waste disposal.

The rate of recycling varies widely across the world, depending on national waste management legislation. With a rate of 59%, Singapore and South Korea had one of the highest rates of urban solid waste recycling. Figures 1.30 [64] and 1.31 [3] depict MSW solid waste recycling rates worldwide in 2017 by region, as well as the number of materials recycled in the United States from 1960 to 2018. The amount of MSW recovery and recycled has risen since then, reaching 69 million tons in 2018 [3].

The bulk of waste processing and recycling takes place at a materials recovery facility (MRF) (Fig. 1.32) [3]. Figure 1.33 shows a statistic for the sum of materials collected from urban waste sources in the United States in 2018, based on material. Rubber and leather products were recovered from urban waste in the amount of 1.67 million metric tons this year [3].

It is a dedicated plant that accepts commingled products and separates and densifies them for sale to end-user manufacturers using a combination of machinery and/or manual labour. Mechanically, using variations in physical characteristics of the waste such as height, density, and magnetic properties, these are carried out. The size of the waste articles is decreased by shredding or pulverizing, resulting in a uniform mass of material. Hammer mills and rotary shredders are used to do this.

Various MSW constituents can be separated and recovered via recycling. The most common ones are paper, plastics (PET and HDPE), glass, aluminium, ferrous

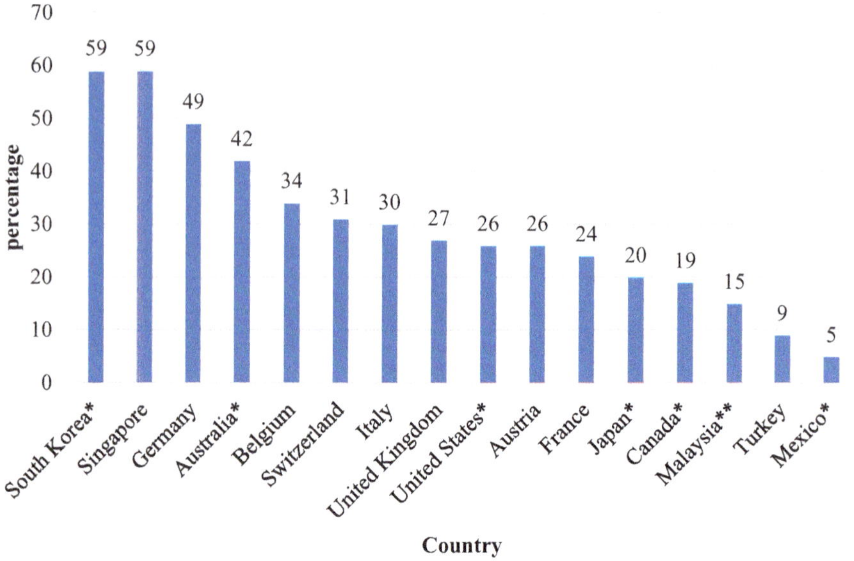

Fig. 1.30 Recycling rates of municipal solid waste in 2017 by country (**Estimated) [58]

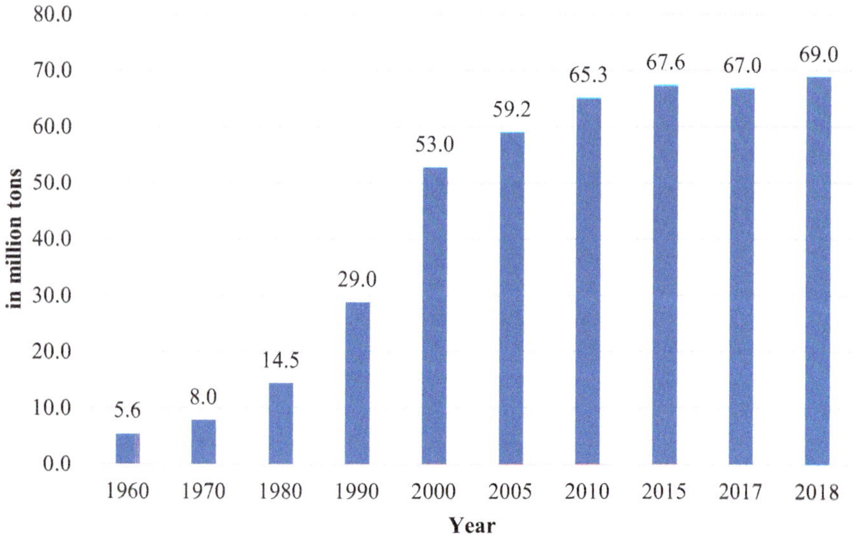

Fig. 1.31 The amount of materials recycled in municipal solid waste in the United States between 1960 and 2018 [3]

Fig. 1.32 An example of a typical MRF facility. (**a**) Commingle waste. (**b**) Manual or automatic sorting process. (**c**) Pulverised and conveyed to next process. (**d**) A baled product

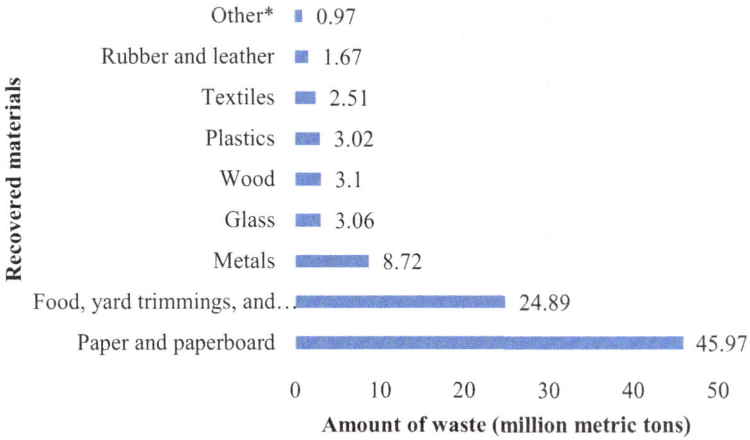

Fig. 1.33 The number of materials recovered from industrial waste in the United States in 2018, based on content [3]

metals, and non-ferrous metals. There are two categories of MRFs: clean and dirty. Clean MRFs manage the contents of commingled recycling bins that are mainly recyclable, while dirty MRFs handle solid waste that contains some salvageable recyclable materials (Table 1.13).

1.5.6 Composting

One of the solid waste treatment techniques is a composting process which has been practised since ancient times. The composting process converts the waste into organics that can be used for enhancing soil quality. In the conversion process, the normal microbes found in soil has been considered in addition to optimizing their carbon-cycling activities. In composting systems, the physical factors in the soil system, including temperature, moisture, and bulk density, are controlled internally.

There are several types of composting, as detailed in USEPA [66], as follows.

1.5.6.1 Basic Composting

Feedstock and nutrient balance, particle size, moisture content, oxygen flow, and temperature are the five key areas that must be 'regulated' during composting. It is important to maintain a proper balance of organic materials and oxygen flow. Grass clippings, food waste, and manure are examples of organic material that contain a lot of nitrogen. Dry leaves, wood chips, and branches are examples of 'brown' organic materials, which contain a lot of carbon but little nitrogen. Experimentation and persistence are needed to find the right nutrient mix. It's all part of the composting art and science. Smaller particles also help maintain optimum temperatures by producing a more homogeneous compost mixture and improving pile insulation. However, if the particles are too small, they can prevent air from freely flowing through the pile. To live, microorganisms in a compost pile need an adequate amount of moisture. Water is a crucial component in the compost pile because it aids in the transport of substances and makes nutrients in organic matter available to microbes. Moisture can be found in organic material in different quantities, but it can also be found in the form of rainfall or deliberate watering. Aerating the pile allows for quicker decomposition than in anaerobic environments. However, be careful not to provide too much oxygen, as this will dry out the pile and slow the composting process. Microorganisms need a specific temperature range to work properly. Temperatures that encourage rapid composting and kill pathogens and weed seeds are ideal. The temperature of the pile's centre will rise to at least 140 °F due to microbial activity. Anaerobic conditions (i.e., rotting) evolve if the temperature does not rise. The proper temperature can be achieved by regulating the previous four variables.

Table 1.13 The available technologies used for solid waste processing, treatment, and disposal [65]

Types of waste and technology available	The processing	Key issues
Paper	Paper recycling	High capital investment is required
	Waste to energy	No auxiliary fuel needed
		High capital investment is required
Plastic	Incineration	High calorific value, no auxiliary fuel needed
		Efficiency is high
		High capital cost
	Recycling	Expand landfill life span
		Needs to identify buyers
Construction waste	Reuse and recycling	Substitute for new products
		Non-recyclables/residuals sent to landfill
Organic and garden	Composting	Improves nutrient quality, thus destroying pathogens and acting as a soil conditioner
		Time-consuming and requiring a substantial amount of land
	Anaerobic digestion and methanation	Generates anaerobic/gaseous fuel
		Reduction in greenhouse gas emission
		Capital-intensive method
		Less effective for lower biodegradable
Inorganic	Sanitary landfills and landfill gas recovery	Cheaper if the land is available
		Potential for energy recovery of landfill gas
		May cause air and water pollution if not designed and maintained properly
		Land requirement is high
	Refuse-derived fuel (RDF) production	A burner made of RDF pellets
		Trained staff are needed
		A large initial capital expenditure is required
Chemical/hazardous	Recycling	Recycled into new products
		Skilled person required
	Incineration, waste to energy	Reduced air pollutants in modern design
		Involves high capital investment
		Requires good air pollution control systems
	Hazardous waste landfill	Secured landfill with extra pollution control
		High capital investment
		Expert requirement
Medical/hospital	Off-site	Proper handling and expert requirements
		Mainly incinerated

(continued)

Table 1.13 (continued)

Types of waste and technology available	The processing	Key issues
E-waste	Recycling	E-waste recycling centres must be devoted
		Electronic devices contain potentially dangerous materials, such as poisonous chemicals
Metal	Recycling	Metal recycling saves landfill space and requires a large initial investment
		Ample space is needed

1.5.6.2 On-Site Composting

Limited quantities of leftover food may be composted. Animal products and significant amounts of food scraps should not be composted on site. Changes in the weather and seasons will have little effect on on-site composting. As things change, such as when the rainy season begins, small adjustments can be made. Food scraps must be handled carefully to prevent odours or to attract unwanted insects or animals. Composting, in this manner, requires very little time and equipment.

1.5.6.3 Vermicomposting

This process employs worms to break down food waste into high-quality compost known as castings. Worm bins are simple to build and can also be purchased. A pound of mature worms (roughly 800–1000 worms) can consume up to half a pound of organic matter per day. The bins can be modified to suit the number of food scraps that will be used to produce castings. The time it takes to make functional castings is normally 3 to 4 months. You may use the castings as potting soil. Worm tea, another by-product of vermicomposting, is a high-quality liquid fertilizer for houseplants and gardens. Temperatures of 55 °F to 77 °F are ideal for vermicomposting. The bin should be put in the shade in humid, arid areas. Many of these issues can be avoided by vermicomposting indoors.

1.5.6.4 Aerated (Turned) Windrow Composting

This form is best suited to large volumes, such as those produced by entire communities and collected by local governments, as well as high-volume food-processing operations. It will create a significant amount of compost, which could be sold as a finished product. This method of composting entails sorting organic waste into rows of long piles known as 'windrows' and aerating them on a regular basis by turning the piles manually or mechanically. The ideal pile height is 4 to 8 ft tall, with a width of 14 to 16 ft. This size pile will produce enough heat and keep

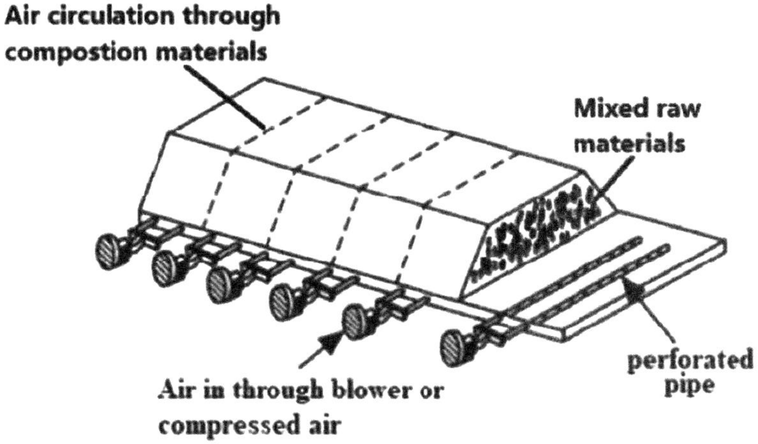

Fig. 1.34 Examples of an aerated static pile composting [67]

temperatures steady. It's small enough to allow oxygen to flow to the windrow. Windrows are often covered or placed under a shelter in a humid, arid environment to prevent water from evaporating. During rainy seasons, the pile's shapes can be changed such that water flows off the surface instead of being drained by the pile. This form can generate odours, which will need to be handled.

1.5.6.5 Aerated Static Pile Composting

This method of composting (Fig. 1.34) creates compost in a short time (between 1 and 6 months). It works well for larger volume generators of yard trimmings and compostable urban solid waste (e.g., food scraps, paper products), like local governments, landscapers, or farms, and is ideal for a relatively homogeneous mix of organic waste [67]. However, this approach does not work well for composting animal waste or grease from food processing industries; in this case, organic waste is mixed in a large pile. Layers of loosely stacked bulking agents (e.g., wood chips, shredded newspaper) are added to aerate the mound, allowing air to flow from the bottom to the top. The piles may also be placed over a network of pipes that supply or extract air from the pile. A timer or temperature sensors can be used to trigger air blowers. To avoid water from evaporating in a hot, arid environment, the pile may need to be covered or placed under a shelter. The pile will keep its warm temperature in the cold. Since passive airflow rather than active turning is used, aeration can be more difficult. It's also possible to bring the aerated static piles indoors with adequate ventilation. To buy, mount, and maintain equipment such as blowers, pipes, sensors, and fans, this method can necessitate a substantial financial investment as well as technical assistance.

1.5.6.6 In-Vessel Composting

This system can treat large quantities of waste with a smaller footprint as the windrow method, and it can handle almost any form of organic waste (e.g., meat, animal manure, biosolids, food scraps). Organic materials are fed into a drum, silo, concrete-lined trench, or other similar device using this process. This allows for precise monitoring of environmental factors, including temperature, humidity, and airflow. To ensure that the material is aerated, it is mechanically turned or blended. The size and capability of the vessel may differ. This process yields compost in a matter of weeks. Since the microbial activity must be balanced and the pile must cool, it will take a few more weeks or months until it is ready to use. This form produces no odour or leachate and can be used in extremely cold weather. However, this approach is costly, and proper operation can necessitate technical expertise.

1.5.7 Thermal Treatment Methods

Incineration is the heat treatment of solid wastes by controlled and total combustion. Solid waste incineration is a viable option in densely populated areas where landfills are unavailable. It results in energy recovery and hazardous waste degradation at high temperatures (between 980 °C and 2000 °C). The ability to reduce the original amount of combustible solid waste by 80–95% is one of the most appealing features of the incineration process. It also eliminates pathogenic bacteria. However, this process produced a high amount of air emissions which contains some invisible hazardous air pollutants that have a high potential health risk if not properly designed and operated. Hence, incinerators should be operated with care and in a proper way to minimize possible pollution.

An incineration is a form of producing steam via pyrolysis, gasification, and plasma and gasification. Several technologies have been established to make the processing of MSW with energy recovery cleaner and more cost-effective than ever before. Although older waste incineration plants generated a substantial amount of contaminants, recent regulatory changes and new technology have greatly reduced this issue. New and recent incinerators have been well built to minimize dioxin emissions from waste-to-energy plants. With the installation of sophisticated scrubbing and cleaning system, the waste-to-energy unit is also now ‘clean’ with less environmental impact

Incineration is costly, but it offers high energy returns while still being low on environmental impact and energy required to process MSW when properly built. It also necessitates a small footprint.

There are several types of incinerators, such as:

- Rotary kiln
- Fluidized bed
- Liquid injection

Table 1.14 Waste-to-energy facility in the US in 2018 [68]

Total capacity (by energy)					
Daily throughput		Gross electric capacity		Equivalent combined heat and power (CHP) capacity	
94,243 tons/day		2,534 MW		2,725 MW	
Number of facilities					
No. of operating facilities in the US		Ownership		Operation	
Operating facilities	75	Private	41	Private	65
States with waste to energy (WTE)	21	Public	34	Public	10
No. of facilities (by technology)			*No. of facilities (by offtake)*		
Mass burn	58		Electricity generation	58	
Refuse-derived fuel (RDF)	13		Steam export	3	
Modular	4		Combined heat and power (CHP)	14	

- Multiple hearth
- Catalytic combustion
- Waste-gas flare
- Direct flame

The first three types, namely, rotary kiln, fluidized bed, and liquid injection, are typically used in industry because of their large size use and their versatility. Furthermore, all three forms of incinerators can be run in pyrolysis or with very little oxygen.

Table 1.14 displays the number of waste-to-energy facilities in the United States' leading states in 2018. In Minnesota, there were eight waste-to-energy plants this year. The majority of facilities combust municipal solid waste without pre-processing using mass-burn technology [68].

Waste-to-energy plants in the United Kingdom are responsible for generating an estimated 7.77 terawatt-hours of electricity. Between 2015 and 2019, the estimated gross electricity generation of energy-from-waste (EfW) of waste-to-energy (WTE) plants increased by some 2.3 terawatt-hours (Fig. 1.35). There were 48 operational EfW power plants in the country as of 2019 [69]. A typical WTE plant is shown in Fig. 1.36.

1.5.8 Final Disposal by Landfilling

There are several disposal options for solid waste based on their composition and sources. These include the following:

- Direct burning of solid waste and dumping in the sea is not advisable.
- Dumping on land in landfills or dumpsites.

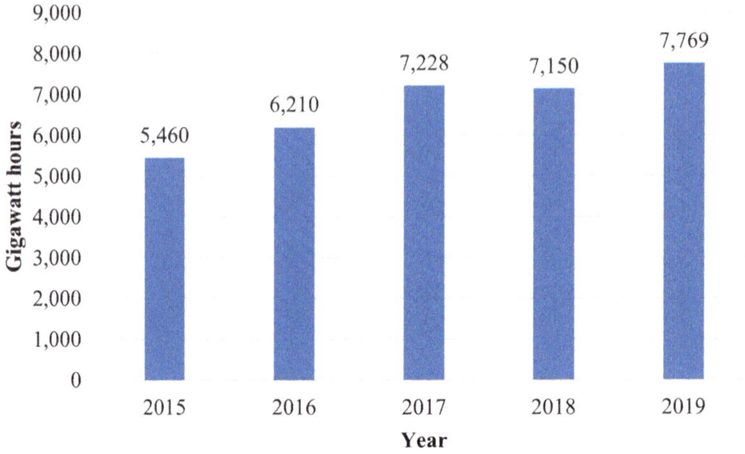

Fig. 1.35 Estimated gross electricity generation of energy-from-waste (EfW) incinerators in the United Kingdom (UK) from 2015 to 2019 (in gigawatt hours) [69]

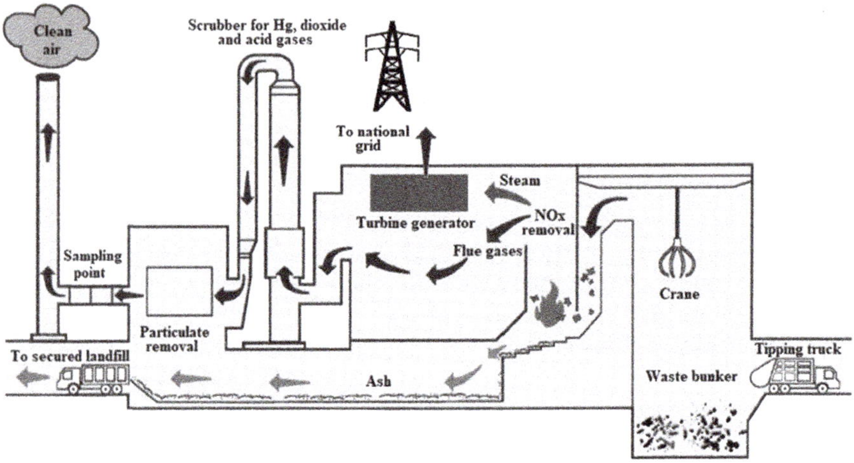

Fig. 1.36 A typical waste-to-energy incinerator

- Heat treatment – incineration, pyrolysis, gasification.
- Composting and reuse process in agricultural activities.
- Biological fermentation and digestion.

1.5.8.1 Introduction

Sanitary landfill is the most typical options for final waste disposal, especially in developing countries. It is the most basic, inexpensive, and cost-effective form of waste disposal. Many countries still consider landfills to be the oldest and most realistic method for disposing of solid waste. Many landfill sites are rapidly filling up due to the increase in waste volume produced daily, and many do not have a sufficient 3R facility. Due to land scarcity and increasing land prices, it is becoming more difficult to acquire new land for landfilling purposes, especially in urban areas.

Modern municipal waste disposal sites are well engineered with facilities to operate waste disposal. Nonetheless, this technique can be correlated to serious environmental problems if the disposal site is not adequately managed. The most popular form of land dumping is the mass dumping of waste into a designated area, normally a hole or a sidehill. Big machines compact the waste after it has been dumped. When the dumping cell is completed, a plastic sheet or soil is used to 'seal' it. A landfill allows solid waste to decompose before converting into a relatively inert and stable material. In fact, landfilling is an essential stage in waste management practices. However, the recycling process also generates non-recyclable products and residuals that require final disposal in a landfill. However, proper landfill design and monitoring after their closure are important issues for better and safe disposal and management of solid waste. This includes landfill gas control systems and leachate collection and treatment systems.

1.5.8.2 Landfills in the World

According to Worldatlas [70], nearly half of the world's population lacks access to basic waste collection and disposal facilities. More than 70% of MSW in the world is disposed of in landfills, with the majority of waste produced in low- and middle-income developing countries going to landfills. However, a substantial portion of the world's waste is still illegally disposed of on open dumpsites, which are particularly prevalent in low-income countries. These unregulated landfills are often located near cities. Pollution from open burning and pollution of groundwater are also common complaints from unregulated sites.

As of 2019, this figure gives a ranking of some of the world's largest dumpsites (Fig. 1.37). The Apex Regional Landfill in Las Vegas, Nevada, covered around 2200 acres of land this year. As the largest landfill in the United States, it is expected to last 250 years and contains close to 50 million tons of waste [70].

Figure 1.37 shows the world biggest landfill in 2019. The waste generation and number of a landfill in EU and other disposals/treatment method are given in Table 1.15 [71]. Table 1.16 presents the waste treatment methods used in ASEAN.

The majority of waste is actually discarded or disposed of in landfills around the world (Fig. 1.38) [70]. About 37% of waste is disposed of in a landfill, with 8% of that going to sanitary landfills equipped with gas collection systems. Around 31% of waste is discarded publicly, while 19% is recovered by recycling and composting,

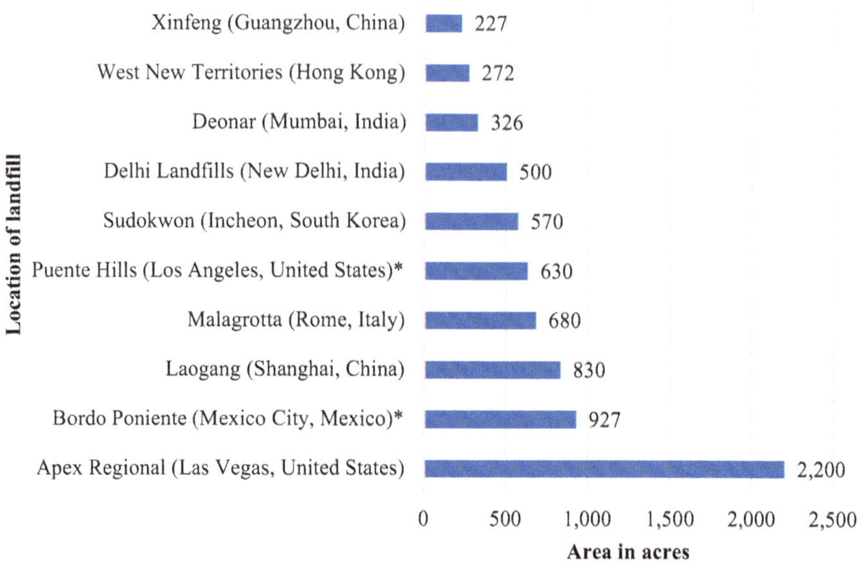

Fig. 1.37 Size of largest landfills globally as of 2019 [70]

and 11% is incinerated for final disposal. High- and upper-middle-income countries are almost entirely responsible for appropriate waste management or treatment, such as managed landfills or more stringently run facilities. Open dumping is widespread in low-income countries; 93% of waste is dumped in low-income countries, while only 2% is dumped in high-income countries. The Middle East and North Africa, Sub-Saharan Africa, and South Asia are the three regions that publicly spill more than half of their waste. The highest percentage of waste in landfills (54%) is found in upper-middle-income countries. In high-income nations, this figure drops to 39%, with 36% of waste diverted to recycling and composting and 22% to incineration. Incineration is mainly used in countries with high energy, high income, and limited property.

1.5.8.3 Categories of Landfill

Landfill sites can basically be divided into five categories, as being practised in Malaysia and in many countries. Its functional details are explained in Table 1.17. There are anaerobic landfills, anaerobic sanitary landfills, improved anaerobic landfills, semi-aerobic landfills, and aerobic landfills.

Table 1.15 Waste generation and treatment facilities (kg/capita) [71]

Country	Year	Source	Population (000s)	Waste Generated	Landfill	Incineration	Other recovery	Material recycling	Composting and digestion	Recycling rate
1. Germany	2015	OECD	81,202	628.6	59.3	196.7		300.7	114.6	66.10%
2. Wales	2016/17	Welsh Government	3100	512.8	48.7	125.8	0.3	327.2		63.80%
3. Singapore	2016	Singapore Government	5607	1394.6	543.1			850.5		61.00%
4. South Korea	2014	OECD	50,424	361.3	60.9	91.5		209.9	3.3	59.00%
5. Taiwan, ROC	2016	Taiwan EPA	23,492	317.6	3.9	127.4	2.6	159.7	24.5	58.00%
6. Netherlands	2016	Netherlands Government	16,981	560.6	12.5	209.7		171.5	145.7	56.60%
7. Austria	2015	OECD	8538	566.4	16.8	214.6		145.3	177	55.90%
8. Slovenia	2015	OECD	2067	448.1	101.7	76.5	14.3	208.1	34.3	53.90%
9. Belgium	2015	OECD	11,369	414.1	78.4	179.7		142.1	79.2	53.50%
10. Switzerland	2015	OECD	8129	741.8		350.6		236.7	154.5	52.70%
11. Italy	2016	Italy EPA (ISPRA)	60,656	496.2	122.5	96.7		153.4	107.4	52.60%
12. Luxembourg	2015	OECD	563	632.7	111.5	215.7		180.4	125.1	48.30%
13. Sweden	2015	OECD	9799	446.6	3.6	228.7		144.6	69.8	48.10%
14. Denmark	2015	OECD	5611	799.3	9.1	420.4		217.9	151.8	46.30%
15. Scotland	2015	SEPA	5400	457.2	213.1	41.9		202.1		44.20%
16. United Kingdom	2015	UK Government – DEFRA	64,532	489.2	115	153.5		133.3	79.4	43.50%
17. Norway	2015	OECD	4904	446	15.1	233.5	6.3	116.5	74.4	42.80%

(continued)

Table 1.15 (continued)

Country	Year	Source	Population (000s)	Waste Generated	Landfill	Incineration	Other recovery	Material recycling	Composting and digestion	Recycling rate
18. England	2016/17	UK Government – DEFRA	55,268	476.2	74.8	184	13.8	127.1	76.5	42.80%
19. Poland	2015	OECD	38,016	285.7	129.7	37.9		75.4	46	42.30%
20. Northern Ireland	2015/16	NI Government – DAERA	1900	510.1	205.4	90	1.4	122.3	91.1	41.80%
21. Australia	2015	OECD	23,941	557.2	260.6	64.9		231.7		41.60%
22. Finland	2015	OECD	5493	498.5	57.3	238.9		140.2	62.1	40.60%
23. France	2015	OECD	66,498	502.3	134.8	174.4		111.8	86.7	39.60%
24. Hong Kong, China	2014	Hong Kong Government	7240	776.2	492.7			283.6		36.50%
25. United States	2014	OECD	318,857	735.3	386.7	94.3		188.9	65.5	34.60%

Note: *OECD* Organisation for Economic Co-operation and Development, *EPA* Environmental Protection Agency, *ISPRA* Istituto Superiore per la Protezione e la Ricerca Ambientale (Italian National Institute for Environmental Protection and Research), *SEPA* Scottish Environment Protection Agency, *DEFRA* Department for Environment, Food and Rural Affairs, *DAERA* Department of Agriculture, Environment and Rural Affairs

Table 1.16 MSW management and disposal in ASEAN [9]

No	Country	Segregation at source (%)	Collection rate (urban) (%)	Recycling rate (%)	Management/treatment/disposal method				
					Composting	Incineration	Sanitary landfill	Open dumps	Open burning
1	Brunei Darussalam	<50	90	15			√	√	
2	Cambodia	<50	80	<50	√		√	√	√
3	Indonesia	<50	56–75	<50	√	√	√	√	√
4	Lao PDR	<50	40–70	<50					
5	Malaysia	<50	>70	50–60 (metal, paper, plastic)	√		√	√	√
				<50 (others)					
6	Myanmar	<50		70 (plastic, paper, metal)		√	√	√	
7	Philippines	50–70	40–90	20–23 (paper)	√		√	√	
				30–70 (Al)					
				20–58 (other metals)					
				23–42 (plastic)					
				28–60 (glass)					
8	Singapore	70	>90	50–60 (paper, horticulture)		√	√	√	
				>90 (Fe, CanD, used slag)					
				>80 (scrap tyre)					
				>80 (wood)					
				>50 (others)					
				60 (overall)					

(continued)

Table 1.16 (continued)

No	Country	Segregation at source (%)	Collection rate (urban) (%)	Recycling rate (%)	Management/treatment/disposal method				
					Composting	Incineration	Sanitary landfill	Open dumps	Open burning
9	Thailand	<50	>80	>90 (metal)	√	√	√	√	
				50–60 (paper, construction)					
				<50 (others)					
10	Vietnam	<50	80–82	>90 (metal)	√			√	
				>70 (plastic, e-waste)					
				50 (paper)					
				<50 (others)					

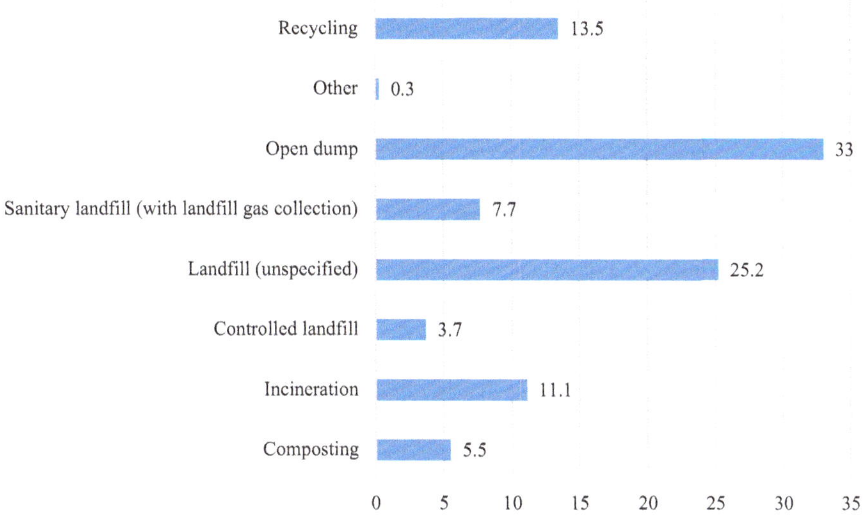

Fig. 1.38 Global treatment and disposal of waste (%) [8]

1.5.8.4 Landfill Leachate and Its Treatment

The accumulated municipal solid waste in landfill sites is exposed to several physicochemical processes, which result in the generation of highly polluted, dark-black liquid with a bad scent known as the leachate. Leachate is the fluid that disintegrates solid waste into leachate originates from the natural moisture and water in organic matter residue (product of the biologically decayed organic matter), as well as rainwater which percolates into the internal layers of the landfill and further enhances the solubility of suspended materials (Fig. 1.39).

The generation of leachate in landfill sites is also highly dependent on several main parameters such as the composition of MSW, site topography, area hydrogeological condition, age of solid waste, climate variations, humidity, and landfill site operation. Hence, it is important to protect the sustainability of the ecosystem by exploring appropriate environmentally friendly and effective treatment systems that could treat the leachate to such a degree that it is safe to be discharged into surface water resources (Fig. 1.40). A typical characteristic and classification of landfill leachate are shown in Table 1.18 [72–75].

The benefits and drawbacks of each physical and chemical treatment process are clearly displayed in Table 1.19.

Biological treatment is also only effective on young and intermediate leachates that have a high content of biodegradable organic matter ($BOD_5/COD > 0.5$). Physical–chemical approaches are more fitted to treat strong contaminant landfill leachate with a low biodegradability index of less than 0.1 and a high concentration of ammoniacal nitrogen. Both physical and chemical treatments were proved to be highly effective in treating old or stabilized leachate, which is hard to degrade [74].

Table 1.17 Types and characteristics of conventional landfills

Types	Characteristics	Illustrations
Anaerobic landfill	Solid waste is dumped into a dug area or a valley, and water is merged to allow for the anaerobic process to take place. This basic landfill has caused many serious environmental and human health problems by producing hazardous leachate	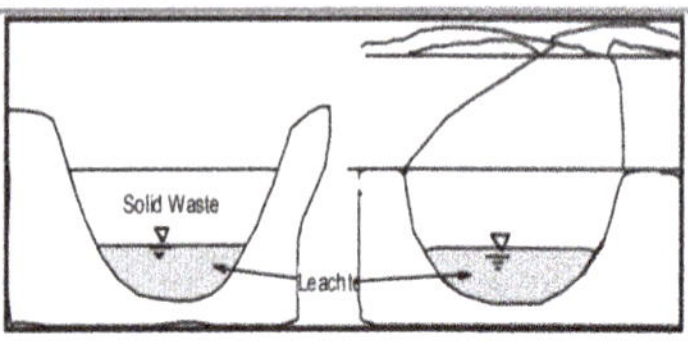
Anaerobic sanitary landfill	This method layers solid waste with soil (sandwich form). Other characteristics are similar to those of anaerobic landfills	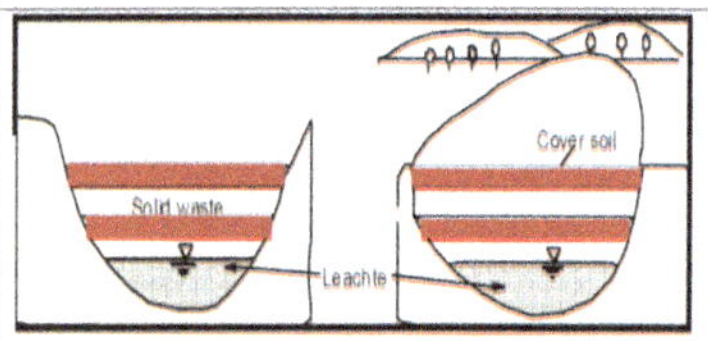
Improved anaerobic landfill	Improvement was made to this design by adding a leachate collection system at the bottom of the pond. Other characteristics are similar to that of anaerobic landfills, except for the moisture content that is notably low	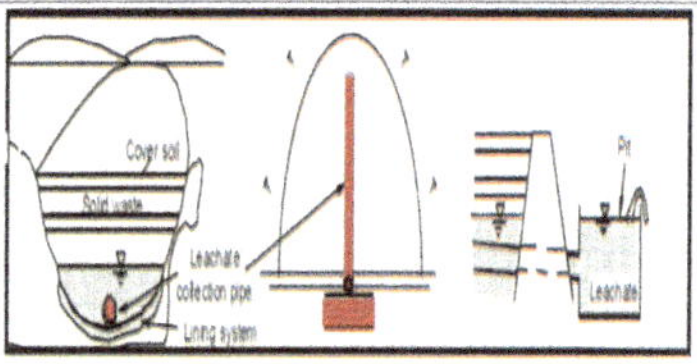
Semi-aerobic landfill	In this model (Fukuoka method), O_2 is supplied spontaneously through the collection pipe to stabilize the solid waste. Therefore, the collection pipe is designed to be bigger than the previous model so that it can function to collect leachate and provide O_2. The aerobic process occurs here and increases the decomposition rate of solid waste	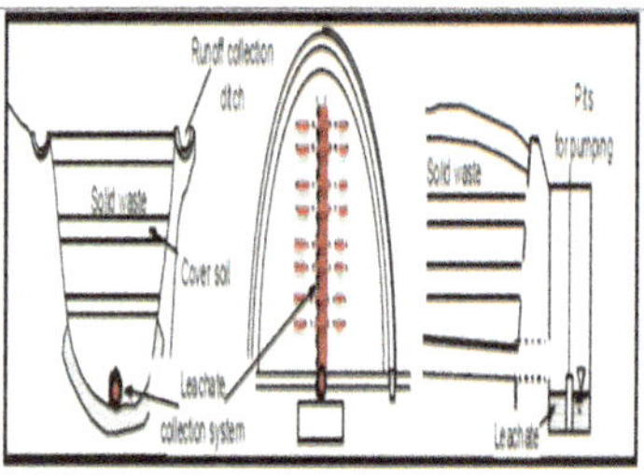
Aerobic landfill	This method is designed to enhance the aerobic process of landfill systems since semi-aerobic landfills have performed well in terms of biodegradation and stabilization of landfills. Air and recirculation leachate systems are also installed in order to increase and maintain the humidity as well as to supply nutrients for the microorganisms present in the water sample	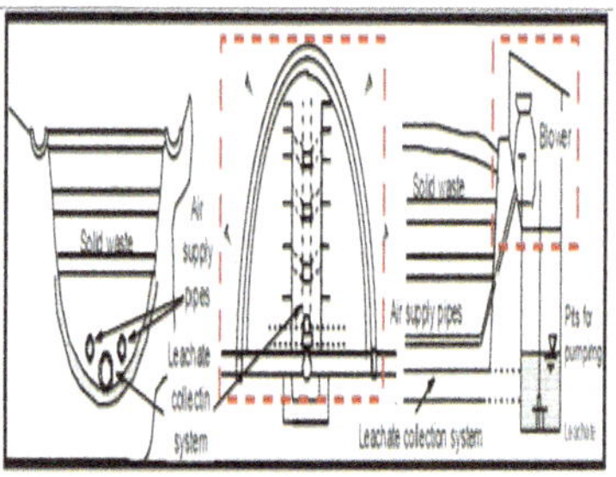

1.5.8.5 Design, Operation, and Challenges for Landfilling

Landfill siting: Site selection for landfill construction is one of the main challenges facing the local authority for managing the waste due to the difficulty in identifying a suitable location. One of the greatest challenges is the public concern who normally would object to sitting the landfill near their neighbourhood.

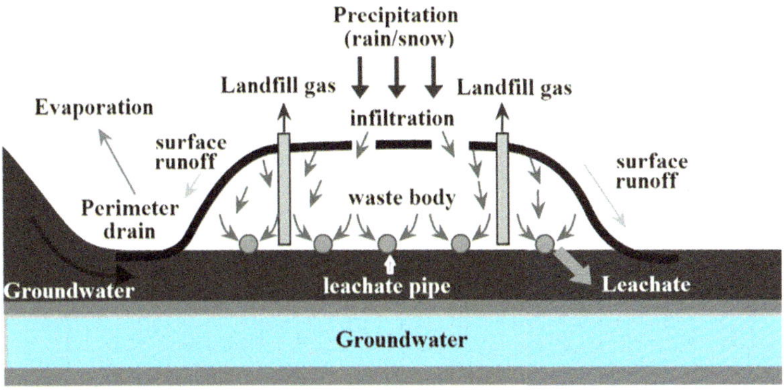

Fig. 1.39 Leachate formation in landfill

Landfill design: The landfill should be properly built to account for all various features, such as a suitable baselining system, leachate collection system, daily soil covering and final top liner, surface run-off collection and discharge system, gas emission system, and appropriate access facilities to the landfill, according to engineering principles and environmental codes of practice. The materials used in landfill construction and lining systems should have no negative impact on the environment, particularly groundwater. The landfill baseliner is made up of two upper and lower liners that compacted the soil. Furthermore, the primary aim of a baseliner is to reduce the risk of groundwater contamination due to leachate infiltration.

Gas emission: The natural decomposition of waste in the landfill involving microorganisms to break down the waste usually happens. The rate of degradation and decomposition of waste depends on the amount of water in and the temperature of the waste. During this process, the organic fraction of the wastes turns into CH_4 and CO_2. Moreover, some organics can be directly transformed into gas, such as cleaning materials waste. Typical constituents in the gas produced by municipal solid waste landfill are given in Table 1.20 [76].

Operation and maintenance for landfill: For the public's and environmental health's sake, the landfill should be constructed to meet safety requirements. The operation and maintenance should be managed well, including the following:

1. The waste in landfill should be identified by the operators to be non-hazardous, safe, and acceptable for disposal.
2. Daily covering the waste in the landfill.
3. Control the surface runoff to keep the waste decomposition in the landfill and not increase the leachate production.
4. The equipment used in landfill should be protective and should implement safety operation to the site workers and operators.

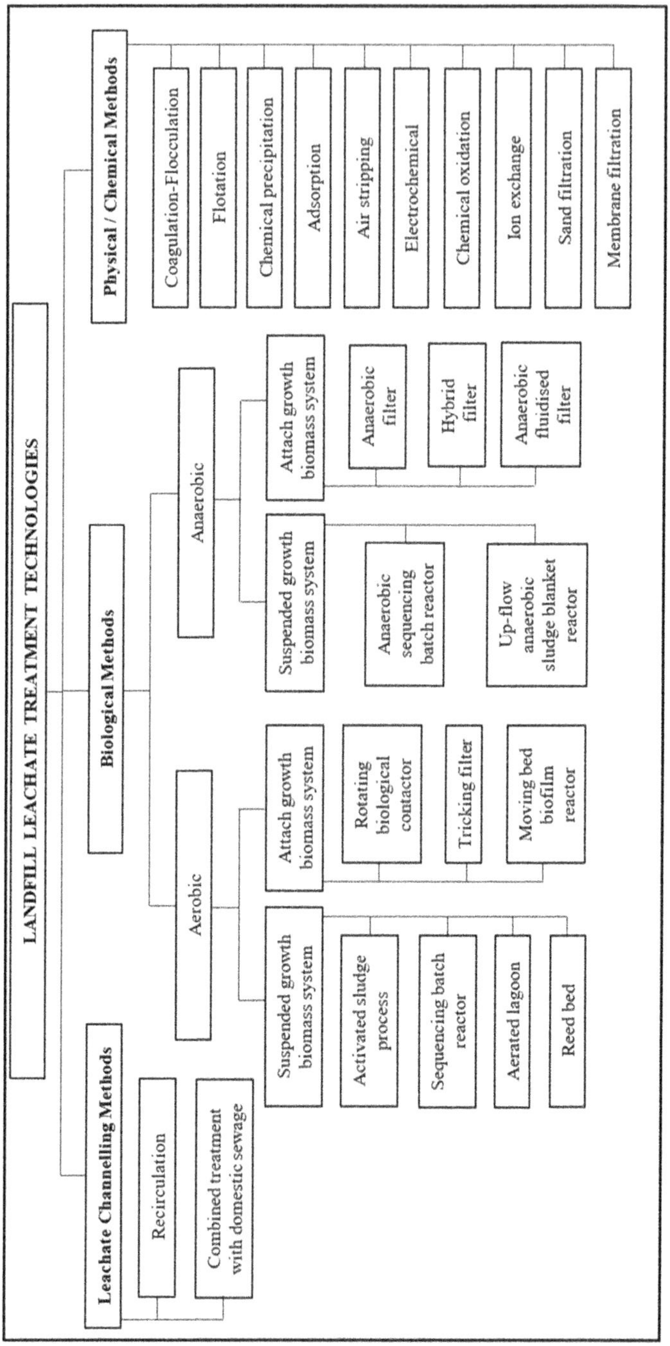

Fig. 1.40 Schematic diagram of landfill leachate treatment technologies

Table 1.18 Typical characteristics and classification of landfill leachate

No.	Parameter	Unit	Category of landfill leachate		
			Young	Intermediate	Stabilized
1	Age	Year	<5	5–10	>10
2	pH	–	<6.5	6.5–7.5	>7.5
3	COD	mg/L	>10,000	5000–10,000	<5000
4	BOD_5/COD	–	>0.5	0.5–0.1	<0.1
5	Organic	–	80% (VFA)	5–30% (VFA + HFA)	HFA
6	NH_3-N	mg/L	<400	–	>400
7	Colour	PtCo	<1000	–	1500–7000
8	TOC/COD	–	<0.3	0.3–.05	>0.5
9	Conductivity	μs/cm	15,000–41,500	6000–14,000	–
10	Heavy metal	mg/L	Low	Low	Low
11	Biodegradability	–	Important	Medium	Low

Note: *VFA* volatile fatty acids, *HFA* humic and fulvic acids
Source: [72–75]

Table 1.19 Comparison of the physical and chemical methods in leachate treatment [71]

No.	Method	Leachate			Cost	Remark
		Y	M	O		
1	Coagulant – flocculation	Poor	Fair	Fair	Low	High sludge production
2	Air stripping	Poor	Fair	Fair	High	Air pollution
3	Chemical precipitation	Poor	Fair	Poor	Low	Disposal of hazardous waste
4	Adsorption	Poor	Fair	Good	Low	Carbon fouling
5	Chemical oxidation	Poor	Fair	Fair	High	Toxic by-product
6	Electrochemical	Poor	Fair	Fair	High	High energy usage
7	Membrane filtration	Good	Good	Good	High	Membrane clogging
8	Ion exchange	Poor	Fair	Fair	High	High anion/cation
9	Flotation	Poor	Fair	Fair	High	High capital cost

Legend – *Y* young, *M* medium, *O* old

1.6 Integrated Solid Waste Management

Integrated solid waste management (ISWM) is a holistic procedure in managing solid waste that includes waste prevention, reduction, recycling, reuse, treatment, and disposal. How waste is handled from source to source is referred to as management (a cradle-to-grave approach). It also takes into account the waste management hierarchy, taking into account both direct and indirect impacts of waste transportation, processing, treatment, and disposal. From planning to design, commissioning, and service, solid waste management must be sustainable. ISWM may be implemented to create a long-term solid waste management system that is both environmentally and economically feasible also as socially acceptable.

For solid waste management, and ISWM solution is currently the most suitable scheme. ISWM promotes waste minimization by waste recovery, reuse, and

Table 1.20 Typical parameters found in landfill gas [76]

Component	Percentage (dry volume basis)
Methane	45–60
Carbon dioxide	40–60
Carbon monoxide	0–0.2
Nitrogen	2–5
Oxygen	0.1–1.0
Sulphides	0–1.0
Ammonia	0.1–1.0
Hydrogen	0–0.2
Trace constituents	0.01–0.6
Non-methane organic compounds	0.01–0.6

recycling through enhancing the quality of the total management system for all types of wastes, composting, incineration, and landfilling, in addition to waste treatment using conventional or advanced methods.

In waste management, a hierarchy (a structure in order of importance) may be used to prioritize actions for implementing initiatives in the community. Source reduction, recycling, waste transformation/processing, and landfilling are typically in the ISWM hierarchy. As shown in Fig. 1.14, a greater emphasis should be placed on source reduction and the least on final disposal. Prevention, minimization, reuse, recycle, energy recovery, and landfill disposal are the most widely sought solid waste management approaches.

Source reduction (reduce): At the top of the ISWM hierarchy, source reduction (reduce) entails lowering the volume and/or toxicity of wastes currently generated. Waste reduction can be accomplished by designing, processing, and packaging goods that have a low toxic content, a limited amount of material, or a longer useful life. Selective purchasing habits and the reuse of goods and resources can also help to reduce waste in the home, business, or industrial setting.

Reuse and recycling is the second-highest level in the hierarchy, and it entails (1) waste separation and collection; (2) preparation of waste materials for reuse, reprocessing, and remanufacture; and (3) reuse, reprocessing, and remanufacture of waste materials. Recycling is a vital part of reducing resource demand and the amount of waste that must be disposed of in landfills.

By putting in a Materials Recovery Plant, you will get a lot of recyclables back (MRF). This facility will sort all recyclable materials in a systematic manner, allowing them to be recycled and converted into new goods. This will also extend the landfill's useful life.

A proper waste segregation system is expected to enhance the overall recycling process and extend the landfill's life. At the same time, a 3Rs (recover, reuse, recycle) campaign should be prioritized. Residents should be reminded of the value of sorting their trash at a source. As soon as possible, a proper recycling programme must be devised. In the long run, education is extremely important. The 3Rs campaign is critical, and it is a difficult challenge that can only be accomplished through education. This could take a long time, but it should begin as soon as possible. Residents would be required to sort their trash at the point of collection. It is necessary to provide facilities. It will be important to assess performance and failure. It is hoped that by integrating these two methods (recycling at the point of usage by MRFs and recycling through education), the recycling rate will increase dramatically in the near future.

Waste processing/transformation: Waste transformation, the third rank in the ISWM hierarchy, entails the physical, chemical, or biological transformation of wastes. Physical, chemical, and biological transformations of MSW are usually used to (1) increase the performance of solid waste management operations and systems, (2) recover recycled and recyclable materials, and (3) recover conversion products (e.g., compost) as well as energy in the form of heat and combustible biogas. In most cases, waste transformation results in less landfill space being used. One well-known example is the reduction of waste volume by incinerator heat treatment.

Disposal by landfilling: In the end, something must be done with (1) the non-recyclables and are of no further use; (2) the residuals after solid wastes have been separated at a materials recovery facility; and (3) the residuals after the waste-to-energy facility. Landfilling, which is ranked fourth in the ISWM hierarchy, includes the controlled disposal of wastes on or in the earth's mantle, and it is by far the most typical method of final disposal for waste residuals.

However, in order to implement a good integrated solid waste management system, various factors need to be investigated. This includes forecasting the waste amount and undertake various studies in terms of Environmental Impact Assessment (EIA), Socio-economic Impact Assessment (SoEIA), Sustainable Assessment (SA), Risk Assessment, and Life Cycle Assessment (LCA). On top of that, a Cost–Benefit Analysis and a feasibility study are also necessary.

1.7 Legislative Aspects of Solid Waste

Waste management laws regulate the transportation, handling, storage, and disposal of different forms of waste. Waste laws are usually enforced to discourage contamination by restricting or preventing the unregulated diffusion of waste materials into the atmosphere. They also contain legislation aimed at reducing waste generation and promoting waste recovery and recycling.

The method of classifying a substance as a 'waste' subject to regulation is known as waste identification. For example, in the United States and many other countries, non-hazardous municipal solid waste may be disposed of in landfills, whereas some metal scrap is considered hazardous and cannot be disposed of in landfills, but must instead be handled, stored, treated, and disposed of according to stricter regulations. The disposal options for a specific waste are governed by disposal requirements. Littering is the most popular and prevalent of these standards in many countries. Some waste should be handled in a certain way before being disposed of at a disposal site. The United States Environmental Protection Agency's Land Disposal Restrictions under the Resource Conservation and Recovery Act Subtitle C hazardous waste management programme, for example, prohibit hazardous waste from being disposed of on land (primarily in landfills) without prior approval.

Specific requirements for the construction and operation of a landfill may also be adopted, particularly with regards to the need to adhere to location restrictions in order to avoid surface and ground water contamination. It also manages operation policies to eliminate dust and other annoyances (leachate and gases), as well as environmental control programmes that ensure compliance.

There is also international law, which includes agreements on international hazardous waste transportation and disposal. European Agreement Concerning the International Carriage of Dangerous Goods by Inland Waterways (AND), Geneva, 2000; Convention to Ban the Importation of Hazardous and Radioactive Wastes into Forum Island Countries and to Control the Transboundary Transport and Management of Hazardous and Radioactive Wastes, Geneva, 2000; Convention to Ban the Importation of Hazardous and Radioactive Wastes into Forum

The Resource Conservation and Recovery Act (RCRA), which was passed in 1976, is the primary federal legislation regulating the management of solid waste and hazardous waste in the United States. The federal Superfund program in the United States was gazetted by the Comprehensive Environmental Response, Compensation, and Liability Act of 1980 (CERCLA), which is managed by the United States Environmental Protection Agency (USEPA). The program's aim is to inspect and clean up hazardous substance–contaminated areas. The United States Environmental Protection Agency (USEPA), which regulates the hazardous waste generation and disposal; the United States Department of Transportation (DOT), which regulates hazardous waste transportation; and the Nuclear Regulatory Commission (NRC), which regulates nuclear waste, are among the regulatory bodies in the United States [78–81].

Waste legislation in the United Kingdom is primarily derived from EU governance and transposed into UK law through Statutory Instruments. In response to the rising waste amount disposed of in landfills, the government levied a Landfill Tax on some forms of waste deposited in landfills starting in October 1996. Landfill operators who were authorized under the Environmental Protection Act (EPA) or the Pollution Control & Local Government Order 1978, for example, had until 31 August 1996 to register their tax liability. The tax is seen as a critical means for the UK to attain its objectives for landfilling biodegradable waste specified in the Landfill Directive. Other advanced waste management systems with higher tipping charges are more financially attractive by raising the cost of landfill. The Landfill Directive, also known as Council Directive 1999/31/EC of 26 April 1999, is a European Union directive that governs landfill waste management in the EU. By 16 July 2001, all of its member states had adopted it. The Directive's ultimate goal is to 'prevent or mitigate as much as possible negative environmental impacts from waste landfilling, as well as any resulting danger to human health'. This legislation has far-reaching consequences for waste management and disposal. Following that, in 2000 and 2002, waste targets for England and Wales were adopted, focusing on recycling, composting, and energy from waste (EFW) technologies for MSW recovery [11].

The new European waste strategy is based on the Landfill Directive [31], the Waste Incineration Directive, and the Packaging and Waste Packaging Directive [31, 77]. The obligatory targets set out in the EU Landfill Directive, which was adopted on 26 April 999, and entered into effect on 16 July 1999, would require the UK and other EU countries to lessen the biodegradable portion of municipal waste disposed of to landfill to 75% of the amount in 1995 by 2010. Similarly, by 2013, this will have to be reduced to 50%, and by 2020, it will have to be reduced to 35%. In 2002, the Welsh Assembly Government released the 'Wise About Waste' National Waste Strategy for Wales, which aims to ensure compliance with European waste management directives. According to the goals, a minimum of 15% of MSW must be recycled or composted by 2003/2004, with a 5% minimum goal for each group. By 2006/2007, the target has risen to 25%, with a minimum of 10% for each group. By 2009/2010, the overall goal is set at 40%, with a minimum goal of 15% for each group [77].

The majority of ASEAN countries have already enacted environmental legislation as well as other green growth, sustainable development, climate change policies, regulatory frameworks, and strategies. Waste management legislation exists in Indonesia, Malaysia, the Philippines, and Thailand. The Ministry of the Environment is largely responsible for waste management policy. Other related ministries are also in charge of particular waste sources (for instance, the Ministry of Health for Hospital Waste, Ministry of Local Government for Domestic Waste). Municipalities and state or local governments are directly responsible for waste management systems at the local level.

The Local Government Act of 1976, the Environmental Quality Act of 1974, the Town and Country Planning Act of 1976, and the Streets, Drainage and Construction Act of 1976 are the four subsidiary laws that regulate solid waste management in Malaysia. Local governments are currently the most influential institution active in solid waste management. As a result, the Technical Section of the Local Government Department in Malaysia's Ministry of Housing and Local Government proposed a National Solid Waste Management Action Plan, also known as the Action Plan for a Beautiful and Clean Malaysia, in 1988. (ABC). The Solid Waste and Public Cleansing Management Act 672, 2008, and the Solid Waste and Public Cleansing Management Corporation Act 673, 2008, were both published in the *Gazette* this year.

1.8 Concluding Remarks

This chapter has endeavoured to provide an overall image of the nature of the solid waste crisis, both qualitatively and quantitatively, as well as the issues surrounding its management. The rapid growth of waste volumes and a diverse waste composition of new and emerging waste sources are among the various environmental challenges in waste management. In both developed and developing countries, proper solid waste management is essential for maintaining human health and the environment. Full compliance with waste management hierarchy systems is needed for integrated municipal solid waste management. Minimizing and reducing waste is a critical component of effective urban solid waste management. Recycling and reuse of waste were considered an important option for improving the economy and reducing the quantity of final waste disposal. Despite the numerous disposal options, landfilling is commonly regarded as a viable choice for the disposal of urban solid waste around the world. Many local governments are facing a major challenge: Growing waste collection volumes and the need to meet more rigorous regulatory requirements in disposal operations necessitate increased capital and operating revenue reserves.

In the overall management system, both short- and long-term options are important. Many countries face major challenges in terms of infrastructure, technology, funding, governance, and stakeholder involvement. These difficulties, on the other hand, could become opportunities if we change our perspective of waste as a resource. Although improving waste recycling rates and waste to energy (WTE) technologies and methods, front-end solutions, such as frameworks for waste reduction/prevention through sustainable consumption and resource management, must also be considered. For a successful recovery system, all countries should encourage segregation at the source. Certain incentives, such as a recycle for life card that rewards people who send recyclables to recycling centres with money, maybe devised to encourage people to recycle. A model that works A proper value chain in the entire solid waste management system (including waste generation, segregation, collection, transfer, treatment, and disposal, as well as resource recovery via the 3Rs) should be designed so that waste can be converted into income and the new digital economy cycle can be accelerated. Furthermore, every country should look into producing co-benefits from the waste sector, such as reduced greenhouse gas (GHG) emissions, which facilitates the achievement of sustainable development goals (SDGs), and so on. Greater coordination between public and private organizations in waste value chains would help pool resources and gather shared responsibilities for waste management, particularly in terms of selecting and implementing environmentally sound technologies (ESTs) that are suitable for the local waste characteristics [78–81].

Furthermore, the design and implementation of the right mix of legislative, economic, and social instruments, as well as incentives for strong enforcement monitoring by all relevant stakeholders, are critical. Improving operational performance and encouraging interdepartmental/agency collaboration is also critical. Alternative and creative financing methods, such as public–private partnerships (PPPs), public funding initiatives (PFIs), and the implementation of the polluter pays concept, will bolster existing revenue streams. Companies that are doing ‘healthy’ in terms of waste recycling should be given tax breaks.

Glossary

US Environmental Protection Agency (USEPA) is the United States federal government agency whose mission is to protect human and environmental health.

Environmental Impact Assessment (EIA) is the process of examining the anticipated environmental effects of a proposed project from consideration of environmental aspects at the design stage

American Society of Mechanical Engineers (ASME) is an American professional association that promotes the art, science, and practise of multidisciplinary engineering and allied sciences around the world.

Cost–benefit analysis (CBA) is a systematic process that businesses use to analyse which decisions to make.

Life Cycle Assessment (LCA) is a methodology for assessing environmental impacts associated with all the stages of the life cycle of a commercial product, process, or service.

Material Flow Analysis (MFA) is an analytical method to quantify flows and stocks of materials in a system.

Socio-economic Assessment (SoEA) is the analysis of social, cultural, economic, and political conditions of individuals, groups, communities and organizations.

Risk Assessment (RA) is the process of identifying and analysing potential events that may negatively impact individuals or the environment and making judgements on the tolerability of the risk on the basis of a risk analysis.

Strategic Environmental Assessment (SEA) is a systematic decision support process aiming to ensure that environmental and possibly other sustainability aspects are considered effectively in policy, plan, and programme making.

Resource Conservation And Recovery Act (RCRA) is the principal federal law in the United States governing the disposal of solid waste and hazardous waste, which was enacted in 1976.

United Nations Environment Programme (UNEP) is the leading environmental authority in the United Nations system.

Volatile fatty acids (VFA) are short-chain fatty acids composed mainly of C2–C6 carboxylic acids produced in the anaerobic digestion process, which does not need sterilization, additional hydrolysis enzymes, or high-cost pre-treatment step.

Air Pollution Control Residues (APCr) is typically a mixture of ash, carbon, and lime.

Brominated flame retardants (BFR) are mixtures of man-made chemicals that are added to a wide variety of products, including for industrial use, to make them less flammable.

Chlorofluorocarbons (CFC) are fully or partly halogenated paraffin hydrocarbons that contain only carbon (C), hydrogen (H), chlorine (Cl), and fluorine (F), produced as a volatile derivative of methane, ethane, and propane.

Hydrochlorofluorocarbons (HCFC) are compounds containing carbon, hydrogen, chlorine, and fluorine.

References

1. USEPA-US Environmental Protection Agency. (2013). *Waste-hazardous waste-waste minimisation* [Online]. Available from http://www.epa.gov/waste/hazard/wastemin/index.htm. Accessed 22 Mar 2021.
2. USEPA-US Environmental Protection Agency. (2021a). *Criteria for the definition of solid waste and solid and hazardous waste exclusions* [Online]. Available from https://www.epa.gov/hw/criteria-definition-solid-waste-and-solid-and-hazardous-waste-exclusions. Accessed 22 Mar 2021.
3. USEPA-US Environmental Protection Agency. (2020). *Advancing sustainable materials management: 2018: Tables and figures*, November.
4. HKEPD-Hong Kong Environmental Protection Department. (2020). *Monitoring of solid waste in Hong Kong Waste statistics for 2019*, Statistics Unit, Environmental Protection Department, Hong Kong, December 2020.
5. IPCC Guidelines for National Greenhouse Gas Inventories, the 2019 Refinement was considered in May 2019 during the IPCC's 49th Session (Kyoto, Japan) and adopted/accepted on 12 May 2019 (Decision – IPCC-XLIX-9 – Adoption and Acceptance of 2019 refinement). Accessed online on 26 July 2020.
6. OECD-Organisation for Economic Co-operation and Development. (2021). *Municipal waste generation and treatment* [Online]. Available from. https://stats.oecd.org/Index.aspx?DataSetCode=WSECTOR. Accessed 22 Mar 2021.
7. Eurostat. (2021). *Municipal waste statistics* [Online]. Available from https://ec.europa.eu/eurostat/statistics-explained/index.php?%20title=Municipal_waste_statistics&oldid=343958%20. Accessed 22 Mar 2021.
8. Kaza, S., Yao, L., Bhada-Tata, P., & Van Woerden, F. (2018). *What a waste 2.0: A global snapshot of solid waste management to 2050* (Urban Development Series). World Bank. https://doi.org/10.1596/978-1-4648-1329-0. License: Creative Commons Attribution CC BY 3.0 IGO.
9. UNEP-United Nations Environment Programme. (2017). *Highlights Asia waste management outlook-Summary for decision maker*. Nairobi, Kenya.
10. Abdel-Shafy, H. I., & Mansour, M. S. M. (2018). *Solid waste issue: Sources, composition, disposal, recycling, and valorization*. Egyptian Journal of Petroleum. (in press).
11. Buttol, P., Masoni, P., Bonoli, A., Goldoni, S., Belladonna, V., & Cavazzuti, C. (2007). LCA of integrated MSW management systems: Case study of the Bologna District. *Waste Management, 27*(8), 1059–1070.
12. Hui, Z., AiHong, M., YanQiu, L., QingHai, L., & YanGuo, Z. (2014). An overview of characteristics of municipal solid waste fuel in China: Physical, chemical composition and heating value. *Renewable and Sustainable Energy Reviews, 36*, 107–122. https://doi.org/10.1016/j.rser.2014.04.024
13. USEPA-US Environmental Protection Agency. (2021b). *Wastes-non-hazardous waste-industrial waste* [Online]. Available from https://archive.epa.gov/epawaste/nonhaz/industrial/special/web/html/index-5.html. Accessed 22 Mar 2021.
14. Pashkevich, M. A. (2017). Chapter 1 – Classification and environmental impact of mine dumps, author links open overlay panel. In *Assessment, restoration and reclamation of mining influenced soils* (pp. 1–32). Academic.
15. Collins, R. J., & Ciesielski, S. K. (1994). Recycling and use of waste materials and by-products in highway construction. In *National cooperative highway research program synthesis of highway practice 199*. Transportation Research Board.
16. USEPA-US Environmental Protection Agency (1985). Report to Congress on wastes from the extraction and beneficiation of metallic ores, phosphate rock, asbestos, overburden from uranium mining, and oil shale. Report No. EPA/530-SW-85-033,
17. Bowdish, L. (2016). *Trash to treasure: Changing waste streams to profit streams, U.S* (p. 21). Chamber of Commerce Foundation.

18. Food & Water Watch. (2021). *Factory farm map: What's wrong with factory farms?* [Online]. Available from http://www.factoryfarmmap.org/problems/. Accessed 22 Mar 2021.
19. Palese, A. M., Persiani, A., D'Adamo, C., Pergola, M., Pastore, V., Sileo, R., Ippolito, G., Lombardi, M. A., & Celano, G. (2020). Composting as manure disposal strategy in small/medium-size livestock farms: Some demonstrations with operative indications. *Sustainability, 2020*(12), 3315. https://doi.org/10.3390/su12083315
20. Quest. (2021). [Online]. Available from https://www.questrmg.com/2019/08/08/food-waste-statistics-the-reality-of-food-waste/. Accessed 22 Mar 2021.
21. Simão, L., Hotza, D., Raupp-Pereira, F., Labrincha, J. A., & Montedo, O. R. K. (2018). Wastes from pulp and paper mills – A review of generation and recycling alternatives. *Cerâmica, 64*(371).
22. Labfresh. (2021). [Online]. Available from https://labfresh.eu/pages/fashion-waste-index. Accessed 22 Mar 2021.
23. USEPA-US Environmental Protection Agency. (2021c). *Facts and figures about materials, waste and recycling- wood: Material-specific data* [Online]. Available from https://www.epa.gov/facts-and-Fig.s-about-materials-waste-and-recycling/wood-material-specific-data. Accessed 22 Mar 2021.
24. Ed Suttie. (2004). Wood waste management-UK update, final workshop COST Action E22 '*Environmental optimisation of wood protection*', Lisbon-Portugal, 22nd – 23rd March.
25. FPA. (2019). *State of industry report, flexible packaging association*, September 2019 [Online]. Available from https://www.packworld.com/home/news/15693115/fpa-publishes-2019-state-of-the-flexible-packaging-industry-report. Accessed 22 Mar 2021.
26. Scotia Bank. (2021). Global economics|Global Auto Report, January 27, 2021 [Online]. Available from https://www.scotiabank.com/ca/en/about/economics/economics-publications/post.other-publications.autos.global-auto-report.january-27-2021.html. Accessed 22 Mar 2021.
27. Statisca. (2019). *U.S. paper products market value 2015–2025*, Statisca, November 2019 [Online]. Available from https://www.statista.com/statistics/1066665/us-paper-products-market-value/. Accessed 22 Mar 2021.
28. Statisca. (2021). *Waste statistics in Abu Dhabi Emirates*, July 2020 [Online]. Available from https://www.statista.com/statistics/895550/uae-total-solid-waste-generated/. Accessed 22 Mar 2021.
29. Poon, C. S., Yu, A. T. W., Wong, A., & Yip, R. (2013). Quantifying the impact of construction waste charging scheme on construction waste management in Hong Kong. *Journal of Construction Engineering and Management, 139*(5), 466–479. https://doi.org/10.1061/(ASCE)CO.1943-7862.0000631. hdl:10397/6714. ISSN:1943-7862.
30. Wang, Y., Zhang, X., Liao, W., Wu, J., Yang, X., Shui, W., Deng, S., Zhang, Y., Lin, L., Xiao, Y., Yu, X., & Peng, H. (2018). Investigating impact of waste re-use on the sustainability of municipal solid waste (MSW) incineration industry using energy approach: A case study from Sichuan Province, China. *Waste Management, 77*, 252–267. https://doi.org/10.1016/j.wasman.2018.04.003
31. European Commission. (2018). *EU construction and demolition waste protocol and guidelines*. Internal Market, Industry, Entrepreneurship and SMEs-European Commission [Online]. Available from https://ec.europa.eu/growth/content/eu-construction-and-demolition-waste-protocol-0_en. Accessed 22 Mar 2021.
32. CIS-Construction Industry Scheme. (2021). GOV.UK [Online]. Available from https://www.gov.uk/what-is-the-construction-industry-scheme. Accessed 22 Mar 2021.
33. Yu, A., Poon, C., Wong, A., Yip, R., & Jaillon, L. (2013). Impact of construction waste disposal charging scheme on work practices at construction sites in Hong Kong. *Waste Management, 33*(1), 138–146. https://doi.org/10.1016/j.wasman.2012.09.023. hdl:10397/6713. S2CID 20266040.
34. CIDB. (2008). *Guidelines on construction waste management*. Construction Industry Development Board Malaysia.

35. Forti, V., Baldé, C. P., Kuehr, R., & Bel, G. (2020). *The Global E-waste Monitor: Quantities, flows and the circular economy potential*. United Nations University (UNU)/United Nations Institute for Training and Research (UNITAR)-Co-Hosted SCYCLE Programme, International Telecommunication Union (ITU) & International Solid Waste Association (ISWA), Bonn/Geneva/Rotterdam.
36. Cho, R. (2018). *What can we do about the growing e-waste problem*? General Earth Institute, August 27, 2018 [Online]. Available from https://blogs.ei.columbia.edu/2018/08/27/growing-E-waste-problem/. Accessed 22 Mar 2021.
37. Agence nationale pour la gestion des déchets radioactifs. (2020). *Édition 2018 de l'inventaire national des matières et déchets radioactifs l'Essentiel 2020* (p. 15). Agence nationale pour la gestion des déchets radioactifs, April 2020.
38. JAEA – The Japan Atomic Energy Agency. (2019). *White paper on nuclear energy 2019* (p. 276). JAEA, August 2020 [Online]. Available from http://www.aec.go.jp/jicst/NC/about/hakusho/index_e.htm. Accessed 22 Mar 2021.
39. World Nuclear Association. (2021). *Storage and disposal of radioactive waste* [Online]. Available from https://www.world-nuclear.org/information-library/nuclear-fuel-cycle/nuclear-waste/storage-and-disposal-of-radioactivE-waste.aspx. Accessed 22 Mar 2021.
40. Marine Conservation Society. (2019). *Great British Beach clean 2019 report*, November 2019.
41. Mohajerani, A., Burnett, L., Smith, J. V., Markovski, S., Rodwell, G., Rahman, T., Kurmus, H., Mirzababaei, M., Arulrajah, A., Horpibulsuk, S., et al. (2020). Recycling waste rubber tyres in construction materials and associated environmental considerations: A review. *Resources, Conservation & Recycling, 155*, 104679.
42. Morin, J. E., & Farris, R. J. (2000). Recycling of 100% cross linked rubber powder by high temperature high pressure sintering. *Encyclopedia of Polymer Science and Engineering, 37*, 95–101.
43. ISRI. (2019). *Recycling industry yearbook 2019.*
44. Macie, S., Justyna, K. L., Helena, J., & Adolf, B. (2012). Progress in used tyres management in the European Union: A review. *Waste Management, 32*(10), 1742–1751. https://doi.org/10.1016/j.wasman.2012.05.010
45. Fazli, A., & Rodrigue, D. (2020). Waste rubber recycling: A review on the evolution and properties of thermoplastic elastomers. *Materials, 13*, 782. Imbernon, L., & Norvez, S. (2016). From landfilling to vitrimer chemistry in rubber life cycle. *European Polymer Journal, 82*, 347–376.
46. Saha, P. (2018). *Rubber recycling: Challenges and developments*. Royal Society of Chemistry.
47. Chatziaras, N., Psomopoulos, C. S., & Themelis, N. J. (2016). Use of waste derived fuels in cement industry: A review. *Management of Environmental Quality: An International Journal, 27*(2), 178–193.
48. Mangi, S. A., Wan Ibrahim, M. H., Jamaluddin, N., Arshad, M. F., & Mudjanarko, S. W. (2019). Recycling of coal ash in concrete as a partial cementitious resource. *Resources, 8*, 99. https://doi.org/10.3390/resources8020099
49. Dwivedi, A., & Jain, M. K. (2014). Fly ash-waste management and overview: A review. *Recent Research in Science and Technology, 6*(1), 30–35.
50. Hannan, J. (2021). *Chemical makeup of fly and bottom ash varies significantly; must be analyzed before recycled. ThermoFisher Scientific* [Online]. Available from https://www.thermofisher.com/blog/mining/chemical-makeup-of-fly-and-bottom-ash-varies-significantly-must-be-analyzed-before-recycled/#:~:text=The%20European%20Coal%20Combustion%20Products,of%20fly%20ash%20at%2043%25.&text=The%20chemical%20makeup%20of%20fly,of%20the%20coal%20being%20burned. Accessed 22 Mar 2021.
51. Airquality news.com. (2021). *Rule change sought on Air Pollution Control residues* [Online]. Available from https://airqualitynews.com/2016/04/14/rule-change-sought-on-air-pollution-control-residues/#:~:text=What%20is%20Air%20Pollution%20Control,to%20a%20non%2Dhazardous%20landfill. Accessed 22 Mar 2021.

52. Karagiannidis, A., Samaras, P., Kasampalis, T., Perkoulidis, G., Ziogasa, P., & Zorpas, A. (2011). Evaluation of sewage sludge production and utilization in Greece in the frame of integrated energy recovery. *Desalination and Water Treatment, 33*, 185–193.
53. IWA. (2021). *Sludge production* [Online]. Available from https://www.iwapublishing.com/news/sludge-production. Accessed 22 Mar 2021.
54. Yang, G., Zhang, G., & Wang, H. (2015). Current state of sludge production, management, treatment and disposal in China. *Water Research*. https://doi.org/10.1016/j.watres.2015.04.002
55. Chandler, A. J., Eighmy, T. T., Hartlen, O., Kosson, D., Sawell, S. E., van der Sloot, H., & Vehlow, J. (1997). *Municipal solid waste incinerator residues* (Studies in Environmental Science, International Ash Working Group) (Vol. 67). Elsevier Science.
56. UK Department for Business. (2020). *Energy and industrial strategy, digest of UK energy statistics: renewable sources of energy (DUKES 6.4)*, July.
57. Tinmaz, E., (2002). *Research on integrated solid waste management system in Corlu Town*. Master thesis, Istanbul Technical University, Istanbul, Turkey.
58. Recyclingbin.com. (2021). *Recycling facts* [Online]. Available from https://www.recyclingbin.com/Recycling-Facts#:~:text=Recycling%20Facts-,Paper,from%20the%20air%20each%20year. Accessed 22 Mar 2021.
59. Tchobanoglous, G. H., & Theisen, S. V. (1993). *Integrated solid waste management: Engineering principle and management issue. International Ed*. McGraw – Hill Book Co.
60. Wilson, D. C., Cowing, M. J., Oelz, B., Scheinberg, A., Stretz, J., Velis, C. A., Rodic, L., Masterson, D., Vilches, R., & Whiteman, A. D. (2014). 'Wasteaware' benchmark indicators for integrated sustainable waste management in cities. *Waste Management, 35*, 329–342.
61. Waste Collection Report. ISWA Working Group on Collection and Transportation Technology. (2007) [Online]. Available from https://www.semanticscholar.org/paper/Underground-Automated-Vacuum-Waste-Collection-for-(-Dixit-Rastogi/d8a662a748b2e1d19cb8f1962ef9d44b9ea21325. Accessed 22 Mar 2021.
62. Chafer, M., Sole-Mauri, F., Sole, A., Boer, D., & Cabeza, L. F. (2019). Life cycle assessment (LCA) of a pneumatic municipal waste collection system compared to traditional truck collection. Sensitivity study of the influence of the energy source. *Journal of Cleaner Production, 231*, 1122–1135. with permission from Elsevier.
63. Recycling Magazine. (2020). Best practices for construction waste management-trends, analyses, opinions and facts for the recycling industry. *Recycling magazine* 04/2020 [Online]. Available from https://www.recycling-magazine.com/ausgabe/recycling-magazine-04-2020/. Accessed 22 Mar 2021.
64. OECD. (2021). *Municipal waste, generation and treatment*. OECD-The Organisation for Economic Co-operation and Development [Online]. Available from https://stats.oecd.org/Index.aspx?DataSetCode=MUNW#. Accessed 22 Mar 2021.
65. Parvez, N., Agrawal, A., & Kumar, A. (2019). Solid waste management on a campus in a developing country: A study of the Indian Institute of Technology Roorkee. *Recycling, 2019*(4), 28. https://doi.org/10.3390/recycling4030028
66. USEPA-US Environmental Protection Agency. (2021d). *Sustainable management of food-types of composting and understanding the process* [Online]. Available from https://www.epa.gov/sustainable-management-food/types-composting-and-understanding-process. Accessed 22 Mar 2021.
67. Odeh, A., & Al-Sa'ed, R. (2018). *Evaluation of windrow composting pilots for domestic organic waste amended by horse manure and biosolids*. In 6th Balkans Joint Conference and Exhibition, 7–9 November 2018, Expocity Albania, Tirana, Albania.
68. Michaels, T., & Krishnan, K. (2018). *Directory of waste to energy facilities*. Energy Recovery Council US, October 2018.
69. Tolvik Consulting Limited, UK. (2020). *Energy from waste statistics*-2019, May.
70. Worldatlas. (2019). *Largest landfills, waste sites, and trash dumps in the world* [Online]. Available from https://www.worldatlas.com/articles/largest-landfills-waste-sites-and-trash-dumps-in-the-world.html. Accessed 22 Mar 2021.

71. Eunomia. (2021). *Recycling-who really leads the world? Identifying the world's best municipal waste recyclers* [Online]. Available from https://resource.co/sites/default/files/World%20Recycling%20League%20-%20Full%20Report%20-%20FINAL.pdf. Accessed 22 Mar 2021.
72. Shehzad, A., Bashir, M. J. K., Sethupathi, S., & Lim, J. (2015). An overview of heavily polluted landfill leachate treatment using food waste as an alternative and renewable source of activated carbon. *Process Safety and Environmental Protection, 6*, 1–42. https://doi.org/10.1016/j.psep.2015.09.005
73. Aziz, H. A., & Ramli, S. F. (2018). Recent development in sanitary landfilling and landfill leachate treatment in Malaysia. *International Journal of Environmental Engineering, 9*, 201–229. https://doi.org/10.1504/ijee.2018.10018737
74. Costa, A. M., de Souza Marotta Alfaia, R. G., & Campos, J. C. (2019). Landfill leachate treatment in Brazil – An overview. *Journal of Environmental Management, 232*, 110–116. https://doi.org/10.1016/j.jenvman.2018.11.006
75. Pasalari, H., Farzadkia, M., Gholami, M., & Emamjomeh, M. M. (2019). Management of landfill leachate in Iran: Valorization, characteristics, and environmental approaches. *Environmental Chemistry Letters, 17*, 335–348. https://doi.org/10.1007/s10311-018-0804-x
76. WHO. (2014). *Safe management of wastes from health-care activities*. WHO.
77. Emery, A., Davies, A., Griffiths, A., & Williams, K. (2007). Environmental and economic modelling: A case study of municipal solid waste management scenarios in Wales. *Resources, Conservation and Recycling, 49*(3), 244–263.
78. Wang, L. K., & Pereira, N. C. (1980). *Handbook of environmental engineering, Volume 2, Solid waste processing and resources recovery* (480 pages). The Humana Press.
79. Hung, Y. T., Wang, L. K., & Shammas, N. K. (2014). *Handbook of environment and waste management: Land and groundwater pollution control* (Vol. 2, 1091 pages). World Scientific.
80. Wang, L. K., Wang, M. H. S., Hung, Y. T., & Shammas, N. K. (2016). *Natural resources and control processes* (493–623 pages). Springer.
81. Hung, Y. T., Wang, L. K., & Shammas, N. K. (2020). *Handbook of environment and waste management: Acid rain and greenhouse gas pollution control* (Vol. 3, 770 pages). World Scientific.

Chapter 2
Legislation for Solid Waste Management

Siti Hafizan Hassan, Azhar Abdul Halim, Mohd Suffian Yusoff, Lawrence K. Wang, and Mu-Hao Sung Wang

Abstracts The law and institutional structure provide the basis for formulating a policy framework aimed at improving the future practice for solid waste management (SWM). SWM has recently become a worldwide concern for metropolitan environments, which could have a negative impact on the economy. The authors have reviewed many regulations, which applied in many countries like the US, Europe, Korea, and Japan. Related international standards and principles have also been reviewed at the federal, state, and local levels. This chapter covers the legislations based on the countries mentioned and also includes Malaysian legislations such as the Federal Constitution of 1957, the Act for Local government (Act 171) 1976, the Town and Country Planning Act (Act 127) 1976, the Act for the Environmental Quality (Act 127) 1974, the Act for Street, Drainage and Building Act (Act 133) 1974, and the Act for Solid Waste and Public Cleansing 2007. An overview of SWM implementation in Malaysia and the rest of the world will be given through this subject. The most recent (updated to 2020) US Federal Acts affecting solid and hazardous waste management are also covered in detail: the Solid Waste Disposal Act (1965), Resource Recovery Act (1970), the Resource Conservation and Recovery Act (1976), the Comprehensive Environmental Response, Compensation, and Liability Act (1980), the Solid Waste Disposal Act

S. H. Hassan (✉)
School of Civil Engineering, College of Engineering, Universiti Teknologi MARA, Cawangan Pulau Pinang, Permatang Pauh, Pulau Pinang, Malaysia
e-mail: sitihafizan741@uitm.edu.my

A. A. Halim
School of Environmental and Natural Resource Sciences, Faculty Science and Technology, Universiti Kebangsaan Malaysia, Bangi, Selangor, Malaysia
e-mail: azharhalim@ukm.edu.my

M. S. Yusoff
School of Civil Engineering, Universiti Sains Malaysia, Nibong Tebal, Pulau Pinang, Malaysia
e-mail: suffian@usm.my

L. K. Wang · M.-H. S. Wang
Lenox Institute of Water Technology, Latham, NY, USA

L. K. Wang et al. (eds.), *Solid Waste Engineering and Management*, Volume 1, Handbook of Environmental Engineering 23,
https://doi.org/10.1007/978-3-030-84180-5_2

Amendments (1980), the Used Oil Recycling Act (1980), and the Hazardous and Solid Waste Amendments (1984); and some acts' recent amendments up to 2020.

Keywords Solid waste management · Legislation · Environmental standards · Regulation

Acronyms

CAA	Clean Air Act
CERCLA	Comprehensive Environmental Response, Compensation, and Liability Act
CWA	Clean Water Act
DOE	Department of Environment
EFTA	European Free Trade Association
EIA	Environment Impact Assessment
EPSM	Environmental Protection Society of Malaysia
ISWM	Integrated Solid Waste Management
LA	Local Authorities
MENGO	Malaysian Environment Non-Governmental Organizations
METI	Ministry of Economy, Trade and Industry
MHLG	Ministry of Housing and Local Government
MOE	Ministry of Environment
MSW	Municipal Solid Waste
MSWM	Municipal Solid Waste
NCLG	National Council for Local Government
NGOs	Non-Governmental Organizations
NSP	National Strategic Plan
NSWMD	National Solid Waste Management Department
RCRA	Resource Conservative and Recovery Act
RCRA	Resource Conservation and Recovery Act
RRA	Resource Recovery Act
SDBA	Streets, Drainage and Building Act
SDWA	Safe Drinking Water Act
SWDA	Solid Waste Disposal Act
SWDA	Solid Waste Disposal Act
SWM	Solid Waste Management
SWMPC	Solid Waste Management and Public Cleansing
SWPCMC	Solid Waste and Public Cleansing Management Corporation
TCPA	Town and Country Planning Act
TrEEs	Treat Every Environment Special
TSCA	Toxic Substances Control Act
UORA	Used Oil Recycling Act
USEPA	US Environmental Protection Agency

UST	Underground Storage Tank
WQA	Water Quality Act

2.1 Introduction to Solid Waste Management Legislation

The legislation and institutional structure provide the basis for formulating a policy framework to promote future practice in solid waste management, as well as international developments in solid waste management (SWM). Recently, SWM has become a global concern for urban areas such as the US, Europe, Japan, and Korea, which can have a detrimental effect on the economy.

2.1.1 Highlights of the Scope and Focus of the Chapter

This chapter discusses the Solid Waste Management (SWM) regulations internationally, including Malaysia. SWM laws offer a summary of the sources of solid waste in their relevant legislations. In addition, the phases of solid waste management laws are discussed, with a focus on the structure, requirements, and specifications. The Solid Waste Management Act illustrates collections of SWM laws that have applied internationally, including in the USA, Europe, Japan, and Korea. The operation of the Solid Waste Compliance Act in Malaysia is also discussed in this chapter. In addition, an overview of the legal structure of Malaysian Integrated Solid Waste Management (ISWM) as well as the institutional framework of the federal, state, local governments, and private sector are also presented. This chapter discusses the concerns and consequences of waste reduction legislations and the future reform of solid waste legislation.

2.1.2 The Trends of Waste Management Controls

The objective of moving forward waste administration is to decrease waste era and the sum of materials arranged in landfills. This point is part of the overarching objective of improving natural supportability and advancing feasible strategy. The essential objective of great squander administration strategy to utilize materials and, eventually, decrease the sum of squander within the last transfer process. A case of a great waste administration strategy that can be utilized by buyers is the application of the squander chain of command concept, such as killing, reusing, reusing, recuperating, and arranging. Appropriately, the application of best activities in waste administration ought to incorporate understanding of the progression of squander. In this way, it diminishes the sum of squander created in ventures and hence arranged in landfills [1].

The Malaysian Solid Waste and Public Cleansing Administration Act 2007 (MSWPCM) Act [2] characterizes 'waste hierarchy' as the positioning of squander administration operations in understanding with their natural benefits within the taking after arranging of need: (1) lessening of squander era, (2) reuse of squandered materials, (3) reusing of squandered materials for comparable or unused items, (4) utilization of vitality from squander, and (5) transfer of squander from landfilling. As presented within the chain of command, the foremost, naturally inviting choices are positioned to begin with (at the beat), whereas the foremost undesirable choice is positioned final.

2.1.2.1 Sources Reduction

With regard to material waste, avoidance is the best approach; in most cases, however, this may not be possible or practical [1]. The elimination of sources is the next best solution to controlling waste generation. 'Source reduction is defined as the design, manufacture, purchase or use of materials to reduce their quantity or toxicity until they enter the waste stream' [3]. According to Vesilind [4], three strategies can be used to minimize waste: (a) reduce the quantity of material used per product, (b) increase the life of the product, or (c) remove the need for the product. Reducing waste often implies reducing waste at the consumer level. The USEPA indicates that the reduction of sources can be accomplished by the following activities:

1. Conserve the capital of nature.
2. Save resources.
3. Decrease emissions.
4. Elimination of the toxicity of our waste.
5. Save money for both customers and corporations.

2.1.2.2 Reuse of Existing Materials

Reuse involves using the components nearly as they were initially meant to be used. It is the most valuable recycling form. The reuse of waste materials is considered an efficient technique to improve the use of resources prior to disposal [1].

2.1.2.3 Recycling and Composting

The US Environmental Protection Agency (USEPA) [3] pointed out that there is a range of recycling operations, including the storage of waste, reused or discarded items otherwise known as waste; the processing and treatment of recyclable products in the manufacture of raw materials; and replication of new products made of recycled commodities. Consumers deliver the final link to recycling by buying recycled material items. Recycling, in particular, avoids the release of many greenhouse gases and contaminants, saves electricity, provides vital raw material to the industry, generates employment, encourages greener technology growth, retains capital for our children's future, and decreases the need for new landfills and combustion plants. Guerrero [5] found that effective recycling is dependent not only on levels of participation but also on the efficiency of facilities and equipment.

2.1.2.4 Regeneration of Resources

The United States Environmental Protection Agency [6] described waste energy regeneration as a conversion of non-recyclable waste material through a range of processes such as combustion, gasification, pyritizations, anaerobic digestion, and waste gas recovery into wastable heat, electricity and power. This is often called the 'waste-to-energy' process.

2.1.2.5 Refining and Disposal

Landfills are the most common waste disposal method and form a significant component of an integrated waste management system. MSW-accepting landfills are mostly controlled by state, tribal, and local governments. Methane gas can be obtained and used as a fuel to generate energy, a by-product of decomposing waste. Land, such as parks, golf courses, and ski slopes, can be used as recreation areas before the landfill is closed. Guidelines for the management of solid waste and landfills identify and explain concerns that should be addressed in the solid waste disposal plans and the construction of landfills for urban and hazardous waste. Specific thoughts need to reflect on the main topics. The issue identified in these guidelines provides guidance on the establishment of facilities for solid waste disposal and landfill sites. Guerrero [5] stressed that inadequate waste collection systems due, inter alia, to a lack of facilities or municipal inefficiencies, allow individuals to seek other waste disposal options, including domestic incineration (combustible materials) and composting of a rotten fraction of waste.

The concepts of the management hierarchy should be applied with the implementation of the MSWPCM Act. For all the types of waste referred to in the Act, the enforcement of the Act should be carried out. In order to align Malaysia with future sustainable development, waste management is crucial.

2.1.3 Movement of Legislation to Eliminate Waste

Waste management is primarily the responsibility of state and territorial governments that, in accordance with applicable legislation, policies, and programmers, monitor and manage waste. Local authorities, as provided for in the legislative system of each state or territory, are responsible for waste management within their local areas. In providing household waste collection and recycling facilities, the conservation and operation of landfill sites, the introduction of education and awareness initiatives, and the provision and protection of litter instruments, local governments play an important role. The law governs the management of waste; waste management plans and programmers; the rights and obligations of legal agencies and individuals dealing with waste management; the manner and conditions of collection, transport, care, treatment, storage, and disposal of waste; the manufacture, export, and transit of waste; monitoring; the information system and funding to fix current issues.

Resource quality is the current subject of the regulatory debate on urban solid waste management. Waste managers see themselves as having progressed beyond merely handling municipal waste to resolve human health and environmental concerns and are now focused on waste reduction and recovery from waste generated. Control of waste is not just the task of governments. Waste management and recovery include a number of sectors and firms, as well as towns, households, and individuals.

2.2 Solid Waste Management Legislation Overview

SWM has recently become a global concern in urban environments like Malaysia, which could have a negative effect on the economy. A review of the legislation on solid waste is covered in this section.

2.2.1 Identify the Types of Solid Waste and the Specific Legislation

Municipal Solid Waste (MSW) contains packaging, materials for herbal cutting, furniture, clothes, bottles, leftover food, newspapers, instruments, paint, and batteries. Municipal Solid Waste (MSW) is made up of daily items used and then thrown away, the Environmental Protection Agency of the United States [8] said. Homes, schools, hospitals, and factories are the sources of such waste. MSW is generally a term used to describe a heterogeneous collection of waste produced in urban areas that varies between regions [7].

Solid waste is described as managed solid waste by the Malaysian Solid Waste and Public Cleansing Management Act [2], which includes commercial solid waste, household solid waste, institutional solid waste, and public solid waste.

(a) Any scrap material or other unacceptable surplus substance or rejected product resulting from the use of any process.
(b) Any material to be disposed of as broken, worn out, contaminated or otherwise damaged.

(c) Any other material necessary to be disposed of by the authority in accordance with this Act or any other written legislation. The Act also classifies managed solid waste as follows: (3a) Commercial solid waste: Any solid waste produced from any commercial activity such as supermarkets, store lots, etc.; (3b) Solid waste construction: Any solid waste created from any construction or demolition operation, including improvement, preparation, repair or alteration work; and (3c) Solid waste from the household: Any solid waste created by a household and of a nature that, when occupied as a dwelling house, is or is filthily generated or produced by any premises and includes garden waste.
(d) Solid industrial waste – solid wastes treated: (4a) Wastes obtained from the production process (product rejects, trimmings, surplus etc.); (4b) Wastes produced from the packaging of goods, raw materials (wood pallets, carton boxes, plastic sheets etc.); (4c) Quite particular and homogeneous forms.
(e) Solid industrial waste – non-process solid wastes: (5a) Wastes developed from food courts/canteens (food wastes, plastic bottles, tin cans etc.); (5b) Office-generated waste (computer papers, magazines, plastic bags, aluminium cans, etc.).
(f) Institutional solid waste: Any solid waste which is created by (6a) Any premises licensed for use wholly or partly for religious worship or for charitable purposes under any written law or by the State Authority; (6b) Any premises used by any agency of the Federal or State Government, any municipal authority or legislative body; (6c) Every school premises; (6d) All health services; (6e) any premises which are used as public zoos, museums, libraries, orphanages.
(g) Imported solid waste: Any solid waste produced in other nations and imported for processing or disposal to Malaysia.
(h) Public solid waste: Any solid waste created by public places under any local authority's oversight or control. This includes any open field, parking space, garden, leisure and pleasure ground or square, etc.

2.2.2 The Solid Waste Management Legislation Overview

There are three levels of solid waste management legislation:

(a) Structure regulations, such as the definition of waste and the criteria for permits.
(b) Highly qualified criteria to guarantee a high degree of environmental protection for waste facilities management.
(c) Criteria for particular waste sources, such as recycling or hazard mitigation measures.

2.3 Implementation of Solid Waste Legislation Act

One of the key mechanisms for improving the delivery of public administration and public services is regulation. An enactment provides for and governs the management and public cleaning of managed waste material to ensure adequate hygiene and for incidental problems.

2.4 Introductory to Solid Waste Legislation Enforcement Act

To ensure that solid waste disposal functions effectively, there are several kinds of solid waste law enforcement acts set up to ensure that solid waste disposal works effectively. Many public policy scholars have drawn attention to the issue of insufficient implementation of solid waste management policies in developing countries. Indeed, poor solid waste management will have an adverse impact on the development of human health, society, and the environment. Researching solid waste patterns and taking steps to reinforce the introduction of solid waste policy is therefore very important.

Under the long-term national development plan, environmental security and restoration are becoming more important. In this regard, Malaysia expressed its commitment to environmental protection and conservation [9]. A strong legal structure needs to be developed for pollution reduction activities, as such activities include legal authority, competence, and punishment to ensure successful compliance [10]. Malaysia's legal and institutional structure is the basis for analysing the current environment and SWM framework in Malaysia. This structure is an indication of the situation we are now pursuing as the legal and structural framework for policy and practice is the basis for its development [11].

2.5 United States of America

International developments in the management of solid waste provide a benchmark for the assessment and potential course of the integrated management of solid waste in Malaysia [12]. Other countries are more advanced in solving this specific problem than Malaysia. As such, it is possible to model relevant development efforts, technologies, facilities, and programmes from the experience of other countries. This chapter analyses applicable policies of several countries to benchmark solid waste management initiatives. For the US, all types of Federal Acts related to solid and hazardous waste management are covered in this chapter. They are mainly the Solid Waste Disposal Act (1965), Resource Recovery Act (1970), the Resource Conservation and Recovery Act (1976), the Comprehensive Environmental Response, Compensation, and Liability Act (1980), the Solid Waste Disposal Act

Amendments (1980), the Used Oil Recycling Act (1980), and the Hazardous and Solid and Hazardous Waste Amendments (1984).

2.5.1 *Solid Waste Disposal Act (1965) and Resource Recovery Act (1970)*

The US federal solid waste law has gone through four major phases. The Solid Waste Disposal Act (SWDA) became the first phase law on 20 October 1965. It was a great endeavour in its original nature to address the solid waste problems the country was facing through a series of in-depth projects, investigations, trials, training, demonstrations, surveys, and studies. The decade preceding its passage confirmed that the SWDA was not adequately designed to fix the country's rising amount of waste disposal problems. This first US federal law discusses MSW management was passed by Congress in 1965 and focused on research, demonstrations, training and also promoted increased state-level activity. The SWDA authorized the Department of Health, Education, and Welfare should provide state and local governments with technical and financial assistance in preparing and implementing environmental sustainability and solid waste disposal strategies. Thus, just as the states established MSW management strategies under the stimulation of federal grants given by the SWDA, with the enactment of waste management guidelines, the federal government became more actively involved in waste management [13].

The US federal government believed that with the introduction of new technologies along with the implementation of SWDA, they had found the ideal formula for a management system that would be able to solve the MSW problem. However, along with the growth in the urban population, the amount of MSW increased. After the passage of SWDA in 1965, the quantity of MSW generated continued to increase. There was a decline in supply and a rise in demand to bring this change in MSW in simple economic terms. The number of suitable methods of disposal decreased while the demand for a place to put MSW grew. This move caused about 90% of the waste in this country to be disposed of on the ground [13]. In a second phase, the Resource Recovery Act of 1970 emphasized reclaiming energy and materials from solid waste instead of dumping. Specifically, the Resource Recovery Act of 1970 provides state and local governments with technical and financial help in planning and developing resource recovery and waste disposal systems.

2.5.2 Resource Conservation and Recovery Act (1976), Comprehensive Environmental Response, Compensation, and Liability Act (1980), Solid Waste Disposal Act Amendments (1980), Used Oil Recycling Act (1980), and Hazardous and Solid Waste Amendments (1984)

In a third phase, the Resource Conservation and Recovery Act (RCRA) of 1976, which became law on 21 October 1976, was enacted. There were major amendments to the act. The Resource Conservation and Recovery Act, generally referred to as RCRA, is the primary legislation of the nation controlling the disposal of solid and hazardous waste, according to the United States Environmental Protection Agency (USEPA). The US Congress approved RCRA to address the rising challenges threatening the nation from the quantity of municipal and industrial waste. The RCRA, amending the Solid Waste Disposal Act of 1965 and the Resource Recovery Act of 1970, sets national objectives for the protection of human health and the environment from potential hazards of waste disposal, the conservation of energy and natural resources, the reduction of the quantity of waste produced and the sustainable management of waste [14]. RCRA made open dumping illegal and also instituted the first federal permitting programme for hazardous waste. RCRA focuses only on active and future facilities and does not address abandoned or historical sites, which are managed separately under the Comprehensive Environmental Response, Compensation, and Liability Act (CERCLA) of 1980, which is commonly known as Superfund [15]. The Solid Waste Disposal Act Amendments of 1980 targeted hazardous waste dumping. The Used Oil Recycling Act of 1980 defined the terms used oil, recycled oil, lubricating oil, and re-refined oil, and encouraged the state to use recycled oil. In the USA, the solid and hazardous waste legislation has been constantly strengthened and improved by the introduction of amendments to the above major laws and other specific related laws.

The fourth phase of basic major solid waste environmental laws began when the US Congress reauthorized and strengthened RCRA through the Hazardous and Solid Waste Amendments (HSWA) of 1984 [14]. The 1984 HSWA Amendments suggested a policy shift away from land disposal and towards more preventive solutions. RCRA has been amended on two occasions since HSWA: (a) the Federal Facility Compliance Act of 1992, which strengthened enforcement of RCRA at federal facilities; and (b) the Land Disposal Program Flexibility Act of 1996, which provided regulatory flexibility for land disposal of certain wastes.

In order to protect the environment, the amended RCRA set national objectives [14]: (a) protecting public health and the environment from the possible risks of disposal of waste; (b) energy and natural resources management; (c) decrease in the rate of waste generated; and (d) waste management in an environmentally responsible manner.

The US Environmental Protection Agency (USEPA) has established comprehensive regulations specifying what items are categorized as solid waste and hazardous waste. RCRA-governed materials are known as solid wastes. Only materials which comply with the RCRA definition of solid waste can be categorized as dangerous waste, subject to additional regulations.

Based on the United States Environmental Protection Agency, RCRA [14] has established three independent but interrelated programmes, which are:

(a) Under RCRA Subtitle C programmed hazardous waste offers a structure for hazardous waste management from the moment it is created to its ultimate disposal, essentially from 'cradle to grave'.
(b) Under RCRA Subtitle D, the Solid Waste Policy requires states to develop comprehensive policies for the management of non-hazardous industrial solid waste and municipal solid waste, to create requirements for municipal solid waste landfills and other solid waste disposal facilities, and to prohibit open dumping of solid waste.
(c) RCRA Subtitle I's programmed underground storage tank (UST) tracks underground storage tanks containing hazardous materials and petroleum.
(d) Subtitle J was added by Congress after medical waste washed up on Atlantic beaches in the summer of 1988, revealing the inadequacy of medical waste management procedures. Subtitle J urged the USEPA to establish a 2-year demonstration programmed to monitor medical waste from production to disposal in demonstration countries.

2.5.3 Continuous US Legislation for Solid and Hazardous Waste Management

Solid and hazardous waste legislation has been constantly strengthened and improved by the introduction of amendments to the major laws mentioned above and other specific laws. RCRA has been amended on many occasions since the Hazardous and Solid Waste Amendments of 1984 (HSWA) in the fourth phase [16–20]:

1. Superfund Amendments and Reauthorization Act (SARA) of 1986 has amended the CERCLA of 1980, has increased state involvement in the Superfund program and has encouraged greater citizen participation in decision-making [25].
2. Medical Waste Tracking Act of 1988 has created RCRA Subtitle J, which expired on 22 March 1991 [26].
3. Ocean Dumping Ban Act of 1988 has prohibited all municipal sewage sludge and industrial waste dumping into the ocean [27].
4. Pollution Prevention Act of 1990 has required the USEPA to establish an Office of Pollution Prevention and the owners and operators of manufacturing facilities to report annually on source reduction and recycling activities [28].
5. The Federal Facility Compliance Act of 1992 has strengthened enforcement of RCRA at federal facilities [14].

6. The Land Disposal Program Flexibility Act (LDPFA) of 1996 has directed the USEPA Administrator to provide additional flexibility to approved states for any landfill that receives 20 tons or less of municipal solid waste per day. The additional flexibility applied to alternative frequencies of daily cover, frequencies of methane monitoring, infiltration layers for the final cover, and means for demonstrating financial assurance. The additional flexibility allows owners and operators of small municipal solid waste landfills (MSWLFs) the opportunity to reduce the costs of MSWLF operations while still protecting human health and the environment. This direct final rule (29 June 1997) recognizes that these decisions are best made at the state and local level and, therefore, offers this flexibility to approved states [29].
7. The RCRA Expanded Public Participation Rule of 1996 has encouraged communities' involvement in the process of permitting hazardous waste facilities and has expanded public access to information about such facilities [30, 31].
8. The Federal Insecticide, Fungicide, and Rodenticide Act (FIFRA) of 1996 has provided for federal regulation of pesticide distribution, sale, and use. All pesticides distributed or sold in the United States must be registered (licensed) by USEPA. Before USEPA may register a pesticide under FIFRA, the applicant must show, among other things, that using the pesticide according to specifications 'will not generally cause unreasonable adverse effects on the environment' [32].
9. The Hazardous Waste Combustors Revised Standards Final Rule – Part 1 of 1998 has provided for a conditional exclusion from RCRA for fuels that are produced from hazardous waste and promotes the installation of cost-effective pollution prevention technologies [34].
10. The RCRA Cleanup Reforms I&II of 1999 and 2001 have accelerated the clean-up of hazardous waste facilities regulated under RCRA, and the Used Oil Management Standards of 2003 defined used oil management standards in a more exact manner [34].
11. Alternative Liner Performance, Leachate Recirculation, and Bioreactor Landfills, 6 April 2000: USEPA has considered revisions to the Criteria for MSWLs (40 CFR part 258) regarding the use of alternative liners when landfill leachate is recirculated and allowing the operation of landfills as more advanced bioreactors. USEPA has requested more information on these types of landfill processes to proceed with any revisions [35].
12. MSW Landfill Location Restrictions for Airport Safety – Technical amendment, 8 October 2002: USEPA has amended the location restriction section in the criteria for municipal solid waste landfills (MSWLFs) under Resource Conservation and Recovery Act (RCRA) to add a note providing information about landfill siting requirements enacted in the Wendell H. Ford Aviation Investment and Reform Act for the twenty-first century (Ford Act). The amendment does not change existing criteria under RCRA with respect to siting MSWLF units. Background information for this notice is available through Regulations.gov using docket number EPA-HQ-RCRA-2002-0034. More

information can be located using 67 FR 45948 and 67 FR 45915 at FederalRegister.gov. [35].

13. Lead-Based Paint Rule and Supporting Materials, 18 June 2003: They include (a) the USEPA Criteria for Classification of Solid Waste Disposal Facilities and Practices and (b) Criteria for Municipal Solid Waste Landfills (MSWLFs): Disposal of Residential Lead-Based Paint Waste; Final Rule [35].

2.5.4 Clean Air Act (1970) Related to Solid and Hazardous Waste Management

The Resources Conservation and Recovery Act (RCRA) is the US public law that creates the framework for theproper management of both hazardous solid waste and non-hazardous solid waste [14]. The law describes the waste management programme mandated by the US Congress that gave USEPA authority to develop the RCRA programme. This section introduces additional important US legislation indirectly related to solid and hazardous waste management. They are the Clean Air Act (1970); Clean Water Act (1972); Toxic Substances Control Act (1976); and Federal Insecticide, Fungicide, and Rodenticide Act (1996) related to solid and hazardous waste management. They are often used interchangeably to refer to the US environmental laws, regulations, and USEPA policies and guidance.

The Clean Air Act (CAA) was initially enacted in 1963 and later amended in 1965, 1967, 1970, 1977, and 1990 is the comprehensive US federal law that regulates air emissions from stationary and mobile sources [21]. Among other things, this law authorizes USEPA to establish National Ambient Air Quality Standards (NAAQS) to protect public health and public welfare and to regulate emissions of hazardous air pollutants. Today, the major regulatory programmes under the Clean Air Act are partially related to solid and hazardous waste management, such as incineration facility installation, incinerator emission control (greenhouse gases, carbon dioxide, methane, heavy metals, hazardous air pollutants, etc.), municipal landfill installation, bioreactor landfill installation, bioreactor landfill operation, bioreactor landfill emission control, solid and hazardous waste transportation, transportation station operation, hazardous solid waste landfill, landfill emission control (greenhouse gases, carbon dioxide, methane, etc.), permit application, state collaboration, etc. USEPA began regulating greenhouse gases (GHGs) from mobile and stationary sources of air pollution under the Clean Air Act for the first time on 2 January 2011, after having established its first auto emissions standards in 2010. USEPA regulation of greenhouse gas emissions by the USA includes carbon dioxide and methane, which are both generated from incinerators and landfills [36].

(a) National Ambient Air Quality Standards (NAAQS), which legally decides how much ground-level ozone (O_3), carbon monoxide (CO), particulate matter (PM_{10}, $PM_{2.5}$), lead (Pb), sulphur dioxide (SO_2, and nitrogen dioxide (NO_2) are allowed in the outdoor air;

(b) National Emissions Standards for Hazardous Air Pollutants (NESHAPs), which legally decides how much of 187 toxic air pollutants are allowed to be emitted from industrial facilities and other sources. Under the CAA, hazardous air pollutants (HAPs, or air toxics) are air pollutants other than those for which NAAQS exist, which threaten human health and welfare. The NESPHAPs are the standards used for controlling, reducing, and eliminating HAPs emissions from stationary sources such as industrial facilities.
(c) New Source Performance Standards (NSPS): Rules for the equipment required to be installed in new and modified industrial facilities, and the rules for determining whether a facility is 'new'. The 1970 CAA required USEPA to develop standards for newly constructed and modified stationary sources (industrial facilities) using the 'best system of emission reduction which (taking into account the cost of achieving such reduction) the USEPA determines has been adequately demonstrated'. USEPA issued its first NSPS regulation the next year, covering steam generators, incinerators, Portland cement plants, and nitric and sulphuric acid plants.
(d) Acid Rain Program (ARP): An emissions trading programme for power plants to control the pollutants that cause acid rain. The 1990 CAA Amendments created a new title to address the issue of acid rain, and particularly nitrogen oxides (NO_x) and sulphur dioxide (SO_2) emissions from electric power plants powered by fossil fuels and other industrial sources. The Acid Rain Program was the first emissions trading programme in the United States, setting a cap on total emissions that was reduced over time by way of traded emissions credits rather than direct controls on emissions.
(e) Mobile Source Programmes: Rules for pollutants emitted from internal combustion engines in vehicles. Since 1965, Congress has mandated increasingly stringent controls on vehicle engine technology and reductions in tailpipe emissions. Today, the law requires USEPA to establish and regularly update regulations for pollutants that may threaten public health, from a wide variety of classes of motor vehicles, that incorporate technology to achieve the 'greatest degree of emission reduction achievable', factoring in availability, cost, energy, and safety.
(f) Fuel Controls: USEPA has regulated the chemical composition of transportation fuels since 1967, with significant new authority added in 1970 to protect public health.
(g) State Implementation Plans (SIPs) (40 CFR 51, 40 CFR 52): Since its earliest version in 1963, the Clean Air Act has set up a cooperative federalist programme for developing pollution control standards and programmes. Rather than create an entirely federal system, the CAA imposes responsibilities on the US states to create plans to implement the Act's requirements. USEPA then reviews, amends, and approves those plans.
(h) Title V Permitting. The 1990 amendments authorized a national operating permit programme, covering thousands of large industrial and commercial sources. It required large businesses to address pollutants released into the air, measure their quantity, and have a plan to control and minimize them as well as to peri-

odically report. These consolidated requirements for a facility into a single document. In non-attainment areas, permits were required for sources that emit as little as 50, 25, or 10 tons per year of volatile organic compounds (VOCs), depending on the severity of the region's non-attainment status. Most permits are issued by state and local agencies. If the state does not adequately monitor requirements, the USEPA may take control.

2.5.5 Clean Water Act (1972) and Water Quality Act (1987) Related to Solid and Hazardous Waste Management

The Clean Water Act (CWA) establishes the basic structure for regulating discharges of pollutants into the waters of the United States and regulating quality standards for surface waters. The basis of the CWA was enacted in 1948 and was originally called the Federal Water Pollution Control Act, but the Act was significantly reorganized and expanded in 1972. 'Clean Water Act' became the Act's common name with amendments in 1972. Major changes have subsequently been introduced via amendatory legislation, including the Clean Water Act of 1977 and the Water Quality Act (WQA) of 1987 [37].

Under the CWA, USEPA has implemented pollution control programmes such as setting wastewater standards for industry and municipalities. USEPA has also developed national water quality criteria recommendations for pollutants in surface waters. The CWA made it unlawful to discharge any pollutant from a point source into navigable waters unless a permit was obtained under the USEPA's National Pollutant Discharge Elimination System (NPDES) permit program or under a specific State's State Pollutant Discharge Elimination System (SPDES) that has jurisdiction to control local discharges. The CWA has six major statutory provisions as follows:

(a) Title I – Research and Related Programs
(b) Title II – Grants for Construction of Treatment Works
(c) Title III – Standards and Enforcement: (a) Discharge Permits Required; (b) Technology-Based Standards Program; (c) Water Quality Standards Program; (d) National Water Quality Inventory; (e) Enforcement; (f) Federal Facilities; (g) Thermal Pollution; (h) Non-point Source Management Program; (i) Military Vessels
(d) Title IV – Permits and Licenses: (a) State Certification of Compliance; (b) NPDES Permits for Point Sources; (c) Dredge and Fill Permits, and (d) POTW Biosolids Management Program
(e) Title V – General Provisions: (a) Citizen Suits; and (b) Employee Protection
(f) Title VI – State Water Pollution Control Revolving Funds

The Clean Water Act was one of the United States' first and most influential modern environmental laws which are primarily administered by the USEPA in coordination with state governments, though some of its provisions, such as those involving filling or dredging, are administered by the US Army Corps of Engineers. For solid and hazardous waste management, an environmental engineer or manager is concerned about point discharge from landfill, non-point source from the farm, thermal pollution from incinerators, biosolids management, and dredge and fill operations in lakes and rivers, so CWA must be carefully examined and in compliance.

The CWA introduced the NPDES (or SPDES, if authorized by USEPA), a permit system for regulating point sources of pollution. Point sources mainly include: (a) industrial facilities (including manufacturing, mining, shipping activities, oil and gas extraction and service industries); (b) municipal governments (particularly sewage treatment plants) and other government facilities (such as military bases); and (c) agricultural facilities, such as animal feedlots.

Non-point source pollutants, such as sediments, nutrients, pesticides, fertilizers, and animal wastes, account for more than half of the pollution in the US waters. Agricultural stormwater discharges and irrigation return flows were specifically exempted from permit requirements. The US Congress, however, provided support for research, technical, and financial assistance programmes at the US Department of Agriculture to improve run-off management practices on farms.

The CWA Section 404 requires that a discharger of dredged or fill material obtain a permit unless the activity is eligible for an exemption. Essentially, all discharges affecting the bottom elevation of a jurisdictional water body require a permit from the US Army Corps of Engineers (USACE). These permits are an essential part of protecting streams and wetlands, which are often filled by land developers. Congress placed the so-called recapture clause limitation on these new project exemptions. Under CWA section 404(f)(2), such new projects would be deprived of their exemption if all of the following three characteristics could be shown: (a) a discharge of dredged or fill material in the navigable waters of the United States; (b) the discharge is incidental to an activity having as its purpose the bringing of an area of navigable waters into a use to which it was not previously subject; and (c) where the flow or circulation of navigable waters may be impaired, or the reach of such waters may be reduced. To remove the exemption, all of these requirements must be fulfilled – the discharge, the project purpose of bringing an area into a use to which it was not previously subject, and the impairment or reduction of navigable waters.

The 1987 WQA created a programme for the management of biosolids (sludge) generated by publicly owned treatment works (POTWs). The WQA Act instructed USEPA to develop guidelines for the usage and disposal of sewage sludge or biosolids. The USEPA regulations: (a) identify uses for sewage sludge, including disposal; (b) specify factors to be taken into account in determining the measures and practices applicable to each such use or disposal (including publication of information on costs), and (c) identify concentrations of pollutants which interfere with each such use or disposal. The term *biosolids* are used to differentiate treated sewage sludge that can be beneficially recycled. Environmental advantages of sewage sludge consist of application of sludge to land due to its soil condition properties and nutrient content. Advantages also extend to a reduction in adverse health effects of incineration, decreased chemical fertilizer dependency, diminishing greenhouse gas emissions deriving from incineration and reduction in incineration fuel and energy costs. Benefits of reusing sewage sludge from the use of organic and nutrient content in biosolids include: (a) improving marginal lands and serving as supplements to fertilizers and soil conditioners; (b) increasing forest productivity, accelerated tree growth, revegetation of forest land previously devastated by natural disasters or construction activities; and (c) aiding the growth of final vegetative cap for municipal solid waste landfills is enormously beneficial.

In the Water Quality Act (WQA) of 1987, US Congress defined industrial stormwater dischargers and municipal separate storm sewer systems (often called 'MS4') as point sources and required them to obtain NPDES (or SPDES, if authorized by USEPA) permits, by specific deadlines. The permit exemption for agricultural discharges continued.

2.5.6 Toxic Substances Control Act (1976) Related to Solid and Hazardous Waste Management

The Toxic Substances Control Act (TSCA) of 1976 provides USEPA with authority to require reporting, record-keeping, and testing requirements, and restrictions relating to chemical substances and/or mixtures. Certain substances are generally excluded from TSCA, including, among others, food, drugs, cosmetics, and pesticides. TSCA addresses the production, importation, use, and disposal of specific chemicals, including polychlorinated biphenyls (PCBs), asbestos, radon and lead-based paint. Therefore, the amount of PCB, asbestos, lead-based paint going into a landfill, an incinerator, or any other solid and hazardous waste treatment facilities must be examined carefully. Various sections of TSCA provide authority for the following [38]:

(a) Require, under Section 5, pre-manufacture notification for 'new chemical substances' before manufacture.
(b) Require, under Section 4, testing of chemicals by manufacturers, importers, and processors where risks or exposures of concern are found.
(c) Issue Significant New Use Rules (SNURs), under Section 5, when it identifies a 'significant new use' that could result in exposures to, or releases of, a substance of concern.
(d) Maintain the TSCA Inventory, under Section 8, which contains more than 83,000 chemicals. As new chemicals are commercially manufactured or imported, they are placed on the list.
(e) Require those importing or exporting chemicals, under Sections 12(b) and 13, to comply with certification reporting and/or other requirements.
(f) Require, under Section 8, reporting and record-keeping by persons who manufacture, import, process, and/or distribute chemical substances in commerce.

(g) Require, under Section 8(e), that any person who manufactures (including imports), processes, or distributes in commerce a chemical substance or mixture and who obtains information which reasonably supports the conclusion that such substance or mixture presents a substantial risk of injury to health or the environment to immediately inform USEPA, except where USEPA has been adequately informed of such information. USEPA screens all TSCA (e) submissions as well as voluntary 'For Your Information' (FYI) submissions. The latter is not required by law but are submitted by industry and public interest groups for a variety of reasons.
(h) Require inspections and investigations under Toxic Substances Control Act Compliance Monitoring regulations.
(i) Require compliance under waste, chemical, and clean-up enforcement regulations.

2.5.7 Safe Drinking Water Act (1974) Related to Solid and Hazardous Waste Management

The US Federal legislation is very complex. In the case of a new bioreactor landfill engineering project, its solid waste examination, solid waste collection, bioreactor landfill siting, bioreactor design, landfill construction, landfill operation, air emission monitoring, landfill leachate collection, leachate treatment, air emission discharge of methane and carbon dioxide, bioreactor liquid effluent discharges to a river, liquid effluent discharge to a groundwater source, leachate discharge to a publicly owned treatment works (POTW), bio-solids management, permit application, etc. will involve the examination and compliance of many previously introduced federal rules and regulations. In addition to required RCRA compliance for bioreactor landfill itself, more must be considered: (a) CAA compliance will be required for air emission discharge of methane and carbon dioxide; (b) CWA compliance will be required for leachate discharge to a POTW or a pre-treatment facility; (c) WQA compliance will be required for biosolids management; (d) Safe Drinking Water Act (SDWA) of 1974 compliance along with RCRA, CERCLA (Superfund) compliance will be required if the possibility of bioreactor landfill siting and/or liquid effluent discharge to a groundwater source is considered. In case the collected solid waste contains a certain amount of hazardous substances, then TSCA compliance shall be examined. Yes, it is hard to believe that sometimes even the SDWA of 1974 is related to solid and hazardous waste management.

The Safe Drinking Water Act (SDWA) of 1974 is the principal federal environmental law in the United States intended to ensure safe drinking water for the public. Pursuant to the act, the US Environmental Protection Agency (USEPA) is required to set standards for drinking water quality and oversee all states, localities, and water suppliers that implement the standards. In the United States, the SDWA applies to currently all 151,000+ public water systems (PWS) but does not cover private wells; even about 13% of US households were served by private wells in 2020. The SDWA does not apply to bottled water because it is regulated by the Food and Drug Administration (FDA) under the Federal Food, Drug, and Cosmetic Act. The following is an oversimplified SDWA content in which the portions related to solid and hazardous waste management are identified later [39]:

(a) National Primary Drinking Water Regulations: (1) microorganisms, (2) disinfectants, (3) disinfection by-products, (4) inorganic chemicals, (5) 'lead-free' plumbing requirements, (6) organic chemicals, (7) radionuclides
(b) National Secondary Drinking Water Standards
(c) Health advisories
(d) State standards
(e) Future standards: (1) unregulated contaminants, (2) perchlorate, (3) perfluorinated alkylated substances, and (4) non-community water systems
(f) Monitoring, compliance, and enforcement
(g) Protection of underground sources of drinking water: (1) underground injection control (UIC) program and hydraulic fracturing exemption, (2) wellhead protection areas, (3) emergency power, (4) judicial review and civil actions
(h) Related programmes: (1) airline water supplies, (2) source water assessment, (3) whistle-blower protection
(i) History: (1) prelude; (2) 1974 Act; (3) 1986 Amendments; (4) 1996 SDWA Amends; (5) 2005, 2011, 2015, 2016, and 2018 Amendments
(j) Environmental justice

It is noted above that the federal drinking water standards are organized into six groups [40]: (a) microorganisms, (b) disinfectants, (c) disinfection by-products, (d) inorganic chemicals, (e) 'lead-free' plumbing requirements, (f) organic chemicals, and (g) radionuclides. The regulations include both mandatory requirements (maximum contaminant levels or MCLs; and treatment techniques or TTs) and non-enforceable health goals (maximum contaminant level goals, or MCLGs) for each included contaminant. As of 2019, USEPA has issued 88 standards for microorganisms, chemicals, and radionuclides. For some contaminants, USEPA establishes a TT instead of an MCL. TTs are enforceable procedures that drinking water systems must follow in treating their water for a contaminant. MCLs and TTs have additional significance because they can be used under the RCRA/CERCLA (Superfund) law as 'Applicable or Relevant and Appropriate Requirements' in clean-ups of contaminated sites on the National Priorities List, in case the site is chosen as a site for construction of a future bioreactor landfill, for instance.

Protection of underground sources of drinking water is enforced by USEPA under the SDWA because any groundwater contamination can have disastrous consequences for drinking water source and, in turn, for human health. the pollutants are hard to trace once it enters the ground, and polluted aquifers are hard to remediate. An underground source of drinking water (USDW) means an aquifer with sufficient quality and quantity of groundwater to supply a public water system now or in the future.

The SDWA's Underground Injection Control (UIC) Program prohibits any underground injection which endangers drinking water sources. Even clean water can result in the illegal movement of a fluid containing a contaminant into a USDW because injections of clean water into the ground can cause the movement of contaminants, and may dissolve into clean water as the injected water passes through the soil on its way to an aquifer. The SDWA and its implementing regulations are not concerned with whether an injected fluid is itself contaminated. Rather, they are concerned with the result of injection activity. A permit applicant must show that the proposed activity will not allow 'the movement of fluid containing a contaminant'.

The 1974 SDWA authorized USEPA to regulate injection wells in order to protect underground sources of drinking water. The UIC permit system is organized into six classes of wells: (a) Class I: Industrial waste (hazardous and non-hazardous) and municipal wastewater disposal wells; (b) Class II: Oil and gas-related injection wells (except wells solely used for production; (c) Class III: Solution mining wells; (d) Class IV: Shallow hazardous and radioactive waste injection wells (no longer permitted); (e) Class V: Wells that inject non-hazardous fluids into or above underground sources of drinking water; and (f) Class VI: Geologic sequestration wells for carbon dioxide. A bioreactor landfill's liquid effluent, for example, is either an industrial waste or municipal wastewater within Class I, depending on the ownership (industry or municipality). This portion of the SDWA must be considered for landfill siting or leachate management.

2.6 Europe

This chapter analyses applicable policies in several countries to benchmark solid waste management initiatives. For the European, there are five types of acts cover in this chapter. They are Waste Framework Directive (75/442/1975), Landfill of Waste Directive 1999, Directive 2008/98/EC (2008), Directive 2006/12/EC (2006) and Regulation (EC) No 1013/2006 of the European Parliament and of the Council of 14 June 2006.

2.6.1 Waste Framework Directive (75/442/1975)

The adoption, in part, in 1975, of Waste Framework Directive 75/442/EEC was a reaction to the enactment by some Member States of legislation meant to provide for a national waste policy framework and attempted to create a coherent collection of measures to be implemented in all Member States (Article 6) [22]. It offers a simple description of concepts relevant to waste and waste and a waste management system. The Directive's core responsibilities are as follows:

1. *Waste management strategies*: In Article 7, in order to achieve the goals set out in the Structure Directive, responsible authorities are obliged to draw up waste management plans. These agreements can also include products, but they must include details on products in accordance with the provisions of the Directive: (a) The sort, quantity, and source of waste to be recovered or discarded; (b) General technical specifications; (c) All basic transparent waste arrangements; and (d) Relevant disposal sites or facilities. The Directive also emphasizes the need to work together on waste management strategies according to Article 7. The Member States must ensure that in compliance with the Framework Directive, each waste producer has its own waste treated or disposed of by a collector of private or public waste, which is stated in Article 8.
2. *Reporting*: The public should also be able to access all of this material. Furthermore, all institutions or undertakings carrying out 'disposal' or 'recovery' operations are needed to maintain a log of the amount, form, source, and, where appropriate, location, frequency of collection, method of travel, and method of treatment for all of the waste referred to in Annex I and for all of the operations referred to in Annexes II A and II B as mentioned in Article 14. Member States are required to report on all waste referred to in Annexes II A and II B and must be submitted every 3 years to the Commission as claimed in Article 16.
3. *Permits and Inspection*: Any establishment or association engaged in recovery or disposal operations shall obtain from the competent authority the authority referred to in Article 10 as a permit, as set out in Annexes II A and II B. Such permits shall be given for the transport of waste to any private undertaking operating in the local territory. In order to ensure compliance with the municipal standards for the prevention of harm to the health of local citizens and the environment, all disposal or treatment plants shall be inspected and revised (Article 13).

2.6.2 Landfill of Waste Directive 1999

Waste must be disposed of and sent to landfills that comply with the waste landfill requirements of Directive 1999/31/EC. Dumping is the least appropriate alternative and should be limited to the minimum possible. The Directive aims to avoid or minimize the harmful effects of waste disposal on the general environment of surface water, groundwater, soil, air, and public health to the fullest degree possible by establishing strict technical requirements for waste and landfills. The purpose of the Directive is to protect human and environmental health by means of consistent licensing, operation, inspection, and aftercare procedures for new and existing landfills. The Directive also provides a process by which the Commission will, from time to time, lay down and implement new technical requirements.

The different types of waste are specified by the Landfill Directive and relate to all landfills, defined as dumpsites for waste disposal on or on land. Hazardous, non-hazardous, and inert waste are categorized into three groups of landfills, based on Article 4. The Directive also aims to reduce the volume of emissions of methane gas from landfills by setting limits on the amount of biodegradable waste to be disposed of in landfills (Article 5). Subsequent waste, such as liquid waste, fuel waste, toxic or oxidizing waste, medical and other infectious waste, tyres used, and, in some cases, any other form of waste which does not comply with the acceptance criteria set out in Annex II shall not be allowed to enter a landfill, including a declaration prohibiting the mixing of different types of waste in a landfill in Article 6 [23].

On the basis of Article 6, a standard procedure is laid down for the acceptance of waste in a landfill in order to avoid any hazards that are as follows:

1. Non-hazardous waste landfills need to be designated for agricultural and other non-hazardous waste.
2. Inert waste landfill areas must be used exclusively for inert waste.
3. The requirements for the admission of waste into each landfill class must be accepted by the Commission in compliance with the general principles of Annex II.
4. Waste must be processed before being landfilled.
5. In the sense of the directive, hazardous waste must be assigned to a hazardous waste landfill.

The operation and control of landfills also need to be taken care of, which is landfill users, whether public or private enterprises, the procedures laid down in the annexes to the Directive for acceptance shall be followed, management and monitoring of waste (Article 11 and Article 12). In addition, the operator shall continue to take aftercare steps until the responsible authority finds that the landfill is no longer a threat to physical or environmental health and ensures that the landfill is safely and properly closed (Section 13). Member States are then obligated to provide the Commission with reports every 3 years on the application of this guideline (Article 15) [23].

The aim of the Directive, as regards recovery costs, is to ensure that the true cost of landfilling is reflected in the cost to the waste producer so that it is not viewed as an inexpensive alternative to landfilling. The Directive requires the Member States to take steps to ensure that the price charged by the landfill operator is compensated by all costs related to the establishment and maintenance of a landfill site, including expected closure and aftercare costs for a period of at least 30 years (Article 10) [23].

2.6.3 *Directive 2008/98/EC (2008)*

Basic waste management concepts and interpretations include waste, recycling, recovery, etc. It explains when waste ceases to be waste and becomes a secondary raw material and how to distinguish between waste and by-products. 'Waste' is any material or object that the holder discards or needs to discard or must be discarded by the holder [24]. Directive 2008/98/EC of the European Parliament and of the

Council of 19 November 2008 concerning waste and repealing those directives, referred to as the Waste System Directive in its short title, entered into force on 12 December 2008. This Directive lays down steps to protect human health and the environment by minimizing or avoiding the production of waste, reducing the negative impacts of waste generation and management, reducing the overall impacts of the use of resources and improving the quality of the use needed for the transition to a circular economy, and ensuring the long-term stability of the use of waste. It imposes provisions on hazardous waste and waste oils (the old Hazardous Waste and Waste Oils Directives are repealed with effect from 12 December 2010) and includes two new goals for recycling and recovery to be reached by 2020, which are 50% preparation of some household waste items and other household-like sources for reuse and recycling and 70% preparation for recycling [26].

The main provisions of the Waste Framework Directive cover the meaning and classification of the main regulatory terms of waste law, the establishment of a current five waste management hierarchy (prevention, reuse planning, recycling, other conservation, such as energy recovery and disposal) to replace the previous three-level hierarchy (prevention, recovery, disposal) regulations [26]. The 'waste hierarchy', which stipulates a priority order from top to bottom of the least favoured 'prevention' choice, has a major influence on existing practices on waste management. The 'waste hierarchy' collects different steps before and/or after it has become waste that could be applied to substances, products, and/or items. In the Directive, prevention is specified as steps are taken to 'reduce the amount of waste'. These programmes include the reuse of goods and the increase of the product life cycle. Recovery operations are ready to inspect, disinfect, or repair waste items and parts of products so that they can be reused in preparation for reuse without any other pre-treatment. Any recycling operation involving the reprocessing of waste products into articles, materials, or substances for original or other uses shall be carried out for recycling purposes. This involves the reprocessing of organic material but does not involve the recovery of energy or the reprocessing of material used as fuels or as service backfill. Recovery is considered to be the opposite of 'disposal' because it consists of waste treatment. The key feature in Article 3(15) distinguishing 'recovery' from 'disposal' is that 'waste' must be used for a useful function of 'recovery' as a part of the process of 'recovery'. Any activity that meets the concept of 'recovery' in compliance with the Waste Directive 2008/98/EC is a further recovery. Last but not least, any operation not recovered, even if the recovery of substances or energy is claimed as a disposal stage within the hierarchy as a secondary result.

The Directive also requires the introduction of waste management policies and programmes for the avoidance of waste by the Member States [23]. Member States shall describe the preventive measures in place and evaluate the efficacy of, or other applicable measures, the examples of measures referred to in Annex IV to the updated Waste Framework Directive. The aim of such goals and initiatives is to break the connection between economic growth and the environmental impact of waste generation. Such plans will be checked and revised at least every sixth year as needed. They are either incorporated, as the case may be, into waste management plans or other environmental policy schemes or function as separate programmes. Where every such programme is introduced in the context of the waste management plan or other programmes, the preventive measures for waste management shall be clearly defined.

2.6.4 Directive 2006/12/EC (2006)

The aim of Directive 2008/98/EC and Directive 2006/12/EC (originally 75/442/EEC) is to provide a basis for consistent action by the Member States in order to solve the waste management problem. The scope of the Directive excludes such types of waste, such as gaseous effluents and where other legislation applies, hazardous waste, mining waste, wastewater, and certain agricultural waste [17]. Under the Directive, four general obligations are placed on the Member States to take the following measures:

(a) Establish an interconnected and appropriate network of disposal facilities taking into account the latest technology and without additional costs. The goal of the scheme is to make it possible for the community to be self-sufficient in disposing of waste.
(b) Promoting avoidance or reductions in the production and harm of waste, in particular by the use of safe technology and techniques for the final disposal of hazardous materials in waste for recovery and the creation and sale by the design of manufacture, use, or final disposal of products intended to have a minimal impact on the environment.
(c) Confirm that waste is recovered or disposed of without prejudice to the health of the population, without the use of technologies or methods capable of affecting the atmosphere, in particular without prejudice to the air, soil, plants, and animals, without prejudice to noise or odour, and without prejudice to the environment or places of special interest. Waste recycling, waste processing, or unauthorized disposal must be forbidden.
(d) Promoting waste recovery, including waste recycling, reuse, or recovery and the use of waste as an energy source

The Advisory Committee shall consist of representatives of the Member States under this Directive and shall be chaired by the Commission delegate to assist; for example, the Commission is amending its Annexes [17]. In accordance with their responsibilities, Member States shall inform the Commission of the actions they intend to take. In December 2010, Directive 2006/12/EC, repealed in accordance with Directive 2008/98/EC (Directive Hazardous Waste and Waste Oils), then takes on the role of the EU's central control measure in the waste legislation.

2.6.5 Regulation (EC) No 1013/2006 of the European Parliament and of the Council of 14 June 2006

Law (EC) No 1013/2006 of the European Parliament and of the Council of 14 June 2006 on waste shipments. No 1013/2006 of Regulation (EC) lays down systems for tracking and monitoring waste shipments inside, into, and out of the Society, generally known as the Waste Shipment Regulation. It states that the primary aim of the regulation is to protect the environment and that its effects on foreign trade are just accessory. This law lays down protocols and control schemes for the shipping of waste, depending on the origin, destination, and path of the shipment; the type of waste transported; and the handling of the waste at the destination of the shipment. The legislation only allows for shipments of waste within the Community, with or without transit through third countries, exporting to third countries from the Community, being transported to the Community from third countries, and in transit through the Community [22].

Regulation (EC) No 1013/2006 simplifies the earlier procedures for inspecting waste inside, into and outside of the European Community, implements reforms of international law and improves guidelines on compliance and cooperation between the Member States for illicit transportations. In this regulation, there is some forbidden part that must give attention to. Exports of hazardous waste intended for recycling are forbidden, with the exception of those intended for use in countries protected by the OECD decision and in third countries that are party to the Basel Convention. Moreover, exports of waste destined for recycling to third countries are prohibited, with the exception of countries that, under the European Free Trade Agreement, are parties to the Basel Convention (EFTA). The year 1992 saw the entry into force of the Basel Convention on Transboundary Movements of Risky Wastes. The Basel Convention's primary objective is to protect human health and the environment by environmentally sustainable management by minimizing, wherever possible, the production of hazardous waste. The Convention provides an integrated life cycle strategy for managing hazardous waste production that includes rigorous controls from generation to storage, transportation, treatment, reuse, recycling, and final disposal. The bilateral agreement with the Community has been concluded. Imports of waste intended for recycling or recovery from third countries are subject to the same regulations as exports [22].

The legislation has two methods for regulating waste shipments:

(a) Previous written notice and consent protocol (for both shipments of disposable waste and hazardous and semi-hazardous waste for recovery).
(b) A process by which such knowledge (applicable to recovery-oriented non-hazardous waste) is followed by waste shipments.

Waste shipping could result from a contract with the individual responsible for supplying or distributing the waste, and the recipient must be notified, and the contract must provide financial protection. Article 22 lays down the criteria for comprehensive notification of various options relating to the return or appropriate waste management of such waste. If it is not possible to complete a duly approved shipment, the competent shipping body is expected to ensure that the notifier returned the waste to the shipping State within 90 days, if it is satisfied; otherwise, the waste may be handled in an environmentally responsible manner. The prototype of the notification document is provided by Annex IA, and the 'amber list' (Annex IV) comprises waste subject to notice and approval, while the 'green list' (Annex III) contains waste referred to for information purposes only. Annex V then lists waste that is forbidden from shipping [17].

Member States shall, in accordance with Waste Directive 2006/12/EC, provide for inspections of facilities and undertakings and spot checks for waste shipments or related recovery or disposal, in accordance with Directive 2006/12/EC (Article 50). Then the Member States were also expected to set down the rules applicable to punishments for violations of the provisions of the Regulation. The Regulation also requires the Member States to work together to facilitate the prevention and detection of illegal shipments by recognizing and informing the Commission of its permanent staff members involved. Next, the Commission is required to produce a 3-year implementation report. Member States shall prepare before the end of the year an annual report in compliance with Article 13(3) of the Basel Convention on the additional reporting criteria referred to in Annex IX and transmit it to the Secretariat of the Commission and to the Basel Convention[22].

2.7 Japan

The Waste Management and Public Cleaning Act, enacted in December 1970, provides that municipal waste generated within its jurisdiction should be collected, transported, and disposed of by each municipality, according to Borongan and Okumura [23]. In addition to providing technical and financial assistance to municipalities, the government is introducing a fundamental waste reduction policy and a waste management planning development plan. The Ministry of the Environment is responsible for the management of MSW at the national level (MOE). Certain types of municipal waste are recycled according to relevant recycling laws, such as the following recycling legislation:

(a) Container and Packaging Recycling Law (enacted in June 1995, administered by the MOE and the Ministry of Economy, Trade and Industry (METI))
(b) Home Appliances Recycling Law (enacted in June 1998, administered by the MOE and the METI)
(c) Food Waste Recycling Law (enacted in June 2000, administered by the MOE and the Ministry of Agriculture, Forestry and Fishery)
(d) End-of-Life Vehicle Recycling Law (enacted in July 2002, administered by the MOE and the METI)

In line with 'Mottainai', Japan illustrates best practices through the 3R (reduction, reuse and recycling) principle in the creation of a sound material cycle society.

Mottainai is a long-lived Japanese word, meaning something lost without fully exploiting its potential is a shame [23].

2.7.1 Urgent Measures Law on Capacity Increasing of Waste Management Facilities 1963

The 'Law on Emergency Steps for the Development of Environmental Sanitation Facilities' was declared in 1963, and on that basis, the Ministry of Health and Welfare developed the First Five-Year Plan for the Development of Environmental Sanitation Facilities in 1965, outlining a strategy that as a general rule, the disposal of urban waste would be carried out by incineration, and any residual would be disposed of. Therefore, due to the huge increase in the quantity of plastics, progress has been made in the construction of waste incineration plants in each city. However, the calorific value of waste has also increased dramatically, and the construction of incineration plants has not been able to keep up with the increase in the calorific value produced by each local public authority [24].

2.7.2 Air Pollution Control Law 1968

Air pollution has been a problem in Japanese culture since the 1930s. In 1932, pollution resistance protests in the prefecture of Osaka led to the first legislative cap for particulate matter in the world. Rapid economic growth following the war became a cause of extreme air pollution many decades later. The Air Pollution Control Act was adopted in 1968, and the first SO2 standards were set a year later [25]. The aim of this Act is to protect people's health and the living environment from air pollution by controlling, inter alia, emissions of soot and smoke, organic volatile compounds and particulates associated with commercial operations at plants and workplaces, and demolition facilities, etc. [26]. The development of a soot- and smoke-emitting facility includes prior approval by the governor of the prefecture. If deemed necessary, governors of prefectures may order modifications or the abolition of the plans submitted. This law also includes that aggregate volume levels for sulphur oxide and nitrogen oxide can be set in a region with significant emissions, and the aggregate volume per factory unit can be calculated by plans drawn up by the governor of the prefecture.

There are also provisions for implementing controls in the Air Pollution Control Legislation. Mainly from facilities such as factories and mobile sources such as motor vehicles, pollutants are produced. As far as factories are concerned, national emission standards are laid down by the Minister for the Environment, but prefecture governors may lay down more stringent standards in areas under their authority. For specific substances, certain concentration requirements have been developed, and emissions of these materials from facilities cannot surpass these standards. The tasks of regulation are performed by prefectures and large cities.

There are provisions in the Air Pollution Control Legislation to reduce roadside air pollution from mobile sources and the use of other regulations to protect air quality [27]. Under the law, major actions include the following:

(a) The Environment Minister shall set overall allowable limits for the quantity of exhaust gases from motor vehicles. The Minister of the Environment shall also set the maximum allowable limits for exhaust gas amounts emitted by special motor vehicles produced by non-road engines such as building machines.
(b) The Minister of Property, Housing, Transport and Tourism, under the Law on Road Transport Vehicles, shall decide the required questions in relation to the controls of motor vehicle exhaust emissions by ensuring the minimum allowable limits set by the Minister for the Environment.
(c) The Minister of the Environment shall, where necessary for the prevention of air pollution caused by automotive exhaust gases, set maximum permissible limits on the quality of automotive fuel.

2.7.3 Water Pollution Control 1970

Enactment of water pollution legislation in 1970 and enforced in 1971; establishment of public water bodies for public purposes, including rivers, reservoirs, wetlands, ports, coastal areas, and their connection channels for unique inspection areas. The objective of this act is as follows[28]:

(a) Preventing contamination in drinking water, protecting human health, and maintaining the living environment by controlling effluent from plants and companies discharged into public water.
(b) Protecting the victims of this discharge by determining the responsibility of the factory owners and the company owners who are responsible. For people suffering from discharge-related health conditions, compensation may be assessed against factories and enterprises.

Specified Facility and Company Identification

The Water Pollution Control Act defines the method for mitigating water pollution [29]: (a) Define the characteristics of wastewater that causes water contamination from public water sources; (b) Define the facilities specified for the discharge of wastewater from these types; and (c) Identify and list factories and companies manufacturing offending types of wastewater processes as listed facilities.

The water pollution control law controls effluent wastewater from specified factories where specified facilities have been identified. The quality standard of effluent water for the preservation of the living environment is applied to factories and enterprises that discharge more than 50 m^3/a of wastewater volume. All factories and businesses that discharge wastewater are covered by standards for the protection of human health. For several types of industries, provisional quality standards of effluent water for the preservation of the living environment were applied, which required:

(a) Major investment in the installation of wastewater treatment (food processing, textile printing, tannery, processing of fur, slaughterhouses, dead animals).
(b) Additional time because treatment methods (starch processing, coal mining, non-metal mining, and processing of non-ferrous metals) did not meet the standard specifications.
(c) Postponement until it was better to alter wastewater facilities and manufacturing processes (pulp and distilleries).

After a given grace period, the above-mentioned industries were added to national effluent standards. Shortly after the enactment of the standard, the effluent standard for human health safety was extended to all sectors. The classification of facilities began in 1972 as a result of the water pollution control legislation. Under the statute, the livestock industry, hospitals, research and education institutions, waterworks, industrial waste treatment and disposal facilities, such as the stated facilities and enterprises, were also designated.

Stringent Effluent Standards
In the case of public water areas where the common national minimum effluent standards are reorganized to be inadequate to protect human health, the prefectural government can enforce stricter effluent standards to protect the living environment in order to protect human health. Through the establishment of strict regulations on wastewater disposal, the extension and expansion of wastewater systems, the improvement of river water flow and an effective environmental impact assessment of the various wastewater projects in public water areas, national and prefectural governments control water pollution. Standards of effluent water quality for the safety of human health (heavy metals, PCBs, etc.) are 10 times stricter than standards of environmental water quality. Based on its maximum daily value, the quality of effluent water is regulated. Via data monitoring that exceeds expectations, violations of effluent water quality standards are easily established. Nationwide quality requirements of effluent water for the storage of suspended solids (SS) etc., in the living environment, are planned to resolve effluent from household septic tanks. This system is used because it is assumed that companies in unregulated areas will discharge effluent to at least the same quality as that from household septic tanks. Standard values are given for BOD, COD, and SS as the maximum and average values, respectively. A fine is subject to breach of either the average or maximum rule. As stated earlier, the Prefectural Government can set more stringent quality standards for effluent water for public water bodies where the common national minimum effluent standards are inadequate to protect human health. In addition, a significant source of public water pollution is waste from household preparing food, cleaning, and grooming. Therefore, in the Water Pollution Control Act, the Ministry of the Environment lays down the provisions for promoting systematic countermeasures against domestic waste as follows [30]:

(a) Clarification of the administrative obligation for domestic effluent countermeasures: These provisions mean that cities, towns, and villages have been called upon under the Water Pollution Control Law framework to participate in activities to enforce countermeasures against domestic waste. These municipalities are expected to take the initiative to upgrade and educate citizens on countermeasures to reduce domestic effluent levels and their impact on effluent treatment facilities. At the same time, regional countermeasures are to be coordinated on a broader scale by the prefectures against domestic effluents, while the central government is to disseminate similar information and fund local government countermeasures.
(b) Clarification of citizens' responsibilities in relation to domestic waste countermeasures: These laws encourage the general public to dispose of food waste, cooking oil, and other waste properly, and also to educate fair use of detergents. It is also expected that the general public will comply with measures by the central and local governments to ensure the quality of public water. It is anticipated that individuals discharging domestic effluent would keep their treatment facilities in good order.
(c) Systemic promotion of household waste countermeasures: With regard to systematic promotion, the Water Pollution Control Law specifies that a region where countermeasures against domestic effluents need to be implemented due, for example, to excess environmental quality requirements, such as the Field of Emphasized Countermeasures against Domestic Effluents, should be designated by the governor of the prefecture. It is important for the municipalities in the area to compile a plan for the promotion of countermeasures against domestic effluents for the establishment of treatment facilities, the dissemination of countermeasures, etc. The municipalities are implementing the countermeasures under the plan and encouraging them.
(d) Extension of regulation in areas subject to Areawide Total Pollutant Load Control to such effluent-generating facilities: For 201–500 persons, on-site sewerage facilities were designated as 'listed facilities in the specified areas', which are managed only in the Areawide Total Pollutant Load Control Areas to improve effluent facility control.

Flexible infrastructure Constructing a variety of local, appropriate effluent treatment facilities (e.g. municipal sewage systems, sewage treatment tanks, rural community sewage systems for agricultural areas, community sewage systems for urban areas, and individual sewage treatment tanks) is vital to the solution in order to improve policies to resolve domestic effluent issues.

Consequences for Breach of Legislation

The Water Pollution Control Act forbids the discharge of wastewater that does not comply with effluent water quality requirements and punishes those who break the standards [28].

2.8 Korea

Borongan and Okumura [23] argued that the 3R definition is agreed upon and obeyed by all municipalities in South Korea. The core principles of Korea's waste management are to ensure a healthy environment by reducing waste generation, maximizing waste recycling, and managing waste created in an environmentally sustainable manner for people and the natural world. As the central indicator for achieving targets in the field, a strategic approach to promoting the 3R concept is assured. Therefore, given their limited facilities in their respective areas for waste segregation/sorting, collection, and care, both municipalities understand and attempt to enforce this strategy. The laws used in South Korea contain the following:

2.8.1 Waste Management Act (1986)

Enacted in 1986 and amended in 2007, the Waste Control Act and the Resource Saving and Recycling Promotion Act, enacted in 1992 and amended in 2008, primarily govern waste management in South Korea. There are also specific waste management laws covering the disposal of electrical/electronic devices and vehicles, the recycling of building waste, the construction of waste disposal facilities, and hazardous waste management.

According to Yang [31], by implementing several specialized, autonomous acts under this basic law and introducing increased producer liability and a volume-driven garbage rate scheme based on the polluter payment principle, waste management has become more effective for both general household waste and industrial hazardous/massive waste. Management mainly requires not only reducing waste generation but also careful management and total recycling of waste. The system established that the management of waste was not only about containment but also about eliminating waste. Since its introduction in 1986, the practice of systematic and integrated waste management has been increasing in South Korea.

South Korea's fundamental waste management system is laid out in the Wastes Control Act, including the definition of waste, national/local government and citizens' obligations, waste discharge standards and rules, and treatment procedures, etc. The Act also requires the Minister of the Environment to prepare a master plan for the proper management of waste created across the country every 10 years.

Based on the Waste Management Report (2015), South Korea has seen its landfill rates decline from over 90% to under 10% since the mid-1980s, while its recycling rates have risen from under 10% to over 80%. In a paper published in the *Journal of Material Cycles and Waste Management*, Professor Yong-Chil Seo of the Department of Environmental Engineering at Yonsei University and associates argued that the law offered a context in which waste management was not only about containment but generally about waste reduction. It implemented what became a general waste management standard: the form of the 3Rs: reducing, reusing, recycling. Because of limited land with no waste disposal locations and the high land price, the 3Rs were more important in Korea as the priorities, which could encourage the more successful push of such 3R practices [32].

2.8.2 Act on the Control of Transboundary Movement of Hazardous Wastes and Their Disposal (1994)

In the 1970s and 1980s, the awakening of environmental concern and the subsequent tightening of environmental laws in the developing world led to increased public opposition to hazardous waste disposal. In the late 1980s, the Basel Convention was signed, and its goal at the time of its adoption was to combat toxic trade. In February 1994, Korea acceded to the Basel Convention and the Basel Act.

Implementation of the agreement took place in March 1994. The aim of this Act is to avoid any environmental contamination, to track the movement of hazardous waste generated by the transboundary movement of waste, to encourage international cooperation and to contribute to the protection of the environment, and to the qualitative improvement of the lives of people through the implementation of the Basel Convention on the Monitoring of Transboundary Movements [33].

Centred on the Basel Convention International, Article 2 of the Presidential Decree on the Management and Disposal of Transboundary Movements of Hazardous Wastes describes hazardous waste as follows [34]:

(a) Waste is referred to in Annex I or Annex VIII, with some of the dangerous characteristics referred to in Annex III.
(b) Waste, as alluded to in Annex II.
(c) Waste that has been identified by Korea to the Secretariat of the Convention as hazardous under Paragraphs I, II, III, and III of Article III and Article XI of Article III.

The Convention's regulations focus on the following key objectives [35]:

(a) Reducing the production of hazardous waste and supporting environmentally friendly hazardous waste management wherever it is disposed of.
(b) Limit the transboundary movement of hazardous waste, except where the principles of environmentally sustainable management are considered to comply with it.
(c) A regulatory framework that extends to situations where it is appropriate to travel across borders.

The goal is to address a range of general requirements requiring States not to export hazardous waste to the Antarctic, to a State that is not a Party to the Basel Convention or to a Party that has prohibited the import of hazardous waste, and to comply with the basic principles of environmentally sustainable waste management (Article 4). However, because such agreements are no less environmentally friendly than the

Basel Convention, the Contracting Parties can enter into bilateral or multilateral agreements with other Contracting Parties or non-Contracting Parties on the management of hazardous waste (Article 11). The authorities of the exporting State should communicate the expected movements to the authorities of the prospective importing and transiting States and provide them with accurate details. The campaign will only continue if and when all the participating States have given their written consent (Articles 6 and 7). Cooperation between the parties is also provided for by the Basel Convention, ranging from the exchange of information on issues relating to the application of the Convention to technical assistance, in particular to developing countries (Articles 10 and 13) [35].

The Secretariat is obligated to foster and sustain this relationship, functioning as a clearing house (Article 16). Where the transboundary transport of hazardous waste has been carried out illegally, such as in breach of the provisions of Articles 6 and 7, or is incomplete as foreseen, the Convention assigns responsibility to one or more of the States concerned and imposes an obligation to ensure that hazardous waste is disposed of safely, whether or not it is transported back into the State of generation (Articles 8 and 9). The Convention also provides for regional or subregional training and technology transfer centres to be set up to deal with and mitigate the production of hazardous waste and other waste in order to meet the particular needs of different regions and subregions (Article 14) [35].

Until prior written approval has been obtained from the competent authority, export, manufacture and transit of hazardous waste are strictly prohibited pursuant to Articles 6, 10, and 16 of this Act. The exportation of hazardous waste for final disposal is, in principle, prohibited. Exceptions could, however, be made where Korea does not have adequate facilities and/or technology for the proper disposal of hazardous waste, which can be disposed of in an environmentally sound manner by importing countries. The Ministry of the Environment (MOE) has the power to regulate or prohibit the export and import of particular waste that, if and when considered necessary, may have a significant effect on human health or the environment [35].

2.8.3 Act on the Promotion of Saving and Recycling of Resources

The Promotion of Saving and Recycling of Resources Act sets out the framework for waste recycling, such as basic methods of recycling, the responsibilities and duties of organizations and individuals in promoting waste recycling, and provisions for waste reduction. The Act also mandates the Minister of the Environment, in consultation with relevant bodies, such as local government chief executives, to create a fundamental plan for resource recycling every 5 years.

As of 24 December 2018, the Act on the Promotion of Energy Conservation and Recycling, which will come into force on 25 December 2019, has been amended by the Republic of Korea. The amended Act facilitates the rating and inspection of packaging products and structures according to their recyclability. In order to facilitate the creation of packages that are easily recyclable, they should also be identified on the product labels once the grades have been determined. The grades should be labelled on each product, but they have nothing to do with the sales ban [36].

The Promotion of Saving and Recycling of Resources Act sets out the framework of waste recycling, such as basic recycling methods, organizations' and individuals' functions and obligations in promoting waste recycling, and waste reduction provisions. The Act also mandates the Minister of the Environment (MOE) to create a fundamental resource recycling plan every 5 years in collaboration with related bodies, such as local government chief executives.

According to the 3R Portfolio of the Republic of Korea [37], the reduction of packaging waste was included in major activities in the Promotion of waste generation reduction, which are reduction of packaging waste and restraints on the usage of disposable items. The Act for the Promotion of Resource Saving and Recycling and the Packaging Methods and Materials Ordinance Specifications were introduced in 1993. The key goal is to reduce the production of packaging waste, which accounts for the largest proportion (37%) of municipal solid waste, and to regulate some forms of hard-to-recycle packaging materials. The following steps have been taken by the Ministry of the Environment (MOE): (i) Regulation of packaging materials such as expanded polystyrene such as Styrofoam since 1993 and separate PVC packaging materials since 2001; (ii) regulation of packaging methods to reduce overpackaging; and (iii) incremental reduction of annual plastic packaging materials. Consequently, despite the increase in economic size and population, the total amount of packaging waste produced in 2002 (49,902 tons/day) decreased by about 20% from that of 1993 (62,940 tons/day).

Since 1994, MOE, based on the Saving and Recycling of Resources Promotion Act, has been implementing regulatory policies on the use of disposable goods. In restaurants, bathrooms, retail outlets, and other businesses, the use of disposable items such as disposable cups, bowls, reusable plastic bags, and paper bags is restricted. Offering such products to customers free of charge is illegal. As the use of plastic shopping bags decreased, the production of plastic waste gradually declined. Plastic containers and cups have also been replaced by paper items in fast-food restaurants [37].

The Act for the Promotion of the Recycling of Construction Waste came into force in 2005 to encourage recycling. More than a certain amount of recycled aggregate must be used in any building work contracted by a public entity. The Minister of Construction and Transportation is obliged to set quality requirements for recyclable aggregates by their use and has the authority to certify goods. South Korea has based the legislation on the Promotion of the Saving and Recycling of Resources since 1996 with regard to the promotion of product design for 3Rs to enhance the content and product structure. In the design process, producers and importers of both automotive and electronic devices, such as televisions, refrigerators, washing machines, and air conditioners, are expected to consider ways to use less types of materials, opt for recyclable materials, eliminate the use of toxic chemicals, reduce the device's weight, and make it easier to uninstall items. The manufacturers and importers of the products referred to above should assess the recyclability of materials on the basis of the requirements set out in the Ministry of Electricity and Energy (MOEE) [37].

2.8.4 Guidelines on the Reduction of Industrial Wastes (2001)

In Korea, waste is loosely divided according to its source into two categories: municipal household waste and industrial waste from commercial sites or large-scale factories (waste generation above 300 kg/day). In addition, industrial waste is classified into two categories: ' general industrial waste' consisting of slag, ash, dust, and building waste, 'defined waste' consisting of hazardous waste such as waste acid, waste alkali, waste oil, organic solvent waste, and so on. In Korea, in a dual system, waste is handled. The final disposal of urban waste is the responsibility of the local government, with the final disposal of municipal waste being the responsibility of the discharger of industrial waste. The total amount of waste produced has risen steadily since 1993 [38]. Industrial waste, as it turns out, can be used in various industries as a secondary resource, especially in the manufacture of building materials, goods and structures. Improving current procedures, designing and applying the new technologies should be the key tasks in the field of industrial waste processing. Measures aimed at the incorporation of the waste directly into the production process need to be created and implemented [39].

In South Korea, there are six successful waste disposal methods that foreign and domestic companies tend to venture into the following [39]:

(a) Waste prevention or disposal: Widespread use of new or unnecessary goods is the main reason why untested waste is produced. Rapid population growth involves the reuse of resources or the judicious use of existing items since there is a possible risk that people would otherwise be harmed by the adverse effects of radioactive waste. The management of waste can also take on a formidable form. A deliberate decision must be taken on a personal and technological level to curb the detrimental rise in waste.
(b) Recycling: Serves to turn waste into objects of its kind through industrial recycling. Paper, glass, aluminium, and plastic are typically recycled. The reuse of waste, instead of returning it to nature, in an environmentally friendly way. However, the processing technology is very expensive.
(c) Incineration: Incineration involves waste incineration, converting it into essential components and then producing electricity using the heat produced. For products, distinct gases and inert ash are popular. Depending on the quality of the burning waste and the configuration of the incinerator, pollution is caused to varying degrees. Hydroponic solutions can sustain the nutrient-rich ash obtained from burning organic waste. It is easy to get rid of hazardous and toxic waste using this process. The energy gathered can be used to cook, heat, and supply

the turbines with energy. However, in looking for the unintended leakage of micro-level pollutants, strict caution and due diligence should be practised.

(d) Composting: Helps bacteria to decompose organic waste, allowing waste to be collected for a long period of time in the pit. The method, however, is slow and takes up a large amount of soil. Biological processing greatly increases the fertility of the soil.

(e) Sanitary landfill: This is a landfill for waste. The base consists of a protective lining that serves as a barrier between groundwater and waste and prevents the leakage of harmful pollutants into the water region. It compacts the layers of waste, and then a soiled sheet covers them. In order to decrease susceptibility to accidental leakage of hazardous chemicals, non-porous soil is favoured.

2.9 Malaysia

SWM typically includes different government agencies, from the federal government to the state of Malaysia and the local government (LA). The SWM participates directly or indirectly in these government departments. Malaysia has three levels of government as a representative democracy, namely the Federal Government, the Government of the State, and LA.

2.9.1 Overview of Malaysian Legal Framework in Integrated Solid Waste Management (ISWM)

It has 14 states and 14 LAs. LAs are composed of the city mayor, council, town hall, and council [12]. LAs include the city council. The development of the Malaysian Solid Waste and Public Cleansing Management Act 2007 (MSWPCM) is considered informal due to indirect regulations on SWM enforcement before the MSWPCM Act is implemented. It takes a significant amount of time to grow. Directly or indirectly, the previous SWM law is derived from Federal Constitution 1957.

The Federal Constitution's Ninth Schedule [11] divides the federal state's legislative authority into three lists:

(a) Federal List (contains matters based on which the Parliament may make laws)
(b) State List (contains matters based on which State Legislatures may make laws)
(c) Concurrent List (contains common subject matters on which both the Parliament and State Legislatures have competence)

The Federal List consists of matters under the control of the federal government only, while the State List consists of matters under the jurisdiction of state governments only. The supreme law of the nation from which all other laws are derived is the Federal Constitution. The authority is defined by the Ninth Schedule of the Constitution between the federal and state governments. The constitutional summary is summarized below:

(a) The term solid waste or urban waste does not exist in the Constitution, nor is there an explicit reference to solid waste. Solid waste management, however, is implicitly referred to under the Concurrent List of the Constitution in the fields of public health, sanitation, and disease prevention. The Constitution also authorizes all issues relating to local governments under the State list, except for matters relating to the Federal territories.
(b) Article 95A: The National Council of Local Government (NCLG) has been established. In consultation with the Federal Government and the State Governments, it is the responsibility of the National Council for Local Government to formulate, from time to time, a national policy for the promotion of the establishment and rule of local government across the Federation and the administration of any legislation relating thereto, and to formulate the policy in such a manner that the Federal and the Federal.
(c) Article 76: Authority of the Parliament in some situations to legislate for States. (1) With regard to any matter enumerated in the State List, Parliament can make laws, but only as follows, that is to say. This also allows the Parliament to make laws under the jurisdiction of the State to promote uniformity of the laws of two or more states through the concern.
(d) Local government issues such as sanitation and solid waste management usually fall under the authority of the state.
(e) The provisions of Article 95A of the NCLG and the fact that sanitation is included in the Concurrent List suggests that the management of solid waste will indirectly fall under the control of federal and state governments.
(f) The provisions of Article 76(1) mean that the federal government may enact solid waste legislation for the entire country in order to promote uniformity in management.

Conceptually, legislation on all subject matters contained in the Federal List is constitutionally appropriate for the federal government to enact. Those issues include national policies, taxes, education, national defence, and internal security.

The governments of states are responsible for the issues listed in the list of States. Religious matters, property, rivers, forests, and waterways could involve such matters. These laws have minimal applicability in terms of implementation. In comparison to federal law, state laws apply only within a single state's limits and territories (Article 74i). With regard to the Concurrent List, federal and state governments will share legal powers and authority over issues such as history, recreation, antiques, neighbourhood and village planning, and wildlife protection [40].

Victor [12] states that there is no statutory expression of 'solid waste' or 'municipal waste', and there is no clear reference to solid waste available. Yusof [40] endorsed this argument, stating that the word 'climate' is curiously not included in any of the Federal Constitution's statutory lists. Moreover, in the Federal List, the State List, or the Concurrent List, the word is not included. The state of Malaysia as a whole is possibly responsible for this absence, which in previous years was largely underdeveloped when Malaya, Malaysia's former name, was used. However, under the Concurrent List of the Constitution, solid waste is implicitly included in public health, healthcare, and disease prevention [12].

Victor [12] stated that the state is responsible for local municipalities' general issues, such as sanitation and SWM. However, Article 95A (National Local Government Council) and the fact that sanitation is referred to in the Concurrent List indicates that federal and state governments will indirectly be responsible for the SWM. The provisions referred to in Article 76(1) also mean that the Federal Government may enact legislation on the solid waste throughout the country in order to encourage consistency in its management, but that legislation must be adopted by each government.

2.9.2 Local Government Act (Act 171) 1976

The 1976 Local Government Act [41] was passed in March 1976 and only applied to Malaysia. The Act lists the competencies and duties of the Local Authorities (LAs). In general, the LAs have a responsibility under the legislation for environmental control and conservation, sanitation and public health, epidemic disease prevention, and the general welfare of citizens of LAs. The LAs will, however, expand their roles to other areas. Summary of the act are as follows:

(a) To consolidate all other laws relating to the administration of local authorities, the Local Government Act was enacted.
(b) The management of solid waste was indirectly mentioned through Sections 9, 69–72, 84, and 102.
(c) Section 69 – Committing stream nuisances, etc.: Any person who, within the local authority jurisdiction, commits a nuisance or deposits any filth in or on the bank of any stream, canal, public drain, or another watercourse shall be guilty of an offence and shall be liable to a fine on conviction.
(d) Section 70 – Contamination of streams with commercial refuse: Any person who knowingly causes, within or without the limits of the flow of the local authority area, to fall or flow into any stream, any solid or liquid waste, any solid or liquid waste, any solid or liquid refuse of any manufacture, manufacturing process or any other waste or any putrid matter, any other person to interfere with the flow or to pollute its waters.
(e) Section 71 – The local authority can recover for work performed where the local authority has incurred any expenses in carrying out any work as a result of the offences referred to in Sections 69 and 70, the local authority shall certify the costs of the work to defaulting persons, and the local authority's certificate shall constitute conclusive evidence of the amount due. This sum is called a debt owed to the local authority and can be collected for the recovery of unpaid rates in the manner provided for in this Act.

(f) Section 72 – The local authority shall have the power to do all or more of the following, namely, to create, maintain, and carry out such sanitary facilities for the care, control, and management of waste.

In Sections 9, 69–72, 84, and 102, Victor [12] suggested that the Act applies implicitly to the SWM. The Act typically requires state governments to provide input on the LAs' strategy and organizational management. The following are the areas where the legislation empowers the local authorities to perform their SWM duties:

i. The Act forbids any waste of any manner from being disposed of in any drain, stream, or watercourse within the jurisdiction of the LAs.
ii. The Acts empower LAs in areas under their administration to protect public health. The LAs are also expressly empowered to include sanitary facilities, such as waste disposal and other forms of waste.
iii. The Act authorizes the LAs to create, amend, or revoke by-laws for the proper management and operation of sanitary facilities, such as the disposal of waste and other types of waste.
iv. The Act allows the LAs to take legal action against any person who might cause a nuisance deliberately or through negligence within the region of the LAs. Generally, the word nuisance may be interpreted to include the indiscriminate dumping of waste or waste.
v. The Act provides for the regulation and control of dangerous or nuisance-causing sectors or activities.

The additional section, such as Section 42 of the Act, for example, forbids anyone from moving construction material onto the road without the prior consent of the LAs. Indeed, the Act demands that the occupant of an office cleans the way in front of the premise entrance. Generally speaking, discarded content may be littered or disposed of in a public place within any LA area and punishable by fines or incarceration, or both, by the Local Government Act. Perhaps the most critical rule of the SWM. It enables the LAs to take responsibility for SWM generally in their areas of competence. SWMs, organizational activities, and the creation of legislation and prosecution guidance seem to have broad and detailed empowerment areas. This Act tends to be largely based on the existing functioning of LAs in terms of SWM.

2.9.3 Town and Country Planning Act (Act 127) 1976

The City and Country Planning Act [42] Malaysia Act 172 was passed in 1976 to ensure the local authority's proper management and governance of urban planning and development. The description of this Act is as follows:

(a) The Local Authorities are determined by the Act as the local planning body for specific areas. It was implemented to coordinate and control planning and development in the area of local authorities.

(b) The Act provides for the planning and growth of an area by establishing frameworks and municipal plans.
(c) These strategies are management support plans that decide when and how to improve an area by analysing the normal social, environmental, and economic problems of a region.
(d) The Act requires the Local Authorities to review and accept or deny the request or withhold approval for the planning and development of the submitted area.
(e) It also provides the framework for structured environmental planning and development, including solid waste disposal systems, such as landfills, transfer stations, and incinerators, which are planned and supplied.

Zubaidah and Mustafa [10] explained that this act imposes a legal obligation on the local planning authority to carry out effective planning controls within the local authority. Sinha and Kew [43] endorsed the account, adding that the draught structural plan, including steps to reinforce physical conditions and cooperation of planning authorities, will be required by the Act to be prepared. The act designating LAs as local planning authorities in their respective regions was summarized by Victor [12]. The Act uses the institutional system and local strategies to prepare and develop an environment that is provided for by the statute. These are development control plans that specify which fields and how to research physical, social, economic, and environmental issues. These development plans are legally binding documents that provide general guidance on a geographer's development and strategy.

2.9.4 Environmental Quality Act (Act 127) 1974

To prevent, mitigate, and regulate emissions and to improve the overall quality and ensure effective environmental management, the Environmental Quality Act was passed in 1974 [43]. The following summaries are located below:

(a) Enacted to avoid and monitor pollution from the surroundings, as well as to ensure proper environmental management.
(b) In Section 3, the Act empowers the Director-General of the Department of the Environment (DOE) to facilitate, encourage, coordinate, and carry out environmental preparation, waste management, and planning for pollution control.
(c) In Section 2, wood waste can be defined here as planned waste or other solid, semi-solid, and liquid substances that could lead to pollution.
(d) It is also specified in Section 24 that soil pollution involves waste dumps, waste dumps or other solid waste as a form of waste because of its harm to humans, including open burning, which is also known as solid waste.
(e) An Environmental Impact Assessment (EIA) is required under this Act for prescribed operations before any such action could be carried out.

(f) The Act also empowers the Minister in Section 51 to impose laws to limit, forbid, and enhance the overall environmental quality that may restrict the formation of solid or liquid waste disposal sites.
(g) Although the overall Act contents it is not into solid waste management, the Act shows that coordination and planning of waste management come under its purview.
(h) Both solid waste management services, such as landfills and incinerators, are also introduced by means of EIA documents. It also mentioned that in the region, the Minister might establish solid disposal locations.
(i) This Act can be used in the construction of a solid waste management framework since this Act has the provisions to regulates such issues.

As set out in Section 2, the wood waste can be defined here as planned waste or other solid, semi-solid, and liquid substances that could lead to pollution. The Act offers for the monitoring or protection of environmental damage. The presence of soil contamination in refuse dumps, garbage dumps or other solid waste on the basis of its adverse effects on humans is also stated in section 24. Section 29 of the Open Burning of Waste, which typically includes solid waste, is also governed by Section 29 of the Open Burning Act.

According to Victor [12], a significant aspect of this act is that, before such activities can be carried out, it needs an Environmental Impact Assessment (EIA) for prescribed activities. The list of prescribed operations includes the installation of infrastructure for municipal solid waste, such as incineration plants, composting plants, recovery/recycling plants, and solid waste landfill facilities. Finally, the Act requires the Minister to issue legislation to govern, prevent, and enhance the overall environmental standard (Section 51). In compliance with this clause, the Minister may provide for the establishment of sites for the disposal of solid or liquid waste on or in any territory.

2.9.5 Street, Drainage and Building Act 1974

On 13 June 1974, the Lane, Drainage and Building Act [44] was enacted. To deal with drainage and development in local authority areas, the Streets, Drainage and Building Act (SDBA) were passed to consolidate the laws of the highway. Victor [12] explained that solid waste management in public areas is empowered by this Act. Although this may sound excessive, the act focuses primarily on the cleanliness of public places, as opposed to solid waste management in general. As follows, the summaries of the act:

(a) Enacted to consolidate laws in local government areas regulating highways, drainage, and houses.
(b) The Act requires the preservation, repair, extension, and widening of roads in terms of routes. With the approval of local authorities, either local authorities or private individuals can build the streets. The Act also lays down guidelines for

the construction of trees, lamps, water pipes, and other services under or along the streets. Furthermore, the Act provides that local authorities are responsible for the development and maintenance of drains and watercourses, as well as back lanes.

(c) There are also regulations banning waste disposal from domestic or industrial sources to any public area within the local authority.

(d) Usually, this Act promotes good waste management in public areas, but in comparison to solid waste management in general, the cleanliness of public places is stressed.

2.9.6 Solid Waste and Public Cleansing Management Act 2007

This Act was gazetted to provide to govern and manage public cleaning of solid waste to ensure adequate hygiene and for incidental problems. There have been several challenges to the history of the SWMPC Act as a major solid waste law [45]. According to Yahaya and Larsen [46], the government agreed in September 1995 that it was necessary to privatize solid waste management. After 10 years of privatization, the government approved the National Strategic Strategy for Waste Management (NSP) in July 2005. In the same year, the Ministry created a Solid Waste Management Division to prepare for the implementation of the programme. The government adopted the National Solid Waste Policy less than a year after the implementation of the NSP, which provides the overall course of solid waste management. Law 672, or the SWMPC Act, was passed by Parliament in July 2007 and published in August 2007 in the National Gazette. The Ministry of Housing and Local Government administered this Act, and it was implemented for every state in Peninsular Malaysia except Penang and Selangor on 1 September 2011. Provision and control of public cleaning and management of managed solid waste in order to ensure proper sanitation are the main objectives of the Act.

Act 672 requires all suppliers of solid waste and public cleaning management and operators of facilities to apply for a licence from the Director-General of the National Solid Waste Management Agency. Before any construction, modification, or removal of prescribed solid waste management facilities, the Act also allows the Director-General of the National Solid Waste Management Agency to give prior written approval. In addition, Act 672 includes the following:

(a) Regulation of solid waste generators and people who are in possession of managed solid waste
(b) Enforcement
(c) Reduction of managed solid waste and recovery

Act 672 requires the Federal Government in Peninsular Malaysia and the Federal Territories of Putrajaya and Labuan to help solid waste management and public cleaning from the LAs. A significant improvement in the modus operandi of solid waste and public cleansing management in Malaysia is expected to be driven by the introduction of Act 672 and its related regulations on solid waste and public cleansing management. Upon implementation of Act 672, the Solid Waste and Public Cleansing Management Company (SWPCMC) [45] will completely handle all work on solid waste and public cleansing management. Such enforcement includes the successful assumption of all solid waste and public cleaning management functions by the LAs, as the LAs would lose power over the collection and disposal of waste, as well as the cleaning of public roads, public places, public toilets, and drains. SWPCMC will then assume the management and regulation of all landfill sites previously under the LAs, including the identification and assessment of sanitary landfills, transfer stations, incineration methods, and plant management [45].

The Federal Government shall, upon the entry into force of this Act, have executive power with regard to all matters relating to solid waste management and public cleansing. The two previously existing Acts which gave legal authority to the LAs on solid waste management need to be amended with the enactment of the SWMPC Act 2007 [2]. Therefore, revisions to the applicable provisions were accepted by both the Local Government Act of 1976 and the Street, Drainage and Building Act of 1974, thus eliminating any reference to solid waste management.

The Town and Country Planning Act (TCPA) of 1976 is another significant piece of legislation that was also revised. A new provision was introduced by the TCPA in order to clarify that the provisions of the 2007 SWMPC Act should be regarded by the local planning authority when processing the planning permission application. This provision is a long-term measure to ensure a coordinated plan for Solid Waste Management. Additionally, some regulations have been established by the National Solid Waste Management Department to implement the power granted to it on 1 September 2011, by the SWMPC Act of 2007. These laws include:

(a) Solid Waste and Public Cleansing Management (Manner of Appeal) Regulations 2011
(b) Solid Waste and Public Cleansing Management (Prescribed Solid Waste Management Facilities and Approval for The Construction, Alteration and Closure of Facilities) Regulations 2011
(c) Solid Waste and Public Cleansing Management (Compounding of Offences) Regulations 2011
(d) Solid Waste and Public Cleansing Management (Licensing) (Management or Operation of Prescribed Solid Waste Management Facilities) Regulations 2011
(e) Solid Waste and Public Cleansing Management (Licensing) (Undertaking or Provision of Collection Services for Household Solid Waste, Public Solid Waste, Public Institutional Solid Waste and Solid Waste Similar to Household Solid Waste) Regulations 2011
(f) Solid Waste and Public Cleansing Management (Licensing) (Undertaking or Provision of Transportation Services by Long Haulage) Regulations 2011
(g) Solid Waste and Public Cleansing Management (Licensing) (Undertaking or Provision of Public Cleansing Management Services) Regulations 2011
(h) Solid Waste and Public Cleansing Management (Scheme for Household Solid Waste and Solid Waste Similar to Household Solid Waste) Regulations 2011

Other proposals to be included in the compliance are as follows:

1. The public will be given a 2-year grace period to get used to the procedure before fines are levied on anyone who has not done so.
2. The three concessionaires responsible for solid waste management in the nation will have to raise awareness of waste separation at the source over the next 2 years.
3. Before dumping it, homeowners have to know their responsibility to distinguish their own garbage.

A review of the final results from the latest information [47] on the amendment that resulted in the 2007 SWMPC Act being enacted is given below.

1. All references to Solid Waste Management were omitted in the 1976 Local Government Act and the 1974 Street, Drainage and Building Act. These adjustments are necessary to ensure the smooth implementation of the new act and to avoid any misunderstandings during implementation.
2. A new regulation was adopted by the Town and Country Planning Act of 1976 (TCPA) to ensure that local planning authorities comply with the requirements of the 2007 SWMPC Act while processing an application for a planning permit.
3. With the implementation of the SWMPC Act of 2007 in Malaysia, the modus operandi of solid waste management and public cleaning has been changed. Upon enactment of Act 672, all work on solid waste and public cleansing management will be taken over by SWPCMC, which requires the immediate takeover of solid waste and public cleansing management duties from all LAs. This move is because it is no longer possible for the LAs to collect and dispose of waste and clean public roads, public places, public toilets, and drains.

All solid waste management regulations, such as the Local Government Act 2007 (Act A1311) (as amended), the Lane, Drainage and Building Act 2007 (Act A1312) (as amended), and the Town and Country Planning Act 2007 (Act 1313) are also subject to the Federation's Act 672. (as amended). In Peninsular Malaysia and the Putrajaya and Labuan Federal Territories, the uniformity of these Acts is applied [48].

In the SWMPC Act, there is a well-developed waste management act that creates substantial compliance and control of waste collection, transport, and processing. A significant principle in the SWMPC Act continues to be the waste hierarchy. This implies that the SWMPC Act offers systematic enforcement of waste management from the initial generation of waste to its ultimate elimination, with the ability to reduce waste at each stage of the hierarchy. The description and area of jurisdiction of the legislation are presented in Table 2.1.

Table 2.1 Summary and area of jurisdiction for the legislation modified from Victor [11]

Legislation	Area of jurisdiction
Federal Constitution, 1957	The foundation for the formulation of the NCLG and other related solid waste management legislation
Local Government Act, 1976	Authorizes the state government and the LAs to be responsible for solid waste disposal
Town and Country Planning Act, 1976	Authorizes LAs to prepare for environmental growth, which also includes solid waste management, especially at the sites of waste management plants
Environmental Quality Act, 1974	Regulates solid waste pollution and, by requiring an EIA, the implementation of waste treatment facilities
Streets, Drainage, and Buildings Act, 1974	Authorizes the LAs in its territory to control and prohibit solid waste disposal
Solid Waste and Public Cleansing Management Act 2007	Providing and controlling public cleaning and the disposal of managed solid waste in order to ensure proper sanitation
	Regulation of generators of solid waste and persons in possession of managed solid waste
	Enforcement
	Managed solid waste reduction and recovery

2.9.7 Overview of Institutional Framework in ISWM

This segment provides a summary of the key parties involved in the management of solid waste in Malaysia. Pursuant to Victor [11], these organizations constitute the key process by which the regulatory structure for the operation and management of solid waste on the ground is translated.

The Cabinet, consisting of different ministers who are elected representatives of the Malaysian people, leads the overall institutional structure of the Malaysian government. By formulating policies and providing directions for growth, many councils help the Cabinet to guide the country's development. These councils are made up of different members of the Cabinet, as well as state government representatives. Such boards serve as liaisons between state and federal governments.

The administration of municipal solid waste management (MSWM) under the Federal Government is the responsibility of the Ministry of Housing and Local Government (MHLG). In this Ministry, there are two divisions directly involved with MSWM, namely the National Department of Solid Waste Management (NSWMD) and Company Solid Waste and Public Cleaning Management (SWPCMC). The Federal Government's positions in MSWM are largely advisory and coordinating in nature. The roles of NSWMD and SWPCMC as the key federal agencies responsible for this matter are presented in this section.

This section addresses the summary of the implementation of the Act in Malaysia.

2.9.8 National Solid Waste Management Department (NSWMD)

Under the SWMPC Act of 2007, the NSWMD was formed to integrate the solid waste management system at the national level. This Act grants the Federal Government the administrative power to carry out the duties of solid waste and public cleaning management [12].

The role of the SWM falls within the Engineering Division of Environmental Health and Project Implementation Division, Local Government Department, prior to the establishment of NSWMD. This role was transferred to NSWMD and the Solid Waste and Public Cleaning Management Company when Act 672 and Act 673 were passed (SWMPC) [12]. Such functions have characteristics:

1. Proposing solid waste and public cleansing management policies, plans and strategies.
2. Formulate solid waste management plans, including the location, characteristics of new treatment facilities, target areas for solid waste disposal facilities, solid waste management schemes for the demand of controlled solid waste to solid waste management facilities, and time scales for the implementation of plans.
3. Set standards, specifications, and codes of practice for all aspects of solid waste and public cleaning services management.
4. To perform the regulatory function laid down in Act 672 and any regulation made under the Act.
5. To issue licenses and approvals pursuant to Act 672.
6. The success of those additional activities for the enforcement of the Act.

2.9.9 Solid Waste and Public Cleansing Management Corporation (SWPCMC)

To complement and ensure the successful implementation of the National Solid Waste Management Policy, the SWPCMC was set up. In general, the policy is intended to provide a comprehensive, integrated, cost-effective, and efficient system of solid waste management that satisfies the environmental and public well-being requirements. The company was formed under the 2007 Solid Waste Management and Public Cleansing Corporation Act (Act 673) and started operations under the MHLG on 1 June 2008 [46].

The organization is allowed to administer and execute laws and matters relating to solid waste and the management of public cleaning. The duty of the Company is to ensure effective and integrated solid waste management and public cleaning systems and meet consumer expectations [46]. Companies' functions include the following:

(a) To propose to the Department of National Solid Waste Management policies, proposals, and strategies for solid waste management and public cleaning services.
(b) Implementing and encouraging public engagement and understanding through different solid waste management and public cleansing initiatives.
(c) Establish systemic practices for recycling and waste segregation, focusing on sustainable growth and integrating economic, political, and social aspects.
(d) Processing applications for new licences and renewals, including revocation or suspension of licences for providers of solid waste management services and submission of recommendations to the Department of National Solid Waste Management.
(e) Track the efficiency of concessionaires engaged in the provision of facilities for solid waste management and public cleansing.
(f) To find safe and cost-effective environmentally friendly solutions and to perform research on the advancement of solid waste management and public cleansing.
(g) Implement and execute the Solid Waste Management and Public Cleaning Acts and Regulations.

2.9.10 Private Waste Manager

In 1983, as a national strategy, privatization in Malaysia was introduced to tackle the economic problems at that time. *The Economic Times* defines privatization as a transfer of ownership, property, or industry from the government to the private sector. During privatization, the government ceases to be the owner of the agency or company. Three key areas are involved in this process, namely, transparency, properties, and personnel management. Malaysia's Economic Planning Unit lists the national privatization policy priorities as follows:

1. To promote the country's economic growth.
2. Reducing the administrative and financial burden on the government.
3. Reducing the government's presence in the economy.
4. Reducing the level and extent of public expenditure.
5. Enabling market forces to control economic activities and increase quality and productivity, in line with the National Development Policy.

Local government sector privatization entails the provision of public services such as sewerage, transport, and solid waste management. In particular, the privatization of urban solid waste management was implemented in 1993 by the Malaysian government. Transportation, collection, transport, processing, and disposal of solid waste is part of this privatization. The aim was to have, as part of Vision 2020, an integrated, sustainable, safe, and technologically advanced solid waste management system. The main purpose of this step was to eliminate and recycle waste, thus reducing the need for final waste disposal.

In addition, privatization was intended to restructure the LAs' existing solid waste management system into a system incorporating successful measures in recycling and environmental management. Privatization was also required to overcome and fix the problems facing LAs in solid waste management, such as finance, lack of expertise, illegal dumping, open burning, and lack of appropriate solid waste sites [48].

The Solid Waste and Public Cleansing Services Corporation was the government-appointed regulatory authority. Three private waste management consortiums were then designated for the entire zone, that is:

1. Alam Flora Sdn Bhd, which is responsible for WP Kuala Lumpur, WP Putrajaya, and Pahang.
2. SWM Environment Sdn Bhd, which is responsible for Johor, Melaka, and Negeri Sembilan.
3. Environment Idaman Sdn Bhd, which is responsible for Kedah and Perlis.

However, several states have refused the collection of concessionaires, such as Penang, Selangor, and Perak. This crisis has occurred due to the lack of satisfaction of the local authorities with the quality of services and operating costs [48].

2.9.11 Non-Governmental Organizations (NGOs)

In the solid waste management process, non-governmental organizations such as Malaysian Environmental NGOs (MENGO), Treat Every Environment Unique (TrEEs), the Environmental Protection Society of Malaysia (EPSM), the Lions Club, and the Rotary Club play a major role. At the community level, these NGOs are also involved and organize programmes such as recycling and environmental awareness campaigns that ultimately drive community engagement in solid waste management.

2.10 The Issues and Effect of Legislation Towards Waste Reduction

Globally, waste volumes are rising rapidly, much quicker than the speed of urbanization. The world's cities are delivering almost 1.3 billion tonnes of heavy waste per year right now. It is predicted that this sum will grow to 2.2 billion tonnes by 2025. Their financial success rises as nations urbanize. The consumption of goods and services increases as living and disposable income levels increase, resulting in a corresponding increase in the amount of waste produced [49].

Roughly US$25 billion per year is as of now invested by nearby governments in Asia on urban strong waste administration. This is used to collect more than 90% of the waste in high-income countries, 50–80% in middle-income nations, and just 30–60% in low-income nations. In 2025, investing in strong waste management programmes is estimated to extend by 200% in 2015 [50]. Based on the enactment, most nations recommend decreasing waste dumping at a landfill, burning, produce and promote 3R, which is reduced, reuse, and reuse. Although the necessity may offer assistance to reach the objectives on waste reduction, certain issues and impact may appear up after enactment constrained.

Malaysian laws were too common and were far from satisfactory due to a shortage of funding and local budget constraints due to the problems Malaysia faced with legislation. According to the municipality's scale, the waste collection budget ranged from 20% to 70% [51]. In open fields and rivers, waste dumping is still normal today, and a Kuala Lumpur waste disposal operation study found that 31.9% of waste was disposed of by open combustion, while 6.5% was dumped into the river system [52]. A common problem in Malaysia is the lack of concern about the causes and contributory variables of the waste produced [53].

Subsequently, the environmental protection problem in Malaysia was secondary, and most districts had a difficult time looking for new transfer locations as the current transfer locations were about to be depleted [53]. Due to its population density, Kuala Lumpur has an urgent need to reduce its reliance on landfills; however, an elective arrangement such as an incinerator is difficult to be implemented. A major challenge in Malaysia remains in terms of tackling problems in improving administration and waste management services provided by the respective local authorities. In the late 1980s, 3Rs was generally centred on reusing exercises, but tragically it fizzled to upgrade the current practice of waste administration. The open mindfulness of the 3Rs is still poor, despite the Malaysian government's support for various campaigns. Also, there is an inadequate concrete policy on 3Rs to get full public involvement.

In the US, many recyclables become contaminated when items are recycled in the wrong bin or when a dirty food container gets into the recycling bin. Contamination can prevent the recycling of large batches of material. In those facilities, the manufacturing of other products is not feasible. In addition, many items collected are also not recyclable, such as plastic straws and packets, eating utensils, yoghurt, and take-out containers. In landfills, they usually end up being incinerated, discarded, or washed into the ocean. While incineration is mostly used for electricity generation, waste-to-energy plants have in the past been linked to toxic emissions. One of the early aspects of the EU environmental requirement was legislation to restrict the environmental effects of waste. Based on Farmer [37], policies are aimed at reducing waste generation, preventing waste generation, reducing waste disposal levels and growing levels of reuse, recycling, restricting pollutants entering waste streams resulting in less hazardous waste and product recovery, restricting the export to third countries of hazardous waste and waste for disposal, and setting environmental standards for the disposal of waste to third countries.

In developing countries, solid waste management is compounded by unsustainable practices that increase atmospheric degradation and disease dissemination. In particular, open dumping at unregulated sites, open combustion of waste fractions and mismanagement of leachate generated at final disposal sites are key observable issues [53]. The situation in slum areas is worsening with the additional issues of high-density neighbourhoods, noise, air, and water pollution. In these cases, unregulated dumping in open spaces near water sources is a common concern that contributes to public health concerns [54]. For example, the landfills in Korea moreover postured a genuine issue due to the natural contamination. Deposits also led to the pollution of the wastewater, soil, and surface water, as well as the leakage of hazardous contaminants, aggressive odours, and unforeseen explosions. An agent of landfill-induced natural pollution was the landfill location that fit most of the waste generated in Seoul from 1978 to 1992. Moreover, the problem of landfills is at the core of Japan's garbage problem. Typically, 10% of the initial weight remains in the form of ash that must be buried on the side of non-burnable waste in landfills while the waste collected by each district is burned. Inventive modern strategies for trash cremation and the utilize of the coming about fiery debris can offer assistance to ease the issue of rubbish. However, in order to produce as little waste as possible, it would also be important for individuals to change their way of life and for recycling to be further encouraged to increase the usage time of each landfill [55].

Solid waste burning, especially for cities within the Global South, is frequently depicted as a 'quick-fix' arrangement for rising waste volumes for producing power. Burning, be that as it may, is among the most noticeably awful measures that cities can take to realize both waste lessening and vitality targets. It is exorbitant, wasteful, and postures dangers to the environment. By constraining them to proceed, creating tonnes of waste to fuel the incinerator, undermining endeavours to play down waste era or raise reusing rates, it locks towns into high-carbon pathways. Building and running cremation offices are expensive, which is the foremost costly frame of producing power. The tall forthright taken a toll implies that open segment financing is frequently required to guarantee private speculations. Waste cremation makes destructive emanations, requiring strict natural controls to maintain a strategic distance from their discharge into the environment, moreover the costliest working cost. Offices regularly cut back on these natural controls when budgets are strained, with critical emanation results. The vitality made from the incineration of waste isn't secure or feasible since it burns materials such as plastics that are inferred from fossil fuels. It discharges gasses from nurseries [56].

2.11 Improvement in Future of Solid Waste Legislation

There are various highlights to a suitable waste organization: political, social, common, budgetary, and mechanical. In show disdain towards the reality that the waste organization course of action targets moves little from country to country, the methods utilized to achieve them must be balanced in each country to the winning circumstances. These components join the availability of and competition for inventive, monetary, and human capital from another national course of action necessities, particularly in making countries. There's no single altered procedure for finishing suitable organization of waste. In any case, there are common needs that all countries that need to appropriately control their waste must reply. These criteria will incorporate satisfactory understanding of the sorts of waste to be arranged of how much there's where it exists, who makes it, and what happens to it.

Waste administration may be an energetic, multidisciplinary operation, and an adequate supply of adequately qualified and experienced people is nearly the foremost vital asset required. The capabilities change, agreeing to whether the person is a mechanical handle director, a waste treater or disposer, a controller, organizer, or plan design since a distinctive adjustment of abilities and mastery is required for each viewpoint of waste administration. The waste director is likely to be a completely prepared individual in one of the normal sciences or building areas which has extended his skill to regions such as well-being and security, fund, administration, and natural sciences, such as affect investigation, arranging, and chance appraisal.

Policies that can lead to a more prosperous environment have to be introduced by all sectors of society, including waste management, in order to step in a more sustainable direction. Waste manufacturing and management are focused on what is happening in society and how public bodies control these activities. Decision-making bodies shall, for the purpose of monitoring operations, adopt particular policy instruments and issue documents setting out general policy objectives. To ensure sustainability, techniques to manage the production of waste material must be among the best and most suitable approaches.

The lowest level is recycling in the commodity recovery hierarchy, as claimed by Amelia [48]. Introducing a recycling programme to decrease waste disposal problems. It uses an alternative that is affordable and better for the environment than looking for a new location for landfills and can increase the life of the existing landfill. In addition, by replacing raw materials with used materials, the programme is more affordable, conserves resources, and generates employment opportunities [57]. Controlling urban waste development will help minimize the shipping of municipal waste to landfills. Municipal waste generation may also be tackled by various means, such as compliance with waste regulation, recycling, waste management at source, the design of a smart public waste composition control framework, and an ongoing awareness campaign on waste-related issues [58]. Improvement can be suggested on each legislation according to a current global situation which are, considering society, politics and economy.

When COVID-19 swept the world starting March 2020, the total number of infected persons increased and increased, which seems to have not reached its current level. This has led to the production of waste and different stages of problems in waste management practices. Impacts that affect handling and treatment practises include changes in the waste amount, composition, timing (temporal), distribution (spatial), and risk. The new impacts, problems, and innovations of the COVID-19 waste management response have been analysed in this update. The wearing of protective masks, particularly at the beginning of the pandemic, was a typical controversy. It is recommended in certain countries only for people who are ill or highly vulnerable. Owing to the lack of proper storage and disposal and the lack of healthcare services, the use of masks for ordinary individuals has been problematic [59]. During this crisis, whether to stop or to continue, for example, to recycle, is a difficult decision. The worker could, by recycling, be exposed to the risk of infection. Resource Recycling, Inc. [60] announced that since mid-March, when widespread stay-at-home orders and other legislation began to change everyday life around the world, the coronavirus pandemic had had an effect on the US recycling programmes.

At this current point, where the understanding of COVID-19 is still limited, the approach chosen has been better safe than sorry for most of the time. However, even if it is not possible to follow the highest preferential options for waste management and the hierarchy of the circular economy, it is still important to minimize the environmental impact of waste management to the greatest extent possible. Changes in waste content, the volume of waste, frequency and timing of disposal (temporary), distribution (spatial), and risk of infection are included [61]. Dynamic and sensitive steps are needed to overcome unprecedented challenges. With the improvement of legislation in the reduction of waste in each country, the government should play a major role to access the recent events that occur, which can enhance the policies and requirements suitable for the application of current use.

2.12 Conclusion

The solid waste management challenges and issues that should be considered while framing solid waste policy include proper waste generation; segregation or separation at sources; collection, transportation, and disposal methods; landfill management; hazardous and other toxic material management; treatment, incineration, recycling, and other technology standards; monitoring, evaluation, and continuous improvement methods. The most important challenges are the change the mindset of the community to reduce the waste at sources. In addition to these issues, policy needs to address the short- and long-term economic, environmental, and social costs and benefits, funding methods, and roles of various stakeholders.

The joint projects between municipalities and stakeholders bring forward in terms of implementing recycling concepts. This initiative may reduce the amount of waste that goes to landfills. Partnerships and the continued cooperation of various stakeholders in municipal waste management efforts should be enhanced in the provision of resources for the effective implementation of waste management.

Effective management of solid wastes is one of the prerequisites for any countries in achieving a successful landmark. It is generally recognized that there is a strong relationship between effective management of solid wastes and good quality of life and a healthy environment. The attractiveness of the country to foreign visitors and investments is very much influenced by a clean and healthy environment. Solid and hazardous waste legislation in all countries has been constantly strengthened and improved by the introduction of new amendments to their major environmental laws introduced in this chapter. The US legislation is the most complex which changes whenever the presidency changes. A practising solid and hazardous waste engineer must pay attention to these changes from time to time in order to manage his/her solid and hazardous waste projects properly and legally.

Glossary

Department of Environment (DOE) An organization tasked with implementing, conserving, and maintaining sound environmental management as part of the nation-building process, as well as ensuring that the environment is still clean, stable, and safe for the people's well-being.

European Free Trade Association (EFTA) An intergovernmental organization formed to promote free trade and economic integration for the benefit of its four member countries – Iceland, Liechtenstein, Norway, and Switzerland – as well as their global trading partners.

Environment Impact Assessment (EIA) A method of assessing the possible environmental consequences of a proposed project or production, taking into account interconnected socio-economic, cultural, and human-health effects, both positive and negative.

Environmental Protection Agency (EPA) A federal government agency whose mission is to protect human and environmental health. The EPA is in charge of developing standards and laws that promote individual and environmental health.

Environmental Protection Society of Malaysia (EPSM) A membership-based national association administered solely on a volunteer, non-profit basis by an elected Executive Committee. Its goals are to avoid environmental degradation as a result of human activities, monitor human activities that lead to environmental degradation, implement environmental improvement initiatives, and raise public awareness about the state of our environment.

Integrated Solid Waste Management (ISWM) A comprehensive waste prevention which includes waste reduction, recycling, composting, and disposal. It considers how to avoid, recycle, and handle solid waste in ways that protect human health and the environment as effectively as possible.

Local Authorities (LA) An association that is legally in charge of all public services and facilities in a specific region.

Malaysian Environment Non-Governmental Organizations (MENGO) A grouping of Malaysian Environmental NGOs (MENGO) was formed under the DANIDA-supported programme for environmental assistance to Malaysia.

Ministry of Economy, Trade and Industry (METI) A ministry of the Japanese government. It has jurisdiction over a broad policy area, including Japan's economic and industrial policy, trade policy, energy security policy, intellectual property policy, industrial technology and innovation policy, control of arms exports, etc.

Ministry of Housing and Local Government (MHLG) A Malaysian government ministry in charge of housing, local government, town planning, country planning, fire and rescue, landscape, solid waste management, strata management, moneylenders, and pawnbrokers.

Ministry of Environment (MOE) The Ministry of the Environment plays a central role in government environmental policy. The ministry exclusively handles the planning and formulation of all government environmental policy and planning and all waste and recycling measures.

Municipal Solid Waste (MSW) Commonly known as trash or garbage in the United States and rubbish in Britain, it is a waste type consisting of everyday items that are discarded by the public.

National Council for Local Government (NCLG) The commission was established under the federal constitution to coordinate policies and laws between the federal, state, and local levels of government.

Non-Governmental Organizations (NGOs) Non-governmental organizations or non-profit organizations are establishments that are not governed by the government.

National Strategic Plan (NSP) A strategy for resource planning and allocation focused on national priorities and consensus.

National Solid Waste Management Department (NSWMD) Formed to integrate the solid waste management system at the national level.

Resource Conservative and Recovery Act (RCRA) The primary legislation of the nation controlling the disposal of solid and hazardous waste, according to the United States Environmental Protection Agency.

Streets, Drainage and Building Act (SDBA) Act focuses primarily on the cleanliness of public places, as opposed to solid waste management in general.

Solid Waste Management (SWM) The method of collecting and handling solid wastes is referred to as solid waste management. It also provides recycling options for things that do not belong in the garbage or trash.

Solid Waste Management and Public Cleansing (SWMPC) A Malaysian law enacted to provide for and govern the management of managed solid waste and public cleaning for the purpose of maintaining proper sanitation, as well as other matters.

Solid Waste and Public Cleansing Management Corporation (SWPCMC) A organization that implements policy, plans, strategies with the standards, specifications, and codes of practice and enforce the laws and regulation.

Town and Country Planning Act (TCPA) An Act for the proper control and regulation of town and country planning in Peninsular Malaysia and for purposes connected therewith or ancillary thereto.

Treat Every Environment Special (TrEEs) Programmes are primarily based in urban areas as the lifestyles of the urban community have a tremendous impact on Malaysia's natural resources, yet this community is the most disconnected from the natural environment.

Underground Storage Tank (UST) A tank and any underground piping connected to the tank that has at least 10% of its combined volume underground.

References

1. Wimalasena, B. A. D. (2011). *A planning tool for construction waste management*. PhD thesis, University of Calgary.
2. Malaysia Act 672. (2007). *Solid waste and public cleansing management act*. Ministry of Housing and Local Government, 1–7, Malaysia.
3. USEPA. (n.d.) *EPA history; resources conservation and recovery act*. United States Environmental Protection Agency, Washington, DC, USA. www2.epa.gov/aboutepa/epa-history-resource-conservation-and-recovery-act
4. Vesilind, P. A., Worrell, W., & Reinhart, D. (2002). *Solid waste engineering*. Brooks/Cole Thomson Learning.
5. Guerrero, L. A., Maas, G., & Hogland, W. (2013). Solid waste management challenges for cities in developing countries. *Waste Management, 33*(1), 220–232.
6. USEPA. (n.d.). *Source reduction program potential manual*. United States Environmental Protection Agency, Washington DC, USA. www.epa.gov/osw/conserve/downloads/source.pdf
7. UNEP Annual Evaluation Report 2005. (n.d.). *UN programme*. Retrieved from https://www.unenvironment.org/resources/synthesis-reports/unep-annual-evaluation-report-2005
8. USEPA. (2013). *Municipal solid waste*. United States Environmental Protection Agency, Washington, DC, USA. United States Environmental Protection Agency, Washington, DC, USA. Retrieved from https://archive.epa.gov/epawaste/nonhaz/municipal/web/html/. Raya, J. K. (2008). *A practical guide for environmental protection & enhancement works, Arahan Teknik (Jalan)* 16/03; (Pindaan 2008) JKR 21300-0002-09.
9. Zubaidah, S., & Mustafa, M. (2008). *Legal and institutional framework for pollution prevention in Malaysian rivers*. Malayan Law Jur.
10. Malaysia Federal Constitution 1957, Malaysia.
11. Victor, D. J.. (2001). *Policy for integrated solid waste in Malaysia*. Master thesis, University of Malaya, Malaysia.
12. National solid Waste Management Department. http://www.kpkt.gov.my/jpspn_en_2013/main.php?Content=sitemap on November 2020, Japan.
13. Solid Waste Disposal Act. (2020). http://definitions.uslegal.com/s/solid-waste-disposal-act-swda/ on the 5 November 2020.
14. USEPA History: Resource Conservation and Recovery Act, United States Environmental Protection Agency, Washington, DC, USA. https://www.epa.gov/history/epa-history-resource-conservation-and-recovery-act on November 2020.
15. USEPA. (2019). *Superfund Task Force Recommendation 13 examine opportunities to achieve protective cleanup at NPL-caliber sites without listing on the NPL summary of findings*. US Environmental Protection Agency.
16. Wang, L. K., & Wang, M. H. S. (1988). *Guidelines for disposal of solid wastes and hazardous wastes* (Vol. 1) (Technical Report No. PB88178066. OCLC No. 18438829). Department of Commerce, National Technical Information Service, VA, USA.
17. Wang, L. K., & Wang, M. H. S. (1988). *Guidelines for disposal of solid wastes and hazardous wastes* (Vol. 2) (Technical Report No. PB88178074. OCLC No. 18438829). Department of Commerce, National Technical Information Service, VA, USA.
18. Wang, L. K., & Wang, M. H. S. (1988). *Guidelines for disposal of solid wastes and hazardous wastes* (Vol. 3) (Technical Report No. PB88178082. OCLC No. 18438829). Department of Commerce, National Technical Information Service, VA, USA.
19. Wang, L. K., & Wang, M. H. S. (1988). *Guidelines for disposal of solid wastes and hazardous wastes* (Vol. 4) (Technical Report No. PB88178090. OCLC No. 18438829). Department of Commerce, National Technical Information Service, VA, USA.
20. Wang, L. K., & Wang, M. H. S. (1988). *Guidelines for disposal of solid wastes and hazardous wastes* (Vol. 5) (Technical Report No. PB88178108. OCLC No. 18438829). Department of Commerce, National Technical Information Service, VA, USA.

21. USEPA. (2019). *Implementing the RCRA/CAA air emission controls compliance exemption/election provisions through RCRA permits*. US Environmental Protection Agency.
22. USEPA. (2018). *CERCLA and EPCRA reporting requirements for air releases of hazardous substances from animal waste at farms*. US Environmental Protection Agency.
23. USEPA. (2020). *Medical waste tracking act 1988*. US Environmental Protection Agency.
24. USEPA. (2020). *United States federal strategy for addressing the global issue of marine litter*. US Environmental Protection Agency.
25. USEPA. (2020). *Pollution prevention act of 1990*. US Environmental Protection Agency.
26. USEPA. (2021). *Municipal solid waste landfill*. US Environmental Protection Agency.
27. USEPA. (2020). *RCRA public participation manual tools RCRA Community Engagement Initiative (CEI) case study*. US Environmental Protection Agency.
28. USEPA. (2020). *RCRA public participation checklist: Community engagement assessment worksheet*. US Environmental Protection Agency.
29. USEPA. (2019). *Fact sheet: Interim recommendations for addressing groundwater contaminated with PFOA and PFOS*. US Environmental Protection Agency.
30. USEPA. (2018). *Superfund, RCRA CA, and brownfields sites*. US Environmental Protection Agency.
31. USEPA. (2018). *Incomplete oversight of state hazardous waste rule authorization creates regulatory gaps and human health and environmental risks*. US Environmental Protection Agency.
32. USEPA. (2017). *Self-insurance for companies with multiple cleanup liabilities presents financial and environmental risks for EPA and the public*. US Environmental Protection Agency.
33. USEPA. (2019). *Office of land and emergency management Fy 2020–2021 national program guidance*. US Environmental Protection Agency.
34. USEPA. (2017). *National Program Manager (NPM) guidance fiscal years 2018–2019 Office of Enforcement and Compliance Assurance Draft*. US Environmental Protection Agency.
35. USEPA. (2019). *Office of Enforcement and Compliance Assurance National Program guidance fiscal years 2020–2021*. US Environmental Protection Agency. 41 Pages.
36. USEPA. (2019). *Clean air act permitting for greenhouse gases*. US Environmental Protection Agency.
37. USEPA History: Clean water act. United States Environmental Protection Agency, Washington DC, USA. https://www.epa.gov/laws-regulations/history-clean-water-act
38. USEPA. (2020). *Summary of the toxic substances control act*. United States Environmental Protection Agency, Washington, DC, USA.
39. Wikipedia Contributors. (2021, March 27). *Safe drinking water act*. Retrieved June 2, 2021, from https://en.wikipedia.org/wiki/Safe_Drinking_Water_Act
40. Galal-Gorchev, H. (1993). WHO guidelines for drinking-water quality. *Water Supply, 11*(3–4), 1–16.
41. Farmer, A. M. (Ed.). (2012). *Manual of European environmental policy*. Routledge. 1043pp.
42. Hansen, W., Cristopher, M., & Verbuecheln, M.. (2002). *EU waste policy and challenges for regional and local authorities*. December 2002 Ecologic, Institute for International and European Environmental Policy. Retrieved from https://www.ecologic.eu/sites/files/download/projekte/1900-1949/1921-1922/1921-1922_background_paper_waste_en.PDF
43. Gharfalkar, M., Court, R., Campbell, C., Ali, Z., & Hillier, G. (2015). Analysis of waste hierarchy in the European waste Directive 2008/98/EC. *Waste Management, 39*, 305–313.
44. Directive 2008/98/EC of the European Parliament and of the Council. (2008). Legslation.gov.uk. from https://www.legislation.gov.uk/eudr/2008/98/article/1 in November 2020.
45. Federal Ministry for The Environment, Nature Conservation, and Nuclear Safety. (2008) *Directive 2008/98/EC on waste and repealing certain Directives*. Waste Framework Directive. Retrieved from https://www.bmu.de/en/law/directive-200898ec-on-waste-and-repealing-certain-directives/ in November 2020.
46. European Commission. (2020). *Directive 2008/98/EC on waste (Waste Framework Directive)*. Retrieved from https://ec.europa.eu/environment/waste/framework/ in November 2020.

47. Borongan, G., & Okumura, S. (2010). *Municipal waste management report*. Status-quo and Issues in Southeast and East Asian Countries, AIT/UNEP Regional Resource Center for Asia and the Pacific (RRC.AP).
48. Japan Waste Management & 3Rs Research Foundation. (2017). *State of the 3Rs in Asia and the Pacific*. United Nations Centre for Regional Development (UNCRD), Japan. November 2017 https://www.uncrd.or.jp/content/documents/5690[Nov%202017]%20Japan.pdf
49. Japan Air Quality Standard. (n.d.). *Transport policy*, Japan. https://www.transportpolicy.net/standard/japan-air-quality-standards/
50. Air Pollution Control. (1968). *Japanese law translate*, Japan. Retrieved from http://www.japaneselawtranslation.go.jp/law/detail/?printID=&id=2146&vm=02
51. Japan's Environmental Policy. (2010). *Research Institute of Economy, Trade and Industry*, Japan. https://www.rieti.go.jp/en/special/policy-update/039.html
52. Water Environment Department, Environmental Management Bureau. (n.d.). *Water pollution control law*, Japan. Retrieved from http://www.wepa-db.net/policies/law/japan/wpctop.htm
53. Water Environment Department, Environmental Management Bureau. (n.d.). *Enforcement of effluent water quality standards*, Japan. Retrieved from http://www.wepa-db.net/policies/law/japan/wpc3.htm
54. Water Environment Department, Environmental Management Bureau. (n.d.). *Promotion of countermeasures against domestic effluents*, Japan. Retrieved from http://www.wepa-db.net/policies/law/japan/wpc5.htm
55. Yang, W. S., Park, J. K., Park, S. W., et al. (2015). Past, present and future of waste management in Korea. *Journal of Material Cycles and Waste Management, 17*, 207–217. https://doi.org/10.1007/s10163-014-0301-7
56. South Korea Legislates Towards a Zero Waste Society. (2015, July 17). *Waste management review*, Korea. Retrieved from https://wastemanagementreview.com.au/south-korea-legislates-towards-a-zero-waste-society/
57. Basel Convention. (2002). *Country fact sheet 2006*. Republic of Korea.
58. Overview; Basel Convention. (n.d.). *UN programme*. Retrieved from http://www.basel.int/the-convention/overview/tabid/1271/default.aspx
59. South Korea – Sub-Act on the Promotion of Saving and Recycling of Resources. (n.d.). *TUV Rheinland*, Korea. Retrieved from https://www.tuv.com/regulations-and-standards/en/south-korea-sub-act-on-the-promotion-of-saving-and-recycling-of-resources.html
60. 3R Portfolio Good Practices to Promote the 3Rs -Country: Republic of Korea, Ministry of Environment (Japan). (n.d.). https://www.env.go.jp/recycle/3r/en/info/05_12.pdf
61. Policies for Sustainable Resources Management in the Republic of Korea. (n.d.) *International Policy Research Center Korea Environment & Resources Corporation*, Korea. http://me.go.kr/home/web/main.do

Chapter 3
Waste Transportation and Transfer Station

Mohd Azhar Abd Hamid, Siti Fatihah Ramli, Hamidi Abdul Aziz, and Yung-Tse Hung

Abstract Municipal solid waste includes various types of complex waste predominantly from the residential, commercial, institution, and industrial collections. Waste collection and transport operation has now become a primary challenge requiring a regular response from the waste service provider. Transfer stations usually act as spaces for the waste collector to discharge their load before reloading it into a large vehicle for long-haulage waste transportation. Besides, a transfer station also plays a critical role that directly links the waste collection stream to the waste treatment facilities or final disposal site. The main issues and challenges to strengthen the efficiency of municipal solid waste transport and transfer operation are discussed in this chapter. These elements could aid to improve the sustainable and efficient total waste management system that could induce the balances of anaesthetic and a pleasant environment.

Keywords Solid waste · Municipal Solid Waste (MSW) · Waste transportation · Transfer station · Waste collection

---

M. A. A. Hamid · S. F. Ramli
School of Civil Engineering, Engineering Campus, Universiti Sains Malaysia, Nibong Tebal, Pulau Pinang, Malaysia

H. A. Aziz (✉)
School of Civil Engineering, Engineering Campus, Universiti Sains Malaysia, Nibong Tebal, Pulau Pinang, Malaysia

Environmental Engineering, Solid Waste Management Cluster, Engineering Campus, Universiti Sains Malaysia, Nibong Tebal, Pulau Pinang, Malaysia
e-mail: cehamidi@usm.my

Y.-T. Hung
Department of Civil and Environmental Engineering, Cleveland State University, Cleveland, OH, USA

L. K. Wang et al. (eds.), *Solid Waste Engineering and Management*, Volume 1, Handbook of Environmental Engineering 23,
https://doi.org/10.1007/978-3-030-84180-5_3

Acronyms

KRI	Khazanah Research Institute
MCO	Movement Control Order
MSW	Municipal Solid Waste
RORO	Roll-on–roll-off

Nomenclature

B	= time to remove and replace each loaded vehicles (min),
B_C	= the total cycle time for clearing each pushpit and compacting waste into the trailer (min)
F	= peaking factor (the ratio vehicles received of average 30 min to peak 30 min)
H_t	= hours per day used to load vehicles
H_W	= hours per day that waste is delivered
L	= the total length of dumping space (m)
L_n	= the length of each hopper
L_p	= the length of each pushpit (m)
N	= number of transfer vehicles loading simultaneously
N_n	= number of hoppers
N_p	= number of pushpit
P_C	= collection vehicle payloads (tons)
P_t	= the transferred vehicle's payload (tons)
T_C	= time to unload each collection vehicle (min)
T_t	= time to load each transfer vehicle (min)
W	= the width of each dumping space (m)

3.1 Introduction

According to the United Nations, up to 50% of the world population will live in cities area in the year 2050. This fact necessarily makes the sustainable development of cities areas amongst the most critical agenda for the next century. One of the adverse effects of this urbanization activity and great economic growth is the high generation of municipal solid waste (MSW). The design and application of collection and transportation operation for an effective solid waste management system is an essential step for the ecological protection and sustainable growth of a country [1]. Thus, the primary purpose of waste collection transport is not only to collect recyclable solid waste materials; it also plays a role in transferring the waste from the source to a site where the loaded vehicle can be dumped to safe waste treatment facilities [2]. Organized and comprehensive waste collection and treatment are important for social well-being and health purposes, such as for disease prevention or odours mitigation in the public area. In recent years, waste effects have been attributed to environmental issues like global warming, ecosystem pollution or depletion of resources [3]. The association between waste operation practice and the effects of greenhouse gases causes high environmental impacts due to methane effluents contributing to climate change and stratosphere ozone depletion. At the same time, inadequate or ineffective recycling operation results in an increased scarcity of the useful resource. Because of this, legislation across the globe has attempted to better regulate waste management practices in order to facilitate more effective waste operation systems. Building a solid waste transfer station or recycling centre becomes a practical solution to this waste problem. Besides, the transfer station facilities may also allude to alternative options to sort in homogeneous waste with cost-effective transportation. This could lead to minimize landfills requirements and promote sustainable waste management practice. Hence, the basic principle of waste management concepts would also include the various associated operation of collection, transportation, and disposal as well as steps to minimize the waste that is much wider than the term of waste treatment [4].

3.2 Municipal Solid Waste (MSW)

3.2.1 Introduction

Rapid urbanization and modernization in most developing countries have resulted in the massive production of MSW. This condition poses a serious concern among the government as well as the public as it may cause and lead to many problems, such as deterioration of the environment and other natural resources. In addition, the rapid growth of population, industrialization, and improper planning and execution of waste disposal can be a major cause for the degradation of the environment. Generally, poor solid waste management leads to two main problems, which are the loss of resourceful materials and the social cost due to the health impact on the public [5]. Basically, MSW can be categorized into industrial, commercial, and household waste. Waste such as durable and non-durable goods, food waste, yard waste, inorganic waste from residential, commercial, and industrial sources are all classified as MSW. However, waste coming from sources such as municipal sludge, combustion ash, and industrial non-hazardous process waste is not included as the MSW. Depending on the municipalities, the quantity and the composition of MSW vary greatly. Factors such as climate change, social customs, per capita income, living standards, geographical locations, cultural habits of individuals, type of housing, degree of urbanization and industrialization affect the characteristics of MSW.

In Malaysia, for example, the main sources of MSW come from household waste. According to Khazanah Research Institute (KRI), about 44.5% of total solid waste is generated from the household sector, which attributed to 6.1 billion tonnes per year. Among the 20 listed categories of MSW, food waste made up about 50%

of the total waste composition [6]. Food waste sources in Malaysia were derived from residential, commercial, institutional, commercial, and city areas. Generally, food waste is defined as food that is safe to consume, but due to the spoilage, it needs to be discarded. Different countries will produce different levels of food waste. A much lower-income country usually wastes food during the early phase, which is the production phase. Meanwhile, food-wasting is observed at the last stage of the household for middle- and higher-income countries [7]. Normally, food-wasting increase due to overbuying, which leads to the expiration of the product even before consumption.

Seasons/cultural habits are one of the factors that affect the composition and production of the MSW. For example, in Malaysia, during the fasting month (Ramadhan) and Hari Raya celebration, the production of solid waste tends to increase due to overbuying or higher consumption of food. The untouched or unused food is usually thrown away and ends up as waste.

In addition, just recently, Malaysia had introduced and undergo the Movement Control Order (MCO) due to the outbreak of the COVID19 pandemic. This restriction was effective from 18 March 2020, which required the citizen to stay at home to isolate and break the outbreak chain. This has exaggerated the waste amount as most of the time people stayed at home. However, a study by Ismail et al. (2020) [8] on the effect of food waste generation during MCO in one of the towns in Malaysia state (Klang Valley) contradicted this expectation. It is expected that the generation of household waste, especially food waste, will be increased due to the restriction and higher frequency of activity during the stay-at-home order. This phenomenon leads to more overbuying/panic buying of food supplies. However, the food waste data after the study showed a decrease of 15.1% during the MCO compared to during the normal condition.

3.2.2 Collection of Municipal Solid Waste

Basically, there are four phases involved in the solid waste collection and transportation, which are as follows:

i) Storage at the generation and pickup points
ii) Waste pickup by the workers
iii) Trucks driving around the neighbourhood
iv) Transportation of waste to a transfer station/disposal point

The waste collection is quite costly, difficult, complex, and time-consuming. An improvement in terms of waste collection should be implemented to overcome these issues. Typically, the collection cost of solid waste requires 60–80% of the municipality's expenses budget. Generally, the collection of waste involves these two steps:

i) Onsite Storage and Handling

The commonly used container by a single-family residential area is either plastic, galvanized metal containers, disposable paper, or plastic bags. It is normally handled by the residents or the tenants themselves. Generally, single-use plastic bags are used by the homeowner only when the curb service is provided. For the high-rise building residents, the picked up is normally conducted by the building maintenance personnel, or special vertical chute are provided. The function of the vertical chutes is to deliver the waste directly to the central location for storage, processing, or resource recovery. Apartments utilize stationary container systems into which the residents drop the solid wastes. Solid wastes from commercial buildings are collected in large containers that may be stationary or transportable [9].

ii) Collection

For residential area, the collection method is either curb or alley, set out–set back, and backyard carry. In curb/alley service, the waste is brought to the curb/collection point by the residents by carrying the single-use plastic bags and container which contain the waste. Then, the empty container is taken back to its original spot after the pickup. For the set out–set back method, the collection crew returns the empty containers. Meanwhile, in the backyard carry service, the transfer of solid waste into the wheeled barrel is conducted by the collection crew and then unload into the collection truck [9].

3.3 Waste Transportation

Advanced solid waste management normally applies the integrated approach for a sustainable waste operation system. The system encompasses activities of generation, collection, transportation, segregation, transfer, recovery, treatment, and disposal with focusing on optimizing the efficiency of resources usage [10, 11]. Thus, waste transport is a critical task in the overall waste management approaches. This operation begins from the sources of waste generation at residential, commercial, institutional, and industrial areas until they are transferred to the waste treatment facilities or disposal site [2]. However, the challenges of waste transportation may occur in terms of low frequencies, route distance, inappropriate vehicle, and transport time [12]. These factors may have an impact on the quantity of waste transported and the total level of operation services. Care must be taken not only in terms of the environmental issues associated with emissions of greenhouse gas and usages of resources like energy, food, and materials but also related to waste transporting and its generation [13]. Several critical aspects, such as types of transport vehicle, waste amount, and operation costs of managing waste significantly influence the level of waste collection services [14]. At present, waste generation remains one of the major global issues that need to be solved. The raising in the numbers of the population has led to an increase in the waste amount. Hence, the operation of urban waste management planning should be a key focus of the municipalities authority [15]. Solid waste transportation is a part and parcel of the challenges and complications faced by the waste authority because of the rising waste generation at different sources [16, 17]. Ristić et al. (2015) [18], Haerani and Budi (2019) [12] reported that approximately waste transportation constitutes about 50% to 70% of the overall waste management costs. Furthermore, inefficient waste transport would have serious effects on public health, increase operational cost, low sweeping efficiency, and clogging of the drainage system [12].

Figure 3.1 depicts the anticipated relationship between the main element involved in the waste transportation framework. Usually, waste is sorted at transfer stations after collection by collection vehicles. The waste segmentation is categorized based on recyclable, organic, low and high energy, which will be directed to facilities of recycling, composting, landfill, and incinerator plant [15]. The ash of by-product from waste produces by the incinerator plant also finally goes to a disposal site.

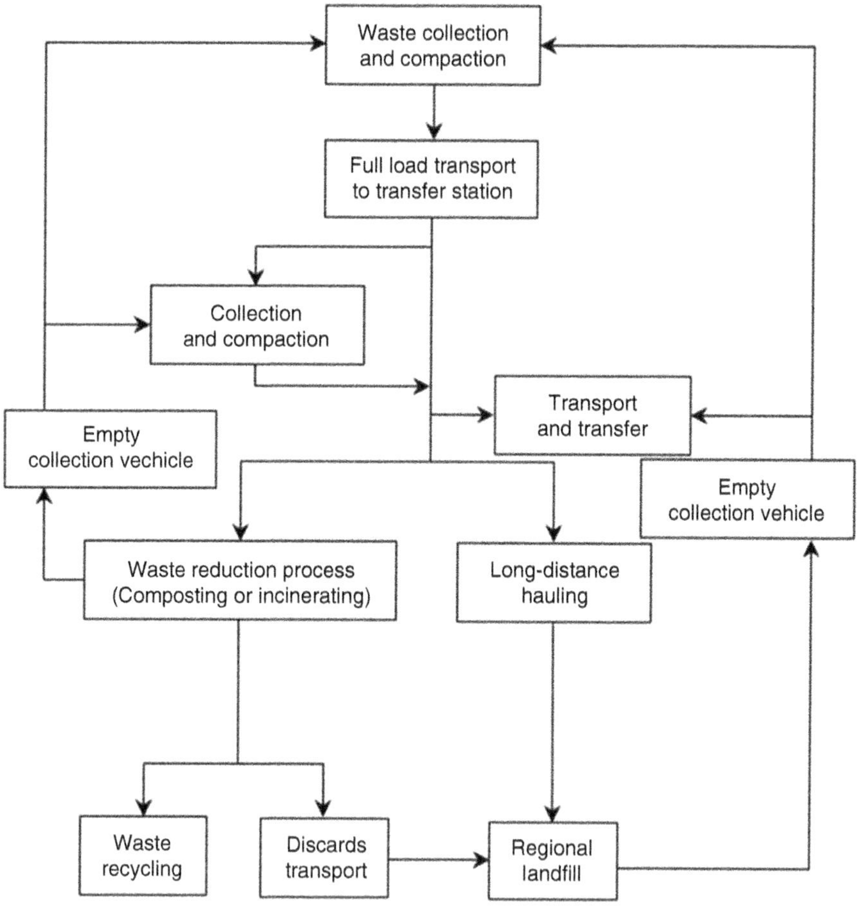

Fig. 3.1 Waste transportation framework [15]

3.3.1 The Importance of Solid Waste Transportation

Solid waste management involves several stages such as generation control, storage, collection, transfer and transport, processing, and ends with the disposal of solid waste wastes [5]. However, in most developing countries, unfortunately, the solid waste management faces various kind of issues such as lack of coverage of the collections and irregular collections, crude open dumping and burns without air and water pollution control, fly and vermin cultivation, and informal waste picker treatment and control [19]. Improper management of waste collection or transportation might also lead to a serious problem. The occupational and environmental health problems as well as a lack of proper training of workers during the collection and transportation will cause the injury issues at the site. Moreover, the uncollected waste led to drain clogging, which will cause stagnant water. Hence lead to the breeding of the mosquito vector is also one of the environmental health issues. Besides, the poor solid waste management also affected the nearby environment as the downstream ground and surface water quality will be deteriorated by the contamination of leachate from the land disposal facilities. With poor control, volatile organic compounds and dioxins in the air emissions may increase cancer incidence and psychological stress for those living near the land disposal facilities.

To date, landfill is a commonly used waste disposal method, and it is generally the cheapest and easiest way for waste disposal. However, the lack of proper management of the site will contribute a detrimental effect to the environment. Disposal of food waste at the landfill site tends to create many adverse effects as the food waste contains high organic content. Through landfilling, food waste is estimated to increase the emission of methane gases up to 50% by 2020 [20]. The success of the disposal systems is very much influenced by a good transportation system

3.3.2 Role of Waste Transportation

Generally, an integrated municipal solid waste management operation can possibly be divided into four main phases, which involve the collection, transfer, transport, and treatment [21, 22]. From the environmental perspective, the efficacy of waste management operation depends heavily on the degree of segregated waste collection. The effectiveness of this operation could lead to the minimal deposited waste into landfill, which in less release of carbon gases [23]. The sustainability of a waste management system is of concern to researchers around the globe [24], particularly in terms of energy usage and its returns outputs [25]. Most of the focus in the waste management chain, however, heavily tends to be based either on the initial step of waste collection or the final ones of waste treatment. Besides the environmental impact imposed by those facilities, the transport of waste from the source of collection to the waste disposal site requires special concern to better determine the entire responsibility of the waste management operating systems [26]. Waste transportation is all procedures for moving the collected waste from one geographical point to another. This involves the transportation of loaded waste from a source of the collection area to discharge location, treatment facilities, or processing industries for materials recovery [14]. This waste handling and segregation are the main steps for final disposal. Handling involves the shifting of loaded containers to the site of the collection, followed by transportation of the waste. Waste segregation is a significant process for the collection, handling, and proper storage of solid waste at the source level. Transportation involves two necessary steps: the waste is moved from normal collection vehicles to a long-haul vehicle where the waste is then transported to a landfill site far away from the urban settlement for final disposal [27].

The optimization of waste transport routes is an important instrument for minimizing waste collection and transport costs [28]. The waste transportation cost includes two main elements: capital and operating costs. These elements make waste transportation services expensive relative to other waste treatment operations costs due to labour, fuel, vehicle, container, and maintenance expenses [29]. The average capacity for waste collection transport worldwide is between 100 m^3 and 300 m^3/day/mil people served, especially in developing countries [30]. In most countries, transportation cost is the highest components of the total cost of a waste management system. Hence, emphasis must be given to optimizing the waste transport efficiency to improve the total amount of transported waste in a vehicle per day. Traffic congestion is another important aspect that needs to be considered, as it disrupts the fleet stream will reduce the effectiveness of this operating system [31, 12]. In addition, traffic flow disruptions may also occur in the cities with a large number of intersections and circulation on the route from the urban centre to the disposal site [32]. As a result, this causes an increase in pollution emission, resulting in a decrease in the air quality, especially in the urban area. Waste transport is indeed a very critical segment in the overall waste management system, which needs to be adequately considered in the waste management plan.

3.3.3 Selecting the Waste Transport

Solid waste transportation is one of the central components of the waste management system. The waste collection initially starts by picking up the waste from their source via a waste collection transport. This is the most expensive part of the waste management system. Hence, in order to minimize the operation costs, it is important to analyse the technical and rational factors to select the types of waste transportation vehicle to be used [33]. Many technical components have significant effects on the choice of operating system and transport vehicle for a specific situation. In most situations, the selection of vehicles and storing system are closely related. There are several key criteria that must be taken into account before choosing the desired type of waste transportation vehicle. In this case, the main purpose of waste transportation operation is to collect and transport waste from the generation point at optimum intervals to discharge to the waste treatment facilities with minimal costs [34].

(a) Waste Generation Rate

Increased population, industrial, urban development, economic, and living standards have led to a huge rise in the production of solid waste. Generally, richer countries and their societies produce more solid waste. In developing countries like Malaysia, the average generation rate is generally 1.17 kg/capita/day [35]. While the average waste production rates in most industrialized countries like the USA is about 2.1 kg/capita/day, the real trends of waste production for lower-middle-income countries such as Armenia and Philippines respectively are 0.33 kg/capita/day and 0.73 kg/capita/day [36]. The waste production rate has a critical impact on the selection of waste transport vehicle.

The waste generation rate has a critical influence on the selection of collection system and vehicle. For example, in the case of a curb side collection in two communities where the rate of waste generation in the second is twice that in the first, it is proven that more stops are needed in the case of the first community to transport the same final payload to the disposal site. Vehicle productivity will, therefore, be less, and alternative systems, such as the use of portable community containers, can prove more productive in such a case.

(b) Waste Density

The weight and density of waste are critical in order to accurately estimate the amount of solid waste and vehicle type and size that is needed [33]. The density of waste is different based on communities affluence, composition, moisture content, physical shape, degree of compaction, and storage [10]. Before beginning to design any transportation for waste collection operation, it is important to have appropriate data on the density of solid waste. If the density of the waste is low, it must be compacted in order to raise its density and reduces its volumes, thus improve its capacity and more cost-effective. This is why vehicles equipped with compaction feature are widely used in most countries. For example, in low-income countries like India, the application of compaction vehicles could obtain a final waste density of approximately 400 kg/m^3 with a ratio for compaction up to 4:1. In an industrial country, it is normal to achieve about a compaction ratio of 1.5:1 or less [37]. The waste density may also vary based on its location and the type of bin used. This difference is attributed to several main factors such as living standard, compression, decomposition, evaporation, air entrainment, loading, and travelling method [2]. Generally, the uncompacted waste density is about 150 kg/m^3, while the collected waste density is 235 to 350 kg/m^3 [38]. However, in certain situations, it is common for the density value of waste to decrease during the transfer process from waste containers to collection vehicles. During the waste transferring process, the air adsorption may reduce the overall density of the waste. The application of communities' bins may increase the efficiency of the waste collection vehicle. Figure 3.2 shows the possible estimation of waste density at the different collection stages, which can be used to estimate the capacity of containers or collection vehicles required.

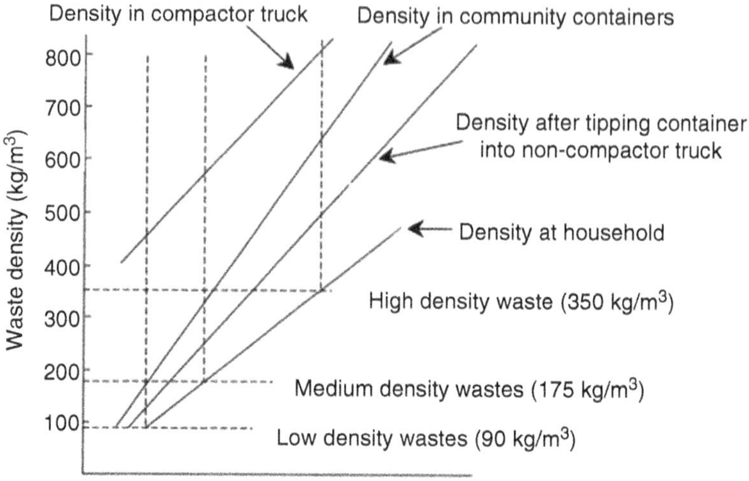

Fig. 3.2 Change in waste density during handling [38]

(c) Waste Volume Per Capita

The integration of lower rates of waste production and higher density would give a significant change in the volumes of the total waste. Per capita waste generation is a useful measure for evaluating the intensity of waste generation over time that can contribute to better waste management planning [39]. The generation and density of this waste vary between countries. The per capita waste production degree typically rises linearly with the improvement of the quality of living standard. Generally, the daily average production rates in kg/capita/day ranged from 0.8 to 1.2 kg/capita/day in large cities, whereas from 0.35 to 0.5 kg/capita/day in small cities [37]. In contrast, the average rate of waste generation is higher in industrialized cities such as in Japan at 1.64 kg/person/day [2]. Bangladesh with low incomes has average lower rates of waste generation of around 0.387 kg/person/day [40]. Thus, any waste transport and transfer planning must take into account any expected rise in the waste amount which could be produced in the futures. This condition can be attributed to population development, additional service areas, rising in waste per capita production and a decrease in waste density due to improvement in standard living.

(d) Waste Constituents

MSW is usually composed of biodegradable, non-biodegradable, and inert waste. Up to 4.15% weight of inert wastes may be present, which comprise of coarse substances like ashes, stones, and sand as well as dust [2]. This kind of substances in waste may break sliding components and wear out the hydraulic parts, which are the main elements in waste collection trucks. The compaction vehicles are equipped with characteristics that need a high degree of maintenance, especially for a replacement of sliding and rotating parts. These parts usually get in contact with solid waste containing a huge amount of inert substances. Generally, solid waste also usually contains large percentages of biodegradable materials (around 50% to 66%) [37]. Acids produced from the process of organic matter decomposition could pose significant corrosion issues in the body and storages area of the vehicle. The design and choice, as well as the operation, should consider these problems. The generation of waste from commercial areas is expected to have low biodegradable waste. Thus, the wastes from this area are likely to highly differ from the common household, which mainly contains mixed wastes. Depends on the waste storage method and degree of prosperity, the wastes from specific communities' area could consist of loose items or big bulky waste. Air is also frequently caught in disposal bags to make wastes more voluminous than expected [10].

(e) Transport Distance

Fuel consumption, traffic conditions, and hauling distances significantly impact the selection of waste transportation and decisions on the use of the transfer station [41]. Thus, the service area and the final disposal site, such as the landfill, should be thoroughly assessed. There is little point in investing in expensive high-speed waste transports where hauling distances are very short and slow traffic speed. The option of any waste transportation should be a compromise in terms of effective operation in the waste collection and transferred to the transfer station, treatment facilities, and landfill. The distance from a collection area and discharge location influence the waste transport size and speed. Small transport and the simple-to-load vehicle should be compared to large transport that runs slowly and hard to load; however, it requires fewer journeys to waste treatment facilities. Besides, the total hauling distances also define the requirement for the transfer stations to transferring the waste from the normal collection vehicle to long-haul transfer vehicle, flatbed railcar, or ship containers for transporting to remote waste treatment facilities [42]. According to Pires et al. (2018) [41], waste collection operation vehicle can be reduced by up to 20% of the total cost depends on engine vehicles' efficiency and driving characteristics.

(f) Loading Heights

The difference in vehicle loading heights has an important influence on the loading speed since it affects operational productivity. Generally, loading height for waste collection vehicles is usually determined by the distance from the ground where a container has to be lifted to empty. If the waste container can be lowered down, it is much easier by lifting a door so that the loading height can be lowered for effective loading. On the other hand, if the height for loading is too high, it can increase loading time and injury risk, which in turn would also increase the pressure and exhaustion of workers [10]. Workers must not be asked to load the waste over their shoulder level as it may pose excessive workload and the health risks associated with falling waste [37]. The traditional sideloading waste transport vehicle has high-level bodies positioned where the waste collector unload the containers above the head height (Fig. 3.3). The waste transport with higher loading heights may need additional crews inside of the vehicle to unload the container. This process is quite

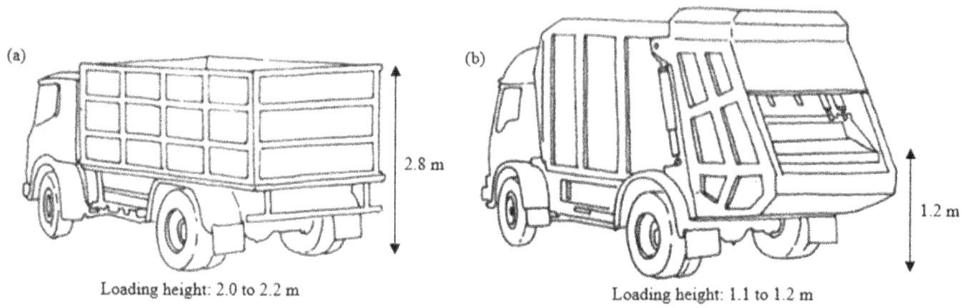

a) Open top and (b) compactor

Fig. 3.3 Vehicle loading height vehicle [30]

risky and dangerous. They can handle large bulky items, which allow very low loading heights and are then easy and fast to load [41].

(g) Traffic Impacts

An increase in traffic density affects the collection of waste transportation, resulting adversely in the productivity rate of waste generation [43]. Besides, the amount of fuel consumption operation efficiency could also be influenced by traffic conditions [44]. Most delays in waste transport operation can be attributed to high traffic flow, which coupled with the poor design of local road [45]. In a certain city centre, it has a restriction on the use of heavy vehicles in the daytime or during times of high traffic flow. Besides, a heavy vehicle might be prevented to use certain roads or bridges due to weight limitations, traffic jams, or pollution emissions.

(h) Vehicle Maintenance

The critical factors for a sustainable waste management operation are the fast availability of replacement parts and accessibility to the vehicle facilities for the maintenance process [46]. Advanced waste vehicle may eventually cause a long-time delay and high expenses when the vehicle breakdowns. It often happens that the daily operation of solid management has failed due to scarcity and long wait for receiving replacements parts. This may result in lengthy delays in the operating systems and a shortage of vehicles associated with inadequate preventive maintenance [37]. Thus, in cases of modern imported waste transport, the procurement of replacement parts from abroad could be very difficult; therefore, it is critical to choose the available parts locally. Generally, the performance of waste operators is evaluated by their abilities in performing operational activities, which also includes vehicles preparation, operational monitoring, and vehicle maintenance planning [47]. Thus, all financial aspects and operation facilities of waste transportation vehicle, including waste containers, should be taken into accounts while selecting an appropriate waste transport vehicle. In addition, the operational budgets should also include the overall operating and maintenance costs and the estimation of loading speed by crews and standby or replacement of transportation vehicles.

(i) Level of Service

Solid waste services are very important in keeping the locality clean and safe. Unfortunately, it has been given unfair attention by some municipalities. This can be shown by the limited allocation of funds for operating, maintenance, and investment in the waste management system. If the whole expense of the solid waste management system is covered by a service charge, the willingness of the householders to pay for the service would decide the quality of service that can be offered [37]. Hence, a more effective waste transportation system based on service quality can be promised to the community. A high degree of service will require a direct waste collection from door-to-door for each day that has greatly impacted the costs of capital and operation [48]. A viable waste management system may require the waste generators from the residential, commercial, institution, and industrial should be prepared to pay any incurred costs either directly (direct billing) or indirectly (service tax).

(j) Participation of Private Sector

It is quite common practice that when a new waste management system is going to be developed/set up in a country, an external advisor is hired by the consultants to advise on the operating systems that need to be used. They normally involve in planning the technology, managing skills, operational, customer service, as well as providing resources to assist the government in its waste management efforts [37]. Local data like waste density, labour supply, hauling distance, and future expected changes should be detailed. However, in a certain condition, these consultants draw on their expertise without taking into consideration characteristics of local conditions such as various wastes properties, road network, labour cost, capital expenses, and maintenance ability [2].

(k) Variations in Transportation Mode

In selecting a suitable waste vehicle type, the modes of other transportation sectors like public transport, goods freight, and agriculture haulage, which is used locally, are also beneficial to observe [42]. Waste transport like animal and human carts, trishaw, or modern vehicles can be applied for freight transportation and also suitable for transporting recyclables items but will rely on the needs of operating countries [37].

(l) Computer Software

Computer software can be used to aid in selecting the most efficient waste transport and transfer system by comparing the suitable operation vehicles. The application of a computer programme can be set up to link with the inputs which can correspond with the local condition, the density of population, waste characteristics, travelling distances, road condition, traffic information, rates of waste generation and density, etc. [49]. In addition, cost information in managing this operation should also include labour cost, vehicle characteristics, fuel consumption, transport equipment, and economic cost also could be synergized [18]. From this overall data and information, a sophisticated waste management computer system can be built through a local database.

3.3.4 *Type of Waste Transportation Vehicle*

The use of waste transportation vehicle must be suitable for their operation purpose. There are several main types of waste transport vehicles, such as small, non-compaction, semi compaction, and full compaction [41]. In general, most of the collection routes covered the whole service area. Data on the quantity of waste variation and generation are very useful in planning for a collection vehicle and disposal system. For example, bulky waste requires a hook lift, while smaller items are collected by compactor vehicles.

3.3.4.1 Small Vehicles

Waste transportation to waste treatment facilities is needed when storage containers are full. Small waste transports powered by an animal (Fig. 3.4) and human as well as micro-vehicle are ideal for transporting the waste within small areas like a market and short distance. Furthermore, this type of vehicle is also likely to be commonly used locally. In the normal operating system, handcart containers would be utilized in a short-range distance from the transfer station facilities, while trishaw and micro-vehicle respectively would be effective to use in middle- and wider-range distances areas [50]. However, the use of human and animal-powered waste transport leads to some disadvantages which are seen like old-fashioned and shameful, issues with temperament, exposure to a health problem, limited travelling range, and greatly slow compared to motorized vehicles [51].

Fig. 3.4 Animal transport [30]

(a) Handcarts and Tricycles

Most of the municipalities also use non-motorized vehicles such as handcarts and tricycles for waste collection and transportation [37]. This type of vehicle might be suitable when short hauling and there is no steep slopes. The biggest advantages of these simple equipment are that they are inexpensive and are easy to operate and maintain. Besides, the utilization of these manual vehicles is to optimize travelling times and vehicle distances [52]. The load-carrying tricycles have a number of handcarts as they can be pedalled to the service area. Figure 3.4 illustrates a simple design of human handcarts and tricycles. Generally, this type of waste transport completes with a bell as an alert system to the resident for them to bring their waste to the waste collector. Hayat and Sheikh (2016) [53] stated that 60% of street waste collection is done with the aid of handcarts, while 10% by auto three-wheelers, and the rest 30% done manually. Meanwhile, the tricycles transport equipped with rear-mounted waste containers is used to transport the wastes from the community's containers and transferred to the small transfer station. This would help the waste collectors by tricycles to collect more waste compared to the handcarts mode. Both handcarts and tricycles can be considered the simplest waste transportation vehicles used to transport small waste quantities (<0.4 m^3) at short distances (<2 km) to the transit point [37]. Thus, in the planning of entire waste transport systems, proper consideration must be provided with the cost-effective and hygienic manner for transporting waste from one vehicle to another. Waste can be transferred by lifting and unloading the containers manually, or by mechanical lifting, or direct unloading into the container at a different level (Fig. 3.5).

(b) Animal Vehicle

In low- or middle-income developing countries, an animal-driven vehicle like donkey or horse cart is used to transport wastes [54, 55]. Their usage can be attributed to a lack of confidence in external aid and problems in obtaining recognition support from public resources [56, 57, 61]. However, the application of animal-powered vehicle requires no fuel usage, low investment and operation cost, low noise than motorized vehicles [46], making it a preferable option. In order to minimize the load for the animal and for easier unloading, attention must be given to design the braces and cart (Fig. 3.6). It is very suitable in the steep sloping area and unpaved road network. Nevertheless, the use of donkey carts is ineffective because

Fig. 3.5 Small waste transport [30]

Fig. 3.6 Animal waste transport [30]

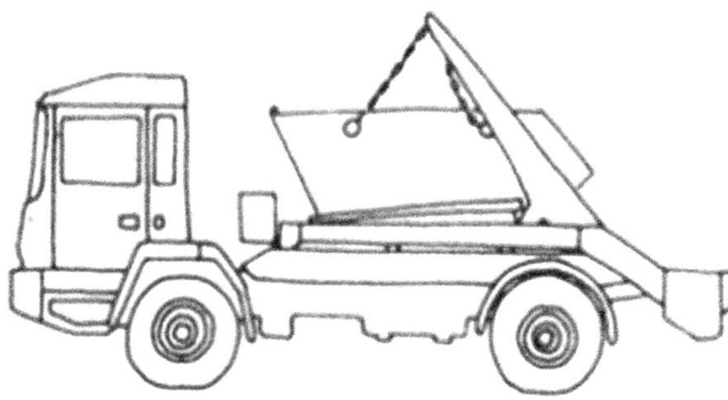

Fig 3.7 Waste micro truck [30]

of limited capacity loading and very slow for unloading. However, there is limitation in the use of donkey carts as it would be hard to pull the loaded waste due to poor balancing and inefficient harnesses. Furthermore, these animals may cause traffic congestion and would be very unpleasant and unsuitable for waste transport near the city centre. Hence, to date, the role of animal-powered vehicles for waste transport has received little attention [54].

(c) Micro Transport

The use of a small or micro truck requires a wider operation service [50]. This micro truck could be equipped with a waste collection container, normally a capacity of about 1.5 m^3 with the open-top or -hinged lids (Fig. 3.7). The micro truck is quite common and conventional and can be in many countries. It is possible to produce a standard tipping body locally. This type of micro truck is proven to be highly effective in an urban area with a high-density population [50]. The characteristics of a universal tipping system make this vehicle very appropriate for use in the small transfer station and allow these trucks to direct unloading into larges containers. The small wheel sizes and low loading height allows a quick loading rate; however, they are not ideal with bad road condition. This type of small vehicle is suitable to enter many narrow areas when the large vehicle is unable to enter. It is also very useful in high traffic density area. These features make the use of this vehicle is very profitable compared to other large sizes of the modern transportation system [58].

3.3.4.2 Non-compaction vehicle

The non-compaction vehicle is an alternative waste transport to motorized compacted vehicles. The loaded waste is still in loose form. If the loaded waste is highly dense, then the wastes loads can be obtained without compaction. Besides, this type of vehicles can also be suitable for very wet waste, limited access to skilled maintenance, and long collection routes [51]. However, several evaluations are needed to select the best types of non-compaction bodies to optimize vehicle utilization and efficiency. The vehicle body design has an important influence on the speed at which waste is loaded and thus could enhance the waste amount to be collected. The bodies vehicle for the non-compaction system needs a greater load space to allow the vehicle to bear its maximum weight. A non-compaction body, however, usually has low capital and maintenance cost and could be operationally effective to suit local conditions [59].

(a) Open Tipper Vehicles

The open-top vehicle is traditional dumping vehicles that can stretch sides upwards to raise its load capacities (Fig. 3.8). Transporting waste using this vehicle type is quite slow and unhygienically since the wastes should be transferred to the upper level by the crew inside the vehicle. When unloading, it is only raised at an angle, causing the door to open and solid waste to slide out. Due to the least cost-effective and versatility, these vehicles are extensively used to transport various types of MSW in many municipalities in Asia regions [37]. The physical features like high-sided bodies fitted, which are often in excess of 2.5 m, make it practically useful for loading bulky waste and can commonly be loaded up to 11 m^3 [60]. Besides, physical appearances like small trucks, easy operation, faster load have excessive loading heights make these types of vehicles widely utilized. However, the slow loading speed caused by the high loading heights necessitates the use of an enormous number of these vehicles. There is also a propensity to lose the waste by

Fig. 3.8 Open tipper vehicle [30]

blowing off the top of the vehicle during travelling unless the driver use tarpaulin to shield the loading.

(b) Roll-on–roll-off

Roll-on–roll-off (RORO) vehicle could lift their container to the deck by raising and dragging the front of the container (Fig. 3.9). Initially, the RORO vehicle utilizes a lifting frame system with a metals cable in order to wire up the container by the crews. However, this system has evolved and, in most cases, have been upgraded and overtaken with the hook-lift system, which enables a single driver could operate it without leaving the vehicle. The hook-lift system is mounted with a hydraulic-powered arm coupled with a steel hook in front of the container, which engages a loop [61]. The container rear-ends rolls is done via the metal rollers, and the container is emptied by a tipping system. RORO vehicle could collect the container with a complete width without having crossbars or barriers to restrict the load height. Therefore, it will collect containers that are much bigger than the open tipper system and further appropriate to handles urban wastes. Typically, these RORO containers can hold up to 20m^3 and used to transport bulky waste and other waste like construction, remoulding, garden waste, etc. [48]. However, it is possible to lift the 30 m^3 capacity container with more than three-axle vehicles, yet this system is very expensive and often used for long-distance hauling transport [62]. There are different types of hook-lift containers to handle various container types with a single vehicle only. The open-top containers with slides more than 2 m in height could be loaded by wheel loaders at the different floor level in the transfer station. Meanwhile, the container that is used for long-haul transport may be loaded using a stationary compactor system which compacts the wastes to the RORO containers. The open-top container must be a shield with a tarpaulin during travelling to prevent the waste from dropping.

(c) Crane Vehicle

Initially, this crane tipper system for waste collection vehicle was established in developing countries (Fig. 3.9); they use a hydraulically powered crane to reach out the container with the curb side collection [42]. The tipper crane could lift

Fig. 3.9 Roll-on–roll-off waste vehicle [30]

Fig. 3.10 Crane tipper system with container [30]

containers of 1.0 m^3 within distances 8 m from the vehicle, and thus the container can be placed at any type of roadside, either pave or unpaved [41]. This collection waste vehicle is mounted with special features that allow the collector to pull the containers into the open-top containers to empty. This vehicle is quite rapid and hygienic during loading, while by tipping technique for unloading. The collection operation by the hydraulic system can cover the vehicle body during transportation to prevent the wastes from being flying. This vehicle can be safely used when the containers are not placed under the cable lines. The vehicle body can be used as an open tipper truck, but the hydraulic crane should specially designed. The capacity container up to 1.5 m^3 could be mounted to the rear-end of the vehicle and loaded from the residential container, which is at the curb side. It can be emptied in a similar manner as the other container. The collection system by using this vehicle can be in conjunction with a small open-top vehicle that is equipped with a rear-end lifter (Fig. 3.10).

3.3.4.3 Semi-Compaction Vehicle

The waste collection vehicle with semi-compaction is a midway compaction system between vehicle mounted with non-compaction and full-compaction equipment. Most waste reduction volumes can be achieved through this type of vehicle, which is more appropriate for generated wastes in low- to medium-density population area. Besides, these semi-compaction vehicles only require a simple additional compaction system and less demand in maintenance compared with a full compaction system.

(a) Sideloading Vehicle

This type of vehicles has open space on either side of their bodies for waste loading (Fig. 3.11). For easy loading purpose, this opening is higher up close to the front end. During the loading process, the filled waste is pushed into the back of the vehicle body by a packing steel plate. This is possibly the most popular form of specialized waste collection vehicle found in developing countries. This is because its main benefits are low capital costs and simple local manufacturing. Nevertheless, the opening door is slightly small, resulting in a speed rate reduction for loading waste, and the crews have to wait as the system moves the waste into the rear. Furthermore, the opening loading only can handle a single loaded at one time. In certain cases, it is normal to observe an extra crew inside the vehicle bodies to increase the loading process to clear the waste from the loading opening [41]. Some types of waste collection vehicles are equipped with a container lifter system that can lift and emptying the resident and communal containers [63]. For this reason, extra capacities are available by this type of semi-compaction vehicle, which are beneficial against non-compaction vehicles.

(b) Fore and Aft Vehicle

This type of vehicle body has a good design that combines waste low loading heights features, easy loading, rear-loaded with body capacities of up to 12 m^3 [30]. The loading operation system filled the wastes in the back hopper. This collection vehicle works by tilting it in both directions to the frontward to move the waste from aft to fore, thus making the space for the next loading [46]. The steel panel swings downwards from the top of the roof vehicle and then compress the waste to the fore of the vehicle body (Fig. 3.12). This is a very basic semi-compaction vehicle system without any sliding components to compact the waste within it. The hydraulic arm is mounted on the rooftop of the vehicle to compress the wastes and is thus shielded from contact with the waste. In general, this type of vehicle is suitable for handling wastes with densities above 250 kg/m^3 [30].

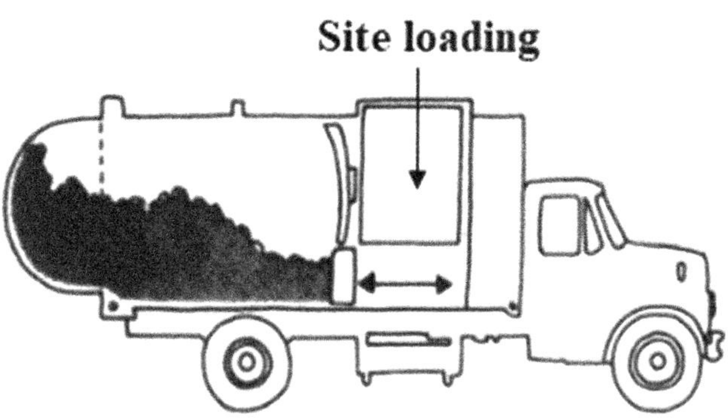

Fig. 3.11 Sideloading semi-compaction vehicle [30]

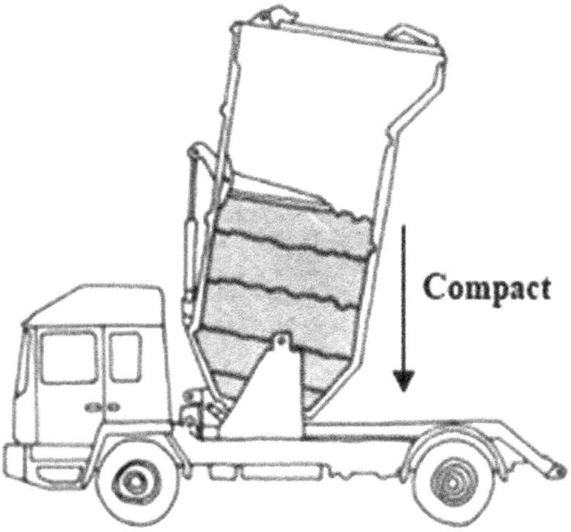

Fig. 3.12 Fore and aft semi-compaction vehicle [30]

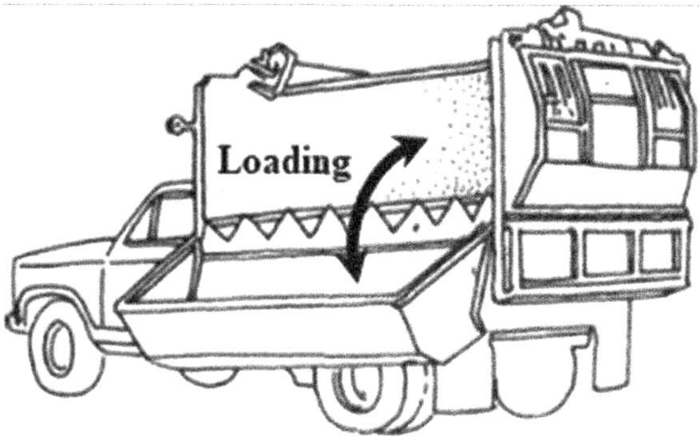

Fig. 3.13 Sideloading hopper semi compactor vehicle [30]

(c) Sideloading with Hopper

Compared with full-load compaction vehicles, semi-compaction of sideloading with hopper have simpler systems for compaction and less force to compress the waste [64]. This type of vehicle is operated by the one-side loading with a wide loading hopper with the aid of a hydraulic system. Then the waste is loaded into the bodies and compacted towards the different sides of the vehicle (Fig. 3.13). Once the vehicle body is full load, the accumulated wastes are transferred to the backside of the vehicle body. The drawback of this type of semi-compaction vehicle is that it requires the use of large roads to operate, which is only appropriate for a one-side loading vehicle. These waste collection vehicles are typically used in a rural and suburban area as there has much space to operate and when the waste collector is not obstructed by many parked vehicles [65]. Besides, it has a relatively sideloading cycle. Albeit this vehicle is still being used in several industrial countries. However, the use of sideloading of the semi-compaction system has some restricted application, particularly when the access is limited. Similar to other semi-compaction vehicles, this type of vehicle is only appropriate to load a moderate dense waste [64].

3.3.4.4 Full-Compaction Vehicle

The full-compaction waste vehicle is well recognized as mostly used municipal solid waste collection transports systems in various municipalities area. The utilization of full-compaction vehicles for waste transport is aimed to minimize the volume of waste and could also increase the waste density initially from 200 to 250 kg/ m^3 on average to 450 to 520 kg/m^3 [66]. Waste density is among the key elements in deciding the best waste collection vehicle to be used. The factors in Table 3.2 demonstrate the importance of densities of waste to be taken into account to illustrate the response of the full-compaction vehicles that can be efficient if the waste density is low and high at allowable payloads. Full-compaction vehicle bodies could accomplish a reduction in waste volume ratio of about 4:1 or 5:1 in low waste density, while in high waste density, the reduction ratio can reach up to 1.5:1 only [37]. In this case, the overall dimensions of the non-compaction body must be no more than those of the compactor truck to achieve the same load, resulting in a volumetric capacity of 20 m^3 for the open truck [64]. When the density of waste surpassed 250 kg/m^3, then the full-compaction vehicles are rarely be accepted [2, 50]. However, the use of this type of waste transport also gives many advantages and disadvantages, as listed in Table 3.1 below.

(a) Screw Compaction

This compaction vehicle system is slightly different from the rear-loading compaction plate vehicle. A rotating screw compaction vehicle is used to drive the wastes to their bodies so that they can be compressed and fragmented (Fig. 3.14). It consists of a rear spiral compacting screw system to transport the wastes into the vehicle bodies so that it continually continuously compressed and fragmented during collection [41]. The screw compaction waste system is the only running parts that are contacting with the loaded wastes and can be removed easily for maintenance. This type of full-compaction vehicle with a screw compactor system is much lighter compared to other hydraulic mechanisms. Thus it allows higher payloads up to 1000 to 1500 kg waste weight that is greater compared to rear-loaded compactor vehicle [30]. It is also ideal for handling loose and coarse wastes but is unable to handle bulky materials because of its limited screw opening. This type of waste

Table 3.1 Comparison waste density between non-compactor and compactor vehicle [30]

			Type of vehicle	
No	Item weight (kg)	Waste density (kg/m^3)	Non-compactor	Compactor
1	Permissible weight		13000	13000
2	Body Weight		1500	3500
3	Chassis		4500	4500
4	Max waste load		7000	5000
5	Max waste payload (Density)	100	2000	5000
		250	5000	5000
		400	7000	5000

Table 3.2 Advantages and disadvantages using full compaction waste transport [51]

No	Advantages	Disadvantages
1	Able to carry payloads with the wastes have a density which is too low	Compacting bodies are expensive
2	The vehicle is readily available in large quantity	The sophisticated system required special training
3	The vehicle has a sophisticated and modern appearance	High maintenance cost
4	The waste and liquid are largely enclosed	The rear loading hopper and compaction are so heavy
5	Loading is fast and more convenient.	Compaction operation quite noisy
6	More hygiene and less smell	High fuel consumption

Fig. 3.14 Rear screw compactor vehicle [30]

transport is compact and light, but it needs special equipment to manufacture the screw compactor and gearbox hydraulic.

(b) Rotating-Drum Compaction

This type of compaction vehicle is also ideal for managing dense solid waste and small materials. The wastes are loaded in the big rotating cylinders drum at the back of the vehicle, which is driven by helical blades to convey the wastes to the front bodies as it rotates (Fig. 3.15). The volume reduction wastes can be accomplished by the fragmenting effects of the waste during the rolling process with fixed capacities that, in turn, can facilitate the full waste compaction [41]. The loaded waste is unloaded by raising the back gate and further changing the drum rotation direction. However, this unloading process requires a longer time than other compaction vehicles, and some large compactor vehicle may have to move ahead a few times. Besides, these types of vehicles may cause breakdown issues due to connection with coarse waste materials, and irregular vehicle body shape could decrease the total waste volume capacities. These factors could be the reason that the operation and maintenance cost of this compactor vehicle system is quite higher than the rear loader [67].

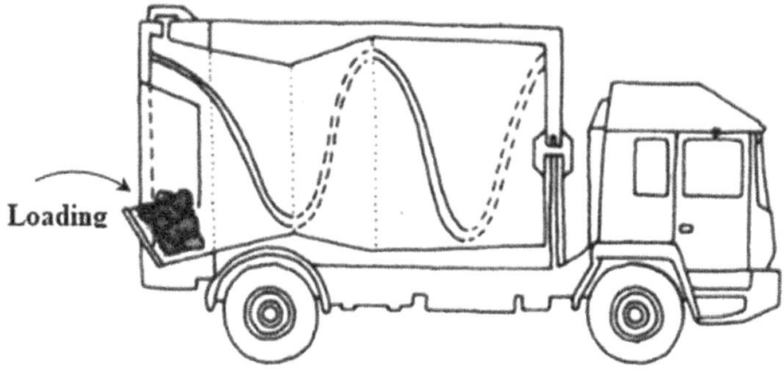

Fig. 3.15 Rotating-drum compaction vehicle [30]

Fig. 3.16 Rear-loading compaction vehicle [30]

(c) Rear-loader compaction

This types of rear-equipped full-compaction system with a compactor plate and big hoppers are the most popular used in many countries (Fig. 3.16). A sliding steel plate transfers the wastes from the rear-loading space during the compacting process to a running wall so that keeping the steady forces with leachate storage underneath the vehicle body [41]. For the unloading process of the compacted wastes, the rear-mounted compaction is raised up by the hydraulic system, and the ejection steel plate pushes the wastes out of the back of the vehicle. Usually, this compaction system with this type of vehicle body have weights between 8 and 12 m^3 [60]. However, more waste can be loaded if the waste density is high. The steel hopper and other related equipment should be mounted at the rear of the wheels to lower the loading heights so that the compactor lift weight can be distributed to the rear of the vehicle. If the compacted wastes have a density nearly double that of the compactor trucks' original design, the trucks will need to be equipped with very small bodies to avoid overloading. The benefit of this rear equipped hydraulics systems is that it is capable of handlings bulky wastes. Also, it allows slow loading heights, which in turn make it easier for rapid loading [64]. However, this system has a short lifespan due to the sliding plate, and the hydraulic components are exposed to coarse wastes and caused wear and tear. In addition, they are also often susceptible to anaerobic reaction, which may lead to high corrosion in many parts of vehicle bodies. The rear-loading waste full-compaction vehicles are typically mounted with a hydraulic lifts system which could lift and empty containers using a rear hopper. Some vehicle systems can lift small or large container individually or together, and some can manage different types of containers at the same operation time. This vehicle also has a design with a big rear loading system for lifting and emptying large communal containers of 10 m^3 capacity [68]. This rear compaction type of vehicle, however, should be suitably utilized together with other small-capacity vehicles since it can be employed to service different operation areas with a narrow road condition and transferring collected waste to temporary transfer station facilities [68].

(d) Front Loaders

Another solution for the collection and transportation of waste using a compaction system is a front-loading truck system that provides advantages for the urban waste transportation process [69]. This type of waste collection vehicle is operated by lifting the waste containers with a capacity of up to 1.5 to 3 m^3 by using reinforced-steel forks mounted at the front end of the vehicle. The single driver could lift up the containers above the driver cab up to a 90° angle with the ground, while keeping the open rear and tipped into the top vehicle bodies that the wastes are compacted backwards (Fig. 3.17). The front-loading compaction vehicle could be used on a multiple shifts basis, high performances, flexible exchanges containers, and does not need an additional crew to aid the driver [70]. However, this vehicle is not suitable for manual loading and waste collection at the curb side. Furthermore, the front-loading vehicle also uses dual rear axles, often very expensive, that need direct loading access. Space restrictions with narrow road conditions can inhibit its efficient application in most of the cities areas [48]. Nevertheless, this sophisticated vehicle is used for high productivity operation, which makes them very suitable to further enhance the overall level of an integrated waste management system. In addition, this type of compactor vehicle works faster at each collection spot compared to RORO and other compaction vehicles [33].

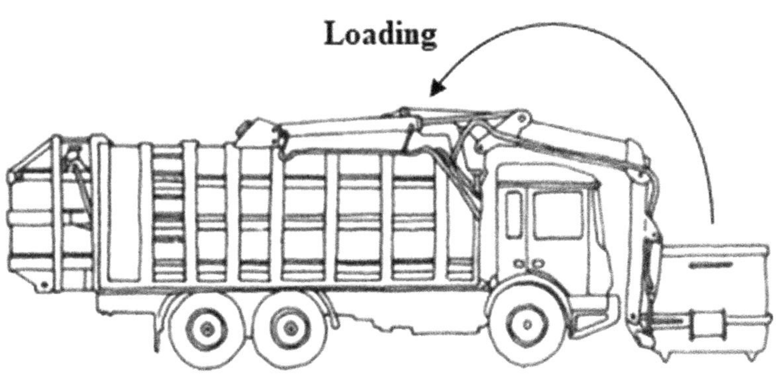

Fig. 3.17 Front-loading compaction vehicle [30]

3.3.5 A Pneumatic Waste Transportation System

The modern and efficient waste transportation system through the underground to the central processing facilities using vacuum extraction and hydraulic flushing are two main elements in the pneumatic system [29]. Compared to conventional waste transportation, the pneumatic system could minimize energy usage up to 60% and reduce over half of the carbon gases emissions [71]. With the pneumatic waste transporting system, wastes are moved through the buried pipeline network starting at the waste generation source. The operation is by using negative air forces to draw the wastes via a pipeline to the collection point before waste is compressed and transported to waste treatment facilities [71]. Waste is dumped into gravity-fed inlets (either indoor garbage chutes or outdoor litter bins), where it collects (either within the chute or in a tank underneath the litterbin) before the valves connecting the inlets to the tube transport network are opened. When the material arrives at the terminal, it passes through a cyclone separator, which spirals the heavier-than-air waste down into a compactor while the air in which the waste was entrained rises through a fabric filter. Before the air is pumped through the exhausters and then out through the stacks, the fabric filter eliminates dust and impurities. Figure 3.18 illustrates the basic concept of a pneumatic system for waste transport. The pneumatic waste collection transport has two types of systems: stationary and mobile. In a stationary method, the waste collection loops start either inside the buildings or the roads. The plastic bags wastes are put inside the sluice on the top of the vertical pipe that running downwards. The bags fall by the gravity forces effects through a valve in the basement that is properly closed and stored. The pneumatic operational valves are connected to a horizontal pipeline network that transports the waste to the drop off terminal [72]. In contrast, a mobile pneumatic method requires a special vacuum waste collection vehicle to suck the wastes through connector stations linked to the pipeline networks [53]. The vehicle can serve several networks by built-in a vacuum system to compact the wastes and transferred them to the waste treatment facilities. Both network operation types may be used to collect multiple waste sources or fractions isolated from the source.

3.3.6 Waste Transportation System

The waste transportation operation not only includes waste collection from different generation waste, but also moves the wastes to the unloaded point and continue the waste collection operation to reload again. The critical factors for the optimal design of the waste management collection and transport system include several factors like the waste, waste airspace, traffic conditions, and the distance to waste treatment facilities [21]. Haerani and Budi, 2019 [12] describe that transportation of solid waste could be performed by two main waste transportation container modes: hauled and stationary systems.

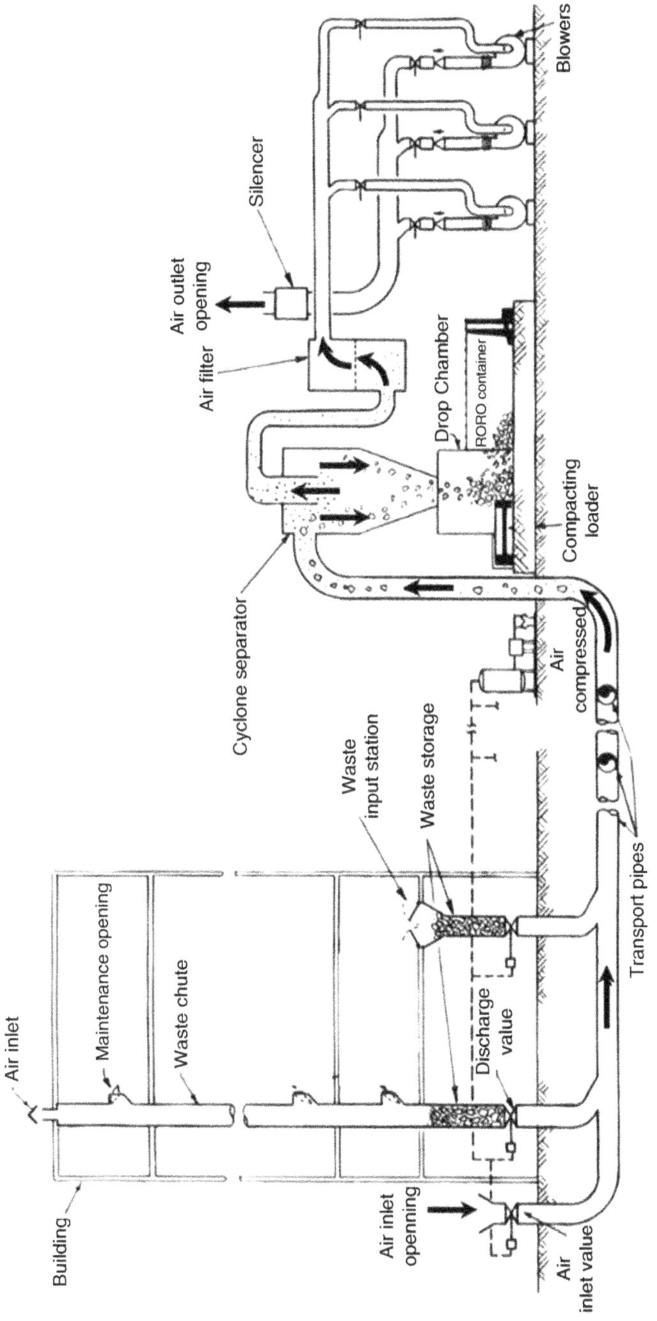

Fig. 3.18 Basic concept of waste transport using a pneumatic system [73]

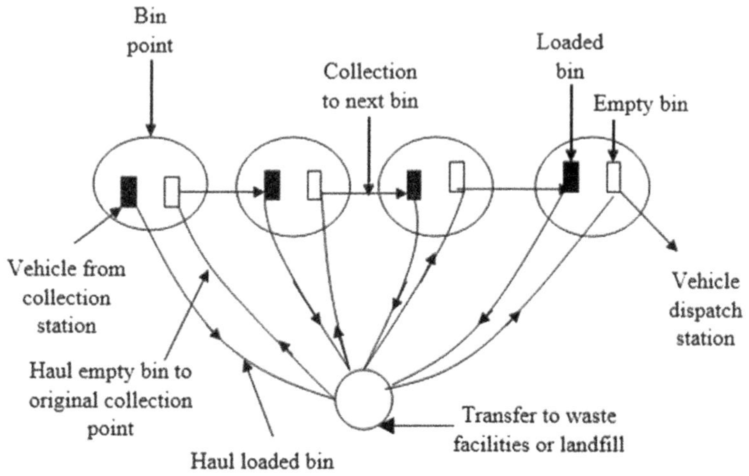

Fig. 3.19 Hauled container mode [53]. Reprinted with permission from Hayat S, Sheikh SH. Municipal Solid Waste: Engineering Principles and Management. 2nd ed. Farhan K, editor. 2016. 1–234 with permission from The Urban Unit, Lahore, Pakistan

3.3.6.1 Hauled Container Mode

Hauled container mode (Fig. 3.19) or moved containers system is the waste transport operation by transporting the wastes using containers throughout the collection areas that can then be transported to waste treatment facilities [12]. The system typically uses bins as a collection container, and it is widely applied in commercial and residential areas. This hauled container mode can be subdivided into two main operation modes such as conventional and exchange containers.

3.3.6.2 Stationary Container Mode

Stationary container mode or static waste containers system is the operation of transporting waste where the collection containers are stored at the places [12]. Usually, the containers are easy to lift and are widely used in household areas (Fig. 3.20). The stationary container system can be subdivided into two main operation modes, such as mechanically and manually loaded vehicles. The comparison between hauled and stationary container system are summarized in Table 3.3.

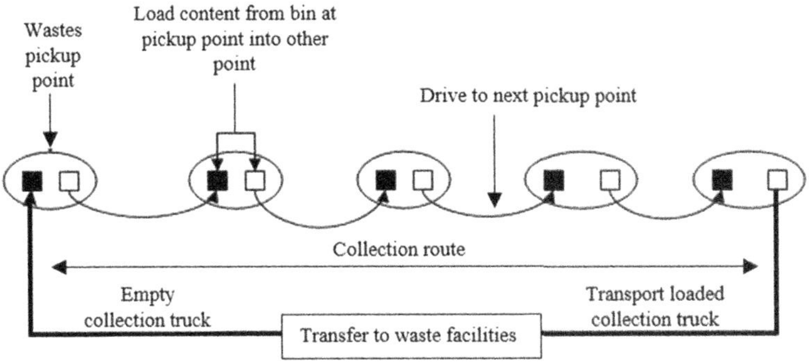

Fig. 3.20 Stationary container system [53]. Reprinted with permission from Hayat S, Sheikh SH. Municipal Solid Waste: Engineering Principles and Management. 2nd ed. Farhan K, editor. 2016. 1–234 with permission from The Urban Unit, Lahore, Pakistan

Table 3.3 Operation mode for hauled and stationary container system [59, 53, 12]

Operation	Description
i) Hauled container system	Containers used for the storage of wastes are hauled to an MRF, transfer station, or disposal site, emptied and returned to their original location
a) Conventional mode b) Exchange container mode	Containers used for the storage of wastes are hauled to waste treatment facilities to empty and return to a different location for exchange. In the exchange model, the driver must give the collection route an empty container to be deposited at the first collection site
ii) Stationary container system a) Mechanical mode b) Manual mode	A variety of container sizes can be utilized with waste transport. This stationary mechanical system is very usual for waste collection in the residential area. However, it is not suitable for waste collection in industrial and construction areas A manual-type stationary system is usually located for waste collection transport in a residential area. This system can be competed with the mechanical type due to a lot of residential areas, or point of the collection might be not accessible using mechanized waste transport

3.3.7 Waste Transportation Routes

3.3.7.1 Introduction

The waste collection process is the transportation of waste from the source of generation to the point of waste treatment facilities (processing or disposal) [2]. Thus, the primary purpose of waste collection is not only to collect solid waste materials but also to transfer the waste from the source of generation to a site via a good routing system for efficient waste transportation operation [17]. When the equipment and workers needs are defined, the transports route for wastes collection operation should be established, such that the service and those requirements can be performed in an effective manner. Generally, the route design for collection operation usually includes a number of tests. However, waste transportation routes should be designed within the minimal cost, effective distance, and minimize the probability of accident [74] by evaluating the expenses and revenues in the waste management systems. The optimization of the route for waste transportation is an efficient way to minimize the overall cost of transportation, fuel consumption, and save the time of service operation [75, 28]. The transportation cost always constitutes more than half of the total expenditure; hence, the determination of transportation route limitation is very critical for the optimal waste management system [29]. The routing design of waste transportation can be subdivided into three primary sustainability pillars related to economic feasibility, environmental impact, as well as social relations [41]. These are illustrated in Table 3.4.

3.3.7.2 Factors for Routing Planning

An effective waste management expenses could be achieved by optimizing the routing transport system [76]. Therefore, several primary considerations such as waste composition, population, collection point, road capacity, segregation level, route length, and time consumed need to be considered (Table 3.5). Besides, a good waste transportation system should also be designed to have effective time management, less manpower, fewer emissions, fuel efficiency, and maximum collection points [2]. In addition, an optimal route is also required to be cross-validated and continuously monitored to achieve a sustainable, cost-effective operation system.

Table 3.4 The elements for waste transportation planning route

No	Perspective	Description
1	Economic	a) Reducing time, distances, cost are the priorities when designing waste transportation routes. Thus, a close review should also analyse cost-effectiveness during day and night-time operation b) Optimizing the waste transport vehicles is one of the goals in the planning of waste transportation routes. However, low numbers of vehicles might lead to employees' overtime work c) Maximizing routes sections is the best waste transport management planning. Not only is it a cheap and efficient compact route, it also makes the entire route planning and operation more efficient
2	Environment	a) Effective routes planning not only reduce costs, but it also contributes to positive environmental impact such as reduction in gases emissions b) Minimization needs of energy due to active waste transportation attributed to fuel consumptions. These are influenced by vehicle load, road gradient, speed, engine and air density.
3	Social	a) The social aspect is often not viewed into consideration, specifically waste transportation planning. Balance of work between employees and safety concerns during operation work should be well addressed

Table 3.5 Factor for routing planning [77, 75, 78]

No	Factors	Descriptions
1	Waste composition	Depending on the type of waste, such as dry or wet; thus, it should be collected separately. Hence, the waste transportation system can be designed based on its composition
2	Population density	Population density is linearly proportional to waste generation. Hence, the population level and the rate of waste generation must be taken into account while designing any waste collection route
3	Collection points	Collection points must be created depending on the population density and the level of waste generation. Thus, collection points should be easy access for waste transportation operation
4	Road capacity	They are various vehicle types and sizes available for waste collection and transport. Proposal of vehicles used for waste transportation system should be suited with road capacity
5	Segregation level	Waste is required to segregate and disposed of, which is dependent on the behavioural patterns of the population. Thus, the level of segregation, community bins, and access to a waste vehicle can be designed
6	Route length	The waste transportation route needs to be optimized so that most of the vehicles do not travel more than 15 km for an effective routine waste operation system
7	Route timing	Time-zone planning is necessary for the efficient designing a waste transportation system for the appropriate collection process. Hence, the shortest and most efficient path must be created for this mechanism

3.4 Transfer Station

Solid wastes must be transported from generation points and moved to a transfer station, processing facilities, or landfills [42]. The transfer station is a transportation hub linking waste gencrated with waste treatment facilities. Thus, the needs for waste collection vehicles are somehow different from transporting them from the transfer station to a waste treatment facility. The vehicle should be compact and easily manoeuvrable for the waste collection process in the condition of high traffic density, which has an acceptable load height for efficient and sanitary loading [17]. It should also provide a secure loader standing platform for the crew. When operating in the city areas, they need to comply with a maximum allowable speed; thus, a high-speed waste collection vehicle may not be necessary [79]. In a low, dense area, a vehicle with a compaction mechanism might be appropriate to minimize the overall waste volume so that it can carry the maximum loads [33]. For a long-haul waste transport to a waste processing facility or final disposal, a larger waste vehicle is needed with a high-powered engine for easy and fast travelling time [68]. The cabin crew are not required, and either a loading or compaction system is also not needed. A vehicle's specifications for collecting wastes vary significantly from those for transporting the wastes collected to a transfer station, disposal site, or processing facility.

The waste collection works usually operate in two shifts. In order to optimize the vehicle's usage, the operation in household areas is normally done during the day due to less traffic and fewer problems with parked cars. In contrast, the collection process at the commercial areas is mostly performed at night time until early morning to avoid heavy traffic congestion [59]. Thus, the transfer station facilities allow the opportunity to temporarily store waste that is collected before it is transported in larger quantities to the waste treatment facilities at a predetermined time. In the city areas, when the hauling distances from the waste collection areas to the landfill is close, collection transport might be suitable for the direct discharge of the waste. However, large collecting vehicles may cause huge traffic issues in the service areas, and waste transport must be carried out concurrently with the collection operation [10]. This condition could lead to large collection vehicles causing traffic congestion in collection areas, and transportation must occur concurrently with the collection. Thus, more waste collection vehicle is needed to ensure the effectiveness of the operation system. This can be overcome if the haul distances to transit sites like transfer stations are very short before it is transferred to other waste treatment facilities. There are many options to transfer the waste from the collection vehicle to a long-haul transport vehicle at the transfer station facility. The majority of the requirements for choosing a suitable location for waste facilities such as large transfer stations are high costs and a large area away from public areas to prevent nuisances [80, 81]. Modern pit types of a small transfer station only need a smaller area and could be placed near to the source of the waste; this allows the use of a low-cost waste collection system [82]. A number of small transfer stations in the service areas can be replaced with a single large station which could decrease the travelling time and distances.

3.4.1 Role of Transfer Station

A waste transfer station is deemed helpful alternatives to promote holistic, solid waste management operation in various settlements and different population densities [82]. These facilities are part and parcel of the contemporary urban waste management systems. However, the transfer station should be financially viable [83]. The role of transfer stations in waste management has become even more prominent with the increase in the number of regional landfills, which are often located in remote areas [84]. This station is typically located close to population centres. It is used to serve as nodes where collected waste streams can be consolidated for long-distance transportation to which may cut costs [42]. Integrating smaller waste loads from primary collecting vehicles into the large transport vehicles could lower long-hauling transportation costs. This could hinder direct discharge into the landfill, which allows waste collectors to spend more time than effective collection operation [85, 89]. This factor could also reduce fuel usages, operation costs, pollution issues, traffic congestion, and road wear [10]. According to Pichtel (2014) [10], a transfer station can be economically justified; the disposal site must be at least 16–25 km away from the generation area. The transfer stations are also very useful facilities for small-sized communities where waste generation is not sufficient to support direct discharge into waste treatment facilities. In addition, a transfer station may be used to aid in the recycling programme by providing the option of waste screening before final disposal. This, in turn, could be cost-effective and reduce workload operation in a solid waste management system [29].

In most transfer station operation, operators screen the loaded wastes on the conveyor or tipping floors or platform pits. The wastes screening process involves two main elements, namely the isolation of recyclables materials from waste streams and the detection of waste that may not be suitable for disposal. Identifying and removing recyclables materials helps to reduce the volume and weight of waste quantity significantly before final disposal. Solid waste inspection is an important opportunity to inform the public about the benefits of recycling at a transfer station rather than a landfill [86]. Transfer stations also offer much flexibility for more disposal options. Operators may have options to choose the most economically advantageous landfill sites even though they are further away. In addition, they could also consider various facilities for waste disposal, competitive fees, and choose a suitable disposal method. In transfer stations, the unnecessary wastes or recyclables items can be transferred from waste collection vehicles to bulk transportation vehicles that could transport the materials by road, ship, or rail [69]. Therefore, the use of a transfer station with efficient transportation infrastructure is highly required to enhance the economic effect. This is more critical in the waste collection operation system when the treatment facilities are a distance from waste generation sources, and the quantity of waste to be transported is enormous [29] (Fig. 3.21).

3.4.2 Planning of Transfer Stations

The transfer station facility is an integral part of integrated solid waste management operation. Traditionally the planning of transfer stations was motivated by the need to reduce the waste volume, minimize transportation costs, increase collection frequency, and flexibility in siting treatment facilities [87]. Besides, transfer station provision also could reduce transportation cost and management expenses up to 18% of energy consumption along with a reduction in environmental pollution [80]. If the waste treatment facilities are distances from the city centre, the required time by the collectors for travelling become unproductive and would disturb the collection service. According to Schneider (2014) [68], the requirement of a transfer station needs to be determined when the travelling distance to waste treatment facilities is more than 15 km or over 30 min in one way travel. Waste collection transport is designed purposely for waste collection service; meanwhile, for long-haul distances transport, the trailer vehicle is normally used. Singh et al. (2014) [38] reported that the use of trailer vehicle for long-distance hauling is cheaper when the average haul trip distances are over 50 km. As a result, it might not be cost-effective to use waste collection transport for travel to the waste treatment facilities far from the collection area. A transfer station is generally located near the population centres since it is more economical to transport a bulk quantity of wastes using long-haul

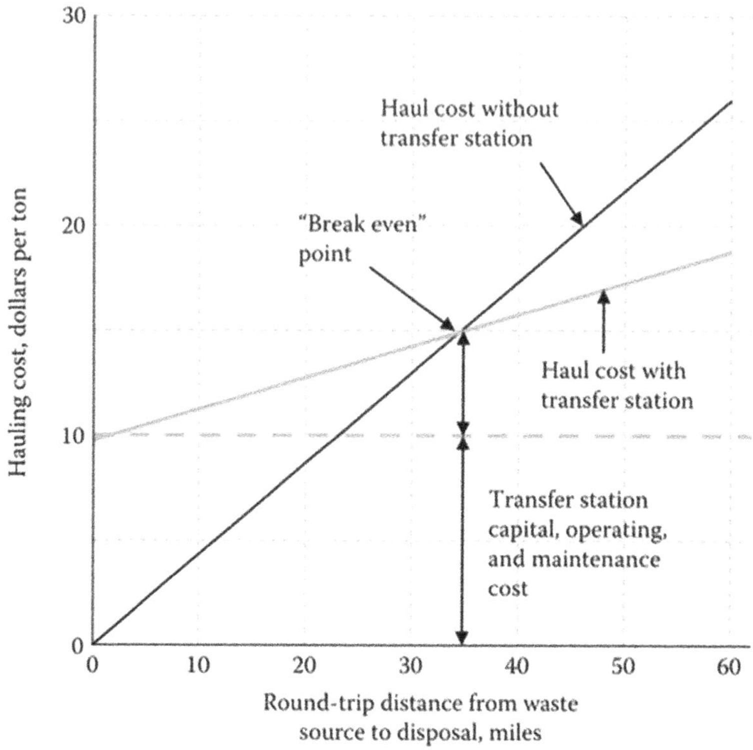

Fig. 3.21 Comparison of waste hauling costs with and without a transfer station [10]

transportation mode compared to small ones. Usually, a personal trailer or transport vehicle that has a large capacity between 27 and 46 m^3 or more are used to move bulk wastes from the transfer station to waste treatment facilities [38]. If the destination of the material collected is far off the transfer station, the use of more effective transport vehicles may be applied like long-trailer or ships or trains [42]. Among the important considerations in planning are types of waste typically accepted, uses of the transfer station, the capacity of a transfer station, location of the transfer station, and public involvement [10]. Partially or complete waste-processing activities like segregation, crushing, balers, compaction, or composting can be offered at the waste transfer station facility. The goal of this process is to minimize the amount of wastes and transforms them into other physical structure as well as recover the materials used (Fig. 3.22).

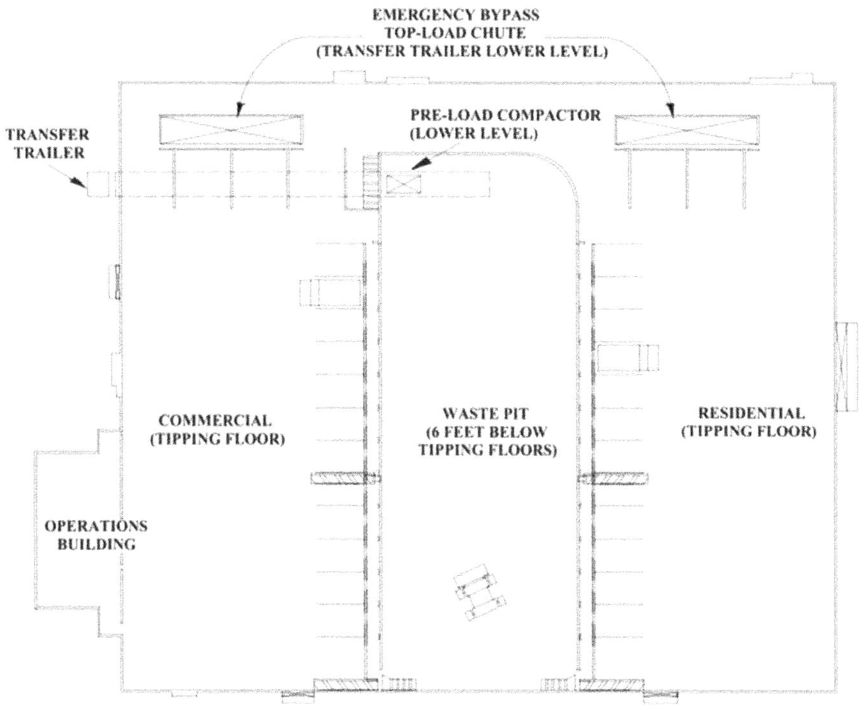

Fig. 3.22 Example of facilities layout in transfer station design [10]

3.4.2.1 Wastes Handled in Transfer Stations

The generation of municipal solid waste typically sources from residential, commercial, institutional, and industrial [88, 89, 91]. The characteristics of solid waste are highly heterogeneous, dynamic, which influenced by socio-geographical factors containing a variety of recyclables and organic waste [90, 91]. In order to process these mixed wastes, several transfer stations provide an operation that can separate, specifically, recyclables materials. Thus, it can divert the waste from being directly disposed of, which in turn could support the recycling initiatives [86]. The materials that are processed often varies depending on the location of a transfer station facility. Nevertheless, not all wastes are acceptable in this facility for numerous reasons. Some wastes are forbidden by certain countries due to difficult and expensive to process while other has a high potential of health or fire hazard [45]. The very large waste type which could damage trucks or operating equipment at a transfer station is also prevented. The main types of wastes that commonly acceptable and unacceptable at the transfer station are described in Table 3.6. This is a generic list; however, some transfer stations may be set up to specifically handle these types of wastes, while other facilities may also have a different list. Even though this unacceptable waste constitutes a small amount of the existing waste stream, but proper

Table 3.6 Types of municipal solid waste are acceptable and unacceptable [86, 48, 45]

Conditions	Types of waste
Acceptable	a) Garden waste: leaves, grass clippings, tree trimmings, and agriculture b) Household waste: cleaning products, cooking products, loose furniture, automotive products, brake fluid, antifreeze, and paint c) Recyclables waste: paper, newsprint, ferrous metals, plastic, glass containers, aluminium cans, motor oil, and tires d) Construction and demolition: concrete, brick, wood, masonry, roofing materials, sheetrock, plasters, and metals
Unacceptable	a) Toxic waste: biomedical, pesticides, spent acids, alkalis, etchants, solvents, coolants, and waste oils b) Flammable waste: ammunition, dry and wet carbide waste, fireworks, and self-igniting waste c) Hazardous waste: chlorinated herbicides, insecticides, and fungicides d) Bulky waste: tree stumps, large furniture, or objects

management by the transfer station operators and cooperation of local authorities with other agencies may be required to handle these types of waste.

3.4.2.2 Determination of Transfer Station Capacity

The determination of transfer station size is usually dependent on the capacity of facilities based on the volume of generated waste within an area to be serviced [45]. There are several factors that are considered to determine the facility capacity and size required to accommodate waste deliveries, as explained in Table 3.7. Besides, the type of waste delivery vehicles may also be compared and will, as well, determine the land area needed for the facility. In addition, the maximum rate at which waste is delivered is a crucial consideration to avoid a long waiting time for the collection vehicles [53]. In general, it is best to design a facility to accommodate the present and future projection waste load together with proper planning for forthcoming facility expansion. The arrangement and the capacity planning of these transfer stations are strategic decision-making steps that should be thoroughly thought of. This is because the planning has a lasting effect on the system's behaviour due to the transport flow network and the operation cost that will be borne in future [1]. The following Equation [1] to [5] describes the design of the main element for determining transfer station capacity, as explained by Pichtel (2014) [10]

Determining transfer station capacity:

a) Surge pits	
i) Unload rate (tons/day) = $P_C x \left(\frac{L}{W}\right) x \left[(60 x H_W) / T_C \right] x F$	(1)
ii) Loaded rate (tons/day) = $\frac{P_t x N x 60 H_t}{T_t + B}$	(2)

Table 3.7 Determination of transfer station capacity [92]

No	Factors	Descriptions
1	The service area of waste collection	All municipal solid waste generated in the governorate or collected by the residential, commercial, and industrial waste contractor
2	The generated amount MSW at the designated area	This could be including the growth of population and recycling activities as well as seasonal programme in localities
3	The number of collection vehicles delivering MSW	Involvement in overall traffic system management
4	Types of MSW to be delivered	Includes solid waste generated from commercial, industrial, garden, agriculture, construction, compacted or loose
5	Patterns of arrival MSW	Tendency MSW arrival depending on the daily, hourly or peak time
6	The availability of transfer vehicles	The efficiencies of transfer operation system such as load and unload MSW to treatment facilities
7	Forecasting increases in MSW generated during facility services	Projection of 20 years' operating planning service to be designed for twice capacity in early operation years
8	The correlation to other MSW facilities	This included the existing or proposed MSW facilities management such as MRF, RDF, incinerator, composting, and landfills
9	Amount of lane for queuing vehicles	During peak times operation vehicles required to check-in; thus, it is critical to ensures a good traffic control system
10	Size and number pits	It depends on the corresponding number of transfer vehicles used at the loading or unloading positions
11	Temporary MSW processing area	For routine operation such as holding MSW to reloaded into vehicles

b) Direct dump	
Station capacity (tons/day) = $\dfrac{N_n x P_t x F x 60 x H_W}{\left[\left(\dfrac{P_t}{P_C}\right) x \left(\dfrac{W}{L_n}\right) x T_C\right] + B}$	(3)
c) Hopper compaction	
Station capacity (tons/day) = $\dfrac{N_n x P_t x F x 60 x H_W}{\left(\dfrac{P_t}{P_C x T_C}\right) + B}$	(4)
d) Push pit compaction	
Station capacity (tons/day) = $\dfrac{N_P x P_t x F x 60 x H_W}{\left[\left(\dfrac{P_t}{P_C}\right) x \left(\dfrac{W}{L_P}\right) x T_C\right] + B_C + B}$	(5)

Where, P_C is collection vehicle payloads (tons), L is the total length of dumping space (m), W is the width of each dumping space (m), H_W is hours per day that waste is delivered, T_C is time to unload each collection vehicle (min), F is peaking factor (the ratio vehicles received of average 30 min to peak 30 min), P_t is the transferred vehicle's payload (tons), N is number of transfer vehicles loading simultaneously, H_t is hours per day used to load vehicles, B is time to remove and replace each loaded vehicles (min), T_t is time to load each transfer vehicle (min), N_n is number of hoppers, L_n is the length of each hopper, L_p is the length of each pushpit (m), N_p is number of pushpits and B_C is the total cycle time for clearing each pushpit and compacting waste into the trailer (min).

3.4.2.3 Location of Transfer Station

Identifying a suitable location for a waste transfer station can be a challenging process for local authorities [81, 86]. The potential location should be considered on several factors such as health hazards, bad odour, stray animals, and machine noise [93]. Besides, the distance from the transfer station to waste treatment facilities could also be one of the important indicators in the planning [80]. When selecting a location, a balance needs to be achieved among multiple criteria that might have competing objectives. If the location of the transfer station is large enough to accommodate all required functions, it might not be able to be centrally located in the area where waste is generated [94]. Likewise, in densely developed urban areas, ideal locations that include effective natural buffers may not be available. Less than ideal location may still present the best option due to transportation, environmental, and economic considerations. Several issues relating to whether the transfer station location is in an urban, suburban, or rural setting will also play a role in the final selection. There are many benefits that can be obtained by having a transfer station facility for waste processing. This includes environmental and resources conservation, ultimate energy production and land use, reduction of emission, and increase in social health [81]. Thus, it is necessary for the transfer station facilities to be located in suitable areas in order to obtain these advantages. Several critical criteria for the selection of the transfer station location are explained in Table 3.8.

3.4.2.4 Siting of Transfer Station

A decision-maker for siting of the transfer station may be the municipality or commercial entity. However, there are many stakeholders that could influence this decision [95]. For sure, a siting of transfer station may require continuous resident's involvement which is a critical process for developing this facility. Local residents are most likely would be able to accept a transfer station facility if the site is carefully designed with buildings that are fitted with a good landscaping area that is appropriate for the modern transfer station [45]. To avoid the unsuitable siting of this transfer station and to alleviate its adverse environmental effects, there are many other criteria that may be evaluated in the decision analysis as tabulated in Table 3.9. The public should be engaged at the earlier stage of this facility siting

Table 3.8 Selection criteria for transfer station location [81]

No	Criteria	Descriptions
1	Central location for collection vehicles routes	To maximize waste collection efficiency, transfer stations should be located centrally to waste collection routes
2	Easy entry to main transportation routes	The location of the transfer station should have direct and convenient access to all vehicles used like highway, expressway, or railways. It is preferable to avoid routing traffic through residential areas
3	Requirements for location area	The location area required for specific transfer stations varies significantly depending on the volume of MSW to be transferred, frequency of delivery, and operation to be carried out
4	Sufficient space for the traffic system	Transfer station vehicles require a convenient traffic flow system to move vehicles around various section by considering onsite roadways, vehicles queuing, and temporary parking
5	Compatibility of vehicles and traffic system	Transfer stations receive surges of traffic when collection vehicles finish routes simultaneously. Thus, the traffic flow in the transfer station will vary at the normal time but tends to peak at first and last trips of waste collection
6	Capability for future expansion	This is necessary to allow for a daily increase amount of MSW to enhance processing capabilities. Hence, it is less expensive to expand an existing transfer station than to develop a new site
7	Area for receiving recyclables	A sustainable transfer station could also be conducive to a recycling and composting programme. Sufficient space is requiring compensation for these activities
8	Buffer area	In order to mitigate the impact on the surrounding community, a transfer station should be located in an area that provides separation from residences or sensitive areas
9	Utility accessibility	The general transfer station is required utilities to operate like electricity, water, Internet, cleaning, restrooms, sanitary sewer system, etc.

process to incorporate the needs and concerns of the communities. It is also an important matter to address all public concerns to build integrity and establish good communications with the community [10]. Establishing a good reputation and high public confidence is as essential as addressing concerns of social, environmental, and economic relating to waste facilities [29]. Therefore, it is important to developing and implementing a siting transfer station process by integrating public inputs. Other than the criteria that have been outlined, several main public outreach initiatives could also be considered, including special targeted group meetings, stakeholder engagements, interviews with media, public conferences, education programmes, and community workshops. In addition, beyond this public outreach initiative, the potential host community and identification of resident's conditions requirements should also be addressed. Thus, community members may then become supporters of the proposed facility if properly approached.

Table 3.9 Siting criteria for transfer station [45]

No	Criteria	Description
1	Proximity to collection operation	Helps to increase savings from reduced transportation time and distance as well as operation costs
2	Ease in the accessibility of haul routes	Facilitates transfer vehicles to enter highways or other major routes, hence reducing haul times and possible impacts on close-by residences and businesses
3	Site zoning and design	This requirement is to be confirmed by residents of the responsible community whether the use meets the zoning requirements of the site
4	Visual impacts	This effect should be oriented so that the operations during waste transfer are not visible to area residents. However, the visibility of the required area will depend on traffic created by haulage trucks, storage capacity, allowed buffer areas, and station design

3.4.3 Types of Transfer Station

Transfer stations should be designed to reduce transportation time, provide effective equipment facilities, less fuel consumption, and low waste collection operation costs [68, 94]. The transfer stations may also be fitted with different material recovery systems for sorting the recyclables before sending them for further processing to recyclable plants. These facilities typically have a large size of waste containers between the range of 15 to 25 m^3 to transport loaded waste. A basic ramp facility could be provided in order to facilitate the collection vehicle for unloading and direct loading into large container or vehicles at a lower level. Several transfer stations may also provide unloading facilities like a hopper. This facility is used to transfer the waste into the static compactor and then push driven into the large long-haul container vehicle. A common transfer station is generally designed based on the amount of daily receiving solid waste, which depends on its capacity of small (less than 100 tons/day), medium (between 100 to 500 tons/day), and large (more than 500 tons/day) [53]. A typical unloading operational procedure by collection vehicles in a transfer station is illustrated in Fig. 3.23.

3.4.3.1 Small and Medium Size

A small and medium-sized transfer station is typically using a direct unloading waste station that does not have any temporary storage area [10]. This type of facilities usually provides a drop-off platform for public use, which are devoted to processing municipal and private collection waste [45]. Depends on the location aesthetics, environmental concerns, and weather conditions, the daily operation of a transfer station can take place either indoors or outdoors (Fig. 3.24). A small transfer station is normally more complex. This is because it needs to attend a long period of waste transfer operation since it has only a few basic waste processing facilities. Generally, the direct unloading operation system has two operating platforms where a large open-top container or static compactor is located at a low level.

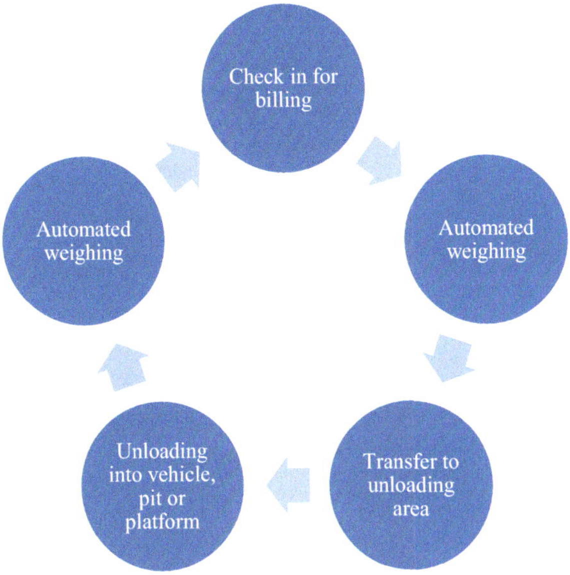

Fig. 3.23 Operation procedure for a large transfer station [53]

User will unload wastes from the top level into hoppers directly to those containers. This type of transfer station uses the drop-off system for the unloading method, which involves a number of open-top large containers provided by the station user. Then, these large containers are transported with a transport vehicle and hauled to the processing facilities or landfill. For this reason, the requirement for the total capacity of the station highly depends on the population, area density, and operating frequency [96]. As reported by Coffey and Coad (2011) [30], the common operation of small and medium transfer station depends on several main principles as follows:

a) An electric hoist deposits a large open top container into a ground pit so it can be easily direct discharged.
b) Collection vehicles of all types can tip their loads directly into the containers.
c) The capacity of the container is matched with secondary transportation vehicle capacity.
d) Weigh cells at the ground pit are used to determine the weight of waste in a container.
e) There is space at each pit for temporary storing empty or full containers of wastes.
f) The hydraulic hoist is used to lift the full or empty containers onto the secondary transport vehicles.
g) Small space is required around 20 m × 10 m for a double and 12 m × 8 m for a single pit system.
h) This transfer station can be located close to the waste generation point due to the small space requirement.

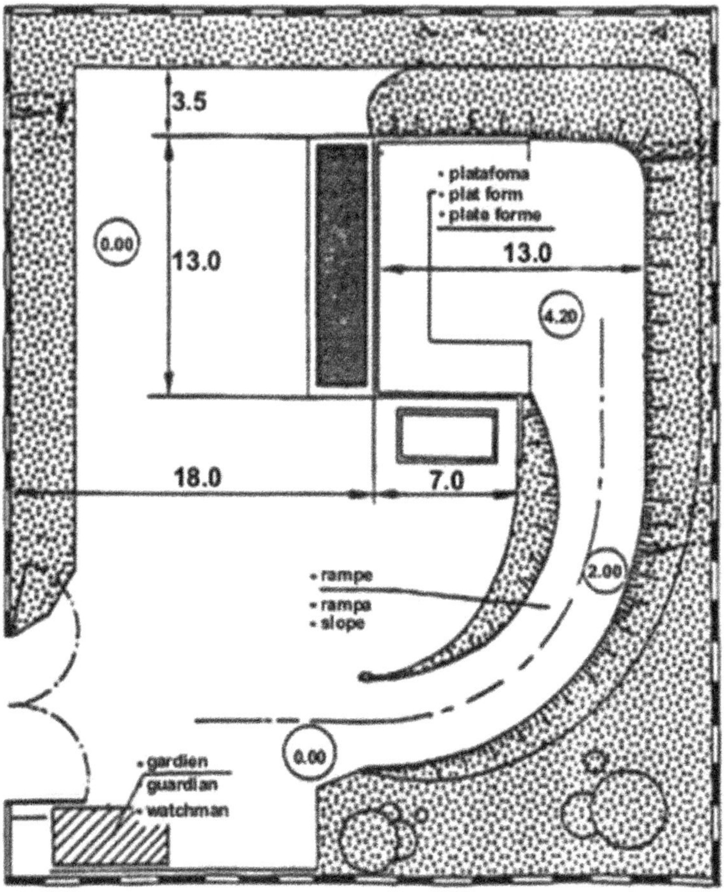

Fig. 3.24 Typical small and medium transfer station [53]. Reprinted with permission from Hayat S, Sheikh SH. Municipal Solid Waste: Engineering Principles and Management. 2nd ed. Farhan K, editor. 2016. 1–234 with permission from The Urban Unit, Lahore, Pakistan

i) The transfer station needs to be cleaned daily to ensure its hygiene and free from unpleasant odours.

3.4.3.2 Large Size

Larger-scale transfer stations are normally planned for commercial use with collection vehicles by a private company or local authorities [10]. In several cases, the communities sometimes have access to the facilities. For such a case, the appropriate facilities are required to be included in the early stage of transfer station planning. The design of this transfer station generally includes a floor for tipping the waste, after which bulldozers are used to push the waste into transfer trucks or a

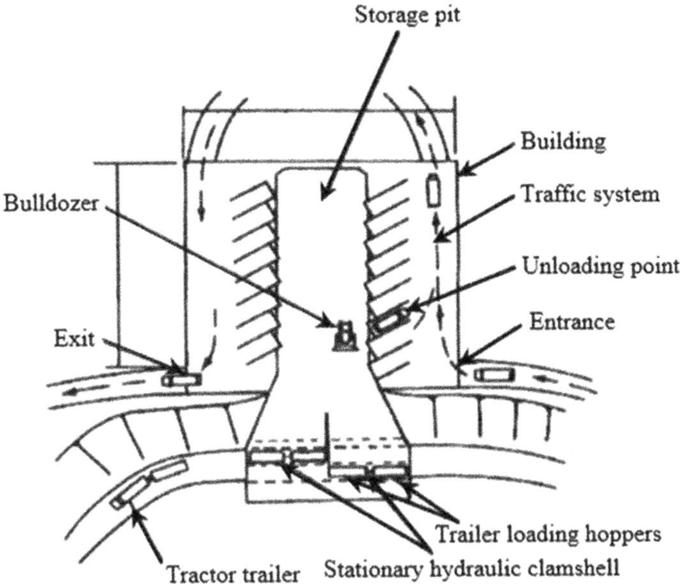

Fig. 3.25 Large-capacity transfer station [53]. Reprinted with permission from Hayat S, Sheikh SH. Municipal Solid Waste: Engineering Principles and Management. 2nd ed. Farhan K, editor. 2016. 1–234 with permission from The Urban Unit, Lahore, Pakistan

compacting chamber. Next, the waste is packed into trucks or compacting the waste into a high-density bale that is wrapped in wire mesh. In comparison, recyclables materials are increasingly being temporarily sorted and processed at transfer stations before transferred to waste treatment facilities. The typical layout of a larger transfer station is illustrated in Fig. 3.25. The large-scale transfer stations are usually located at a farther distance from the residential area to avoid noise, odours, leachate from waste, and vehicular traffic, but closer to the generation points [45]. On the basis of secondary loading from collection vehicles to a long-distance vehicle, the transfer station may be classified into three main categorized discharge load, which are direct, storage, and combine [87].

(a) Direct Unloading Station

A direct unloading waste of a non-compaction station is usually operated with two main platforms (Fig. 3.26). During the transfer process, wastes are unloaded directly from the top level by the collection vehicle into an open-top large container vehicle on the low-level platform [10]. In this operation, the transfer trailers vehicles are placed on scales such that when the full payload is achieved, discharge can be stopped. The waste collectors themselves load the waste into the transport trailer, and once the vehicle is loaded, it leaves to the waste treatment facility; meanwhile, the remaining collectors will have to wait for the next truck to deliver their load [64]. Upon a loading process, a plastic cover or tarpaulin canvas is placed on top of the loaded vehicle. This station is highly efficient due to wastes is only processed

Fig. 3.26 A typical direct discharge station invert [10]

once time. This system does not use any sophisticated infrastructure that needs high capital. However, some provisions for waste storage during peak time or system interruptions need to be further developed. The excess waste may be emptied and temporarily stored on the part of the tipping floor.

(b) Storage Unloading Station

In the platform station, collection vehicles temporarily discharged the wastes on a floor where segregation and sorting of recyclables or unrecyclable products could be performed, as needed [10]. The wastes are next transferred to open-top container trailers, normally via front-end loaders (Fig. 3.27). As similar to direct loading stations, this operating system has a two-level platform. However, the station would have three platform levels if the pit is utilized. In this station, waste from the storages pit is driven into the transfer vehicle or may be transported via a conveyor system to the processing or compaction facilities [29]. The main advantage of this type of transfer station is that it allows temporary storages that give a longer period of time for peak inflow waste. However, infrastructure costs for this type of facility are typically greater due to increased platform area. The opportunity to store waste on a temporary basis require fewer vehicles to be purchased, thus, enable the operators to transfer the waste in suitable traffic condition or at night. Usually, the design of this station facilities has waste storage between 0.5 and 2 days.

Fig. 3.27 Platform station invert [30]

(c) Combine Unloading Station

In a combined unloading transfer station, both types of method of direct and storage unload are used for waste processing and sorting recyclables in a single facility [29]. A combined transfer station can also be a facility for material recovery operations. Wastes that require a segregation process can be stored in the storage unloading transit facility; on the other hand, wastes that do not require recovery can be directed to the unloaded section. Hayat and Sheikh [53] stated that usually a commonly combined transfer station requires a stationary compactor system (Fig. 3.28). Waste from collection vehicles is dumped into a hopper and compacted by using a static compactor system. The hydraulic ram of the compactor drives the wastes into transfer containers that are normally mechanically connected to the compaction system. The compressed waste is then moved to a large long-haul container and transferred to the waste processing facilities or landfill.

3.4.4 *Operating Systems at the Transfer Station*

The main process involved in managing the wastes at the transfer facility starts at the time it is unloaded by the collection truck till its leaves. This process is fundamental to be considered in designing any transfer station system. The direct unloading system into the tipping floor is a normal practised method. These are the worst

Fig. 3.28 Stationary compactor system in a combine transfer station invert and put [73]

conditions, where the collected waste is directly unloaded into the long-haul container vehicle. However, this method removes the opportunities to segregate or inspect the recyclables items. In several cases, there is also a combination of the waste transfer process between the system of direct, platform, and compaction station [10]. The following sections describe several main basic systems of handling wastes at transfer stations.

3.4.4.1 Open-Top Vehicle System

Waste is directly unloaded onto the open-top vehicle or on the tipping platform to enable the recovery of recyclables items before loaded into a vehicle with non-compacting features [51]. Bulldozers are then used to organize the waste on the tipping floor (Fig. 3.29). The design of this open pit is able to also support a large vehicle on the open tipping platforms. Thus, this system requires higher capital and operation costs as compared to the open tipping floor. A larger vehicle is expected to have a high payload since the waste is uncompacted. This is a basic operating process that does not require any advanced machinery such as a compactor or baler system. Its versatility makes it the preferred selection for operations of a low- or small-volume transfer station.

3.4.4.2 Surge-Pit System

The surge-pit operation is not a loading technology, yet it is an interim stage usually by using open-top, pre-compactor, or compactor systems (Fig. 3.30). Aside from the intermediate step, the design of this surge pit system highly require it to handle peak

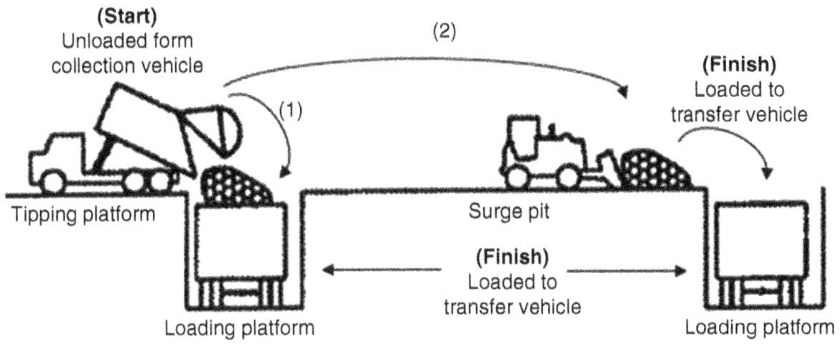

Fig. 3.29 Open-top transfer vehicle system [73]

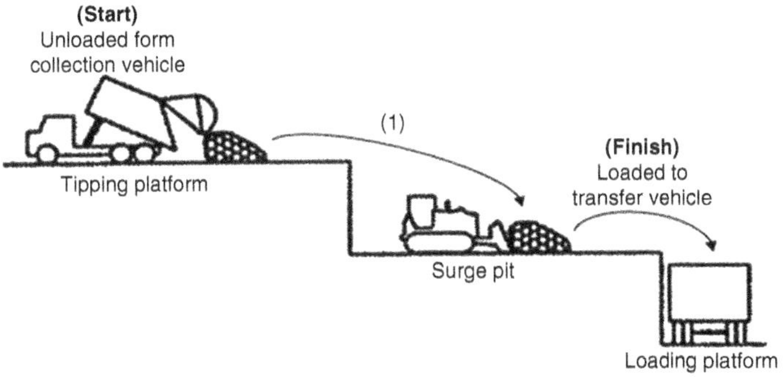

Fig. 3.30 Surge-pit operation process [73]

waste flows from incoming waste collection vehicles [69]. It also provides temporary storage capacity at peak hours to reduce the number of transfer trailers needed to efficiently service the station. A bulldozer is normally used to compact an accumulated waste either by riding on top of it or by pushing it up against a sidewall of the pit, thus increasing the payload. This could increase the waste's density and the resultant payload deposited into the transfer trailer. However, this method will prevent the recovery of recyclables of materials and waste screening efforts due to waste is always directly unloaded into the pit.

3.4.4.3 Compaction System

A compaction system at the transfer station facility is used to compress the wastes using a mechanical force before being transferred. Wastes are fed into the static compactor either from a direct collection vehicle or by pit after storage (Fig. 3.31). The hydraulic-powered ram drives the waste into a transfer vehicle that is mechanically attached to the compactor system. Therefore, the vehicle containers made of reinforced steel should be designed to withstand the compacted force. Normally, the heavy vehicle and onboard unloading ram are used to reduce the available waste

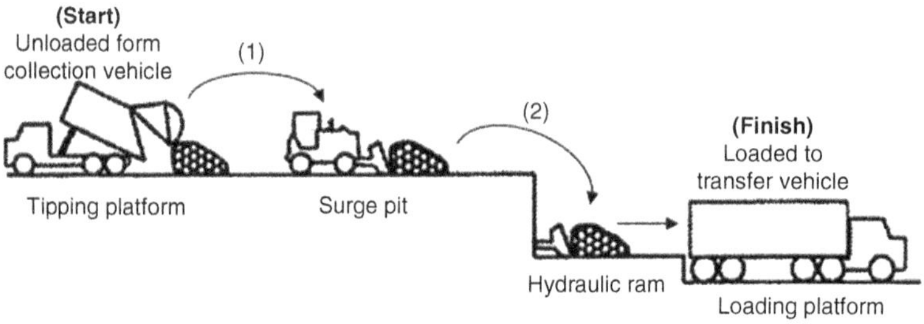

Fig. 3.31 Compaction operation process [73]

payload. Because the compactor is separated from the container, liquid wastes and residues may spill, which cause odours and attract pests [10]. In response to this practical problem, the self-contained compactor was developed. These units comprise a compactor and roll-off box housed within the same unit.

3.4.4.4 Pre-compaction System

In this pre-compaction system, the loaded waste falls via a hopper chamber. The compaction pressure is performed in a chamber that is away from the long-haul vehicle, which eventually would receive and transferred the compacted waste. The pre-compaction system is positioned on the lower-level platform of the transfer station, which contains a mechanical hydraulic ram (Fig. 3.32). This equipment employs a heavy hydraulic ram inside a tube to produce a dense waste know as a log. The log is transferred into a transfer vehicle by using the walking floor method for unloading or relies on a tipper at the landfill site to be unloaded with the aid of gravity [10]. Due to the transfer trailer vehicle has no compaction forces; thus, there is no requirement to have a heavily reinforced container body. The design of a vehicle used for this system is quite similar to the vehicle with a non-compaction system. Most of the pre-compaction system is mounted with two units in the case of one unit needs maintenance. Though this system has relatively high investment costs, the excellent payload will compensate for these initial costs.

3.4.4.5 Balers System

Balers system is a technology that compacts the waste into a dense, self-contained bale. Typically, the balers system has a wide range of capacities and various levels of sophistication. Some balers can operate by a completely automated system, while others need significant operator input. During the baling process, a wire strap might be applied to keep the bales intact. The baler provides a means to achieve high levels of waste density without the need for heavily reinforced truck bodies [69]. This equipment compacts the waste into the form of high-dense bricks, logs, and self-contained bales. They are normally moved using forklifts and transferred with the

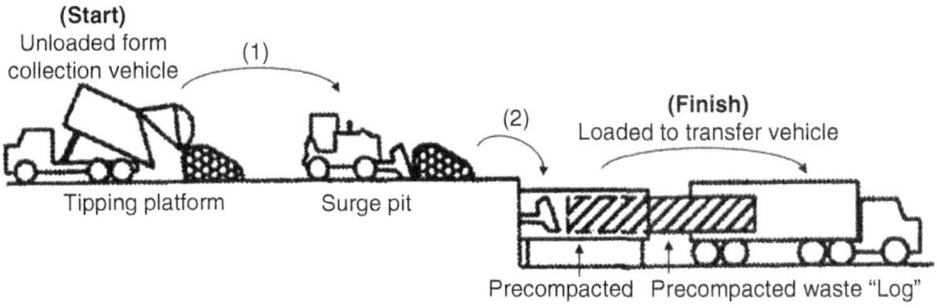

Fig. 3.32 Pre-compacter operation process [73]

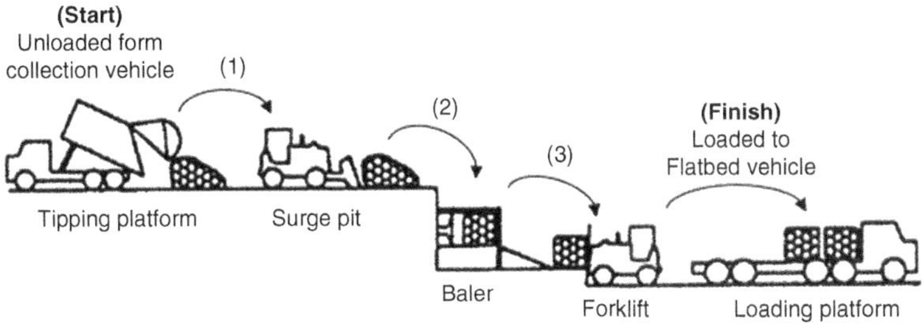

Fig. 3.33 Baler operation process [73]

trailer-flatbed vehicle (Fig. 3.33). The baler equipment is also suitable for baling recyclables items like paper, metal, and plastic. This system makes a high payload, so its capital costs are also high. Most of the balers system is mounted with two units in the case of one unit needs maintenance. This high-tech equipment is typically used when a high-volume operation; however, for unloading, it may require a special device or area at the final processing waste.

3.4.4.6 Intermodal Container System

Transfer stations serve as nodes where collected waste streams can be consolidated for long-distance transport. They do not just load in trailer trucks but also various transportation mediums. At transfer stations, waste can be transferred to other forms of intermodal transportation such as road trailers, rail cars, and barges or ships, depending on the station's location and regional transport infrastructure [69]. Waste loaded onto the intermodal system typically has a mechanism to control moisture and odour, as well as the container, suitable for both flatbed trailers and railroad. The loading containers could be directly loaded into railcars or transferred by a trailer vehicle to a train or ship terminal (Fig. 3.34). The packed container could be stored temporarily on sites till a sufficient number of containers filled to allow economical transportation to the waste treatment facilities. This alternative enables the overall transfer of vehicle traffic in local roads to be minimized and could make the distance site of waste treatment facilities economically viable.

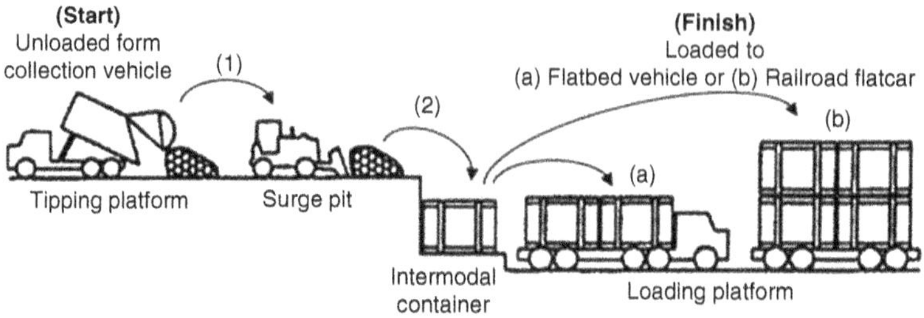

Fig. 3.34 Intermodal transportation process [73]

3.4.5 Advantages and Disadvantages of Transfer Station System

The transfer station is a superior facility in an integrated solid waste management system. Owing to the higher efficiency in this operation system approach, transfer stations presents more benefits than the direct transfer of solid waste to a disposal site. However, every technology must have its own advantages and disadvantages, including the system used in the transfer station. Several transfer stations use integration technology with other systems to minimized some of the drawbacks of a particular design. For instance, they may use the top-loading system as a contingency method if the pre-load compactor is under repair. It also shows that various interrelated considerations have to be addressed when making decisions for the best technology use on a transfer station. The main factors to be taken into account include design capability, shipping distance, cost, reliability, safety, unloading method, and transportation type [29]. Table 3.10 summarizes the advantages and disadvantages of the main transfer station system.

3.4.6 Types of Bulk Transportation

Waste transportation systems involved all procedures that start after the completion of waste collection and end at treatment facilities or disposal sites [94]. The primary transportation starts after the collection of waste at the source of generation to a transfer station. Meanwhile, secondary transportation is long-distance waste transport after the waste transit at transfer stations to the waste processing facilities or disposal site. At the transfer station, waste will undergo several sorting activities and are reloaded to long-distance various transportation vehicles. There are different kinds of transport like full-load, intermediate, and empty. The choice solid waste transportation method is determined by local conditions and should satisfy several main requirements such as minimum hauling costs, covering of wastes during haulage, designed according to authorities, distances to treatment facilities, allowable weight limits, and simple unloading methods [53]. Countries with a limited road

Table 3.10 Advantages and disadvantages of transfer station system

Systems	Advantages	Disadvantages	Applications
Open top vehicle			
Waste is directly unloaded into transfer vehicles from collection trucks	a) Simple equipment used and low potential to breakdown b) Minimizes waste handling and less housekeeping c) Inexpensive capital and operation costs d) Vehicles movement can be conveniently arranged	a) Direct dumped bulky waste could be damaged vehicles b) No temporary storage during peak times c) No potential for waste screening and recovery	Ideal for low waste volumes and short transmission distances
Surge pit			
Waste is unloaded on a platform and loaded onto the vehicles by using the equipment	a) Provides temporary waste storage and acts as an intermediate operation b) Allows to compact and break the bulky waste c) Prevent the risk of collisions with facilities machinery	a) Higher costs needed in terms of structure and equipment b) Unloaded waste on the pit can be messy and harmful c) Double handling to reload waste into vehicles d) Unacceptable waste on the pit may be hard to remove	Suitable for large transfer stations with high-density waste
Compaction			
MSW is dumped from the collection vehicle by a hopper and transfer into a large vehicle using a hydraulic ram	a) Compaction of MSW typically reaches high densities b) Compaction can enable optimum usage space c) The compactors can be mounted according to the transfer vehicle suitability	a) Potential to shut down operation if compactor fails b) High capital and operation costs using a compaction system c) Hydraulic system equipment might be very noisy	Not widely applied in a new transfer station
Pre-compactor			
Pre-compactor MSW by hydraulic ram then pushed into a transfer vehicle	a) Usually uses much smaller vehicles than uncompacted type b) Containers could be completely covered to avoid intrusion of rainwater c) Compacted waste payload can be easily measured	a) Equipment capital expenses are significant high b) Not suitable for some types of MSW c) High electrical usages to operate d) Equipment used fairly complex	Most ideally for high waste volume transit station with long operation transfer

(continued)

Table 3.10 (continued)

Systems	Advantages	Disadvantages	Applications
Balers			
Compress MSW to dense bales form	a) Enables an effective transportation system due to compressed waste b) Vehicle could be completely covered to avoid intrusion of rainwater c) Could be used to manage recyclables items	a) Compress system equipment might be very noisy b) Higher capital cost is needed c) A special tool is required at the landfill	Suitable for a high-density station, especially which have a long-distance waste transfer
Intermodal			
Tipper vehicle mounted with containers for transfers is used	a) Allows optimizing payloads using a lightweight vehicle b) Suitable for flatted container system c) Extremely fast and easily handle large volume loading	a) If a tipper fails very difficult to unload in the landfill b) Tippers are very unstable in the landfill site c) Required fixed point for loading and unloading operation	Ideal for fixed waste treatment facilities

system for transporting waste to remote facilities may depend on railways or shipping and non-coastal transport. Meanwhile, countries may have a good road system use transport by vehicles. Thus, the efficient transport system for a transfer station usually can integrate the use of the roadway, railway, and waterway as a waste transportation medium [97, 29].

3.4.6.1 Road Transport

When the destination of final waste processing or disposition may be reached by transport road, the right option method to transport the waste from the transfer station is by using trucks, trailers, and semitrailers. Transfer trailers come in many sizes, typically 15 to 22 m long, and the capacities can be excess up to 77 m^3 with a weight-load capacity between 15 to 25 tons [69]. Collection vehicles generally transport the waste from a service area to a facility in which the waste is discharged. Road transfer vehicles like trailers are used for long-hauling operating from transfer stations to other waste treatment facilities. This is because of their greater capacity that could reduce the total operating costs of overland bulk transport. Economical feasibilities also show that it is less expensive to transport large quantities of waste long-distance using a large trailer than smaller quantities of waste in smaller trucks. In term of fuel usages for the road transport system using a long-haul trailer, it depends on various considerations like axles amount, carrying ability, capacity utilization, and driving style. In general, a vehicle used during waste transport on the highways should consider several criteria such as minimal expenses, waste covered, suitable for highway use, adequate weight limits, easy unloading and reliability [98].

3.4.6.2 Rail Transport

The transport by rail mostly happens with compacted waste in compactor containers. While the railroads were widely employed in the past to transport waste, however, only a few communities still use them to date. Presently, the renewed interests in using railroads for waste transport is increasing again, particularly in remote areas when railway lines still exist and travelling with the roadway is very difficult due to congestion, etc. Trains comprise one or multiple locomotives driving a long chain of a heavy flatbed railcar. Normally it carries containers for long-haul transport of large quantities from a transfer station to waste processing facilities or landfill site. About 90 tons of compacted waste could be transported in 18.28 m long containers rail relative to a transfer trailer vehicle which normally transports 20 to 25 tons only [10]. The independence from weather and discharge of the road network made this transportation method give advantageous for waste transfer. If there is no railway siding at transfer stations or at treatment facilities, further transfers alternative is necessary which results in higher transportation costs. Since the high quantities and masses can be carried per train unit hence, the compacted waste is highly suitable for this method. Besides, waste transport by rail is also environmentally friendly, free from the road network, low energy consumption, higher transport safety, and shorter distances [98]. However, the new development of rail at the transfer station is typically costly compares to a roadway. This can be related to the requirement of building new rail lines, installation of special loading and unloading equipment [10].

3.4.6.3 Water Transport

At transfer stations, waste can be loaded loosely in barges or with containers on container ships. Using water transport to the final destination points is often possible without another transfer vehicle. Therefore, transport on the waterway is rare. In some countries, like Bangladesh, the use of water transport for transporting waste from the transfer stations is cheapest than other transportation modes via road and rail [99]. Besides, the main advantages of freight transport using waterways are high energy efficiency, low emissions, lowest noise level, and low infrastructural costs [100]. Usually, barges, scows, special boats, and ship are used to transfer the waste to the treatment facilities or disposal sites. Even though several self-propelled ships are used, but the most normal practice is the use of barges pulled by tugs or particular boats. In contrast, ship transport is a form of transportation for inter the regional or around the globe. Normally, a mode of ship transport is filled with large containers containing various kinds of mixed waste. The overall waste load that can be carried in a single trip depends on the ship's size. Generally, the ship is divided into small ships with a weight between 2k to 50k tons and large ships in the range of 50k to 300k tons. One of the major problems encountered using ocean vessels is moving barges and ships during heavy seas [101]. Besides, other influences like floods, low tide, or ice drift could also affect transport [29]. In such an instance, the wastes must be kept and require costly facilities for storage. Furthermore, long transport times are expected; thus, only waste which will not decompose can be transported.

3.5 Issues and Challenges of Waste Transportation

According to Themelis (2019) [42], the waste collection system involves the process of gathering and transporting waste from the generation sources to the place where the content carried till is discharged. Instinctively, due to high labour costs and extensive use of waste transport make this component is the most expensive process of overall solid waste management [74]. Several main problems encountered in the process include insufficient funds, lack of priority, poor operational scheduling, poor road network, and inadequate technical expertise [79].

3.5.1 Constraint in Budget

Effective and sustainable municipal solid waste management needs sufficient funding or financial allocation. Guerrero et al. (2013) [102] has reported that increasing waste production greatly burdens the municipalities financial budget because of the high costs of operation for collection and transporting waste activities. Owing to the enormous expense necessary to provide this service, most municipalities in developed countries have often neglected the efficient management of solid waste [79]. The funding provided for managing solid waste is dwindling across different regions around the world [103]. In addition, the provision of sufficient waste collection operation and disposal facilities is further hindered by insufficient financial aid, inadequate resources, consumers' inability to pay, and the absence of adequate use of economic incentives [104]. As a result, poor funding for waste management contributes to inadequate service of collection and transport, not in tandem with high waste production. As a result, illegal dumping sites may crop up that pose a significant social problem and environmental hazards [105, 61]. In general, municipalities allocate a budget in the range of 20% to 50% of their revenues on waste management in most countries. Due to the limited budget, only about 30% of the city's residents have access to adequate and proper service of this waste collection system [79]. Sufficient funds and strong budgetary support is very critical to establish adequate waste management services for the collection operation, transportation process, and disposal activities

3.5.2 *Poor Operational Scheduling*

Aside from the aspect of inadequate financing, another limiting factor is the optimization of operational scheduling and vehicular routing [2]. Drivers should be scheduled and supervised in their particular operating service areas for the regular transportation of waste collection. However, ineffective waste transport processes can be attributed to ineffective operation schedules, inadequate waste facilities, and a limited vehicle number which leads to a heap of waste that pose environmental issues [79]. Sometimes the routes used by collectors are decided by them, which may put aside operational cost reductions [106]. This condition results in higher transportation cost as well as serious environmental pollution. Therefore, effective and sustainable waste transport operation must consider a scientific approach to determine efficient scheduling and systematic routes that can achieve a reduction in overall costs and environmental protection [107].

3.5.3 *Lack of Priority*

The municipalities in many developing countries are known for their lack of funds to effectively manage transportation and waste disposal [79]. This condition is also aggravated by inconsistencies in the financing abilities assessment and regulation of private partnerships waste management programmes [47]. The deficiency of a political will leads to less or no commitment to further enhances the waste collection and transportation system in between high competition demands for the development of road, healthcare, education, etc. [108]. This results in ineffective service delivery to the deterioration of public and environmental health. Waste operators are often correlated with conditions of low social class and less education, which results in less motivation for them to offer their good service. Politicians typically also offer low priority to waste management service in comparison with other municipal operations, which limits skilled workers in this sector [102]. Therefore, in order to provide good solid waste management services, professionalism should transcend favouritism and will. Besides, a high commitment and dedication to ensuring the efficiency of the overall waste management system from all sides are highly required.

3.5.4 Poor Road Network

The necessity of road networks in many countries for any transport service cannot be compromised. Most developing countries have fewer sufficient road transport routes. According to Guerrero et al. (2013) [102], the bad quality of roads network, amongst many factors, adversely affects the waste operation of collection and transport. The limited access to a road network particularly in the suburban area for waste collection service is the main problem faced by local authorities [61]. There are also cases where waste collection vehicles could not access some of the routes due to small and poor or unpaved road conditions; hence, it very difficult for the waste collection vehicle to operate, which leads to a pile of waste generation in a long time [109]. This low-grade road would have a significant direct effect on the fuel consumption of waste transportation vehicles [110]. This condition gives opportunities to illegal waste collectors operators. Bad road conditions may also result in frequent vehicles failure, thus growing the maintenance costs and expenses of incomes. Guerrero et al. (2013) [102] have stated that neither the design of roads as well as also the inefficient routing planning during urban development could have an impact on the overall successful collection system. The urban planning process shall consider the ease of transportation of large waste operation vehicles. This could ensure the facilities of the built route effective for the waste collection and transport service.

3.5.5 Inadequate Technical Expertise

The waste management system is often influenced by the technical knowledge or abilities of operational personnel who handle solid waste service [111,112]. Besides insufficient funds as the main factor in managing solid waste management, other factors that are no less important include poor skills, less community involvement, and insufficient technologies [113]. These disadvantages hinder the effective waste management operation of transfer and transportation. The developed countries commonly have the expertise, technologies, welfare, ability, and facilities for a good waste management system [104]. However, these are mostly absent in many developed countries. They also face inadequate technical skilled, financial issues and the absence of a sustainability system [109]. The recruitment of operational personnel shall be based upon professional expertise and learning, as well as the development programs are required to ensure the workforce is aware of current emerging problems [79].

3.6 Conclusion and Prospects

Municipal solid waste management requires the right selection of waste transportation vehicles, optimization of the route planning, and effective operation of a transfer station. It is a complex process that often requires the application of multiple criteria. Decision-making in an uncertain condition is often a difficult task that needs good management and proper attention. The aspect of waste collection, transfer, and transportation operating systems can easily take up more than 70% of total waste management budgets; hence, these are an important issue to address. Therefore, optimization of waste transportation vehicles and routing systems to reduce the high fuel consumption and vehicle transportation expenditures would not only reduce operation costs on manpower, time, and maintenance but also environmental benefits such as a reduction in greenhouse gas emissions and social issues. A recent experience suggests that small transfer stations are encouraged to be used with a small collector vehicle, particularly in the urban and commercial area. In several cases, an effective and cost-effective collection operation only requires small motorized or non-motorized vehicle to transport the waste into a small transfer station and vice versa also for large vehicles. The need for an integrated approach to the waste transportation system should be strategized to suit the local conditions. This will further increase the effectiveness of the implementation of strategies and will ultimately reduce the overall waste management costs in the interest of all stakeholders. Such an approach will facilitate information flow and enable more efficient planning of transport stream distribution with the optimization links between waste transport and transfer. However, the choice of any waste transportation and transfer station operation must address the main purposes for sustainable integrated solid waste management systems, following the waste management hierarchy.

Glossary

Bulk density is the ratio of mass to volume.

Compactor collection vehicle is a large vehicle with an enclosed body having special power-driven equipment for loading, compressing, and distributing wastes within the body.

Compactor is any power-driven mechanical equipment designed to compress and thereby reduce the volume of wastes.

Communal collection is a system in which individuals bring waste to a predetermined collection point, from which it is collected for further processing and disposal.

Composition Quantitative description of the materials that are found within a particular waste stream in the form of a list of materials and their absolute quantities per day or per year, or as a percentage of total materials.

Curb-side collection Programmes in which recyclable materials are collected at the curb, often from special containers, and then taken to various processing facilities.

Food waste is food that is not eaten. The causes of food waste or loss are numerous and occur throughout the food system, during production, processing, distribution, retail, and consumption. Global food loss and waste amount to between one-third and one-half of all food produced.

Facility is any location wherein the process incidental to the collection, reception, segregation, storage, dismantling, treatment, and disposal of e-waste are carried out.

Geographic information system (GIS) is a conceptualized framework that provides the ability to capture and analyse spatial and geographic data.

High-income countries is defined by the World Bank as a country with a gross national income per capita of US$12,536 or more in 2019.

Household container The vessel or basket used by a household or commercial generator to store and set out the waste materials, commonly made of metal, plastic, rubber, or wood.

Hazardous wastes A material that poses substantial or potential threats to public health or the environment and generally exhibits one or more of these characteristics like ignitable, oxidant, corrosive, radioactive, explosive, toxic, carcinogenic, disease vector.

Heavy metal Metals of high atomic weight and density that are toxic to living organisms, such as mercury, lead, and cadmium.

Integrated sustainable waste management (ISWM) is a systems approach to waste management that recognizes three important dimensions of waste management included stakeholders, waste system elements, and sustainability aspects.

Incineration The process of combusting solid waste under controlled, approximately stoichiometric conditions to reduce its weight and volume and often to produce energy. In combustion chemistry, the condition whereupon the quantity of oxygen provided to the combustion process is exactly that needed to completely oxidize all carbon in the fuel to carbon dioxide.

Low-income countries Is defined as those with a gross income per capita, calculated using the World Bank Atlas method, of $1,025 or less in 2019.

Landfill An engineered method of disposing of solid wastes on land, in which, at a minimum, there is perimeter fencing, gate control and the waste is covered every day. Some form of reporting is usual, often in the form of a weighbridge (scale house), and some form of tipping fee is usually charged. A landfill differs from a sanitary landfill in that it is not necessarily sealed from below and does not necessarily have a leachate collection system.

Leachate The liquid generated from municipal solid waste that seeps through solid waste or other medium and has extracts of dissolved or suspended material from it.

Materials recovery facility (MRF) An industrial facility of moderate scale that is designed for post-collection sorting, processing, and packing of recyclable and compostable materials. It is usually of moderate technical complexity with a combination of automated and hand sorting.

Municipal authority is a form of a special-purpose governmental unit. The municipal authority is an alternate vehicle for accomplishing public purposes without the direct action of counties, municipalities, and school districts.

Municipal solid waste (MSW) Commonly known as trash or garbage in the United States and rubbish in Britain, is a waste type consisting of everyday items that are discarded by the public. 'Garbage' can also refer specifically to food waste, as in a garbage disposal; the two are sometimes collected separately.

Movement Control Order (MCO) is a cordon sanitaire implemented as a preventive measure by the federal government of Malaysia in response to the COVID-19 pandemic in the country.

MSW generation per capita represents the intensity of waste generation and can be used to assess progress in waste prevention activities (reducing and reusing).

Municipal Solid Waste Management (MSWM) The collection, segregation, storage, transportation, processing, and disposal of municipal solid waste, including reduction, reuse, recovery, recycling in a scientific and hygienic manner.

Organic content refers to the large source of carbon-based compounds found within natural and engineered, terrestrial, and aquatic environments.

Refuse-Derived Fuel (RDF) Fuel in the form of pellets or fluff produced by shredding and dehydrating combustible components of municipal solid waste.

Roll-off container A large waste container that fits onto a tractor trailer that can be dropped off and picked up hydraulically.

Sanitary landfill An engineered method of disposing solid wastes on land in a manner that protects human health and the environment. The waste is compacted and covered every day. The landfill is sealed from below, and leachate and gas are collected, and there is a gate control and a weighbridge.

Stakeholder Individual or institution (public and private) interested and involved in related processes and activities associated with a modernization process, plan, project goal, or desired change.

United Nations (UN) is an intergovernmental organization that aims to maintain international peace and security, develop friendly relations among nations, achieve international cooperation, and be a centre for harmonizing the actions of nations.

Waste transportation is the movement of waste over a specific area by trains, tankers, trucks, barges, or other vehicles.

Waste generation is includes all materials discarded, whether or not they are later recycled or disposed of in a landfill.

References

1. Kůdela, J., Šomplák, R., Nevrlý, V., Lipovský, T., Smejkalová, V., & Dobrovský, L. (2019). Multi-objective strategic waste transfer station planning. *Journal of Cleaner Production, 230*, 1294–1304.
2. Kumar, S. (2016). Municipal solid waste management in developing countries. *Municipal Solid Waste Management in Developing Countries*, 1–176.
3. Dobrowolski, D., Droździel, P., Madleňák, R., Rybicka, I., & Siłuch, D. (2018). Daily Kilometrage Analysis for Selected Vehicle Groups. *Advances in Science and Technology Research Journal, 12*, 39–46.
4. Stalmašeková, N., Genzorová, T., Čorejová, T., & Gašperová, L. (2017). The Impact of Using the Digital Environment in Transport. In *TRANSCOM 2017: International scientific conference on sustainable, modern and safe transport* (pp. 231–236). Elsevier Ltd.
5. Eusuf, M.A., Ibrahim, M., Shamzani Affendy, M.D., Islam, R. (2011). Solid waste generation characteristics: The Malaysian local authorities' outlook. Plan Malaysia.
6. Sundaram, J. K., Gen, T. Z., & Jarud, R. K. (2019). *Achieving food security for all Malaysians* (pp. 1–78). Khazanah Res Inst Kuala Lumpur.
7. Agriculture. USDO. (2019). *The impact of food waste*. United States Department of Agriculture.
8. Ismail, M. H., Ghazi, T. I. M., Hamzah, M. H., Manaf, L. A., Tahir, R. M., Nasir, A. M., & Omar, A. E. (2020). Impact of Movement Control Order (MCO) due to Coronavirus Disease (COVID-19) on food waste generation: A case study in Klang Valley, Malaysia. *Sustainability, 12*, 1–17.
9. Singh, G. K., Gupta, K., & Chaudhary, S. (2014). Solid waste management: Its sources, collection, transportation and recycling. *International Journal of Environmental Science and Development, 5*(4), 347–351.
10. Pichtel, J. (2014). *Waste Management Practices: Municipal, Hazardous and Industrial* (2nd ed., pp. 1–676). Taylor & Francis Group.
11. Põldnurk, J. (2015). Optimisation of the economic, environmental and administrative efficiency of the municipal waste management model in rural areas. *Resources, Conservation and Recycling, 97*, 55–65.
12. Haerani, D., & Budi, S. S. (2019). Review modeling of solid waste transportation routes using Geographical Information System (GIS). *Web Conference, 6*, 1–5.
13. Amaral, A. R., Rodrigues, E., Gaspar, A. R., & Gomes, Á. (2020). A review of empirical data of sustainability initiatives in university campus operations. *Journal of Cleaner Production, 250*, 1–41.
14. Kumar, S. (2016). *Municipal Solid Waste Management in Developing Countries* (1st ed.). CRC Press, Taylor & Francis Group.
15. Khan, M., Vaezi, M., & Kumar, A. (2018). Science of the Total Environment Optimal siting of solid waste-to-value-added facilities through a GIS-based assessment. *Science Total Environmental, 610*, 1065–1075.
16. Karak, T., Bhagat, R. M., & Bhattacharyya, P. (2012). Municipal solid waste generation, composition, and management: The world scenario. *Critical Reviews in Environmental Science and Technology, 42*, 1509–1630.
17. Malakahmad, A., Bakri, P., Mokhtar, M. R., & Khalil, N. (2014). Solid waste collection routes optimization via GIS techniques in Ipoh City, Malaysia. *Procedia Engineering., 77*, 20–27.
18. Ristić, G., Djordjević, A., Hristov, S., Umičević, P., Petković, A., & Milošević, L. (2015). Methodology for route optimization for solid waste collection and transportation in urban areas. *Working and Living Environmental Protection, 12*(2), 187–197.
19. Jereme, I. A., Siwar, C., & Alam, M. M. (2014). Waste recycling in Malaysia: Transition from developing to developed country waste recycling in Malaysia: Transition from developing to developed country. *Indian Journal of Education and Information Management, 4*, 1–14.

20. Lim, W. J., Chin, N. L., Yusof, A. Y., Yahya, A., & Tee, T. P. (2016). Food waste handling in Malaysia and comparison with other Asian countries. *International Food Research Journal, 23*, S1–S6.
21. Monzambea, G. M., Mpofua, K., & Daniyan, I. A. (2019). Statistical analysis of determinant factors and framework development for the optimal and sustainable design of municipal solid waste management systems in the context of industry 4.0. *Procedia CIRP., 84*, 245–250.
22. Durán-Romero, G., López, A. M., Beliaeva, T., Ferasso, M., Garonne, C., & Jones, P. (2020). Bridging the gap between circular economy and climate change mitigation policies through eco-innovations and Quintuple Helix Model. *Technological Forecasting and Social Change, 160*, 120246.
23. Calabrò, P. S., Gori, M., & Lubello, C. (2015). European trends in greenhouse gases emissions from integrated solid waste management. *Environ Technology (United Kingdom), 36*, 2125–2137.
24. Rigamonti, L., Sterpi, I., & Grosso, M. (2016). Integrated municipal waste management systems: An indicator to assess their environmental and economic sustainability. *Ecological Indicators, 60*, 1–7.
25. Tomić, T., & Schneider, D. R. (2017). Municipal solid waste system analysis through energy consumption and return approach. *Journal of Environmental Management, 203*, 973–987.
26. Peri, G., Ferrante, P., La Gennusa, M., Pianello, C., & Rizzo, G. (2018). Greening MSW management systems by saving footprint: The contribution of the waste transportation. *Journal of Environmental Management, 219*, 74–83.
27. Troschinetz, A. M., & Mihelcic, J. R. (2009). Sustainable recycling of municipal solid waste in developing countries. *Waste Management, 29*, 915–923.
28. Das, S., & Bhattacharyya, B. K. (2015). Optimization of municipal solid waste collection and transportation routes. *Waste Management, 43*, 9–18.
29. Zhou, M. C., Chang, N.-B., & Pires, A. (2016). Sustainable Solid Waste Management: A Systems Engineering Approach. In T. Samad, G. W. Arnold, V. Lumelsky, & L. Shafer (Eds.), (Vol. 4, pp. 1–932). Wiley.
30. Coffey, M., & Coad, A. (2011). Collection of municipal solid waste in developing countries. In D. Vilnersson (Ed.), *United Nations Human Settlements Programme* (Vol. 70, pp. 1013–1014). United Nations Human Settlements Programme.
31. Ismail, A. H., Usman, Y. V., Hidayah, N. Y., & Chairani, L. (2012). Metropolitan cities' waste transportation model. *Procedia – Social and Behavioral Sciences, 65*, 1046–1053.
32. Corriere, F., Rizzo, G., & Guerrieri, M. (2013). Estimation of air pollutant emissions in "turbo" and in conventional roundabouts. *Applied Mechanics and Materials, 394*, 597–604.
33. Jaunich, M. K., Levis, J. W., DeCarolis, J. F., Gaston, E. V., Barlaz, M. A., Bartelt-Hunt, S. L., et al. (2016). Characterization of municipal solid waste collection operations. *Resources, Conservation and Recycling, 114*, 92–102.
34. Kallel, A., Serbaji, M. M., & Zairi, M. (2016). Using GIS-based tools for the optimization of solid waste collection and transport : Case study of Sfax City, Tunisia. *Journal of Engineering, 2016*(2), 1–8.
35. Lacovidou, E., & Ng, K. S. (2020). Malaysia versus waste. *Chemical Engineer, 8*(949), 1–5.
36. Aleluia, J., & Ferrão, P. (2016). Characterization of urban waste management practices in developing Asian countries: A new analytical framework based on waste characteristics and urban dimension. *Waste Management, 58*, 415–429.
37. Periathamby, A., & Tanaka, M. (2014). *Municipal Solid Waste Management in Asia and the Pacific Islands: Challenges and Strategic Solutions* (pp. 1–382). Springer.
38. Kawai, K., & Tasaki, T. (2016). Revisiting estimates of municipal solid waste generation per capita and their reliability. *Journal of Material Cycles and Waste Management, 18*, 1–13.
39. Abedin, M. A., & Jahiruddin, M. (2015). Waste generation and management in Bangladesh: An overview. *Asian Journal of Medical and Biological Research, 1*, 114–120.
40. Pires, A., Martinho, G., Rodrigues, S., & Gomes, M. I. (2018). Sustainable solid waste collection and management. *Sustainable Solid Waste Collection and Management*, 1–368.

41. Themelis, N. J. (2019). Recovery of Materials and Energy from Urban Waste. In R. A. Meyers (Ed.), *Recovery of Materials and Energy from Urban Wastes* (Vol. 1, 3rd ed., pp. 1–551). Springer.
42. Budhiarta, I., Siwar, C., & Basri, H. (2012). Current status of municipal solid waste generation in Malaysia. *International Journal on Advanced Science, Engineering and Information Technology, 2*, 129.
43. Höke, M. C., & Yalcinkaya, S. (2020). Municipal solid waste transfer station planning through vehicle routing problem-based scenario analysis. *Waste Management & Research, 2*, 1–12.
44. Christian, B. (2014). *Multi-Criteria Selection of Waste Transfer Stations in the Kumasi Bosompem Christian*. Kwame Nkrumah University of Science and Technology.
45. EPA.. (2020). Best practices for solid waste management : Best practices for solid waste management : A guide for decision-makers in developing countries (1st ed.). Washington, DC.
46. Oduro-kwarteng, S. (2011). *Private Sector Involvement in Urban Solid Waste Collection*. Erasmus University.
47. Yanke D, Gregory D. Solid Waste Assessment and Management Study. 2014.
48. Lella, J., Mandla, V. R., & Zhu, X. (2017). Solid waste collection/transport optimization and vegetation land cover estimation using Geographic Information System (GIS): A case study of a proposed smart-city. *Sustainable Cities and Society, 35*, 336–349.
49. Scheinberg, A., Wilson, D. C., & Rodic-Wiersma, L. (2010). Solid Waste Management in the World's Cities. In L. Rodic-Wiersma (Ed.), *Solid Waste Management in the World's Cities* (pp. 1–257). Earthscan.
50. Chandrappa, R., & Bhusan, D. D. (2012). Solid Waste Management: Principles and Practice. In R. Allan, U. Forstner, & W. Salomons (Eds.), (pp. 1–428). Springer Nature.
51. Le Son, H., & Louati, A. (2016). Modeling municipal solid waste collection : A generalized vehicle routing model with multiple transfer stations, gather sites and inhomogeneous vehicles in time windows. *Waste Management, 5*, 1–16.
52. Hayat, S., & Sheikh, S. H. (2016). In K. Farhan (Ed.), *Municipal Solid Waste: Engineering Principles and Management* (2nd ed., pp. 1–234). The Urban Unit.
53. Shah, S. Z. A., Nawaz, Z., Nawaz, S., Carder, G., Ali, M., Soomro, N., et al. (2019). The role and welfare of cart donkeys used in waste management in Karachi. *Pakistan. Animals., 9*(4), 159.
54. Kunz, J. R., Granella, M. C. S., Mendes, R. P., Müller, T. R., Kau, S., & Fonteque, J. H. (2020). High prevalence of orodental disorders in South Brazilian Cart Horses: Walking a Tightrope between animal welfare and socioeconomic inevitability. *Journal of Veterinary Dentistry, 10*, 1–10.
55. Rivero-Moreno, A. (2014). Proceedings of the 7 th international meeting on Narrow Gap semiconductors. In J. F. Wade (Ed.), *Proceedings of the 7th International Colloquium on Working Equids* (p. 154). World Horse Welfare.
56. Warboys, D., Robles, M., Chapman, A., & Lagos Blessing, O. G. (2014). Cross-sectional survey on the importance of the role of working equids in Honduras. In J. F. Wade (Ed.), *Proceedings of the 7th International Colloquium on Working Equids, Royal Holloway* (p. 7). University of London.
57. Sunaryo, S. (2020). Break-even analysis of clay transport as raw materials in small bricks industry using Grandong based appropriate technology in rural areas. *International Journal of Scientific and Engineering Research, 11*(2), 384–392.
58. Olukanni, D. O., & Adeleke, O. (2016). A review of local factors affecting solid waste collection in Nigeria. *Pollution., 2*, 339–356.
59. Basnayake, B. F. A., & Sustainable, V. C. (2014). In Pariatamby & M. Tanaka (Eds.), *Solid Waste Management in India* (pp. 1–18). Springer-Verlag.
60. Amoah, S. T., & Kosoe, E. A. (2015). Solid waste management in urban areas of Ghana : Issues and experiences from Wa. *Journal of Environment Pollution and Human Health, 2*, 110–117.

61. Hillary. (2016). *Comprehensive solid waste management and recycling plan*. Maryland.
62. Sandhu, G. S., Frey, H. C., Bartelt-Hunt, S., & Jones, E. (2016). Real-world activity, fuel use, and emissions of diesel side-loader refuse trucks. *Atmospheric Environment, 129*, 98–104.
63. García, A. M. L. (2017). *A Technological Approach Towards Integrated Solid Waste Management*. Jyväskylä University.
64. Kogler, T. (2007). *Waste Collection*. European union.
65. Rada, E. C. (2016). *Solid Waste Management: Policy and Planning for a Sustainable Society. Apple Academic Press* (265 p). CRC Press.
66. Nakou, D., Benardos, A., & Kaliampakos, D. (2014). Assessing the financial and environmental performance of underground automated vacuum waste collection systems. *Tunnelling and Underground Space Technology, 41*, 263–271.
67. Schneider, S. H. (2014). *Municipal Solid Waste Management Manual* (pp. 1–933). Central Public Health and Environmental Engineering Organisation.
68. Trotti, J., Duncan, B., & Rachford, D. (2016). Municipal solid waste management. *SWANA., 6*, 1–118.
69. Bucknell, P. A. V., Southern, S. M. M., & Heine, L. G. (2012). Introduction to Environmental Engineering. In J. Dinsmore (Ed.), *Introduction to Environmental Engineering* (3rd ed., pp. 1–628). Cengage Learning.
70. Douglass L, Ross B. A Study of the Feasibility of Pneumatic Transport of Municipal Solid Waste and Recyclables in Manhattan Using Existing Transportation Infrastructure. 2013.
71. Teerioja, N., Moliis, K., Kuvaja, E., Ollikainen, M., Punkkinen, H., & Merta, E. (2012). Pneumatic vs. door-to-door waste collection systems in existing urban areas: A comparison of economic performance. *Waste Management, 32*, 1782–1791.
72. Christensen, T. H. (2010). Waste transfer stations: A manual for decision-making. *Solid Waste Technology and Management, 1*, 311–318.
73. Beliën, J., De Boeck, L., Van Ackere, J., Beliën, J., De Boeck, L., & Van Ackere, J. (2014). Municipal solid waste collection and management municipal solid waste collection and management problems : A literature review. *Transportation Science, 2*, 1–26.
74. Bhambulkar, A. V., & Khedikar, I. P. (2011). Municipal Solid Waste (Msw) collection route for Laxmi Nagar by geographical information system. *International Journal of Advanced Engineering and Technology, 2*(4), 48–53.
75. Singh, S., & Behera, S. N. (2019). Development of GIS-Based Optimization Method for Selection of Transportation Routes in Municipal Solid Waste Management. In A. S. Kalamdhad (Ed.), *Advances in Waste Management* (pp. 1–13). Springer.
76. Tavares, G., Zsigraiova, Z., Semiao, V., & Carvalho, M. G. (2009). Optimisation of MSW collection routes for minimum fuel consumption using 3D GIS modelling. *Waste Management, 29*, 1176–1185.
77. Mrówczyńska, B. (2014). Route planning of separate waste collection on a small settlement. *Transport Problems, 9*(1), 61–68.
78. Sulemana, A., Donkor, E. A., Forkuo, E. K., & Oduro-kwarteng, S. (2018). Optimal routing of solid waste collection trucks : A review of methods. *Journal of Engineering, 2018*, 1–12.
79. Jaiswal, A., & Bharat, A. (2015). Identification of criteria and Indicators for locating waste transfer station sites in urban areas by Delphi consensus. *International Journal of Scientific Research, 4*(2), 318–323.
80. Azadeh, A., Ahmadzadeh, K., & Eslami, H. (2018). Location optimization of municipal solid waste considering health, safety, environmental, and economic factors. *Journal of Environmental Planning and Management, 65*, 1–12.
81. Ağaçsapan, B., & Çabuk, S. N. (2020). Determination of suitable waste transfer station areas for sustainable territories: Eskisehir case. *Sustainable Cities and Society, 52*, 1–39.
82. Yadav, V., Bhurjee, A. K., Karmakar, S., & Dikshit, A. K. (2017). A facility location model for municipal solid waste management system under uncertain environment. *Science of the Total Environment, 603*, 760–771.

83. EPA.. (2014). *Municipal solid waste landfills: Economic impact analysis for the proposed new subpart to the new source performance standards*.
84. Garibay-Rodriguez, J., Laguna-Martinez, M. G., Rico-Ramirez, V., & Botello-Alvarez, J. E. (2018). Optimal municipal solid waste energy recovery and management: A mathematical programming approach. *Computers and Chemical Engineering, 119*, 394–405.
85. Herrera HE Consultants. Optimized Transfer Station Recycling Feasibility Study Task 6 : Strategy Evaluation and Recommendations. 2013.
86. Yadav, V., Karmakar, S., Dikshit, A. K., & Vanjari, S. (2016). Transfer stations siting in India: A feasibility demonstration. *Waste Management, 47*, 1–5.
87. Bhalla, B., Saini, M. S., & Jha, M. K. (2013). Characterization of Leachate from Municipal Solid Waste (MSW) landfilling sites of Ludhiana, India : A comparative study. *International Journal of Engineering Research and Applications, 2*(6), 732–745.
88. Johari, A., Alkali, H., Hashim, H., Ahmed, S. I., & Mat, R. (2014). Municipal solid waste management and potential revenue from recycling in Malaysia. *Modern Applied Science, 8*, 37–49.
89. Fazeli, A., Bakhtvar, F., Jahanshaloo, L., Che Sidik, N. A., & Bayat, A. E. (2016). Malaysia's stand on municipal solid waste conversion to energy: A review. *Renewable and Sustainable Energy Reviews, 58*, 1007–1016.
90. Kamaruddin, M. A., Yusoff, M. S., Rui, L. M., Isa, A. M., Zawawi, M. H., & Alrozi, R. (2017). An overview of municipal solid waste management and landfill leachate treatment: Malaysia and Asian perspectives. *Environmental Science and Pollution Research, 24*, 1–35.
91. Yadav, V., Karmakar, S., Dikshit, A. K., & Vanjari, S. (2016). A feasibility study for the locations of waste transfer stations in urban centers: A case study on the city of Nashik, India. *The Journal of Cleaner Production, 126*, 191–205.
92. Steve-Harold, W. K., Qin, X., Sha, C., & Song, C. (2020). Greenhouse gas reduction and cost-benefit through improving municipal solid waste management in Ouagadougou. *African Journal of Environmental Science and Technology, 14*, 222–230.
93. Cirucci, J. F., Miller, D. A., & Blanford, J. I. (2015). Retrospective GIS-based multi-criteria decision analysis: A case study of California waste transfer station siting decisions. *Proceedings of the International Symposium Sustainable Systems and Technologies, 3*, 1–16.
94. Phinyoyang, A. (2016). *Waste transportation management based on transportation cost and environmental impact of sites*. Suranaree University of Technology.
95. Economopoulou, M. A., Economopoulou, A. A., & Economopoulos, A. P. (2013). A methodology for optimal MSW management, with an application in the waste transportation of Attica Region, Greece. *Waste Management, 2*, 1–11.
96. Dolman G. Why short-haul intermodal rail services succeed. 2016.
97. Islam, M. S., Pervin, L., & Muyeed, A. A. (2012). Feasibility study for using the water ways to transport of solid waste of Dhaka city and waste transport route preparation using GIS for DCC solid waste management. *Canadian Journal of Environmental Construction of Civil Engineering., 3*(1), 52–57.
98. Pfoser, S., Jung, E., & Putz, L.-M. (2018). Same river same rules? – Administrative barriers in the Danube countries. *Journal of Sustainable Development of Transport and Logistics, 3*, 27–37.
99. Miloslavskaya, S., Plotnikova, E., & Myskina, A. (2019). Inland waterways transport in large cities transport system: Practice and prospects. *Transport Means – Proceedings of International Conference, 10*(12), 639–643.
100. Guerrero, A. L., Maas, G., & Hogland, W. (2013). Solid waste management challenges for cities in developing countries. *Waste Management, 33*, 220–232.
101. Sanjeevi, V., & Shahabudeen, P. (2015). Development of performance indicators for municipal solid waste management (PIMS): A review. *Waste Management & Research, 2*, 1–14.
102. Akhtara, M., Hannan, M. A., Basri, H., & Scavinob, E. (2015). Solid waste generation and collection efficiencies: Issues and challenges. *Journal Teknologi, 11*(75), 41–49.

103. Ababio, M. O. (2011). Missing links in solid waste management in the Greater Accra Metropolitan Area in Ghana. *GeoJournal, 76*, 551–560.
104. Alam, O., & Qiao, X. (2020). An in-depth review on municipal solid waste management, treatment and disposal in Bangladesh. *Sustainable Cities and Society, 52*, 101775.
105. Hannan, M. A., Begum, R. A., Al-Shetwi, A. Q., Ker, P. J., Al Mamun, M. A., Hussain, A., & TMI, M. (2020). Waste collection route optimisation model for linking cost saving and emission reduction to achieve sustainable development goals. *Sustainable Cities and Society, 62*, 102393.
106. Abalo, E. M., Peprah, P., Nyonyo, J., Ampomah-sarpong, R., & Agyemang-duah, W. (2018). A review of the triple gains of waste and the way forward for Ghana. *The journal, Renewable Energy, 2018*, 1–13.
107. Mmereki, D., Baldwin, A., & Li, B. (2016). A comparative analysis of solid waste management in developed, developing and lesser developed countries. *Environmental Technology Reviews, 5*, 120–141.
108. Lopp, S., Wood, E., & Duran, A. (2015). Evaluating the impact of road grade on simulated commercial vehicle fuel economy using real-world drive cycles. *SAE International, 1*, 1–6.
109. Achi, H., Adeofun, C., Ufoegbune, G., Gbadebo, A., & Oyedepo, J. A. (2012). Disposal sites and transport route selection using geographic information system and remote sensing in Abeokuta, Nigeria. *Global Journal of Human Social Sciences, 12*(12), 1–11.
110. Aja, O. C., Oseghale, S. D., & Al-Kayiem, H. H. (2014). Review and Evaluation of Municipal Solid Waste Management Practices in. *The Journal of Solid Waste Technology and Management, 8*, 1–18.
111. Pamnani, A. M. S. (2014). Municipal solid waste management in India: A review and some new results. *International Journal of Civil Engineering and Technology, 5*(2), 1–8.

Chapter 4
Characterization and Measurement of Solid Waste

Mohamad Anuar Kamaruddin, Faris Aiman Norashiddin, Mohamad Haziq Mohd Hanif, Abdubaki Mohamed Hussen Shadi, Mohd Suffian Yusoff, Lawrence K. Wang, and Mu-Hao Sung Wang

Abstract Development in the economy, urbanization, and industrialization has resulted in an increase in the amount and variety of solid waste. Ineffective and inefficient waste management may cause greenhouse gas and toxic emissions, as well as the loss of valuable materials and services, in addition to negative health effects and increased pollution of the air, soil, and water. Waste prevention and minimization need to be promoted to increase reuse, recycling, and energy recovery. Solid wastes must be classified according to their origins, generation rates, the types of wastes produced, and their compositions. These are important data used in the formulation of economic, regulatory, and institutional policies. Hence, waste characterization measurement needs to be properly evaluated based on standard protocols. Hence, this chapter presented the measurement of the common physical and chemical data of waste. These include, but are not limited to, quantity, density, size distribution, moisture content, and ash. Leachate quantity, its composition, and gas evaluation were also discussed.

Keywords Municipal solid waste · Waste characterization · Waste composition · Landfill leachate · Leachate quantification · Proximate analysis · Leachate composition

---

M. A. Kamaruddin (✉) · F. A. Norashiddin · M. H. M. Hanif · A. M. H. Shadi
School of Industrial Technology, Universiti Sains Malaysia, Minden, Pulau Pinang, Malaysia
e-mail: anuarkamaruddin@usm.my

M. S. Yusoff
School of Civil Engineering, Engineering Campus, Universiti Sains Malaysia, Nibong Tebal, Pulau Pinang, Malaysia
e-mail: suffian@usm.my

L. K. Wang · M.-H. S. Wang
Lenox Institute of Water Technology, Latham, USA

L. K. Wang et al. (eds.), *Solid Waste Engineering and Management*, Volume 1, Handbook of Environmental Engineering 23,
https://doi.org/10.1007/978-3-030-84180-5_4

Nomenclature

ADEME	Agency for Environment and Energy Management, France
ASEAN	Association of Southeast Asian Nations
ASTM	A Standard Test Method
ASTM	American Society for Testing and Materials.
BOD	Biochemical oxygen demand
COD	Chemical oxygen demand
EC	Electrical Conductivity
ED	Electrodialysis
ERT	Electrical Resistivity Tomography
FID	Flame ionization detector
FILL	Flow Investigation for Landfill Leachate
FW	Food waste
HELP	Hydrologic Evaluation of Landfill Performance
HHV	Higher heat value
LFG	Landfill gas
LHV	Lower Heating Value
MC	Moisture content
MFA	Material Flow Analysis
MRF	Material Recovery Facility
MSW	Municipal solid waste
RDF	Refuse-derived fuel
RIVM	National Institute of Public Health and Environmental Protection
SPLP	Synthetic Precipitation Leaching Procedure
SWCORP	Solid Waste and Public Cleansing Management Corporation
SWPCM	Solid Waste and Public Cleansing Management
TCLP	Toxicity characteristics leaching procedure
UNEP	United Nation Environment Programme
USEPA	United States Environmental Protection Agency
VFA	Volatile fatty acid
WBM	Water Balance Method
WET	Waste extraction test
WGR	Waster generation rate
WHO	World Health Organization
WTE	Waste-to-energy

4.1 Introduction

Waste materials are produced by human activities and are often discarded because they are deemed useless. These wastes are usually solid, and the word waste implies that the material is unnecessary and useless [1]. However, if properly treated, many of these waste materials can be reused and thus become a resource for industrial production or energy generation. Waste materials are produced by human activities and are often discarded because they are deemed useless.

These wastes are usually solid, and the word waste implies that the material is unnecessary and useless. However, if properly treated, many of these waste materials can be reused and thus become a resource for industrial production or energy generation. Industry, private citizens, and state governments are all looking for ways to minimize the increasing amount of waste that households and businesses generate and reuse or dispose of it in a sustainable and cost-effective manner.

More laws dealing with solid waste management have been passed by state legislatures in recent years than any other subject on their legislative agendas [2]. An understanding of the waste stream's characteristics is needed regardless of the form of solid waste management is considered or introduced. Long-term trends in waste stream characteristics are also critical in good planning, which goes beyond developing a snapshot of current waste composition [3]. If potential waste stream amounts and components are underestimated or overestimated, facilities can be over- or undersized, affecting project revenues and costs [4]. A typical view of waste refuse in a dumping site is shown in Fig. 4.1. Typically, the dumping site receives a wide spectrum of MSW, and unfortunately, a certain amount of waste compositions which having recyclable items potential are dumped together in the refuse cell.

Fig. 4.1 Mixture of MSW

4.1.1 *Waste Quantity*

The types and quantities of waste produced in any area are determined by the citizens' lifestyles and living standards, as well as the region's natural resources. Excessive amounts of waste are generated by society as a result of inefficient manufacturing processes, poor product durability, and unsustainable resource consumption [5]. It is difficult to generalize or standardize solid waste management in the private sector due to the varying degrees of growth in different countries. Solid waste management entails both an understanding of current waste management techniques and the introduction of innovative approaches to counteract them. Higher-income regions are notorious for their 'use and throw' habits, which generate massive amounts of waste. Lower-income regions maximize the use and reuse of available capital, resulting in less waste generation. Environment, economy, disaster frequency, people's mentality, and all the other variables contribute to varying waste quantities and qualities. Solid waste collection ditches were the key cause of epidemics in Europe between 1348 and 1665, which can now be seen in developing countries. When the technology of waste disposal evolve from century to century, many waste components were diverted and recovered for recycling purpose. Some of the recovered wastes were reproduced or even repurposed to fit the needs of a selected class of populations. Figure 4.2 typically shows the scavenging activity carried out at an open dump site in Malaysia. Interestingly, scavengers able to recover around 45% to 65% worth of recyclable items and sold them to a vendor for their ends meet.

Fig. 4.2 Scavenging activity in Malaysia's dump site

Most municipal governments have waste management as one of their key services. Solid wastes must be classified by sources, generation rates, types of wastes generated, and composition to track and control current waste management systems while improving the existing system. Financial, regulatory, and institutional decisions will be aided by this information. However, the rapid growth of the human population and the discovery of new materials have caused the quantities and characteristics to change on a daily basis. According to a conservative World Bank estimate from 1999, municipal solid waste (MSW) from Asian cities will increase from 760,000 tonnes per day in 1999 to 1.8 million tonnes per day in 2025. With that income in Asian countries, solid waste management will become more difficult in the coming years on the continent. New goods wrapped in new packaging materials, new living standards and expectancies, and changes in wealthy people's income and lifestyle have all contributed to a rise in global waste volume. However, as mentioned in the following pages, after reaching a limit that is unique to a country or area, the per capita waste quantity begins to decline. This may be due to a shift in technology that reduces waste generation, a shift in people's attitudes, a shift in purchasing power, or a reduction in product demand. The United States, which has the most vehicles per thousand people in the world, will obviously produce fewer cars and waste from car manufacturing than India and China, where people have a greater desire to own private transportation.

Commercial, residential, manufacturing, institutional, demolition, renovation, and urban wastes are all included in MSW. However, the data on MSW varies greatly between waste studies. Household waste, which makes up a small portion of the overall waste stream, is usually the basis for waste management decisions. In addition, businesses and economic operation conceal details in order to escape legal obligations.

4.1.2 MSW Source

MSW is defined as waste from residential, commercial, institutional, and some industrial sources, according to this definition. The definition of MSW [6] is as follows:

- The amount of materials and items in MSW that reach the waste stream before being recycled, composted, or burned is referred to as a generation.
- Recovery is the process of removing products from the waste stream so that they can be recycled or composted. Recovery does not always imply recycling.
- The MSW that remains after recovery is referred to as discards. The waste is usually burned or buried, but it can also be littered, stored, or disposed of on-site, particularly in rural areas.

Table 4.1 shows the MSW sources according to established reference [6]:

However, Malaysia has a specific definition of MSW, which describes the term as follows:

Table 4.1 Waste source

Source	Examples
Residential	Single-family homes, duplexes, town houses, apartments
Commercial	Office buildings, shopping malls, warehouses, hotels, airports, restaurants
Institutional	Schools, medical facilities, prisons
Industrial	Packaging of components, office wastes, lunchroom and restroom wastes (but not industrial process wastes)

The Solid Waste and Public Cleansing Management (SWPCM) Act 2007 defines solid waste and controlled solid waste as follows:

'Solid waste' includes:

a) Any scrap material or other unwanted surplus substance or rejected products arising from the application of any process
b) Any substance required to be disposed of as being broken, worn out, contaminated, or otherwise spoiled; or
c) Any other material that according to this Act or any other written law is required by the authority to be disposed of, but does not include scheduled wastes as prescribed under the Environmental Quality Act 1974 [Act 127], sewage as defined in the Water Services Industry Act 2006 [Act 655] or radioactive waste as defined in the Atomic Energy Licensing Act 1984 [Act 304]

'Controlled solid waste' means any solid waste falling within any of the following categories:

- Commercial solid waste
- Construction solid waste
- Household solid waste
- Industrial solid waste
- Institutional solid waste
- Imported solid waste
- Public solid waste
- Solid waste which may be prescribed from time to time

Table 4.2 further divides the SWPCA 672 for the coverage of disposal of controlled waste, and public cleansing roles undertake by the Solid Waste Management and Public Cleansing Corporation (SWCorp).

Almost all anthropogenic practises, as well as waste management, would have an effect on the environment. While proper waste management reduces the severity of the effect, it does not completely eradicate it. It is essential to determine the environmental impacts to protect the climate. Any stage of waste management can have an effect on the environment. Figure 4.3 shows the cattle wandering near an excavator for spreading the fresh waste. This scene is quite a typical view in certain dumping sites in Malaysia, which has or limited environmental control, especially entrance and hoarding gate.

Table 4.2 Controlled waste and public cleansing coverage

Collection and disposal of controlled solid waste	Public cleansing
• Household	• Streets, public areas, public toilets, public drainage
• Commercial	• Markets and hawker centres
• Construction	• Illegal dumping
• Industry	• Beaches
• Institutions	• Public area & roadside grass cutting
• Public	• Removal of carcasses
• Imported	
• As determined by the Minister from time to time	

Fig. 4.3 Cattle wandering at a dump site

4.1.3 Waste Generation Status

Human activities generate solid waste, and inadequate solid waste management leads to major public health issues. Solid waste portion quantification and characterization is a critical phase in solid waste management procedures [7]. In the sampling process and solid waste segregation operations, the World Health Organization (WHO) and a few other well-known countries used a system for municipal solid waste (MSW) quantification and characterization that took into account seasonal differences and socio-economic status. Poor solid waste management practices pose a significant threat to public health and the environment because they can result in contamination of the air, soil, and water. Unfortunately, only 60% of the waste produced in developed countries is collected and disposed of properly. Due to a lack of complete information on the volume of municipal solid waste discards, collection and disposal of municipal solid waste are often insufficient [8]. For proper solid waste preparation and management, an accurate estimate of the city's solid waste quantity is critical [9].

The underestimation of the amount of solid waste produced has been a major stumbling block for most waste management plans. This leads to erroneous design calculations, which have a negative impact on waste management system performance. It is only by providing accurate and sufficient information on the rate of generation and composition of wastes produced that successful management can be planned. Since factors such as the degree of industrialization, the environment, and the existence of socio-economic development affect waste generation rates and composition, they vary from one country to the next and also between cities within a country. In developing countries, it is normal for the daily amount of waste collected to be less than the actual amount of waste generated by households.

Inadequate waste collection services and informal waste-picking practices are to blame. As a result, basic data on waste characteristics, which are required for the design and planning of solid waste management facilities, are available [10]. Figure 4.4 shows the typical failure or overcapacity of a dumping site that was not well planned and caused severe environmental deterioration as the percolating water leaching into water that covers with refuse. It mixed with water and brought further downstream to receiving water body.

Data on waste composition, production, and recycling are critical for designing collection routes, deciding bin placements, managing collection crews, and

Fig. 4.4 Flooded of refuse in perimeter ditches

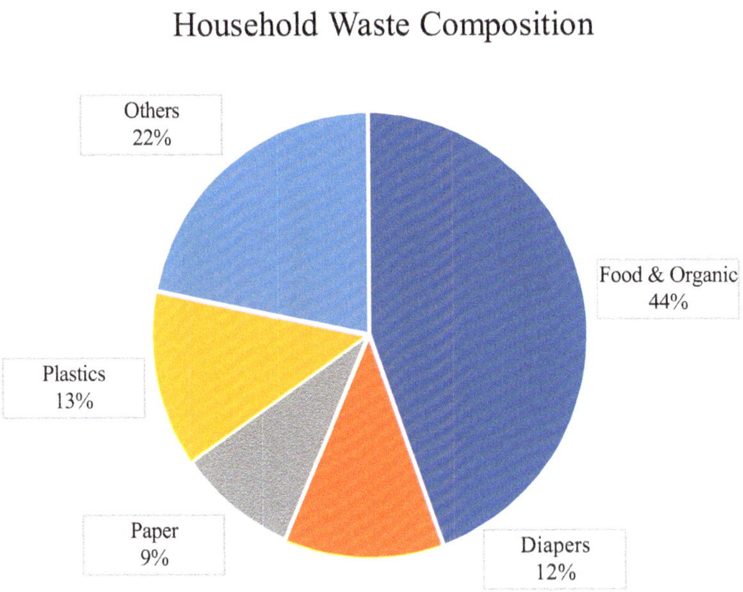

Fig. 4.5 Typical MSW composition from a landfill in Malaysia [14]

selecting suitable solid waste disposal options, as well as determining whether alternative waste management schemes, such as material recycling or biological treatment, should be developed. Several studies on the characterization of solid waste were conducted in developing countries to assess their physical, chemical, and thermal properties [1, 3, 5, 6, 11]; however, many countries lack reliable and up-to-date data on solid waste due to a lack of technological and financial resources. These studies revealed that waste characteristic analysis is important for a variety of reasons, including the need to recognise the potential for material recovery from waste mixtures, assess waste generation sources, and simplify treatment facility planning [11].

Figure 4.5 shows typical waste composition from Malaysia landfill. Typically, Malaysia, like any Asian countries, generates more than 40% of organic fraction consists of food waste and other carbon and nitrogen elements [12]. Asian population comprising of community that celebrate food more than any other community in the world. As a result, food waste which carries more than 50% moisture content, is the main reason for poor selection by favours more on the landfilling method, which is not sustainable [13].

4.2 Waste Measurement

The treatment and disposal facilities, such as recycling plants, composting plants, incineration plants, and landfills, are the primary sources of waste treatment and disposal data [15]. The use of administrative data collected for licencing and

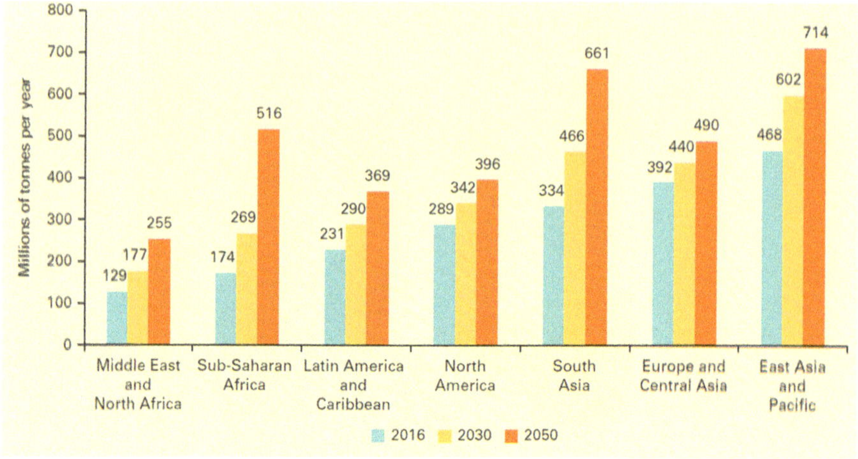

Fig. 4.6 Waste generation rate across the region (https://datatopics.worldbank.org/what-a-waste/trends_in_solid_waste_management.html)

tracking purposes, such as facility registers, consignment notices, or waste management reports, is heavily reliant on data collection on waste treatment and disposal [16]. For the collection of facility-related information and data on treated amounts, comprehensive waste facility registers are needed. When data on waste management and disposal isn't available from administrative sources, surveys are commonly used [17]. The majority of waste treatment and disposal statistics are derived from surveys of all waste treatment and disposal facilities that are subject to applicable obligations. Because of the broad range of waste management operations and waste streams, data is often gathered from several sources, necessitating the harmonization of definitions, classifications, and reporting requirements. Figure 4.6 shows the potential of waste generation rate across the region.

4.2.1 Landfill Infrastructure

Solid waste generation can be calculated or assessed in a variety of locations, including the point of generation, collection point, and disposal site, all of which can influence the amount of waste identified by sources [18]. These wastes are often characterized by the manner in which they are obtained. Mixed household waste is typically collected by trucks designed specifically for that purpose, while recyclables are collected either alongside the mixed waste in a separate compartment or by other vehicles designed specifically for that purpose [19]. Yard waste may be collected along with household garbage or stored in a separate truck. Big containers are used for commercial waste, which is poured into specially designed trucks. Demolition and construction waste is stored in stationary roll-off containers that

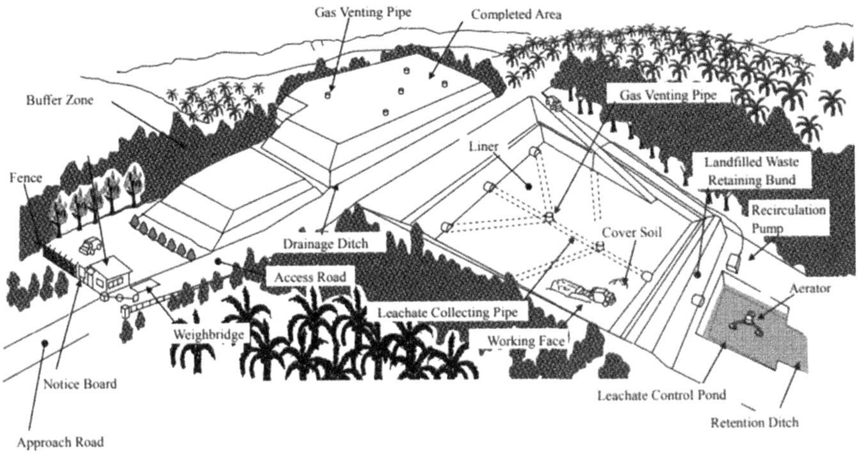

Fig. 4.7 Typical landfill infrastructure

stay on the job site until they are completed. Larger trucks capable of carrying large objects are often used to collect bulky items on an 'as required' basis.

As long as the MSW is going to be disposed of in a landfill, there's no need to do anything more than calculating the tonnes of waste produced and consider the issues with special (hazardous) materials. If, on the other hand, the aim is to extract gas from a landfill and put it to good use, the amount of organic material is critical [20]. When recycling is planned, or if material or energy recovery by combustion is the goal, it is important to consider the following factors. Figure 4.7 illustrates a typical schematic of landfill infrastructure and its component.

Any municipality must determine how much and what kind of waste is generated in its jurisdiction. This entails calculating the current amounts of waste generated, recycled, treated, and disposed of. The most common method for calculating waste amounts is to weigh them (kilograms or tons). For the following purposes, this data is required:

- Ensure appropriate resource planning for waste management services.
- Government-prioritized expansion of waste collection systems to previously unserved areas.
- Creating waste recycling programmes, such as buyback centres, that can help reduce the amount of recyclables that end up in landfills, and waste management businesses that can encourage people to start their own businesses.
- The provision of a certain number of waste receptacles and the determination of collection needs, that is, how many skips should be installed in communal areas such as shopping malls, hospitals, and other public areas such as sports clubs and the implementation of separation at source initiatives

4.2.2 Weighbridge

Making predictions about potential waste volumes to ensure that enough money is set aside for future waste services and infrastructure growth. Specifically, the data will reveal whether new waste disposal facilities or transfer stations are needed for faster and more efficient collection rates. It is important to provide easy access to a disposal site from existing roads. At the very least, the access road should allow two trucks to pass in either direction. Depending on the amount of waste collected at the site, a two-lane road with passing places sections will accomplish this. Path conditions must permit permanent entry, and the road must be protected so that garbage trucks are not impaired by an excessively rough surface.

A reception area should be clearly defined and large enough for incoming vehicles to be stopped and inspected by workers. The reception area should have an entry gate or fence, as well as a gate house to store waste records and documents, as well as provide shelter for the landfill workers from inclement weather. Two (or more) trucks should be able to park in the reception area without interfering with vehicle movement in and out of the site. A water supply and sanitation facilities must also be available. Electricity should be available at large landfill sites that receive 100 tons a day or more. Figure 4.7 shows a typical landfill infrastructure, which consists primarily of the following:

- Notice board
- Access road
- Weighbridge
- Drainage ditch
- Leachate collection
- Gas venting pipe
- Recirculation pump
- Leachate pond

At the landfill site's entrance, a scale or weighbridge (Fig. 4.8) device weighs waste vehicles when they arrive, records and verifies load data, and reweighs the vehicle once the load has been deposited [21]. Data are processed and can be moved to back office systems to meet legislative reporting standards and increase performance. The operator must use a weighbridge to record the amount of waste entering its waste disposal facility by measuring the vehicles at the point of entry and exit. The mass of the waste is determined by the difference in vehicle mass between the 'in' and 'out' points. The types of waste disposed of must be accurately identified by a weighbridge operator. The information is collected using weighbridge software that is confided with spreadsheet software like Microsoft Excel or a personalized weighbridge software that can provide billing information depending on the type of waste and the vehicle's size. This information is then saved in specialized software designed for weighbridge operations, and it is accessible through software export functions or waste report printing.

Fig. 4.8 A typical weighbridge system

The concept of the weighbridge system is as follows:

- On their way in and out, all vehicles passing through the site pass through the weighbridge. The waste mass is determined by the difference in vehicle mass between the 'in' and 'out' points.
- The weighbridge operator records the type of waste in the vehicle. The price paid to the disposer is usually determined by the type of waste.

When there is a lack of a weighbridge system at certain landfills, municipalities have unreliable solid waste data, which leads to both uninformed and misinformed planning processes, as well as the formulation of policies that are meaningless to municipalities [22]. Figure 4.9 shows a typical dump site without having a weighbridge system.

Whether paid or not, all waste entering a landfill site must pass through the weighbridge. Certain pollutants, such as garden waste (organics) and construction debris, are often accepted at landfills for free and are sometimes diverted across the weighbridge. This must not be permitted as unrecorded waste will lead to reducing the lifespan of a landfill. A weighbridge system is applied to record the amount of waste being carried and disposed of at the landfill [23]. If there is no weighbridge system provided at the landfill, a density-volume estimation can be a simple indicator for the operator to measure the amount of waste dumped at the landfill. To calculate the mass of waste entering the landfill, waste mass estimation systems are normally used with a formula that includes waste volume, waste density, and waste loading. This is shown in Eq. 4.1.

Fig. 4.9 Typical view of a dump site without having weighbridge system

$$waste\ mass(kg) = vehicle\ volume(m^3) \times loading\ factor \times waste\ density(kg/m^3) \quad (4.1)$$

The volume of a vehicle can be calculated in a variety of ways:

• The vehicle specification may be used to estimate the number of municipal vehicles.

• The number of regular vehicles that use landfills has been calculated and registered.

• When entering the landfill, the vehicle can be weighed. The following formula can be used to calculate a vehicle's volume:

$$Volume\ m^3 = length(m) \times breadth(m) \times height(m) \quad (4.2)$$

4.2.3 Waste Quantification

In order to minimize food waste, it is important to first identify the issue that needs to be addressed. One of the common method determining the amount of waste being disposed of is by using the waste quantification method [24]. However, the lack of a shared standard for quantifying and disclosing food waste is a barrier to performing quantifications, making it impossible to analyse reports from various organizations. It is arguable that due to differences in building methods, work processes, and traditional standards, waste rates cannot be strictly comparable between countries [25]. The quantification of waste needed to assess the true size of waste produced. There are many methods for calculating the volume or rate of waste generation at the generation stage. Waste generation rates are commonly studied using two approaches: soft and hard steps. Interviewed and questionnaire, as well as an estimation based on statistical data, are examples of soft measure methods. Though hard measurement methods such as Material Flow Analysis (MFA) Approach and Sorted and Weighed Waste Materials are used to quantify the waste. There are plenty of methods and technics involves in waste quantification [26–30].

Waste quantification can be carried out at three major point sources, which are as follows:

- Measurements at the point of generation and waste quantification by inspection of records at the point of generation (industries, hospitals, commercial, institution, etc.)
- Quantification of waste based on documents examined at a waste disposal facility
- Quantification of waste based on a sample of waste removal vehicles a

As long as the MSW is going to be disposed of in a landfill, there's no need to do anything more than calculating the tons of waste produced and consider the issues with special (hazardous) materials. If the aim is to extract gas from a landfill and put it to good use, however, the amount of organic material is critical. Besides, it is relatively crucial to have a clearer picture of the solid waste when recycling is planned, or if materials or energy recovery through combustion is the goal [31]. The following are some of the characteristics of interest:

- Composition by identifiable items (steel cans, office paper, cardboard, glass, etc.)
- Moisture content
- Particle size
- Chemical composition (carbon, hydrogen, etc.)
- Heat values
- Density
- Mechanical properties
- Biodegradability

The following is the flow in determining the composition of waste as shown in Fig. 4.10:

4.2.4 Composition by Identifiable Items

Data from published industry output statistics can be used to estimate waste composition on a national level. The input method of estimating solid waste production is the name for this method. Generally, industry output for all materials for consumption is considered as the total used for the domestic. Likewise, the estimation of

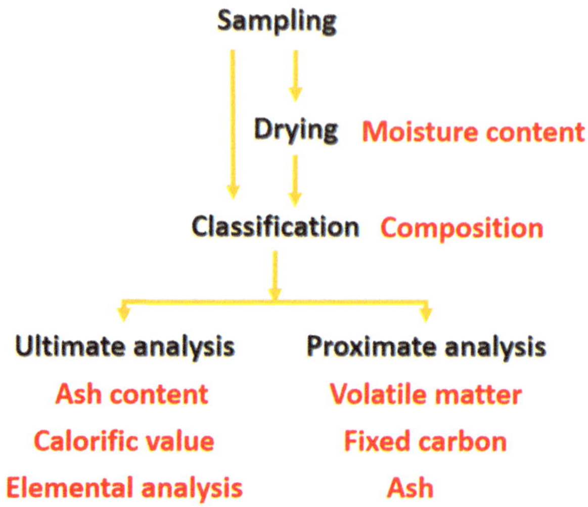

Fig. 4.10 Flow of composition and quantification of waste

selected materials to be as an indicator of waste production is very difficult to be obtained unless specific and coordinated efforts are implemented to determine the total amount of production.

Typically, waste generated is not in tandem with the amount of materials produced as it may be lost in the other waste streams [32]. However, we can simply use the rule of thumb to obtain the total waste generated by multiplying the population by the average waste generation rate. For example, 32.2 million of the population with an average waste generation rate of 2.8 kg/cap will generate around million 77 million of waste. When the input data can be collected from specialist organizations that regularly collect and publish industry-wide data, the input method of estimating solid waste generation can be used. Since collecting data is costly, this method also allows for frequent updates of waste generation estimates. Furthermore, since the data obtained by the same agencies include potential forecasts, future solid waste production may be estimated [33]. Sampling studies for characterizing waste must be structured to yield the most valuable and reliable data for the least amount of money and effort. The sample size and the method of characterizing the refuse are two important variables to consider when conducting such a report. Measuring the composition of a completely heterogeneous substance, such as mixed municipal waste, is rather a difficult task, but it is important if the different fractions are to be separated and recovered. However, some authorities argue that (because establishing the composition with fair precision requires a significant amount of effort), it is sometimes not worth the difficulty and cost and that a national average should be used instead [34]. When precise data is needed for estimating the economics of potential solid waste management alternatives, composition studies should be used. According to United Nation for Education Program, ASEAN member countries' municipal solid waste has become a major issue in recent years, as waste generation has risen dramatically as a result of rapid urbanization and industrialization, population growth, and improved lifestyles. As per se, Table 4.3 summarises the total

Table 4.3 Total waste generation

No.	Country	Annual MSW generation ('000,000)	Per capita MSW generation
1.	Indonesia	640.0	0.70
2	Thailand	267.7	1.05
3.	Vietnam	220.02	0.84
4.	Philippines	146.6	0.69
5.	Malaysia	128.4	1.17
6.	Myanmar	84.15	0.53
7.	Singapore	75.15	3.763
8.	Cambodia	1.08	0.55
9.	Brunei Darussalam	0.210	1.05
10.	Laos	0.077	0.69

MSW generation for the 2017 year. MSW is mainly generated by households, but it also includes waste from offices, hotels, shopping malls/stores, schools, and organizations, as well as municipal services such as street sweeping and parks maintenance.

The waste composition analysis is driven by the sampling plan and it is done manually. Even if the sampling process is carried out correctly, the plan will be useless unless the results are accurate. The waste must first be correctly interpreted by proper load selection to avoid biassing the final analysis. The truckload that will be analysed must reflect (as closely as possible) the community's average refuse production.

Following the selection of the load, a technique for generating a sample small enough to be analysed but large enough to be statistically representative of the MSW must be created. The American Society for Testing and Materials (ASTM) Standard Test for Determination of the Composition of Unprocessed Municipal Solid Waste is the most commonly used technique for calculating the number of samples necessary to achieve statistical validity (ASTM designation D 5231-92). This approach includes a statistically dependent method for calculating the number of samples needed to classify the waste, as well as a script to follow while performing a waste composition analysis. The number of samples needed to achieve the desired level of measurement precision depends on the component(s) in question and the desired level of confidence. The calculations are an interactive operation, starting with a suggested sample mean and standard deviation for waste components. For most tests, a level of trust of 90% is appropriate. Sorting and evaluating more than 200 lb (90 kg) in each study, as a rough first guess, would provide little statistical benefit. The question is how many of the 200-pound samples are needed for the testers to be confident in the results statistically. Quartering and coning are recommended by ASTM. After thoroughly mixing the contents with a front-end loader, a truckload of waste is quartered into successive quarters. The samples are then coned and quartered once more until they weigh about 200 pounds (90 kg). The number of 200 lb samples analysed would increase as the target precision increases. Figure 4.11 shows the sampling of MSW (90kg), while Fig. 4.12 illustrates the quartering and coning method.

The steps of quartering and coning are shown in Fig. 4.13, and the explaining of the steps are described in Table 4.4.

Fig. 4.11 Collecting of MSW sample

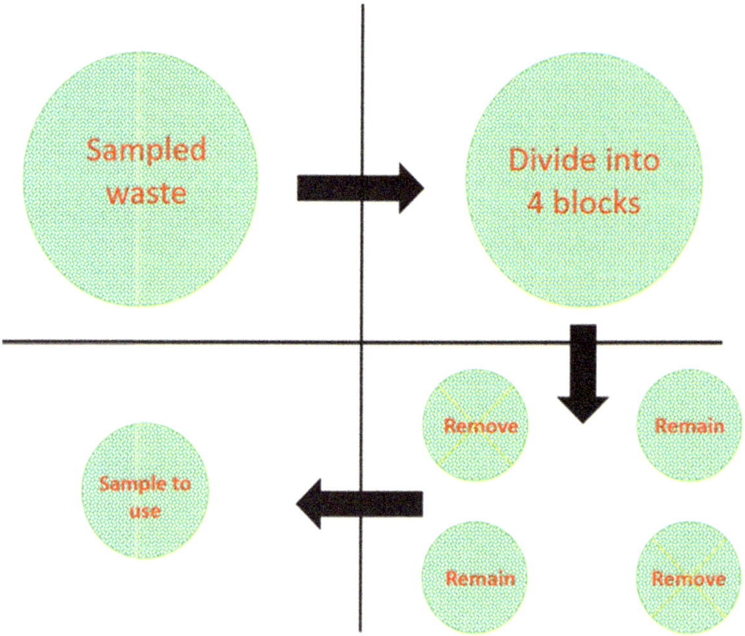

Fig. 4.12 Quartering and coning method

Fig. 4.13 Stepwise of waste composition study based on ASTM

Table 4.4 Quartering and coning method

Step	Description
1	Prepare 200 kg of waste from waste pit/collection vehicle
2	Mix and flatten the waste
3	Divide into four blocks
4	Remove two blocks of waste diagonally opposite ends
5	Another diagonal opposite ends waste are remained
6	Mix and flatten the waste
7	Divide into four blocks
8	Tear the content of waste from waste plastics
9	Pick up recyclable items and place
10	Measure the weight to check the density

Equation 4.3 can be used to quantify the density of the waste. Weighing inert waste in a landfill shown in Fig. 4.14:

$$\text{Density}\left(\text{kg}/\text{m}^3\right) = \frac{\text{Mass}\left(\text{kg}\right)}{\text{Volume}\,\text{m}^3} \tag{4.3}$$

If the waste analysis is being used to design a waste incineration system, the waste would be dried to make the next steps easier. However, in the event of improved collection/final disposal, a wet base waste portion analysis will be conducted (without drying). Because of the composition of refuse, even careful sampling studies produce imprecise results. Some objects are difficult to categorize into the desired components. A tin can with an aluminium top and a paper wrapper, for example, has four parts: steel, tin, aluminium, and paper. Inaccuracies are inserted into the final values regardless of the item's final classification [3]. Moisture, food,

Fig. 4.14 Weighing inert waste (organic)

and dirt are all common pollutants in waste. Despite the fact that these products are common in the waste stream, they are problematic because they greatly increase the weight of paper, plastic film, yard waste, and containers.

Contamination increases as waste are compressed in collection vehicles, allowing materials to smear or stick together and pushing moisture from food and other wet wastes onto other absorbent products. Contamination can also happen during sampling and storage as a result of mixing and/or bad weather.

The samples should be taken to a laboratory after sorting to be measured, cleaned of contaminants, and air-dried. Durable products (such as glass and plastic containers) can be washed before being air dried, and filled containers can be drained. Each category should be adequately calibrated for the weight of contamination if the contaminant category can be detected (e.g. food). Waste composition studies are critical methods in the management of urban solid waste. However, the data given is often unreliable and imprecise due to a lack of consistent procedure and underfunding of studies. Too frequently, not enough samples are collected, sampling activities are not indicative of seasonal and economic changes, pollution is not taken into account, and the research is not replicated in response to community changes. If the numbers collected are to be used for design purposes, a mediocre study could be worse than none at all.

4.2.5 Moisture Content

The amount of water found in a substance, such as soil (also known as soil moisture), rock, ceramics, fruit, or wood, is referred to as moisture content. Water content is a ratio that can range from zero (completely dry) to the value of the materials' porosity at saturation. It is used in a wide range of scientific and technological fields. It can be given in terms of volume or mass (gravimetric). Biodegradable waste fractions decompose more quickly in wet environments than in dry conditions. It also makes the waste unsuitable for thermochemical conversion (incineration, pyrolysis/gasification) for energy recovery because heat is needed to extract moisture first.

This property is particularly important because it has an effect on the waste's unique energy and ignition characteristics. Paper products, plastics, organics, textiles, and other residue are the most vulnerable to changes in moisture content. Since the moisture content of glass, leather, rubber, wood, and some metals varies very little, previous research can be used to estimate moisture. The moisture content of household solid waste may be influenced by the type of collection process, the time between collections, changes in waste composition, and weather conditions, and these effects should be evaluated. The moisture content must be in the range of 50%–60% in the initial phase of arrival to the landfill. Moisture is transferred between the garbage can and the truck, and the moisture content of different materials varies with time [35]. When it is deposited into the receptacle, the newsprint contains around 7% moisture by weight, but the total moisture content of the newsprint from a refuse truck also reaches 20%. The oven-dry method is used in order to determine the moisture content of compost. All sample is dried in an oven (Fig. 4.15) around 105°C for a day. The sample is placed in a crucible and weight by analytical balance as initial weight, W_1. Then, the sample is dried in an oven at 100°C for a day. After that, the final dry weight of the sample is taken and named W_2. The percentage of moisture content in the sample is determined by Eq. 4.4:

Fig. 4.15 Oven and dried samples

$$\text{Moisture content}\% = \frac{W_1 - W_2}{W_1 - W_o} \times 100 \quad (4.4)$$

where

W_0 = weight of tray (initial)
W_1 = Initial weight of sample with tray
W_2 = Final weight of tray after taken out from the oven

Drying is typically performed in an oven at 77°C (170°F) for 24 hours to ensure full dehydration while avoiding excessive vaporization of volatile content. The moisture content of different types of waste varies greatly. Table 4.5 lists out several items' moisture content [36].

4.2.6 Particle Size

Analytically, any combination of particles of different sizes is difficult to explain. The problem is exacerbated if the particles are irregularly formed. Municipal solid waste (MSW) is probably the worst material for particle size analysis, but most MSW

Table 4.5 Moisture content of several items [31]

	Moisture content	
Component	Range	Typical
Residential		
Aluminium cans	2–4	3
Cardboard	4–8	5
Fines (dirt, etc)	6–12	8
Food waste	50–80	70
Glass	1–4	2
Grass	40–80	60
Leather	8–12	10
Leaves	20–40	30
Paper	4–10	6
Plastics	1–4	2
Rubber	1–4	2
Steel cans	2–4	3
Textiles	6–15	10
Wood	15–40	20
Yard waste	30–80	60
Commercial		
Food waste	50–80	70
Mixed commercial	10–25	15
Wood crates and pellets	10–30	20
Construction (mixed)	2–15	8

processing technology relies on accurate particle size descriptions. A mixture of particles cannot be accurately defined by a single value [31]. The size and distribution of waste components are critical for material recovery, particularly when mechanical methods such as trommel screens and magnetic separators are used. The best attempt in that direction is to characterise the mixture using a curve that plots percent of particles (by number or weight) versus particle size. Particle size is possibly one of the most significant to evaluate among these properties, as it has a direct effect on the efficiency of many processes, including sorting and size reduction operations [37].

Caputo and Pelagagge [38], for example, compared the mass recovery and the lower heating value (LHV) of various sequences of a refuse-derived fuel (RDF) production facility. They demonstrated that adding a trommel and a shredder to a simple series increases the LHV of the resulting RDF while decreasing mass recovery. In MSW processing, such as in MRF plants, these types of equipment, which are mostly dependent on the particulate properties of the waste materials, are extremely important. However, MSW has a lot of form and size variability, which makes describing particle size difficult [39, 40]. Furthermore, the wide range of ductility of MSW particles has been described as a challenge in predicting flow properties [40]. These concerns highlight the need for further research into MSW particle size properties.

The sample size chosen was a compromise between sample representativeness and the time taken to sort the sample manually. These samples, known as M-samples, were manually sieved to distinguish the particle size categories of >100 mm, 40–100 mm, 20–40 mm, and 20 mm. To make comparisons between manual and mechanical operations, sieves with mesh sizes of 100 and 40 mm were chosen to reflect the openings of the drum screen of the full-scale mechanical equipment. The finest fraction was separated using a 20 mm sieve. With the exception of the 20 mm content, each particle size category was manually divided into seven fractions: metals, plastics, paper and cardboard, textiles, soil, wood, and others. Figure 4.16 demonstrates the process of screening inert materials as the final stage after manual sorting. The following is the steps that can be used to calculate the size of waste components:

- $Sc = L$
- $Sc = (L + w)/2$
- $Sc = (L + w + h)/3$
- *Sc*: size of component, mm
- *L*: length, mm
- *W*: width, mm
- *h*: height, mm

4.2.7 Ash Content

The ash content of MSW may be used to determine if it needs to be classified for better use as a fuel or other use [36]. However, some MSW samples are obtained from real waste in the thermochemical study, while others are extracted from clean

Fig. 4.16 Typical screening process for inert waste

or pure materials. These two collection methods have a large difference in moisture and ash content. The moisture content of MSW in Asian countries is usually much higher than that of MSW in European and American countries, which ranges from 10% to 30% [41].

It may be attributed to climatic and lifestyle variations [42]. As a result, torrefaction of MSW from Asian countries is critical before incineration, pyrolysis, or gasification. MSW had an average ash content of 43.57%, which came from glass, ash, dust, ceramics, and tiles. The ash content of MSW, on the other hand, varied greatly, from 20.56% to 76.76%, owing to local economics and heating systems.

The chemical composition of biodegradable compounds will also change with time. Typical measurement of ash is as follow:

- The sample is weighed into a clean and dry tarred porcelain crucible in a range of 1.5 to 2.5 g. Then, the sample is placed in the muffle furnace at 760 for 1.5 hours.
- After that, the sample is cooled in the desiccator and the sample re-weighed. The ash content of a given sample is calculated as follow:

$$\text{Ash content}\% = \frac{C-A}{B-A} \times 100 \tag{4.5}$$

where

A = Initial weight of the crucible
B = Initial weight of crucible with sample
C = Final weight of crucible with sample after heating

4.2.8 Calorific Value

The calorific value of calorific fractions was determined as part of the sample characterization [42]. A calorimeter, a system in which a sample is combusted and the temperature rise is registered, is commonly used to calculate the heat values of refuse and other heterogeneous materials [42]. In resource recovery, the heat values of refuse are significant. Table 4.6 shows some published values for a variety of fuels to demonstrate how the fuels vary depending on how they are derived [43].

Organic materials, inorganic materials, and water are all components of refuse. The calorific value is usually expressed in terms of all three components (Btu/lb), with the inorganics and water included in the sample weight. However, the heat value is often represented as moisture-free, with the water portion removed from the denominator. A third method of determining calorific value is to remove the inorganics, resulting in a Btu that is moisture and ash free, with the ash defining the inorganic during combustion [43]. Figure 4.17 shows a typical bomb calorific metre used to measure the calorific value of a given sample.

Separate recycling of municipal solid waste (MSW) allows for the reuse of usable materials and a reduction in MSW management's environmental effects. Waste-to-energy (WTE) plants may be used to dispose of residual waste (MSW not intercepted by separate collection). WTE plants generate energy in the form of electricity and/or heat.

The heat value of refuse where the fraction of components is known can be calculated using those estimated heat values for the different components, just as it can

Table 4.6 Typical calorific value for MSW [44]

	Calorific value		
Source	As collected	Moisture-free	Ash-moisture free
Cardboard	7040	7400	7840
Food waste	1800	6000	7180
Magazines	5250	5480	7160
Newspapers	7980	8480	8610
Paper (mixed)	6800	7570	8050
Plastics (mixed)	14,100	14,390	16,020
HDPE	18,700	18,700	18,900
PS	16,400	16,400	16,400
PVC	9750	9770	9980
Steel cans	0	0	0
Yard waste	2600	6500	6580

Fig. 4.17 Bomb calorimeter

be done with moisture content [31]. The distinction between higher and lower heat values is a significant feature of calorimetric heat values. The gross calorific energy is also known as the higher heat value (HHV), whereas the lower heat value (LHV) is also known as the net calorific energy. In the design of combustion units, this distinction is critical.

4.2.9 Elemental Analysis

The percentages of C (carbon), H (hydrogen), O (oxygen), N (nitrogen), S (sulphur), and ash are calculated. The proximate analysis aims to determine the percentage of volatile organics and fixed carbon in the fuel, while the ultimate analysis is focused on elemental compositions. An elemental analyser can be used to calculate the amount of carbon, hydrogen, nitrogen, sulphur, and oxygen in organic, inorganic, and polymeric materials by weight percent [45]. The sample size for the determination of CHNS shall be less than 180 μm. Figure 4.18 shows the typical setup of an elemental analyser for CHNS measurement.

If such equipment is not available, the chemical formula for solid waste can be determined as shown in the following section. Solid waste is made up of a variety of different materials, each of which has its own chemical structure and chemical formula. However, estimating oxygen requirements and other possible emissions during natural degradation or waste treatment would be easier with an estimated formula. The following example demonstrates how to derive a chemical formula:

Chemical formula for solid waste: $C_{98.26}\,H_{1.57}\,O_{173.96}\,N_{1.05}S$

Fig. 4.18 An elemental analyser

Components	Wet mass (kg)	Dry mass (kg)	Moisture (kg)	Composition (kg)					
				C	H	O	N	S	Ash
Food	16	5	11	2.40	0.32	1.88	0.13	0.02	0.26
Paper	46	43	3	18.70	2.58	18.92	0.13	0.08	2.58
Cardboard	11	10	1	4.40	0.59	4.46	0.03	0.02	0.51
Plastic	11	10	1	6.00	0.72	2.38	0	0	0
Total	84	68	16	31.5	4.21	27.64	0.29	0.12	4.35

Step one: Derive ultimate analysis and moisture content of individual solid waste components

Step two: Convert moisture content into Hydrogen and Oxygen element

- Hydrogen: (2/18)*16 = 1.78 kg
- Oxygen: (16/18)*16 = 14.22 kg

Step three: Revise composition of an element in kg

Element	C	H	O	N	S	Ash
Mass (kg)	31.50	5.99	41.86	0.29	0.12	4.35

Step four: Compute molar composition of the waste

Step five: Compute normalised mole ratio

The chemical formula for solid waste: $C_{98.26}\,H_{1.57}\,O_{173.96}\,N_{1.05}S$

Element	C	H	O	N	S
Mass (kg)	31.50	5.99	41.86	0.29	0.12

Element	C	H	O	N	S
Kg/mol	12.01	1.01	16.00	14.01	32.06
Moles	378.32	6.05	669.76	4.06	3.85
Moles ratio	98.26	1.57	173.96	1.05	1.00

4.2.10 Proximate Analysis

The characterization of municipal solid waste is an essential analysis needed to design effective solid waste management and to implement waste-to-energy systems. The proximate analysis consists of the determination of ash content, moisture content, volatile matter, and fixed carbon [46–48]. This analysis is carried out, which measures, in percentage, the potential as an energy source of solid waste.

A study carried out by assessed the proximate analysis of wastes from Saggian landfill site, Lahore, Pakistan. Relatively high moisture content was measured in the food waste of all towns ranging from 57.3% to 66.2%. The paper and textile waste also contained high moisture contents ranging from 26.1% to 58.8% and 60.3% to 78.5%, respectively. The levels of VOCs were found higher in paper waste (30.4–54.7%) than food (20.2–27%) and textile (20.2–38.5%) waste of all three towns. The average value of fixed carbon in food, paper, and textile wastes of all towns was 2.1%, 5.8%, and 0.9%, respectively. The highest ash contents were measured in food waste (11.8–12.8%).

Firstly, the moisture content (MC) of the targeted sample. The percent moisture of the MSW samples is determined by weighing 100 g of the sample into a pre-weighed dish and drying the samples in an oven at 105°C to a constant weight. The percent of moisture content can be calculated as a percentage loss in weight before and after drying using Eq. 4.6 [49]:

$$\%\text{Moisture content} = \left[\frac{(\text{Wet weight} - \text{Dry weight})}{\text{Wet weight}}\right] \text{X}100\% \tag{4.6}$$

It is noted that for each sample, triplicate samples were carried out to ensure the replication and reproducibility of the result is obtained.

Next, analysis of the volatile matter content can be determined by the method of ignition of the sample at 550°C. The samples are then weighed, then dried using a muffle furnace for 24 hours at 550°C. After combustion, the samples were weighed to determine the ash dry weight, with volatile solids being the difference between the dried solids and the ash as shown in Eq. 4.7 [50]:

$$\%\text{VS} = \left[\frac{(\text{Dry sample weight} - \text{Ash weight})}{\text{Dry sample weight}}\right] \text{X}100\% \tag{4.7}$$

The ash content of the samples waste is determined by heating the samples in an oven at 750°C [51]. The residue left after combustion represents the ash content. Whereas fixed carbon is determined by the following Eq. 4.8:

$$\%FC = 100 - \text{weight}\left(\%\text{moisture content} + \%\text{ash} + \%\text{volatile matter}\right) \quad (4.8)$$

4.2.11 Heavy Metals in Waste

Heavy metals are commonly and naturally found in the environment, such as soil and food. They are broadly used in manufacturing processes and consequently are transferred to and become ever-present in composted organic residuals [52]. Chemicals have become a significant aspect of human existence and have become impossible to live without. In addition, these chemicals and their products have their benefits, but at the same time, exposure to them during production, usage and uncontrolled discharge into the environment has caused lots of hazards and harm to humans, animals, and the environment.

Heavy metals (Table 4.7) are any metallic chemical element characterized by their relatively high density and potential to be toxic at low concentrations. These include mercury (Hg), cadmium (Cd), arsenic (As), chromium (Cr), thallium (Tl), and lead (Pb). While heavy metals such as copper (Cu), selenium (Se), and zinc (Zn), are referred to as trace elements, which are essential contributors to various metabolic pathways within the human body, making small amounts of these chemical elements essential for survival.

Heavy metals possess a variety of physical and chemical properties which have been utilized in consumer, automotive, and medical supplies and application throughout history. They can be divided into two categories: natural and man-made heavy metal sources. Natural pollution occurs as a result of metal processing, industrial effluents, and solid waste disposal are examples of man-made sources; for example, Pb and Ni come from coal and gasoline, while Mn, Cd, and Zn come from discarded batteries.

When disposed of in landfills, municipal solid waste (MSW) is made up of paper, plastics, textiles, food wastes, yard wastes, and other organic materials, with the addition of inorganic materials. The component of MSW contain heavy metals and are declared toxic when achieving certain concentrations. This is especially true when batteries, paints, dyes and inks in paper, pesticides, and fertilizers in yard waste are present, which are some examples containing high quantities of heavy metals [53].

Leachate is generated by percolating water through the solid waste matrix and is the by-product of landfilling. This is added with precipitation on-site, which becomes runoff or infiltrates into the site itself, contributing to the generated leachate. This infiltrating water ultimately will form leachate, which through leaching, picks up soluble and suspended contaminants before discharging into a collection

Table 4.7 Sources of toxicological effects of some heavy metals

Heavy metals	Sources	Effects
Copper	Water pipes, copper water heaters, frozen greens and canned greens using copper to produce an ultra-green colour, alcoholic beverages from copper brewery equipment Instant gas hot water heaters, hormone pills, pesticides, insecticides, fungicides, copper jewellery, copper cooking pots	Mental disorders, anaemia; arthritis/rheumatoid arthritis; hypertension, nausea/vomiting, hyperactivity, schizophrenia, insomnia Autism, stuttering, postpartum psychosis, inflammation and enlargement of the liver, heart problem, cystic fibrosis
Chromium	Steel and textile industry	Skin rashes, respiratory problems, haemolysis, acute renal failure, weakened immune systems, kidney and liver damage, alteration of genetic material, lung cancer, pulmonary fibrosis
Nickel	Effluents of silver refineries, electroplating, zinc base casting and storage battery industries	Dermatitis, myocarditis, encephalopathy, pulmonary fibrosis, cancer of lungs, nose and bone, headache, dizziness, nausea and vomiting, chest pain, rapid respiration
Lead	Industries such as mining, steel, automobile, batteries and paints. Pollutants arising from increasing industrialization	Nausea, encephalopathy, headache and vomiting, learning difficulties, mental retardation, hyperactivity, vertigo, kidney damage, birth defects, muscle weakness, anorexia, cirrhosis of the liver, thyroid dysfunction, insomnia, fatigue, degeneration of motor neurons, schizophrenic-like behaviour
Mercury	Industries like chloro-alkali, paints, pulp and paper, oil refining, rubber processing and fertilizer, batteries, dental fillings adhesives, fabric softeners, drugs, thermometers, fluorescent light tubes and high intensity streetlamps, pesticides, cosmetics and pharmaceuticals	Tremors, birth defects, kidney damage, nausea, loss of hearing or vision, gingivitis, chromosome damage, mental retardation, tooth loss, seizures, cerebral palsy, blindness and deafness, hypertonia – muscle rigidity, Minamata disease

pond. As discussed in the previous section, heavy metals are present in the solid wastes disposed of at sites, which will be passed on as water percolates through the refuse.

There are many adverse effects of heavy metals on human life and the environment as a whole if not treated or managed sustainably. This could occur if leachate contaminates surface water or groundwater due to treatment failure. When water supplies are contaminated with leachate containing heavy metals, the mechanism leading to health hazards is bioaccumulation. Many living organisms, including humans, are known to accumulate specific toxicants (chemical pesticides, industrial organics, heavy metals) from the environment. This capability is widespread, and the amount accumulated may range from barely detectable concentrations to concentrations that greatly exceed the amount present in the environment, depending on the contaminant and organism involved. The adverse effects of heavy metals on human and animal health include birth defects, kidney damage, nausea, loss of hearing or vision, mental disorders, dermatitis, and many forms of cancer [54].

4.2.11.1 Measurement of Heavy Metal

Generally, heavy metals are quantified by traditional methods of analysis such as atomic absorption spectroscopy (AAS), inductively coupled plasma optical emission spectrometry (ICP-OES), and inductively coupled plasma mass spectrometry (ICP-MS) due to the high sensitivities of these instruments.

Atomic absorption spectrometry (AAS) is useful due to its quick analysis of the unknown materials and their composition. Overall, the available techniques of analysing atomic spectrometry are able to detect atoms or ions of analyte elements only in gaseous phases. Thus, the drawback of this method is that it requires the transformation from its condensed state into the atomic vapour state of the sample prior to analysis. There are two types of atomizers that are commonly applied: high temperate flames and electrothermal atomiser.

To analyse the contents or traces of heavy metals present in the leachate sample, FAAS is the quickest and most convenient method [55]. Firstly, the pressures of acetylene, N_2O, and air were adjusted. The fan, spectrometer, computer, and AAS programme for the AA initialization part was started. The AA instrument was turned on, with the lamp controlled and background correction measured, if any. The flame control window was opened, and the oxidant was chosen before adjusting the flow rate with flame ignition. The air/acetylene, N_2O/acetylene, and burner position were optimized. This is the most critical and hazardous action during FAAS use. The method of analysis was chosen between either manual or automated analysis. The calibration curve was checked with a standard solution for each desired metal. Once the calibration curve of the desired metal was achieved, analysis of the sample was started. When the analysis is completed, the flame, air lines, acetylene cylinders, spectrometer, fan, and computer were turned off. The results of the analysis were printed in chronological orders for all iterations. SW-846 is an EPA publication titled Test Methods for Evaluating Solid Waste: Physical/Chemical Methods. Inductively Coupled Plasma-Optical Emission Spectrometry (SW-846 Test Method 6010D) is a test method for SW-846 (ICP-OES).

The spectrometric technique of inductively coupled plasma – optical emission spectrometry (ICP-OES) is used to assess trace elements in aqueous solutions. ICP-OES involves constantly aspirating (or nebulizing) a sample solution into an inductively coupled argon plasma discharge, where analytes of interest are transformed to excited state, gas-phase atoms, or ions. As excited-state atoms or ions return to their

Table 4.8 Heavy metals analysis with respect to Chemical Abstract Service Registry Number

Element	Symbol	CASRNa	Element	Symbol	CASRNa
Aluminium	Al	7429-90-5	Mercury*	Hg	7439-97-6
Antimony	Sb	7440-36-0	Molybdenum	Mo	7439-98-7
Arsenic	As	7440-38-2	Nickel	Ni	7440-02-0
Barium	Ba	7440-39-3	Phosphorus	P	7723-14-0
Beryllium	Be	7440-41-7	Potassium	K	7440-09-7
Boron	B	7440-42-8	Selenium	Se	7782-49-2
Cadmium	Cd	7440-43-9	Silica	SiO2	7631-86-9
Calcium	Ca	7440-70-2	Silver	Ag	7440-22-4
Chromium	Cr	7440-47-3	Sodium	Na	7440-23-5
Cobalt	Co	7440-48-4	Strontium	Sr	7440-24-6
Copper	Cu	7440-50-8	Thallium	TI	7440-28-0
Iron	Fe	7439-89-6	Tin	Sn	7440-31-5
Lead	Pb	7439-92-1	Titanium	Ti	7440-32-6
Lithium	Li	7439-93-2	Vanadium	V	7440-62-2
Element	**Symbol**	**CASRNa**	**Element**	**Symbol**	**CASRNa**
Magnesium	Mg	7439-95-4	Zinc	Zn	7440-66-6
Manganese	Mn	7439-96-5			

ground state, they release energy in the form of light with wavelengths unique to each particle. The sum (concentration) of that element in the analysed sample is proportional to the strength of the energy emitted at the chosen wavelength. Thus, the elemental composition of a sample relative to a reference norm can be quantified by assessing which wavelengths it emits and their respective intensities. Direct ICP-OES analysis can only be performed on reasonably safe, aqueous matrices (e.g. pre-filtered groundwater samples) for reliable results. Acid digestion is needed for other, more complex aqueous and/or solid samples prior to analysis; the analyst should choose a sample digestion method that is suitable for each analyte and the intended use of the data (Table 4.8).

4.3 Landfill Gas

Landfill gas (LFG) are the by-products of landfilling under anaerobic conditions [55]. LFG consists of mainly methane and carbon dioxide with traces of its constituents, transferred by vaporization from solid and liquid waste. Due to the biological degradation processes occurring throughout the landfill, LFG contains mostly such as ammonia (NH_3) or hydrogen sulphide (H_2S) at low concentrations. Some trace gases cause odours; thus, to avoid these gaseous odours, landfill gases, especially when emitted from the early phases of degradation, has to be diluted the odour can no longer be detected. LFG quality may also change with time due to the different degradation phases and waste quality disposed at the landfills.

Methane and carbon dioxide makes up the majority of LFG, so methane being more stable. It is also composed of constituents that were previously part of solids and liquids, which were previously vaporized during wastewater treatment, which is expanding. Other by-products of biological degradation are present in low concentrations, such as ammonia (NH_3) or hydrogen sulphide (H_2S).

4.3.1 *Landfill Gas Collection*

Gas collection can be implemented by using an active or passive system. When operated as an active system, gas is extracted out of the landfill using blowers, which provides vacuum pressures in the landfill body in the range of −1 to −30 kPa. Passive gas collection systems use positive pressure in the landfill to transport gas out under semi-controlled conditions. Passive systems are only feasible, given methane gas production is less than 0.5 L/m^2.h and has installed a gas and airtight top cover (Table 4.9).

Furthermore, LFG contains hundreds of trace constituents. These compounds may be organically deposited in the landfill. Some of these gases are naturally generated in the landfill while some are anthropogenic. Some of the gases are acute

Table 4.9 Oxygen-containing constituents in landfill gas

Compound	Concentration range (mg/m^3)
Ethanol	16–1450
Methanol	2.2–210
1-Propanol	4.1–630
2-Propanol	1.2–73
1-Butanol	2.3–73
2-Butanol	18–626
Acetone	0.27–4.1
Butanone	0.078–38
Pentanal	0.8
Hexanal	4.04
Acetic ester	2.4–263
Butyric ester	<0.9–350
Acetic butyl ester	60
Butyric propyl ester	<0.1–100
Acetic propyl ester	<0.5–50
Acetic acid	<0.06–3.4
Butyric acid	<0.02–6.8
Furan	0.01–2.4
Methylfuran	0.06–170
Tetrahydrofuran	<0.5–8.8

Table 4.10 Phases of LFG generation

Phase	Description
Phase I–III	LFG production starts developing
Phase IV	Stable gas starts production, pores and small voids in the landfill fill up with LFG. At the surface, gas emissions can be measured
Phase V	The gas production decreases but is stable. Easy degradable waste components will be biologically reduced. The LFG emissions will be lower. The composition of the gas changes because the easily degradable fraction has almost been degraded, and more difficult organics are present
Phase VI	Due to a decrease in gas production, air can migrate into the landfill body. This process happens from the surface into deeper layers. Aerobic processes will start. The LFG emissions are reduced and can be detected in some areas but only in very low concentrations
Phase VII	As a result of aerobic processes, carbon dioxide will be generated. Residual amounts of methane will be degraded due to microbiological processes. There will be only very low LFG emissions
Phase VIII	The landfill body is almost in an aerobic condition. As some organic material is still left in the landfill, low concentrations of carbon dioxide will be produced
Phase IX	The waste in the landfill is almost inert. The gas quality in the pores will be similar to in natural soil (air and some carbon dioxide)

toxic; some are carcinogenic and genetically harmful; thus, that is why LFG is classified as a hazard. The utilization of LFG has caused damages such as clogging or corrosion in gas engines and other facilities. This is mainly due to the nature of the gases containing sulphur, chlorine, fluorine, and siloxane. Therefore, gas treatment or increased maintenance of LFG may become necessary. Therefore, it is quite common and necessary to control the cumulative concentrations of these compounds. In the early phase of gas generation, oxygen-containing gases will occur, as shown in Table 4.10.

Table 4.10 describes the LFG composition, which changes with time and is divided into nine phases [57].

It is standard practice in most landfills that each well is connected via a pipe to a distribution station, where the transportation pipe of several wells is connected to the main header pipe that transports the gas to the blower station. In the distribution station, a valve can adjust the flow rate with technical facilities for measuring the pipe pressure, flow rate, gas composition, and temperature. In the gas distribution station, it is advised to distinguish between lean gas and normal gas as it may occur in landfills, which consist of an older and a younger part. In this case, the distribution station may be connected to two header pipes; as a result, two separate gas collection systems exist. In most cases, this is not necessary as long as the gas collection system can be adjusted in a way that high methane concentrations can be reached.

As the LFG is, in general, water-saturated and cools down when it leaves the landfill, it is necessary to remove the condensate from the pipes at their deepest points. These may be located at the gas wells, the distribution stations, or the

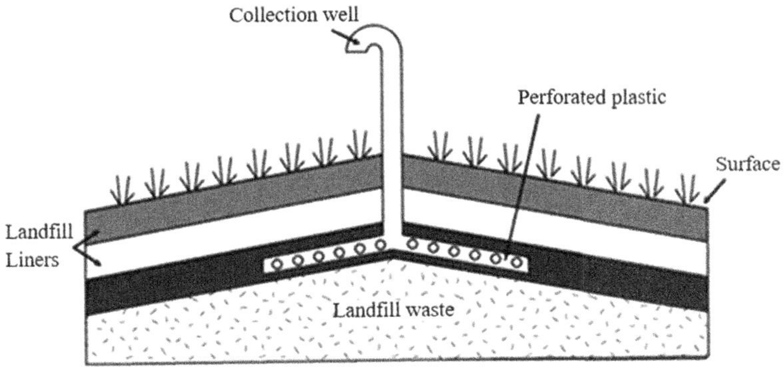

Fig. 4.19 Passive gas venting using methane oxidation processes in the top cover

utilization unit. In cases where there is more gas extracted than utilized, the gas has to be flared. Flaring may also be necessary when the gas utilization plant is experiencing a breakdown or is in maintenance. As gas extraction rates vary with time, utilization units may not use all the extracted gas. Lean gas that is not used for energy production should also be flared until low methane concentrations is being achieved. Standard flares are able to burn gas until methane concentrations reach 20–25%, with recently developed technology able to operate gas with methane concentrations of 8–10% or even lower. If the gas is not feasible through thermal treatment, it is then emitted into the atmosphere and causes odour problems, and the released methane gas becomes a contributor to global climate change. For all these reasons, flares should be standard at all landfills. Figure 4.22 shows the passive gas venting using methane oxidation processes in the top cover [153] (Fig. 4.19).

4.3.2 *Measurement of Landfill Gas*

By now, LFG extraction and utilization must be a standard procedure at all landfills due to the effects of these gases on the environment, especially air quality. Emissions these gases are influencing the global climate can be significantly reduced, energy can be utilized, odours can be avoided, and risks of explosions can be reduced. Under the Environmental Protection Agency Guideline (EPA), there is a standard method to measure landfill gas (Method 2E). Note that this approach does not provide all of the necessary requirements (e.g., equipment and supplies) and procedures (e.g., sampling and analysis). In this section, some material is integrated by analogy from other approaches. This approach is used to determine the flow rate of non-methane organic compounds (NMOC) from landfills and to assess landfill gas (LFG) output flow rate from urban solid waste landfills. Extraction wells are located in the landfill in either a cluster of three or five scattered locations. LFG is extracted from the landfill using a blower. The landfill gas production, as well as the LFG composition, landfill pressures, and orifice pressure differentials from the wells, are all measured.

One approach to detect gas emissions is by using the flame ionization detector (FID) method. The device monitors the entire area of the landfill by placing a little cap on top of the surface at different points and pumping small amounts of gas into the FID detector. Generally, the concentration range that can be measured with the detector is between 0 and 10,000 ppm of CH_4. This method allows detecting spots emitting gas. To reduce uncontrolled gas emissions, the gas extraction rates have to be adjusted, more wells may be installed, and/or the areas where the gas is emitting may be sealed. An alternative detection method is the control of diffusive emissions, which measures methane concentrations outside of the landfill and the meteorological data to calculate the emission rate using gas distribution models. The disadvantage of this method is identifying the hot spots where the main emissions occur.

Another method is by using flux chambers, a popular method utilized for several years, from which from soi l surfaces is able to measure the flux of various gases. This method traps the gas as it leaves the soil surface, either by gas build up in a closed enclosure (closed or static chamber technique) or by releasing and simultaneously measuring the gas as it leaves the enclosure (open or dynamic chamber technique) [59].

Lastly, the mass-balance method is also able to measure the LFG released from landfills [60]. This method utilises the vertical methane and carbon dioxide concentration profiles, measured along with a wind-velocity profile, using sampling points in a pole up to 10, 15, or 25 meters high. Profiles are interpreted, and emissions from the region upstream of the pole are obtained by the equation below:

$$J = \frac{\int_{z=0}^{z=l} u_z \left(c_z - c_l\right) d_z}{x}$$

In which J (m^{-2} s^{-1}) is the methane flux through the surface of the landfill; uz (m s^{-1}) the wind velocity at height z; cz (m^{-3}, vol%, or 1000 ppm) is the concentration at height z; cl (m^{-3}) is the background concentration of methane; l (m) is the length of the pole; x (m) is the fetch (the upstream length from the pole to the landfill slopes).

4.4 Landfill Leachate Measurement

4.4.1 Leachate Flow

Traditionally, landfill leachate is measured either continuously or once in a certain period [61]. The continuous method applied the method of multiplying the number of leachate pump being switched on with the volume between upper and lower switch points. Another method practices the filling of vessels with leachate to obtain the time taken. The exact quality and quantity of landfill leachate remain uncertain as various consideration needs to be addressed. Generally, the landfill is considered the ultimate disposal method, assembling waste from different sources. The complex and heterogeneous mixing of waste forming a multiplex internal geometry which is influenced by spatial and temporal changes [62].

Leachate is defined as the liquid percolation via solid waste matrix as a result of chemical and physical changes from the degradation process of solid waste refuse and soil matrix once the landfilling process is accomplished. The generation of landfill leachate from the solid refuse accelerates by rainwater percolation, biochemical, chemical, and physical reactions, thus increasing the quality and quantity of it [63]. Environmental and landfill conditions involving pH, moisture, precipitation, weather variations, age, waste type, and composition pose a significant impact on leachate quality and quantity. The tendency of having unauthorized waste such as industrial, medical, and hazardous wastes contribute to high toxicity and accumulation of heavy metal that may deteriorate the surface water bodies and groundwater aquifer. The water cycle in the landfill process creates a hydrological system that is complex and consists of various subsystems.

4.4.2 Hydrological System of the Landfill

The landfill is exposed to various incoming water from multiple sources into the landfill body. Precipitation occurring in landfill sites, especially in a country that received rains throughout the year, would increase the water input. Malaysia, as an example, receives 2420 mm considered the country situated near the equator, having a hot and humid climate throughout the year. Figure 4.20 shows the average monthly rainfall in different regions in Malaysia (Fig. 4.20).

Water stored in the landfill is not only caused by rainfall. The waste dumped at the landfill itself could act as a water source due to the nature of the waste itself that contains high moisture content. Unstrategic geographical landfill sites could lead to the overflowing of surface water or groundwater into the landfill cells. A small portion of water is contributed by the product of the anaerobic process that occurs during the decomposition of waste but is often neglected due to the small value [64]. The water cycle in the landfill requires consideration of atmospheric fluxes, precipitation, evapotranspiration, and overland flow to obtain the actual amount of leachate generated. Water that diffuses and percolates downwards until the root zone normally will form leachate.

Typically, young landfill leachate only accumulates, and only a small volume will infiltrate and forming leachate. The structure of organic material changes as a result of biological decomposition, which is necessary for water retention and storage. The degrading conditions of organic materials caused various leachate flow in the landfill, including flow in channels, flow in saturated and in unsaturated parts, thus complexing the calculation for quantification and flow of leachate. Computing the existing mathematical models to obtain the landfill leachate flow is very difficult and resulting in a different value compared to the actual value. According to [61], leachate flow in landfills should be best carried out schematically. The hydrological

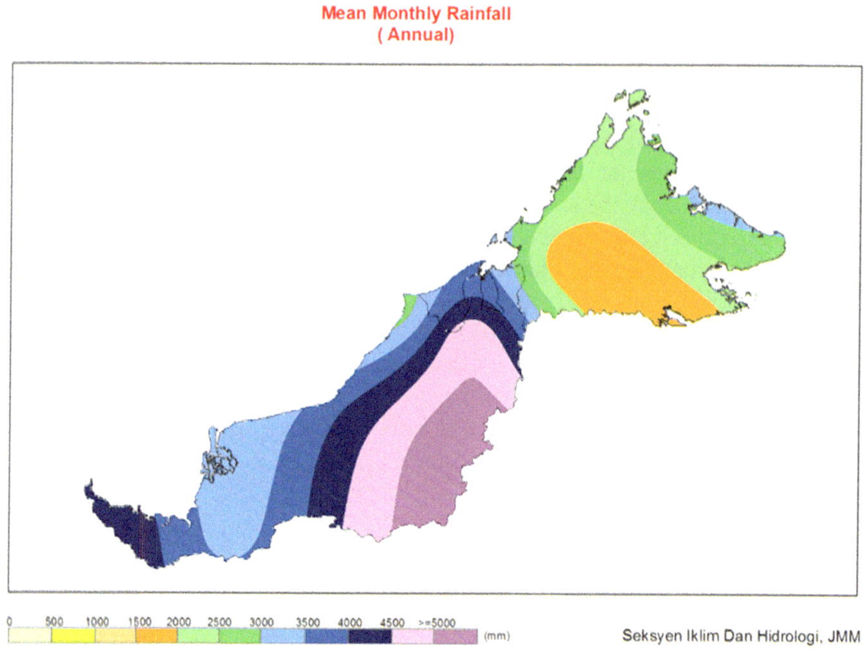

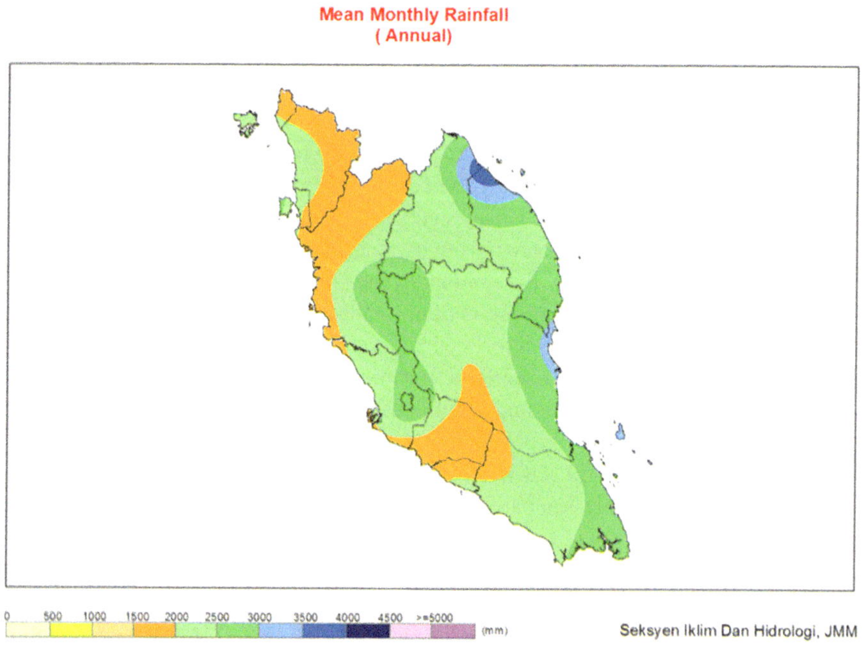

Fig. 4.20 Average monthly rainfall in different regions in Malaysia

system in the landfill could be either linear or non-linear, time-variant or time-invariant, lumped or distributed, deterministic or stochastic. A system is linear if it satisfies the properties of proportionality and superposition; it is time variant if the input–output relationship varies over time; it is lumped if spatial heterogeneity is ignored; it is deterministic if the system uniquely determines its output for a given input and initial and boundary conditions, and it is stochastic if its behaviour is regulated by laws of probability.

The continuity equation was established to determine the spatial lumped hydrologic system as in Eq. 4.9:

$$I - Q = \frac{dS}{dt} \tag{4.9}$$

I= Rate of water input
Q= *Runoff flux*
S= Storage

S can be discarded and (1) can be solved for Q if the relationship S – f (Q) is understood. Figure 4.21 shows the water cycle in a landfill.

4.4.3 Leachate Production and Water Budget

The water balance approach is used to determine when and by how much field capacities are exceeded and thus to calculate the amount of leachate generated by landfills.

The water balance in the natural environment is different compared to the landfill due to the different properties of the soil, such as size and organic matter amount. Landfill soil poses a coarse-grained structure that inhibits uniform water distribution until the degradation of organic matter is completed and improves the soil properties. The water budget of the landfill focusing on the surface of the landfill,

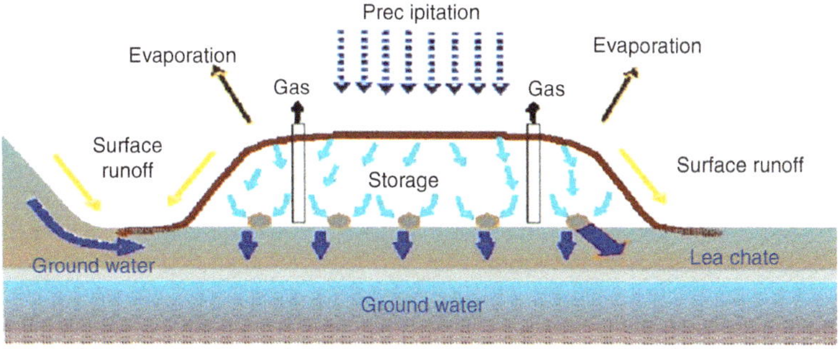

Fig. 4.21 The water cycle in the landfill

especially the soil cover and method of compaction. The soil cover provides smooth distribution of the precipitation period due to hydraulic resistance and soil cover storage capability and water depletion due to evapotranspiration that draws soil water. Absolute evaporation may be considered to be equivalent to regional evaporation for a landfill with a well-vegetated soil cover. The vegetated soil cover eventually absorbs water from the soil and maintaining the field capacity below the maximum amount. During the operating process of an uncovered landfill, the circumstances are very different. Infiltration of water for the uncovered landfill soils is about 35%–80% of precipitation. If percolation exceeds evapotranspiration over an extended period, the soil's ability to carry water (its field capacity) would be surpassed. Commonly, the initial moisture content of municipal solid waste is lower than field capacity causing only a small amount of leachate flow during compaction. In any cases of exceeded storage capacity, the amount of infiltration must be equal to the amount of leachate flow. The leachate production rate depends on the landfill operation technique involving compaction, either using a crawler tractor or steel wheel compactor. [61] provides a general leachate flow in the landfill, as stated in Table 4.11 below.

Water balance takes into consideration the surface water run-off to estimate the amount of water refuse. Different surface soils and influence of slope being expressed in coefficient surface water run-off in grass-covered soils, as stated in Table 4.12, is used in calculation of water balance [65].

Field capacity or also known as soil capacity, is vital for determining the water-holding amount in the landfill. The field potential of soil (or any other substance such as refuse) is the maximum amount of moisture that can be retained without constant downward percolation caused by gravity. Different materials carry different properties of field capacity, as stated in Table 4.13 [66]. The following Eq. 4.10 can be used to approximate the field capacity, where it varies with the overburden weight.

$$\mathrm{FC} = 0.6 - 0.55\left(\frac{W}{10{,}000 + W}\right) \tag{4.10}$$

where FC = field capacity (i.e., the fraction of water in the waste based on the dry weight of the waste)

W = overburden mass calculated at the mid-height of the waste in the lift in question, lb

A one-dimensional model of water movement through soil and compacted refuse is used to calculate leachate production. Figure 4.22 illustrated the mass balance occurring in a landfill environment, and Eq. 4.11 provides the mass balance of the

Table 4.11 General leachate flow in the landfill

Compaction	Leachate Flow
Crawler Tractor	25%–50% of annual precipitation
Steel Wheel Tractor	15%–25% of annual precipitation

Table 4.12 The coefficient of surface water run-off in grass-covered soils [65]

Surface soils	Run-off coefficient
Sandy soil, flat to 2% slope	0.05–0.10
Sandy soil, 2% to 7% slope	0.01–0.15
Sandy soil, over 7% slope	0.15–0.20
Heavy soil, flat to 2% slope	0.13–0.17
Heavy soil, 2% to 7% slope	0.18–0.22
Heavy soil, over 7% slope	0.25–0.35

Table 4.13 The field capacity of different materials [66]

Material	Field capacity, mm water/m of soil
Fine sand	120
Sandy loam	200
Silty loam	300
Clay loam	375
Clay	450
Solid waste	200–350

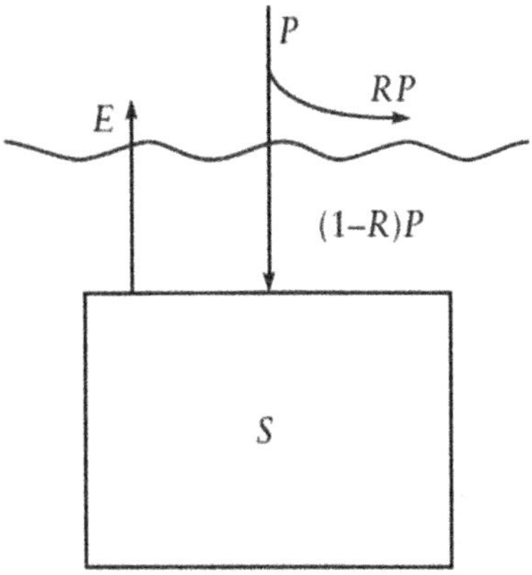

Fig. 4.22 Mass balance occurring in a landfill environment

equation. These equations are predicated on the theory that moisture travels in a frontal wetting fashion. Sometimes, the waste contains channels and cracks that allow for preferential flow. Additionally, some leachate is generated when precipitation runs over the landfill surface. As a result, leachate can be formed (and often is) much earlier than these calculations suggest.

$$C = P(1 - R) - S - E \tag{4.11}$$

Where

C = total percolation into the topsoil layer, mm/year

P = precipitation, mm/year
R = run-off coefficient
S = storage within the soil or waste, mm/year
E = evapotranspiration, mm/year

[67] describes and elaborates extensively on the water balance of each component using Eq. 4.12. Each component, either gain or loss, is described thoroughly in Table 4.14. Fig. 4.23 indicates the mass balance of a typical landfill extensively.

$$S_{SW} = W_{SW} + W_{RS} + W_{CM} + W_{A(R)} - W_{LG} - W_{WV} - W_{E} - W_{B(L)} \qquad (4.12)$$

Where,

S_{SW} = change in the amount of water stored in solid waste landfill, lb/yd^3

Table 4.14 The details and description of each component

Component	Description
Water in Solid Waste	The solid waste itself contains a high moisture content either readily available or absorbed from the atmosphere due to precipitation. However, moisture content can be lost due to the dry season. Typically, the moisture content in residential and commercial MSW range from 15 to 35 percent.
Water in cover material	Cover material is best described by field capacity.
Water from above	Water from above refers to snow that has percolated into the cover material in the top layer of the landfill. Water from above refers to water that has percolated into the solid waste above the layer in question for the layers below. When leachate is recirculated in landfills, the water from above would also contain the recirculated leachate. One of the most important aspects of preparing a landfill's water balance is determining the volume of rainfall that ultimately percolates into the landfill cover sheet. In the absence of a geomembrane, the amount of rainfall that percolates through the landfill cover can be measured using the most recent iteration of the Hydrologic Evaluation of Landfill Performance (HELP) model.
Water lost in the formation of landfill gas	Anaerobic decomposition of organic material in MSW consumes water as part of the reaction. Water consumption per cubic foot of gas generated is usually between 0.012 and 0.015 lb H_2O/ft^3 of gas.
Water lost as water vapor	Landfill gas typically is rich in water vapor. The amount of water vapor emitted from the landfill is calculated by assuming that the landfill gas is filled with water vapor. The density of water vapor present in a cubic foot of landfill gas at 90 degrees Fahrenheit is approximately 0.0022 lb H_2O/ft^3 landfill gas.
Water lost due to evaporation	Moisture would be lost to evaporation when the waste is landfilled. The amounts are insignificant and sometimes neglected. Whether or not these factors are used in the water balance study would be driven by local conditions surrounding the landfill.
Water leaving from below	Leachate is the term used to describe the water that drains from the bottom of the landfill's first cell. As previously mentioned, water exiting the bottom of the second and subsequent cells correlates to water entering from above the cell below.

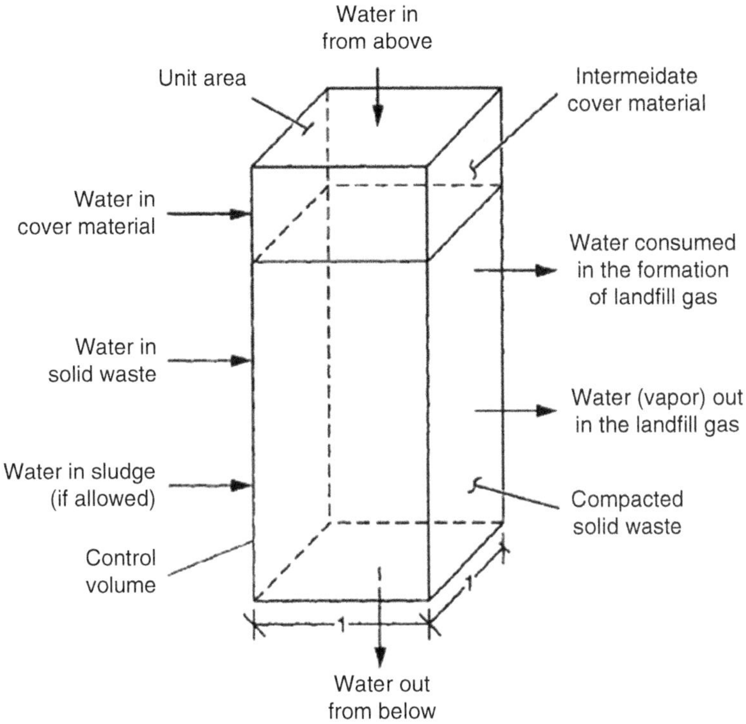

Fig. 4.23 Detailed mass balance occurring in a landfill environment

W_{SW} = water (moisture) in incoming solid waste, lb/yd^3
W_{RS} = water (moisture) in incoming treatment plant sludge, lb/yd^3
W_{CM} = water (moisture) in cover material, lb/yd^3
$W_{A(R)}$ = water from above (for upper landfill layer water from above corresponding to rainfall or water from snowfall), lb/yd^2
W_{LG} = water lost in the formation of landfill gas, lb/yd^3
W_{WV} = water lost as saturated water vapour with landfill gas, lb/yd^3
$W_{E=}$ water lost due to surface evaporation, lb/yd^2
$W_{B(L)}$ = water leaving from bottom of element (for the cell placed directly above a leachate collection system; water from bottom corresponds to leachate), lb/yd^3

4.4.4 *Leachate Flow and Transport Process*

Hydrological and landfill characteristics are vital for modelling purposes. Channel flow, internal geometry, spatial and temporal variability are the list of factors required in modelling. The internal geometry of landfills is strongly stratified and exhibits anisotropy in the vertical plane as a result of the compaction process. Field findings indicate that the landfilled waste medium is discretely hierarchical and consisting of porous lens-shaped components with a partly impermeable base. These are compacted refuse bags or sacks represented by these lenses, and some are probably partially torn due to the compaction process. The structural spaces between the lenses form a network of streamlines or channels into which water is thought to flow as a thin viscous film, propelled by gravity along the channel's solid surfaces. Capillary potentials in landfilled waste are low, particularly in fresh waste due to its course structure. As a result, the capillary forces were thought to have a static effect in the channels so that their effect can be accounted for as a loss term in the channel flow equations.

Gaps in the underlying layer, such as seams and hinges, or defects or cracks in a lining significantly increase leachate flux. To assess the condition, it is critical to collect hydrogeological details and knowledge about the structure and status of potentially formed lining structures both underneath and downstream of landfills.

The flow pattern of landfill leachate follows Darcy's Law as stated in Eq. 4.13 below.

$$q = ki * A. \tag{4.13}$$

k = hydraulic conductivity at places where the flow has been measured
i = hydraulic tilt at the same spot
q = flow at the desired place
A = cross-sectional surface via which flow is happening at the desired place

4.4.5 Quantification of Landfill Leachate Using the Mathematical Model

Evaluation of the quality and quantity of landfill leachate is critical for determining its environmental effects, as well as for treatment and maintenance purposes. The initial attempts to calculate the leachate production utilizing a simplified version of the water balance method (WBM) that was previously developed by the US Environment Protection Agency [68]. Several attempts had been carried out from previous decades using various mathematical models and equations for forecasting the generated leachate amount by varying in prediction and accuracy. The commonly used model is the Hydrologic Evaluation of Landfill Performance (HELP). HELP performs landfill's hydrological performance forecasting in the long term [69, 70] [71]. HELP model was used alone or paired with conventional WBM for comparison and verification purposes as carried by [72]. According to [73], their work suggested two patterns, Deterministic Multiple Linear Reservoir Model (DMLRM) and Stochastic Multiple Linear Reservoir Model (SMLRM), to better imitate creation at active landfills. Numerous computer programs have been created to estimate leachate generation, including Hydrology Evaluation Leachate Performance (HELP), FILL, and SOILINER. [74] in his study introduced three models for understanding the leachate flow over and through the liners of solid-waste landfills for steady-state, quasi-steady state, and transient state. These trends were all developed using the US Environmental Protection Agency's Water Balance Method (WBM). Some programs use finite margin tools to implement 1D scalar solutions. Each software, regardless of the ultimate constraints, has critical benefits. These limits are as follows:

1. The cells are not fully used at the same time, and sometimes cells are not closed with daily soil cover.
2. The contact between cells that is imposed by the structure of adjacent cells to form strips and/or the structure of other cells on top to form layers is excluded.
3. The assumption for space and time discharge for leachate during the operation and after cell closure cannot be made.
4. The calculation occurs on a single level, disregarding changes caused by matters on top or by solid waste as the depth or height of the landfill grows.

HELP pattern is a quasi-2D hydrological model for water equilibrium analysis and cycle in the landfill and other solid waste pollutant equipment [75]. This pattern allows for the incorporation of temperature, soil, and plan data and employs solution methods that account for the effects of surface storage, snowmelt, drainage, percolation, evapotranspiration, plant formation, soil humidity storage, lateral underlying sewerage, leachate recirculation, un-soaked vertical sewerage, and penetration to the soil, geo-film, or composite liners. Combinations of various landfill components such as vegetation, soil cover, cells, lateral drain layers, synthetic geomembrane liners, and barrier soils can be simulated. This pattern enables rapid estimation of the amount of drainage, evapotranspiration, sewerage, leachate accumulation, and liner infiltration that may occur during the execution of various types of landfill plans. HELP utilizes various empirical and numerical equations such as conservation service (CS) method and curve number (CN) method for run-off estimation soil, modified Penman method for evapotranspiration.

Several studies have indicated that the result obtained from HELP is close to WBM but higher than the actual amount [71]. However, some indication can be predicted for any different result between HELP, WBM, and the actual amount of leachate generated. The lack of a traditional method to quantify exact leachate volume by using human observation and mathematical estimation contributes to a different value of leachate amount. A portion of the produced leachate was accumulated and percolated into the subsurface soil layers through the lining system. As a result, the observed leachate levels at the site were lower than those predicted by the model.

SOILINER was developed to simulate the dynamics of a percolation case involving the incongruity of compressed soil liner systems and the relationship between liner features and the degree of saturation. Percolation of various soil/clay liners can be exactly determined. According to [76], SOILINER poses the potential to determine multi-layered mechanisms, changing initial humidity and changing the situation on the margins of the compressed soil liner flow area that are vital for the planning of liner forms, especially depth and liner conductivity.

Flow Investigation for Landfill Leachate (FILL) is another method used to determine leachate flow [74]. FILL is a 2D unsteady state moisture flow model. The FILL model routed moisture first into the soil cover and then through the compacted solid waste underneath. Additionally, the formulation and solution methods were used to calculate the leachate mound head in a landfill's saturated region [77]. A model was also used to analyse the flow of contaminants from the landfill using the moisture routing technique as stated by [78].

4.4.6 Leachate Plume

Leaching of organic and inorganic pollutants could deteriorate the surface and underground water system seriously for the long term, thus exposing it to high risk and for health and environment [79]. Landfill leachate is considered a point source contamination and potentially generates contaminated leachate plumes [80, 81]. The landfill waste materials decay and dissolve over time, and as water percolates into landfill leachate, inorganic and organic constituents are formed. Mostly, older landfills lack a leachate disposal system or liners under the landfill, resulting in leachate polluting groundwater downstream from the landfill. The interaction between surrounding aquifers and the contaminant plume from the point source requires a better understanding by carried out field investigations of the contaminant plume. The geological and hydrogeological assessment was previously used by gaining the information for boreholes and data on the chemical properties of water and soil. According to [82], the data-driven method is considered incomplete as limited spatial information is achieved and could lead to incomplete site assessment.

Geophysical methods such as electrical resistivity tomography (ERT) have been utilized widely to study the landfill and its contaminants. The geophysical method is considered non-invasive and minimizes the spatial information gap as this method has broad lateral coverage as well as high-resolution data. The ERT assesses the migration of contaminants by observing the groundwater ion concentration changes which being expressed in 2D and large-scale 3D detailed images. The 3D layout incorporates a vast number of electrodes, allowing for completely undiscovered spatial coverage while retaining high resolution. The presence of inorganic materials will increase the electrical conductivity (EC) of the polluted groundwater, resulting in a resistivity contrast between the contaminated region and the host aquifer that can be detected by surface resistivity assessments. Multichannel calculations have made the ERT process more stable, simpler, and easier to perform in the field [83]. 2D inversion codes able to provides high-resolution subsurface resistivity images. However, for the assessment of landfill leachate plume, the 3D resistivity method needed to obtain a larger set of data coverage, data processing, and includes data inversion. Broad areal coverage requires more electrodes. The 3D modelling and inversion are best described by finite different and finite element methods that utilize algorithm via optimization of Gauss–Newton technique [84–86]. Figure 4.24

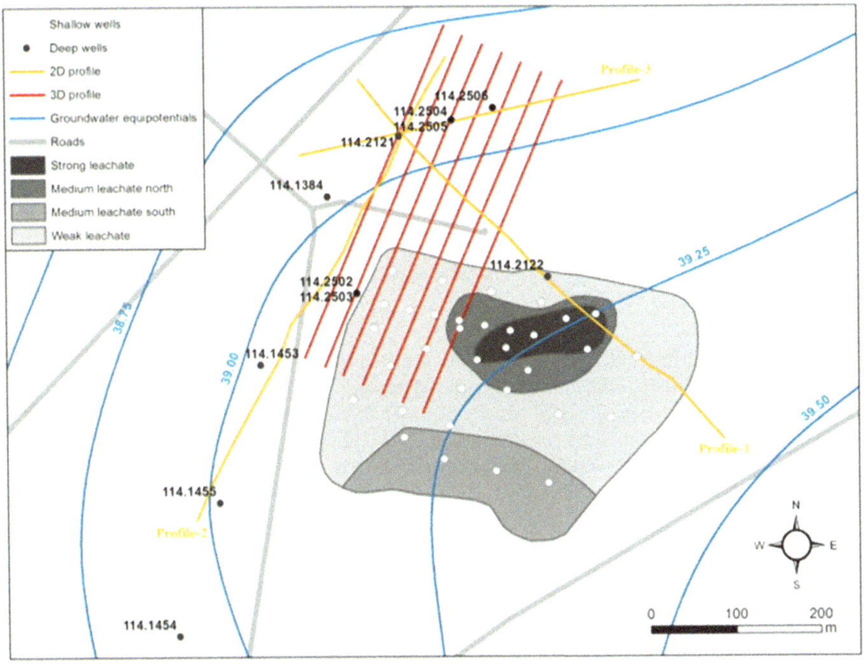

Fig. 4.24 A typical iso-potential map carried out by [87]

is an example of work carried out for iso-potential map carried out by [87]. The flow field exhibits some temporal variation, which may aid in pollution distribution [88].

Reprinted from P.K. Maurya, V. K Rønde, G. Fiandaca, N. Balbarini, E. Auken, P.L. Bjerg, A.V. Christiansen.

Detailed landfill leachate plume mapping using 2D and 3D electrical resistivity tomography with correlation to ionic strength measured in screens. *Journal of Applied Geophysics*, 138 (2017), 1–8, with permission from Elsevier

The 3D ERT modelling used the 2D profile to obtain from the iso-potential map in Fig. 4.24. The data collected from the resistivity meter transferred to the computer automatically. The principle of 2D inversion follows the AarhusInv inversion code [89]. The coding mechanism is based on an approximated covariance analysis that utilizes real model output from the inversion and data standard deviation [90]. Another potential feature of the code is being able to obtain the depth of investigation (DOI), which is being improvised by [91]. Various methods are available for 3D inversion methods, which are commonly based on the Gauss–Newton approach that is able to reduce the misfit between data and model response [92]. Solving the inverse problem requires proper planning involving the selection of a proper earth model for the forward problem, such as practised by [84]. The author practised the code based on the triple-grid inversion technique, applying the unstructured tetrahedral meshes and incorporates topography in the process.

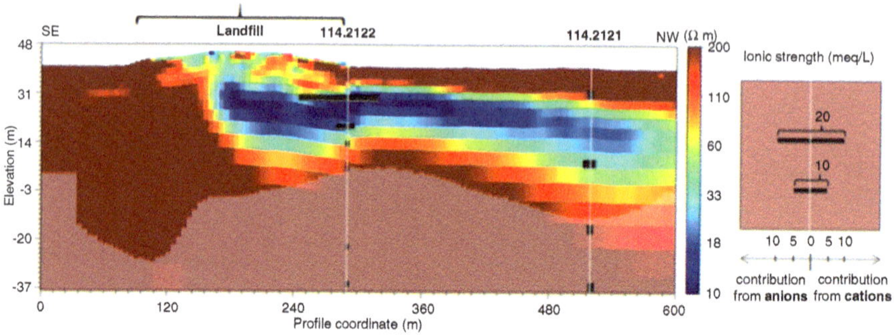

Fig. 4.25 Inverted resistivity model (2D) along the profile crossing and approximately in the direction of groundwater. Reprinted from P. K. Maurya, V. K Rønde, G. Fiandaca, N. Balbarini, E. Auken, P. L. Bjerg, A. V. Christiansen. Detailed landfill leachate plume mapping using 2D and 3D electrical resistivity tomography with correlation to ionic strength measured in screens. *Journal of Applied Geophysics*, 138 (2017), 1–8, with permission from Elsevier

Based on the study carried out by [82], the ionic strength and resistivity responses in the landfill source and landfill leachate plume compared. Figure 4.25 shows a 2D inverted resistivity model obtained from the landfill. The result shows the presence of a low-resistivity layer at a certain depth and thickness, which is known as leachate plume. The plume originates from the landfill, refuses beneath the landfill in the flow direction, and causes the electrical conductivity to decrease gradually due to the increasing ion content generated in the landfill leachate. Based on the resistivity model, the researcher could predict the plume direction.

The landfilling of various waste materials from different sources creates different sub-plumes. The 3D inversion enables visualization of plumes sources and their flow [93]. Figure 4.26 is an example of 3D inversion results obtained from the landfill. As shown in the figure, there are two distinct plumes originating from different landfill locations that contain different waste sources. This is supported by the identification of two separate low resistivity anomalies shown by the red arrows and separated by a dashed yellow line in Fig. 4.26.

4.4.7 *The Decomposition Process in Landfills*

Each day, various types of municipal solid refuse are delivered to landfill sites, and the decomposition process differs greatly between sites. The decomposition method is highly complex since it is influenced by a variety of factors, including the solid waste structure, changes in climate, landfill operations, how old the site is, level of moisture, and levels of pH [94]. These differences have a major effect on the method of leachate treatment in terms of how it is designed and operated [63].

From the beginning of the process of decomposition at the landfill site, complicated reactions occur biologically and chemically [95]. In consequence, it is

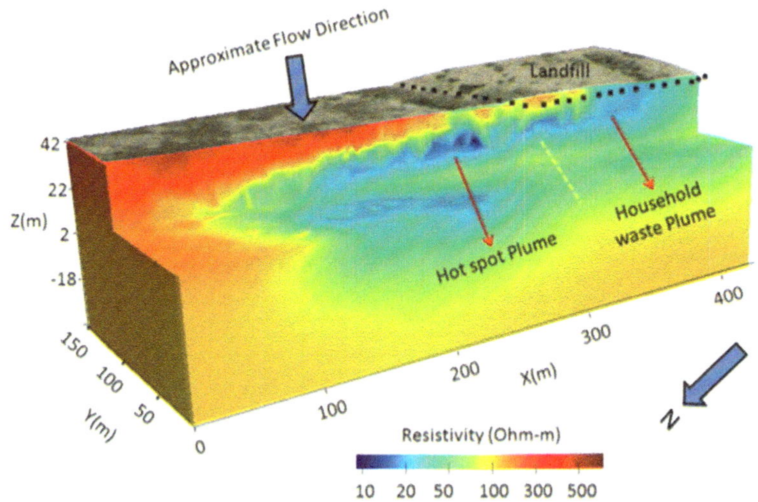

Fig. 4.26 The 3D cube of resistivities is sliced vertically along a line approximately perpendicular to the groundwater flow direction and horizontally at 25 m elevation. Reprinted from P. K. Maurya, V. K Rønde, G. Fiandaca, N. Balbarini, E. Auken, P. L. Bjerg, A. V. Christiansen. Detailed landfill leachate plume mapping using 2D and 3D electrical resistivity tomography with correlation to ionic strength measured in screens. *Journal of Applied Geophysics*, 138 (2017), 1–8, with permission from Elsevier

possible to identify five stages of these changes: firstly, initial adjustment (Phase I); secondly the phase of transition (Phase II); thirdly, the acidogenic stage (Phase III); followed by the fermentation of methane (Phase IV); and lastly, maturation (Phase V). At each landfill site, the rates of decomposition within all the phases depend on the particular chemical, physical, and microbiological features[96].

Phase I – Initial Adjustment Within a landfill, confined air and microbes cause biodegradable organic matter to decompose. At the initial adjustment point, this generally occurs in an aerobic state. Only small amounts of leachate are produced at this stage, yet the pollutant concentration levels are high [97].

Phase II – Transition A matter that is organic and biodegradable undergoes a microbial decomposition process. The first stage produces a complex solution due to the creation of leachate in aerobic conditions; the level of pH is nearly identical to neutral. Once the waste materials have been shut off inside the landfill, they are unable to access any supply of oxygen. This causes the phase of microbial decomposition to perpetuate to the point at which the total depletion of oxygen remaining is complete. The aerobic degradation produces heat, which means the leachate may reach approximate temperatures of 80–90°C. In terms of the heat retained, such temperatures can increase the level of leachate produced in subsequent phases. In this stage, leachate processing happens due to the confined refuse, which produces moisture discharge during its compaction and because its access to rainfall has been short-circuited [98].

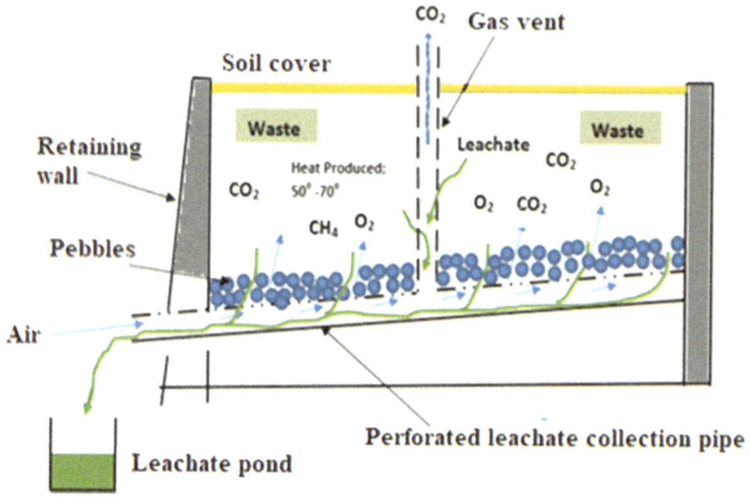

Fig. 4.27 Semi-aerobic landfill leachate mechanism

Phase III – Acidogenic Once the depletion of oxygen within the confined waste in the landfill is complete, the anaerobic phase starts. The beginning of this acidogenic phase sees the production of soluble degradable organic matter in high concentrations, as well a slightly to strongly acid pH level. The acidity of the pH is strengthened due to CO_2 being present. The pH leachate level reduces to 5 or less because acidic leachate and organic acids are produced. The reduced level of pH level means that vital nutrients within the leachate are removed, and the heavy metals disintegrate. Meanwhile, in this phase, the concentrations of ammonium and metal rise, but complex molecule numbers reduce. The entire process is completed in around a third of a year, but stabilising the levels of gas generated in the landfill takes from 1 to 2 years [99].

Phase IV – Methane Fermentation In methanogenic conditions, leachate becomes neutral or marginally alkaline, a process that generally happens within several months or possibly years. Methanogens generate methane and CO_2. Once the methanogenic state in landfill sites stabilizes, the composition of these gases is 55–60% methane and 40–45% CO_2 (other gases are present in trace forms) [100]. The consumption of CO_2 and acetate is performed by two types of bacteria. One type is mesophilic bacteria, thriving at heat levels between 30°C and 35°C. The other is thermophilic bacteria, growing at heat levels between 45°C and 65°C (see Fig. 4.27). Although the process occurs slowly and consumes time, a beneficial effect is that the pH level of the leachate becomes established at 7 to 8, which causes the amount of heavy metals in the leachate to reduce [101].

Phase V – Maturation When the process of turning the waste into CO_2 and methane is complete, aerobic conditions might be reintroduced due to new forms of aero-

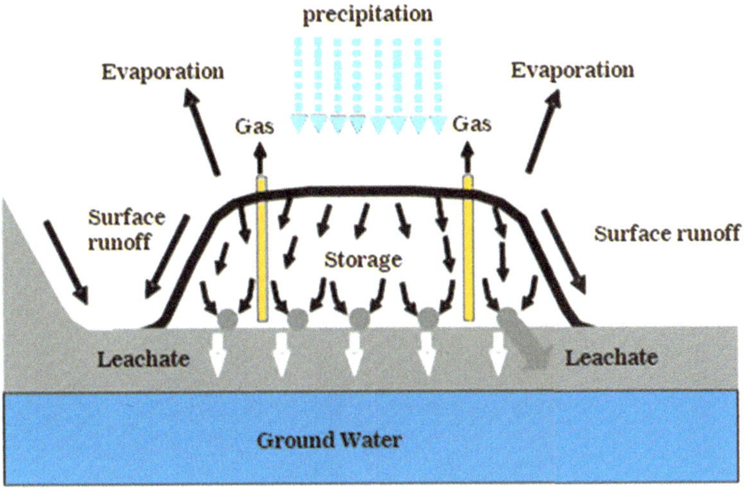

Fig. 4.28 Water movements in the landfill

bic microorganism production. These microorganisms replace the anaerobic types, leading to the aerobic state being re-established [101].

4.4.8 *Characterization of Landfill Leachate*

Generally, various forms of contaminant are released by the municipal solid refuse at landfills into the nearby environment. These include the emission of gases, leachate in liquid form, and solid material that is not degradable. Leachate from landfills combines contaminants of various kinds: toxic, organic, and inorganic. The production of leachate is generally dependent on various factors, including levels of rain, the solid waste composition, the size of the particles, the site hydrology, the compaction degree, the conditions of heat and moisture, the availability of oxygen, and the landfill age. Meanwhile, the variations in the composition of leachate and the amounts of pollutants take from the refuse are sometimes explained by the amount of water infiltration into a landfill site. Variations are also in direct relation to the processes of nature that occur within the landfill (Fig. 4.28). Landfills can function in a similar way to large anaerobic reactors if conditions allow. Suitable conditions are generally created if the moisture needed to support the activity of microbes is present in sufficient quantities [76, 102].

Sanitary landfills experience four steps in terms of the performances of their internal biological processes. Each of these states might influence the composition of leachate in landfills. It is easy to identify leachates according to landfill age, which was classified by Kamaruddin [76] (Table 4.15). Young leachate is normally a few weeks old (aerobic phase). Medium age occurs from 2 to 10 years (acidic

Fig. 4.29 Image of column leaching test for landfill leachate

phase), while old age involves it being more than 10 (methane phase). Leachate has a certain set of parameters; for example, the BOD/COD ratio, the chemical oxygen demand (COD), and the biochemical oxygen demand (BOD) and level of pH vary widely with landfills' increasing age [103]. New, younger landfills contain huge amounts of biodegradable organic material, which stimulates faster anaerobic fermentation and causes volatile fatty acids (VFA) to be produced [104]. In such sites, the organic contaminant COD volume is over 10,000mg/l, in comparison to that of landfills over 10, in which the volume is only 3,000 mg/l [105]. The COD levels in leachates fall within older landfill sites, but concentrations of the amounts of ammonia nitrogen (AN) rise [106]. Hydrolysis and nitrogen fermentation in biodegradable refuse substrates cause higher amounts of the contaminant ammonia nitrogen (AN). If the level increases, this results in more damaging environmental effects [107]. Phase four can be termed the 'humic phase'. Limited knowledge is available about this stage since only a minority of landfills observed have reached it. This phase is anticipated to happen at least a century after a sanitary landfill has been closed, or possibly hundreds of years later [108].

Many alterations in the composition of leachate that would occur in the future have not been identified through observation [109]. Instead, chemical knowledge is the basis for each analogy and rational hypothesis. The discussion section that follows will refer to the initial three landfill stages, while a particular focus will be devoted to phases two and three [110]. It is important to note that the 'phasing'

Table 4.15 Characterization of landfill leachate based on age

Landfill age (years)	< 2	2–10	> 10
Stabilization status	Young (fresh)	Intermediate	Mature (stabilized)
BOD_5	2000–30,000	N.A.	100–200
COD	3000–60,000	3000–15,000	100–2800
TOC	1500–20,000	N.A.	80–160
BOD_5/COD	0.5–1.0	0.06–0.5	< 0.1
TOC/COD	< 0.3	0.3–0.5	> 0.5
Total Kjeldahl nitrogen	100–2000	N.A.	N.A.
Ammoniacal nitrogen	10–800	30–1800	20–900
Organic nitrogen	10–800	N.A.	80–120
Nitrate	5–40	N.A.	5–10
pH	4.5–7.5	6.5–7.5	6.6–7.5
Alkalinity as $CaCO_3$	1000–10,000	N.A.	200–1000
Total hardness as $CaCO_3$	300–10,000	N.A.	200–500
Total suspended solids	200–2000	N.A.	100–400
Heavy metals	> 2.0	< 2.0	< 2.0
Total phosphorus	5–100	N.A.	5–10
Orthophosphate	4–80	N.A.	4–8
Calcium	200–3000	N.A.	100–400
Magnesium	50–1500	N.A.	50–200
Potassium	200–1000	N.A.	50–400
Sodium	200–2500	N.A.	100–200
Chloride	200–3000	N.A.	100–400
Sulphate	50–1000	N.A.	20–50
Total iron	50–1200	N.A.	20–200

concept in each consecutive landfill process relates to various conditions. These include:

- The speed of each aerobic/anaerobic reaction is influenced by the composition of solid refuse, particularly when this solid waste comprises organic matter in significant or minimal amounts, as it is relatively readily degradable[111].
- Leachate formation and the capacity it has to move matter inside the landfill.
- Ambient temperatures and climate conditions.
- Landfill arrangement: anaerobic conditions are more likely to accelerate when the landfill arrangement involves relatively smaller deposit cells, which take just 1–2 years to close and seal [112]. The cells in the landfill would, therefore, be converted into anaerobic reactors.

4.4.9 Leaching Procedure

The intended purpose of tests of leaching is the assessment of the levels of toxic constituents released into the nearby environment from liquid, solid, and multiphasic refuse [113]. Tests may involve using leachate to make contact with the form of waste involved, in precise conditions and over a defined time span. After this, effluents are retrieved, and the levels of extracted contaminant concentrated in the material are ascertained [114]. The solid refuse matter might contain waste products from industry, soil containing contaminants, or refuse management materials [115]. On the other hand, forms of wastewater (including industrial) and leachate from landfills are found in liquid waste. Tests of leaching are categorized into two forms. One type uses the batch method, and control over pH is maintained [116]. The aim of this set of tests is to achieve conditions of equilibrium when the leaching experiments have been concluded. Controls in the second group are in the form of diffusion tests, while the column method is used. Consideration must be given to numerous factors in the construction of leaching tests [117]. These include assessments of character, how compliant the material is, and verification on-site. Nevertheless, leaching is influenced by various factors; the most significant of these are as follows: ratio of L/S, exchange of water, biologically influenced degradation, leachate recirculation, particle size, and temperature [118]. This profusion of factors might explain the discrepancies that have arisen in the monitoring of wells and the laboratory results of investigations into leachate characterization.

- L/S ratio

  This parameter, the ratio of liquid/solid in landfills, offers the optimal description of the water quantity that flows across a small-scale disused refuse disposal site [119]. The resulting value established the relationship between the amount of water that infiltrates into landfill sites and the waste body's dry mass [120]. The ratio involves a direct relationship with the site climate conditions, the system of surface cover, the landfill height, and each stage on landfill sites, the mathematical calculation of the parameter involves the following equation: [121].

$$\frac{L}{S}\left(\frac{1}{Kg}\right)=\frac{\left[I_0*\left(a_0-f_i\right)\right]+\left[I_r*\left(a_r+f_i\right)\right]}{m_{DM}}*f_g \tag{4.14}$$

where:

I_0 is the infiltration into the uncovered landfill during the operation
I_r is the infiltration into the closed and or recultivated landfill during
a_0 is the number of years of waste disposal
a_r is the number of years in which the landfill is closed
f_i is a factor considering the presence of intermediate coverage during landfill operations
f_g is a factor considering the influence of groundwater level on the disposal site

m_{DM} is the weight expressed in kg of the dry matter of the landfill section of 1 m^2 multiplied with the estimated average height of the landfill body.

The progressive calculation of the ratio of L/S is possible in tests of leaching. It involves totalling the freshwater quantity that enters the body of waste and dividing it by the waste body dry mass total, as measured by the test [122]. The amount of water introduced into the system has a relationship with the duration. In a landfill, the true ratio of L/S ratio can be ascertained if the time-dependent rainfall variations at the landfill site are known. These are divided by the complete volume of waste disposed of, the calculation for which derives from the waste density and the characteristics of the location geometrically, and produces its final volume [123]. Hence, any interconnection between different types of leachate can be determined: those developed in laboratory tests and those forming on actual landfill sites. This improved design for conducting tests would enable the generation of leaching test outcomes covering broader L/S intervals while at the same time considering biological activities.

- Leachate composition

  The composition of leachate depends on the environmental conditions used to store the wastewater and other forms of waste. Furthermore, it is important to evaluate the wastewater pH because this is affected by the conditions experienced by the wastewater. Experiments that simulate scenarios of waste disposal generally employ organic acids [124]. Leachate such as these might make contaminant complexation more likely.

- Biological activity

  The quality of leachate is also affected by the key factor of biological degradation. The impacts on the environment can be understood and controlled by recognizing the processes of degradation that occur within the landfill [96]. Processes involving chemical, physical, and microbial reactions occur within the refuse, resulting in gas-based and dissolved compounds being released in the form of leachate and gases from landfill sites (Christensen and Kjeldsen) [125]. Landfills' decomposition processes take a long time to complete. The early stages, the initial aerobic phase, and the subsequent anaerobic phase have gotten a lot of attention up until now. The humic phase, which occurs after perhaps a century, is the next phase. It is last a very long time, thousands of years, perhaps. In the literature, the humic process is underrepresented, with only a few quantitative explanations of processes occurring during this period.

  Leaching is influenced by more than just the presence or absence of biological activity. The type of biological activity has a significant impact on leachate consistency. There are two main degradation phases of anaerobic degradation: acidogenic and methanogenic [126]. The organic content of the first process (acidogenic step) is considerably very high; however, the organic content of the methanogenic leachate is much lower. Additional technical information can be found from the literature [56, 58, 122, 150–152, and 154].

- Recirculation

Since water aids biological processes, leachate recirculation is a basic approach derived from bioreactor practices that aim to monitor and improve landfill stabilization [127]. In reality, leachate recirculation promotes biological activity by increasing and equalizing moisture content, allowing good contact between microbes, substrate, and nutrients, and transporting degradation products away [128]. The advantages of leachate recirculation include the delivery of nutrients and enzymes, pH buffering, dilution of inhibitory compounds, processing and distribution of methanogens, liquid storage, and evaporation opportunities. Many research using lysimeters, model cells, and full-scale experiments have shown the efficacy of leachate recirculation [129].

Many researchers have documented the benefits of leachate recirculation, including the acceleration of organic compound biodegradation as well as the reduction of the time needed for stabilization from decades to 2 to 3 years [130]. The rate of recirculation is a significant factor in the acceleration of degradation. The recirculation of leachate into a fresh waste can inhibit methanogenesis due to the accumulation of volatile fatty acids (VFAs) and low pH in effluent leachate, as well as the accumulation of ammonia nitrogen [131].

- Preferential pathways

Another significant factor that affects leachate quality variation is the preferential routes of leachate in landfills [132]. Water movement in restricted channels and voids is facilitated by the extremely heterogeneous physical composition of the solid waste material that composes a landfill. Several field tests, as well as laboratory studies, have revealed the presence of rapid flow along preferred flow paths in solid waste media [133, 134].

Prediction models focused on representing solid waste as a homogeneous porous medium, which is a popular method for modelling water flow and solute transport in solid waste, are ineffective. The effects of rapid water flow in preferential flow paths must be considered and quantified in order to enhance long-term predictions of leachate quality [135].

Because of the existence of these preferential pathways, findings from water flow in field-scale landfills and laboratory reactors can differ. For example, it has been discovered that in laboratory-scale studies, about 40% of pore water participates in advective solute transport, while this fraction is less than 0.2% in the investigated full-scale landfill, resulting in differences in moisture distribution and water flow [119]. The increased biological activity has the potential to alter water flow routes. If the waste degrades, the structure of the waste weakens, the waste's channels fail, and the water seeks new routes. Waste that was not originally included in the leaching process can now be used. Also present in the landfill body are plastic bags containing waste, which are effective barriers for water movement in the waste body: They may be able to force water flow and avoid water from coming into contact with waste.

- Aeration

The development of a long-term landfill is a key objective in waste management around the world. In relation to this, landfill aeration aids in the rapid, regulated, and long-term conversion of traditional anaerobic landfills to a biologically stabilized state, with reduced emission potential [136]. The composition of leachate is affected by the aeration of MSW landfills. The emission behaviour should be significantly improved in accordance with the ultimate goal of landfill aeration. Compounds like dissolved organic carbon and ammonium-nitrogen, for example, should have considerably lower concentrations due to efficient conversion of biodegradable organic compounds in the first case, and nitrification and simultaneous denitrification in the second case. Several experiments have yielded the previous findings [137]. Currently, landfill aeration methods aren't well established, and they depend on a variety of pressures, air injection systems, and off-gas disposal systems. Although high-pressure aeration is typically used to reduce the risk of explosion and odour irritation during landfill excavation or mining projects, low-pressure aeration as well as the semi-aerobic approach have been recognized for its potential in landfill remediation aimed at accelerated biological waste stabilization [138].

Based on the results of lab-scale and full-scale preliminary tests, the effectiveness of landfill in situ aeration is dependent on proper control of oxygen distribution, waste temperature, and moisture content; proper management of air flow and water inlet in the landfill body is also needed [55]. When choosing an aeration system for a specific disposal site, local climatic conditions must be considered.

Many studies were performed on waste samples taken from the landfill site prior to full-scale aeration to assess the possible emission reduction of the landfilled waste and its long-term emission behaviour after aeration completion and state the positive effects of aeration on leachate quality [139].

- Temperature

Tests, the resultant solubility, and the rates of adsorption and diffusion can be continuously affected by the temperature, which has a subsequent effect on the results of leaching (Fleming et al., 2011). Testing is generally conducted at room temperature.

4.4.10 Leaching Test

The term 'leaching' refers to elements being extracted due to a solvent's influence; this is generally water. The waste management context is important to consider: The body of waste is placed at landfill sites, where it experiences water percolation due to surface runoff, rain, or groundwater. The effect is that pollutants pass into the water from the refuse; the water may then enter and affect the surrounding environment [140]. How waste behaves in terms of leaching is linked to the manner in which it releases, or does not release, its constituent parts when influenced by its exposure conditions. Such behaviour might be anticipated when conducting

Fig. 4.30 Image of batch leaching test for landfill leachate

experiments involving tests of leaching. Hence, the characterization of leachate may be possible in the future. Such testing involves loading liquid with elements of chemical and bacteriological materials that are produced when waste is degraded as water circulates within the refuse. Tests for characterization and compliance must be distinguished in terms of the methods of conducting leaching tests. The former provides a better understanding of the stages involved in the control of materials released from particular stores of stabilized waste. In addition, they form the foundation for developing acceptance criteria that can be used in particular release situations. The latter tests provide standardized waste verification of waste that use particular thresholds of reference [141]. Figures 4.29 and 4.30 show the differences between the column and batch leaching test in landfills waste.

4.4.11 Available Tests

There is a need to differentiate between characterization and compliance testing in the field of leaching test methods. The first provides an understanding of the processes that govern the release of particular stabilized waste, as well as the foundation for developing standards for acceptance in a suitable release scenario, while the second is used for routine waste verification with specific reference thresholds [142]. Various tests of leaching are available. They are often categorized into two: equilibrium test (statistic tests) and dynamic tests.

a. Equilibrium (statistic) tests produce detail about the mechanisms controlling the solubility of species, such as sorption, complexation, and discrete phase precipitation. Static testing aims to reach equilibrium for certain parameters, for example, attaining a pH in the leaching test for pH dependence. Leaching tests of pH dependence might provide a measurement of the material of interest's chemical speciation and capacity for acid–base neutralization [143].
b. Dynamic tests involve the interpretation of processes of leaching, such as diffusion, dissolution, and surface wash-off. Experts have devised numerous tests that measure and analyse the contaminant leachability originating in leachate wastewater from landfills and industry, among other sources. The literature provides more information about many forms of measurement tests. Several tests offer details about contaminant leachability from waste. The following section outlines the test that is most commonly used [144].

The leaching experiments serve as a baseline to which findings from other tests can be compared and contrasted. Usually, the waste was subjected to leaching tests in order to obtain information on the waste's characteristic properties and short and long-term behaviours under the conditions defined by the scenario: anaerobic and aerobic conditions.

4.4.12 *Methods of Leaching Tests*

- Toxicity Characteristic Leaching Procedure (TCLP) is utilized for purposes of regulation in nations such as the United States and Australia; Method 1311 is a single batch form of testing. In this process, waste is mixed with acetic acid (pH 2.88 or 4.93) at a ratio of 20/1 L/S over an 18-hour period. The leachant selection depends on the alkalinity of the waste.
- Method of California Waste Extraction Test (WET) involves tumbling waste in a 0.2-M sodium citrate solution for 48 hours at an L/S ratio of 10/1. In compliance with Californian toxicity guidelines, it is used as a complement to the TCLP.
- Method 1312 of the Synthetic Precipitation Leaching Procedure (SPLP) is a single batch test used in the United States for regulatory purposes. It uses 60% of sulphuric acid/40% of nitric acid leachant at pH 4.2 or 5.0 to imitate acid rain leaching. The mixture is agitated for 18 hours and use a 20/1 L/S ratio.
- JLT-13 is a single batch test for granular waste used in Japan's regulatory system for landfill waste disposal. It takes 6 hours to shake 50 g of waste in 500 ml distilled water (at pH 5.8–6.3).
- A common regulatory column test used in the Netherlands to model leaching from mineral wastes in short to medium term (50 years) is NEN 7343 column up. Over a period of 21 hours, distilled water (pH 4 with HNO_3) is passed through waste in a column of specific dimensions, and seven eluate fractions in the L/S range 0.1/1–10/1 are obtained.

- Waste with a minimum particle size of 40 mm is fully submerged in water (pH 4) at a 5/1 L/S ratio with no agitation. This called NEN 7345 tank leach test. At intervals of 0.25 to 64 days, the liquid is updated and analysed. The test is intended for situations where contaminant diffusion is more relevant than convectional transport.

One particular aspect of leaching is addressed by each stipulated procedure of the test. It is often difficult to determine which test is best suited to the scenario being considered [145]. Four main factors should be incorporated if a leach testing procedure is followed:

1. Capacity to represent the actual conditions in the field.
2. How compatible each test is when examining various types of waste.
3. How reproducible tests are if repeatedly applied.
4. As a function of the testing time, stability needed to provide consistent results.

4.4.13 Risk Analysis

The main issue with long-term leachate pollution is the temporal variation in the concentration of contaminants present in the leachate released by the landfill and how these concentrations can be characterized using laboratory tests. For use in risk assessments, leaching tests and associated modelling techniques have been developed in the last decade to resolve long-term release characteristics of pollutants from wastes [146].

Risk assessment is a useful management method for safeguarding the environment from the dangers of landfills. All potential hazards must be recognized and the threats associated with them evaluated in order to achieve optimum environmental protection against the hazards associated with landfill sites. The issue is that an integrated risk management approach does not exist, despite the fact that it is fraught with uncertainty and necessitates a high level of expertise. The same could be said for a comprehensive knowledge-based computer model that could handle anything from risk management to risk quantification and hazard indices for landfills [147].

The aim is to assist government agencies and researchers in determining the environmental and health risks associated with polluted sites. Risk is more than just the presence of a hazard; it also includes its composition, pathway, and goal. The concentration of a given hazard that exceeds a given target crosses the target limits, and the safe and appropriate levels of hazard concentration for the given target all influence the degree of risk. Risk analysis may be defined as direct if the risk is measured from available data on concentrations or inverse, if reasonable risk limits are determined a priori and concentrations within these limits are calculated from these values [148]. The following are the measures in conducting a risk analysis:

- Identifying the cause of pollution, which in the case of a landfill is the landfill itself and its pollutants, which are primarily leachate and biogas.

- Characterization of exposure paths, mostly through the atmosphere, surface water, and groundwater in the case of a landfill.
- Characterization of the target site, which could be humans or the environment.
- Calculation of exposure, taking into account a variety of exposure pathways and scenarios, like ingestion, dermal touch, indoor and outdoor inhalation, and contact or ingestion with surface or groundwater.
- The risk is characterized by comparing threshold concentrations to hazards concentrations.

4.4.14 Applications and Limitations

Tests of leaching tests can be utilized for various purposes during their major role in the context of regulations. A more accurate representation of the environment of landfills is provided with dynamic leaching tests compared to batch tests. However, in the laboratory context, such tests remain limited in their applicability since they are time-consuming, and it may not be possible to rely on the results if the tests are not kept under close control.

Channelling will occur if the column is not packed properly, for example. Furthermore, if liquid flow rates are too high, the local equilibrium assumption is invalidated, resulting in poor reproducibility of performance. The equilibrium assumption states that the liquid must be in contact with the solid for a sufficient amount of time to achieve equilibrium in the extraction of metal ions from the solid by the liquid. Single batch experiments are suitable for regulatory purposes because of their shorter time frames and high reproducibility. For regulatory purposes, it is widely agreed that the most significant factor in the disposal of highly toxic wastes is a repeatable examination. As a result, batch procedures are often used to assess the leachability of pollutants from industrial wastes.

The most common criticism of batch tests is their inability to accurately simulate the landfill environment, particularly over time. Single-step batch experiments are the worst examples of laboratory-scale tests that cannot accurately replicate the field environment. Furthermore, batch experiments do not reveal the speed and kinetics of metal release in real-world situations.

Tests such as the sequential extraction procedure can be conducted to investigate the partitioning of metals between phases. The process, however, may suffer from interference, selectivity, and sensitivity issues, according to [149]. The non-specificity of the extractants and re-adsorption of metals after extraction were also listed as limitations of the sequential extraction method. As compared to batch tests, column tests are designed to be a more accurate representation of field conditions, where both dissolution/precipitation and diffusion processes are relevant at low L/S ratios. The focus of the field studies is on the composition of the leachate, with an emphasis on hydrology and complex processes. Laboratory leaching tests provide valuable information on leachate characteristics such as pH and conductivity, but they are unable to replicate the difficulty of field conditions, according to a discussion of laboratory versus field research [144].

Although the different leaching procedures mentioned could all be flawed, the absence of an acceptable test for a particular application often leads local authorities to use an incorrect test, amplifying the degree of misrepresentation of the results. Tests are often used arbitrarily, regardless of a material's ultimate fate. As a result, the purpose of a test must be clearly defined before deciding which test is most suitable in a given situation.

4.4.15 *Analytical Procedures for Landfill Leachate*

The term 'landfill leachate' refers to the pore water that collects in landfills. In terms of how they're leached and the complexity of the leachate matrix, this form of leachate differs significantly from laboratory-prepared leachates.

There are a variety of laboratory leaching methods that are designed to replicate a variety of environmental leaching events, but they all have the same goal: To obtain data on analyte leachability, which corresponds to a conservative determination of associated risk. This leachability can be used in the polluted land industry to model natural attenuation, assess the risk of contaminant leaching and reaching natural groundwater, and even decide the best waste disposal route if the risk is too great.

Since all leachate experiments are basically empirical, the analytical results collected can be suspect unless the process protocol is strictly followed. Because of the numerous and dynamic interactions that can occur between the solute and the solvent, it is critical that every approach be applied consistently.

The following table shows the simplified list of leachate analytical procedure under the Standard Methods for Water and Wastewater Analysis published by the American Public Health Association (APHA) (Table 4.16).

Chemical Oxygen Demand. COD (Titrimetric, Mid-Level) Under particular conditions of the oxidising agent, temperature, and time, the Chemical Oxygen Demand (COD) method calculates the amount of oxygen needed to oxidize the

Table 4.16 Standard methods for leachate parameters analysis

No.	Parameters	Method number
1.	Chemical oxygen demand (COD)	Method 5220, APHA
2.	Biochemical oxygen demand (BOD)	Method 5210, APHA
3.	Dissolved oxygen (DO)	Method 422F, APHA
4.	Total organic carbon (TOC)	Method 5310, APHA
5.	Nitrogen ammonia	Method 4500, APHA
6.	pH	Method 4500-H
7.	Trace metal	EPA-600/4-79-020

organic matter in a waste sample. Since the test uses particular chemical oxidation, the result has no clear association with the waste's Biochemical Oxygen Demand (BOD) or Total Organic Carbon (TOC) level. The test result should be regarded as a separate calculation of organic matter in the sample rather than a replacement for the BOD or TOC tests. The process can be used for effluent containing more than 50 mg/L of organic carbon. The Low-Level Modification can be used for lower carbon concentrations, such as those found in surface water samples. When the sample's chloride concentration reaches 2000 mg/L, saline water modification is needed.

The procedure in determining the COD concentration is as follow:

- Fill the reflux flask halfway with boiling stones, then add 50.0 mL of sample or an aliquot diluted to 50.0 mL and 1 g of $HgSO_4$. (6.5). Swirl in 5.0 mL concentrated H_2SO_4 (6.8) until the mercuric sulphate is fully dissolved. Place the reflux flask in an ice bath and slowly pour in 25.0 mL of 0.025 N $K_2Cr_2O_7$ while swirling (6.2). Slowly pour 70 mL of sulphuric acid–silver sulphate solution (6.3) into the cooled reflux flask, swirling continuously. Caution: Make sure the contents of the flask are thoroughly mixed before proceeding. If not, superheating will occur, and the mixture will be blown out of the condenser's open end.
- If the sample contains volatile organics, use condenser to apply the sulfuric acid–silver sulphate solution when cooling the flask to minimize loss due to volatilization.
- Reflux for 2 hours after applying heat to the flask. The 2-hour reflux cycle is not needed for certain waste waters. A shorter duration of refluxing can be permitted if the time needed to achieve full oxidation for wastewater of constant or known composition can be calculated.
- Allow the flask to cool completely before cleaning the condenser with around 25 mL distilled water. If you used a round bottom flask, move the mixture to a 500 mL Erlenmeyer flask and wash the reflux flask three or four times with distilled water. Enable the acid solution to cool to room temperature after diluting it with distilled water to around 300 mL. Titrate the excess dichromate with 0.25 N ferrous ammonium sulphate (6.4) solution to the end stage, adding 8 to 10 drops of ferroin indicator (6.6) to the solution. The transition from a blue-green to a reddish hue will be abrupt.
- Simultaneously perform a blank determination using low COD water.

Determine the COD in the sample in milligrams per litre (mg/L) as follows:where

A is the number of millilitres of Fe (NH_4)
B = millilitres of $Fe_2(SO_4)_2$ solution needed for titration of the blank $(NH_4)_2(SO_4)_2$ solution needed for the sample
N = $Fe(NH_4)_2(SO_4)_2$ solution normality
and *S* = millilitres of sample used for the test

Biochemical Oxygen Demand (BOD) The relative oxygen requirements of the landfill leachate can be measured using the biochemical oxygen demand (BOD)

Fig. 4.31 Winkler bottle

test. The test's application to organic waste discharges allows for the measurement of the discharges' effect on the receiving water's oxygen supplies. Technology standards for the design of wastewater treatment plants were developed using data from BOD studies.

The BOD test (Fig. 4.31) is an analytical bioassay-type technique that determines how much dissolved oxygen microbial life consumes when assimilating and oxidizing organic matter. Dark incubation at 20°C for a given time span is part of the normal test procedure (often 5 days). Temperature, biological population, water flow, sunlight, and oxygen concentration in the laboratory cannot be reliably replicated. When comparing BOD findings to stream oxygen demands, all of the above considerations must be taken into account.

The waste sample, or an adequate dilution, is incubated in the dark for 5 days at 20°C. The biochemical oxygen demand is calculated from the decrease in dissolved oxygen concentration during the incubation cycle.

1. A wastewater sample is needed to ensure proper biological activity during the BOD test and follows the following characteristics:
   - It must be chlorine-free. If the sample contains chlorine, a de-chlorination chemical (such as sodium sulphite) must be applied until testing.
   - The pH level should be between 6.5 and 7.5. If the sample falls outside of this range, acid or base must be applied to compensate.
   - Requires the presence of a sufficient microbiological population.
   - If the microbial population is insufficient or uncertain, a 'seed' solution of bacteria is applied, along with an essential nutrient buffer solution, to ensure that the bacteria population is healthy.
2. Specialized 300 mL BOD bottles are used, which are designed to allow for complete filling with no air space and an airtight seal. The bottles are filled with

either the sample to be checked or diluted (distilled or deionized) water, with varying quantities of wastewater sample added to represent different dilutions. As a monitor or 'blank', at least one bottle is filled entirely with dilution water.

3. Each bottle's initial dissolved oxygen concentration (mg/L) is measured using a DO metre, which should be at least 8.0 mg/L.
4. After that, each bottle was put in a dark, 20°C incubator for 5 days.
5. The DO metre is used again after 5 days (3 hours) to calculate the final dissolved oxygen concentration (mg/L), which should be reduced by at least 4.0 mg/L.
6. The BOD concentration (mg/L) is calculated by subtracting the final DO reading from the initial DO reading. The BOD concentration reading is multiplied by the dilution factor if the wastewater sample needed dilution.

Dissolved Oxygen Electrochemical reactions are used in the most popular instrumental probes for determining dissolved oxygen in the water. The current or potential can be associated with DO concentrations under steady-state conditions. An example of a dissolved oxygen meter is shown in Fig. 4.32.

Probe response is influenced by interfacial dynamics at the probe–sample interface, and a significant amount of interfacial turbulence is needed. Turbulence should be constant for precise results.

- There is no evidence that dissolved organic materials interact with the production of dissolved oxygen probes.
- The efficiency of a dissolved oxygen probe is affected by dissolved inorganic salts.

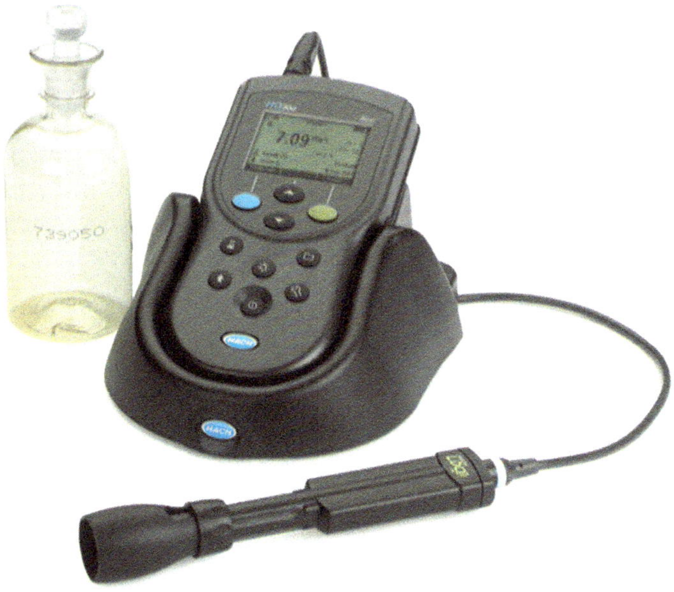

Fig. 4.32 Dissolved oxygen meter (HACH)

- Membrane-based probes react to oxygen partial pressure, which is a feature of dissolved inorganic salts. Data on dissolved oxygen saturation versus salinity can be used to measure conversion factors for seawater and brackish waters. Experiments may be used to develop conversion factors for particular inorganic salts. The use of a membrane probe can be difficult due to large differences in the types and concentrations of salts in samples.
- The production or efficiency of dissolved oxygen probes may be hampered by reactive compounds.
- Reactive gases passing through the membrane probes may cause problems.
- Chlorine, for example, can depolarize the cathode and result in high probe performance. Long-term chlorine contact coats the anode with the anode metal's chloride, gradually desensitizing the probe.
- Free chlorine is not present in alkaline samples, so they would not intervene.
- If the applied potential is greater than the sulphide ion's half-wave potential, hydrogen sulphide can interfere with membrane probes. An intervening reaction will not occur if the applied potential is less than the half-wave potential, but the anode may be coated with the sulphide of the anode metal.
- Temperature compensation is usually given by the manufacturer because dissolved oxygen probes are temperature sensitive. Temperature coefficients for membrane probes range from 4% to 6%/°C, depending on the membrane used.

Total Organic Carbon (TOC) TOC test procedures are straightforward and simple, but they are dependent on the type of carbon-analysing instrument used in the lab. As a result, there is no such thing as a 'typical' TOC method. The manufacturer's procedures should be strictly followed to produce the best performance. The TOC test, like the COD test, can be used to quickly estimate BOD concentration once a stable level has been established.

The TOC to BOD ratio is determined for a specific wastewater stream. A carbon analysing tool, which tests total organic carbon in a wastewater sample, is at the core of the TOC test. Different heat and oxygen, ultraviolet radiation, and chemical oxidant-based methods for measuring TOC are available, depending on the carbon analysing instrument used. The TOC test converts organic carbon to carbon dioxide (CO_2), which is then analysed with an infrared analyser.

Nitrogen, Ammonia An ion-selective ammonia electrode and a pH metre with an extended millivolt scale or a specific ion metre are used to calculate ammonia potentiometrically (Fig. 4.33). The sample solution is separated from an ammonium chloride internal solution by a hydrophobic gas-permeable membrane in the ammonia electrode. Ammonia in the sample diffuses through the membrane, causing the internal solution's pH to change, which is detected by a pH electrode. A chloride selective ion electrode, which serves as the reference electrode, detects the constant amount of chloride in the internal solution.

pH The pH of a sample is determined electrometrically using either a glass electrode in combination with a reference potential or a combination electrode

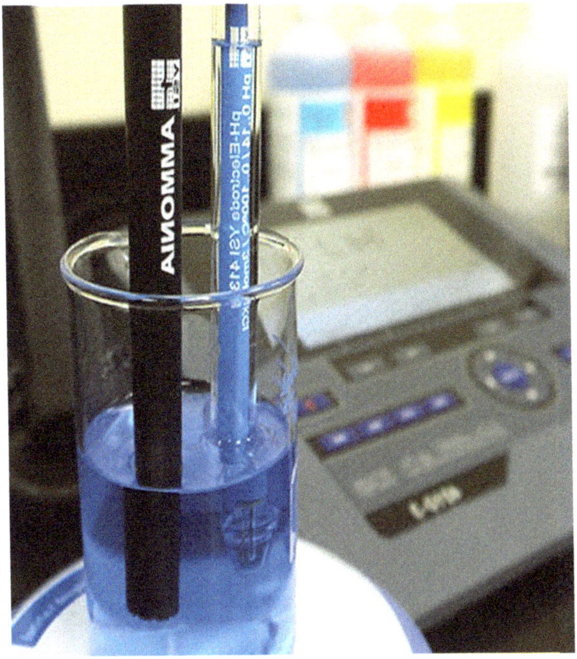

Fig. 4.33 Ammonia ISE (YSI)

(Fig. 4.34). Samples should be examined as soon as possible, preferably while they are being collected in the field. Since high-purity waters and waters that are not at equilibrium with the atmosphere will alter when exposed to the atmosphere, sample containers should be fully filled and sealed before the examination. The following are the procedures in determining the pH of a given sample:

- Place the sample or buffer solution in a clean glass beaker with enough volume to cover the electrode sensing elements and allow clearance for the magnetic stirring bar.
- If field measurements are being taken, the electrodes can be submerged to a reasonable depth in the sample stream and relocated in such a way that sufficient sample movement through the electrode sensing element is ensured, as indicated by drift-free (0.1 pH) readings.
- The calculated pH values must be adjusted if the sample temperature varies by more than 2°C from the buffer solution. Temperature variations are electronically compensated by automatic or manual compensators in instruments. Consult the instructions given by the manufacturer.
- Immerse the electrodes in the sample beaker or sample stream after rinsing and gently cleaning them if possible, and stir at a constant rate to ensure homogeneity and solids suspension. The air transfer rate at the sample's air–water interface should be as low as possible. Take note of the pH and temperature of the sample.

Fig. 4.34 pH meter (EUTECH)

Measure successive volumes of the sample until the differences in pH are less than 0.1 pH units. Typically, two or three volume changes are necessary.

- During the calculation, the air–water interface should not be disrupted. As dissolved gases are absorbed or desorbed, pH values will change if the sample is not in equilibrium with the atmosphere. Take note of the pH and temperature of the sample.
- pH meter show readings in pH units; report pH and temperature to the nearest 0.1 unit and degree °C, respectively.

Trace Element Analysis The method defines a technique for determining multiple trace elements in a solution simultaneously or sequentially. The approach is based on using an optical spectroscopic technique to calculate atomic emission. The samples are nebulized, and the resulting aerosol is transferred to the plasma torch, where it is excited.

A radio frequency inductively coupled plasma (ICP) produces distinctive atomic-line emission spectra. A grating spectrometer disperses the spectra, and photomultiplier tubes track the intensity of the lines. A computer device processes and controls the photocurrents from the photomultiplier tubes. To compensate for the variable background contribution to trace element determination, a background correction technique is necessary.

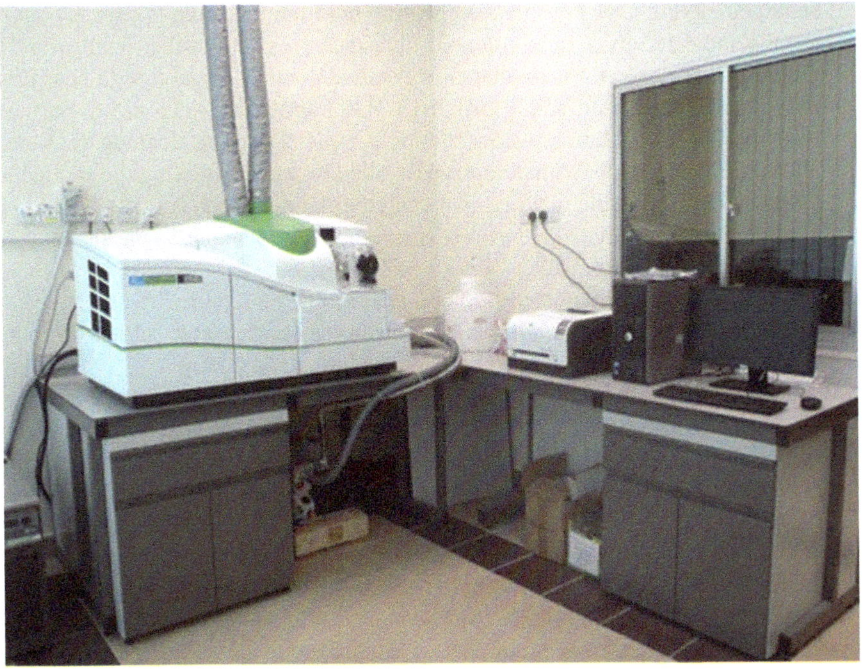

Fig. 4.35 ICP-OES

During the examination, the background must be measured adjacent to analyte lines on samples. The complexity of the range adjacent to the analyte line will determine where the background intensity measurement will be taken on one or both sides of the analytical line. The measurement location must be free of spectral interference and represent the same change in background intensity as the wavelength of the analyte being measured. In cases of line broadening, where a background correction calculation will degrade the analytical outcome, background correction is not needed. Figure 4.35 shows a typical of Inductively Coupled Plasma – Optical Emission Spectrometry (ICP-OES).

The filtered, preserved sample can also be analysed as obtained for dissolved element determinations. Samples and calibration requirements must have the same acid matrix and concentration. If a precipitate form during acidification, transit, or storage of the sample, it must be redissolved before analysis by adding more acid and/or heat. The following is the procedures in analysing trace elements:

- Switch the membrane filter containing the insoluble content to a 150 mL Griffin beaker and add 4 mL concentrated HNO_3 to determine the suspended components. Warm the beaker 3 gently by covering it with a watch glass. The membrane would be dissolved quickly by the warm acid.
- Increase the hot plate's temperature and digest the material. Cool the beaker and watch glass after the acid has nearly evaporated, then add another 3 mL of con-

centrated HNO_3. Cover 3 and keep heating until the digestion is over, which is usually indicated by a light-coloured digestate. Heat the beaker gently for 15 minutes to remove any precipitated or residual content. Evaporate to near dryness (2 mL), cool, and add 10 mL HCl (1+1) and 15 mL deionized, purified water per 100 mL dilution. Allow cooling completely before cleaning the watch glass and beaker walls with deionized distilled water and filtering the sample to eliminate any insoluble material that can clog the nebulizer. Adjust the volume according to the predicted element concentrations. Depending on the elements to be calculated, the volume can vary.

- The concentrations calculated in this manner must be reported as 'suspended'. Instead of filtering, the sample can be centrifuged or allowed to settle by gravity overnight to extract insoluble content after diluting and mixing.
- Choose a measured volume of the well-mixed acid preserved sample suitable for the estimated amount of elements and transfer to a Griffin beaker for total element determination. For 3 mL concentrated HNO_3, place the beaker on a hot plate and carefully evaporate the sample to near-dryness, ensuring that the sample does not boil and that no part of the beaker's bottom becomes dry. Cool the beaker before adding another 5 mL of concentrated HNO_3. Return the beaker to the hot 3 plates after covering it with a watch glass. Increase the hot plate's temperature before a gentle reflux action occurs.
- Continue to heat, adding more acid as needed, until the digestion is finished (generally indicated when the digestate is light in colour or does not change in appearance with continued refluxing). Cool the beaker after evaporating to near-dryness.
- Heat the beaker gently for 15 minutes after adding 10 mL of 1+1 HCl and 15 mL of deionized, distilled water per 100 mL of the final solution to remove any precipitate or residue leftover from evaporation. Allow to cool completely before cleaning the beaker walls and watch glass with deionized distilled water and filtering the sample to eliminate any insoluble material that can clog the nebulizer. Based on the predicted concentrations of elements present, adjust the sample to a predetermined amount. The sample has now been prepared for the examination. The concentrations calculated in this manner must be stated as 'absolute'.

Colour, ADMI The Tristimulus Filter System is extended in this method. Using the Adams Nickerson Colour Difference Calculator, tristimulus values are transformed to an American Dye Manufacturers Institute (ADMI) single number colour difference of the same magnitude as platinum-cobalt standards (DE).

Turbid samples must be screened prior to analysis because even minor quantities of turbidity interfere with the determination. There is yet to be discovered the best filter media for removing turbidity without removing colour. It is simple to use membrane and glass fibre filters with practical pore sizes of about 0.45. Other methods, such as centrifugation and/or filter aids, can be employed. The following are the procedures in measuring colour based on ADMI:

Fig. 4.36 A typical spectrophotometer (HACH)

- Prepare two 100 mL volumes of the sample by keeping one aliquot at its original pH and changing the second aliquot to pH 7.6 with sulphuric acid or sodium hydroxide as required.
- Using a 0.45 membrane filter, glass fibre filter, or other appropriate media, filter samples to eliminate turbidity.
- Set the transmittance to 100% with distilled water, then check the transmittance of the clarified sample or norm with each of the three Tristimulus filters. It is recommended to use calibration standards ranging from 25 to 250 units.
- ADMI values can be calculated using the Tristimulus values obtained using Spectrophotometric Method 204B1 (Fig. 4.36).

Using the equations below, calculate intermediate Tristimulus values for samples and standards based on the transmittance data:

$Xs = (T3 \times 0.1899) + (T1 \times 0.791)$
$Ys = T2$
$Zs = T3 \times 1.1835$

Where

T1 = transmittance value in % using filter number 1
T2 = transmittance value in % using filter number 2
T3 = transmittance value in % using filter number 3

4.5 Concluding Remarks

Integrated waste management includes the safe and effective disposal of municipal solid waste (MSW) and solid waste residues. Solid waste residues are non-recyclable waste components that remain after processing at a materials recovery facility or after the recovery of conversion products and/or resources. Solid waste has traditionally been collected on or in the earth's surface soils or in the oceans. An understanding of the waste stream's characteristics is needed regardless of the form of solid waste management is considered or introduced. Long-term trends in waste stream characteristics are also critical in good planning, which goes beyond developing a snapshot of current waste composition. If potential waste stream amounts and components are underestimated or overestimated, facilities can be over- or undersized, affecting project revenues and costs. Landfills are infrastructure ventures that necessitate a unique combination of technological and public-relations expertise, with the latter often outweighing the former. The nontechnical issues associated with the preparation, construction, and operation of landfills are barely discussed here, but that is not to say that they are unimportant.

Additional important technical information concerning the characterization, measurement, and glossary of solid waste can be found from the literature [155–165].

Glossary

Aerobic composting A method of composting organic wastes that involve the use of bacteria that require oxygen. This necessitates exposing the waste to sunlight, either by turning it or pushing air into pipes that pass through it.

Anaerobic digestion A form of composting that does not necessitate the use of oxygen. Methane is generated by this composting process. Anaerobic composting is another name for it.

Ash Solid by-products of incineration or other burning processes that are non-combustible.

Autoclaving A pressurised, high-temperature steam process is used to sterilise the products.

Baghouse An emission control system for a combustion plant that consists of a series of fabric filters that carry flue gases via an incinerator flue. Particles are suspended, preventing them from entering the atmosphere.

Basel Convention The Basel Convention is a treaty that was signed by over 100 countries have signed an international agreement on the management of transboundary movements of hazardous wastes and their disposal, which was drafted in March 1989 in Basel, Switzerland.

Biodegradable material Any organic material that microorganisms can break down into simpler, more stable compounds. The majority of organic wastes (such as food and paper) are biodegradable.

Bottom ash The incinerator residue that accumulates on the grate of a furnace is relatively coarse, non-combustible, and generally toxic.

Bulky waste Big wastes, such as machinery, furniture, and trees and branches, cannot be processed using standard MSW methods.

Cell The fundamental building block of a landfill. It's where incoming waste is flipped, scattered, compacted, and sealed.

Cleaner production Processes that aim to reduce the amount of waste produced during processing.

Co-disposal Generation of both electricity and steam from the same fuel source in a single plant.

Collection Paper, plastics, wood, and food and garden wastes are all combustible materials in the waste stream.

Combustion Materials are burned in an incinerator.

Commingled After being isolated from mixed MSW, mixed recyclables are collected together.

Communal collection A waste disposal system in which people carry their trash to a central location where it is processed.

Compactor vehicle To minimize the amount of solid waste, a recycling vehicle with high-power mechanical or hydraulic equipment is used.

Composite liner A landfill liner system made up of an engineered soil layer and a synthetic sheet of material.

Compost The content that results from a composting. Compost, also known as humus, is a soil conditioner that can also be used as a fertilizer in some cases.

Composting Biological decomposition of solid organic materials into a soil-like substance by bacteria, fungi, and other species.

Construction and demolition debris Waste includes bricks, asphalt, drywall, lumber, miscellaneous metal parts and sheets, packaging products, and other materials.

Controlled dump A proposed landfill with some of the characteristics of a sanitary landfill: hydrogeological suitability, grading, compaction in some cases, leachate control, partial gas management, frequent (but not always daily) cover, access control, simple record-keeping, and managed waste picking.

Curb side collection Compostable, recyclables, and garbage are collected at the edge of a sidewalk in front of a home or business.

Disposal Following collection, sorting, or incineration, the final handling of solid waste. The most common method of disposal is to deposit waste in a landfill or a dump.

Emissions Gases that have been emitted into the atmosphere.

Energy recovery The method of extracting useful energy from waste, usually by using the heat produced by incineration or landfill methane gas.

Environmental impact assessment (EIA) An assessment aimed at determining and forecasting the effect of a policy or project on the climate, human health, and well-being. Risk evaluation, as well as economic and land use assessments, are all possible components.

Environmental risk assessment (EnRA) A study of the relationships between agents, humans, and natural resources. Usually assessing the probabilities and magnitudes of harm that may be caused by environmental pollutants, it is made up of human health risk assessment and ecological risk assessment.

Flaring At a landfill, methane released from storage pipes is burned.

Fluidized-bed incinerator The stoker grate is replaced by a bed of limestone or sand that can withstand high temperatures in this form of incinerator. The word 'fluidized' comes from the fact that the bed is heated and high air velocities are used, causing the bed to bubble.

Fly ash The extremely toxic particulate matter captured by an air pollution control device from an incinerator's flue gas.

Groundwater Water that fills underground pockets (known as aquifers) and supplies wells and springs under the earth's surface.

Hazardous waste Reactive, poisonous, corrosive, or otherwise harmful to living things and/or the atmosphere waste. Hazardous manufacturing by-products abound.

Heavy metals Metals with a high atomic weight and density that are poisonous to living organisms, such as mercury, lead, and cadmium.

Household hazardous waste Items used in homes that are harmful to living organisms and/or the atmosphere, such as paints and certain cleaning compounds.

Incineration The method of burning waste by reducing the weight and volume of solid waste while still producing energy under regulated conditions.

Inorganic waste Sand, dust, glass, and a variety of synthetics are examples of waste made up of materials other than plant or animal matter.

Integrated solid waste management Usage of a coordinated collection of waste management strategies, each of which may play a role in a larger MSVVM strategy.

In-vessel composting Composting in a closed vessel or drum with a balanced internal setting, mechanical mixing, and aeration are all options.

Landfill gases Methane, carbon dioxide, and hydrogen sulphide are the main gases produced by the decomposition of organic wastes. Landfills can experience explosions as a result of these gases.

Landfilling The final disposal of solid waste by depositing it in a regulated manner in a long-term location. This concept is used in the Source Book to describe both supervised dumps and sanitary landfills.

Leachate Liquid that has seeped into a landfill or compost pile and has accumulated bacteria and other potentially hazardous dissolved or suspended materials (which may be partly formed by decomposition of organic matter). Leachate can contaminate both groundwater and surface water if it is not properly managed.

Leachate pond A pond or tank built at a landfill to collect leachate from the surrounding area. Typically, the pond is built to handle the leachate in some way, such as allowing solids to settle or allowing for aeration to facilitate biological processes.

Liner A protective layer made of soil and/or synthetic materials that is built along the bottom and sides of a landfill to prevent or minimize leachate from entering the atmosphere.

Materials recovery Obtaining goods that are recyclable or can be reused.

Materials recovery facility (MRF) A facility for manually or mechanically separating commingled recyclables Some MRFs are planned to distinguish recyclables from mixed municipal solid waste. The recovered materials are then baled and sold by MRFs.

Methane Methane is an odourless, colourless, flammable, and explosive gas formed by landfills anaerobically decomposing MSW.

Mixed waste Materials that have been discarded into the waste stream without being sorted.

MSW Municipal solid waste.

MSWM Municipal solid waste management.

Municipal solid waste The term 'municipal solid waste' refers to all solid waste generated in a given region, except industrial and agricultural waste. Construction and demolition debris, as well as other special wastes, can sometimes join the municipal waste stream. Hazardous wastes are generally excluded, except to the degree that they join the industrial waste stream. Occasionally, the term is used to refer to all solid wastes for which a city government takes responsibility in some way.

Open dump An impromptu 'landfill' with few, if any, of the characteristics of a managed landfill. Usually, there is no leachate monitoring, no access control, no cover, no management, and a large number of waste pickers.

Organic waste Is described as carbon-based waste, which includes paper, plastics, wood, food waste, and yard waste. In MSWM practise, the term is often used in a more limited context to refer to material that is derived more directly from plant or animal sources and can be decomposed by microorganisms.

Pathogen Organism is one that is capable of causing disease.

Processing Using processes such as baling, magnetic isolation, grinding, and shredding, MSW materials are prepared for future use or management. Separation of recyclables from mixed MSW is another word for the same thing.

Putrescible Decomposition or decay is a term used to describe the process of decomposition or decay. Food wastes and other organic wastes that decompose easily are often referred to as 'biodegradable'.

Pyrolysis In the absence of oxygen, heat causes chemical decomposition of a material, resulting in various hydrocarbon gases and carbon-like residue.

Recyclables Things that can be reprocessed into new product feedstock. Paper, glass, iron, corrugated cardboard, and plastic containers are all common examples.

Recycling The method of converting materials into raw materials for the production of new goods that may or may not be identical to the original.

Refuse A word that is often interchanged with solid waste.

Refuse-derived fuel (RDF) MSW that has been processed is used to make diesel. Separation of recyclables and non-combustible materials, shredding, size reduction, and pelletizing are all examples of processing.

Resource recovery Utilization of resources and energy from wastes is referred to as resource recovery.

Reuse The use of a commodity in its original form more than once for the same or a different reason.

Rubbish Solid waste is referred to as 'waste' in general. Food wastes and ashes are sometimes excluded.

Sanitary landfill An engineered method of disposing of solid waste on land that meets most of the standard requirements, such as proper siting, comprehensive site planning, proper leachate and gas management, tracking, compaction, regular and final cover, full access control, and record-keeping.

Secure landfill A waste management facility that is built to keep wastes out of the atmosphere indefinitely. This involves burying the wastes in a landfill with clay and/or synthetic liners, leachate collection, gas collection (if gas is produced), and an impermeable cover.

Sewage sludge A semi-liquid residue found at the bottom of canals and pipes containing sewage or industrial wastewaters, as well as the bottom of wastewater treatment tanks.

Site remediation Removing hazardous solids or liquids from a contaminated site or handling them on-site.

Source reduction The process of designing, manufacturing, acquiring, and reusing materials in order to reduce the amount and/or toxicity of waste generated.

Source separation To promote reusing, recycling, and composting, compostable and recyclable materials are separated from the waste stream before being collected with other MSW.

Special wastes Wastes that are preferably kept out of the MSW stream, but which occasionally find their way in and must be dealt with by local governments. Household hazardous waste, medical waste, building and demolition debris, war and earthquake debris, tyres, oils, wet batteries, sewage sludge, human excreta, slaughterhouse waste, and industrial waste are all examples of these.

Tipping fee Unloading or dumping waste at a landfill, transfer station, incinerator, or recycling plant is subject to a levy.

Tipping floor A place, usually on the outskirts of a neighbourhood, where small collection vehicles move waste to larger vehicles for transport to disposal sites.

Vectors Organisms that bear pathogens that cause disease. The key vectors that disperse pathogens outside the landfill site are mice, flies, and birds.

Virgin materials Any raw material for industrial processes that has never been used before, such as wood pulp trees, iron ore, crude oil, and bauxite.

Waste characterization study The analysis of samples from a waste stream to determine its composition is known as waste characterization.

Waste collector An individual hired by a municipality or a private company to collect trash from homes, businesses, and community bins.

Waste management hierarchy A rating of waste management operations based on the environmental or energy benefits they have. The waste management hierarchy was created with the aim of making waste management activities as environmentally friendly as possible.

Waste picker An individual who separates recyclables from mixed waste wherever it is temporarily accessible or discarded.

Waste reduction All methods of minimizing the amount of waste generated at the outset and collected by solid waste authorities. This includes everything from regulations and product design to community-based initiatives aimed at keeping recyclables and compostable out of the final waste stream.

Waste stream A community's, region's, or facility's complete waste flow.

Waste-to-energy (WTE) plant A plant that generates energy from solid waste materials (processed or unprocessed). Incinerators that produce steam for district heating or industrial use, as well as facilities that convert landfill gas to electricity, are examples of WTE plants.

Water table The depth below which the earth's crust becomes filled with water.

Wetland For at least part of the year, an area that is constantly wet or flooded and has a water level that is at or above the ground surface.

Working face The length and width of the waste disposal row at a landfill. The tipping face is another name for it.

Yard waste Yard and garden waste includes leaves, grass clippings, prunings, and other natural organic matter.

References

1. Kamaruddin, M. A., Yusoff, M. S., Aziz, H. A., & Hung, Y.-T. (2015). Sustainable treatment of landfill leachate. *Applied Water Science, 5*, 113–126. https://doi.org/10.1007/s13201-014-0177-7
2. Shadi, A. M. H., Kamaruddin, M. A., Niza, N. M., Emmanuel, M. I., Ismail, N., & Hossain, S. (2021). Effective removal of organic and inorganic pollutants from stabilized sanitary landfill leachate using a combined Fe2O3 nanoparticles/electroflotation process. *Journal of Water Process Engineering., 40*, 101988.
3. Edjabou, M. E., Jensen, M. B., Götze, R., Pivnenko, K., Petersen, C., Scheutz, C., & Astrup, T. F. (2015). Municipal solid waste composition: Sampling methodology, statistical analyses, and case study evaluation. *Waste Management., 36*, 12–23.
4. Niza, N. M., Yusoff, M. S., Zainuri, M. A. A. M., Emmanuel, M. I., Shadi, A. M. H., Hanif, M. H. M., & Kamaruddin, M. A. (2021). Removal of ammoniacal nitrogen from old leachate using batch electrocoagulation with vibration-induced electrode plate. *Journal of Environmental Chemical Engineering., 9*, 105064.
5. Yusoff, M. S., Kamaruddin, M. A., Ijanu, E. M., Jaafar, M. H., Alrozi, R., & Omar, K. F. M. (2021). Synthesis and application of coagulant derived from Durio zibethinus for emerging contaminants removal from landfill leachate. *International Journal of Environmental Engineering., 11*, 39–52.
6. Tchobanoglous, G., & Kreith, F. (2002). *Handbook of solid waste management*. McGraw-Hill Education.
7. Elzaki, A. A. M., & Elhassan, B. M. (2019). Quantification and characterization of solid waste in Alkalakla administrative unit, Khartoum State—Sudan. *International Journal of Waste Resources, 9*.
8. Daura, L. A., Enaburekhan, J., & Rufai, A. I. (2014). Characteristics and composition analysis of municipal solid waste in Kano, Nigeria. *International Journal of Scientific & Engineering Research., 5*, 972–975.
9. Sakawi, Z. (2011). Municipal solid waste management in malaysia: solution for sustainable waste management. *Journal of Applied Sciences in Environmental Sanitation, 6*.
10. Noufal, M., Yuanyuan, L., Maalla, Z., & Adipah, S. (2020). Determinants of household solid waste generation and composition in Homs City, Syria. *Journal of Environmental and Public Health., 2020*, 7460356. https://doi.org/10.1155/2020/7460356

11. Buenrostro, O., Bocco, G., & Vence, J. (2001). Forecasting generation of urban solid waste in developing countries—a case study in Mexico. *Journal of the Air & Waste Management Association., 51*, 86–93.
12. Kamaruddin, M.A., Yusoff, M.S., Adam, N.H., Maz, M.R.R., Abdullah, M.M.A.B., Alrozi, R., & Zawawi, M.H. (2018). Degradation of organic matter from stabilized leachate by using zinc sulphate as coagulant agent. In *IOP conference series: Materials science and engineering*. https://doi.org/10.1088/1757-899X/374/1/012047.
13. Kamaruddin, M. A., Yusoff, M. S., & Ahmad, A. H. (2015). Waste to wealth: Recovery of recyclable items and biodegradable wastes from night markets at Seberang Perai Selatan, Penang. In *Applied mechanics and materials* (pp. 543–548). Trans Tech Publ.
14. Yusoff, M. S., Kamaruddin, M. A., Aziz, H. A., Adlan, M. N., Zaman, N. Q., & Mahmood, N. Z. (2018). Municipal Solid Waste Composition, Characterization And Recyclables Potential: A Case Study Evaluation In Malaysia. *The Journal of Solid Waste Technology and Management., 44*, 330–343.
15. Monavari, S. M., Omrani, G. A., Karbassi, A., & Raof, F. F. (2012). The effects of socio-economic parameters on household solid-waste generation and composition in developing countries (a case study: Ahvaz, Iran). *Environmental Monitoring and Assessment., 184*, 1841–1846.
16. Sankoh, F. P., Yan, X., & Conteh, A. M. H. (2012). A situational assessment of socioeconomic factors affecting solid waste generation and composition in Freetown, Sierra Leone. *Journal of Environmental Protection.*
17. Sujauddin, M., Huda, S. M. S., & Hoque, A. T. M. R. (2008). Household solid waste characteristics and management in Chittagong, Bangladesh. *Waste Management., 28*, 1688–1695.
18. Kamaruddin, M. A., Yusoff, M. S., & Ahmad, M. A. (2012). Optimization of preparation conditions for durian peel-based activated carbon for the removal of COD and Ammoniacal Nitrogen (NH3-N) using response surface methodology. *Kuwait Journal of Science and Engineering., 39*, 37–58.
19. Kamaruddin, M. A., Ibrahim, M. H., Thung, L. M., Emmanuel, M. I., Niza, N. M., Shadi, A. M. H., & Norashiddin, F. A. (2019). Sustainable synthesis of pectinolytic enzymes from citrus and Musa acuminata peels for biochemical oxygen demand and grease removal by batch protocol. *Applied Water Science., 9*, 68.
20. Al-Khatib, I. A., Monou, M., Zahra, A. S. F. A., Shaheen, H. Q., & Kassinos, D. (2010). Solid waste characterization, quantification and management practices in developing countries. A case study: Nablus district–Palestine. *Journal of Environmental Management, 91*, 1131–1138.
21. Nas, S. S., & Bayram, A. (2008). Municipal solid waste characteristics and management in Gümüşhane, Turkey. *Waste Management., 28*, 2435–2442.
22. Mangizvo, R. V., & Wiseman, M. (2012). The management, practice and environmental health implications of the municipal solid waste dump site in Alice, South Africa. *Online Journal of Social Sciences Research, 1*, 125–131.
23. Alrozi, R., Zubir, N. A., Kamaruddin, M. A., Yusof, S. N. F. M., & Yusoff, M. S. (2017). Removal of organic fractions from landfill leachate by casuarina equisetifolia activated carbon: Characteristics and adsorption mechanisms. In *AIP conference proceedings*. https://doi.org/10.1063/1.5002333
24. Eriksson, M., Malefors, C., Callewaert, P., Hartikainen, H., Pietiläinen, O., & Strid, I. (2019). What gets measured gets managed–Or does it? Connection between food waste quantification and food waste reduction in the hospitality sector. *Resources, Conservation & Recycling: X., 4*, 100021.
25. Hassan, S. H., Aziz, H. A., Daud, N. M., Keria, R., Noor, S. M., Johari, I., & Shah, S. M. R. (2018). The methods of waste quantification in the construction sites (A review). In *AIP conference proceedings* (p. 20056). AIP Publishing LLC.

26. Gu, B., Jiang, S., Wang, H., Wang, Z., Jia, R., Yang, J., He, S., & Cheng, R. (2017). Characterization, quantification and management of China's municipal solid waste in spatiotemporal distributions: A review. *Waste Management., 61*, 67–77.
27. Miezah, K., Obiri-Danso, K., Kádár, Z., Fei-Baffoe, B., & Mensah, M. Y. (2015). Municipal solid waste characterization and quantification as a measure towards effective waste management in Ghana. *Waste Management., 46*, 15–27.
28. Sebastian, R. M., Kumar, D., & Alappat, B. J. (2019). A technique to quantify incinerability of municipal solid waste. *Resources, Conservation and Recycling., 140*, 286–296.
29. Lebersorger, S., & Beigl, P. (2011). Municipal solid waste generation in municipalities: Quantifying impacts of household structure, commercial waste and domestic fuel. *Waste Management., 31*, 1907–1915.
30. Ibikunle, R. A., Titiladunayo, I. F., Lukman, A. F., Dahunsi, S. O., & Akeju, E. A. (2020). Municipal solid waste sampling, quantification and seasonal characterization for power evaluation: Energy potential and statistical modelling. *Fuel., 277*, 118122.
31. Worrell, W.A., Vesilind, P.A., & Ludwig, C. (2016). *Solid waste engineering: A global perspective, Cengage learning*.
32. Allesch, A., & Brunner, P. H. (2014). Assessment methods for solid waste management: A literature review. *Waste Management & Research., 32*, 461–473.
33. Dahlén, L., & Lagerkvist, A. (2008). Methods for household waste composition studies. *Waste Management., 28*, 1100–1112.
34. Gidarakos, E., Havas, G., & Ntzamilis, P. (2006). Municipal solid waste composition determination supporting the integrated solid waste management system in the island of Crete. *Waste Management., 26*, 668–679.
35. Ibrahim, N., Yusoff, M. S., & Aziz, H. A. (2011). Food waste characteristics after autoclaving treatment. In *Proceedings of the 2nd international conference on biotechnology and food science* (pp. 54–57).
36. Gawande, N. A., Reinhart, D. R., Thomas, P. A., McCreanor, P. T., & Townsend, T. G. (2003). Municipal solid waste in situ moisture content measurement using an electrical resistance sensor. *Waste Management., 23*, 667–674.
37. Tanguay-Rioux, F., Legros, R., & Spreutels, L. (2020). Particle size analysis of municipal solid waste for treatment process modeling. *Waste Management & Research., 38*, 783–791.
38. Caputo, A. C., & Pelagagge, P. M. (2002). RDF production plants: I Design and costs. *Applied Thermal Engineering., 22*, 423–437.
39. Von Blottnitz, H., Pehlken, A., & Pretz, T. (2002). The description of solid wastes by particle mass instead of particle size distributions. *Resources, Conservation and Recycling., 34*, 193–207.
40. Velis, C. A., Longhurst, P. J., Drew, G. H., Smith, R., & Pollard, S. J. T. (2010). Production and quality assurance of solid recovered fuels using mechanical—biological treatment (MBT) of waste: a comprehensive assessment. *Critical Reviews in Environmental Science and Technology., 40*, 979–1105.
41. Zhou, H., Meng, A., Long, Y., Li, Q., & Zhang, Y. (2014). An overview of characteristics of municipal solid waste fuel in China: Physical, chemical composition and heating value. *Renewable and Sustainable Energy Reviews., 36*, 107–122. https://doi.org/10.1016/j.rser.2014.04.024
42. Kaartinen, T., Sormunen, K., & Rintala, J. (2013). Case study on sampling, processing and characterization of landfilled municipal solid waste in the view of landfill mining. *Journal of Cleaner Production., 55*, 56–66. https://doi.org/10.1016/j.jclepro.2013.02.036
43. Boumanchar, I., Chhiti, Y., M'hamdi Alaoui, F. E., Sahibed-Dine, A., Bentiss, F., Jama, C., & Bensitel, M. (2019). Municipal solid waste higher heating value prediction from ultimate analysis using multiple regression and genetic programming techniques. *Waste Management & Research., 37*, 578–589.
44. Calabrò, P. S. (2010). The effect of separate collection of municipal solid waste on the lower calorific value of the residual waste. *Waste Management & Research., 28*, 754–758.

45. Baawain, M., Al-Mamun, A., Omidvarborna, H., & Al-Amri, W. (2017). Ultimate composition analysis of municipal solid waste in Muscat. *Journal of Cleaner Production., 148*, 355–362. https://doi.org/10.1016/j.jclepro.2017.02.013
46. Sadef, Y., Nizami, A. S., Batool, S. A., Chaudary, M. N., Ouda, O. K. M., Asam, Z.-Z., Habib, K., Rehan, M., & Demirbas, A. (2016). Waste-to-energy and recycling value for developing integrated solid waste management plan in Lahore. *Energy Sources, Part B: Economics, Planning, and Policy, 11*, 569–579.
47. Shen, J., Zhu, S., Liu, X., Zhang, H., & Tan, J. (2010). The prediction of elemental composition of biomass based on proximate analysis. *Energy Conversion and Management., 51*, 983–987.
48. Abu-Qdais, H. A. (2000). Energy content of municipal solid waste in Jordan and its potential utilization. *Energy Conversion and Management., 41*, 983–991.
49. ASTM D3173, Standard Test Method for Moisture in the Analysis Sample of Coal and Coke, ASTM International, West Conshohocken, PA., in: 2017.
50. ASTM D3175, Standard Test Method for Volatile Matter in the Analysis Sample of Coal and Coke, ASTM International, West Conshohocken, PA., in: 2020.
51. ASTM D3174, Standard Test Method for Ash in the Analysis Sample of Coal and Coke from Coal, ASTM International, West Conshohocken, PA., in: 2018.
52. Lineres, M. (1992). *Transfer of inorganic pollution by composts, Composting and Compost Quality Assurance Criteria* (pp. 237–245). Commission of the European Communities.
53. Erses, A. S., Fazal, M. A., Onay, T. T., & Craig, W. H. (2005). Determination of solid waste sorption capacity for selected heavy metals in landfills. *Journal of Hazardous Materials., 121*, 223–232.
54. Karnika Alluri, H., Reddy Ronda, S., Saradhi Settalluri, V., & Singh Bondili, J. (2007). Biosorption: An eco-friendly alternative for heavy metal removal, *African Journal of Biotechnology.* 6 2924–2931. http://www.academicjournals.org/AJB (Accessed 29 Apr 2021).
55. Raga, R., & Cossu, R. (2013). Bioreactor tests preliminary to landfill in situ aeration: A case study. *Waste Management., 33*, 871–880. https://doi.org/10.1016/j.wasman.2012.11.014
56. N.E.E.P, Modern Landfill, (2020). https://www.eia.gov/energyexplained/biomass/landfill-gas-and-biogas.php.
57. Rettenberger, G. (2018). *Collection and disposal of landfill gas, solid waste landfilling* (pp. 449–462). Elsevier.
58. USEPA, Landfill gas collection system, 2018.
59. Perera, M.D.N., Hettiaratchi, J.P.A., & Achari, G. (1991). A mathematical model to improve the accuracy of... , In *Proceedings Sardinia.*
60. Savanne, D., Arnaud, A., Beneito, A., Berne, P., Burkhalter, R., Cellier, P., Tregoures, A., Savanne, D., Arnaud, A., Beneito, A., Berne, P.,.... (1997). Google scholar, In 6th Int. Landfill Symp. S. Margerita Di Pula, Cagliari, Italy. pp. 13–17.
61. Ehrig, H. J. (1983). Quality and quantity of sanitary landfill leachate. *Waste Management & Research., 1*, 53–68. https://doi.org/10.1177/0734242X8300100105
62. Bendz, D. (1998). *Generation of leachate and the flow regime in landfills.* https://www.osti.gov/etdeweb/biblio/646189 (Accessed 20 Apr 2021).
63. Kamaruddin, M. A., Yusoff, M. S., Aziz, H. A., & Hung, Y.-T. (2015). Sustainable treatment of landfill leachate. *Applied Water Science, 5*, 113–126. https://doi.org/10.1007/s13201-014-0177-7
64. Bengtsson, L., Bendz, D., Hogland, W., Rosqvist, H., & Åkesson, M. (1994). Water balance for landfills of different age. *Journal of Hydrology., 158*, 203–217. https://doi.org/10.1016/0022-1694(94)90054-X
65. W.R. Information. (1998). *Service.* Water Atlas.
66. D.G. Fenn, K.J. Hanley, T.V. DeGeare. Use of the water balance method for predicting leachate generation from solid waste disposal sites., , 1975.

67. Tchobanoglous, G., & Kreith, F. (2002). *Handbook of solid waste management*, McGraw-Hill Education, . https://www.accessengineeringlibrary.com/content/book/9780071356237 (Accessed 23 Apr 2021).
68. El-Fadel, M., Findikakis, A. N., & Leckie, J. O. (1997). Modeling leachate generation and transport in solid waste landfills. *Environmental Technology (United Kingdom).*, *18*, 669–686. https://doi.org/10.1080/09593331808616586
69. Pazoki, M., & Ghasemzadeh, R. (2020). *Municipal landfill leachate management, environmental science and engineering*. https://link.springer.com/content/pdf/10.1007/978-3-030-50212-6.pdf (Accessed 23 Apr 2021).
70. Pazoki, M., & Ghasemzadeh, R. (2020). Leachate quality. In *Environmental science and engineering* (pp. 101–127). Springer. https://doi.org/10.1007/978-3-030-50212-6_3
71. Othman, F., & Abualqumboz, M. S. (2017). Quantifying the generated and percolated leachate through a landfill's lining system in Gaza strip, Palestine. *Article in Polish Journal of Environmental Studies*. https://doi.org/10.15244/pjoes/73803
72. Alslaibi, T., Abustan, I., … Y.M.-W.M. &, undefined (2013). Quantification of leachate discharged to groundwater using the water balance method and the Hydrologic Evaluation of Landfill Performance (HELP) model, *Journals.Sagepub.Com*. 31 (2013) 50–59. https://doi.org/10.1177/0734242X12465462.
73. Hatfield, K., & Miller, W. (1994). *Hydrologic management models for operating sanitary landfills*.
74. Demetracopoulos, A. C., Sehayek, L., & Erdogan, H. (1986). Modeling leachate production from municipal landfills. *Journal of Environmental Engineering.*, *112*, 849–866. https://doi.org/10.1061/(ASCE)0733-9372(1986)112:5(849)
75. A.D.C. Schroeder P.R., The hydrologic evaluation of landfill performance (HELP) model: user's guide for version 3., (1994).
76. Kamaruddin, M. A., Yusoff, M. S., Rui, L. M., Isa, A. M., Zawawi, M. H., & Alrozi, R. (2017). An overview of municipal solid waste management and landfill leachate treatment: Malaysia and Asian perspectives. *Environmental Science and Pollution Research.*, *24*, 26988–27020. https://doi.org/10.1007/s11356-017-0303-9
77. Demetracopoulos, A. C., Korfiatis, G. P., Bourodimos, E. L., & Nawy, E. G. (1984). Modeling for design of landfill bottom liners. *Journal of Environmental Engineering.*, *110*, 1084–1098. https://doi.org/10.1061/(asce)0733-9372(1984)110:6(1084)
78. Perrier, E.R., & Gibson, A.C. (1980). *HSSWDS – Hydrologic simulation on solid waste disposal sites*, 129. https://books.google.com/books?hl=en&lr=&id=7UoVRE1EB1YC&oi=fnd&pg=PR8&dq=Perrier+ER,+Gibson+AC+(1980)+Hydrologic+simulation+on+solid+waste+disposal+sites:+EPA-SW-868.+US+EPA,+Cincinnati&ots=n7pmHJbnP8&sig=EHAzCzRIlGflj_vtH8P1m8bhpXM (Accessed 23 Apr 2021).
79. Bjerg, P.L., Albrechtsen, H.-J., Kjeldsen, P., Christensen, T.H., & Cozzarelli, I.M. (2014). The biogeochemistry of contaminant groundwater plumes arising from waste disposal facilities. 573–605. https://orbit.dtu.dk/en/publications/the-biogeochemistry-of-contaminant-groundwater-plumes-arising-fro (Accessed 23 Apr 2021).
80. Barker, J. F., Tessmann, J. S., Plotz, P. E., & Reinhard, M. (1986). The organic geochemistry of a sanitary landfill leachate plume. *Journal of Contaminant Hydrology.*, *1*, 171–189. https://doi.org/10.1016/0169-7722(86)90015-X
81. Cozzarelli, I. M., Böhlke, J. K., Masoner, J., Breit, G. N., Lorah, M. M., Tuttle, M. L. W., & Jaeschke, J. B. (2011). Biogeochemical evolution of a landfill leachate plume, Norman, Oklahoma. *Ground Water.*, *49*, 663–687. https://doi.org/10.1111/j.1745-6584.2010.00792.x
82. Maurya, P. K., Rønde, V. K., Fiandaca, G., Balbarini, N., Auken, E., Bjerg, P. L., & Christiansen, A. V. (2017). Detailed landfill leachate plume mapping using 2D and 3D electrical resistivity tomography – with correlation to ionic strength measured in screens. *Journal of Applied Geophysics*, *138*, 1–8. https://doi.org/10.1016/j.jappgeo.2017.01.019

83. Dahlin, T., & Zhou, B. (2006). Multiple-gradient array measurements for multichannel 2D resistivity imaging. *Near Surface Geophysics., 4*, 113–123. https://doi.org/10.3997/1873-0604.2005037
84. Rücker, C., Günther, T., & Spitzer, K. (2006). Three-dimensional modelling and inversion of dc resistivity data incorporating topography – I. Modelling. *Geophys. J. Int., 166*, 495–505. https://doi.org/10.1111/j.1365-246X.2006.03010.x
85. Loke, M. H., & Barker, R. D. (1996). Practical techniques for 3D resistivity surveys and data inversion 1. *Geophysical Prospecting., 44*, 499–523. https://doi.org/10.1111/j.1365-2478.1996.tb00162.x
86. Yi, M. J., Kim, J. H., Song, Y., Cho, S. J., Chung, S. H., & Suh, J. H. (2001). Three-dimensional imaging of subsurface structures using resistivity data. *Geophysical Prospecting., 49*, 483–497. https://doi.org/10.1046/j.1365-2478.2001.00269.x
87. Orbicon, 3 dybe boringer ved Grindsted (Data report in Danish), Region, Syddanmark., (2013).
88. Kjeldsen, P., Bjerg, P. L., Rügge, K., Christensen, T. H., & Pedersen, J. K. (1998). Characterization of an old municipal landfill (Grindsted, Denmark) as a groundwater pollution source: Landfill hydrology and leachate migration. *Waste Management and Research., 16*, 14–22. https://doi.org/10.1177/0734242X9801600103
89. Fiandaca, G., Ramm, J., Binley, A., Gazoty, A., Christiansen, A. V., & Auken, E. (2013). Resolving spectral information from time domain induced polarization data through 2-D inversion. *Geophysical Journal International., 192*, 631–646. https://doi.org/10.1093/gji/ggs060
90. Fiandaca, G., Christiansen, A. V., & Auken, E. (2015). Depth of investigation for multi-parameters inversions. In *Near surface geoscience 2015 – 21st european meeting of environmental and engineering geophysics, European association of geoscientists and engineers* (pp. 666–670). EAGE. https://doi.org/10.3997/2214-4609.201413797
91. Vest Christiansen, A., & Auken, E. (2012). A global measure for depth of investigation. *Geophysics, 77*. https://doi.org/10.1190/geo2011-0393.1
92. Park, S. K., & Van, G. P. (1991). Inversion of pole-pole data for 3-D resistivity structure beneath arrays of electrodes. *Geophysics., 56*, 951–960. https://doi.org/10.1190/1.1443128
93. Kjeldsen, P., Bjerg, P. L., Rügge, K., Christensen, T. H., & Pedersen, J. K. (1998). Characterization of an old municipal landfill (Grindsted, Denmark) as a groundwater pollution source: landfill hydrology and leachate migration. *Waste Management & Research: The Journal for a Sustainable Circular Economy., 16*, 14–22. https://doi.org/10.1177/0734242X9801600103
94. Guo, J. S., Abbas, A. A., Chen, Y. P., Liu, Z. P., Fang, F., & Chen, P. (2010). Treatment of landfill leachate using a combined stripping, Fenton, SBR, and coagulation process. *Journal of Hazardous Materials., 178*, 699–705. https://doi.org/10.1016/j.jhazmat.2010.01.144
95. Faitli, J., Magyar, T., Erdélyi, A., & Murányi, A. (2015). Characterization of thermal properties of municipal solid waste landfills. *Waste Management., 36*, 213–221. https://doi.org/10.1016/j.wasman.2014.10.028
96. Kjeldsen, P., Barlaz, M. A., Rooker, A. P., Baun, A., Ledin, A., & Christensen, T. H. (2002). Present and long-term composition of MSW landfill leachate: A review. *Critical Reviews in Environmental Science and Technology., 32*, 297–336. https://doi.org/10.1080/10643380290813462
97. Salem, Z., Hamouri, K., Djemaa, R., & Allia, K. (2008). Evaluation of landfill leachate pollution and treatment. *Desalination., 220*, 108–114. https://doi.org/10.1016/j.desal.2007.01.026
98. Ghosh, S.,& Hasan, S.E., SANITARY LANDFILL, (n.d.). https://www.eolss.net/Sample-Chapters/C09/E6-65-02-05.pdf (Accessed 24 April 2021).
99. Cossu, R., & Raga, R. (2008). Test methods for assessing the biological stability of biodegradable waste. *Waste Management., 28*, 381–388. https://doi.org/10.1016/j.wasman.2007.01.014
100. Bove, D., Merello, S., Frumento, D., al Arni, S., Aliakbarian, B., & Converti, A. (2015). A critical review of biological processes and technologies for landfill leachate treatment. *Chemical Engineering and Technology., 38*, 2115–2126. https://doi.org/10.1002/ceat.201500257

101. Williams, P. (2005). *Waste treatment and disposal*. https://books.google.com/books?hl=en&lr=&id=vSp0hqqU4e4C&oi=fnd&pg=PP11&dq=Williams,+P.T.,+Waste+treatment+and+disposal.+2005:+John+Wiley+%26+Sons.&ots=l_hNfjfbQJ&sig=Hv5VUpGvnK2x4MLZ-mbhQOqOiCo (Accessed 24 Apr 2021).
102. Kamaruddin, M. A., Yusoff, M. S., Aziz, H. A., & Basri, N. K. (2013). Removal of COD, ammoniacal nitrogen and colour from stabilized landfill leachate by anaerobic organism. *Applied Water Science., 3*, 359–366. https://doi.org/10.1007/s13201-013-0086-1
103. Adhikari, B., & Khanal, S. N. (2015). Qualitative study of landfill leachate from different ages of landfill sites of various countries including Nepal reconstruction project view project anaerobic digestion of waste view project qualitative study of landfill leachate from different ages of landfill sites of various countries including Nepal. *IOSR Journal of Environmental Science., 9*, 23–36. https://doi.org/10.9790/2402-09132336
104. Zhang, S., Jiang, L., Liu, D., & Li, B. (2010). Organic contamination and biodegradability of landfill leachate. In *2010 4th international conference on bioinformatics and biomedical engineering*. ICBBE. https://doi.org/10.1109/ICBBE.2010.5517898
105. Wu, J. J., Wu, C. C., Ma, H. W., & Chang, C. C. (2004). Treatment of landfill leachate by ozone-based advanced oxidation processes. *Chemosphere., 54*, 997–1003. https://doi.org/10.1016/j.chemosphere.2003.10.006
106. Deng, Y., & Englehardt, J. D. (2007). Electrochemical oxidation for landfill leachate treatment. *Waste Management., 27*, 380–388. https://doi.org/10.1016/j.wasman.2006.02.004
107. Bhatt, A. H., Karanjekar, R. V., Altouqi, S., Sattler, M. L., Hossain, M. D. S., & Chen, V. P. (2017). Estimating landfill leachate BOD and COD based on rainfall, ambient temperature, and waste composition: Exploration of a MARS statistical approach. *Environmental Technology and Innovation, 8*, 1–16. https://doi.org/10.1016/j.eti.2017.03.003
108. Shadi, A., Niza, N. Mohamed, Suffian, Y. Mohd., Shadia, A. Mohamed H., Kamaruddina, M. A., Nizaa, N. Muhammad, Emmanuelaa, M. Ijanu, Shaaha, M. A., Yusoffb, M. S., & Allafia, F. A. (2020). Characterization of stabilized leachate and evaluation of LPI from sanitary landfill in Penang, Malaysia Landfill characterization in Malaysia View project Solid Waste Management View project Desalination and Water Treatment Characterization of stabilized leachate and evaluation of LPI from sanitary landfill in Penang, Malaysia, Researchgate.Net. https://doi.org/10.5004/dwt.2020.25468.
109. S.M.- Vann, undefined 2007, Landfill leachate, generation, composition, and some findings from leachate treatment at Swedish plants, (n.d.). https://vannforeningen.no/wp-content/uploads/2015/06/2007_30327.pdf (Accessed 24 Apr 2021).
110. Ziyang, L., Youcai, Z., Tao, Y., Yu, S., Huili, C., Nanwen, Z., & Renhua, H. (2009). Natural attenuation and characterization of contaminants composition in landfill leachate under different disposing ages. *Science of the Total Environment., 407*, 3385–3391. https://doi.org/10.1016/j.scitotenv.2009.01.028
111. Lesteur, M., Bellon-Maurel, V., Gonzalez, C., Latrille, E., Roger, J. M., Junqua, G., & Steyer, J. P. (2010). Alternative methods for determining anaerobic biodegradability: A review. *Process Biochemistry., 45*, 431–440. https://doi.org/10.1016/j.procbio.2009.11.018
112. Selamat, M., H.A.-S.I.B. in, undefined 2020, Items of Consideration in the DESIGN of a Malaysian Landfill, Igi-Global.Com. (n.d.). https://www.igi-global.com/chapter/items-of-consideration-in-the-design-of-a-malaysian-landfill/240871 (Accessed 24 Apr 2021).
113. Tiwari, M. K., Bajpai, S., Dewangan, U. K., & Tamrakar, R. K. (2015). Suitability of leaching test methods for fly ash and slag: A review. *Journal of Radiation Research and Applied Sciences., 8*, 523–537. https://doi.org/10.1016/j.jrras.2015.06.003
114. Çoruh, S., & Ergun, O. N. (2010). Use of fly ash, phosphogypsum and red mud as a liner material for the disposal of hazardous zinc leach residue waste. *Journal of Hazardous Materials., 173*, 468–473. https://doi.org/10.1016/j.jhazmat.2009.08.108
115. Abdel-Shafy, H. I., & Mansour, M. S. M. (2018). Solid waste issue: Sources, composition, disposal, recycling, and valorization. *Egyptian Journal of Petroleum., 27*, 1275–1290. https://doi.org/10.1016/j.ejpe.2018.07.003

116. Galvín, A. P., Ayuso, J., Jiménez, J. R., & Agrela, F. (2012). Comparison of batch leaching tests and influence of pH on the release of metals from construction and demolition wastes. *Waste Management., 32*, 88–95. https://doi.org/10.1016/j.wasman.2011.09.010
117. Ruggieri, F., Gil, R. A., Fernandez-Turiel, J. L., Saavedra, J., Gimeno, D., Lobo, A., Martinez, L. D., & Rodriguez-Gonzalez, A. (2012). Multivariate factorial analysis to design a robust batch leaching test to assess the volcanic ash geochemical hazard. *Journal of Hazardous Materials., 213–214*, 273–284. https://doi.org/10.1016/j.jhazmat.2012.01.091
118. Kylefors, K., Andreas, L., & Lagerkvist, A. (2003). A comparison of small-scale, pilot-scale and large-scale tests for predicting leaching behaviour of landfilled wastes. *Waste Management., 23*, 45–59. https://doi.org/10.1016/S0956-053X(02)00112-5
119. Fellner, J., Döberl, G., Allgaier, G., & Brunner, P. H. (2009). Comparing field investigations with laboratory models to predict landfill leachate emissions. *Waste Management., 29*, 1844–1851. https://doi.org/10.1016/j.wasman.2008.12.022
120. Ritzkowski, M., Heyer, K. U., & Stegmann, R. (2006). Fundamental processes and implications during in situ aeration of old landfills. In *Waste management* (pp. 356–372). Pergamon. https://doi.org/10.1016/j.wasman.2005.11.009
121. Scott, J., Beydoun, D., Amal, R., Low, G., & Cattle, J. (2005). Landfill management, leachate generation, and leach testing of solid wastes in Australia and overseas. *Critical Reviews in Environmental Science and Technology., 35*, 239–332. https://doi.org/10.1080/10643380590917969
122. Polato, F. (2013). Use of leaching tests to characterize landfill leachate in the long term period. *Application in risk analysis*.
123. Ramos-Ruiz, A., Wilkening, J. V., Field, J. A., & Sierra-Alvarez, R. (2017). Leaching of cadmium and tellurium from cadmium telluride (CdTe) thin-film solar panels under simulated landfill conditions. *Journal of Hazardous Materials, 336*, 57–64. https://doi.org/10.1016/j.jhazmat.2017.04.052
124. Tatsi, A. A., & Zouboulis, A. I. (2002). A field investigation of the quantity and quality of leachate from a municipal solid waste landfill in a Mediterranean climate (Thessaloniki, Greece). *Advances in Environmental Research., 6*, 207–219. https://doi.org/10.1016/S1093-0191(01)00052-1
125. Hans, M., & Kumar, S. (2019). Biohythane production in two-stage anaerobic digestion system. *International Journal of Hydrogen Energy., 44*, 17363–17380. https://doi.org/10.1016/j.ijhydene.2018.10.022
126. Andersson, S., Nilsson, S. I., & Saetre, P. (2000). Leaching of dissolved organic carbon (DOC) and dissolved organic nitrogen (DON) in mor humus as affected by temperature and pH. *Soil Biology and Biochemistry., 32*, 1–10. https://doi.org/10.1016/S0038-0717(99)00103-0
127. Bilgili, M. S., Demir, A., & Özkaya, B. (2007). Influence of leachate recirculation on aerobic and anaerobic decomposition of solid wastes. *Journal of Hazardous Materials., 143*, 177–183. https://doi.org/10.1016/j.jhazmat.2006.09.012
128. Kumar, S., Chiemchaisri, C., & Mudhoo, A. (2011). Bioreactor landfill technology in municipal solid waste treatment: An overview. *Critical Reviews in Biotechnology., 31*, 77–97. https://doi.org/10.3109/07388551.2010.492206
129. Reinhart, D. R., & Basel Al-Yousfi, A. (1996). The impact of leachate recirculation on municipal solid waste landfill operating characteristics. *Waste Management & Research: The Journal for a Sustainable Circular Economy., 14*, 337–346. https://doi.org/10.1177/0734242X9601400402
130. Huang, Q., Yang, Y., Pang, X., & Wang, Q. (2008). Evolution on qualities of leachate and landfill gas in the semi-aerobic landfill. *Journal of Environmental Sciences., 20*, 499–504. https://doi.org/10.1016/S1001-0742(08)62086-0
131. Jiang, J., Yang, G., Deng, Z., Huang, Y., Huang, Z., Feng, X., Zhou, S., & Zhang, C. (2007). Pilot-scale experiment on anaerobic bioreactor landfills in China. *Waste Management., 27*, 893–901. https://doi.org/10.1016/j.wasman.2006.07.008

132. Rosqvist, H., & Destouni, G. (2000). Solute transport through preferential pathways in municipal solid waste. *Journal of Contaminant Hydrology., 46*, 39–60. https://doi.org/10.1016/S0169-7722(00)00127-3
133. Tinet, A. J., Oxarango, L., Bayard, R., Benbelkacem, H., Stoltz, G., Staub, M. J., & Gourc, J. P. (2011). Experimental and theoretical assessment of the multi-domain flow behaviour in a waste body during leachate infiltration. *Waste Management., 31*, 1797–1806. https://doi.org/10.1016/j.wasman.2011.03.003
134. Sormunen, K., Ettala, M., & Rintala, J. (2008). Detailed internal characterisation of two Finnish landfills by waste sampling. *Waste Management., 28*, 151–163. https://doi.org/10.1016/j.wasman.2007.01.003
135. Rosqvist, H. N., Dollar, H. L., & Fourie, B. A. (2005). Preferential flow in municipal solid waste and implications for long-term leachate quality: Valuation of laboratory-scale experiments. *Waste Management and Research., 23*, 367–380. https://doi.org/10.1177/0734242X05056995
136. Ritzkowski, M., & Stegmann, R. (2012). Landfill aeration worldwide: Concepts, indications and findings. *Waste Management., 32*, 1411–1419. https://doi.org/10.1016/j.wasman.2012.02.020
137. Ritzkowski, M., & Stegmann, R. (2013). Landfill aeration within the scope of post-closure care and its completion. *Waste Management., 33*, 2074–2082. https://doi.org/10.1016/j.wasman.2013.02.004
138. Kajiwara, N., Hirata, O., Takigami, H., Noma, Y., Tachifuji, A., & Matsufuji, Y. (2014). Leaching of brominated flame retardants from mixed wastes in lysimeters under conditions simulating landfills in developing countries. *Chemosphere., 116*, 46–53. https://doi.org/10.1016/j.chemosphere.2014.01.025
139. Hrad, M., Gamperling, O., & Huber-Humer, M. (2013). Comparison between lab- and full-scale applications of in situ aeration of an old landfill and assessment of long-term emission development after completion. *Waste Management., 33*, 2061–2073. https://doi.org/10.1016/j.wasman.2013.01.027
140. Fleming, C. A., Mezei, A., Bourricaudy, E., Canizares, M., & Ashbury, M. (2011). Factors influencing the rate of gold cyanide leaching and adsorption on activated carbon, and their impact on the design of CIL and CIP circuits. *Minerals Engineering., 24*, 484–494. https://doi.org/10.1016/j.mineng.2011.03.021
141. Shih, Y. J., Chien, S. K., Jhang, S. R., & Lin, Y. C. (2019). Chemical leaching, precipitation and solvent extraction for sequential separation of valuable metals in cathode material of spent lithium ion batteries. *Journal of the Taiwan Institute of Chemical Engineers., 100*, 151–159. https://doi.org/10.1016/j.jtice.2019.04.017
142. Aarnink, A. J. A., & Verstegen, M. W. A. (2007). Nutrition, key factor to reduce environmental load from pig production. *Livestock Science., 109*, 194–203. https://doi.org/10.1016/j.livsci.2007.01.112
143. Sloot, H. (n.d.). J.D.-E.E.C. of the Netherlands-C-04-060, undefined 2004, Development of Horizontally Standardized Leaching Tests for Construction Materials: a material based or release based approach? Identical leaching mechanisms for, Ecn.Nl. https://www.ecn.nl/publications/ECN-C%2D%2D04-060 (Accessed 24 Apr 2021).
144. Dijkstra, J. J., Meeussen, J. C. L., van der Sloot, H. A., & Comans, R. N. J. (2008). A consistent geochemical modelling approach for the leaching and reactive transport of major and trace elements in MSWI bottom ash. *Applied Geochemistry., 23*, 1544–1562. https://doi.org/10.1016/j.apgeochem.2007.12.032
145. Byrne, B. M., Oakland, T., Leong, F. T. L., van de Vijver, F. J. R., Hambleton, R. K., Cheung, F. M., & Bartram, D. (2009). A critical analysis of cross-cultural research and testing practices: implications for improved education and training in psychology. *Training and Education in Professional Psychology., 3*, 94–105. https://doi.org/10.1037/a0014516
146. Xu, Y., Xue, X., Dong, L., Nai, C., Liu, Y., & Huang, Q. (2018). Long-term dynamics of leachate production, leakage from hazardous waste landfill sites and the impact on ground-

water quality and human health. *Waste Management., 82*, 156–166. https://doi.org/10.1016/j.wasman.2018.10.009
147. Butt, T. E., & Oduyemi, K. O. K. (2003). A holistic approach to concentration assessment of hazards in the risk assessment of landfill leachate. *Environment International., 28*, 597–608. https://doi.org/10.1016/S0160-4120(02)00100-9
148. Laner, D., Crest, M., Scharff, H., Morris, J. W. F., & Barlaz, M. A. (2012). A review of approaches for the long-term management of municipal solid waste landfills. *Waste Management., 32*, 498–512. https://doi.org/10.1016/j.wasman.2011.11.010
149. Marguí, E., Salvadó, V., Queralt, I., & Hidalgo, M. (2004). Comparison of three-stage sequential extraction and toxicity characteristic leaching tests to evaluate metal mobility in mining wastes. In *Analytica Chimica Acta* (pp. 151–159). Elsevier. https://doi.org/10.1016/j.aca.2004.05.043
150. N.E.E.P. Neep, Modern Landfill, (2020).
151. Rettenberger, G. (2018). *Collection and disposal of landfill gas, solid waste landfilling* (pp. 449–462). Elsevier.
152. Shammas, N. K., Wang, L. K., Wang, M. H. S., & Chen, S. L. (2020). Ecological impact and management of solid waste landfill gas. In Y. T. Hang, L. K. Wang, & N. K. Shammas (Eds.), *Handbook of environment and waste management: Acid rain and greenhouse gas pollution* (pp. 397–454). World Scientific.
153. Afor T.S. & D.R. ATSDR (2001). Chapter 5: Landfill Gas Control Measures, in: Landfill Gas Primer – An Overview for Environmental Health Professionals.
154. Rettenberger, G., & Stegmann, R., (1996). 2.2 Landfill gas components, landfilling of waste: Biogas. pp. 51–57.
155. Wang, L. K., & Pereira, N. C. (1980). *Handbook of environmental engineering, Volume 2, solid waste processing and resources recovery* (480 pages). The Humana Press.
156. Hung, Y. T., Wang, L. K., & Shammas, N. K. (2014). *Handbook of environment and waste management: Land and groundwater pollution control, Volume 2* (1091 pages). World Scientific.
157. Wang, L. K., Wang, M. H. S., Hung, Y. T., & Shammas, N. K. (2016). *Natural resources and control processes* (pp. 493–623). Springer Nature.
158. Hung, Y. T., Wang, L. K., & Shammas, N. K. (2020). *Handbook of environment and waste management: Acid rain and greenhouse gas pollution control, Volume 3* (770 pages). World Scientific.
159. USEPA. (2021). 'Test Methods for Evaluating Solid Waste, Physical/Chemical Methods', United States Environmental Protection Agency, Washington DC, USA. USEPA Publication SW-846 Third Edition, (November 1986), as amended by Updates I (July 1992), II (September 1994), IIA (August 1993), IIB (January 1995), III (December 1996), IIIA (April 1998), and IIIB (November 2004), document number 955-001-00000-1), current May 2021.
160. USEPA. (1977). 'Procedures Manual for Groundwater Monitoring at Solid Waste Disposal Facilities', United States Environmental Protection Agency, Washington DC, USA. (EPA-530/SW-611), August 1977. Current May 2021.
161. USEPA. (1983). *'Methods for chemical analysis of water and wastes'*, United States Environmental Protection Agency, Washington DC, USA. National Information Service, Springfield , VA, USA, Tech. Report No. PB84-128677, March 1983.
162. USEPA. (1980) *'Samplers and sampling procedures for hazardous waste streams'*, United States Environmental Protection Agency, Washington DC, USA. (EPA-600/2-80-018), January 1980.
163. USEPA. (1981). *'APTI Course 415: Control of gaseous emissions'*, United States Environmental Protection Agency, Washington DC, USA. Publication No. EPA-450/2-81-005, December 1981.

164. USEPA. (1992). '*Screening procedures for estimating the air quality impact of stationary sources*, Revised', October 1992, Publication No. EPA-450/R-92-019, Environmental Protection Agency, Research Triangle Park, NC., USA.
165. USEPA. (1999). Method 1664, Revision A, n-Hexane Extractable Material (HEM; Oil and Grease) and Silica Gel Treated n-Hexane Extractable Material (SGT-HEM; Non-polar Material) by Extraction and Gravimetry, United States Environmental Protection Agency, Washington DC, USA. February 1999. National Technical Information Service, 5285 Port Royal Road, Springfield, VA 22161, USA.

Chapter 5
Mechanical Volume Reduction

Siti Nor Farhana Zakaria, Hamidi Abdul Aziz, Yung-Tse Hung, Amin Mojiri, and Eugene A. Glysson

Abstract A rise in solid waste generation has become an ever-increasing environmental problem in many parts of the world. Overload of solid waste can cause a lot of problems, especially to the environment and human health, such as water pollution, air and odor pollution by rotting waste, and the spread of diseases. Therefore, to overcome this problem, a good management system in handling the waste is needed. Volume reduction of waste is part of an important element in solid waste management, especially in transportation, storage, treatment, and disposal, which relate very much to the overall operation cost. Thus, in this section, the details of mechanical volume reduction involved in municipal solid waste management are discussed. These include the volume reduction technology, its advantages and limitations, on-site application, and operational and maintenance issues of the systems used.

Keywords Mechanical volume reduction · Municipal solid waste (MSW) · Baling · Shredding · Wet process · Dry process

---

S. N. F. Zakaria
School of Civil Engineering, Engineering Campus, Universiti Sains Malaysia, Nibong Tebal, Pulau Pinang, Malaysia

H. A. Aziz (✉)
School of Civil Engineering, Engineering Campus, Universiti Sains Malaysia, Nibong Tebal, Pulau Pinang, Malaysia

Environmental Engineering, Solid Waste Management Cluster, Engineering Campus, Universiti Sains Malaysia, Nibong Tebal, Pulau Pinang, Malaysia
e-mail: cehamidi@usm.my

Y.-T. Hung
Department of Civil and Environmental Engineering, Cleveland State University, Cleveland, OH, USA

A. Mojiri
Department of Civil and Environmental Engineering, Hiroshima University, Higashihiroshima, Japan

E. A. Glysson
Department of Civil Engineering, University of Michigan, Ann Arbor, MI, USA

L. K. Wang et al. (eds.), *Solid Waste Engineering and Management*, Volume 1, Handbook of Environmental Engineering 23,
https://doi.org/10.1007/978-3-030-84180-5_5

Acronyms

Al	aluminum
BOD_5	biochemical oxygen demand
COD	chemical oxygen demand
CO_2	carbon dioxide gases
CH_4	methane gases
Cu	copper
Cr-Ni	chromium–nickel alloy
Fe	iron
GHG	greenhouse gases
GDP	gross domestic product
MSW	municipal solid waste
NH_3-N	ammonia-nitrogen
N_2O	nitrous gases
rpm	revolutions per minute
TDS	total dissolved solids
WTE	-T

waste to energy technology

3R	reduce, reuse, and recycle

Nomenclature

d	= diameter
fe	= movement of per blade or teeth cutter in mm
N	= number of teeth cutter
Rp	= rpm of the cutter
VR_f	= final volume
VR_i	= initial volume

5.1 Introduction

Solid waste is defined as garbage, trash, and unwanted things that need to be disposed of. In Malaysia, the Solid Waste and Public Cleansing Management Act 2007 has described solid waste as any scrap materials or other unwanted surplus substances or rejected products that arise as a result of human activity, but excluding scheduled wastes, sewage, and radioactive wastes [1].

A rise in solid waste generation has become an ever-increasing environmental problem in many parts of the world. Not only does it affect human life, but an excessive load of solid waste dumped at landfill sites also has a bad effect on water and

Table 5.1 Summary of landfill leachate characteristics in Southeast Asia (Malaysia, Indonesia, and Thailand) [1]

Parameter	Unit	Malaysia	Indonesia	Thailand
pH		8.13	7.42–7.45	8.00
COD	mg/L	3852	291.1–585.0	4300.0
BOD_5	mg/L	196	62.00–218.10	418.00
NH_3-N	mg/L	1241	62–125	1934
TDS	mg/L	6237	1.200–1.263	18,900.000

soil. For example, the by-product produced in a landfill called leachate contains high recalcitrant pollutants and is difficult to be remediated. Table 5.1 shows the characteristic of leachate produced from landfill sites in Southeast Asian countries [2–5]. High concentrations of COD, BOD_5 NH3-N, and TDS make the leachate hazardous to be discharged into water bodies.

5.1.1 General Description

Rapid rise of solid waste production is mostly influenced by the growing population rate as well as aggressive industrial development. Kaza et al. (2018) [6] have projected that municipal solid waste (MSW) generation in the world continuously increased from 2.01 billion tons (2016) up to 3.4 billion tons in 2050, as shown in Fig. 5.1. Malaysia, for example, has produced 33,000 tons/day in 2012 to 38,200 tons/day of MSW in 2016 [7], and the number keeps rising until now.

Table 5.2 shows the statistical data for the gross domestic product (GDP) and solid waste generation in Malaysia for the year 2010 [8]. Kuala Lumpur has recorded the highest GDP/capita and solid waste generation rate (3697.88 tons/day). Sreenivasan et al. (2012) [9] concluded that solid waste generation in Kuala Lumpur is projected to increase up to 9207.84 tons/day in the year 2023, which is in line with the economic growth and human population.

Another country that recorded a high increment of solid waste is New Zealand. This country has shown a rising quantity of solid waste from 2.532 million tons in 2010 up to 3.221 million tons in 2015, as listed in Table 5.3 [10].

Excessive quantities of MSW instantly create a burden in terms of collection, processing, and disposal. Improper management and handling of municipal solid waste (starting from collection until disposal phase) can cause serious problems to the society, such as the spread of diseases, bad odor, and environmental pollution. In addition to this, solid waste is listed as one of the emission sources of carbon dioxide gases (CO_2), methane (CH_4), and nitrous gases (N_2O), or known as greenhouse gases (GHG) in the atmosphere. These gases can cause serious environmental pollution, especially global warming. Solid waste management has contributed between 3 and 5% of GHG, mostly from emissions at the landfill site and during transportation and collection activities [11]. An effective system for each solid waste management phase is therefore necessary. As the MSW production rate is

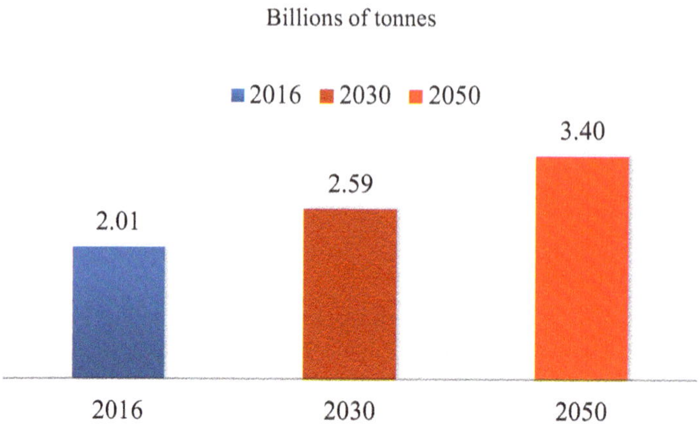

Fig. 5.1 Projection of global waste production [6]

Table 5.2 GDP and solid waste generation for each state in Malaysia, 2010 [8]

State	Population (millions)	GDP/capita at current price (RM)	Waste generation (tons/day)	Waste generation per capita (kg/capita/day)
Johor	3.35	20,911	2800.29	0.83
Kedah	1.95	13,294	1936.66	0.99
Kelantan	1.54	8,273	1512.41	0.98
Melaka	0.82	24,697	752.47	0.91
Negeri Sembilan	1.02	27,485	1106.99	1.08
Pahang	1.5	22,743	1399.59	0.93
Perak	2.35	16,088	2233.09	0.95
Perlis	0.23	15,296	285.9	1.24
Pulau Pinang	1.56	33,456	1590.35	1.01
Sabah	3.21	17,245	1990.91	0.62
Sarawak	2.47	33,307	1889.25	0.76
Selangor	5.46	31,363	4133.21	0.75
Terengganu	1.04	19,225	1290.75	1.24
Kuala Lumpur	1.67	55,951	2679.88	2.21
Labuan	0.09	29,116	95.21	1.06
Putrajaya	0.07	N.A	36.45	0.52
Total (national)	28.3	27,113	26,751.41	

expected to increase every year due to the increase of human population and economic and industrial growth, a good management system is needed in order to control the negative impacts of MSW on our environment and human life.

Table 5.3 Municipal waste of New Zealand [10]

Year	The total amount of waste (millions of tons)
2010	2.532
2011	2.512
2012	2.514
2013	2.684
2014	2.931
2015	3.221

In municipal solid waste management, there are three handling phases: (1) collection and transportation, (2) waste processing and treatment, and (3) disposal. Generally, MSW starts from the waste collection in the domestic area and is usually done manually by workers. After that, waste will be transported to the transfer station, and at this site, some of the waste will be processed prior to further processing at the waste treatment facility. Waste processing is one of the important parts in the solid waste management system to work effectively, especially in recovering material for reuse and conversion into products and energy sources.

Solid waste treatment is a technology to handle solid waste after a long journey of collection and transportation prior to disposal. There are many ways to treat MSW, such as dumping at the landfill site, incinerating, composting, recycling, and others. Although landfill sites offer a systematic place for disposing of MSW, the by-products (leachate) produced in landfills are hazardous and it is therefore not a good choice for a long-term plan.

Fig. 5.2 depicts the percentages of solid waste treatment applied in Italy, 2019 [12]. According to Fig. 5.2, there are 10 methods applied to treat solid waste in Italy.

(a) Biological treatment
(b) Domestic composting
(c) Covering of landfills
(d) Co-incineration
(e) Sorting and biostabilization/intermediate treatment
(f) Incineration
(g) Landfill
(h) Recovery of material
(i) Export and others

As shown in Fig. 5.2, most of the solid waste is converted into reusable material or recycling material by using recovery of material technology (29%). This is a good practice and also has a positive effect on the environment as well as saving our natural resources. Moreover, this method also reduces the quantity of disposal of waste at a landfill and increases the sanitary landfill age. As shown in Fig. 5.2, only 21% of solid waste was dumped at a landfill site in Italy. While 18% incinerated, and the remaining waste was transported to other waste treatment and disposal

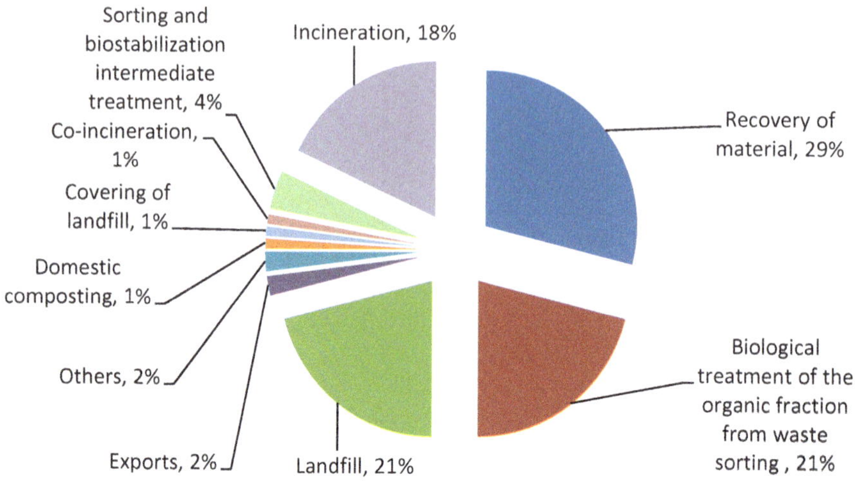

Fig. 5.2 Types of solid waste treatment applied in Italy, 2019 [12]

facilities such as domestic composting (1%), covering of landfills (1%), co-incineration (1%), and biostabilization intermediate treatment (4%).

A part of waste processing is a sorting waste process. This process is done by grouping the waste suited to its material types, such as plastic, glass, and paper. As known, municipal solid waste is abundant with recycling material products. Therefore, sorting and recycling activities can save much energy resources as well as reduce the volume of waste dumps at the landfill site. The characteristic of solid waste is closely related to its origin. In general, there are six main sources of solid waste: domestic waste, institutional waste, commercial waste, industrial waste, street waste, and demolition or construction waste [13].

Scholars have also divided MSW into several types such as paper, textiles, glass, wood, metal, rubber, leather, food scraps, and plastic. In Southeast Asia (Malaysia, Singapore, Indonesia, and Thailand), organic waste, plastic, and paper are dominant in MSW, as shown in Table 5.4 [14–18]. Therefore, the recycling process helps in organizing solid waste as well as makes solid waste treatment more efficient. Moreover, recycling material products are also able to reduce the use of energy resources. For example, recycling paper can reduce the number of trees being cut.

The sorting process often starts from the house by manually sorting the waste to their material group. Japan is one of the countries that has enforced recycling activities to their local society under the legal law of the Basic Recycling Act, 2000 [13]. This act is implemented in society with three mottos (3R): "Reduce," "Reuse," and "Recycle.". Moreover, this law was also developed with five aims: reduction of waste generation, reuse, recycling, thermal resource, and proper disposal of solid waste [13]. Japan has faced the problem of abundant solid waste due to the rapid economic and industrial growth. By implementing this law, it has helped the government to reduce solid waste at the landfill as well as save natural energy resources.

Table 5.4 Composition of municipal solid waste in Southeast Asia (Malaysia, Indonesia, Singapore, and Thailand) [14–18]

	Country			
Composition of MSW	Malaysia	Indonesia	Thailand	Singapore
Organic Matter	45%	55.51%	49.90%	–
Plastic	24%	19.10%	28.50%	10.58%
Paper	7%	10.32%	8.50%	14.86%
Textiles	–	9.57%	5.20%	1.96%
Glass	–	2.02%	4.40%	0.93%
Wood	–	1.84%	–	5.50%
Metal	6%	0.44%	1.40%	–
Rubber/leather	–	0.54%	–	–
Bones	–	0.29%	1.90%	–
Horticultural waste	–	–	–	4.26%
Hazardous	–	–	0.20%	–
Food Waste	–	–	–	10.51%
Construction debris	–	–	–	20.89%
Scrap tires	–	–	–	0.47%
Ash and sludge	–	–	–	3.16%
Others	18%	0.37%	–	26.89%
Total	100%	100.00%	100.00%	100.0%

Some countries have also included a solid waste sorting process by using the municipal waste automatic sorting machine after solid waste collection. The usable waste can be transformed to recycle material and also be able to reduce the volume of dumpling waste in the landfill and increase the age of landfill sites.

Another method applied in waste processing is the volume reduction of waste. Waste volume reduction is defined as a physical alteration of waste in order to make solid waste present in ideal shape and fit for further treatment as well as to optimize the transportation process. Solid waste is usually collected in the mixed type of material, condition, and shape as listed in Table 5.4. Therefore, the effective way to handling this inconsistent waste shape is by reducing the volume of the waste. There are various techniques to reduce the volume, such as mechanical, chemical (thermal), and biological techniques. However, in this chapter, mechanical volume reduction is focused on and is discussed in detail in the next section.

5.1.2 Mechanical Volume Reduction

Volume reduction of solid waste is referring to the reduction of volume or size of the waste to increase the density. There are two common methods applied in physical reduction volume: shredded and compaction processes. In short, solid waste will be shred or compacted into a smaller and uniform form compared to its original. There

Table 5.5 Advantages and disadvantages of volume reduction or compaction of solid waste

Advantages	Disadvantages
1. Solid waste management systems become more effective.	1. Difficult in sorting the waste according to their material type since all the waste has been mixed and compacted.
2. Reduce the volume of waste in the dumping site or landfill site.	2. The quality material of recyclable waste is poor.
3. Increase the life of landfill sites.	3. Wet waste like biodegradable material such as fruit, food, and vegetable will destroy recyclable material (paper and plastic waste).
4. Reduce the operation cost (transportation and storage area).	

are many advantages of doing this method, especially in handling waste, as well as optimizing the operational cost.

Reducing the volume of waste gives more benefit than harm. One of the advantages of this technique is that it is able to optimize the operation, especially during transportation and storage of waste. Compaction of waste during collection can maximize the collection and reduce the number of transportation trips. Transportation costs will be reduced, and space for storage waste at the transfer station will also be increased as the size of waste is reduced. Reduction in space in the storage area will reduce cost. Moreover, applying this technique will reduce the volume of waste dumped at the landfill and may prolong the lifespan of a landfill.

The advantages and disadvantages of a volume reduction of waste are listed in Table 5.5.

Reforming the shape or reducing the volume of waste on a big scale into the required form will need a special machine. Nowadays, there are many machines designed for this purpose, such as hammer mill, rasp mill, and baling. These machines are designed based on the type of material to be processed. For example, Orhorhoro and Oghoghorie (2017) [19] have designed a hammer mill machine for glass waste. This machine managed to effectively crush the glass waste into some useful end products. This technology allows recycling material and will reduce the amount of solid waste as well as saving our earth's resources. Further details of this technology will be discussed in the next section.

5.2 Size Reduction by Shredding Process

Shredding is a process of cutting or destruction of material into smaller pieces or forms by using a mechanical mechanism.

The shredding machine is widely used in various industries and purposes:

(a) Shredder of plastic
(b) Shredder of tires
(c) Shredder of wood
(d) Shredder of electric cables
(e) Shredder of natural or synthetic fibers

Mechanical volume reduction is a process of changing the physical properties of solid waste. Its application requires appropriate machines to employ compression, tension, and shearing force onto the raw waste. Solid wastes collected from households or industries usually come in bulk and in a variety of shapes. Therefore, they will be compacted or shredded first into smaller and uniform forms for easy handling or stored before undergoing another treatment phase. For example, the shredded process is able to improve the workability of the incineration plant compared to nonprocessed waste. This fact is supported by the firing theory, which stated that the efficiency of combustion is increased when solid waste is preshredded.

Also, a shredding machine is important in plastic waste management. A shredding machine is used to process plastic waste prior to recycling it back as new material. Plastic waste needs to be shredded into smaller pieces before melting and converting into a new product. For example, an electrically driven dual shaft multibladed shredder machine with 56° angle cutting and a twin shaft shredder machine with 16 blades on each shaft (the output power required of 2 Hp) can be designed to cut the plastic waste [20].

The shredding process can also be defined as a combined compression process with tension and shearing force together onto waste or cutting waste into smaller forms. The shredding process can be categorized into two types: dry shredded and wet shredded. These two processes are differentiated by physical observation of waste during the shredding process, either in wet or dry conditions. Shredding processes in the dry waste can be done by using hammer mills, large and small grinder, jaw crushes, cutter, clippers, chippers, and shears, while wet shredding machine is done by rasp mills and hydrapulper systems. These are discussed in the next section.

5.2.1 *Dry Processes*

5.2.1.1 Hammer Mills

Dry shredding is normally used on municipal solid waste. The most common machine used in this process is hammer mills. Generally, a milling machine is a machine that functions to cut material by using multiple cutter blades installed in the machine. There are various types of miller cutters designed to fulfill the industry demand, such as plain milling cutters, face milling cutters, side milling cutters, angle milling cutters, end milling cutters, fly cutters, T-slot milling cutters, formed cutters, and metal slitting saw. The properties of some of these milling machine cutters and their applications are listed in Table 5.6 [21]. Also, illustrations of some cutters such as plain milling, angle milling, and straddle milling are illustrated in Fig. 5.3 [22].

Another important factor that influences the efficiency of a milling machine is its spindle speeds. Its speed and also feed rate play a crucial condition in making the milling machine work effectively. Both these factors are related to velocity and occur in different spaces but are important for the cutting process. A milling machine

Table 5.6 Summary types of milling machine cutters [16]

Type of cutter	Properties	Application
Side face	The arrangement of teeth is on the periphery and on both sides Size: Up to 200 mm diameter, 32 mm wide	Steps and slots
Cylindrical	Helical teeth are on the periphery Size: Up to 160 × 160 mm^2	Flat surfaces parallel to the cutter axis
Single angle	The teeth are on a conical surface and flat face Size: 60–85° in 5° steps	Angled surface and chamfers
End mill	Helical teeth at one end and circumferential Size: ≤50 mm	Light work, slots, profiling, facing narrow surface

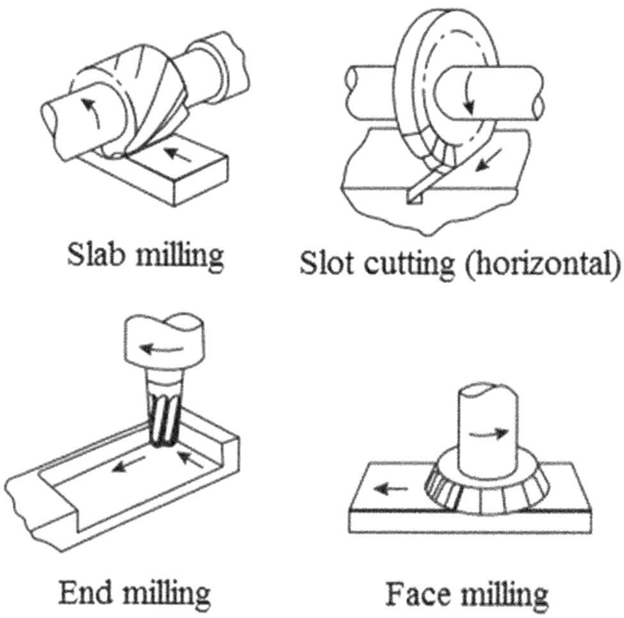

Fig. 5.3 Types of milling cutter applied on-site [22]. Reprinted from Scallan, P. Production equipment and tooling selection, Chapter 5. In Process Planning-The Design/Manufacture Interface, 2003, 171–218, with permission from Elsevier

is a machine that is designed with an electrical motor and subjected to wear and tear. Operation of milling at the right speed condition with the correct feed rate is able to speed up the cutting process and increase the lifetime of the machine.

The spindle speed refers to the velocity or speed of the cutting edge that moves and cuts the material. The speed is defined in revolutions per minute (rpm). The feed rate refers to the velocity at which the cutter is advanced with the workpiece, and the feed rate is usually presented in millimeter per minute (mm/min).

The right amount of spindles, speed cutter, and feed rate used is chosen based on several factors as listed below:

(a) Types of material feed
(b) Workpiece diameter
(c) The strength of the cutter
(d) Diameter cutter
(e) Cutter geometry
(f) Depth of cut
(g) Type of cut
(h) Condition of machine
(i) Power supply on the spindle
(j) Types of finisher desired

The spindle speed and feed rate of the milling machine are mathematically determined by using the following equation:

$$Spindle\ speed\left(rpm\right) = \frac{Cutting\ speed\ of\ the\ material\ x\,1000}{d} \tag{5.1}$$

where rpm is revolutions per minute and d is the diameter of the cutter, and the suitable cutting speed material is given in Table 5.7.

The feed rate is calculated by using the following equation:

$$Feed\ rate\left(mm\,/\min\right) = fe\ x\ N\ x\ Rp \tag{5.2}$$

where *fe* is the movement of per blade or teeth cutter in mm, *N* is the number of teeth cutter, and *Rp* is the rpm of the cutter.

Usually, municipal solid waste is collected in various types and shapes from the site. Hammer mills become the choice because the design of the machine can tolerate any type of material and can work effectively compared with other machines. Hammer mills are not only used in municipal solid waste treatment but also widely used in other industries such as mining, construction, agriculture, forest waste, and milling [23, 24].

Hammer mills work with a set of rotating swing steel hammers or blades, which will crush cut the waste that passes through them. Fig. 5.4 shows a conventional hammer mill used in industry. Hammer mills are designed with a semicircular screen at the bottom of the machine, which allows waste circulation and re-crushing

Table 5.7 Cutting speed of material

Material	High-speed steel	Carbide
Mild steel	25	100
Aluminum	100	500
Hardened steel	–	50

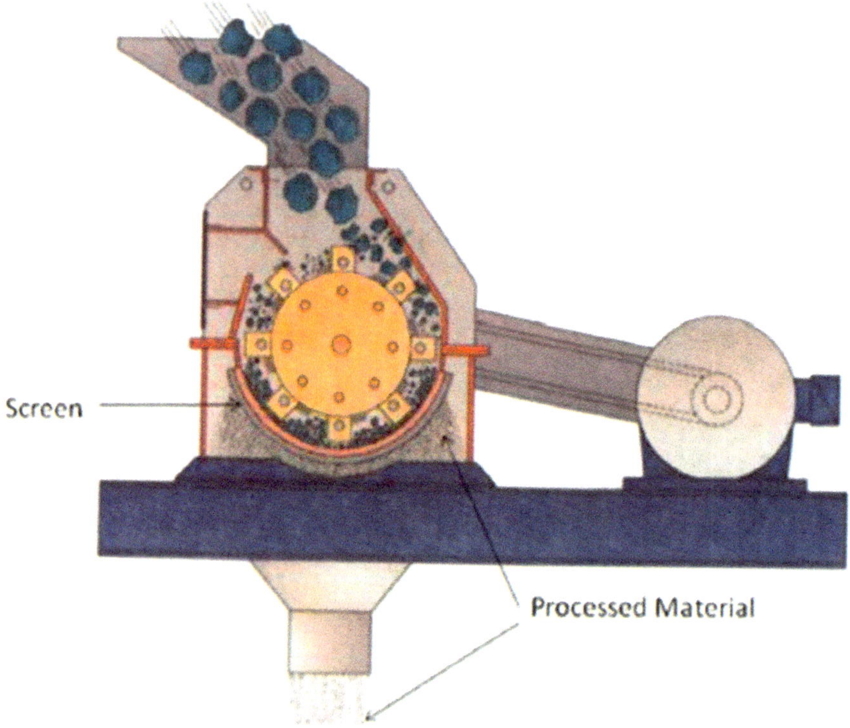

Fig. 5.4 Illustration of conventional hammer mills [23]. Reprinted by permission of Taylor & Francis Ltd.

as well as producing smaller by-products. However, this design has reduced the feed rate of waste and slows the operational process [23]. Therefore, some improvement to overcome this weakness has been made. Ezurike et al. (2018) [23] introduced hammer mills with flat-screen, which resulted in no circulation of waste as well as re-crushing process. Without re-circulation of waste, the rate of feed waste has increased compared with the old design. Their proposed design is illustrated in Fig. 5.5.

Another important part of Hammer mills is the hammer itself. Hammer mills have different shapes of the hammer such as sharp-edged, blunt-edged, two edges, shredder ring, four edges, round-edged, and splitter. The illustration of this hammer is depicted in Fig. 5.6 [25]. Each of the hammers has din weight, starting from a few kilos for a smaller machine to 226.5 kg (bigger Hammer mills) [25].

There are two types of hammer mills: vertical shaft hammer mills and horizontal shaft hammer mills. Between those types of hammer mills, the vertical shaft hammer mills are usually used at the site. Vertical hammer mills are crushing machines with an attached blade to the vertical shaft, as shown in Fig. 5.7. Vertical hammer mills are widely used in the feed production industry, such as corn processing and grains. These machines have become a favorite choice compared to the horizontal

Fig. 5.5 Illustration of proposing hammer mills with flat-screen [23]. Reprinted by permission of Taylor & Francis Ltd.

type because their energy consumption is only 25% compared to horizontal hammer mills; hence, they reduce the cost of operation. Moreover, by using these machines, zero loss of moisture material during shredding is recorded. Another key thing is that these machines produce less noise during operation, which is a maximum of around 83 dB compared to horizontal hammer mills.

In terms of power, vertical hammer mills record between 5 and 550 horsepower per ton of waste per hour [26]. The power of the motor required is usually applied based on three factors as follows:

(a) The size of solid waste
(b) Processing rate required
(c) The size of the end product required

A suitable power source must be supplied to the machine so that the shredder can process waste effectively. Table 5.8 shows the minimum horsepower requirement of the shredder based on the type of solid waste. This data can be used as a guide to manufacturing a shredding machine effectively. Heavy waste or wastes that have dense properties, such as demolition rubble, require bigger shredder horsepower to operate compared to others. It needs a minimum of 2000 horsepower for the shredder to well operate [25]. In contrast, only 250 horsepower is required to shred light waste or domestic waste such as paper, cardboard, bottles, food waste, garbage, and lawn trimmings [25].

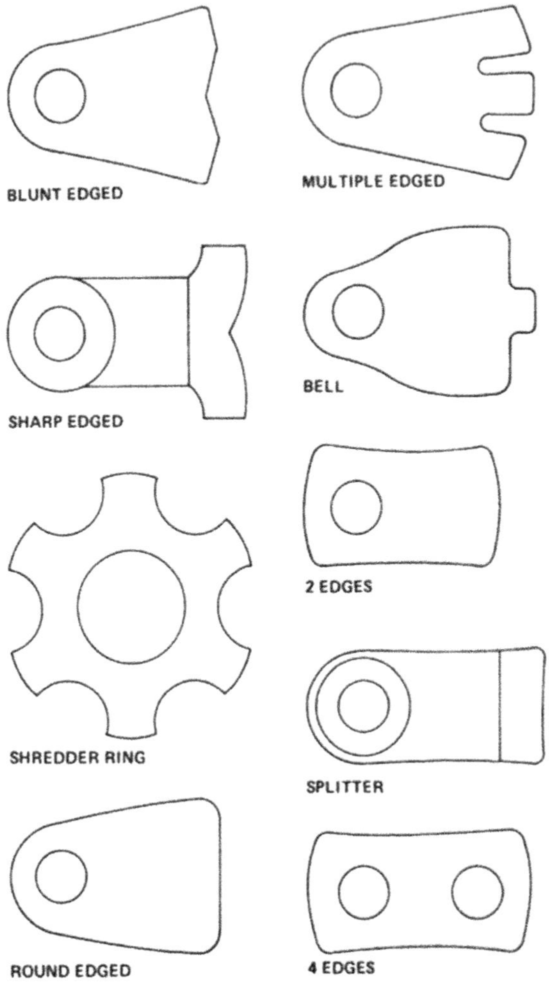

Fig. 5.6 Types of hammers used in hammer mills [25]

The size of the end product is one of the important factors to determine the power needed for operation. Nomographs of power requirement with end size product have been generated as depicted in Fig. 5.8 [25]. This graph can be used as a guide for designing and operating a new shredding machine.

Hammer mill machine is not only applied on MSW but also used widely as a machine crusher in the agriculture industry. Waste from agriculture, for example, stems and leaves, and also animal feed can be reused as organic fertilizer. Hammer mills will cut stems and leaves into proper forms, easy for handling, use, and storage [27]. Another example of a common machine used to process agricultural waste is the reel-type elephant grass chopper machine and bionic saw blade for corn stalk cutting.

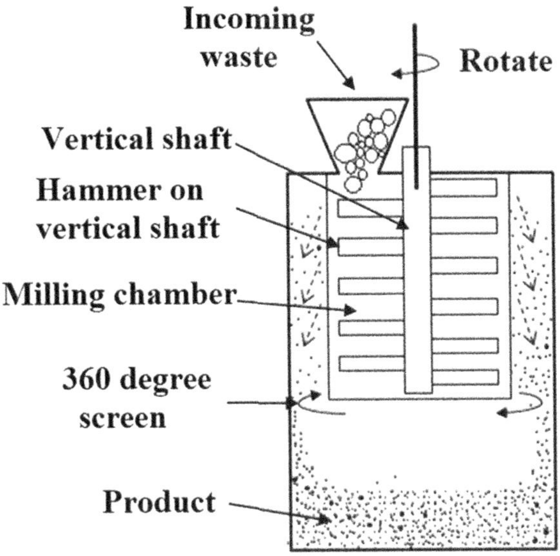

Fig. 5.7 Illustration of vertical hammer mills

Table 5.8 Minimum house power required by shredding machine [25]

Type of solid waste	Composition	Minimum horsepower required
Light	Domestic waste such as paper, bottles, garbage, and lawn trimmings	250
Medium	Normal packer truck waste such as small carting, small appliances, small furniture, bicycles, and occasional auto tires	400
Combined light and medium	Combined light and medium waste as above	600
Bulky	Oversize and bulky waste listed above such as large furniture, springs, washer machine, tree limbs, and truck tires	800
Heavy	Large and dense material such as metal, automobile, and stumps	2000

5.2.1.2 Chipper

Another type of dry shredding machine is called a chipper. Chipper is a crusher machine that is specifically designed for handling wood material such as tree limbs and trunks. There are several types of chippers, such as disc chippers, twin disc chippers, v-drum hop chippers, v-drum hog chippers, and cylinder drum chippers [28]. The common chipper machine used on-site is a drum chipper, as shown in Fig. 5.9. The drum chipper becomes a user favorite because of its capability to dealing effectively with a small size of wood, such as tops and branches.

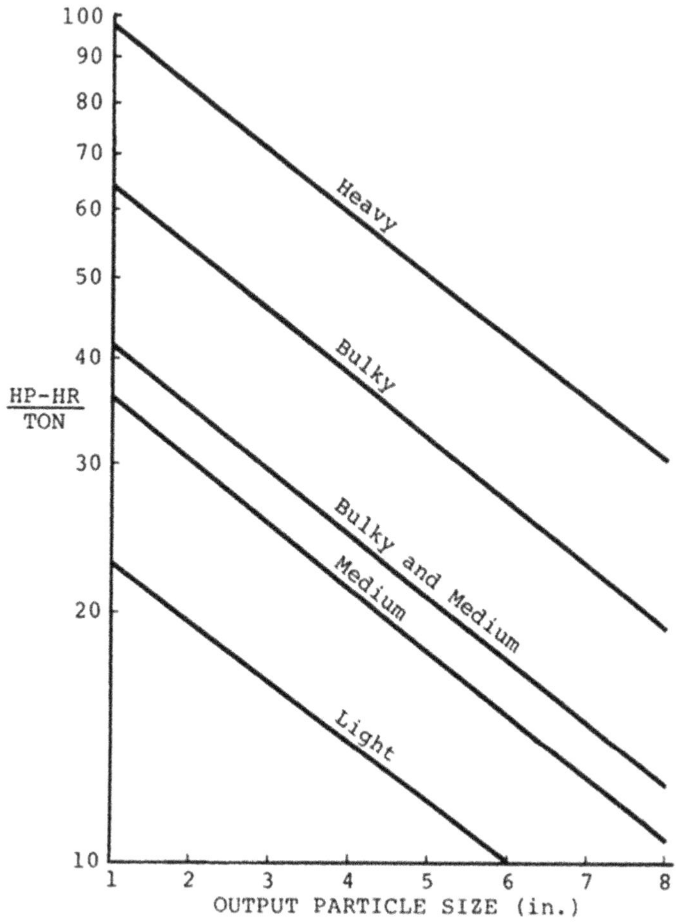

Fig. 5.8 Power requirement for solid waste size reduction with output particle size [25]

One of the important factors that influence the formation of chip wood is the cutting method. Normally, types of cutting method were chosen based on their operational purpose and wood chip uses in the next process. Not only the types and condition of feed material but also the cutting angle plays a crucial role, especially in the thickness of wood chip, removal rate, and cutting force [28, 30]. Usually, the knife angle used in the cutting process is between 30° and 37°. It is important to know the suitable range of cutting angles before starting the process as it can avoid a huge reduction in the thickness of the chip as well as cutting force and the chip damage [30].

On top of that, each type of wood chip formation is also able to explain the cutting mechanism that occurs and is applied to it [31]. There are four types of chip formation, as listed below and illustrated in Fig. 5.10 [32].

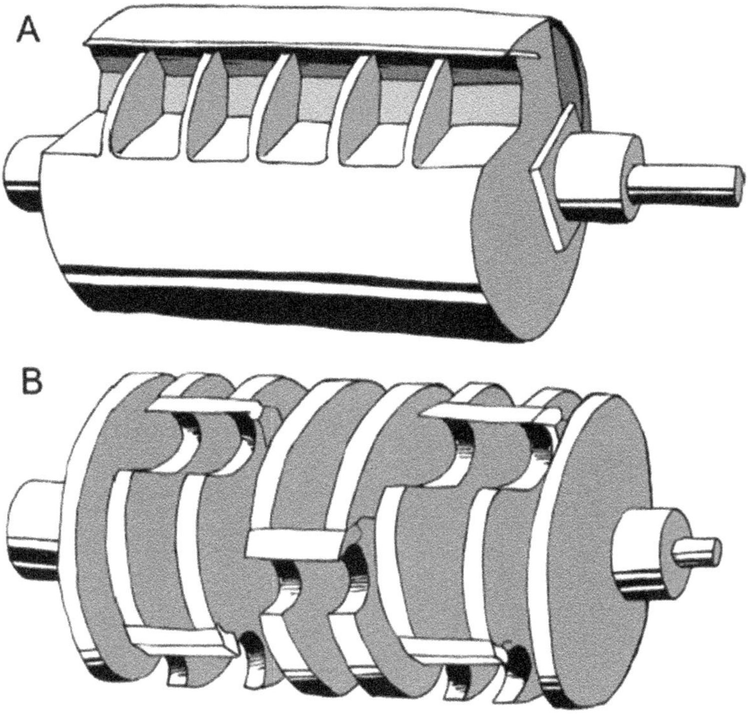

Fig. 5.9 Illustration types of drum chipper machine: closed drum with a full-length knife (A) and open drum with staggered short knives (B) [29]. Reprinted from Spinelli, R., Cavallo, E., Eliasson, L., Facello, A., Magagnotti, N. The effect of drum design on chipper performance, Renewable Energy, 81 (2015), 57–61, with permission from Elsevier

1. Continuous
2. Lamellar
3. Segmented
4. Discontinuous

5.2.1.3 Von Roll Bulky Waste Shear

Shear machines such as Von Roll bulky waste shear and multiblade hopper-type shear are usually used in handling bulky waste like furniture waste and construction waste. As shown in Figs. 5.11 and 5.12, these machines have a series of blades, which will crush the waste placed in between the blades when the blade jaws are closed. The Von Roll bulky shear machine is mostly suitable as a first treatment of waste before it undergoes the incinerator treatment. Moreover, solid wastes like scrap steel and automobile bodies are also cut into shorter lengths by using this machine. However, this machine is not commonly used in municipal solid waste

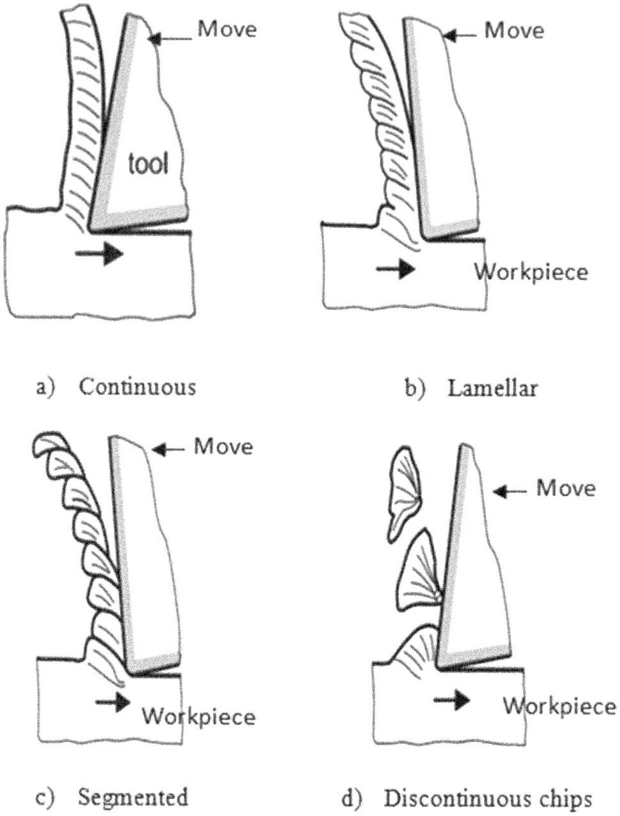

Fig. 5.10 Illustration of the type of chip wood formation [32]

treatment as its operational process is too slow and requires a longer time compared to other machines [33].

5.2.1.4 Ball Mill Machine

Another helpful shredding machine in municipal solid waste treatment is called a ball mill machine or cascade mill machine. This machine has a diameter-to-length ratio of around 3:1 and a shape like a cylinder (Fig. 5.13). This machine functions as a shredded waste machine by rolling action. This machine is filled with a steel ball. During the rotation of the drum at the optimum condition of 14 to 20 rpm, the steel ball will grind the wastes as they get into contact with each other. As a result, the final product of waste becomes smaller than before. After that, the end product with the required size will undergo screening using a screen. The large size waste will further be ground again.

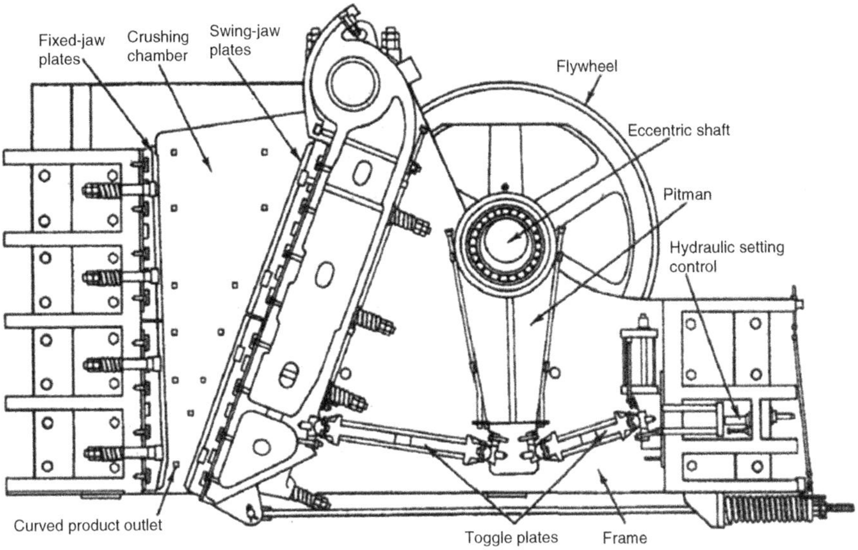

Fig. 5.11 Illustration of bulky waste shear crusher machine [31]. Reprinted from Olawale, J.O., Ibitoye, S.A. (2018). Failure analysis of a crusher jaw, Chapter 10. In Handbook of Materials Failure Analysis-With Case Studies from the Construction Industries, 187–207, with permission from Elsevier

The motion of the steel ball working in the ball mill is illustrated in Fig. 5.14. There are several factors that need to be considered when using this technology. They include the following [35]:

(a) Rotational speed of the drum machine at a constant speed ratio
(b) Milling time
(c) Filling ratio of milling steel balls
(d) Filling ratio of grinding materials

Note that steel balls can also be eroded over time. Therefore, steel balls need to be replaced at intervals.

Also, there are several criteria that have been listed by a researcher as a guide for choosing the right size of shredding machine to be applied on-site. These include the following [36]:

(a) The properties of materials before and after shredding
(b) Size requirements for shredded material by component
(c) Method of feeding the shredders, provision of adequate shredder hood capacity (to avoid bridging), and clearance requirement between feed and transfer conveyors and shredders
(d) Types of operation, either continuous or intermittent

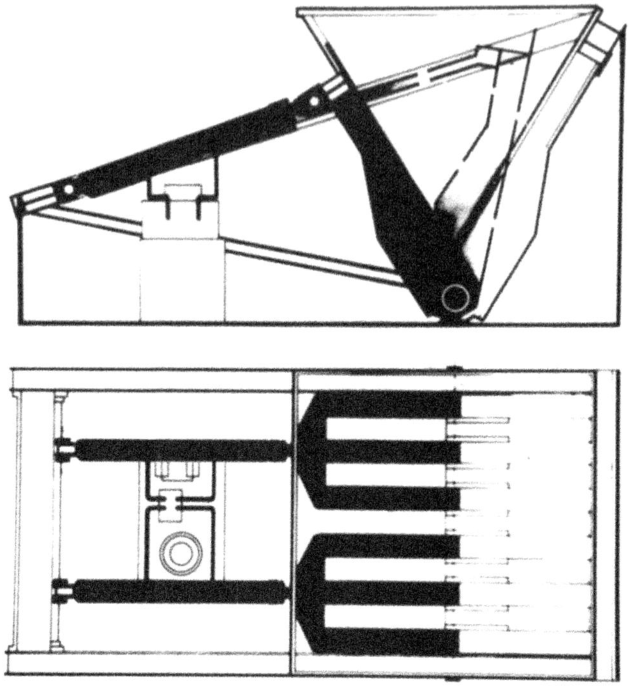

Fig. 5.12 Illustration of multiblade jaws in Von Roll bulky waste shear [25]

(e) Operational characteristics including energy requirements, routine and specialized maintenance requirements, simplicity of operation, reliability, noise output, and air and water pollution control requirements
(f) Site considerations, including space and height, access, noise, and environmental limitations
(g) Metal storage after size reduction for the next operation

5.2.2 *Wet Processes*

Another method of shredding is a wet process. Similar to the dry process, this type of shredder also requires a mechanical structure to complete the operation. The rasp mills and hydrapulper machines are common machines used in this method. These machines are usually effective on organic waste such as paper, agriculture or wood material, and food waste. Thus, these machines are widely used in the paper and cardboard production industries.

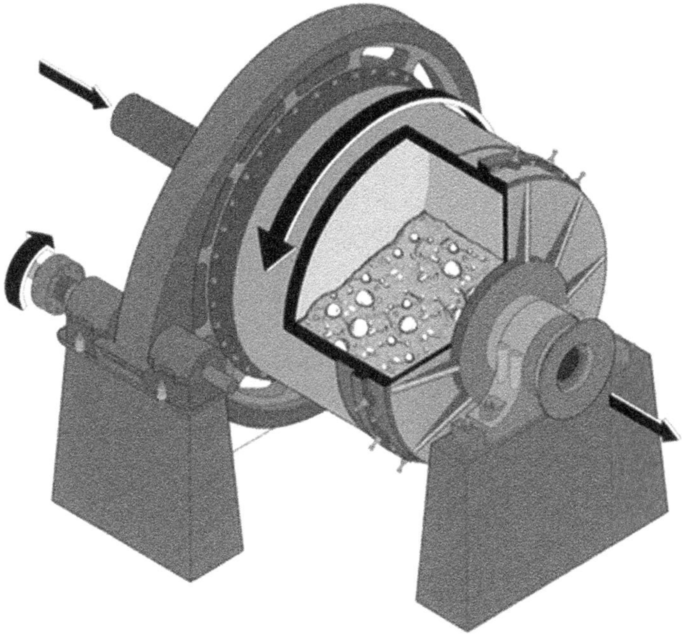

Fig. 5.13 Illustration of ball mill machine [34]

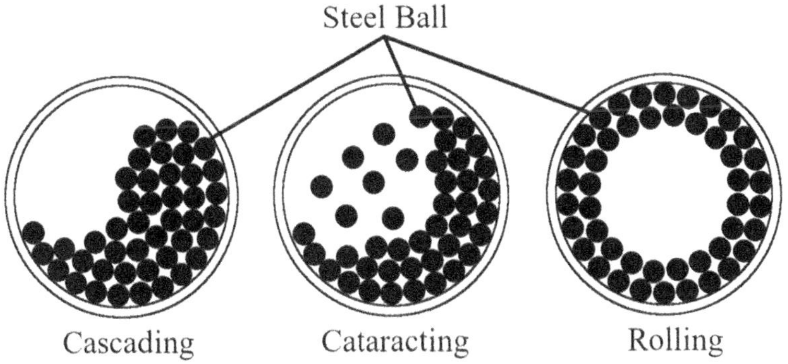

Fig. 5.14 Illustration of types of ball motion during operation in ball mill machine

5.2.2.1 Hydrapulper

Hydrapulper machines are ideally used for shredding organic matter like paper waste and wood as these materials are easy to break in watery condition. Also, this system is widely applied in the recycling fiber industry, especially for the production of paper-based products such as egg tray production lines, paper pulp, and paper cardboard.

Consumption of paper and paperboard is higher and rising in each country as well as their waste. Statistical analysis in 2018 has shown that paper and paperboard have high demand worldwide, as depicted in Fig. 5.16. The largest consumption of paper and paperboard is recorded as China (110 151 thousand metric tons), followed by the United States (70 674 thousand metric tons) and Japan (25 459 thousand metric tons) [37] (Fig. 5.15).

Authorities are concerned about the amount of waste produced as a result of the high demand and use of paper. Therefore, in order to overcome the excessive multi-material waste dumping at the landfill, many countries have implemented various recycling or recovery material technologies, especially for their domestic wastes.

Generally, there are three groups of recycling material in domestic waste: organic recycling, soft plastic recycling, and mixed waste recycling. These are detailed in Table 5.9. However, not all wastes can be recycled for a new product. Some of them still require landfilling and incineration processes. The list of unrecyclable waste materials is as follows:

(a) Broken crockery
(b) Packing straps
(c) Sticky tape
(d) Polystyrene
(e) Glassware (for drinking)

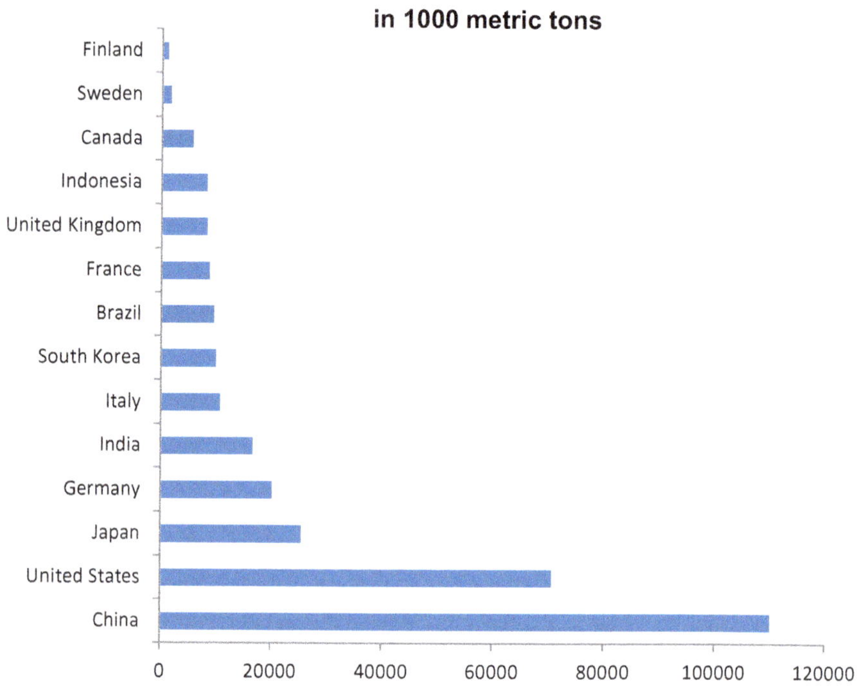

Fig. 5.15 Consumption of paper and paperboard in selected countries around the world [37]

Table 5.9 Types of recycling waste material

Organic waste	Mixed waste	Soft plastic waste
Food	Glass bottles	Food packaging such as bread bag, pasta bag, rice bag, biscuit packets, and cling wrap
Fruit and vegetable	Newspaper	Plastic bag
Meat, fish, and leftovers	Clean cardboard	Bubble wrap
Coffee ground	Plastic	Confectionary bag
Flowers	Office paper	
	Aluminum cans	

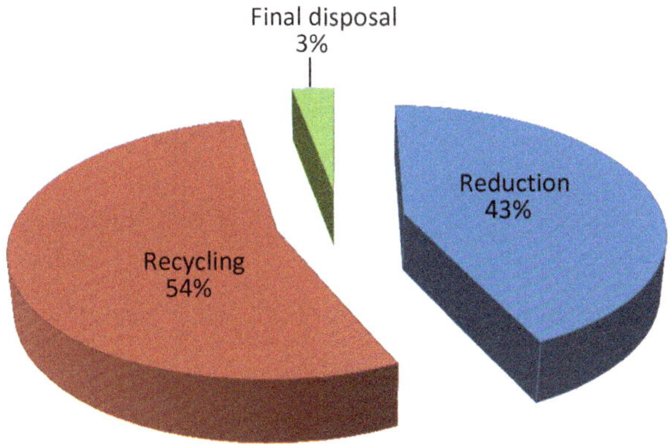

Fig. 5.16 Treatment of industrial waste in Japan, 2014 [36]

Japan and Singapore are an example of Asian countries that have widely implemented recycling technology, starting from the house to the industrial sector. Japan started this recycling program in 2000, and as a result, the volume of waste in landfills has drastically reduced by year (from 2004 until 2014) after that [36]. As listed in Fig. 5.16, various waste treatment methods such as recycling activity could reduce the volume of final disposal waste at the landfill.

According to the data (Fig. 5.16), 54% of waste can be recycled, while 43% of industrial waste undergoes incineration and nonincineration processes such as dewatering, thickening, and drying processes [36]. Furthermore, only 3% of waste remains and is dumped into a landfill after waste treatment. This data shows that waste treatment could increase the lifespan of the landfill and can save natural resources.

Fiber recycling technology, for example, can save a lot of energy sources and reduces the volume of landfill waste. Amemiya (2018) [38] stated that 7000 tons of fiber waste were saved from dumping into the landfill. Not only that, energy needs for the pulping process from waste were also saved at a rate of 11 to 30 kWh per ton energy compared to pulping of virgin wood (1,972 kWh per ton) [37]. Additionally, a combination of waste treatment has also been proven to be able to effectively

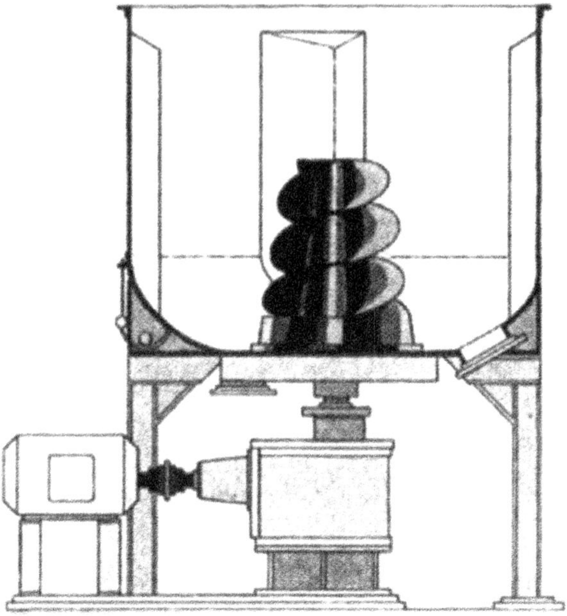

Fig. 5.17 A hydrapulper machine [41]. Reprinted from Batjai, P. (2018). Fiber From Recycled Paper and Utilisation, Chapter 23. Biermann's Handbook of Pulp and Paper (Third Edition), Volume 1: Raw Material and Pulp Making, 2018, 547–582, with permission from Elsevier

manage solid waste, optimize renewable energy, and reduce the waste at landfill sites [39].

In general, a hydrapulper machine is a machine that has a big grinder placed in the center of the tank, as shown in Figs. 5.17 and 5.18. After dumping solid waste (paper) into the tank (filled with water), the blade will rotate, and the rotation force will cut the waste into smaller pieces or fine pulp. Waste or rejected material will be removed separately from the pulp and exit at the side of the tank. The pulp will be dense and settle at the bottom of the tank for collection.

There are two types of blade used in this system: screw pulping and blade pulping. Each of these types has different properties and characteristics, as listed in Table 5.10 and Figs. 5.17 and 18. However, a hydrapulper machine is mostly suitable for paper waste or fiber waste that has less contamination. Contamination here refers to the paper waste or fiber waste that is coated with plastic or foil and potentially obstructs the workability of the machine.

5.2.2.2 Rasp mills

Another wet shredding machine used on-site is rasp mills, as illustrated in Fig. 5.19. The rasp mill machine is designed with a rotating blade at the bottom of a large cylinder. The internal rotor at the bottom of the cylinder usually moves at a speed of 5 to 6 rpm to cut the waste. Similar to the hydrapulper system, water is poured into the drum to allow maximum efficiency of tearing or shredding the waste. Fine pulped sizes of 5 cm will be discharged at the bottom of a cylinder through a sieve

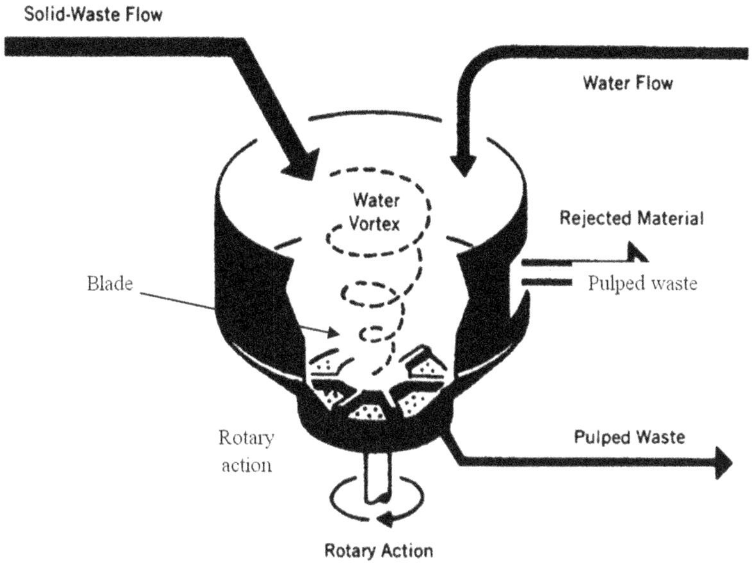

Fig. 5.18 Illustration of blade pulping in a hydrapulper system [25]

Table 5.10 Characteristics of screw pulping and blade pulping in the hydrapulper system

Screw pulping	Blade pulping
1. Volume tank is bigger than 4 m³, depending on productivity.	1. Volume tank is smaller than 4 m³.
2. Place vertically in the center of the tank, as shown in Fig. 5.7.	2. Set up sets of the blade on the bottom of the tank as shown in Fig. 5.8.
3. Able to crush solid waste in bulk effectively.	3. Suitable for a small quantity of process.

plate while the uncut materials, the sizes of which cannot be reduced, will be rejected at the side of the cylinder.

5.3 Size Reduction by Compaction Process

Compaction is a process of reducing the void inside material and size and increasing the density of the subject. During the compaction process, high pressure will be applied at the surface of the subject and the waste is compressed until the maximum size reduction is achieved. The compaction force applied is illustrated in Fig. 5.20 to give a clear understanding of this method.

Compaction of solid waste occurs at an earlier stage, starting from the collection of waste. In terms of their mobility characteristic, the compactor can be grouped into two types: stationary equipment and movable equipment. Stationary equipment

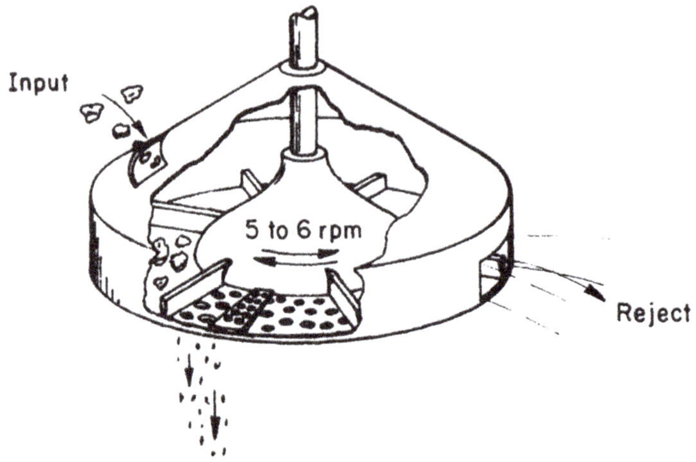

Fig. 5.19 Illustration of a rasp mill machine [25]

Fig. 5.20 Compression force

is equipment that is brought to the collection area, and the waste will be dumped into it by the workers.

A common machine used on-site is a garbage truck. This machine is suitable for lightweight load, commercial, or light industrial waste. The waste is dumped into its steel container and will be compressed into a block before transferring it to the waste treatment station [42]. Compaction of waste can reduce the volume of waste and provide more storage capacity for waste in that truck.

In contrast, movable equipment refers to the truck that is functional to compact waste at a landfill site, and it is usually called a landfill compactor. This machine has four-wheel tractor types and designs with a dozer blade in front of it. The landfill compactor also has a cleat wheel, which is functional as a grinder and crushes the waste as well as compacts it for reducing the volume at the landfill site [40]. The details of these two compactors are explained in Table 5.11.

Table 5.11 Details of stationary and movable compactor equipment

Stationary equipment	Movable equipment
1. Machines where the wastes are brought to and fed into, either manually or mechanically	1. Compactor machines that are used to place or compact solid waste on the soil, such as in a sanitary landfill
2. For example, garbage trucks	2. For example, compactor truck, dozer, and truck loader
3. Ideal for lightweight solid waste such as household wastes or domestic waste (light duty); usually used at the collection stage of solid wastes on-site	3. Suitable for handling a large volume of solid waste such as a sanitary landfill

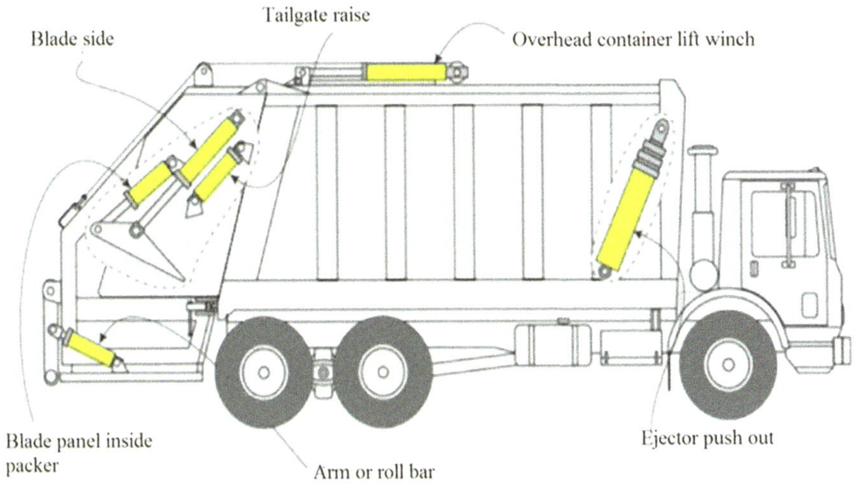

Fig. 5.21 Self-compacted truck [43]. Reprinted from 43. Voicu, G., Lazea, M., Zabava, B. S., Tudor, P., & Moise, V. (2019). Cinematical analysis of the pre-taking and pre-compacting mechanisms of some garbage trucks. Journal of Engineering Studies and Research, 25 [2], 56–62, with permission from the University of Bacau (Alma Mater Publishing House)

5.3.1 Compaction Technology

Normally, municipal solid waste is collected manually by workers and will be thrown inside the garbage truck container equipped with stationary compactor equipment. Each of these trucks has been designed with a compactor system that allows solid wastes to be compressed into smaller forms and instantly provide more space for more trash to be collected. There are several types of garbage trucks, such as rotating drum compactor trucks and self-compacted trucks (Fig. 5.21). They have different compactor systems, and the details are listed in Table 5.12.

Research has been conducted by a few workers to improvise existing waste compactors, such as vibrating compactors and hydraulic garbage compactors. David et al. (2020) [44] have designed and improved a hydraulic garbage compacted machine to become more efficient and steady power in compression compared to

Table 5.12 Details of rotating drum trucks and self-compacted trucks

Rotating drum trucks	Self-compacted trucks
1. The drum is rotating at 4 rpm, and waste will be dropped into that rotating drum by civil workers.	1. Cover a small operating area.
2. Rotation of waste is in a continuous direction with the help of a welded auger, which prevents waste from rotating in the opposite direction.	2. Reduce manpower (can be operated by remote control); affordable and suitable for all kinds of garbage collection, compression, and transport.
3. This method allows compaction, shredding, and mixture effects on solid waste.	
4. Waste collected is suitable for incineration, sorting, and recycling next treatment process.	

others. This machine is designed with high power–weight and high torque–mass ratios, as well as a force–inertia ratio, which allows acceleration and swift response of the hydraulic motors.

The evolution of this solid waste compactor technology is continually done by other researchers until now. Research and development of this technology are important to reduce the solid waste volume as well as reducing the number of manpower in handling it. Many countries (including Malaysia) are currently using garbage compactor trucks for daily domestic waste collection.

Compaction for the volume reduction process could also be applied at different stages of MSW phases, such as in solid waste generation points, processing stations, and disposal sites. Each of these operational points will use different types of compactor equipment based on the solid waste condition and the purpose of the operation. The compactor used in each stage of the MSW process is listed in Table 5.13.

As listed in Table 5.13, there are five types of compactor usually applied in a residential area:

(a) Vertical
(b) Rotary
(c) Bag or extruder
(d) Under-the-counter
(e) Commercial

A stationary compactor becomes a preferred selection at the collection phase, processing-transfer station and a disposal site. It's easy handling, moveability, and big space for waste storage have made garbage trucks mostly used in MSW management.

The compaction process can be done under low and high pressure. Low compactor pressure has less than 7 kg/cm^2 of compaction strength, while high compaction pressure is more than 7 kg/cm^2. The details of these two compaction pressure equipment are listed in Table 5.14.

Also, solid waste reduction volume by compaction can be expressed by percentage reduction and also compaction ratio as presented in Eqs. 5.1–5.2. This

Table 5.13 Type of compactor equipment involved in the MSW process [44]

Location/stage of operation	Type of compactor equipment	Description
Solid waste generation points (residency area or house)	Vertical	Operate by mechanical or hydraulically system, and usually, the wastes are fed into the compactor container. This technique is usually applied in medium- and high-rise apartments.
	Rotary	Waste is compacted by the ram mechanism on rotating platform containers. This technique is usually applied in medium- and high-rise apartments.
	Bag or extruder	Wastes are dumped into the compactor by using a chute, and vertical or horizontal ram can be applied in this compactor. Note that a single bag must be replaced, and continuous bags must be tied off and replaced. This technique is usually applied in medium- and high-rise apartments.
	Under-the-counter	Individual small compactors are used in house, apartment, or residences. Waste is placed in a unique bag for the compaction process.
	Commercial	Compactor with a vertical and horizontal ram. The wastes are compressed into steel containers and compressed wastes are manually tied and removed. This technique is usually applied in medium- and high-rise apartments, as well as in commercial and industrial facilities.
Collection	Stationary	Waste is collected by using vehicles or garbage trucks, which have equipped with a compaction mechanism as explained in Tables 5.3 and 5.4.
Transfer and/or processing station	Stationary	Transfer trucks are equipped with a self-contained compaction mechanism.
	Stationary low pressure	Wastes are compacted into large containers.
	Stationary high pressure	Wastes are compacted into dense bales or other forms.
Disposal site	Movable wheeled or truck equipment	Designed for high or maximum compaction of waste as explained in Table 5.3.
	Stationary	High-pressure movable stationary compactors are used for volume reduction at a disposal site.

Table 5.14 Description of low-pressure compactor and high-pressure compactor

Low-pressure compactor	High-pressure compactor
1. Compaction pressure is less than 7 kg/cm^2	1. Compaction pressure is more than 7 kg/cm^2. Compact systems with a capacity up to 351.5 kg/ cm^2 or 5000 lb/in^2.
2. Solid wastes are compacted in a large container.	2. Solid wastes are compacted by using specialized compaction equipment into blocks or bales of various sizes.
3. Widely used in the industrial sector, especially for recycling waste material like paper and paperboard.	3. The volume reduction was achieved effectively using this method.

measurement is important, especially in making a trade-off analysis between compaction ratio and cost.

$$Volume\,reduction\left(\%\right)=\frac{VR_i-VR_f}{VR_i}x100 \tag{5.3}$$

where VR_i is the initial waste volume before compaction (m^3) and VR_f is the final waste volume after compaction (m^3).

The compaction ratio of the waste is calculated as follows:

$$Compaction\,ratio=\frac{VR_i}{VR_f} \tag{5.4}$$

where VR_i is the initial waste volume before compaction (m^3) and VR_f is the final waste volume after compaction (m^3).

5.4 Size Reduction by Baling Process

The baling process is a process of reducing solid waste volume by using heavy-duty balers. This technique has been applied for a long time, especially in compressing loose materials like hay and cotton in the plantation. There are three types of yarn baling applied on-site: soft bales, pressed bales, and bundling and baling method. Each of these techniques produces a different form of product with respect to the needs.

The baling method has been widely used not only on the farm but also in the recycling industry of solid waste management system. This technology has been applied in solid waste management pioneered in the United States and Japan. It is more convenient and economical when solid waste is compacted prior the transportation to the landfill or recycling center. Usually, metal and paper are common materials processed by this method, as detailed in Table 5.15.

Table 5.15 Application of baling in the recycling industry

Material	Description
Paper	- Waste paper or fiber is usually sorted and separated for recycling, starting from home. After that, the waste will be collected and sent to a recycling center for the next process. - In the recycling center, the paper is pressed into bales and stored before another process of recycling takes place.
Steel or metal	- Similar to paper waste, iron can or metal waste is collected and transported to a recycling center for further process. - Before baling takes place, the waste will be sorted from unwanted material by using a rotary drum magnet. - Only iron material will be stuck to a rotary drum magnet and compressed by a baler machine.

5.4.1 *Baling Process*

Basically, baling is a technique that reduces the void of material and increase the weight as well as the density. As presented in Fig. 5.22, applying forces in all directions toward the subject will reduce the size, will be advantageous in terms of handling and space transportation, and will also cut the cost of storage [46]. The end product of this technology usually comes out as a cubic shape or cylinder shape, which is easier to handle and store. Also, this method promotes fewer gas emissions from the vehicle by reducing the number of transports needed. Obviously, this method has many benefits in solid waste management and it also is environmentally friendly.

During the baling process, a series of forces are applied onto the surface of the subject within the baler as shown in Fig. 5.22. The baling process starts by inserting solid waste from the top into the baler chamber. Then, the top baler is closed, and three-stroke baling is applied sequentially on the side surface (1), top side (2), and smallest surface (3), as shown in Fig. 5.23. These high-pressure forces are transferred by the ram to the surface solid waste and the waste is compressed into a smaller size. After the process is complete, the ram is held in its position for a few seconds for "set." After that, the lid is open, and a block of waste is removed and ready for transportation to the next station or storage. The product from this process becomes a more dense material.

5.4.2 *Baling Technology*

The end product of the baling process is usually present in a large cube. In the earlier days, there are only two types of baling: medium-pressure and high-pressure baling. Medium-pressure baling is a method to reduce the volume of the sample and its ejection bale was tied with a number of wires, while high-pressure baling is self-sustaining bale with no wire attached to the baler and allows its ejection expends.

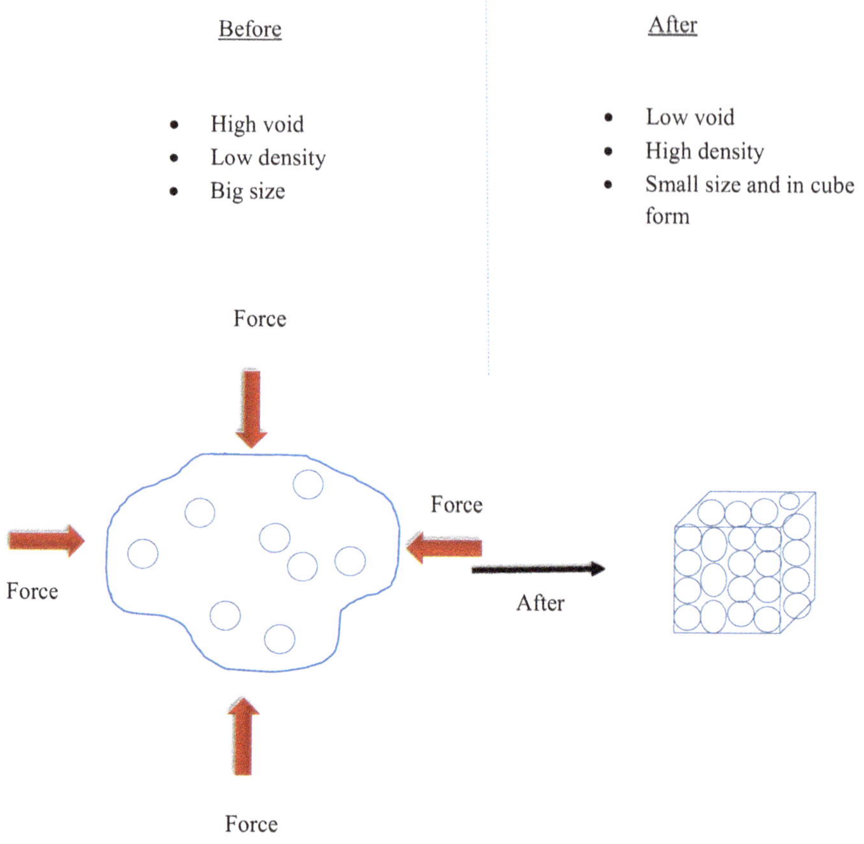

Fig. 5.22 Baling force

However, medium-pressure baling is proven as stronger than high-pressure baling, while in terms of dropping, a self-sustained bale is better. Over time, there are new advances in baling technology to increase the efficiency of this technique. Baling technology can be applied in most material types except for solid or rigid materials such as concrete, steel, beam, and large plastic roll.

Generally, there are three types of baler applied on-site, which are handheld, vertical, and horizontal [36]. Handheld is a common method used in packaging household things such as cloth and sweaters. Usually, these products are packed in a wooden box and strengthen the wrapping by putting wires around it using a hand-held baler. Meanwhile, the vertical and horizontal balers are differentiated by the way the sample is fed into the machines.

Several criteria have been listed by researchers as a guide for choosing the right type of baling. The list is as follows [36]:

(a) The daily volume of waste collected
(b) The density desired from compaction

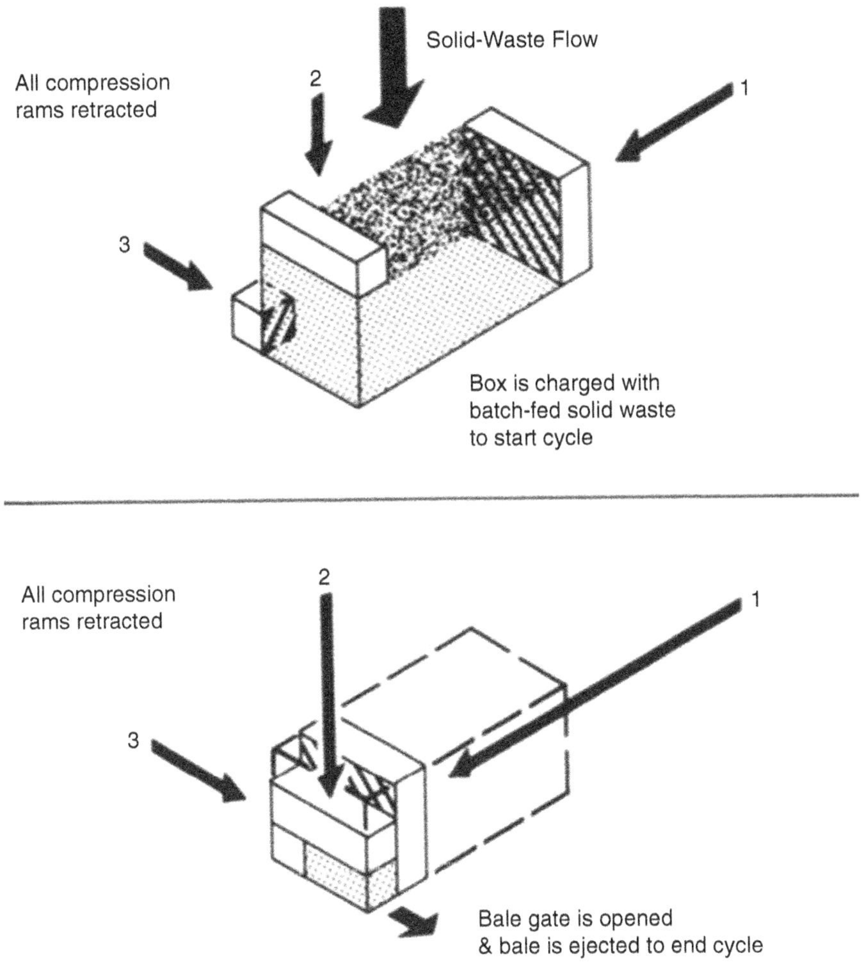

Fig. 5.23 Baling technique [25]

(c) Estimation of future waste volumes
(d) Cost estimation
(e) Shape and size of solid waste
(f) Type material of solid waste
(g) Type of vehicle for transportation

5.5 Application On-Site

As shown in Fig. 5.24, there are three stages of solid waste processes in the MSW management system, which are collection and transfer station, waste processing, and disposal. The collection phase is a stage where domestic waste is collected manually by the worker and transported to the transfer station. For example, in Malaysia, MSW management is under the jurisdiction of the municipality and operated by themselves or by private waste management and public cleansing service companies, as listed in Table 5.16.

Size reduction is applied in most of the solid waste management phases, starting from collection of waste from its source, transfer station, resource recovery, and disposal operation. Each stage has a different characteristic of solid waste and needs. Therefore, a different type of mechanical reduction volume is required in each phase. For example, during the collection of waste at the site, solid waste is received in bulky form. Therefore, it is important to provide a garbage truck with a compactor system.

A compactor system can reduce the volume of solid waste and increase the storage space as well as reduce the collection trip per day. Generally, about 80% of domestic solid waste is easy to be compacted, such as plastic, paper, rags, glassware, and small metal. Therefore, this method has been a favorite choice to be applied on-site. Another 20% of solid waste is difficult to be handled by

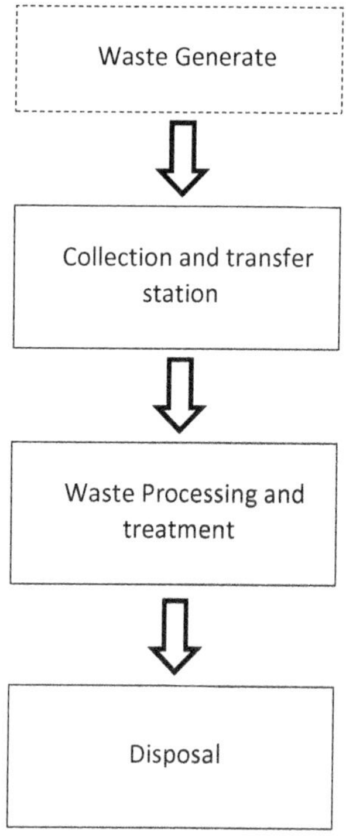

Fig. 5.24 Solid waste management systems

Table 5.16 List of private companies handling MSW in Malaysia

Zone	Company	States covered
Southern zone	SWM Environment Sdn Bhd	Johor, Melaka, and Negeri Sembilan
Northern zone	Environment Idaman	Kedah and Perlis
Central and Eastern zones	Alam Flora	Wilayah Persekutuan Kuala Lumpur/Putrajaya and Pahang

Table 5.17 List of compression pressure applied in different waste materials [31]

Material	Pressure (kg/cm^2)
Cotton and rubber	50
Average sample	65
Filter elements	70
Polyethylene	80
Metallic wastes	110

compaction, such as construction debris like concrete, metal scrap, and soil. It is important to recognize the types of solid waste before undergoing any treatment process. This step can prevent the machine from breaking down and increase the efficiency of the treatment.

Compactor machines can be designed as vertical or horizontal ram depending on the space availability of the operational area and also the position of the feed waste entrance and the exit for the product. Also, each material has different effective work pressure. In short, higher-density material requires more pressure to be compressed. Some experiments have been done by researchers to prove this theory, and the result is depicted in Table 5.17. The result shows that metallic material required high pressure compared to others, while cotton, which is light and less dense, required only 50 kg/cm^2 (smaller). Other materials such as filter components, polyethylene, and commingle samples required 70 kg/cm^2, 80 kg/cm^2, and 65 kg/cm^2, respectively.

There are several common compactors used at the site, such as vibrator compactors, prevailing motorized waste compacting systems, hydraulic compactors, crank mechanism electrical machines, and manual compactors. Usually, suitable compactor machines are chosen based on the type of waste to be processed, the volume of waste, and also the space area available on-site.

5.5.1 *Collection and Transfer Station*

Generally, a transfer station is a place to temporarily store solid waste before transferring it to another place for treatment or disposal. The applications of the mechanical reduction system used in this phase are summarized in Table 5.18. As noted in

Table 5.18 Summary of applications of mechanical volume reduction in transfer station [25]

MSW stage	Technology	Description
Transfer station	Baling	- This method allows a maximum load of solid waste per trip. - Capable of carrying 72,576 kg of waste compared to a normal truck, only 36,288 kg. - Does not require a close car/truck.
	Shredding	- Requires a close car/ truck. - Needs a large truck equipped to shred solid waste and dispose of it at the landfill site.

the table, shredding and baling technologies are applied in this stage. However, the compactor system is most commonly applied and used on-site. For example, most countries, including Malaysia, are using garbage trucks with compactor systems to collect MSW. After that, solid waste is transported to a transfer station or landfill site. This technology has provided a maximum capacity of solid waste during collection and transfer as well as reduces operational costs.

Compaction activities at transfer stations have been applied in several countries around the world. For example, in Japan, a small garbage truck is used to collect and transfer solid waste from its source or residential area to the transfer station. At this station, solid waste will be dumped into a compactor container, and it will be compressed and transferred to the big garbage truck. After that, the waste is transported to another station for further waste treatment. Overall, this system has improved the efficiency of collection and transportation operations in this country. Fig. 5.25 illustrates the collection and transportation of a solid waste system in Japan.

The shredding process in this is a bit challenging task. This process requires a big and closed truck to shred the waste and dump it at the landfill site. This big machine is more costly as well as requiring high maintenance costs.

5.5.2 *Waste Processing and Treatment*

Waste processing and treatment is the second stage in the MSW management system. In this stage, physical, chemical, and biological methods are applied. The methods include mechanical volume reduction, controlled landfill, incineration, gasification and pyrolysis, anaerobic digestion, and aerobic digestion. The suitable waste treatment is selected based on the type of waste, space provided, operational cost, and also product requirement by the next treatment. Some of the wastes are dumped or disposed of at the landfill site after passing this stage. The list of these technologies is listed in Table 5.19.

Nowadays, waste treatment technology is applied to recycle resources, recover materials, and generate energy from solid waste, or called waste to energy technology (WTE-T) [47]. This technology is more sustainable and environmentally friendly. Therefore, some of the researchers have introduced the combination of

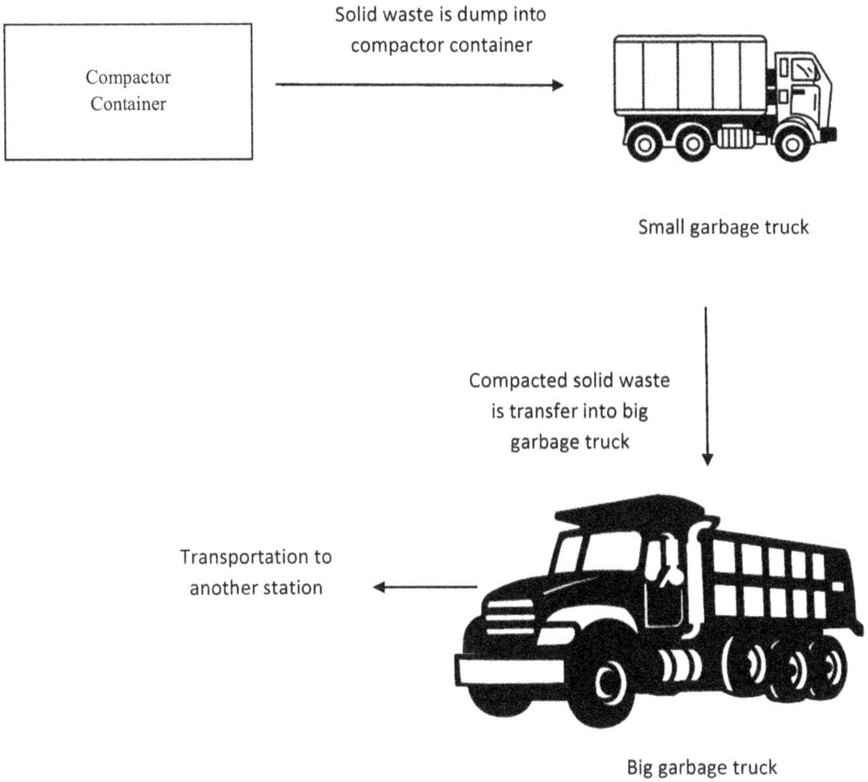

Fig. 5.25 Transfer system of solid waste in Japan

Table 5.19 Types of solid waste treatment in MSW

Type	Treatment
Physical	Recycling Mechanical volume reduction Controlled landfilling
Biological	Anaerobic digestion Aerobic digestion
Chemical/thermal	Incineration Gasification and pyrolysis Open burning

waste treatment such as a mechanical–biological and thermal–chemical method or hybrid treatment in order to achieve the goal.

Generally, waste treatment can be divided into three groups: physical treatment, biological treatment, and chemical treatment. Physical treatment includes recycling activities, grinding, compacting, baling, sieving, and classification for recycling material [48], while biological treatment involves anaerobic and aerobic digestion, which is usually combined with mechanical or physical treatment. The third waste treatment is thermal–chemical treatment. This treatment is usually run as an incineration process, gasification and pyrolysis, and open burning activities. Each of these treatments has advantages and disadvantages. The incineration process and pyrolysis have a negative effect especially on air, water, and land, which may also be harmful to the environment and human health [49].

Table 5.20 shows the mechanical treatment technology applied in this stage, which is a shredding process. This shredding technology is important and helpful in this stage as it provides an efficient condition for solid waste before undergoing further treatment processes. For example, shredding of solid waste prior to combustion will make wastes burning more homogeneous, hence reducing the combustion time as well as operational cost.

Shredding of solid waste before it is disposed of in landfills has also given a big impact on the landfill site and the surrounding as well as on the operational cost. Glisson (1980) [25] has listed several advantages of solid waste shredding prior to disposal at a sanitary landfill.

Advantages of shredding solid waste prior to disposal at the sanitary landfill are as follows:

a. Improved massy condition at the landfill site. Paper is not blowing around by wind since it is shredded and entangling with each other. However, the plastic film may still be a litter problem because this material is less effective in the shredding process.
b. Problem of rodents or flies is controlled by using this method. Some researchers state that fly larvae cannot survive on shredded waste. The garbage content is also well distributed, and rodents cannot find any food supply to survive.
c. Odor problem is also controlled.
d. Shredded solid waste also reduces fire hazards compared to nonshredded waste.
e. Lack of void and easy to be compacted have made the density of solid waste exceed to 710.6 kg/m^3.
f. Researchers have also recorded that loaded trucks can easily pass over the shredded waste even during climate weather (with additional tire wear). This may be

Table 5.20 Summary of application mechanical volume reduction in resource recovery and waste treatment [25]

MSW stage	Technology	Description
Resource Recovery	Shredding	- Shredding is important in this stage. It gives nearly homogeneous characteristics to the solid waste and allows uniform treatment on each solid waste as well as makes each treatment more effective and efficient. - For example, a combustion process like a vortex furnace is required for solid waste to be shredded prior to being fed into the combustion chamber. - Materials such as tires and automobile junk prefer to be shredded for easy handling.

due to the uniform shape of waste after naturally compacted on-site as shredded waste is easy to be compacted.

g. This method also offers little attraction to birds.
h. There is no need for any daily soil earth cover like the conventional method. Thus, landfill operation cost is reduced.

Another effective method in reducing the volume of waste is baling. Baling activities applied here are mostly used to treat or handle heavy and dense wastes such as metal scrap, construction steel waste, aluminum can, and waste copper. Similar to another baling process, this technique provides a smaller and uniform cubic form compared to its original condition. In this shape, waste is easy to handle and also reduces the storage space.

Nowadays, there are many advanced-technology balers designed to fulfill the needs. Most of the balers are designed specifically for their purpose and the types of material that they are handling. Each of the machines is built with a strong body structure and hydraulic system, which makes this baler effectively work in heavy-duty conditions and require less manpower.

The example types of baler machines used in processing dense solid waste are as follows [50]:

(a) Scrap metal copper chips bale machine, specifically design for material like Fe, Al, Cu, Cr-Ni, Ferrous and nonferrous aluminum extrusions.
(b) Waste metal baler machine specifically designed to produce high force with low consumption.
(c) Scrap metal baler packing machine for waste copper, aluminum cans, aluminum and metal shavings.
(d) Hydraulic waste metal baler for aluminum cans.

Advantages of baling solid waste prior to disposal at the sanitary landfill are listed as follows:

(a) Excellent in solving litter and blowing paper problems since waste is well compacted.
(b) Reduces operation cost as no spreading or daily compaction is needed. There is also no daily soil cover, which is usually needed.
(c) No dust problem occurs during discharged waste here.
(d) The baled waste also drives away flies or rodents in a landfill since there is no hole or void for them to access.
(e) Similar to the shredded process, the bailing technique also reduces the odor problem at the landfill.
(f) Fire hazards, which always happen in loose solid waste, are also solved by using this technology.
(g) The most significant benefit of this technique is that the density of waste is increased after the baling process. The dense baler waste is able to fill the landfill and increase the capacity of the landfill by up to 50%.
(h) Due to the maximum density as mentioned previously, the heavy truck can also pass over the bale fill easily without any additional material.

(i) Space or area that has been covered by compacted waste can also be used instantly.

However, there are also disadvantages when applying shredding and baling techniques. For example, the baling and shredded processes require a mechanical structure and manpower to handle. On top of that, these machines also require monthly maintenance and updated system control. The operational cost will be higher compared to a conventional method. Nevertheless, they are still worth considering and are applied on-site as these methods are more environmentally friendly and have many advantages than disadvantages.

5.5.3 *Disposal*

Disposal is the last stage in the solid waste management system. MSW is usually disposed of at landfill sites in some developed countries and in most developing countries, including Malaysia. Combustion and incineration methods are still less applied in these countries. For example, around 94.5% of solid waste in Malaysia is disposed of in landfill sites, while others are transferred for resource recovery and compost practices [51]. There are 150 operating landfill sites, and most of them are classified as nonsanitary or controlled landfills, as presented in Table 5.21. In a sanitary landfill site, solid waste is dumped and compacted layer by layer with soil as a

Table 5.21 Statistics of landfills in Malaysia until 30 June 2017 [52]

State	Close landfill	Operating landfill	
		Nonsanitary	Sanitary
Johor	25	11	1
Kedah	8	6	1
Kelantan	9	11	–
Melaka	7	–	1
Negeri Sembilan	14	4	1
Pahang	20	10	2
Perak	15	15	1
Perlis	2	–	1
Pulau Pinang	1	–	1
Sabah	4	21	1
Sarawak	20	43	3
Selangor	15	2	3
Terengganu	11	9	1
Kuala Lumpur	10	–	–
Labuan	0	–	1
Sum	161	132	18
Total landfill	161	150	

cover material. Compaction machines such as wheel drum trucks or stationary compactors are usually used at the site. This process is rapidly done every day by the worker at the landfill site. This process reduces the overall volume of the waste on-site.

The characteristics of solid waste compaction at landfill sites have been investigated by Hanson et al. (2010) [45] to improve the compaction process on-site. In this study, they investigated the effect of moisture content, degree of compaction, and compaction time. They recorded that adding water or increasing the moisture content in solid waste has increased the effectiveness of compaction and decreased the time of the compaction process [45].

During the experiment, the compaction of solid waste in the hydration condition gives more uniform compacted results compared to the nonhydration condition. This result shows that the workability of the compaction process on-site can be improved by increasing the moisture content and instantaneously will reduce the operational time and increase more storage space at the landfill.

As mentioned previously, shredding solid wastes prior to dumping them at a landfill site has had a good positive effect on a landfill site, its surroundings, and the decomposition of solid waste at this site. Moreover, the application of shredded solid waste improves the effectiveness of the combustion and incineration process, as reported in Table 5.22. However, wet shredding of waste by using hydrapulper or rasp mill machines is not suitable to be applied on waste that will undergo incineration or combustion treatment as wet conditions will decrease the burning process.

Data in Tables 5.18, 5.20, and 5.22 also indicate that shredding is a common process in most of the MSW phases. This technology is very helpful in handling waste as well as improving the efficiency of solid waste treatment. Meanwhile, baling technology is widely used at the earlier stage of the collection and transfer process. Its capability to reduce the volume of waste has effectively saved storage space as well as transportation costs.

Table 5.22 Summary of applications of mechanical volume reduction in disposal stage [25]

MSW stage	Technology	Description
Disposal	Shredding	- Shredding is an important step to manage solid waste in the disposal stage. - Shredded solid waste is needed when using the incineration process as a disposal method. - Similar to the combustion method, shredded waste allows maximum burning of waste in a furnace. - The shredding machine usually used in this process is a hammer mill or jaw-type crusher. - The wet shredding technique is not proper to be applied since that process involves water. It will reduce the efficiency of the incineration process as the waste is in the wet condition.

5.5.4 Renewable Energy Program

Renewable energy program is a program that encourages the usage of green energy such as solar, wind, mini-hydro, and biomass, as well as providing larger access to renewable energy sources by recycling waste products as an energy source. Municipal solid waste has been classified as a potential source for renewable energy, as waste to energy technology (WTE-T) [53]. During this process, the reduced volume of waste also plays an important part, especially in the energy storage section. Energy storage is functional to store the waste for future use. Solid wastes have a good potential as renewable biomass energy that can supply energy or fuel and reduce the amount of waste dumping at landfill sites [53].

Also, there are many energy conversational technologies available to extract energy from solid waste. The selection of these technologies depends on the physicochemical properties of the waste, the type and quantity of waste feedstock, and the desired form of energy [53]. Therefore, the storage energy method is designed for a continuous supply of solid waste as biomass energy sources in the future.

In Sweden, the baling method has been used since 1992 and growing until now. The collection of MSW in this country is recorded as 2.7 million tons and 4.5 million tons for industrial waste per year. However, less than half of it has been disposed of at the landfill, with most of them being recycled as renewable energy [36]. Coupled with that, baling plays an important role in preparing solid waste (a source of renewable energy) for storage and using it (incinerate) during winter. As known, the energy required during this season is higher than that required during another season.

5.6 Operation and Maintenance

Mechanical reduction volume for solid waste generally requires a large mechanical structure to operate. For example, in the shredding process, a hammer mill machine is usually used at the site. A shredding plant is normally set up for two shifts per day with ample time to cater to the maximum load of wastes. It has been compulsory for the manufacturer to put a maintenance slot in the daily schedule for the shredding machine as their job is rugged and in a big scale operation. Extensive and regular maintenance applied will prevent machines from a breakdown and disturb the operational activities. Also, this makes the life span of the machine become longer.

Some researchers have come out with several suggestions and advice in handling and managing these machines. Glisson (1980) [25] suggests that the hammer mill machine can be operated in three shifts per day; one shift should be spared or stopped for maintenance of the mill. This maintenance process will involve component wear on the hammer, mill liner, and grates. Condition of rotors, hammer tipping materials, and also bearing should be checked frequently in order to avoid any clogging or problem during operation.

Similar to a shredding machine, a heavy-duty baler also requires frequent maintenance. Usually, maintenance for the baler is more on the hydraulic systems, equipment lubrication, and motor upkeep. Above all, the feed conveyor system in both machines is most important and must be adequately maintained. Maintenance routine is also a part of the way for us to avoid huge damage or machine breakdown, which will require high repair costs.

Likewise, the compactor machine also requires daily and monthly maintenance service as this machine is doing a rug and heavy work every day. Failure in doing maintenance on the machine may cause not only major breakdowns and additional repair costs but also extended downtime as well as safety concerns, which are also the major issue. Thus, maintenance works of machines cannot be taken for granted. In each purchase of a machine, usually, the manufacturer will provide with the manual book instructions and guidelines for maintenance works. Therefore, the owner should read carefully and follow the instructions.

Some of the general lists to do during the maintenance service of the compaction machine are listed here:

(a) Run the clean cycle at the recommended interval. Usually, some machines have been provided a self-cleaning button on them. So, just press the button to start the cleaning process.
(b) Oil level check.
(c) Grease needs to be applied at each moving part, door hinges, and releases.
(d) Check and change the oil filter as needed (two to three months).
(e) Any scratches on the body of the machine need to be painted to avoid metal rust.
(f) Check and change the hydraulic oil and its filter as necessary.

Similar to the compaction machine, the baler machine is also dealing with heavy-duty stuff every day. Compressing heavy waste into small cubic forms by using high-pressure force definitely requires frequent maintenance. Therefore, general maintenance guidelines on baler machines are listed as follows [54]:

Weekly Maintenance:

(a) All the nuts and bolts need to be checked during the first week of using the machine and repeating this step as monthly routines.
(b) The hydraulic oil level also needs to be checked so that the oil level and cylinder are fully retracted. There is a decal label as "Fill to here" to help the owner maintain the suitable oil level in this machine.
(c) Any leakage or damage at hydraulic hoses also needs to be checked and replaced if necessary.
(d) The power unit should also be in a clean condition and free from any debris that can block the unit's airflow. Suggest always wiping all the grease, dust, and dirt attached to the surface of the control box.
(e) Before starting the operation, make sure all the safety guards and access covers are secure and in good condition.

Monthly Maintenance:

(a) Electrical system connection should be examined by a licensed electrician, including the motor amp draw. Note that successive record readings can help in preventing future failures.
(b) Check hydraulic systems such as the oil level, hoses, and connection. Change any leakage hoses or connections if necessary. Suggest changing the oil filter after the first 50 hours of operation and every 250 hours thereafter.

5.7 Concluding Remark

Mechanical volume reduction is one of the important techniques used in handling, treating, and disposing of solid waste. This technique is beneficial in many angles, especially in transportation, storage area requirements, and operational costs. This method is able to reduce the number of collection trips of garbage track, which makes this operation cheaper than the conventional system. There are several methods used to reduce the volume of waste. The common methods applied on-site are the shredding process, compaction, and baling. Most of these methods are operated by using mechanical machines such as hammer mills, compactors, and balers. Nowadays, there are many improvements and evolutions in the efficiency of the machines, which fit with their functions.

Glossary

Agriculture: A practice of farming or farming industry.
Baling: Applied force at all side surfaces of an object to reduce the void inside it. This process reshapes the object into smaller cubic or cylinder forms (depending on the mold shape) and increases the density.
Combustion: Waste treatment by using a chemical reaction to decompose the waste. This treatment uses heat or flame to burn the waste into ash.
Compaction: Applied force at the surface of an object to reduce the void inside it. This process also reshapes the object into smaller sizes and increases the density.
Compression: Action that reduces the volume of an object.
Incineration: Destruction of waste or material by a burning process.
Landfill: Site to dispose of solid waste.
Mechanical: Operation runs by machines.
Milling: Process of machining using rotary cutters to cut the material into smaller pieces.
Municipal solid waste: Garbage or waste that is commonly thrown every day. This waste is usually collected from houses, schools, hospitals, and business areas.
Recycling: Recycling is a process of transformation and reusing waste material for a new product.

Shredded: Cut or crush something into small pieces.
Solid waste: Garbage, trash, and unwanted things that need to be disposed of.
Transportation: The action of transferring something to another place.
Transfer station: A station that stores or processes solid waste prior to transferring it to the next waste treatment process.
Volume reduction: Making the material physically smaller than its initial form.
Waste processing: Stage where solid waste is being prepared to undergo further waste treatment processes. At this stage, solid waste is usually processed by using mechanical volume reduction technology.

References

1. Agamuthu, P., & Dennis, V. (2011). Policy trends of extended producer responsibility in Malaysia. *Waste Management and Research, 29*, 945.
2. Zin, M. N. S., Aziz, A. H., Adlan, M., & Ariffin, A. (2012). Characterisation of leachate at Matang Landfill site, Perak, Malaysia'. *Academic Journal of Science, 1*(2), 317–322.
3. Zainol, N. A., Aziz, H. A., & Yusoff, M. S. (2012). Characterisation of leachate from Kuala Sepetang and Kulim landfills: a comparative study. *Energy and Environment Research, 2*(2), 45.
4. Zainal, S. F. F. S., & Aziz, H. A. (2017, October). Potential of tin (IV) chloride for treatment in Alor Pongsu as sstabilised landfill leachate. In AIP Conference Proceedings (Vol. 1892, No. 1, p. 040003). AIP Publishing.
5. Nazrieza, N., MY, S. R., Subramaniam, K., & Hazilia, H. (2015). Characterisation of leachate from panchang bedena landfill, Batang Padang Landfill and Matang Landfill: a comparative study. *Malaysian Journal of Science, 34*(1), 69–77. Zakaria (2020). Treatment of stabilised anaerobic landfill leachate by ozonation process with zirconium and tin tetrachloride. phd Thesis.
6. Kaza, S., Yao, L., Bhada-Tata, P., & Van Woerden, F. (2018). *What a waste 2.0: A global snapshot of solid waste management to 2050*. World Bank Publications.
7. Rodzi, R. M. (2019). Analysis of Solid Waste Generation and Composition in Malaysia TVET Campus. *International Journal of Integrated Engineering, 11*(2) Perrot, J. F., & Subiantoro, A. (2018). Municipal waste management strategy review and waste-to-energy potentials in New Zealand. Sustainability, 10(9), 3114.
8. Kamaruddin, M. A., Yusoff, M. S., Rui, L. M., Isa, A. M., Zawawi, M. H., & Alrozi, R. (2017). An overview of municipal solid waste management and landfill leachate treatment: Malaysia and Asian perspectives. *Environmental Science and Pollution Research*, 1–33.
9. Sreenivasan, J., Govindan, M., Chinnasami, M., & Kadiresu, I. (2012). Solid waste management in malaysia–a move towards sustainability. In *Waste management-An integrated vision*. Intech.
10. Perrot, J. F., & Subiantoro, A. (2018). Municipal waste management strategy review and waste-to-energy potentials in New Zealand. *Sustainability, 10*(9), 3114.
11. Abdel-Shafy, H. I., & Mansour, M. S. (2018). Solid waste issue: Sources, composition, disposal, recycling, and valorisation. *Egyptian journal of petroleum, 27*(4), 1275–1290.
12. ISPRA. (December 29, 2020). Distribution of municipal solid waste treatments in Italy in 2019, by method Graph]. In Statista. Retrieved March 30, 2021, from https://www-statista-com.ezproxy.insead.edu/statistics/682944/management-of-solid-urban-waste-in-italy-by-treatment/
13. Mansouri, A & Kacha, L. (2017). *Waste management system in Japan*. Retrieved 06.04.2021 from:https://www.researchgate.net/publication/321214329_Waste_Management_System_in_Japan

14. Davis, M. L., & Corwell, D. A. (2013). *Introduction to environmental engineering* (5th ed., pp. 786–865). McGraw Hill Education.
15. Zainu, Z. A., & Songip, A. R. (2017). Policies, challenges and strategies for Municipal waste management in Malaysia. *Journal of Science, Technology and Innovation Policy, 3*(1).
16. Yanuwiadi, B. (2016). Landfills leachates quality and quantity in Tidal Area in Banjarmasin Landfills, Indonesia. *Resources and Environment, 6*(2), 23–27.
17. Laohalidanond, K., Chaiyawong, P., & Kerdsuwan, S. (2015). Municipal solid waste characteristics and green and clean energy recovery in Asian Megacities. *Energy Procedia, 79*, 391–396.
18. National Environment Agency (NEA). *Waste statistics and overall recycling*. Access on 09.10.2018 https://www.nea.gov.sg/our-services/waste-management/waste-statistics-and-overall-recycling
19. Orhorhoro, E. K., & Oghoghorie, O. (2017). Design of Hammer Mill for Crushing of Glass Waste. *International Journal of Research in Advanced Engineering Technologies, 1*.
20. Nasr, M. F., & Yehia, K. A. (2019). Stress Analysis of a Shredder Blade for Cutting Waste Plastics. *Journal of International Society for Science and Engineering, 1*(1), 9–12.
21. Carvill, J., *5 – Manufacturing technology*, (Eds.),: J. Carvill, Mechanical Engineer's Data Handbook, Butterworth-Heinemann,1993, Pages 172-217, ISBN 9780080511351, https://doi.org/10.1016/B978-0-08-051135-1.50010-8.
22. Scallan, P. (2003). Production equipment and tooling selection, Chapter 5. In *Process planning-the design/manufacture interface* (pp. 171–218). Elsevier.
23. Ezurike, B. O., Osazuwa, O. J., Okoronkwo, C. A., & Okoji, K. I. (2018). Design, construction and performance evaluation of a flat screen hammer mill machine. *African Journal of Science, Technology, Innovation and Development, 10*(6), 683–690. https://doi.org/10.1080/20421338.2018.1491675
24. Hartati, S., Musabbikhah, M., & Suryono, E. (2018, October). Multifunction hammer mill machine design and assembling to destroy an agricultural wastes. In *International Conference on Applied Science and Engineering (ICASE 2018)*. Atlantis Press.
25. Glysson, E. A. (1980). Mechanical volume reduction. In L. K. Wang & N. C. Pereira (Eds.), *Solid waste processing and resource recovery. Handbook of environmental engineering, vol 2*. Humana Press. https://doi.org/10.1007/978-1-4612-5992-3_2
26. USDA. *Agricultural waste management field handbook*. United States Department of Agriculture, Soil conservation Service. Accessed on 21 Apr 2021, from https://www.wcc.nrcs.usda.gov/ftpref/wntsc/AWM/handbook/ch12.pdf
27. Marthiana, W., Arman, R., Mahyoedin, Y., & Wardiyanto, D. (2018). Design and production of rotary type machine for chop up organic waste plantation. In *MATEC Web of Conferences* (Vol. 248, p. 01010). EDP Sciences.
28. Ismail, Z. E., & Ghazy, M. I. (2016). Influence of drum chipper setting on the wood chips performance. *Misr Journal of Agricultural Engineering, 33*(4), 1293–1314.
29. Spinelli, R., Cavallo, E., Eliasson, L., Facello, A., & Magagnotti, N. (2015). The effect of drum design on chipper performance. *Renewable Energy, 81*, 57–61.
30. Tomoiagă, V. B., Popa, M. S., Sattel, S., Conțiu, G., & Bozga, M. (2017). Finite element simulation of the C70 orthogonal cutting process. In *MATEC Web of Conferences* (Vol. 112, p. 06011). EDP Sciences.
31. Olawale, J. O., & Ibitoye, S. A. (2018). Failure analysis of a crusher jaw, Chapter 10. In *Handbook of materials failure analysis-with case studies from the construction industries* (pp. 187–207). Elsevier.
32. Toenshoff, H. K. (2014). Cutting, fundamentals. In L. Laperrière & G. Reinhart (Eds.), *The international academy for production engineering* (CIRP Encyclopedia of Production Engineering). Springer. https://doi.org/10.1007/978-3-642-20617-7_6633
33. Loh, Z. H., Samanta, A. K., & Heng, P. W. S. (2015). Overview of milling techniques for improving the solubility of poorly water-soluble drugs. *Asian journal of pharmaceutical sciences, 10*(4), 255–274.

34. Ball mill. (10 Jun 2021). In *Wikipedia* https://en.wikipedia.org/wiki/Ball_mill
35. Burmeister, C. F., & Kwade, A. (2013). Process engineering with planetary ball mills. *Chemical Society Reviews, 42*(18), 7660–7667.
36. Hogland, W., Marques, M., & Grover, V. (1999). Baling storage method: past, present and Swedish experience. *Linnaeus Eco-Tech*, 203–210.
37. VDP. (August 19, 2020). Consumption of paper and paperboard in selected countries worldwide in 2018 (in 1,000 metric tons). In *Statista*. Retrieved March 27, 2021, from https://www-statista-com.ezproxy.insead.edu/statistics/240565/consumption-volume-of-paper-and-paperboard-in-selected-countries/
38. Amemiya, T. (2018). Current state and trend of waste and recycling in Japan. *International Journal of Earth & Environmental Sciences, 3*, 155.
39. Aue, J. Picard, K. Grabner, K, & Button, A. (2003). Fiber recovery from waste paper: A breakthrough in re-pulping technology. Pulper technology based on patents by Mark Spencer, Inventor.
40. Tan, S. T., Lee, C. T., Hashim, H., Ho, W. S., & Lim, J. S. (2014). Optimal process network for municipal solid waste management in Iskandar Malaysia. *Journal of Cleaner Production, 71*, 48–58. https://doi.org/10.1016/j.jclepro.2013.12.005
41. Batjai, P. (2018). Fiber from recycled paper and utilisation, Chapter 23. In *Biermann's Handbook of Pulp and Paper* (Raw Material and Pulp Making, 2018) (Vol. 1, 3rd ed., pp. 547–582). Elsevier.
42. Guevarra, L. (2012). *Waste management pilot tests a next-gen green garbage truck*. Access on 31 March 2021. https://www.greenbiz.com/article/waste-management-pilot-tests-next-gen-green-garbage-truck
43. Voicu, G., Lazea, M., Zabava, B. S., Tudor, P., & Moise, V. (2019). Cinematical analysis of the pre-taking and pre-compacting mechanisms of some garbage trucks. *Journal of Engineering Studies and Research, 25*(2), 56–62.
44. David, S. K., Biswal, J. N., Muduli, K., Peter, O., & Pumwa, J. (2020). *Design and analysis of an hydraulic trash compactor*.
45. Hanson, J. L., Yesiller, N., Von Stockhausen, S. A., & Wong, W. W. (2010). Compaction characteristics of municipal solid waste. *Journal of geotechnical and geoenvironmental engineering, 136*(8), 1095–1102.
46. Pradel, J., Parsons, P. J., & Malasek, E. (1970). *The volume reduction of low-activity solid wastes*. Report of a panel
47. Zaman, A. U. (2010). Comparative study of municipal solid waste treatment technologies using life cycle assessment method. *International Journal of Environmental Science & Technology, 7*(2), 225–234.
48. Cossu, R. (2018). Mass balance of contaminants: A key for modern landfill design. In *Solid Waste Landfilling. Concepts, processes, technologies*, 73-88.
49. Salman, Z. (2018). *Negative impacts of incineration-based waste-to-energy technology*.
50. Jiangyin Tianfu Technology Co.,Ltd, list of baler product. Access on 1 April 2021: http://www.tianfurecycling.com/y81q-3150-15608510358785356.html
51. Eddine, B. T., & Salah, M. M. (2012). Solid waste as renewable source of energy: current and future possibility in Algeria. *International Journal of Energy and Environmental Engineering, 3*(1), 1–12.
52. MHGL. (2017). *MHGL Selected Statistic 2017. Malaysia: Ministry of Housing and Local Government (MHLG)*. Retrieved 10.10.2018 from: http://www.kpkt.gov.my/resources/index/user_1/GALERI/PDF_PENERBITAN/PERANGKAAN%20TERPILIH/30_Jun_2017[5].pdf
53. Moya, D., Aldás, C., López, G., & Kaparaju, P. (2017). Municipal solid waste as a valuable renewable energy resource: a worldwide opportunity of energy recovery by using Waste-To-Energy Technologies. *Energy Procedia, 134*, 286–295.
54. Harmony. (2021). *Baler and compactor preventative maintenance*, Access on 21 April 2021 at: https://harmony1.com/baler-and-compactor-preventative-maintenance/

Chapter 6
Combustion and Incineration

Ridzuan Zakaria, Hamidi Abdul Aziz, Lawrence K. Wang, and Yung-Tse Hung

Abstract In several countries worldwide, combustion and incineration are the second main option for waste management and disposal. There are different types of incineration systems on the market. This chapter examines the method of urban solid waste mass-burn incineration, from waste collection to bunker and feeding systems, furnaces, and heat recovery systems. Fluidised bed incinerators, starved air incinerators, rotary kiln incinerators, cement kilns, liquid and gaseous waste incinerators, and the waste types incinerated in the various incinerators are also addressed. Particulate matter, heavy metals, toxic and corrosive gases, and incomplete combustion products such as polycyclic aromatic hydrocarbons (PAHs), dioxins, and furans are discussed, as well as pollution creation and control. The wastewater, bottom ash, and fly ash generated by waste incineration are all addressed. The emission dispersion from the chimney stack is defined. The waste-to-energy (WtE) part of incineration is also presented. There are also several case studies mentioned.

R. Zakaria (✉)
School of Chemical Engineering, Engineering Campus, Universiti Sains Malaysia, Nibong Tebal, Pulau Pinang, Malaysia

Solid Waste Management Cluster, Engineering Campus, Universiti Sains Malaysia, Nibong Tebal, Pulau Pinang, Malaysia
e-mail: chduan@usm.my

H. A. Aziz
Environmental Engineering, School of Civil Engineering, Engineering Campus, Universiti Sains Malaysia, Nibong Tebal, Pulau Pinang, Malaysia

Environmental Engineering, Solid Waste Management Cluster, Engineering Campus, Universiti Sains Malaysia, Nibong Tebal, Pulau Pinang, Malaysia
e-mail: cehamidi@usm.my

L. K. Wang
Lenox Institute of Water Technology, Latham, NY, USA

Y-T. Hung
Department of Civil and Environmental Engineering, Cleveland State University, Cleveland, Ohio, USA

L. K. Wang et al. (eds.), *Solid Waste Engineering and Management*, Volume 1, Handbook of Environmental Engineering 23,
https://doi.org/10.1007/978-3-030-84180-5_6

Keywords Municipal Solid Waste · Incineration · Combustion · Waste to Energy

Acronyms

BAT	Best Available Technique (or Best Available Technology, used in the United States)
BREF	Best Available Technique Reference Document
CAR	Clean Air Regulation
Cd	cadmium
CO	carbon monoxide
CO_2	carbon dioxide
DOE	Department of Environment
DRE	destruction and removal efficiency
ESP	electrostatic precipitator
FGT	flue gas treatment
HCl	hydrogen chloride
HF	hydrogen fluoride
ICRF	inductively coupled radio frequency
IPPC	Integrated Pollution Prevention Control
MSW	municipal solid waste
MSWI	municipal solid waste incinerator
NO_x	nitrogen oxides
N_2O	nitrous oxide
ODS	ozone-depleting substances
PCBs	polychlorinated biphenyls
SCR	Selective catalytic reduction
SO_x	sulphur oxide
SO_2	sulphur dioxide
TOC	total organic carbon
VOC	volatile organic compound
WM	waste management
WtE	waste-to-energy

Nomenclature

Δh	enthalpy change
ρ	instantaneous density (g/cm^3)
d_o	particle of initial diameter
f_{CO}	mole fractions of CO
f_{O_2}	mole fractions of O_2
f_{H_2O}	mole fractions of water vapor
k_s	kinetic rate constant for the consumption reaction

k_d	diffusional rate constant
K_p	equilibrium constant
$M\ C_p^0$	molar heat capacity (kcal mol^{-1} °C^{-1})
$M\ C_{p\,avg}^0$	average molar heat capacity (kcal mol^{-1} °C^{-1})
P	absolute pressure (atm)
q	rate of carbon consumption (g cm^{-2} S^{-1}) t
R	universal gas for ideal gases
T	absolute temperature (K)
V	volume (m^3)

6.1 Introduction

In solid waste management schemes, one of the solutions is incineration. It is a mechanism in which combustible waste is combusted, incinerated, or oxidised, producing carbon monoxide (CO), hydrogen chloride (HCl), hydrogen fluoride (HF), nitrogen oxides (NO_x), sulphur dioxide (SO_2), the volatile organic compound (VOC), dioxins and furans, polychlorinated biphenyls (PCBs), heavy metals, and other by-products. As shown in Fig. 6.1, the entire process involves the waste passing through the distribution system, the bunker and feeding system, the boiler, heat recovery systems, and the gas cleaning system, with ash as the end product [1]. This method is appealing because it can dramatically reduce waste volume by up to 80–90% [2] and weight by up to one-third of its pre-burnt weight [3, 4]. It cannot,

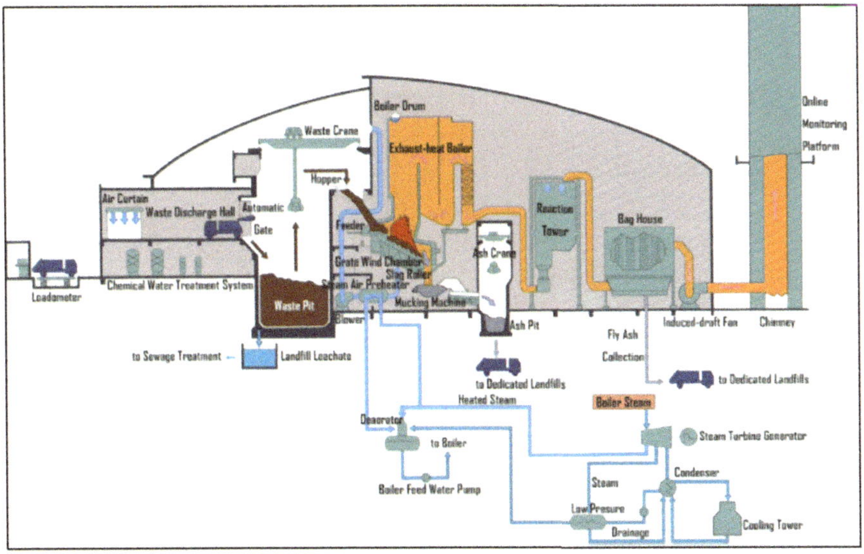

Fig. 6.1 Schematic of a typical solid waste incinerator [1]

however, be a stand-alone operation because its ash or solid residue must be landfilled.

Incineration is also the most practical choice for certain wastes that are too dangerous or expensive to recycle or landfill safely, such as highly flammable materials and radioactive or infectious waste [4]. Furthermore, the energy obtained from waste incineration may be used to generate steam for electricity generation, industrial process heating, or hot water for district heating, thus conserving valuable primary fuel resources [3, 5]. In addition, incinerator bottom ash, a by-product of incineration, may be collected and re-used as secondary aggregates in the building [6].

Incineration, on the other hand, has its own set of drawbacks. The most notable disadvantage is that it has a very high capital cost and a longer payback period [7]. Other disadvantages include operating problems caused by the refuse character, process sophistication, and negative public reactions to the process [8].

However, the benefits of incineration greatly outweigh the drawbacks, and incineration is now accepted as the best environmentally friendly alternative for MSW disposal. The public understanding of the health and environmental risks posed by landfill has resulted in the closure of many landfill sites around the world. In the United States, for example, the figure was over 6,000 in 1990, but by 2017, it had fallen to just 1,270 [9]. Another aspect that favours incineration over landfill disposal is the rising cost of landfill disposal, which includes rises in landfill tax and transportation costs. The landfill tax in the United Kingdom was £7.00 per tonne in 1992, raised to £10.00 per tonne in 1999 [10], and will be £96.70 [11] per tonne in 2021. Thus, the future of waste incineration will be determined by the availability of landfills near densely populated areas, the amount of energy recovered by incinerators, and the extremely high capital and operating costs of incinerator plants [7].

6.2 Municipal Solid Waste Incinerator Plant

6.2.1 Background

In the United Kingdom, the first incinerator was constructed in Paddington City in 1870, while the first MSW incinerator without energy recovery in the United States was installed in New York City in 1885 [12]. However, it was closed due to complaints by the public of smell and smoke production [12]. Another incinerator was built in Manchester in 1876 [13]. It was a cell furnace with batch-wise combustion, in which the waste was manually loaded, and the ash was removed at the high furnace temperature. Many more incinerators have been built since then, but many of them were closed due to fundamental design flaws. As incineration technology improved with provisions for cleaner and efficient combustion, the traditional cell-furnace design was replaced by the shaft-furnace design, and the batch-wise combustion was replaced by continuous combustion using moving grate systems with a capacity between 10 and 50 tonnes per hour [14]. As a result, only moving grate systems are explored in depth in this chapter. A brief description of a smaller incinerator with a waste capacity of 1 to 2 tonnes per hour, such as clinical waste, sewage sludge, and hazardous waste, is given [4]. Typical examples of such devices include fluidised bed, cyclonic, starved air or pyrolytic, rotary kiln, rocking kiln, cement kiln, and liquid and gaseous incinerators.

In general, an incinerator process can be broken down into a few parts, as defined in the previous section: waste delivery, bunker, and feeding systems, furnace systems, heat recovery systems, and gas cleaning systems. Each section of an incinerator will be described in the subsections that follow (please refer to Fig. 6.1).

6.2.2 Waste Delivery, Bunker, and Feeding System

The waste is unloaded and stored in a waste bunker when waste collection vehicles arrive at an MSWI plant. The bunker should be wide enough to hold around 2–3 days' worth of waste equivalent in weight, which will be about 1000–3000 tonnes of waste, to ensure a balance between an uneven waste distribution and the plant's continuous activity [4]. The bunker in some incinerator plants is divided into parts to isolate waste with different calorific values and combustion properties. Until loading them into the feeding hopper, the crane operator can combine them. The crane operator would also remove any bulky or hazardous objects from the refuse for further processing. The waste is then loaded into the feeding machine by the operator. The crane grab has a waste capacity of 6 m^3.

The waste in the steel hopper will flow into the incinerator under its own weight in the feeding device. The material is then conveyed into the grate system through a hydraulic pump or other means. To prevent air leakage into the furnace and ensure uninterrupted feed to the grate, the hoppers are held partially stuffed with waste. A monitor is used to calculate the amount of waste. The furnace entrance is sealed by hydraulic shutters to prevent the fire inside the furnace from spreading to the feeding hopper. The feed chute may also be water-cooled or lined with refractory material.

6.2.3 Furnace System

This is the section where solid waste is incinerated. Commonly, the number of furnaces depends on the capacity of the incinerator, with each furnace having the capacity to burn 10 tonnes/hour of waste [4]. Thus, a typical 60 tonnes/hour incinerator would have six furnaces. The advantage of having multiple furnaces is that if one furnace needs repair, the others can still be operated, which limits its downtime.

Generally, the furnace consisted of a moving grate and incineration chamber. The movement of waste inside this furnace is helped by the individual action of the grate. The primary air is pumped from under the grate, and the secondary air is pumped from just above the waste bed before the radiation shaft. As the solid waste whose orientation differ in size, shape and orientation is transported from the feed input to the ash end of the incinerator through the grate, it will undergo the processes of drying; pyrolysis; solid and gas phase combustion; conductive, convective and radiative heat transfer; and mass transfer. Since the waste stream components differ in moisture content, thermal degradation temperature, volatile composition and ignition temperature, and fixed carbon content, the stages converge in practice [4].

As the raw, wet waste is fed onto the grate, it is initially dried by the heat radiated from the overbed region and from the burning waste. The injection of pre-heated primary air also assists the drying process. At temperature between 50–100°C (373 K), the waste loses most of its moisture content. The amount of water in waste is important because heat is required to evaporate it, which means that more of the waste's usable calorific value is lost in the process of heating up the wet waste, resulting in less energy available. Furthermore, the amount of water in the waste will affect the rate of heating and thus the rate of thermal decomposition. The water content of municipal solid waste can range anywhere from 25% to 60%.

After the moisture is released, the temperature rises to about 260 °C (533 K), and the thermal decomposition and pyrolysis of organic materials such as paper, plastics, food waste, textiles, and so on begins. VOCs, combustible gases, and vapours are produced as a result of the processes. The typical amount of VOC in MSW is between 70% and 90%, and they are formed as hydrogen (H_2), carbon monoxide (CO), methane (CH_4), ethane (C_2H_6), and other higher molecular weight hydrocarbons [4].

In general, waste ignition begins around 316°C (589 K). The waste is then burned above the waste on the grate and in the combustion chamber above the grate until all of the oxygen is consumed, or all of the waste has devolatilised to carbonaceous char. To ensure good mixing and complete combustion of the gases and vapours, a sufficiently high temperature, adequate residence time, and excess turbulent air are needed. Devolatilisation occurs over a temperature range of 200–750 °C, with 425 and 550 °C being the primary release of VOCs. Furthermore, the release of VOCs is affected by the different components present in the waste.

Polystyrene, for example, decomposes at temperatures between 450 and 500°C, yielding approximately 100% volatiles, while wood decomposes at temperatures between 280 and 500°C, yielding roughly 70% volatiles [4]. The rate of thermal decomposition can also be affected by the waste's structure and physical condition. Cellulosic material in thin form, such as paper, decomposes in seconds, while cellulosic material in the form of a large piece of wood will take several minutes to decompose fully.

Since the furnace gas temperature is usually between 750 and 1000 °C but can reach temperatures as high as 1600 °C, the volatile gases and vapours emitted instantly ignite in the furnace. The residual char or partially charred waste can be pyrolysed further, gasified by CO_2 or H_2O to produce CO and H_2, or oxidised by O_2 to produce CO_2, with the ash remaining in the bed at the end of the process. Since

MSW is heterogeneous in nature and the waste elements next to it differ in size and composition, the processes of drying, pyrolysis, and gasification of the waste can occur concurrently in the burning bed.

From the beginning, secondary air is supplied through nozzles above the grate to ensure that there is enough air for combustion and to create turbulence [4]. Excess secondary air is needed to avoid areas where there is no oxygen, which causes the hydrocarbons to pyrolyse rather than burn, potentially resulting in hazardous high molecular weight hydrocarbons and soot. As a result, secondary air circulation and turbulence characteristics play an important role in reducing pollutant formation in the combustion chamber.

From the above, it is stipulated that to design an incinerator plant, a good combination of mechanical and chemical engineering knowledge plays an important role in ensuring that the movement of waste is uninterrupted and combustion is as complete as possible to reduce hazardous emission. The following subsection describes the types of moving grates available in the market.

6.2.3.1 Travelling Grate

As shown in Fig. 6.2, the travelling grate is normally made up of two or more continuous metal-belt conveyors [15]. The waste from the hopper is sent to the first grate. The waste is dried here before starting to volatilise and burn, eventually falling onto the second grate, also known as the burning grate. The bottom ash from the incineration process is collected in an ash hopper at the end of the burning grate.

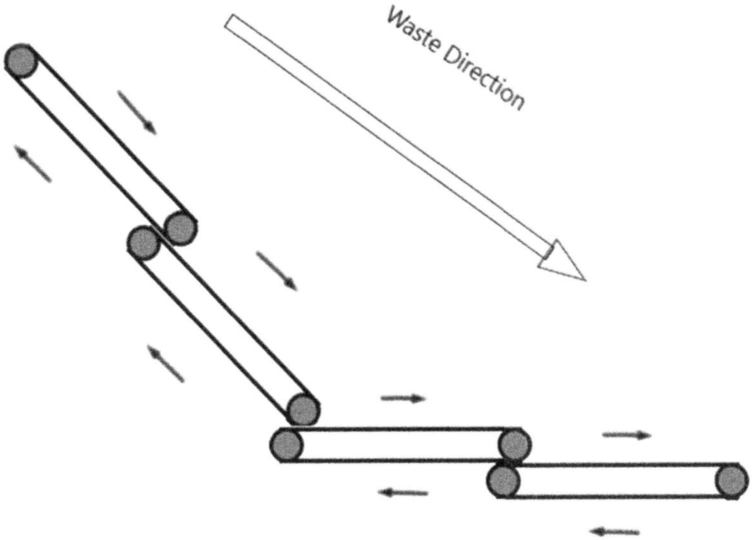

Fig. 6.2 Travelling grate

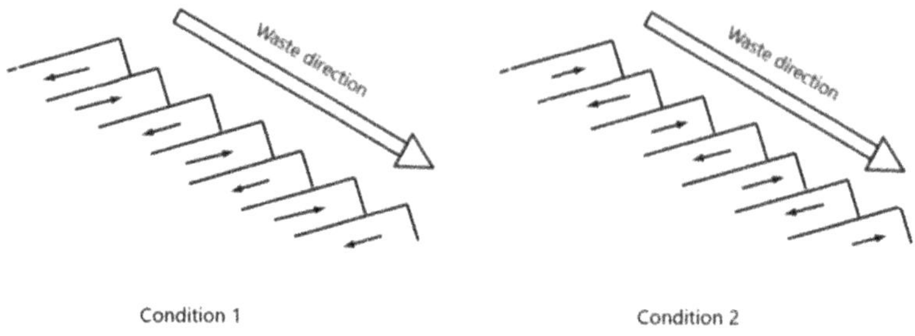

Fig. 6.3 Double motion grate [16]

This system is reliable and relatively cheap. On the other hand, a disadvantage of this system is that the waste does not experience any form of mixing or agitation in the process. The only disturbance to the waste is when it falls from one grate to another, which results in poor waste mixing during combustion. This type of grate thus requires a large amount of air to be provided to improve combustion and consequently produces an extremely large volume of flue gas that needs to be treated. As a result, the particulate removal system can become overloaded, causing high particulate emissions.

6.2.3.2 Double Motion Overthrust Grate

A double motion overthrust grate is shown in Fig. 6.3 [16]. The term "double motion" refers to the arrangement of rows of moving grate bars in opposite directions that are superimposed. When the grate bars adjacent to a stationary bar move away from each other, the grate bars adjacent to the next stationary bar move in the opposite direction. The horizontal structure of the grate, as well as the continuous movement of the rows, enables the waste to be advanced in a controlled manner while preventing sliding. The waste layer loosens as the grate bars move away from each other, and the waste is moved to the next section of the grate before finally falling into the space created by the moving grate bars.

6.2.3.3 Reciprocating Grate

As shown in Fig. 6.4, the reciprocating grate is made up of a series of steps with moving and fixed grate parts angled downwards towards the ash discharge trap. The shifting grates slide back and forth between the stationary grates, agitating and transferring the waste to the ash hopper. The reciprocating grate can also be set up in a multiple-level sequence to provide more agitation to the bed. This grate is ideal for burning wet refuse because it has excellent primary air distribution, resulting in a high-quality burn-out [18]. The biggest drawback of this grate, however, is the bad waste mixing.

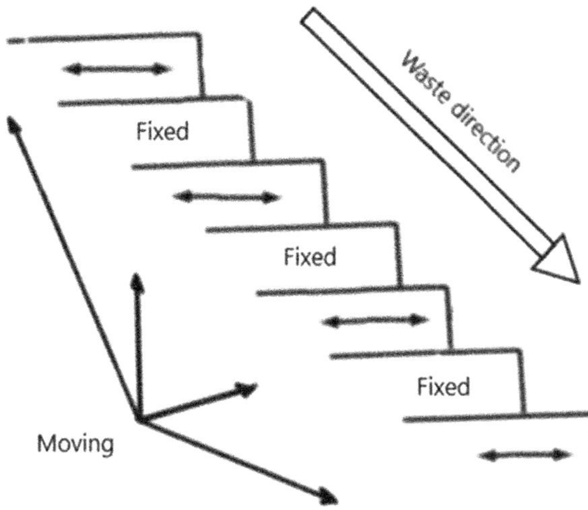

Fig. 6.4 Reciprocating grate

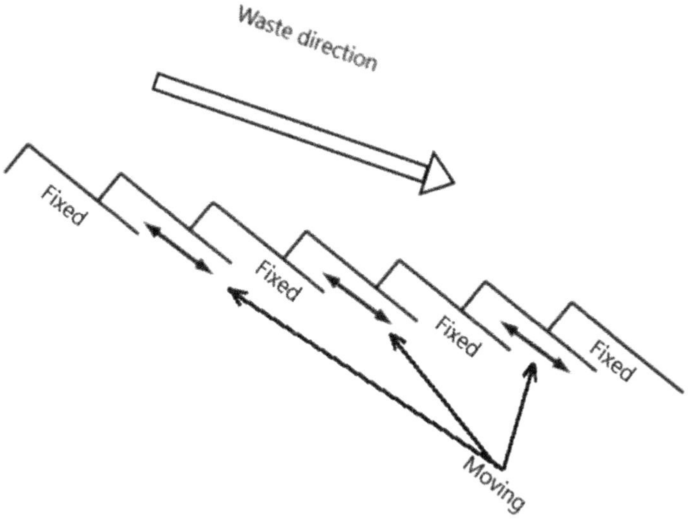

Fig. 6.5 Reverse acting reciprocating grate

6.2.3.4 Reverse Acting Reciprocating Grate

The reverse acting reciprocating grate, like the reciprocating grate mentioned above, is made up of a stack of stepwise configured moving and fixed grate parts that are angled downwards with a steeper angle towards the discharge end, as shown in Fig. 6.5. The key difference being that the grate components reciprocate upwards to the waste's downward progress, allowing the burning content to roll due to the upward reverse thrust. The grate's steplike design provides additional mixing as the waste tumbles from one stage to the next. The reverse acting reciprocating grate has

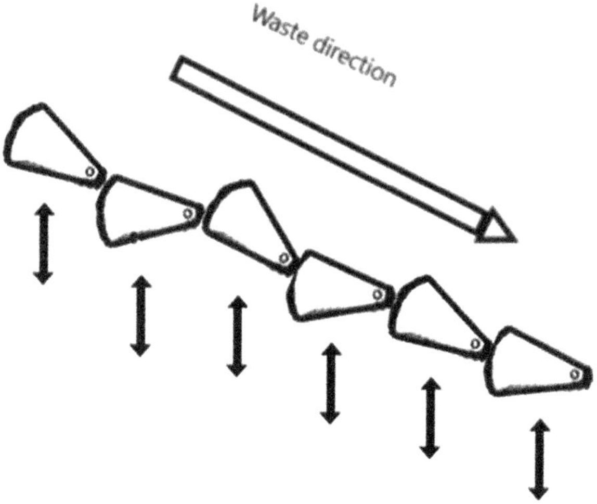

Fig. 6.6 Rocking grate

the advantages of being very effective, producing good burn-out, and being able to manage long operating hours. It is, however, a costly grate with high maintenance costs. Small items being trapped between the grate bars is a common issue with this form of the grate. The Martin reverse-acting reciprocating grate is one of the devices that use the reciprocating grate [19].

6.2.3.5 Rocking Grate

A rocking grate is designed to slope downwards towards the ash discharge end, as shown in Fig. 6.6. It is usually made up of two or three grate parts that are the same width as the furnace. Alternate grate rows are rotated 90 degrees forward around the axis to achieve an upward and forward motion. This movement causes the waste to agitate and step forward. While the alternating grate rows rotate forward, these grates rotate back to their original resting positions. The waste is agitated and pushed forward by the constant back-and-forth movement of alternate grates. This grate is well-known for producing excellent burn-out. It does, however, have drawbacks in that small items may get stuck between the grates, necessitating regular operational maintenance, and in some cases, weekly maintenance. The Esslingen and Nichols systems are two examples of rocking grate systems [4].

6.2.3.6 Roller Grate

This type of grate was developed in the West Germany city of Düsseldorf in 1965 to counteract the high cost of multiple travelling grates. The roller grate is made up of a set of slotted rotating drums, as shown in Fig. 6.7. Each drum rotates forward,

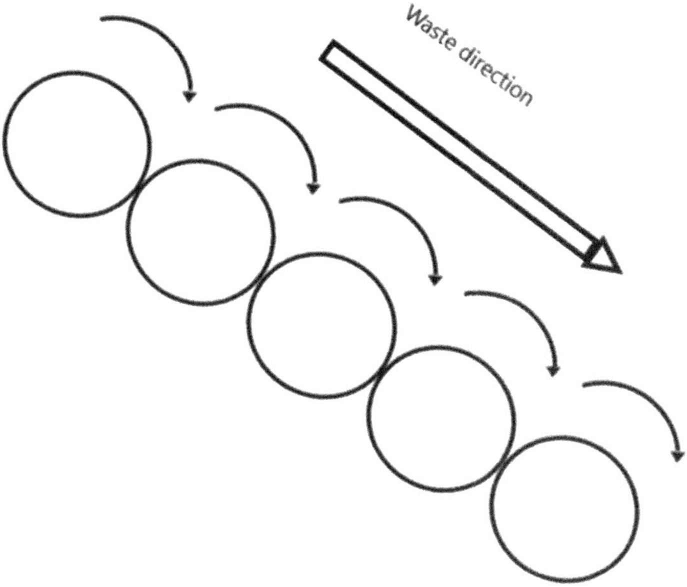

Fig. 6.7 Roller grate

agitating and transporting the waste to the next. The waste is often mixed between the drums as they rotate. Each drum spans the entire width of the furnace and has its own variable speed control, allowing for better grate control during combustion than other grate forms [20]. The major drawback of this grate is that small objects can get stuck between the drums and cause them to stop rotating. The roller grate is exemplified by the Düsseldorf WtE method.

6.2.3.7 Incineration Chamber

The waste's volatile compounds are burned in the combustion chamber, which is situated above the grate. The shapes are to be considered. They are very important as they influence combustion efficiency. The mean residence time of the gaseous volatiles is determined by the size of this chamber, while the form influences the heating pattern of the incoming waste, which receives heat from both the hot flue gases and the furnace wall. Furthermore, the shape of the chamber has an effect on the gaseous flow pattern within it, which affects recirculation and mixing. As shown in Fig. 6.8, the two most common types of combustion chambers are shown.

Vertical shaft combustion chambers are sometimes combined with reverse reciprocating or roller grates. Its architecture allows for efficient gas mixing as well as a long gas residence period. However, the combustion air distribution affects its output. As shown in Fig. 6.8, there are three types of vertical shaft combustion chambers in use today: parallel gas flow designs, contra gas flow designs, and centre gas

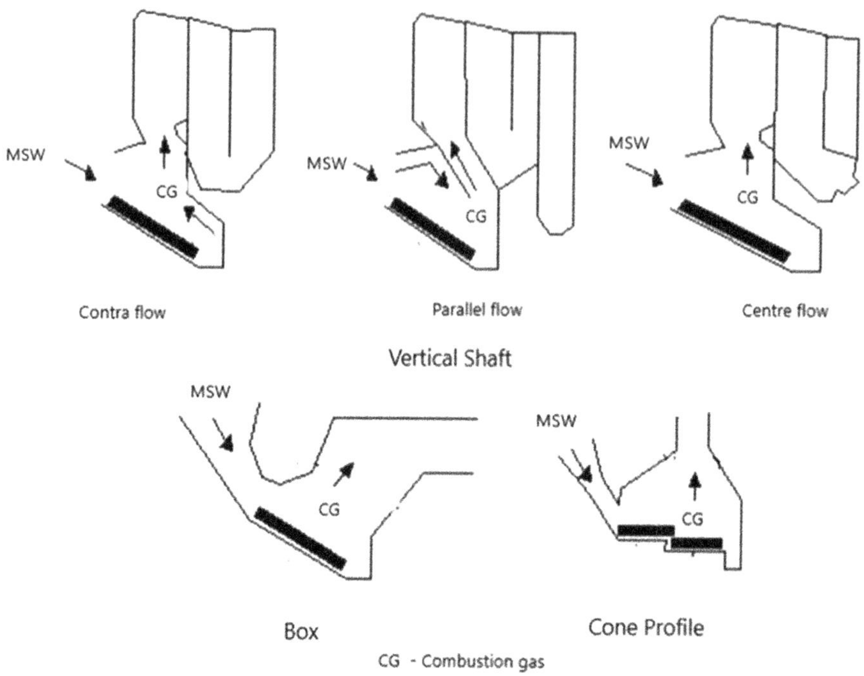

Fig. 6.8 Types of incineration chamber

flow designs. The contra gas flow form is appropriate for wastes that are difficult to burn, such as those with a high moisture content or low volatility [17]. The parallel flow style, on the other hand, favours dry, easily ignited waste [17]. The features of the first two designs are combined in the centre gas flow version.

The rocker and W-grate plants are often combined with the box form of the combustion chamber. As shown in Fig. 6.8, its shape is a variant of a rectangular box that is slightly longer and lower. The downside of this type of combustion chamber is that, due to poor combustion air distribution, its gas turbulence is not as strong as the vertical shaft type.

As shown in Fig. 6.8, the cone profile style combustion chamber is normally used in conjunction with an L-stoker grate. Its architecture does not allow for a long stay. To protect the combustion chamber walls from thermal stresses or deterioration due to high-temperature corrosion and abrasion, they are coated with refractory material. Furthermore, the refractory material emits heat to speed up the drying process, which is then accompanied by the ignition and combustion of the incoming waste. The type of refractory material used is determined by the combustion chamber's predicted strength of combustion [22].

The unburnt gaseous volatiles emerging from the waste bed are first mixed with the secondary air and subsequently burnt. Hot combustion products flow through the radiation shaft to the heat recovery section of an incinerator, as shown in Fig. 6.1.

6.2.4 *Enclosure*

The enclosure that surrounds the grate framework is an essential part of the overall system design. The hot enclosure surfaces not only contain the fuel, but they also radiate heat to the incoming feed, speeding up the drying and combustion of refuse. Furthermore, the shape of the enclosure influences the flow patterns of combustion gases. Furnace enclosures can be made of refractory material or of a "Waterwall furnace," which is made up of a series of water-filled boiler tubes linked by a short metal bridge. At the grating line, refractory material (usually silicon carbide for its abrasion resistance and high thermal conductivity) is often installed, typically reaching one to three meters above the grate in either case.

6.2.5 *Heat Recovery Systems*

Incineration is an extensive heat-generating process. Most of the heat from the incineration process is transferred as flue gases. In a modern incinerator where there is a flue gas treatment system, this high-temperature flue gas, which is usually around 750–1000 °C must be cooled to below 250–300 °C before it can be passed through a flue gas treatment system usually consisted of such as electrostatic precipitators, scrubbers, and bag filters. This process can be achieved through the insertion of a system of boiler tubes at a separate boiler chamber with a specified configuration [5]. Fig. 6.9 shows the plant configuration of the heat recovery system

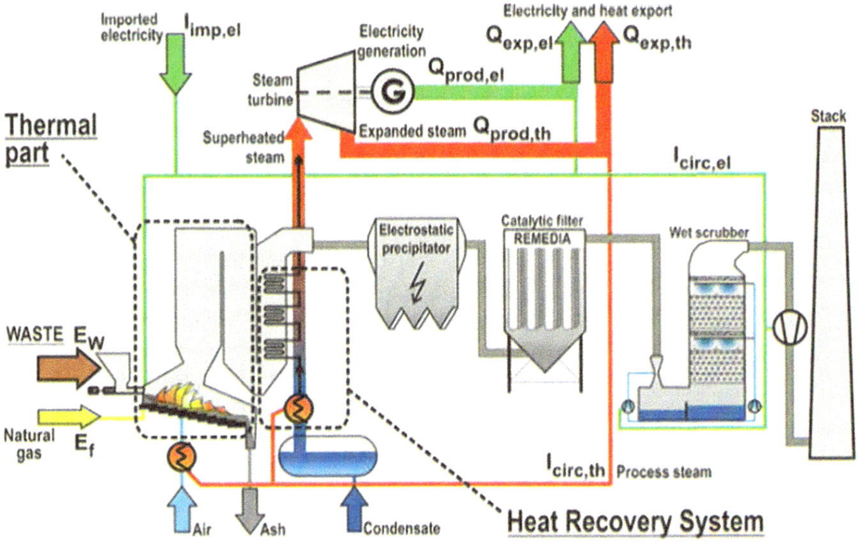

Fig. 6.9 Plant configuration of heat recovery system [21]

[21]. During the cooling process, this boiler device recovers thermal energy by collecting radiant heat from the combustion zone as well as the hot flue gas. Depending on the composition, moisture level, and local activities, this heat content typically ranges from about 2500 kcal/kg [22]. The concentrated heat is used to heat water in a tube bank, resulting in extremely hot steam in the boiler tubes. As a result, this steam is directed to a steam turbine, which generates electricity. Furthermore, hot steam may be used for district heating or for power and space heating inside the plant. The quality of steam is generally determined by temperature, friction, and the water-steam cycle. The natural temperature and pressure are currently around 400°C and 4 MPa, respectively [23]. WtE incineration devices, on the other hand, cannot be equipped for high-temperature and high-pressure systems like power plants because the flue gas contains corrosive gases like hydrogen chloride.

Fouling of the tubes by flue gas deposits, which include fly ash, soot, volatilised metal compounds, and other pollutants, is a major factor in the boiler's performance [4]. The deposits stick to the boiler tubes, limiting heat transfer from the hot flue gases to the water in the steel tubes, lowering steam output, and lowering energy recovery [5]. Flue gas dust filling, fly ash stickiness, which is determined by temperature, flue gas velocity, and tube bank geometry all influence the rate at which tube fouling deposits form. The boiler tubes should be positioned parallel to the gas flow to avoid fouling and corrosion. The presence of molten salts such as calcium, magnesium, and sodium, as well as sulphates, oxides, bisulphates, chlorides, pyrosulphates, and other compounds in the fly ash, as well as SO_3 and HCl, determines the adherence of fly ash to boiler tubes [4]. Soot blowers (which use superheated steam) and shot cleaning (which involves dropping a cast-iron shot on the tubes to shake off the deposits) will also help remove scale deposits (rapping the tube banks to knock off the deposits). Soot blowers are the most common, and they are typically used only once per operational shift. The boiler must be shut down for a thorough mechanical or wet cleaning after 4000 hours of operation when the flue gas outlet temperature reaches a set level [5].

When designing and running incinerator boilers, corrosion is another important factor to remember [5]. Low-temperature acid corrosion can be caused by the formation of HCl from the combustion of chlorine-containing wastes like paper and board, as well as plastics like PVC. At both high and low temperatures, temperature control is critical for avoiding corrosion in the boiler. A series of chemical reactions between tube metal, tube scale deposits, slag deposits, and flue gases in superheater boiler tubes in the boiler chamber at temperatures greater than 450°C causes high-temperature corrosion. Temperature, the presence of low melting phases such as alkali bisulphates and pyrosulphates, acid gases such as HCl and SO_3, the presence of the tube metal, and the frequency of reducing conditions all affect the rate of corrosion. Acid gases such as HCl and H_2SO_4 condense as the temperature drops below the dew point, causing low-temperature corrosion.

As of 2013, the WtE industry is estimated to be worth approximately USD 24 billion [24] and is expected to reach USD 37.6 million by 2020 [25]. As of 2014, at least 80 WtE incinerators were operational in the United States, generating 2769 MWh daily [26]. Many other countries such as Denmark, Sweden, Estonia, Finland, and Japan incinerate at least 50% of their MSW for power generation [27].

Table 6.1 Emission limits for selected countries [28]

Country	Germany -90 Law	UK HMIP IPR 5/3	Sweden -87 Guidelines C ≥ 250 tpd	Netherlands-89 Law New plants	USA-91 Proposal New plants	EC Directives -89, O_2 = 11% New plants C > 5 ton/h
Average	Day	Hourly	Month	Month		Week
	11% O_2 dry	11% O_2 dry	10% CO_2 dry	11% O_2 STP dry	7% O_2 dry	11% O_2 or 9% CO_2 dry
Particulate (mg/m^3)	10	30	20	5	35	30
HCl	10	50	100	10	40 or h = 95%	50
HF	1	2	1	1		2
SO_2	50	300	200	40	85 or h = 85%	300
NO_x as NO_2	70	350	180–300	70	350	200
CO	100			50	500	100
Total C	20			10		20
Dioxin, toxic equivalent (ng/m^3)	–		0.1	0.1		850°C, 2s ≥ 6% O_2
Heavy metals (mg/m^3)						
Total Class I	0.2					0.2
Cd	Cd + TI = 0.05	0.1	0.02	0.05		
Hg	0.05	0.1	0.03	0.05	0.14	
Total Class II	1.0					1.0
As						
Ni						
Total Class III	5.0		0.05 Pb + Zn			5.0 Pb, Cr Mn, Cu
Pb				5.0		
Cr						

6.2.6 *Gas Cleaning System*

Due to the stringent regulations imposed by the introduction of much national and worldwide legislation on MSWI emissions, the control of pollution emission has become a major part of the incineration process. A significant fraction of the total cost of an incinerator plant can be attributed to the provision for pollution control. Table 6.1 shows the emission limit for MSWIs in a number of countries in 1999 [28]. The main directives on emissions are the total particulate or dust in the gas, concentration of acidic gases such as hydrogen fluoride and sulphur dioxide, hydrogen chloride, and heavy metals such as cadmium, mercury, lead, and dioxin. As can also be seen from Table 6.1, these limits are not the same for every country; for example, the emission limit for particulate in the United Kingdom is 30 mg/m^3 while the limit for the Netherlands is 5 mg/m^3. It is the same with the emission limits of NO_x, where the emission limit in the United Kingdom is higher than in the Netherlands. The EC Directives also set a limit for the minimum combustion gas temperature of 850°C (1123 K), a residence time of 2 s, and a minimum oxygen level of 6% to ensure an efficient burn-out [28].

The increasingly stringent emission legislation has forced MSWI operators to install gas treatment plants downstream of their incinerators. The layout of a typical flue gas clean-up system for an MSWI is shown in Fig. 6.10. After the combustion gas from the incinerator exits the heat recovery boiler, the gas often enters a cyclone. This cyclone extracts particles larger than 15 m from the gas as a preliminary collector. The dusty gas stream reaches the cyclone from the side, forms a vortex, and rotates down the cyclone in a helical direction. By using centrifugal force, particles are isolated from the gas and fall to the bottom of the cyclone, where they are deposited. Smaller inlet-orifice cyclones increase collection performance for smaller particles, so they can be used in banks of small units [29].

The gas then flows from the cyclone into an electrostatic precipitator. For municipal waste incinerators, the electrostatic precipitator was once the most common form of particulate removal device. It can remove up to 99.5% of particulates in the flue gas, and it is particularly efficient in removing particles with submicron sizes [29].

The electrostatic precipitator contains an array of wires or thin metal rods with collector plates running between them. Fig. 6.11 shows how the particulates are removed from the flue gas in an electrostatic precipitator [29]. As the gas flows in between the collector plates, the particles are negatively charged by the electrodes. These negatively charged particles are then attached to the positively charged collector plates, where they accumulate and form a dust layer. A rotating hammer system is periodically used to clean the collector plates by "rapping" them to dislodge

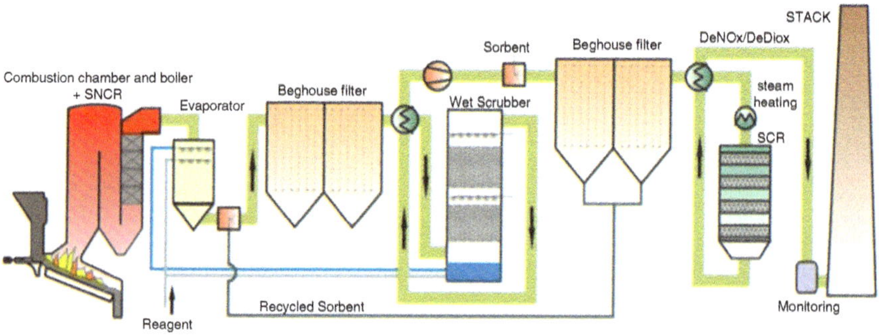

Fig. 6.10 Typical advanced gas clean-up for MSWI [21]

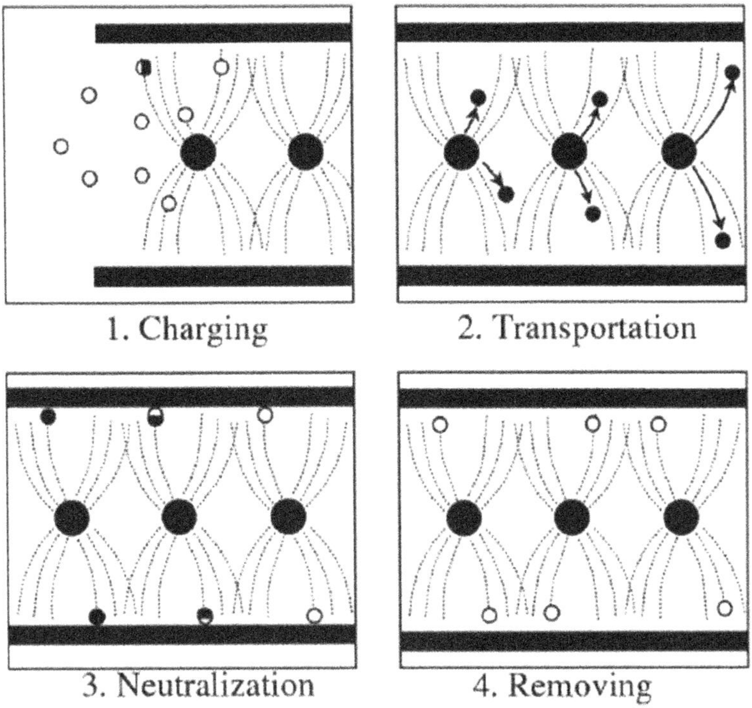

Fig. 6.11 Electrostatic precipitation principle [29]

the layers of accumulated particles. The dust then falls into the collector hopper at the bottom of the precipitator.

The gas then enters a wet, dry, or semi-dry scrubber for removal of soluble acids in it, such as hydrogen fluoride, hydrogen chloride, and sulphur dioxide. It also removes any remaining particulates and heavy metals in the flue gas. Fig. 6.12 shows an example of a wet scrubber.

In a wet scrubber, the gas is firstly cooled to about 60 °C (333 K) in a quench unit. It then enters the first stage of scrubbing, where it is sprayed with water to absorb the hydrogen chloride and hydrogen fluoride. This results in the formation of hydrochloric and hydrofluoric acid. The acid solution also removes the heavy metals in the gas as it flows through the unit. Using an alkaline solution such as lime (calcium hydroxide) or sodium hydroxide, the remaining hydrogen chloride and sulphur dioxide are extracted in the second step. After the de-mister stage, where any liquid carryover is eliminated, the gas exits the scrubber [28].

The dry scrubber system is commonly used in conjunction with fabric filters. As the cooled gas (160 °C) enters the tower, it is sprayed with a dry fine-grained powder, such as dry calcium hydroxide. The reactions between the calcium hydroxide, hydrogen chloride, and sulphur dioxide produce calcium chloride and calcium sulphate, respectively. The dry product is then allowed to drop to the bottom of the

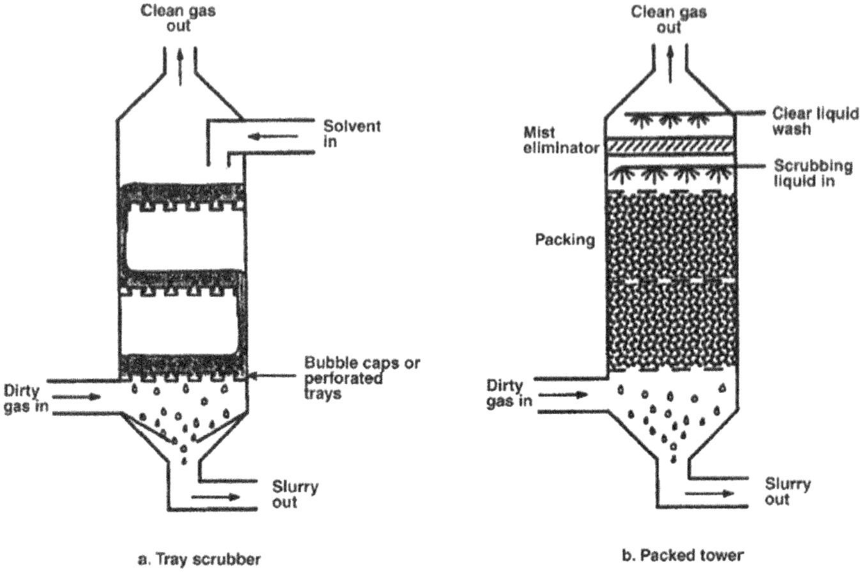

Fig. 6.12 Wet scrubber [29]

tower, where it is collected for further treatment or recycling. This system is also able to remove heavy metals and organic micropollutants by the addition of activated carbon to the calcium hydroxide [21, 23]. The semi-dry scrubber system is also commonly used in conjunction with fabric filters. Its function is the same as the dry scrubber except that droplets of calcium hydroxide solution are sprayed into the gas instead of the dry calcium hydroxide powder, and the water evaporates before the particles reach the wall or fabric filter.

The scrubbed gas then enters a fabric filter where any particulate matter, including fly ash and the activated carbon and lime containing the absorbed pollutant, is removed. After the electrostatic precipitator, scrubber, and injection of additives such as lime or activated carbon, fabric filters are used as the final clean-up stage. They are capable of removing particles as small as submicron from the gas stream, with a particle concentration of less than 10 mg/m^3. A fabric filter is made up of a series of long, permeable fabric bags that are arranged within a baghouse or casing. Fig. 6.13 shows the particulate collection and cleaning processes that occur in a fabric filter. As the gas enters the filter bags, the fine fabric filters out the particulates from the gas stream. The accumulated particulate matter outside the filter bags is then removed by applying a pulse of air into the filter bags. This rapidly expands the bag and releases the accumulated particulates into the hopper located at the base of the baghouse [29].

Before the gas is discharged into the atmosphere, it is treated for the removal of NO_x. Nitrogen oxide (NO_x) emission can be reduced by controlling the combustion conditions and by treating the flue gas. The former method of controlling the NO_x formation is achieved by lowering the combustion temperature and the oxygen

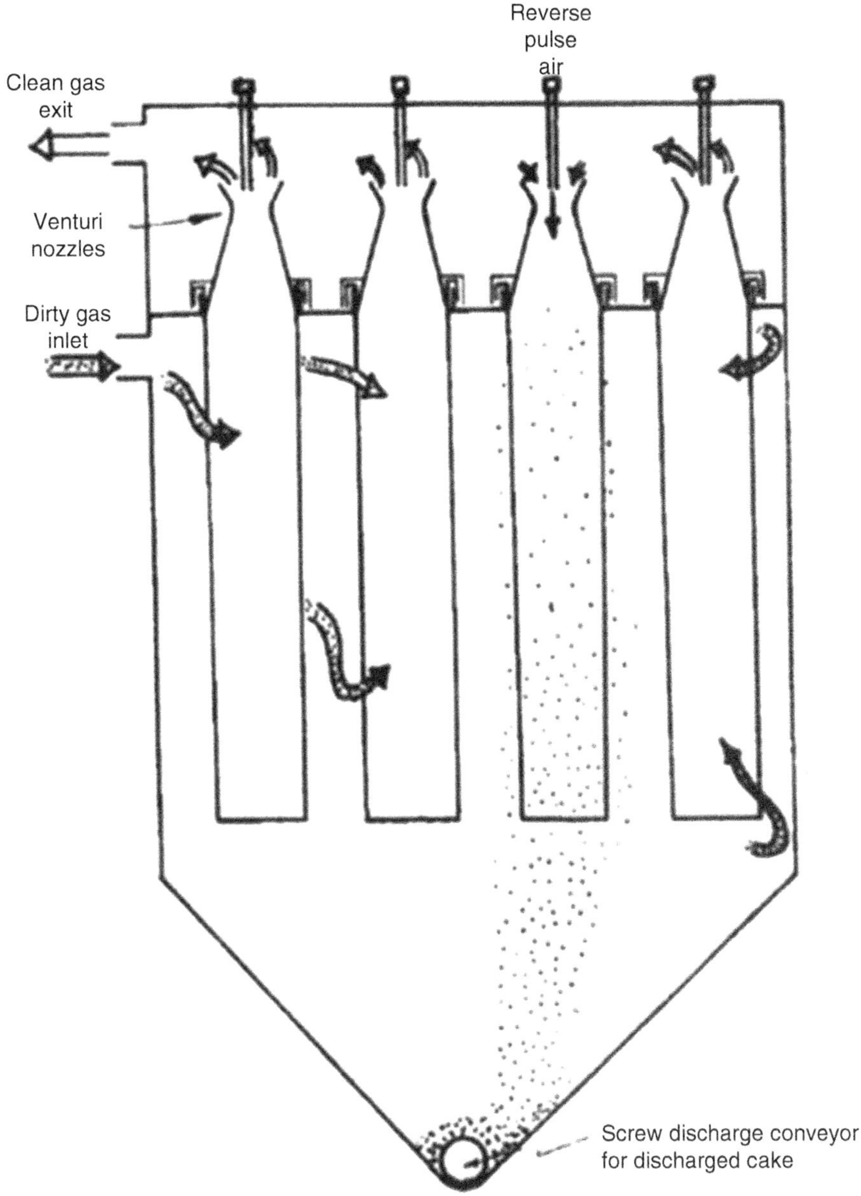

Fig. 6.13 Fabric filter [29]

levels. The latter removal method uses ammonia injection with or without the presence of a catalyst.

The process of ammonia injection where no catalyst is used is called the selective non-catalytic reduction (SNCR) process. This is achieved by injecting ammonia into the gas at temperatures between 870 °C and 900 °C (597 K and 1173 K). This process reduces NO_x to nitrogen and water. SNCR reduces the NO_x level by 75–80% [29]. In the selective catalytic reduction (SCR) process, ammonia is injected into the gas stream in the presence of a catalyst at 300–400 °C (573–673 K) to produce nitrogen and water. Typical catalysts for the SCR process are palladium, platinum, titanium oxide, and vanadium oxide. This method reduces the NO_x level by over 90% [29].

6.2.7 *Residue Processing and Disposal*

Bottom ash and fly ash are the end products of the MSWI plant [23]. Bottom ash and fly ash have different properties depending on the type of incinerator and its location. It must be measured to ensure that the incinerator's combustion efficiency is maintained, as well as to ensure that hazardous substances like heavy metals, dioxins, and furans do not reach regulatory limits. Partially oxidised glass, metal, unburned organic material, inert mineral matter, and char are commonly found in bottom ash [22]. A magnetic separator can be used to recover oxidised metal. Others may be disposed of in landfills or processed for use as cement and building material raw materials [23]. After being treated with cement or a solvent, fly ash is normally disposed of in a managed disposal site [23].

Reburning or landfilling the oversize fraction, which contains the unburned combustible, is an option.

6.3 Other Types of Incinerators

6.3.1 *Fluidised Bed-Type Incinerators*

A fluidised bed incinerator has a layer of sand underneath the combustion chamber, which is blown into by air from the bottom to convert the sand into a fluid. The waste begins to burn on its own on the fluidised bed until the sand layer has been heated. Because of the sand's high heating power, even when waste with high moisture content is applied, it can dry and burn MSW instantly. Fluidised bed incinerators, on the other hand, can be restarted in a limited amount of time after they have stopped operating. However, if the incinerator is not properly built and controlled, incomplete combustion can result in high levels of CO gas due to the high combustion speed. This form of the incinerator is better suited to burning homogeneous materials like sludge than heterogeneous MSW.

6.3.2 Gasification Melting Furnace

The gasification melting furnace is a system that melts bottom ash directly in the furnace to create molten slag. Since molten slag has a higher density than bottom ash, it has greater potential as a construction material. In the gasification melting process, two types of furnaces are used: pyrolysis and gasification melting furnace with a fluidised bed or a kiln and a direct melting furnace with a vertical shaft furnace. Pyrolysis and gasification melting is a process in which waste is thermally decomposed with less oxygen or heated indirectly to induce pyrolysis (partial combustion) at a lower temperature than the combustion temperature, resulting in pyrolysis gas. In fluidised bed and kiln-type furnaces, this process occurs in each furnace that is separate from the melting furnace. In the second step, the pyrolysis gas is fully burned out at high temperatures, and solids are melted using the heat provided by combustion at temperatures of 1,200 to 1,300 °C. Kiln-type furnaces have become increasingly uncommon in recent years.

Waste moves down in a vertical middle pyrolysis layer and lower melting layer in a vertical shaft furnace melting phase. Because of its high costs and operational difficulties, the alternative of gasification melting has not been implemented as quickly as other methods in comparison to traditional incineration methods.

6.3.3 Rotary Kilns

Rotary kilns are exceptionally long-lasting and can burn almost any type of waste, regardless of their shape or composition. Rotary kilns are widely used for the incineration of hazardous wastes. This technology is commonly used to treat clinical waste, but it is not used to treat urban waste. Waste rotary kilns operate at temperatures ranging from 500 °C (as a gasifier) to 1450 °C. When traditional oxidative combustion is used, the temperature is usually above 850 °C. Temperatures of about 1200 °C are common when incinerating hazardous wastes. A schematic drawing of a rotary kiln incineration unit is shown in Fig. 6.14.

The rotary kiln is made up of a cylindrical vessel with a horizontal axis that is slightly inclined. The kiln is normally mounted on rollers, which allows it to rotate or oscillate around its axis (reciprocating motion). As the kiln rotates, gravity transports the waste into it. Rotary kilns can burn solid waste, liquid waste, gaseous waste, and sludges. A post-combustion chamber is normally inserted to maximise the degradation of toxic compounds. The additional firing of liquid waste or additional fuel may be necessary to sustain the temperatures needed for the waste to be incinerated and be destroyed.

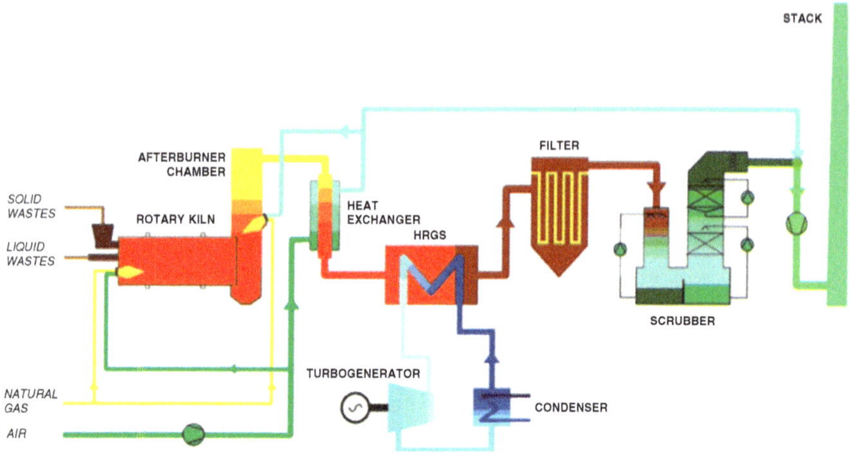

Fig. 6.14 Schematic of a rotary kiln incinerator plant [21]

6.3.4 Plasma Technologies

Electrons, ions, and neutral particles make up plasma (atoms and molecules). The interaction of gas with an electric or magnetic field will produce this high-temperature, ionised, conductive gas. Plasmas contain reactive species, and the high temperatures encourage chemical reactions to occur quickly. To create a plasma, plasma processes use high temperatures (5000 to 15000 °C) that result from the conversion of electrical energy to heat. They entail moving a huge electric current through a gas stream that is inert. Hazardous pollutants such as PCBs, dioxins, furans, toxins, and others are broken down into their atomic constituents by injecting them into the plasma under these conditions. Organics, metals, PCBs (including small-scale equipment), and HCB are all treated using this method. In many cases, waste pre-treatment is needed. The technology's destruction efficiencies are very high, at >99.99%. Plasma technology is a well-established commercial technology, but it can be a difficult, costly, and labour-intensive operation. The following are some examples of plasma technologies.

6.3.4.1 Argon Plasma Arc

The waste reacts with the argon plasma jet directly. Since it is inert and does not react with the torch elements, argon was chosen as the plasma gas. The destruction and removal efficiency (DRE) for destroying ozone-depleting substances (ODS) at 120 kg/h and 150 kW electrical power is stated to be greater than 99.9998%.

6.3.4.2 Inductively Coupled Radio Frequency (ICRF) Plasma

Inductively coupled plasma torches are used in ICRF applications, and energy is transferred to the plasma through the electromagnetic field of the induction coil. Since there are no electrodes, the device can be used for a wide variety of gases, including inert, reducing, or oxidising atmospheres, and it is more reliable than plasma arc methods. The ICRF plasma process has a DRE of over 99.99% and can kill CFC at a rate of 50 to 80 kg/h.

6.3.4.3 Alternating Current (AC) Plasma

The AC plasma is similar to the ICRF plasma in that it is generated directly with 60 Hz high voltage power. The device is believed to be very reliable because it is electrically and mechanically simple. The method does not need argon and can operate with a number of working gases, such as air or steam as plasma gases.

6.3.4.4 CO_2 Plasma Arc

Sending a strong electric discharge into an inert atmospheric gas, such as argon, produces a high-temperature plasma. Depending on the desired process outcomes, the plasma field is maintained with ordinary compressed air or certain atmospheric gases once it has been created. The temperature of the plasma at the point of generation, into which the liquid or gaseous waste is directly pumped, is well over 5000 °C. The temperature in the upper reactor is about 3500 °F, and it gradually drops across the reaction zone to a precisely regulated temperature of around 1300 °F. The use of CO_2, which is produced during the oxidation reaction, as the gas to support the plasma is a unique feature of the process.

6.3.4.5 Microwave Plasma

This method uses microwave energy at 2.45 GHz to produce thermal plasma under atmospheric pressure in a specially built coaxial cavity. The plasma is started with argon, but the process does not need any gas to keep the plasma going. The microwave plasma process is said to have a DRE of over 99.99%, killing CFC-12 at a rate of 2 kg/h. The process's high destruction efficiency is a key benefit. The process is said to be capable of reaching high operating temperatures in a limited amount of time, allowing for greater operational flexibility and less downtime.

6.3.4.6 Nitrogen Plasma Arc

This method produces thermal plasma using a direct current (DC) non-transferred plasma torch with water-cooled electrodes and nitrogen as the working gas. The process was created in 1995, and commercial systems are now available. At a feed rate of 10 kg/h, the process is said to achieve a DRE of 99.99% when destroying CFCs, HCFCs, and HFCs. The equipment is very small, which is a major benefit of this technology.

6.4 Incineration Process

Incineration is a complex process involving the disciplines of chemical and mechanical engineering. To fully understand this process, one must understand the fundamental of incineration, which is its stoichiometry. This involves studying its fundamental relations, material balances, energy balance, equilibrium, and kinetics. In addition, a wide range of waste compositions must also be taken into consideration. In most cases, waste incinerator operators have limited or no control of the precise composition of the incoming waste. However, mass-burn incinerators must be designed to cope with the wide range of waste compositions. Owing to the difficulties ensuring acceptable composition for the incineration process, a triangle diagram was developed, as shown in Fig. 6.15 [30]. The diagram was created to

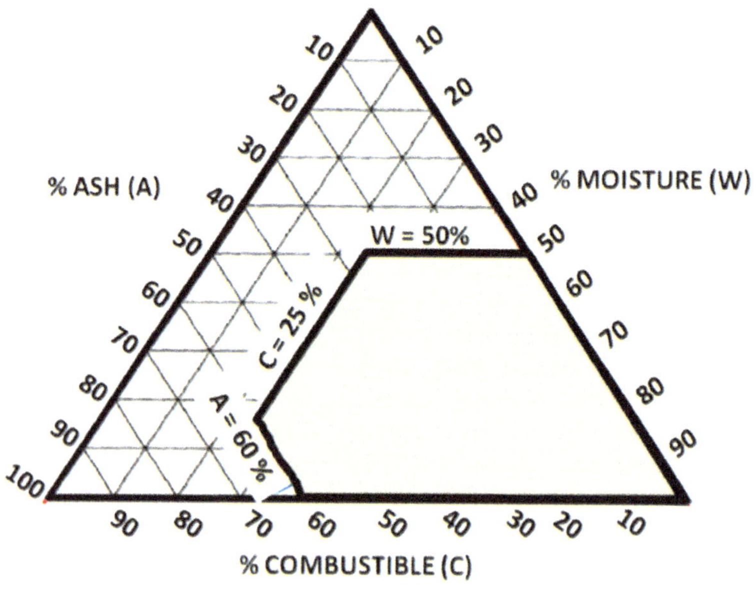

Fig. 6.15 Suitability of MSW composition for incineration [30]

confirm whether the current proximate composition (moisture, ash, and volatile) of municipal solid waste will be appropriate for incineration based on the close relationship between waste composition and calorific value CV. The shaded area depicts a typical municipal solid waste composition that can withstand combustion without the use of auxiliary fuel. The region encompasses the minimum allowable CV as well as the maximum moisture content that can be tolerated. The following parts will go into how to calculate and test the above parameters.

Incineration technology, for example, can be used on waste with a moisture content of 75% or less and volatile content of 20% or more, resulting in a CV of 3,352 kJ/ kg or more. Furthermore, these technologies are typically applicable for waste with a moisture content of 65% or less and volatile content of 30% or more, with a CV of 6,285 kJ/kg or more, in cases where energy recovery is a part of the incineration process.

When looking at the overall composition of waste, the proportion of food and kitchen waste with high moisture levels has the greatest effect. Incineration becomes a viable choice when the volume of organic waste is between 50% and 60%. It is not, however, ideal for energy recovery, which is only possible when the proportion of plastic to paper increases and the proportion of food and kitchen waste falls to about 50 % or less.

6.4.1 Fundamental Relationships

In dealing with the incineration process, one must be familiar with a few chemical and mechanical knowledge related to gas laws, material balances, heat balances, equilibrium, and kinetics. The following subsection will discuss them further.

6.4.1.1 Gas Laws

The ideal gas law is described by Eq. 6.1,

$$PV = nRT \tag{6.1}$$

where P is the absolute pressure, T is the absolute temperature, V is its volume, n is the number of moles of the gas, and R is the universal gas for ideal gases. Table 6.2

Table 6.2 Values of the gas constant R for ideal gases [22]

Pressure	Volume	Moles	Temperature	R
atm	m^3	kg-mol	K	0.08206 m^3-atm kg-mol^{-1} K^{-1}
psia	ft^3	Ib-mol	°R	1543 ft-lb Ib-mol^{-1} °R^{-1}
atm	ft^3	Ib-mol	°R	0.729 ft^3-atm Ib-mol^{-1} °R^{-1}
–	–	kg-mol	K	1.986 kcal kg-mol^{-1} K^{-1}

shows values of the gas constant *r* for an ideal gas [22]. Example 1 will give an overview of how much CO_2 will be produced from an incineration process.

Example 1. An incineration plant is needed to process 6000 kg of waste per day with an average carbon content of 80%, ash content of 7%, and moisture content of 13%. The combustion gases leave the furnace at 1100 °F and pass through a gas cooler before exiting at 90 °F. On a daily basis, how many kilogram-moles and kilograms of CO_2 will be produced? At 1.04 atm, how many cubic meters of CO_2 are emitted per day at the furnace and gas cooler outlets?

The waste contains (6000)(0.80)(1/12) = 400 moles of carbon (atomic weight = 12). In full combustion, each mole of carbon emits 1 mole of CO_2, resulting in 400 mol/d of CO_2. CO_2 (molecular weight 44) has a weight flow of 400(44) = 17600 kg/d.

$$From PV = nRT$$

$$V = \frac{nRT}{P} = \frac{400(0.08206)T}{1.04} = 15.78T$$

At 1100 °C (1373 K), V = 43,334 m^3. At 90 °C (363 K), V = 11,457 m^3.

6.4.1.2 Material Balances

Calculation of material balances is very important in order to know the amount of theoretical air required to completely oxidise carbon, hydrogen, sulphur, etc. It can be represented by the following equation:

$$Input = output + accumulation \tag{6.2}$$

This equation represented a quantitative expression of the law of conservation of matter and is always true for all the elements that pass through a combustion system. However, it is not true for an individual compound that took part in the combustion reaction. The basic data for material balance calculation are the analyses of fuel or waste, gases in the system, reaction rate, proportions in molecules, and heat of reactions. Through this balance on elements in the fuel or waste, the amount of theoretical air required to completely oxidise carbon, hydrogen, sulphur, etc. can be calculated. This theoretical or stoichiometric air requirement is often insufficient in a practical combustor; thus, an excess air must always be supplied. For example, an incinerator operating at 50% excess air denotes a combustion process to which 1.5 times the stoichiometric air requirement has been supplied.

The following is an example of theoretical air calculation for a combustion process.

Example 2. The ultimate analysis of 100 kg/h of waste at an incinerator shows that it has 12.2% moisture, 75% carbon, 5.2% hydrogen, 2.4% sulphur, 2.1% oxygen, 0.5% nitrogen, and 1.6% ash. The combustion air is at 15.5 °C and has 70% relative humidity. Calculate the amount of air needed and products of combustion when it operates at 50% excess air. The sequence of computations is shown in Table 6.3.

Shown below are several elements of the analysis [7]:

(a) Line 1: Carbon is assumed to be entirely converted to carbon dioxide during combustion.
(b) Line 2: The volume of combustion air is increased by hydrogen in the waste (other than hydrogen in moisture), but this is not accounted for in the Orsat study (Lines 16 and 17).
(c) Line 3: Sulphur in the waste, in the form of sulphide or organic sulphur, increases the amount of combustion air needed for SO_2 combustion. Sulphates inorganic can be left as ash or reduced to SO_2. SO_2 (Line 17) is normally listed out as carbon dioxide if the selective analysis is not used.
(d) Line 4: The amount of necessary combustion air is reduced due to the presence of oxygen in the waste.
(e) Line 12: Moisture entering the combustion air appears to be minor and is often overlooked. Despite the fact that this issue only considered waste components of C, H, O, N, and S, the analyst should thoroughly analyse the waste composition and consider the following secondary reactions:
(f) Carbon monoxide is a poisonous gas. CO is generated in significant amounts in solids-burning systems.
(g) Chlorine is a chemical that is used to kill bacteria. Chlorine that appears in waste as inorganic salts will almost certainly remain as salt. Organic chlorides, on the other hand, mainly produce hydrogen chloride.
(h) Metals are a type of metal. While a significant fraction of massive metal feed (e.g., tin cans, sheet steel, etc.) is unoxidised in solid waste burning, metals typically burn to the oxide.
(i) Decomposition due to heat. Some compounds can decompose at high temperatures in the combustor. Carbonates, for example, can dissociate to produce oxide and CO_2, while sulphides can "roast" to produce oxide and SO_2.

Another method of calculation can be done by evaluating an existing MSWI. Here, flue gas composition data can easily be obtained; thus, the operation and the feed waste can be characterised at a much cheaper cost. The percentage of excess air can be calculated using the flue gas composition obtained and the following equation:

$$\text{Percentage excess air} = \frac{\left[O_2 - 0.5\left(CO + H_2\right)\right]100}{0.266N_2 - O_2 + 0.5\left(CO + H_2\right)} \tag{6.3}$$

where O_2, N_2 etc., are the volume percentages of the gases on a dry basis. Example 3 will show how Eq. 6.3 is used.

Table 6.3 Calculations for Example 2

Line	Component	kg	Atoms or moles [a]	Combustion product	Theoretical moles of O_2 required	Moles formed in stoichiometric combustion					
						CO_2	H_2O	SO_2	N_2	O_2	Total
1	Carbon, C	75.0	6.250	CO_2	6.250	6.250	0.0	0.0	0.0	0.0	6.250
2	Hydrogen, H_2	6.2	3.100	H_2O	1.550	0.0	3.1	0.0	0.0	0.0	3.100
3	Sulfur, S	2.4	0.075	SO_2	0.075	0.0	0.0	0.075	0.0	0.0	0.075
4	Oxygen, O_2	2.1	0.066	–	(0.066)	0.0	0.0	0.0	0.0	0.0	0.0
5	Nitrogen, N_2	0.5	0.018	N_2	0.0	0.0	0.0	0.0	0.018	0.0	0.018
6	Moisture, H_2O	12.2	0.678	H_2O	0.0	0.0	0.678	0.0	0.0	0.0	0.678
7	Ash	1.6	N/A	–	0.0	0.0	0.0	0.0	0.0	0.0	0.0
8	Total	100.0			7.809	6.250	3.778	0.075	0.018	0.0	10.121
9	Moles of nitrogen in stoichiometric air[b]			(79/21) (7.809)					29.377		29.377
10	Moles of nitrogen in excess air			(0.5)(79/21)(7.809)					14.688		14.688
11	Moles of oxygen in excess air			(0.5) (7.809)						3.905	3.905
12	Moles moisture in combustion air[c]						0.604				0.604
13	Total moles in flue gas					6.250	4.382	0.075	44.083	3.905	58.695
14	Volume (mole) percentage in wet flue gas					10.648	7.466	0.128	75.105	6.653	100.0
15	Orsat (dry) flue gas analysis, moles					6.250		0.075	44.083	3.905	54.313
16	A. With selective SO_2 testing, volume percentage					11.508	N/A	0.138	81.164	7.180	100.0
17	B. With alkaline CO_2 testing only, volume percentage					11.646	N/A	N/A	81.164	7.190	100.0

[a]The symbol in the component column shows whether these are kg-mol or kg-atom

[b]Throughout this chapter, dry combustion air is assumed to contain 21.0% oxygen by volume and 79.0% nitrogen

[c]Calculated as follows: (0.008/18)[(29.377 + 14.688)(28) + (3.905)(32)] based on the assumption of 0.008 kg water vapor per kg bone-dry air; found from standard psychrometric charts

Example 3. An incinerator that burns a specific amount of waste emits flue gas that contains 11.6% CO_2, 7.2% O_2, and the remaining nitrogen and inerts. Calculate the weight ratio of hydrogen to carbon in the waste, the percentage of carbon and hydrogen in the dry waste, the kilograms of dry air used per pound of dry waste, the percentage of excess air used, and the moles of exhaust gas discharged per kilogram of dry waste burned using these statistics. (It should be noted that this example is based on Example 2.)

As a starting point: 100 mol dry exhaust gas

Component	Moles	Mol O_2
$CO_2 + (SO_2)$	11.6	11.6
O_2	7.2	7.2
N_2	81.2	–
Total	100.0	18.8

If all N_2 comes from the combustion air, a total of 81.2 × (21/79) = 21.6 mol O_2 is entered with the N_2. The difference, 21.6 – 18.8 = 2.8 mol O_2, can be assumed to have been consumed in burning hydrogen.

H_2 burned: 2(2.8) = 5.6 mol	11.6 kg
C burned : 12(11.6) = 139.2 mol	139.2 kg
	150.4 kg

(a) Weight ratio of hydrogen to carbon: (11.2/139.2) = 0.08.
(b) Percentage (by weight) of C in dry fuel: (139.2/150.4)(100) = 92.55.
(c) Kilogram of dry air per kilogram of dry waste.
First, calculate the weight of air resulting in 1 mol dry exhaust gas from a nitrogen balance:
1/100 × (81.2 mol N_2)(1/0.79 mol N_2/mol air)(29 kg air/mol) = 29.81 kg air/mol dry exhaust gas.
Then, 29.81(100/150) mol dry exhaust gas/kg waste = 19.87 kg dry air/kg dry waste.
(d) Percentage of excess air:

The oxygen necessary for combustion is 11.6 + 2.8 = 14.4 mol
The oxygen unnecessary for combustion = 7.2 mol
The total oxygen = 21.6 mol.
The percentage of excess air (or oxygen) may be calculated as

$$\frac{(100)(unnecessary)}{total - unnecessary} = \frac{100(7.2)}{21.6 - 7.2} = 50\%$$

$$\frac{(100)unnecessary}{necessary} = \frac{100(7.2)}{14.4} = 50\%$$

$$\frac{(100)(total - necessary)}{necessary} = \frac{100(21.6 - 14.4)}{14.4} = 50\%$$

(e) Moles of exhaust gas per kilogram of dry waste:

Noting that 5.6 mol water vapour must be added to the dry gas flow, (100 + 5.6)/150 = 0.702 mol/kg fuel.

It is necessary to realise that in calculating the excess air, an engineer needs to acknowledge that [7]

- The results of the waste analysis are crucial in determining the amount of combustion air needed for construction.
- Waste moisture data are needed to calculate total flue gas rates.
- Stack gas analysis may provide information about the waste's composition.
- If data are available, cross-check all data for both fuel and flue gas to ensure accuracy.

6.4.1.3 Heat Balances

A heat balance is a numerical expression of the law of energy conservation. Five energy amounts are of primary importance in waste incineration:

- Chemical Energy. It is a term that refers to the energy that heat produced by chemical reactions, especially combustion.
- Latent Heat. The heat effect of state changes, especially the heat of moisture vaporisation.
- Sensible Heat. The heat content (enthalpy) of materials is proportional to their temperature.
- Useful Heat. The heat that can be used, especially the sensible heat that can be used to generate steam.
- Heat Loss. Through conduction, convection, and radiation, heat is lost through the furnace walls.

The value of heat of combustion and sensible heat are readily available in the literature, usually known as higher heating value (HHV). One can just use it for further calculation. The sensible heat content (Δh) at a temperature T may be calculated relative to the reference temperature T_0 by

$$\Delta h = \int_{T}^{T_0} MC_p^0 dT \,\text{kcal}/\text{kg mol} \tag{6.4}$$

where $M\ C_p^0$ is the molar heat capacity (kcal mol^{-1} °C^{-1}). Another way to calculate Δh is by using $M\ C_{p\,avg}^0$ from Fig. 6.16 and using the following equation:

$$\Delta h = \left(T - T_0\right)\left(MC_{p\,avg}^0\right) \tag{6.5}$$

Example 4. If the 100 kg of waste in Example 2 has a heat of combustion of 7500 kcal/kg (HHV) and the combustion air is pre-heated to 300 °C, what is the tempera-

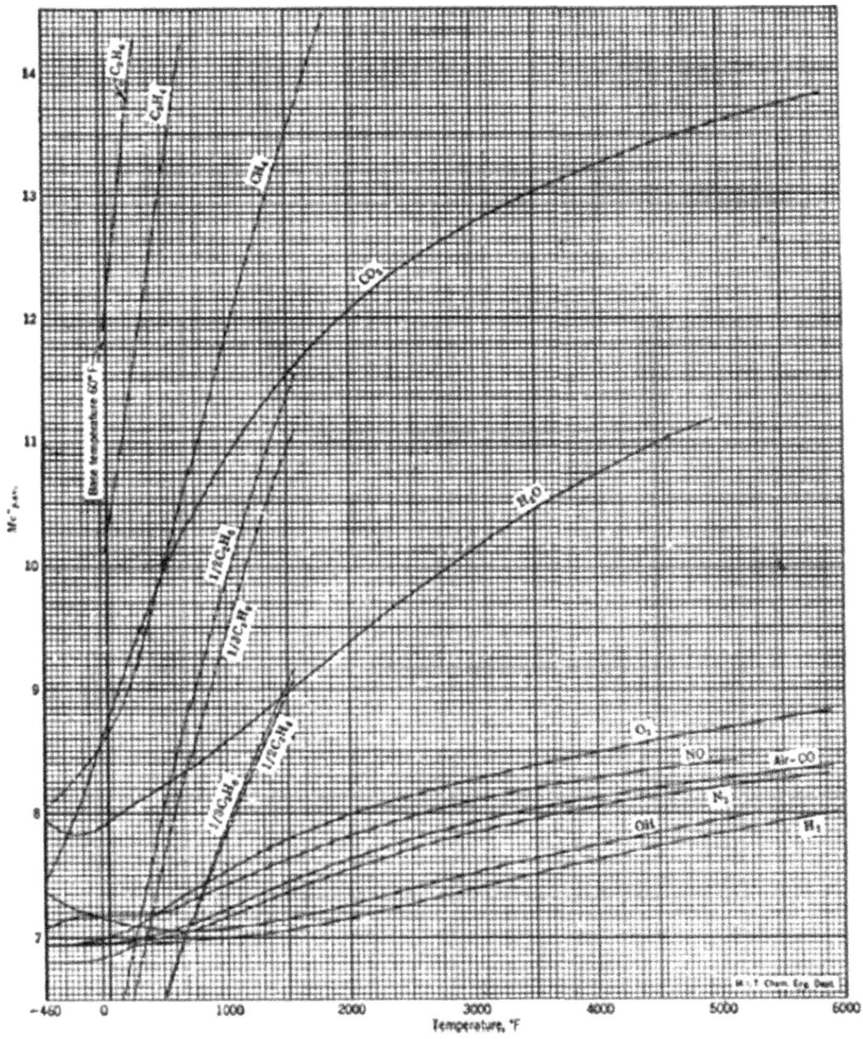

Fig. 6.16 Average molar heat capacity of fuel and combustion gases, using $M\ C^{0}_{pavg}$ at zero pressure between 60°F (15.6 °C) and abscissa temperature [7]

ture of the flue gases? When gases are cooled to 180°C (about 350°F) in a boiler, how much steam can be produced? At 204°C, assume a 5% heat loss in the furnace and a 5% heat loss in the boiler, as well as a 570 kcal/kg enthalpy transition from boiler feedwater to produce steam and 1 h service as a base.

The total combustion air supplied to the system is 29.377 + 14.688 + 3(3.905) + 0.604 = 56.384 mol (see Table 6.3). From Fig. 6.20, the heat content of the preheated air at 300 °C is

$$(56.384)(7.08)(300-15.5)=113,572\,kcal$$

therefore, the total energy impact is

$$7500(100)+113,572=863,572\,kcal\;energy\;addition$$

To determine the combustion chamber's exit temperature and steaming rate, create a plot of the heat content of the gas stream as a function of temperature, as shown in Table 6.4 and shown in Fig. 6.17. Table 6.5 depicts the flow of thermal energy. The enthalpy transition for feedwater (at 100 °C and 15.8 atm) transitioning to saturated steam at 15.8 atm is 567.9 kcal/kg, resulting in the following steaming rate for a burning rate of 1100 kg/h: 661,757/567.9 = 1165 kg /h

For feedwater (at 100 °C and 15.8 atm) changing to saturated steam at 15.8 atm, the enthalpy change is 567.9 kcal/kg, so the resulting steaming rate for a burning rate of 1100 kg/h is

(661,757/567.9) g = 1165 kg/h

Table 6.4 Computation of heat content of flue gases from combustion of benzene waste at 10% excess air

A. Assumed temp., °C	180	500	1,000	1,500	2,000
B. A – 15.5 °C	164.5	484.5	984.5	1,484.5	1.984.5
C. $Mc°p_{,avg}N_2$[a]	7.00	7.13	7.48	7.76	8.00
D. $Mc°p_{,avg}O_2$	7.10	7.50	7.92	8.20	8.40
E. $Mc°p_{,avg}H_2O$	8.10	8.52	9.23	9.90	10.50
F. $Mc°p_{,avg}CO_2$	9.42	10.75	11.90	12.60	13.08
G. Ash[b]	0.2	0.2	0.2	0.2	0.2
H. 44.083(B)(C)	50,750	152,250	324,480	507,925	699,850
I. 3.905 (B)(D)	4,740	14,230	30,565	47,430	65,350
J. 4.382 (B)(E)	5,825	18,125	39,870	64,340	91,265
K. 6.25 (B)(F)	9,680	32,540	73,210	116,910	162,210
L. 1.6(B)(G) + 85[c]	190	290	450	610	770
M. 4.382 (10,595)[d]	46,430	46,430	46,430	46,430	46,430
N. (H + I + J + K + L + M)[e]	117,615	263,865	515,005	783,645	1,065,875
O. kcal/mol gas	2,004	4,495	8,774	13,351	18,160

[a]Source: Fig. 6.16 (kcal/kg mol °C)

[b]Specific heat of the ash (kcal/kg °C) for solid or liquid

[c]The latent heat of fusion of the ash (85 kcal/kg) is added at temperatures greater than 800 °C, the assumed ash fusion temperature

[d]Latent heat of vaporization at 15.5 °C of free water in waste and from combustion of hydrogen in waste (kcal/kg mol)

[e]Total heat content of gas stream (kcal)

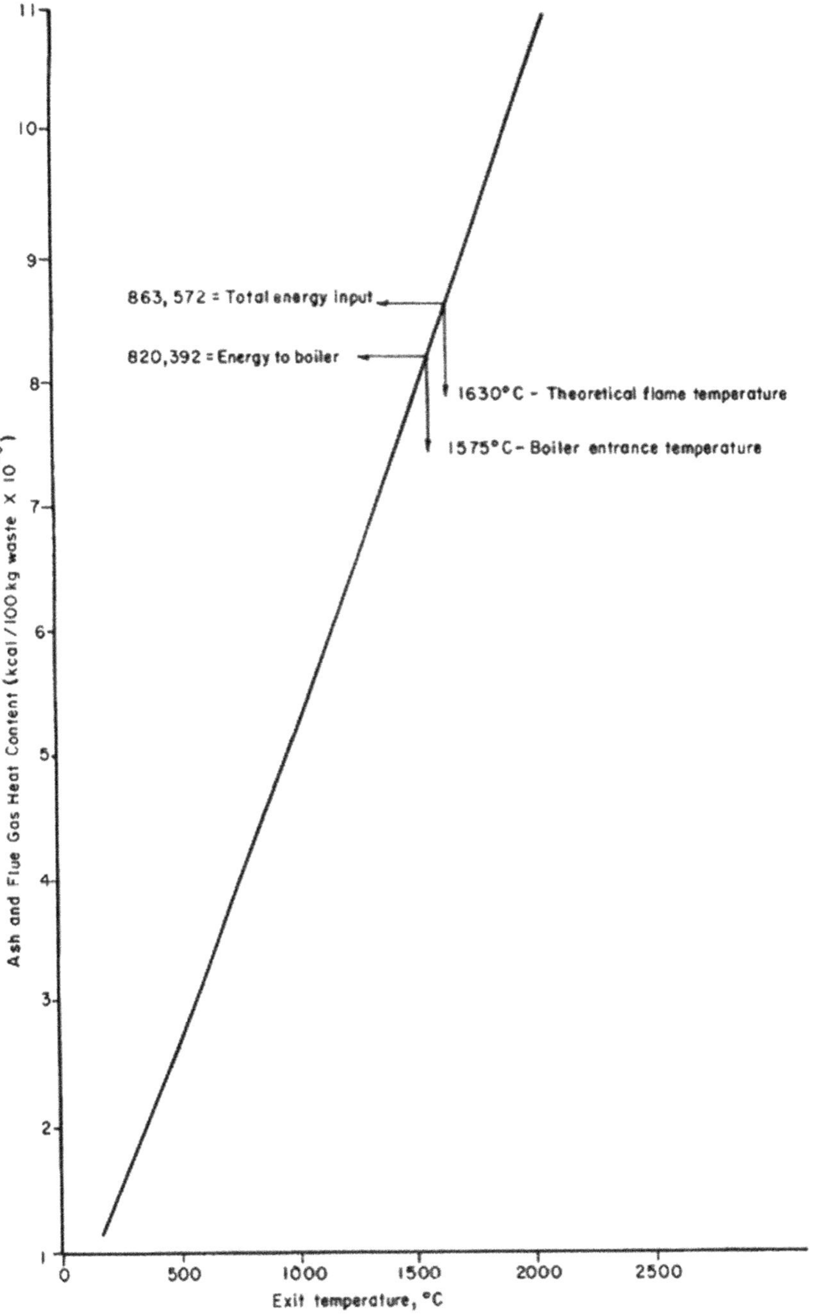

Fig. 6.17 The heat content of exhaust gases relative to 15.5 °C [7]

Table 6.5 Energy flow calculation for Example 4 [7]

Energy flows	kcal	Temperature, °C
Energy into system		
Heat of combustion	750,000	15.5
Air preheat	113,572	300
Total	863,572	1,630°
Heat loss (5%) from combustion chamber	(43,180)	
Energy into boiler	820,392	1,575
Heat loss (5%) from boiler	(41,020)	
Heat loss out stack	(117,615)	180
Net energy into steam	661,757	204

The theoretical (adiabatic) flame temperature for this system (the temperature of the products of combustion assuming no heat loss)

6.4.1.4 Equilibrium

The complete chemical reaction does not exist because some fraction of the reactants remain in the reaction mass. For the gas-phase reaction,

$$aA + bB \sim cC + dD \tag{6.6}$$

where the reactant and product concentrations are expressed as partial pressures and the equilibrium constant K_p, which is a function (only) of temperature, is given by

$$K_p = \frac{p_C^c p_D^d}{p_A^a p_B^b} \tag{6.7}$$

where the units of K_p depend on the stoichiometric coefficients *a, b, c,* and *d* such that if $(c + d - a - b)$ is zero, K_p is dimensionless. If the total is nonzero, K_p will have the units of pressure raised to the appropriate integer or fractional power.

Fig. 6.18 shows the temperature dependence of reactions of interest. Note that when solid carbon is a product or reactant, no partial pressure term for carbon is entered into the mathematical formulation.

Example 5. At the furnace outlet temperature in Example 4 and at a total pressure of 1 atm, what is the emission rate of nitric oxide (NO) formed by the reaction?

$$\frac{1}{2}O_2 + \frac{1}{2}N_2 \leftrightharpoons NO \tag{6.8}$$

From Fig. 6.18 at 1575 °C, log $K_p = 1.9$ ($K_p = 79.43$) where

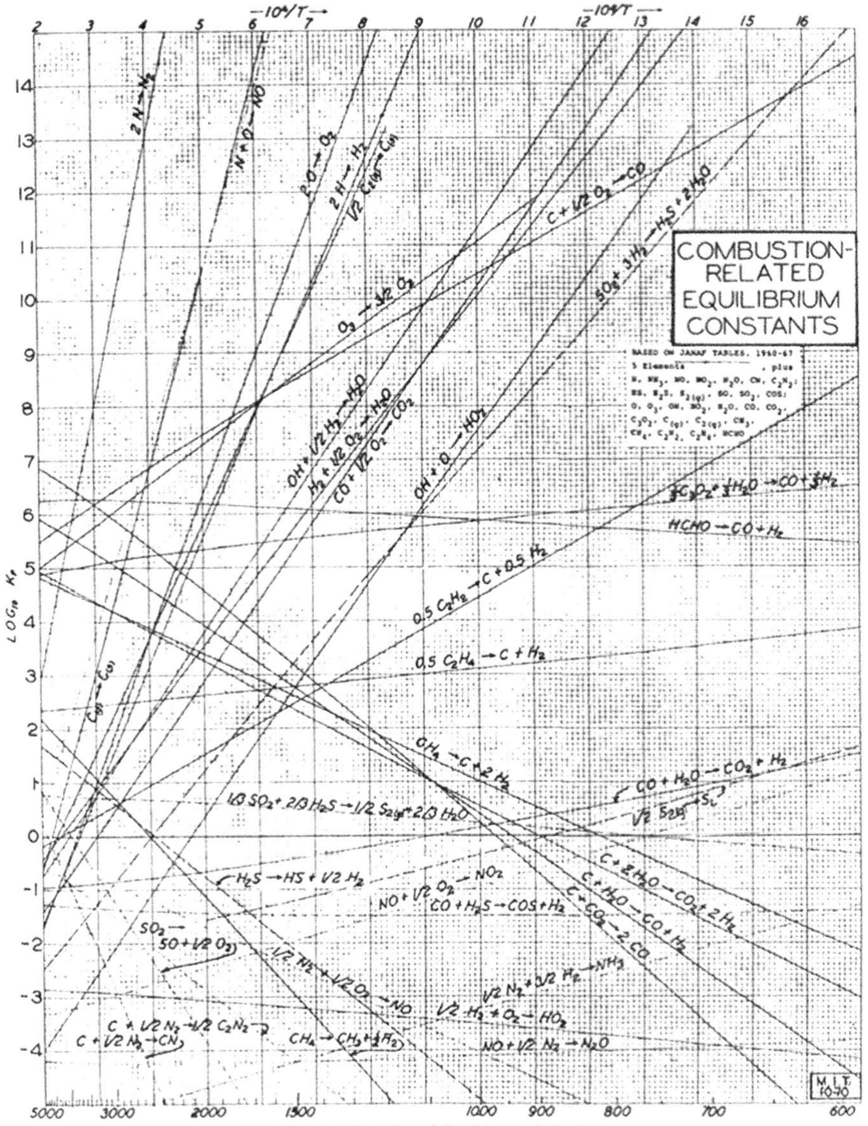

Fig. 6.18 Equilibrium constants of combustion reactions (partial pressure in atm) [7]

$$K_p = p_{O_2}^{1/2} p_{N_2}^{1/2} / p_{NO} \tag{6.9}$$

At equilibrium, then

$$(3.905 - 0.5X)1/2(44.083 - 0.5X)1/2 = 79.43x$$

Solving for x resulted in x = 0.164 mol NO at equilibrium, or 0.279 mol/o or 2794 ppm. Note, however, that kinetic limitations usually result in NO concentrations substantially below those predicted by equilibrium alone [7].

6.4.1.5 Kinetics

Temperature, as well as the concentrations of the reactants and the static pressure (for gas-phase reactions), are all essential factors in chemical reactions. At combustion temperatures, reactions are normally very rapid. The oxidation reactions for carbon monoxide (CO), soot (carbon), and chlorinated hydrocarbons are notable exceptions. The previous publication addressed the reaction rate activity (chemical kinetics) of CO and soot burning [7]. Temperature, as well as the concentrations of the reactants and the static pressure (for gas-phase reactions), are all essential factors in chemical reactions. At combustion temperatures, reactions are normally very rapid. The oxidation reactions for carbon monoxide (CO), soot (carbon), and chlorinated hydrocarbons are notable exceptions. The previous publication addressed the reaction rate activity (chemical kinetics) of CO and soot burning [7].

Kinetics of Carbon Monoxide Oxidation

Carbon monoxide (CO) is an essential air pollutant, a hazardous gas in high concentrations, and represents unavailable combustion energy if present in stack gases. The rate expression for the rate of change of the CO mol fraction (f_{CO}) with time can be expressed by [7]

$$\frac{-df_{CO}}{dt} = 12\,x\,10^{10}\,\exp-\left[\frac{16,000}{RT}\right] f_{O_2}^{0.3} f_{CO} f_{H_2O}^{0.5}\left[\frac{P}{R'T}\right] \tag{6.10}$$

where f_{CO}, f_{O_2}, and f_{H_2O} are the mole fractions of CO, O_2, and water vapor, respectively, T is the absolute temperature (K), P is the absolute pressure (atm), t is the time in seconds, and R and R′ are the gas constant expressed as 1.986 cal g mol^{-1} K^{-1} and 82.06 atm cm^3 g $mol^{-1}K^{-1}$, respectively.

The term (−16,000/RT) is the kinetic expression's core, providing a high-temperature sensitivity by exponentiating the ratio of 16,000 (the Arrhenius "activation energy") to the absolute temperature.

The rate of reaction is affected by the amount of water vapour present, which reflects the position of hydrogen (H) and hydroxyl (OH) free radicals in combustion reactions. In reality, bone-dry CO is extremely difficult to burn, while even a smidgeon of moisture is enough to aid ignition and rapid combustion.

Kinetics of Soot Oxidation

Soot (finely divided carbon) formation is another drawback of burning carbon-bearing wastes. This is the black smoke we see when combustion occurs at a very low oxygen level. It can cause device problems by fouling boiler tube surfaces, reducing the collection efficiency of electrostatic precipitators, and so on, resulting in violations of opacity regulations that apply to stack discharges. In contrast to many other combustion reactions, soot burn-out is relatively slow. The rate of carbon consumption q (g cm^{-2} S^{-1}) to the oxygen partial pressure in atmospheres (P_{O2}) is given by the following equation for spherical particles:

$$q = \frac{P_{O2}}{1/k_s + 1/k_d} \tag{6.11}$$

where k_s is the kinetic rate constant for the consumption reaction and k_d is the diffusional rate constant. For particles of diameter d (cm) at a temperature T (K),

$$k_d = \frac{4.335 x 10^{-6} T^{0.65}}{d} \tag{6.12}$$

$$k_s = 0.13 \exp\left[(-35{,}700/R)\left(\frac{1}{T} - \frac{1}{1600}\right)\right] \tag{6.13}$$

where R is the gas constant (1.986 cal g-mol^{-1} K^{-1}). For a particle of initial diameter d_o and an assumed specific gravity of 2, the time t_b in seconds to completely burn out the soot particle is given by

$$t_b = \frac{1}{P_{O_2}}\left[\frac{d_0}{0.13\exp\left[\left(\frac{-35{,}700}{R}\right)\left(\frac{1}{T} - \frac{1}{1600}\right)\right]} + \frac{d_O^2}{8.67 x 10^{-6} T^{1.75}}\right] \tag{6.14}$$

6.4.2 *Thermal Decomposition (Pyrolysis)*

Pyrolysis is a burning process in the absence of a limited air supply. It generates a low-heat-content stream containing volatilised water, a mixture of CO, hydrogen, and hydrocarbons, and a solid char that is often fully burned in a specialised region of the "pyrolyser."

Pyrolysis starts around 200 °C. It generates complicated partially oxidised tars. If the temperature rises, these materials degrade further, giving way to simpler, more hydrogen-rich gaseous compounds and solid carbon. In terms of chemical composition and physical structure, the solid residue is similar to graphitic carbon.

In pyrolysis, the rate-controlling stage may be either the heat transfer rate into the solid or the chemical reaction rate. For waste pieces less than 1 cm in size, the pyrolysis reactions appear rate-controlling below 500 °C. Pyrolysis reactions are quick over 500 °C, and both heat transfer and product diffusion are rate-limiting. Heat transfer is likely to dominate for all temperatures of practical interest for parts larger than 5 cm.

6.4.2.1 Pyrolysis Time

The rate of heating controlled the time required for the pyrolysis of most wastes. Figs. 6.19 and 6.20 show how long it takes for the centre temperature of plates and spheres to increase by 95% of the initial temperature difference between the specimen and its surroundings. Thermal diffusivity of 3.6×10^{-4} m^2/h has been assumed, which is approximately equivalent to the thermal diffusivity of paper or wood [7]. At infinite cross-flow velocity (V), the heating time corresponds to radiant heating.

6.4.2.2 Pyrolysis Product

In the solid phase, pyrolysis reactions produce ash and carbonaceous char; in the liquid phase, water, various alcohols, ketones, acetic acid, methanol, 2-methyl-1-propanol, l-pentanol, 3-pentanol, 1,3-propanediol, and l-hexanol; and in the gas phase, carbon dioxide, carbon monoxide, hydrogen, and a number of low moles [7]. The distribution of these items is affected by the heating rate, ultimate temperature, and moisture content. Table 6.6 displays the yield of pyrolysis products from various substrates. Tables 6.7, 6.8, 6.9 and 6.10 display the impact of final temperature and heating rate on product mix. Table 6.11 depicts the major variations in gas composition and heat content for various feed materials, while Table 6.12 depicts the distribution of the elements comprising mixed municipal refuse.

6.4.2.3 Decomposition Kinetics

Cellulose pyrolysis appears to be a two-step procedure. The step entails breaking the C–O–C bond to produce a mixture of sugar-like molecules, which are then degraded by breaking the C–O–C bond again. The pyrolysis reaction can be computed using

$$\frac{dp}{dt} = -10^6 (\rho - \rho_c) \exp\left(\frac{-19{,}000}{R_T}\right) \tag{6.15}$$

where ρ is the instantaneous density (g/cm^3) and the subscript c denotes char, t is time (min), R is the gas constant (1.987 cal/mol K), and T is the absolute temperature (K).

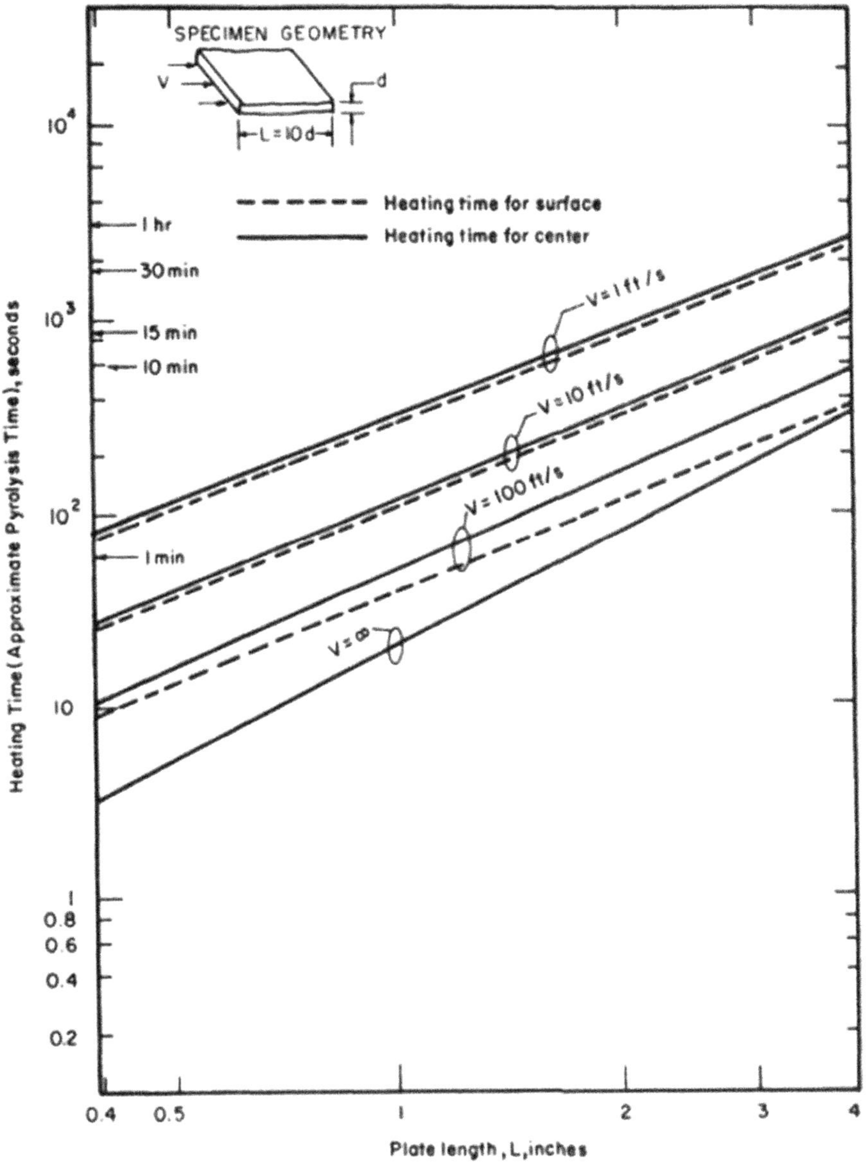

Fig. 6.19 Radiative and convective heating time for a thin plate [7]

6.4.3 Mass Burning

Solid waste is burned in a relatively thick bed at MSWI. Complete combustion occurs at and around the grate in an idealised conceptualisation of the bed processes (after ignition down to the grating line), absorbing the oxygen in the undergrated air to form CO_2 and H_2O.

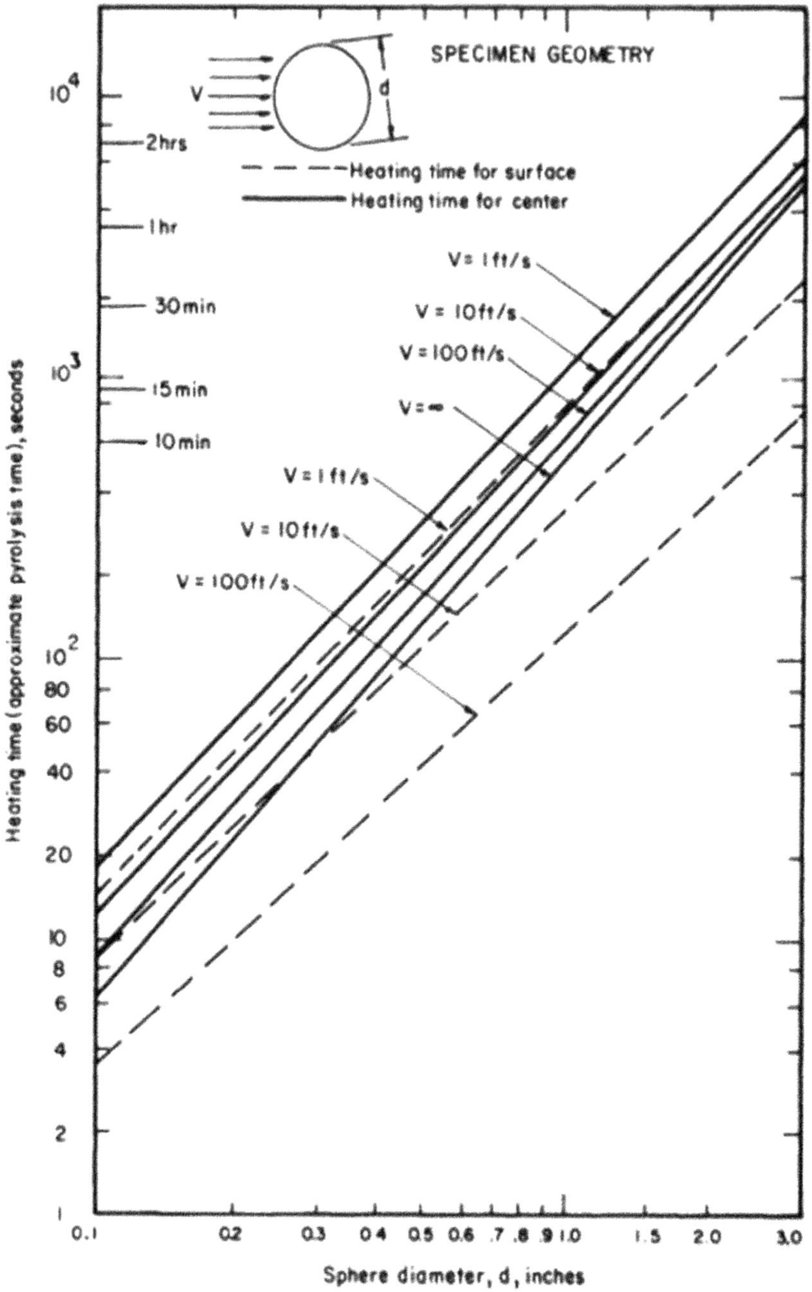

Fig. 6.20 Radiative and convective heating time for a sphere [7]

Table 6.6 Yields of pyrolysis products from different refuse components by weight percentage of refuse[a]

Component	Gas	Water	Other liquid	Char (ash-free)	Ash
Cord hardwood	17.30	31.93	20.80	29.54	0.43
Rubber	17.29	3.91	42.45	27.50	8.85
White pine sawdust	20.41	32.78	24.50	22.17	0.14
Balsam spruce	29.98	21.03	28.61	17.31	3.07
Hardwood leaf mixture	22.29	31.87	12.27	29.75	3.82
Newspaper I	25.82	33.92	10.15	28.68	1.43
II	29.30	31.36	10.80	27.11	1.43
Corrugated box paper	26.32	35.93	5.79	26.90	5.06
Brown paper	20.89	43.10	2.88	32.12	1.01
Magazine paper I	19.53	25.94	10.84	21.22	22.47
II	21.96	25.91	10.17	19.49	22.47
Lawn grass	26.15	24.73	11.46	31.47	6.19
Citrus fruit waste	31.21	29.99	17.50	18.12	3.18
Vegetable food waste	27.55	27.15	20.24	20.17	4.89
Mean Values	24.25	23.50	22.67	24.72	11.30

[a]Refuse was shredded, air-dried, and pyrolyzed in a retort at 815 °C [6]

Table 6.7 Percentage yields of pyrolysis products from refuse at different temperatures by weight of refuse combustibles[a]

Temperature °C	Gases	Liquid (including water)	Char
480	12.33	61.08	24.71
650	18.64	59.18	21.80
815	23.69	59.67	17.24
925	24.36	58.70	17.67

[a]From reference [7]

CO_2 and H_2O react with char to form CO and H_2 in an endothermic reaction mediated by the water–gas change equilibrium as the gases rise.

The only reaction that occurs above this stage is the pyrolysis of refuse in the hot gases from below. The detailed combustion of MSWI has already been discussed.

6.4.4 Suspension Burning

A particle of refuse is unexpectedly thrown into an atmosphere of hot gases and extreme radiation flux in suspension burning. When in the air, the particle rapidly dries and ignites and then burns in an oxygen-rich environment. The particle may be partially or completely burned while still suspended in the gas stream, depending on the particle shape and weight, the velocity of the gas medium, and the geometry and dimensions of the combustion chamber.

Table 6.8 Effect of heating rate on yields of pyrolysis products and heating value of the pyrolysis gas from newspaper[a]

Time taken to heat to 815 °C, min	Yield of air-dried newspaper, wt%				Heating value of gas, kcal/kg of newspaper
	Gas	Water	Other liquid	Char (ash-free)	
1	36.35	24.08	19.14	19.10	1136
6	27.11	27.35	25.55	18.56	792
10	24.80	27.41	25.70	20.66	671
21	23.48	28.23	26.23	20.63	607
30	24.30	27.93	24.48	21.86	662
40	24.15	27.13	24.75	22.54	627
50	25.26	33.23	12.00	28.08	739
60	29.85	30.73	9.93	28.06	961
71	31.10	28.28	10.67	28.52	871

[a]Reference [7]

Table 6.9 Calorific value of pyrolysis gases obtained by pyrolysing refuse at different temperatures[a]

Temperature, °C	Gas yield per kg of refuse combustibles,[b] m^3	Calorific value	
		Gas, kcal/m^3	Refuse combustibles, kcal/kg
480	0.118	2670	316
650	0.173	3346	581
815	0.226	3061	692
925	0.211	3124	661

[a]From reference [7].
[b]At 15 °C, 1 atm.

In general, the furnace's chemistry and heat transfer environment, as well as the specifics of particle characteristics (moisture content, thermal and mass diffusivities, shape factors, and so on), are poorly described, making detailed analysis difficult. Even in the much-simplified case of pulverised coal combustion, many simplifying assumptions must be made in order to predict the flame duration, minimum air requirements, and so on.

For refuse, the second and third stages of the combustion process (heat-up of the dry solid and pyrolysis) may be analysed using Figs. 6.19 and 6.20.

6.5 Economics of Incineration

- There are several conceptual and methodological challenges and pitfalls to be aware of when estimating the MSWI construction and operating costs. First and foremost, it is necessary to differentiate between financial and external costs (the social cost, which is the relevant one for WM policymakers, being the sum of

Table 6.10 Composition of pyrolysis gases obtained by pyrolysing refuse to different temperatures[a]

	Gas composition, volume %					
Temperature, °C	h_2	ch_4	CO	co_2	c_2h_4	c_2h_6
480	5.56	12.43	33.50	44.77	0.45	3.03
650	16.58	15.91	30.49	31.78	2.18	3.06
815	28.55	13.73	34.12	20.59	2.24	0.77
925	32.48	10.45	35.25	18.31	2.43	1.07

[a]From reference [7].

Table 6.11 Produced pyrolysis gas analysis[a]

	Gas analysis (dry basis), volume %								Heating value[f]	
Waste material	H_2	CO_2	CH_4	CO	C_2H_2	C_2H_4	C_2H_6	C_3H_8	BTU/ scf	kcal/ scm
MSW[d]	44.47	15.78	6.96	24.76	4.97	1.49	0.66	0.91	421	6750
Sawdust[c]	29.32	12.13	11.04	43.79	3.12	0.36	0.36	NM[e]	398	6380
Chicken manure	35.91	29.50	8.31	21.37	2.22	NM	0.61	NM	308	4940
Cow manure [d]	31.07	20.60	7.70	38.06	1.86	NM	0.31	NM	328	5260
Animal fat	11.57	27.63	18.12	14.72	25.05	NM	2.91	NM	683	10,950
Tire rubber	33.81	15.33	29.09	5.67	12.94	NM	3.17	NM	661	10,600
PVC plastic	41.02	19.06	14.51	20.76	4.02	0.21	0.43	NM	412	6600
Nylon	45.38	6.03	15.47	34.64	0.0	NM	0.0	NM	403	6460
Bituminous coal	46.88	11.68	16.63	21.72	2.08	NM	1.01	NM	435	6980
Sewage sludge (digested)	47.01	22.88	11.22	15.57	3.12	NM	0.21	NM	360	5770

[a]From reference [7]
[b]Average of five tests
[c]Average of three tests
[d]Average of two tests
[e]NM = not measured
[f]*scf* standard cubic feet (60° F, 1 atm); *scm* standard cubic meter (15 °C, 1 atm)

both). The monetary charges for the design and maintenance of a waste combustion facility are referred to as financial costs. They involve both capital and operating costs, which are likely to be influenced by local conditions and national legislation, though some standardisation is possible.

- Table 6.13 contrasts the findings of some of the most widely cited reference works, which are often published under the auspices of public institutions and national research centres in a variety of countries. All values were translated to € in 2012 using the inflation rate of the country in which the study was conducted, with the assumption that where no information was available, the reference year was the year prior to publication. The $/€ exchange rate is 1.30. The "gross cost" column represents the overall financial cost, while the "net cost" column includes income from energy and by-product sales. The statistics are incongruent; however, if the analysis is limited to the most recent and the EU background, the figures become more comparable. The gross cost of an up-to-date facility that

Table 6.12 Dry-basis yields from pyrolysis of refuse in weight percentage[a]

	C, wt%	H, wt%	O, wty$_0$	N, wt%	S, wt%	Ash, Wt%	Total, Wt%
Feed composition	30.85	3.84	22.32	(0.4)	(0.1)	42.49	100.00
CO	8.01	–	10.68	–	–	–	18.69
CO_2	4.32	–	11.52	–	–	–	15.84
H	–	2.05	–	–	–	–	2.05
CH_4	2.25	0.76	–	–	–	–	3.01
C_2H_2	3.22	0.27	–	–	–	–	3.49
C_2H_4	0.95	0.16	–	–	–	–	1.11
C_2H_6	0.43	0.11	–	–	–	–	0.54
C_3H_6	(0.52)	(0.09)	–	–	–	–	0.61
C_3H_8	(0.35)	(0.08)	–	–	–	–	0.43
Liquids	3.45	(0.32)	(0.12)	(0.1)	–	–	3.99
Ash	–	–	–	–	–	42.49	42.49
Char	7.35	–	–	(0.3)	(0.1)	–	7.75
Pyrolysis product totals	30.85	3.84	22.32	(0.4)	(0.1)	42.49	100.00

[a]From reference [7]. Parentheses indicate estimated values

Table 6.13 Financial cost of MSWI (all values in € as of 2012)

Source	Reference year	Size (kt/yr)	Gross cost (€/t)	Net cost (€/t)	Note
36	2002	648	128	98	Best practice market reference in NL
37	2009	50–500	113–188		Run by private enterprises
	2009	50–500	61–104		Run by municipal associations
38	1987		55–96		WARM model
	2006	80–115	116–126		Configuration typical of SE Asia
		400–600	85–90		Our elaboration assuming economic life = 20 years, r 5%, load factor = 90%
39		300	44–75		
42	2001	200	134		
43	2001	250		79	Based on COWI (2002)
		250		79	

complies with stringent EU regulations and takes advantage of economies of scale is between 100 and 130 € per tonne. The cost of constructing a modern incinerator will range from \$150 million to \$230 million [31].

- The value of recovered capital also affects net costs. These are mostly electricity and ultimately heat in the case of incinerators. Additional revenue can be produced through the recovery of materials (such as metals) and the re-use of ashes as a construction inert.
- Energy prices are subject to their own degree of fluctuation:
- The efficiency of energy recovery is determined by a variety of factors (technology, quality of waste). A plant that treats pre-selected waste will recover 2–3 times more energy and heat than a plant that treats raw waste [32].

- Since the average consumer price in each country is a feature of the technology mix, the market price of electricity can be affected by national market conditions (for example, in Italy, it is significantly higher than in Germany).
- Heat's market value is determined by local conditions, such as the presence of industrial buildings that can easily use heat and/or the viability of district heating. Climate factors clearly matter because they influence heating demand patterns. Many studies support cogeneration, but they assume that recovering heat does not incur additional costs.
- In addition to market rates, potential subsidies must be considered [33, 34]: some countries convert waste to a renewable resource and qualify WtE for green energy subsidies; others impose an incineration levy. In China, the power grid is obligated to buy electricity generated by incinerators at a discounted rate that includes a subsidy [35]. These subsidies should be excluded from the social cost–benefit analysis since they are clearing entries for the collectivity as a whole.

6.6 Case Studies on Incineration Process

6.6.1 Clean Plaza (Yokote City, Japan) [23]

Clean Plaza MSWI was established in March 2016 in Yokote City (population: 90,000). It is a small plant ($47.5 \times 2 = 95$ tonnes per day). The plant was planned to have a power generation efficiency of about 20%. The plant's high efficiency is realised with the application of high-temperature and high-pressure boiler conditions of 400°C and 4 MPa. This incineration plant was designed to recover maximum energy even though incinerator capacity was relatively small. High-temperature and high-pressure boiler conditions of 400°C and 4 MPa are used to achieve the plant's high performance. Even though the incinerator capacity was limited, this incineration plant was built to recover as much energy as possible. In addition, the vacuum degree of condensers must be increased in order to improve boiler performance. Improved quality is also aided by the advancement of materials for machinery and piping. As a result, the designed value of 19.6% for gross power generation efficiency was achieved several years ago.

The relationship between the amount of waste incinerated and the amount of power produced was found to be a linear based on actual plant data obtained three years after the start of plant's operation. The unit power generation value is calculated to be 400 kWh per tonne of MSW incinerated. These figures have risen significantly since the first half of the decade of the 2000s. This is a perfect example of how technological advancements can produce impressive results.

The injection of urea into the furnace, a noncatalytic reduction technique that can also save energy, is used to cool flue gas after the boiler operation. With the use of activated charcoal, dioxins and mercury may be eliminated. In April, July, and October 2018, dioxins in effluent gas were 0.0073, 0.00025, and 0.00087 ng-TEQ/

m3N, respectively, which are significantly lower than Japanese flue gas requirements. In the same time frame, dioxin concentrations in bottom ash were below detection limits, while APC residue (fly ash) concentrations were between 0.20 and 0.58 ng-TEQ/g.

Since Yokote City is situated in a region with heavy snowfall in northern Japan, the energy obtained from waste incineration is often used as fuel to melt snow on roads during the winter.

6.6.2 Joetsu Clean Center, Japan [23]

The LCV (8,100 to 15,900 kJ/kg) of MSW to be incinerated at the incineration facility in Joetsu City (population: 190,000) is higher than that of normal MSW in other parts of Japan. The fact that kitchen waste is collected separately in this city contributes to this benefit. The steam conditions at 5.0 MPa and 420°C are better than at other plants, which is a unique feature of this facility. The generators allow for a power output of 6,290 kW, resulting in a generation efficiency of more than 20%. In addition, the facility uses NO_x reduction technology without a catalyst to improve energy recovery performance.

Before being properly disposed of, the final residue (bottom ash and fly ash) is safely handled. Bottom ash is disposed of without treatment, while fly ash is disposed of after being treated with reagents to prevent heavy metal leaching. Cement may also be made with bottom ash as a raw material.

In Japan, there are a variety of similar examples of modern incineration plants that generate a lot of electricity.

- WtE incineration can be completely realised even if the size of an incineration plant is relatively limited (100 to 200 tonnes/day), as seen in this segment. There are numerous examples of small-scale plants that have achieved high power generation efficiency of about 20%.
- High-performance equipment must be mounted in the incineration plant to achieve high productivity in WtE incineration facilities. Furthermore, solid waste should have an incineration-friendly composition. It is also crucial to have a consistent MSW generation and collection system.

6.6.3 MSW in Phuket, Thailand [23]

In Phuket, there are two incinerators. The Ministry of Interior's Department of Public Works began construction on the first incinerator in 1996, and it has been in service since 1999, with a capacity of handling 250 tonnes of MSW per day. However, since 2012, the incinerator's service has been halted due to facility repairs. A private company (PJT Technology Co., Ltd.) has been operating a second incinerator installed in 2009 with a total capacity of 700 tonnes of MSW per day since 2012.

With a total area of 543 km^2, Phuket Province is the largest island in the Andaman Sea in southern Thailand. In 2017, the population was 410,211 people, with more than 14 million tourists and visitors. The Phuket City Municipality (CM) is in charge of a waste disposal complex that handled 928 tonnes of MSW per day in 2018 and covers an area of approximately 500,000 m^2 (including a landfill area of 214,400 m^2, incinerator plant area of 73,600 m^2, wastewater treatment area of 52,800 m^2, and buffer zone of 124,800 m^2).

PJT Technology Co., Ltd. reported total revenue of THB 545 million and total expenses of THB 275 million to the Ministry of Commerce's Department of Business Development in 2017. With a capacity of 700 tonnes per day, 320 days of service per year, and a storage capacity of 224,000 tonnes per year, revenue per tonne is expected to be THB 2,433 (tipping fees were estimated at THB 520 per tonne, electricity sales at THB 1,913 per tonne, and expenses at THB 1,226 per tonne). In 1994, the Phuket CM held public hearings on the construction of a stoker-type incinerator power plant, going over its historical history and phase-by-phase growth. Construction on the first incinerator with a capacity of 250 tonnes/day, supported by the Ministry of Interior, began in 1996, and an executive committee on waste management in Phuket Province was created. The first stoker incinerator began operating in 1999, generating approximately 2.5 MW of electricity, but the overall amount of waste (approximately 350 tonnes per day) exceeded the incinerator's capacity, resulting in excess waste being deposited in a landfill site since 2003. In 2007, Phuket CM developed a solid waste management (SWM) master plan, held public hearings for the second stoker incinerator power plant, and awarded PJT Technology Co. Ltd. an investment contract in 2009. The decision of the Phuket CM to award a concessionaire was a watershed moment. The second 700 tonne/day WtE incineration plant began operation in 2012, producing 12 MW of electricity, while the first incinerator was shut down for maintenance in 2012. The Phuket CM applied to the central government for a maintenance subsidy, which was denied, resulting in the suspension of operations at the first incineration plant.

In order for WtE incineration facilities to function properly, a specific amount of waste must be collected. Thailand's Pollution Control Department (PCD), the competent authority for municipal waste policies and technologies, recommends that clusters be established among multiple municipalities (PCD 2017). In 1996, the Phuket Governor and Phuket CM created an executive committee on waste management in Phuket Province with about 18 municipalities, local communities, and environmental NGOs (non-governmental organisations).

Pattaraporn [36] claims that SWM in Phuket has grown continuously since the establishment of the executive committee. In 2007, the executive committee suggested creating an SWM master plan for the city. A memorandum of understanding on SWM signed in 2008 specified that municipalities should collect and transport waste to the disposal centre operated by Phuket CM and pay disposal (incineration and landfill) fees of THB 520 per tonne, but the effectiveness of such cooperation was contingent on each municipality's ability and policies. However, the plan's implementation was limited in scope.

According to a public official, the scheme was not carried out because no staff was in charge of its execution. The 2014 Phuket SWM Master Plan, which specifically falls under the purview of the Phuket Office of Natural Resources and Environment, is viewed differently. The executive committee's authority and function have also grown. This enables SWM issues to be handled simultaneously. Nonetheless, the effectiveness of these institutions and policies is dependent on the level of understanding among relevant authorities and the general public of the gravity of the problems, as well as their level of engagement and cooperation in implementing changes.

Understandably, the majority of waste in Phuket is organic, with high moisture content, resulting in LCVs and inefficient incineration. Food waste was historically collected and used as livestock feed, especially for swine, in the early days. However, as tourism and urbanisation increased, piggeries were forced to close, and owners were forced to sell their land or relocate to neighbouring provinces where land was much cheaper. As a result of such social changes, surplus food waste has been shifted into the main waste stream, and the proportion of organic components sent to incinerators has nearly doubled from 34% in 1993 to 64% in 2004.

Following the creation of the Phuket SWM master plan in 2007, the Department of Environment Quality Promotion, in collaboration with local governments and non-governmental organisations, initiated public participation initiatives to encourage waste reduction and separation at the source. The development of an organic waste separation model that uses aerobic composting to produce fertiliser and its successful implementation in pilot communities discovered that removing 15% to 20% of organic waste from the main waste stream would raise the LCV of mixed waste to the designed range, maintain combustion efficiency, reduce incomplete combustion emissions, and increase power generation. Waste separation by societies will help to solve both environmental and energy issues.

6.7 An Approach to Design

The ideas underlying the design of an incinerator were uncomplicated and free of the need to apply both technological and value assessments. This section, however, can only scratch the surface of the system design challenge; we will attempt to structure, if not direct in-depth, the design process in general [37–40].

6.7.1 *Characterise the Waste*

Characterisation of the quantity and composition of the waste is the foremost important factor in deciding whether they are combustible or not. Keep in mind future development as well as the effect of technological and economic developments on organisational trends and decision-making.

6.7.2 *Lay Out the System in Blocks*

Incineration facilities are often built in sections, with inadequate attention paid to the mating of interfaces between different elements of an incinerator. Remember the term "system." It should always begin with waste collection and end with ash disposal.

6.7.3 *Establish Performance Objectives*

Examine current and future regulatory standards for effluent consistency. Determine whether there is a need for volume reduction, residue burn-out, or detoxification. Apply these to the relevant locations in the facility layout.

6.7.4 *Develop Heat and Material Balances*

Determine the material and energy flow in the waste, combustion air, and flue gases using the techniques introduced earlier in this chapter. Take into account the likely building materials and set appropriate temperature limits. Investigate the effect of differences in waste feed composition and quantity from the "average." In reality, these out-of-the-ordinary features would usually best describe day-to-day operating conditions.

6.7.5 *Develop Incinerator Envelope*

The total size of the device can be calculated using heat release rates per unit area and per unit volume. Establish the basic incinerator envelope using burning intensity, flame length and shape, kinetic expressions, and other analysis methods. The final form will be determined by both judgement and these calculations. Make use of the literature as well as the personal experiences of others. Interact with other engineers, manufacturers, technicians, and designers of other combustion systems that have similar operating objectives or physical configurations. Try to strike a balance between being overly conservative at the expense of being overly conservative and the unfortunate fact that a few of the answers are tractable to conclusive analysis and computation. Speak with device operators in particular. Too often, designers only talk to one another, leaving important insights from direct personal experience unheard and, worse, unasked for.

6.7.6 Evaluate Incinerator Dynamics

Apply the jet evaluation methodology, buoyancy measurements, analytical relationships, and traditional furnace draught and pressure drop evaluation techniques to understand, though insufficiently, the system's dynamics.

6.7.7 Develop the Designs of Auxiliary Equipment

Determine the sizes and specifications of the system's burners, fans, grates, materials handling systems, pumps, air compressors, air quality control systems, and numerous other auxiliary equipment. Again, the caution is to be generous, defensive, and tough. The cardinal rule is to plan for when "it" occurs, not when "it" will happen.

6.7.8 Review Heat and Material Balances

This self-explanatory phase will help to strengthen the systems perspective by following the flows through one component element after another.

6.7.9 Build and Operate

Fortunately, in many situations, nature is kind – reasonable engineering designs will work, but maybe not to standards. Plants constructed with the greatest care and attention to detail will fail. This is a lot of staff.

Glossary Stoichiometry is a section of chemistry that involves using relationships between reactants and/or products in a chemical reaction to determine desired quantitative data.

"Developed and developing countries" refers to the classification of economies used by the World Bank in its World Development Indicators report released in 2016. The word "developed countries" refers to high-income countries and regions, while "developing countries" refers to low-income, lower-middle-income, and upper-middle-income countries and regions.

"Best Available Techniques Economically Achievable" (BATEA) means the effective methods to prevent pollution and, where that is not practicable, generally to reduce emissions in the air from the industrial activities and their impact on the environment as a whole. In the United States, BATEA is an abbreviation of "Best Available Technologies Economically Achievable" with the same meaning.

References

1. Ngamsang, T., & Yuttitham, M. (2019). Vulnerability assessment of areas allocated for municipal solid waste disposal systems: a case study of sanitary landfill and incineration. *Environmental Science and Pollution Research, 26*, 27239–27258. https://doi.org/10.1007/s11356-019-05920-6
2. Tyrer, M. (2013). Municipal solid waste incinerator (MSWI) concrete. In F. Pacheco-Torgal, S. Jalali, J. Labrincha, & V. M. John (Eds.), *Eco-efficient concrete* (pp. 273–310). Woodhead Publishing.
3. Swithenbank, J. Nasserzadeh, V., Wasantakoran, A., Lee, P. H., & Swithenbank, C. (1999). 'Future integrated Waste, Energy and Pollution Management (WEP) systems exploit pyrotechnology', In *Proceedings of the 2nd international symposium on incineration and flue gas treatment technology*, Sheffield.
4. Williams, P. T. (2005). *Waste treatment and disposal* (2nd ed., p. 392). John Wiley and Sons.
5. Darley, P. (1999). Energy recovery. In *Short course on incineration of municipal waste with energy recovery*. Department of Fuel and Energy, University of Leeds.
6. Leclaire, T. (1999). Comparison of different treatment methods for residues of MSW-incineration with respect to their ecological impacts. In *Proceedings of the 2nd international symposium on incineration and flue gas treatment technology*, Sheffield, UK.
7. Niessen, W. R. (1980). Combustion and incineration. In L. K. Wang & N. C. Pereira (Eds.), *Handbook of environmental engineering, Vol 2, Solid waste processing and resource recovery* (pp. 151–225). The Humana Press.
8. Tiseo, I., Average cost to landfill municipal solid waste in the United States as of April 2019, by region, Available at https://www.statista.com/topics/2630/waste-management-in-the-united-states/, Access Jan 22, 2021
9. Tillman, D. A., Rossi, A. J., & Vick, K. M. (1989). *Incineration of municipal and hazardous solid wastes* (p. 356). Academic Press.
10. Coggins, C. (1999). Energy from waste: It's role within an integrated waste management strategy. In *Proceedings of the 2nd international symposium on incineration and flue gas treatment technology*, Sheffield.
11. UK HM Revenue & Custom. (2021). Changes to Landfill Tax rates from 1 April 2021, https://www.gov.uk/government/publications/changes-to-landfill-tax-rates-from-1-april-2021/changes-to-landfill-tax-rates-from-1-april-2021, Access Sept 14, 2021
12. Henan Huatai Environmental Equipment Co. Ltd., *The early development of domestic waste incinerator*, http://www.mswpowerplant.com/News/2017/0407/91. html, 2017, Access Jan 26, 2021
13. Costa, J., Santos, R. R., Virgolino, A., & Reis, M. F. (2019). Solid waste incinerators: health impacts. *Encyclopedia of Environmental Health*, 771–784.

14. Fang, Y. Y., Zhou, S. H., & Yan, L. J. (2010). Municipal domestic waste incineration technology and application. *Energy Conservation Technology*, (Issue 1), 76–80.
15. Worrell, W. A., & Vesilind, P. A. (2011). *Solid Waste Engineering* (2nd ed.). Cenage Learning.
16. W+E Umwelt Technik News 1/95 (1995). Information Newsletter by W+E Umwelt Technik, Zurich.
17. Brunner, C. R. (1991). *Handbook of incineration systems*. McGraw-Hill Inc..
18. Dyer K. S. (1996). *A review and assessment of the latest generation pollution abatement equipment for MSW incinerators*, Report ETSU B/RR/00419/REP, Department of Trade and Industry,
19. Martin Biopower. (2013). *MARTIN reverse-acting grate Vario*, Martin GmbH.
20. Düsseldorf, S. (2012). *Waste to energy in düsseldorf*. Company Brochure.
21. Stehlik, P. (2016). *Up-to-Date waste-to-energy approach: From idea to industrial application*. Springer Nature.
22. Niessen, W. R. (2002). *Combustion and incineration processes* (3rd ed.). Marcel Dekker.
23. Liu, C., Nishiyama, T., Kawamoto, K., & Sasaki, S. (2020). *CCET guideline series on intermediate municipal solid waste treatment technologies: Waste-to-Energy Incineration*. United Nations Environment Programme.
24. WEC. (2013). Waste-to-energy 2013. Report. pdf:- World Energy Council, London, UK.
25. Grand View Research, Inc, *Waste to energy market size, share & trends analysis report by technology (Thermal Incineration, Gasification, Pyrolysis), Biological), By region (North America, Europe, APAC, Central & South America, MEA), and segment forecasts, 2020–2027*, Access 28 January 2021.
26. Michaels, T. *ERC directory of waste-to-energy facilities in the United States; 2014*. http://www.wte.org, Accessed 10 Feb 2017
27. Makarichi, L., Jutidamrongphan, W., & Techato, K. A. (2018). The evolution of waste-to-energy incineration: A review. *Renewable and Sustainable Energy Reviews, 91*, 812–821.
28. Wade, J. (1999). Flue gas emissions control, *Short course on incineration of municipal waste with energy recovery*, Department of Fuel and Energy, University of Leeds, UK.
29. Hung Li K., *Air pollution control engineering* in *Handbook of environmental engineering* Lawrence K. Wang, Norman C. Pereira, Yung-Tse Hung, 2004
30. Sebastian, R. M., Kumar, D., & Alappat, B. J. (2020). Demonstration of estimation of incinerability of municipal solid waste using incinerability index. *Environment, Development and Sustainability*, 4821–4844.
31. Seltenrich, N., (2013). Incineration versus recycling: In *Europe, a debate over trash, yale school of the environment*, https://e360.yale.edu/features/incineration_versus_recycling__in_europe_a_debate_over_trash. Accessed 24 Jan 2021.
32. Consonni, S., Giugliano, M., Massarutto, A., Ragazzi, M., & Saccani, C. (2011). Material and energy recovery in integrated waste management systems. Project overview and main results. *Waste Management., 31*(9–10), 2057–2065.
33. Cossu, R., & Masi, S. (2013). Re-thinking incentives and penalties; economic aspects of waste management in Italy. *Waste Management., 33*, 2541–2547.
34. Dubois, M. (2013). Disparities in European taxation of combustible waste. *Waste Management., 33*, 1575–1576.
35. China Research and Intelligence. (2014). *China waste incineration for power generation industry, 2014–2018*, chinamarketresearchreports.com.
36. Pattaraporn, W., (2015). *Evaluation of the local capacity in solid waste management at source, Phuket case study*, Master thesis for Prince of Songkla University, Thailand.
37. Wang, L. K., Shammas, N. K., & Hung, Y. T. (2008). *Biosolids Engineering and Management* (pp. 607–646). Humana Press.
38. Wang, L. K., Shammas, N. K., & Hung, Y. T. (2008). *Biosolids treatment processes* (pp. 613–644). Humana Press.

39. Hung, Y. T., Wang, L. K., & Shammas, N. K. (2014). *Handbook of environment and waste management: Land and groundwater pollution control, Volume 2* (pp. 635–684). World Scientific.
40. Niessen, W. R. (2008). *Combustion and incineration engineering*. In L. K. Wang, N. K. Shammas, & Y. T. Hung (Eds.), *Biosolids engineering and management* (pp. 479–606). Humana Press, NJ, USA.

Chapter 7
Composting Processes for Disposal of Municipal and Agricultural Solid Wastes

Lawrence K. Wang, Mu-Hao Sung Wang, Raul R. Cardenas Jr., Nor Habsah Md Sabiani, Mohd Suffian Yusoff, Siti Hafizan Hassan, Mohamad Anuar Kamaruddin, Fadugba O. George, and Yung-Tse Hung

Abstract Various composting processes have gained a lot of attention in recent years because of pollution concerns and the need of environmentally sound technologies for handling and disposal of municipal and agricultural solid wastes. Composting is extremely significant in terms of its economic viability, ease of operation, ability to recycle nutrients, and waste minimization. This chapter introduces (1) the composting process types: open on-site composting, aerated-turned composting, aerated static pile composting, enclosed mechanical composting, vermicomposting, thermophilic composting, and two-stage in-bin composting; (2) process control parameters: microorganisms, temperature, pH, moisture content, aeration, C:N ratio, and particle size; (3) operational steps: pretreatment prepara-

L. K. Wang (✉) · M.-H. S. Wang
Lenox Institute of Water Technology, Latham, NY, USA

R. R. Cardenas, Jr.
Department of Civil Engineering, Polytechnic Institute of New York, Brooklyn, NY, USA

N. H. M. Sabiani · M. S. Yusoff · F. O. George
School of Civil Engineering, Engineering Campus, Universiti Sains Malaysia, Nibong Tebal, Pulau Pinang, Malaysia
e-mail: cenoor@usm.my; suffian@usm.my; ogfadugba@futa.edu.ng

S. H. Hassan
School of Civil Engineering, Universiti Teknologi MARA Cawangan Pulau Pinang, Permatang Pauh, Pulau Pinang, Malaysia
e-mail: sitihafizan741@uitm.edu.my

M. A. Kamaruddin
School of Industrial Technology, Universiti Sains Malaysia, Minden, Pulau Pinang, Malaysia
e-mail: anuarkamaruddin@usm.my

Y.-T. Hung
Department of Civil and Environmental Engineering, Cleveland State University, Cleveland, OH, USA

L. K. Wang et al. (eds.), *Solid Waste Engineering and Management*, Volume 1, Handbook of Environmental Engineering 23,
https://doi.org/10.1007/978-3-030-84180-5_7

tion, composting-digestion, curing, finishing, and storage; (4) process control; (5) pathogen survival; (6) cost considerations; (7) compost stability and maturity; (8) marketing of compost; (9) compost utilization; and (10) process design considerations and design criteria. Finally, several successful composting schemes, practices, and legal requirements currently in use around the world are also highlighted. A complete two-stage in-bin composting process system is designed and illustrated for handling and disposal of dead animals, poultry, and fish generated from agricultural facilities.

Keywords Composting · Compost · Solid waste · Decomposition · Stability · Maturity · Compost utilization · Municipal solid waste · Biosolids · Dead animal · Poultry · Fish · Case studies · Agricultural solid waste

Acronyms

ADS	Anaerobically digested sludge
ASP	Aerated static pile
BOD	Biochemical oxygen demand
C:N ratio	Carbon to nitrogen ratio
CEC	Cation exchange capacity
CH_4	Methane
CO_2	Carbon dioxide
DASPSS	Dewatered anaerobically stabilized primary sewage sludge
EDS-76	Egg-drop syndrome-76
EU	European Union
GCV	Gross calorific value
HA	Humic acid
HPAI	Highly pathogenic avian influenza
IGES	Institute for Global Environmental Strategies
IR	Infrared
JICA	Japan International Cooperation Agency
KITA	Kitakyushu International Techno-Cooperative Association
MPN	Most probable number
MRF	Material recovery facility
MSW	Municipal solid waste
NGO	Nongovernmental organization
NH_3	Ammonia
NH_4-N	Ammonium nitrogen
NO_3-N	Nitrate nitrogen
O_2	Oxygen
OER	Odor emission rate
OFMSW	Organic fraction of municipal solid waste
OM	Organic matter

OUR	Oxygen uptake rate
rpm	Rotation per minute
RS	Raw sludge
SCADA	Supervisory Control and Data Acquisition
SMC	Spent mushroom compost
SOUR	Specific oxygen uptake rate
TCM	Takakura Composting Method
TMECC	Test Methods for the Evaluation of Composting and Compost
USDA	United States Department of Agriculture
USEPA	United States Environmental Protection Agency
VOCs	Volatile organic compounds
VS	Volatile solids
VSCs	Volatile sulfur compounds
WWTP	Wastewater treatment plant

Nomenclature

$C_aH_bO_cN_dS_eP_f$ =	Empirical mole composition of the organic wastes present at the beginning of the process
$C_uH_vO_wN_xS_yP_z$ =	Empirical mole composition of the organic wastes present at the end of the process
H =	Depth of a single bin (ft)
L =	Length of a single bin (ft)
M =	Percent normal mortality of animals for the entire life cycle expressed as percentage
N =	Head capacity, number of live animals on a farm
PCV =	Primary composter volume (ft^3)
PDADR =	Peak dead animal disposal requirement (lb)
SCV =	Secondary composter volume (ft^3)
T =	Number of days for animals to reach market weight
TFLW =	Theoretical farm live weight (lb)
VF =	Volume factor, between 1.0 and 2.5 cubic feet per pound
VOL =	Volume of a single bin (ft^3)
W =	Width of a single bin (ft)
W_m =	Market weight (lb)

7.1 Introduction

Composting is the decomposition and stabilization of organic matter biologically under conditions that can cause an increase in temperature as a result of biological reactions that produce heat as well as a stable end product, free from pathogens and beneficial to the soil [1]. In general, the composting process is a method of converting organic matter to other materials that are more stable and have many positive effects and can be applied to the soil. The end product is environmentally friendly and clean and has low toxic content [2]. Composting offers several advantages, which include that (i) it produces soil improvement materials that are useful and suitable for agricultural activities, (ii) it produces a cleaner environment in which the composting process can reduce the production of methane gas as occurs in landfills, (iii) it increases the effectiveness of fertilizer application, (iv) it reduces the need for solid waste transportation, (v) it is flexible and can be implemented at various levels from in-house efforts to large-scale levels, and (vi) it can be started with small operating costs and capital costs.

During the composting process, there is an increase in temperature, and the material that has an original identity or character, such as food waste, green waste, sludge, newspaper, etc. turns into a more stable end product (compost) that is dark in color (dark brown or black), with earthy smell and porous. Dark-colored compost has usually reached maturity and stability and contains high organic matter. The texture formed is closely related to the properties of porosity and permeability as well as other parameters that are considered important in plant growth. The presence of an earthy smell odor is the best indication of mature compost. The quality of the final compost is also expected to contain essential nutrients to plants such as nitrogen, carbonaceous matter, and important micronutrients such as phosphorus and potassium. Copper, manganese, iron, and boron are also present in a small percentage.

The finished compost must be of good quality in order for it to be genuinely useful and marketable. To ensure the safety of compost used for agricultural purposes, it must be stable and mature [3]. Because of the many possible negative effects of using immature and unstable compost in agriculture, accurately determining the stability and maturity of compost becomes a priority before it is used in agriculture. The ability to guarantee the safety and effectiveness of compost would certainly help compost gain consumer and regulatory approval for use in a range of applications, while the inability to do so could obstruct compost use.

Compost has long been recognized as a valuable soil conditioner. Its use in a more effective way will increase the quality of crop production, reduce the use of chemical fertilizers, save costs, and indirectly preserve natural resources. Compost has a great potential to improve soil quality and crop yield due to its ability to improve the physical, chemical, and biological properties of the soil. However, the utilization of compost today is not only limited as a plant fertilizer but has also been further developed in controlling soil erosion, as a landfill cover, landscape improvement, chemical fertilizer substitute, bioremediation and pollution prevention, plant disease control, pest control, wetland restoration, and generating renewable energy via compost palletization.

When it comes to selling high-quality compost, there are no fixed rules. The success of a composting project is dependent on a successful marketing strategy. Ineffective marketing strategies have triggered the majority of unsuccessful composting projects in the past [4]. The first step in designing a marketing plan is to determine the value of the current local market. As a consequence, knowing the compost product produced, its uses, and limitations, as well as the ability to estimate the product's value to customers, is an important factor to consider. Furthermore, compost marketing strategies should take into account soil characteristics, types of agricultural activities, weather, and transportation costs in order to meet local needs.

7.2 History of Composting

In the global ecosystem, natural composting is an old process and occurs continuously on the earth's surface. If the natural composting process does not happen, all organic waste resulting from plant waste, grass clippings, and other semi-solid wastes that fall to the earth's surface will not decompose. This method has long been widely practiced over the past few centuries by farmers to produce soil improvement for crops. Farmers, for example, utilized composted human excreta, fruit and vegetable waste, animal feces, and refuse as fertilizer in their agricultural areas in the early days of human civilization. Traditionally, such wastes were piled up in piles or pits located at several suitable locations; they are left to biodegrade because natural conditions allow the material to be decomposed and can be returned to the land by farmers. Good organic humus or soil-like materials that contain organic nutrients suitable for soil replenishment require a long time in the pile or heaps as well as with little or no control.

The process of modern composting known as the "Indore Method" began in the early 1920s in India [5, 6]. This method was developed by a British agronomist, Sir Albert Howard, while he was working at the Indore Institute of Plant Industry, Central India, between 1924 and 1931. In the early stages, the method that combines animal manures, human feces, and garden waste such as leaves, grass, and straw is done anaerobically where the size of the pile is 1.5 m in height and 0.6–0.9 m in depth [7]. The composting process took 6 months or more, during which the waste was only aerated twice. The "Indore method" was well received and widely practiced in the British Empire as it encouraged farmers to treat their crop waste to produce fertilizer rather than previously burned. In 1935, tea plantations in India and Sri Lanka reportedly produced 1,000,000 tons of compost using this method [6]. This technique was later modified, however, so that the aerobic decomposition process could take place by turning the composted waste more regularly to preserve aerobic conditions, thereby ensuring quicker degradation and shortening the composting time [7, 8]. The modified Indore method is known as the "Bangalore process" is widely practiced in Malaysia, East Africa, South Africa, and China.

According to Gotaas and his co-workers [9], the basic principles of composting were first identified in the 1950s. Even since the 1920s, some composting processes have been patented in Europe. Among the earliest to be developed was the Beccari process in Florence, Italy. In the Beccari process, the organic waste is stacked in enclosed cells (during the final aerobic phase) to avoid the production of odors caused by the decomposition of organic matter after going through the anaerobic fermentation phase first. The Beccari cell was a simple cell-type structure with a top-loading hatch and a front unloading door. Air vents were included that allowed composting to continue under partially aerobic conditions when opened for the final stage. Bordas further modified the Beccari process in 1931 to remove the anaerobic stage by adding forced air through a central pipe and along the walls into a fermentation silo. The silo is divided into an upper and a lower segment by a grate. With optimum use of the silo, compost is generated on a batch basis, achieved by lowering the charge through the grate into the lower chamber when it has lost much of its volume through decomposition. In 1939, Earp-Thomas patented a silo-type multiple-grate digester for producing compost in an aerobic state as well as using a rotary plow and forced air for aeration purposes. In this process, special bacterial cultures have been used to aid the composting process. Subsequently, the Frazer process was patented in the United States in 1949. A fully mechanized, sealed aerobic digester is used by the Frazer process in which shredded organic waste is constantly agitated as it travels down from one stage to another and is also placed in direct contact with decomposition gases. When it leaves the bottom of the digester and the tailings of the screen are recycled, the composted product is screened. Later, in line with technological advancements in the field of organic waste composting, the Hardy digester was established. The Hardy digester is a broad round vat containing conveyor screws, placed perpendicular to the ground for aerating and agitating the composting material. The bottom of the vat is designed to be porous to allow air to enter and liquid to flow out. The process of aerobic decomposition takes place continuously in the mixing vat, where the resulting compost is removed while fresh organic waste is added. The Dano process was a method that was widely promoted and used in a variety of countries around the globe. The composting technology that was developed in Denmark conducts operations to separate and reduce the size of waste disposed of. The waste was fed into a slowly rotating cylinder in the process, with the axis tilting slightly downward from the horizontal, where the waste was aerated to eliminate odors, mixed, and partially broken down into smaller particles. When the waste was transported to a grinding and homogenizing unit, the ferrous metals and other recyclable materials were separated by a magnetic separator and hand sorting. Through friction between the waste particles and between the waste and the roughened walls of the rotating cylinder, the size reduction to the desired particle size was achieved. The waste required 4–6 h to travel through the grinding machine. The actual height of the composting pile is 1.5–1.8 m [7]. Meanwhile, large-scale composting facilities in European countries began in 1932 in the Netherlands. The process, which is known as "Van Maanen" and fully operated by the company Vuilafvoer Maatschappij (VAM), is an adaptation of the Indore Method where it is applied to large-scale composting of solid waste [6]. Through this method, the composting period of solid waste is between 4 and 8 months. Monitoring is also done on the moisture content as well as the composted solid waste air requirements. The resulting compost is then filtered, separated, and ground to different sizes before being applied to crops. The composting process in England started in 1906 when the Borough of Southwark has composted street sweepings, market waste, stable manure, and garbage using 70 tons of concrete bins, which is a modified Indore method [10].

During the same era, engineers and scientists in the United States performed basic research on the aerobic degradation of vegetable wastes, stable manure, and crop residue. Among them are studies related to basic principles in the composting process, such as temperature, decomposition rate, the role of microorganisms, and so on. The major urban dilemma concerning solid waste management has grown in the United States since World War II, beginning in the 1950s. The Composting Corporation of America has performed pilot plant operations and researched the financial feasibility of composting municipal and industrial waste from large communities to generate soil improvement [11]. In Oakland, California, a composting facility was built as a private company and was planned to compost 300 tons of mixed waste each day in an 8-h shift, or 600 tons on a two-shift (16 h/day) basis. This composting facility operates aerobically using the windrow method, which is a modern version of the Indore process method. Meanwhile, two demonstration facilities were built by the United States Public Health Service in the 1960s. The facility was intended for the composting of MSW combined with biosolids. According to Breidenbach [12], one plant was built in Johnson City, Tennessee, and another one was situated in Gainesville, Florida. In Florida and Texas, the Metro Waste Conversion System has been practiced for composting municipal solid waste (MSW). This system utilizes forced ventilation through the bottom of the mass incorporate with mixing. During processing, the organic fraction of MSW was placed on a conveyor belt and then dispersed into elongated bins with perforated bottoms. The MSW is propelled using an endless traveling belt along the rails at the top of the elongated bins. Airflow is provided once a day when garbage has filled the bins. The detention time is between 1 and 6 days for composting materials in the bins. To produce a more stable and mature compost, the final product removed from the bins is placed in an open area in the form of a windrow. Table 7.1 summarizes the typical composting process used worldwide, which contains a short overview of the operation and the site where the composting system is located [12].

In 1973, a research program leading to the composting of sludges (biosolids) was conducted by the United States Department of Agriculture (USDA) at the Beltsville (Maryland) Agriculture Research Center, where the study was fully funded by the US Environmental Protection Agency, the Maryland Environmental Service, and the Washington DC Council of Governments. Meanwhile, Epstein et al. [13] reported that the composting method using aerated static pile (ASP) was first introduced in 1975 by a group of researchers in Beltsville. In the same year, several studies related to the composting of biosolids covering process engineering, economic analyzes, and other aspects of composting were conducted by several researchers from Rutgers University [14–16], and a group of researchers from Japan conducted a study related to composting that covered aspects of compost production and utilization in the late 1970s to early 1980s. At the same time, pilot research on plant pathogens and the utilization of compost in controlling plant pathogens was conducted by the University of Ohio [17].

Table 7.1 Typical composting process

Composting process	General description	Location
Bangalore (Indore)	The trench was in-ground, 2–3 ft deep. Material placed in alternate layers of waste, night soil, earth, straw, etc. No grinding. Turned manually as frequently as possible. Detention time of 120–180 days	Common in India
Caspari (briquetting)	The ground material is compacted into blocks and stacked for 30–40 days. Natural diffusion aeration and air passage through stacks. Curing follows initial composting. Blocks are later ground	Schweinfurt, Germany
Dano biostabilizer	Rotating drum, slightly tilted from the horizontal, 9–12 ft in diameter, up to 150 ft long. One to 5 days digestion followed by windrowing. No grinding. Forced aeration into the drum	Predominately in Europe
Earp-Thomas	Silo type with eight decks stacked vertically. Ground waste is moved downward from deck to deck by plow. Air passes upward through the silo. Uses a patented inoculum. Digestion (2–3 days) followed by windrowing	Heidelberg, Germany; Turgi, Switzerland; Verona and Palermo, Italy; Thessaloniki, Greece
Fairfield-Hardy	Circular tank. Vertical screws mounted on two rotating radial arms keep ground material agitated. Forced aeration through tank bottom and holes in screws. Detention time of 5 days	Altoona, Pennsylvania, San Juan, Puerto Rico
Fermascreen	Hexagonal drum, three sides of which are screens. Waste is ground. Batch loaded. Screens are closed for initial composting. Aeration happens when the drum is rotated with screens open. Detention time of 4 days	Epsom, England
Frazer-Eweson	Ground waste is placed in a vertical bin having four or five perforated decks and special arms to force composting material through perforations. Air is forced through a bin. Detention time of 4–5 days	None in operation
Jersey (John Thompson system)	Structure with six floors, each equipped to dump ground waste onto the next lower floor. Aeration effected by dropping from floor to floor. Detention time of 6 days	Jersey, Channel Islands; Great Britain; Bangkok, Thailand
Metrowaste	Open tanks, 20 ft wide, 10 ft deep, 200–400 ft long. Waste ground. Equipped to give one or two turnings during the digestion period (7 days). Air is forced through perforations in the bottom of the tank	Houston, Texas; Gainesville, Florida

Composting process	General description	Location
Naturizer or international	Five 9-ft-wide steel conveyor belts organized to pass material from belt to belt. Each belt is an insulated cell. Air passes upward through the digester. Detention time of 5 days	St. Petersburg, Florida
Riker	Four-story bins with clam-shell floors. Ground waste is dropped from floor to floor. Forced air aeration. Detention time of 20–28 days	None in operation
T.A. Crane	Two cells comprising three horizontal decks. Horizontal ribbon screws extending the length of each deck recirculate ground waste from deck to deck. Air is introduced at the bottom of the cells. Composting followed by curing in a bin	Kobe, Japan
Tollemache	Similar to the Metrowaste digesters	Spain; Southern Rhodesia
Triga	Towers or silos are called hygienizators. In sets of four towers. Waste is ground. Forced air aeration. Detention time of 4 days	Dinard, Plaisir, and Versailles, France; Moscow, U.S.S.R.; Buenos Aires, Argentina
Windrow (normal, aerobic process)	Open windrows, with a "haystack" cross-section. Waste is ground. Aeration by turning windrows. Detention time depends upon the number of turnings and other factors	Mobile, Alabama; Boulder, Colorado; Johnson City, Tennessee; Europe; Israel; and elsewhere
Van Maanen	Unground waste in open piles, 120–180 days. Turned once by grab crane for aeration	Wijster and Mierlo, the Netherlands
Varro conversion	The enclosed digester, eight decks, 160 ft long, 10 ft wide, and approximately 1 ft deep. Continuous flow, mixed by harrows. Forced air aeration. 40 h detention time. Nutrients and water are added to the process	Brooklyn, New York, a pilot plant in Stuttgart, Germany; none in operation

In line with the development of technology, composting systems around the world are also growing rapidly. The compost was once turned manually using conventional equipment, but now it is no longer. With the advancement of technology, equipment such as machines and machinery are used to turn the compost throughout the composting process. In the past, organic waste is deposited in an area and then left to decompose before it is used as fertilizer. However, the development of science and technology in the field of composting causes monitoring such as changes in temperature, pH, moisture content, air requirements, and C:N ratio on the composted organic waste to ensure the quality of the final product. Besides, the final product (compost) produced has also been conducted various tests such as pathogen content, stability, maturity, and phytotoxicity effects to ensure that the compost is safe to use as an organic fertilizer or a soil improvement agent for agricultural activities.

7.3 Microbiology of the Composting Process

Composting is a microbiological process. The first to prove that the self-heating of composts is attributed to biological activity was Browne [18]. Studies related to the microbiology of composting thrived in the 1930s when Waksman conducted research and published papers related to population dynamics [19]. This process is natural where fresh organic waste such as animal manure, food waste, green/yard waste, agricultural waste, etc., are converted into more stable humus-like substances that can be used as organic fertilizer. Microorganisms perform the degradation of organic waste products and are followed by elevations of temperature. Generally, composting microorganisms come from soils that have already mixed with garbage, food waste, or other organic waste.

A large number of bacteria, fungi, and actinomycetes are the most involved microorganisms in the composting process. According to Galitskaya et al. [20], bacteria and fungi have the largest population among all the microorganisms that have been said to be found throughout composting. Apart from bacteria, fungi, and actinomycetes, other microorganisms such as protozoa and yeast also play their role in breaking down organic matter in the composting process. All of these species are active at different times and display different physiological states based on oxygen concentration and temperature [21]. Microorganisms can be divided into three groups in terms of oxygen requirements, namely microorganisms that need oxygen (aerobes), microorganisms that are not exposed to oxygen (anaerobes), and aerobic microorganisms that can survive in the presence of small amounts of oxygen or lack of oxygen (facultative), depending on the environment. In terms of suitability of the temperature range, microorganisms present at temperatures between 10 and 40 °C are known as mesophilic organisms. Meanwhile, thermophilic organisms are present in the temperature range of 40–70 °C. In the temperature range between 20 and 40 °C, the composting process begins in the mesophilic phase with mesophilic organisms decomposing organic matter. According to Hafeez et al. [22], the thermophilic phase is present after the mesophilic phase with an increase in temperature between 40 and 70 °C in which the active decomposition process takes place within this phase. Chennaou et al. [23] reported that during the thermophilic phase, there is an increase in the population and diversity of thermophilic organisms, fungi, and actinomycetes while all mesophilic organisms are killed or inactivated. This phase is known as the active composting phase, where most of the organic matter is degraded, and consequently, most of the oxygen is used in this phase. The curing phase is the second mesophilic phase that occurs after the thermophilic phase. When the compost temperature drops to ambient temperature, the compost maturation phase begins. Mesophilic microorganisms colonize the compost pile at this stage and degrade complex organic compounds such as lignin slowly [24]. At this stage, mature compost will be produced.

Bacteria are microscopic single-cell organisms that develop quickly and can live in both aerobic and anaerobic conditions. They are composed of approximately 80% water and 20% dry matter, with approximately 90% organic and 10% inorganic material [21]. Proteins, carbohydrates, and lipids are organic components present in various parts of the cell. Meanwhile, elements such as phosphorus, sodium, calcium, magnesium, potassium, and iron, as well as trace minerals, are inorganic components found in bacterial cells. At the most active stage of the composting process, bacteria play the most dominant role because of their ability to multiply quickly on soluble proteins and other available substrates [5, 25]. According to Epstein [5], these microorganisms can attack substances that have a more complex structure as well as be able to exploit substances released by less degradable materials due to the extracellular enzyme activities of other organisms. *Bacillus*, *Cellulomonas*, *Pseudomonas*, *Klebsiella*, and *Azomonas* are among the bacterial species that are often involved in the aerobic decomposition of substrates [26]. Meanwhile, *B. subtilis*, *B. licheniformis*, and *B. circulans*, which are bacteria of the *Bacillus* species, are frequently found in the thermophilic phase. This is also supported by Strom [27] that 87% of colonies are from the genus *Bacillus*, resulting from a random selection of colonies during the thermophilic phase of the composting process. Toumela et al. [28] reported that at high temperatures (65–82 °C), thermophilic bacteria species of the genus *Thermus* could be isolated from composting materials. The ammonium-oxidizing and nitrite-oxidizing bacteria found in compost piles are *Nitrosomonas* spp. and *Nitrobacter* spp. [29]. Meanwhile, the presence of denitrifying bacteria in great populations indicates the existence of anaerobic locations in composting piles. Factors such as high content of organic matter and nitrogen in the substrate as well as high initial moisture content (65%) contribute to microbial activity leading to a reduction in oxygen in composting piles. According to Firestone [30], some denitrifying bacteria are facultative and can adapt to the aerobic environment. *Bacillus*, *Flavobacterium*, and *Pseudomonas* are among the denitrifying bacteria commonly found in composting piles [31]. In the early composting stage (<40 °C), mesophilic bacteria such as *Bacillus* spp. and *Azotobacter* spp. play a role in CO_2 evolution [26]. In the temperature range of 40–60 °C, mesophilic bacteria become less active and are partially killed. At higher temperatures (65–80 °C), the most active bacteria in composting are thermophilic bacteria such as *B. schlegelii*, *Hydrogenobacter* spp., and especially from the genus *Thermus* such as *T. thermophilus* and *T. aquaticus* [32]. At this temperature range, mesophilic bacteria play only a very small role in the process of decomposition of organic matter [26].

Fungi are filamentous, spore-forming, nonphotosynthetic, heterotrophic (organic-consuming) microorganisms that, under low moisture and a wide range of pH conditions, have the potential to degrade a wide range of organic compounds [21]. In the early stages of the composting process, there is a competition between fungi and bacteria to obtain available substrates. A good oxygen supply is beneficial for fungi than for bacteria, and transient anoxic conditions can exist even in force-aerated systems [33]. Fungi play an important role throughout the composting process if the substrate used is rich in cellulose and lignin. *Aspergillus*, *Penicillium*, *Rhizopus*, *Fusarium*, *Chaetomium*, *Trichoderma*, *Alternaria*, and *Cladosporium* are the most common cellulolytic fungi species found in compost [24]. There are three types of fungi, namely soft-rot fungi, brown-rot fungi, and white-rot fungi, found in dead woods that can decompose lignocellulose [34]. White-rot fungi are the most effective lignolytic microorganism, for example, *Phanerochaete chrysosporium*, and it is frequently used as a reference. *Coriolus versicolor*, among other well-known white-rot fungi, displays much greater productivity and a broader spectrum of lignolytic functions, along with considerable cellulolytic activity. In paper mill effluents, *Phanerochaete flavidoalba* induces superior lignin loss rather than cellulose and is more effective than *P. chrysosporium* [24]. Lignin is the most difficult plant component to decompose. However, there are several types of fungi and bacteria that are able to decompose lignin; for example, white-rot fungi belong to *Basidiomycetes.* According to Muthukumar and Mahadevan [35], species such as *Polyporus*, *Pleurotus*, *Collybia*, *Poria*, *Fomes*, *Trametes*, *Sporotrichum*, *Cyathus*, and *Coriolus* also have the potential to degrade lignin. Because fungi have lower thermotolerance, the role of these microorganisms is insignificant during the thermophilic phase (above 55 °C). The most important factor influencing the growth of the fungi is the temperature, in addition to several other factors such as carbon, nitrogen, and pH. According to Dix and Webster [36], the majority of fungi belonging to the mesophilic group grow in the optimum temperature range between 25 and 35 °C. However, at high temperatures, there is a small group of thermophilic fungi that play an important role as biodegradation agents. *Taloromyces emersonii*, *T. thermophilus*, *Thermoascus auranticus*, and *Thermomyces lanuginosus* are among the thermophilic fungi found growing on substrates or compounds containing lignocellulose. The process of decomposition of lignin is highest at 50 °C. Meanwhile, Tuomela et al. [28] found that 12% of lignin was solubilized at 75 °C and in alkaline environments.

Actinomycetes are a form of microorganisms that share characteristics with both bacteria and fungi. They resemble fungi in appearance, but they are less filamentous, producing spores and adapting to soil growth [21]. In the deterioration of semi-dry substances in the decomposition process, actinomycetes play a distinct role. Actinomycetes have been observed to have biodegradative activity; they secrete a wide range of extracellular enzymes. They as well have the capacity to metabolize recalcitrant molecules [37]. According to Epstein [5], actinomycetes target polymers such as hemicellulose, lignin, and cellulose. *Micromonospora*, *Streptomyces*, *Nocardia*, and *Thermoactinomyces* are among the most common actinomycetes found in compost. In general, these microorganisms are very active in the final stages of the decomposition process. Compared to bacteria and fungi, actinomycetes undergo a slower colonization process, and this condition is more pronounced if the environment in the composting pile does not receive enough air. Actinomycetes appear in the form of a white film covering the surface of the compost during the thermophilic phase as well as the cooling and maturation phases. Strom [27] reported that *Nocardia*, *Streptomyces*, *Thermoactinomyces*, and *Micromonospora* were among the genera of thermophilic actinomycetes found in compost. Actinomycetes, which tolerate higher temperatures and pH than fungi, are capable of degrading some cellulose and solubilizing lignin. According to Tuomela et al. [28], although their ability to decompose cellulose and lignin is not as high as that of fungi, actinomycetes remain an important agent of lignocellulose degradation during peak heating. The process of hemicellulose decomposition in active composting is performed by actinomycetes during the cooling phase of compost [5]. For fungi and actinomycetes, which are the microorganisms mostly responsible for cellulose degradation, cellulose is not an obligatory carbon source [24]. The introduction of readily metabolizable compounds speeds up cellulose decomposition. The population of cellulose degraders will grow to a large size by initially using the more available carbon sources. When carbon sources become limited, microorganisms will adapt to cellulose, thereby increasing the hydrolysis of cellulose. Yeasts are a fungal growth phase adapted to single-celled, vegetative growth. Yeasts typically favor soluble carbohydrates or substrates of sugar. Protozoa are single-celled microorganisms that eat other microorganisms and may or may not participate in the composting process [21].

Although each microorganism found in all the above groups can decompose all the raw organic matter found in solid waste or other waste, they are more likely to carry out the decomposition process in different compounds or conditions. Celluloses and hemicelluloses are usually favored by fungi, yeasts, and actinomycetes, while bacteria are particularly adept at breaking down simple water-soluble sugars [21]. The predominance of microorganisms differs throughout the composting process, apart from metabolic requirements. In the early stages of the composting process, bacteria are the main microorganisms that dominate the composting pile. According to Ghosh et al. [38], bacteria dominate over fungi during the composting process, and most bacteria are *Bacillus* spp. The increase in temperature is a major factor that allows bacteria to dominate the composting process over fungi. This is because a high-temperature rise is not suitable for fungi growth. Fungi emerged within 7–10 days, and only in the end phase of composting did actinomycetes become visible. According to McGaughey and Goleuke [39], during the composting process using the windrow method, bacteria can be found in all parts of the piles. Meanwhile, actinomycetes and fungi are limited to the outer zone of the composting pile with a thickness of 5–13 cm, starting just below the pile surface. These two classes were likely restricted to the outer region due to temperature and/or aeration. Composting involves the process of decomposing organic waste components biologically and under controlled conditions, so this process is subject to the limitations of all biological activities [21]. Microorganisms need a convenient environment (e.g., temperature, pH, moisture content, and oxygen) and a supply of energy and carbon for the production of their new cellular material [40] to keep growing and functioning properly. For all synthesis, nutrients, such as nitrogen, phosphorus, and other trace elements, are essential.

7.4 Factors Influencing the Composting Process

Composting is the process of decomposition of organic matter that occurs biologically. To produce stable and quality compost while increasing the reaction of microorganisms, factors such as temperature, moisture content, pH, C:N ratio, air requirements, and particle size need to be monitored [41]. The following are the main areas that must be "controlled" during composting.

7.4.1 Temperature

In general, the temperature is one of the parameters that ensures that the composting process can take place more efficiently. Not only does this affect the metabolic process of microorganisms but these parameters are also seen to affect the composition and density of microbes in the compost mass [42]. Microorganisms require a certain temperature range for optimal activity. Certain temperatures promote rapid composting and destroy pathogens and weed seeds. Microbial activity can raise the temperature of the pile's core to at least 140 °F (60 °C). If the temperature does not increase, anaerobic conditions (i.e., rotting) occur. Controlling the previous four factors can bring about the proper temperature.

According to Day and Shaw [43], the population of microorganisms present in the composting system depends on temperature changes and can be classified into three main groups, namely psychrophilic, mesophilic, and thermophilic.

Psychrophilic microorganisms usually exist in the temperature range of −10 to 30 °C [44]. This population is very rare in the composting process, but in Canada and the northern United States, the winter composting process has been successfully carried out where the ambient temperature range is between −27 and 15 °C. Under these conditions, the carbon decomposition process is very slow, and the nitrification process does not occur. Meanwhile, in the commercial composting process, the population of mesophilic and thermophilic microorganisms dominates the compost mass. Mesophilic microorganisms are usually active in the temperature range of 20–50 °C, while thermophilic microorganisms exist in the temperature range of 45–75 °C [44]. Temperature is seen as the best indicator for determining the composting phase. Previous studies have agreed that the optimum temperature range for an efficient composting process is between 50 and 70 °C [2, 44, 45].

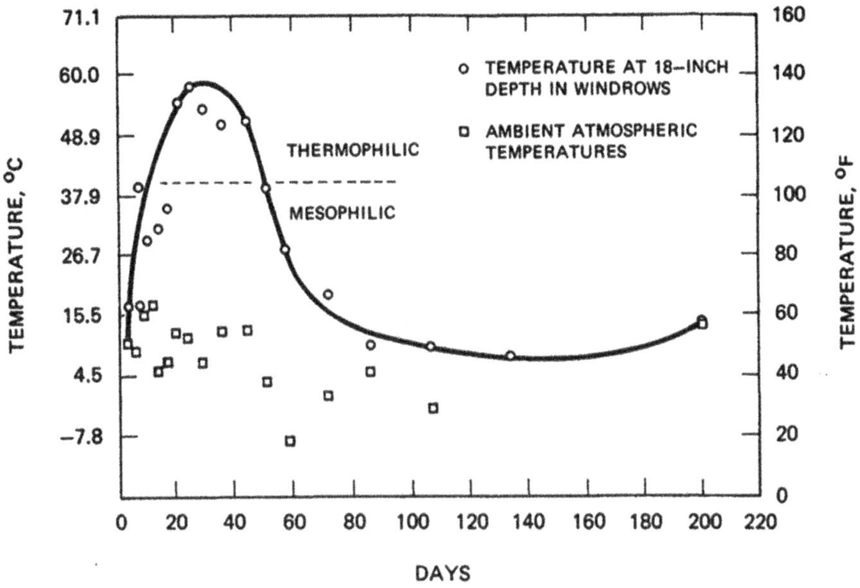

Fig. 7.1 Temperature profile and microbial growth in the compost heap. (Source: USEPA)

Table 7.2 Microbial population during the aerobic composting process

Microorganisms	Mesophilic Initial temperature <40 °C	Thermophilic 40–70 °C	Mesophilic 70 °C to the maturation phase
Bacteria			
Mesophilic	10^8	10^6	10^{11}
Thermophilic	10^4	10^9	10^7
Actinomycete			
Thermophilic	10^4	10^8	10^5
Fungi			
Mesophilic	10^6	10^3	10^5
Thermophilic	10^3	10^7	10^6

Note: The number of microorganisms is based on the mass of 1 g of wet compost [6, 43]

Based on microbial activity, the composting process can be categorized into four main phases, namely mesophilic phase (I), thermophilic, mesophilic (II), and maturation/curing phase. Figure 7.1 shows that the mesophilic phase (I) occurs in the temperature range between 20 and 40 °C. In this phase, the population of mesophilic bacteria and fungi is in the amounts of 10^8 and 10^6 per 1 g of wet compost. Meanwhile, the population of thermophilic bacteria and actinomycetes is 10^4, while the population of thermophilic fungi is 10^3 per 1 g of wet compost, as shown in Table 7.2. An active decomposition of organic matter in this phase will generate heat from the metabolic activity of microorganisms, and indirectly an increase in temperature also occurs in the compost mass. At a temperature range of 35–45 °C, microbial activity increases.

However, the decomposition of organic matter by mesophilic bacteria decreases when the temperature reaches 45 °C. At this point, the thermophilic phase begins, where the decomposition of organic matter will be taken over by thermophilic bacteria. In this phase, the increase in temperature can reach up to 65–70 °C as a result of microbial activity in the compost pile. Based on Table 7.2, the population of thermophilic microorganisms (bacteria, fungi, and actinomycetes) will increase between 10^7 and 10^9, while the population of mesophilic microorganisms will decrease to 10^3–10^6. Meanwhile, the mesophilic phase (II) begins when there is a decrease in compost temperature as well as a reduction in microbial activity. During this phase, the population of mesophilic bacteria and fungi increased again to 10^{11} and 10^5 in the compost mass, respectively. These microbial populations will compete with each other for the remaining food supply to continue the decomposition of the lignin and cellulose found in the compost pile [2]. The maturation phase is the final phase during composting, where the compost temperature has reached ambient temperature, and a more stable compost is produced. This phase can take up to months [43].

Temperatures in normal windrow operations range from 66 to 71 °C. In general, it is possible to sustain temperatures between 60 and 66 °C for a period of around three weeks. In the center of the composting mass, higher temperatures are typically observed. Breidenbach [12] reported that a single weekly temperature reading helps to determine the composting progress in the open windrow system operated in Johnson City, Tennessee. However, there are some composting operations that are reported to reach higher temperatures during the process. Among them is the Metrowaste plant at Gainesville, Florida, and Fairfield-Hardy in Altoona, Pennsylvania, with temperatures reached 82 and 60–71 °C, respectively. In modern composting processes, the temperature ranges designed for mesophilic and thermophilic phases are 10–40 °C and 40–70 °C, respectively. Both environmental experts and engineers believe that the temperature of activity has to be at least 35 °C for effective mechanical composting. For the composting process, the optimal temperature range is about 35–55 °C, which includes the ideal temperatures for the different forms of microorganisms involved throughout the process. For organic decomposition, mesophilic bacteria are more successful than thermophilic bacteria, while thermophilic disintegration happens at a faster rate. Generally, pathogenic microorganisms and weed seeds are destroyed in the thermophilic temperature zone. Thermophilic composting must be implicated at some point in the overall composting process for the prevention of pathogens and weed seeds.

7.4.2 Moisture Content

Microorganisms living in a compost pile need enough moisture to survive. Water is the key element that helps transport substances within the compost pile and makes the nutrients in organic material accessible to the microbes. Organic material contains some moisture in varying amounts, but moisture also might come in the form of rainfall or intentional watering. Moisture content is the most important parameter to ensure that the composting process achieves optimal conditions. During the composting process, water is the medium of nutrient transport for microorganisms to carry out cell metabolic and physiological activities [41–43]. Previous studies have shown that the optimum moisture content is in the range of 50–60% [2, 5, 42, 44, 46]. Excessive moisture content will inhibit the process of oxygen absorption by microorganisms and indirectly create anaerobic conditions in the compost pile as well as produce a foul odor [42]. Also, the excess moisture content will contribute to the problem of loss of nutrients and pathogens in the form of leachate. This condition also causes the airflow inside the compost pile to not be able to occur better. Although 50–60% of moisture content is considered optimal, some researchers consider certain values for different organic matter to be set. For example, the moisture content suitable for domestic waste is between 52–58% and 60% for food waste [43]. The moisture content in compost is usually present from two main sources, namely organic waste used at the beginning of the process and water vapor resulting from the metabolic activity of microorganisms.

Meanwhile, the presence of too low moisture content will inhibit the metabolic process of microorganisms. For example, if the moisture content is 10% or even lower, the microorganisms involved in the process of cellulose decomposition will stop their metabolic activity [2]. Changes in moisture content in compost piles depend on three main factors, namely the organic material used, the bulking agent, and the composting method [43]. According to Epstein [5], the initial moisture content of a mixture of municipal solid waste and biosolid is 67.3%. The windrow method involving the turning process on the mixture has reduced the moisture content to 55% within 15 days and 43% on day 30. Meanwhile, the composting method using positive aeration reduced the moisture content to 48% on day 15 and decreased to 29% after 30 days. Besides, the use of bulking agents such as wood chips and straw is also suitable to reduce the excess moisture content in the compost pile. Environmental factors such as rainfall and evaporation processes will also affect the moisture content in the compost. In general, the moisture content of incoming solid waste to be used in the composting process is highly variable, where most of the moisture content of received solid waste is insufficient to attain high-rate composting. Therefore, moisture additives have to be increased. For adequate composting, the ideal moisture content is between 55% and 65%. The maximum permissible moisture content for different solid wastes is depicted in Table 7.3. At the beginning

Table 7.3 Maximum permissible moisture content for different solid wastes

Type of waste	Moisture content (%)
Municipal solid waste (MSW)	55–65
Residential wet wastes (garbage, lawn clippings, trimmings, etc.)	50–55
Paper	55–65
Animal manure	55–65
Wood (saw dust, other small chips)	75–90
Straw	75–85

Source: [21]

of the process, moisture supplements such as nutrients should be applied. The addition of supplements in liquid form to solid waste must be dependent on calculations and in relation to the material's dry weight. If extra moisture is applied with supplements of nitrogen or phosphorus or is found in sewage or sludge, this must also be included in the measurement of moisture.

Generally, after the composting process is over, this moisture content factor is still taken into account, especially in the screening process and compost packaging before being marketed. Moisture removal is important to ensure that the resulting compost management process is more efficient. For example, the screening process is easier to be conducted on compost that has a moisture content of less than 40% after the active composting process [5].

7.4.3 C:N Ratio

Composting, or controlled decomposition, requires a proper balance of "green" organic materials and "brown" organic materials. "Green" organic material includes grass clippings, food scraps, and manure, which contain large amounts of nitrogen. "Brown" organic materials include dry leaves, wood chips, and branches, which contain large amounts of carbon but little nitrogen. Obtaining the right nutrient mix requires experimentation and patience. It is part of the art and science of composting. The C:N ratio is one of the factors that influence the composting process as well as the quality of the compost produced [47]. In the composting process, carbon is a source of energy for microorganisms, while nitrogen is needed for the construction of cell protoplasm [2, 48]. Most of the absorbed carbon will be converted to CO_2 gas by microorganisms during the cell metabolism process. Meanwhile, the remaining carbon will be exchanged into the form of cell walls (membranes) and also protoplasm. Agamuthu [2] stated that 2/3 of the total carbon used by microorganisms would release CO_2 gas while another 1/3 will combine with nitrogen for the construction of protoplasm cells. In general, carbon is more needed than nitrogen in this process. Previous studies have stated that the optimal C:N ratio for the continuity of the composting process is between 20 and 25 parts carbon compared to 1 part nitrogen [44, 48]. Meanwhile, Huang et al. [47] stated that the appropriate C:N ratio range is 25–30, while Agamuthu [2] suggested that the optimal C:N ratio range is 26–31. However, all these value ranges can still be used because the composting process is also influenced by other factors such as temperature, moisture content, and pH.

In general, a low C:N ratio (<20) in compost mass will cause nitrogen loss in the form of ammonia in addition to odor problems to the environment [1, 48]. Increased temperature, as well as high pH levels in the compost pile, are among the factors that affect the release of ammonia into the environment. A study conducted by Vuorinen and Saharinen [49] stated that nitrogen loss usually occurs through gas release or even in the form of leachate. However, this situation can be overcome by adding materials that have a high carbon content, such as sawdust and straw [48].

Meanwhile, the composting process may be slowed down or take a long time if the organic waste mixture is to be composted with a high C:N ratio (>35) [1]. The addition of substances that have a high nitrogen content, such as sludge and animal manure, can help produce an optimal C:N ratio to enable the composting process to occur more efficiently. Mature and stable compost usually has a C:N ratio of less than 20 and is safe to use on crops without any restrictions [49].

7.4.4 pH Level

pH is a measure to determine the acidity or alkalinity in the compost mass. Control over pH is one of the important parameters during composting in assessing the microbial environment and stabilization of organic waste [44]. According to Agamuthu [2], the optimum pH value for microbial activity is usually between 6.5 and 7.5, while Day and Shaw [43] stated that the optimum pH value for microbial activity is between 6.5 and 8.5. Similar to temperature, the pH value also indicates the variation of fluctuating changes throughout the composting process. In the early stages of the composting process, the pH value will usually decrease to pH 5 or less than that value. The decrease in pH is due to the formation of organic acids in the compost pile. A study conducted by Wong et al. [50] on the soybean and yard waste mixture in Hong Kong also stated that the decomposition of the mixture showed a decrease in pH value from 6.1–6.8 to 5.3–6.1 within 7 days. This decrease is influenced by organic acids resulting from the activity of microorganisms that decompose the wastes. For organic wastes with a pH of 5.5 or less, calcium carbonate can be added to increase the pH in the early stage of the composting process [51]. Moreover, by mixing lime, sodium bicarbonate, caustic soda, or any accessible dilute acid, pH adjustment may also be made. After a few days, the temperature will rise, and the thermophilic phase will result. At this stage as well, the resulting organic acid will act as a substrate that will be used by microorganisms to continue the decomposition of organic matter [52]. Decomposition activity by these microorganisms will increase the pH value to 8.0–8.5.

In general, the process of decomposition of municipal solid waste takes place in a sequential manner in which soluble carbohydrates are decomposed first, and the pH decreases below 7.0. As the composting period increases, these substances are depleted, and proteins start to degrade, allowing the pH to rise steadily. pH, on the other hand, seldom falls below 5.0 or rises above 8.5. The ideal pH range for MSW aerobic composting is between 6.5 and 8.0. The composting process at a plant in Johnson City, Tennessee, for instance, reported a pH value of between 5.0 and 7.0. The pH pattern of the composting mechanism using the windrow system at the facility is shown in Fig. 7.2. It can be seen from Fig. 7.2 that at the beginning of the decomposition phase, the pH initially decreases to 5.0, then leveled off around pH 8.0. As long as the system is in an aerobic state, this value will remain constant.

Temperature is also one of the factors that influence the increase in pH. Thermophilic composting (45–75 °C) usually occurs in the pH range between

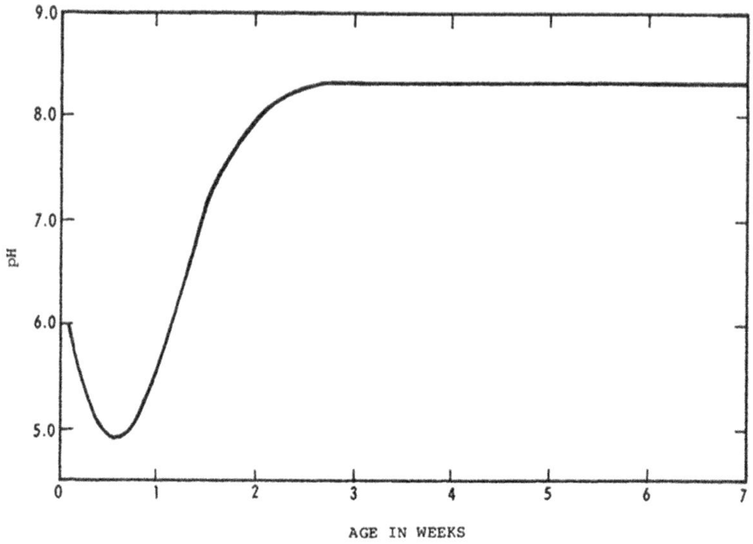

Fig. 7.2 The pH profile obtained in the windrow process from composting facility in Johnson City, Tennessee. (Source: USEPA)

7.5 and 8.5 [5]. Besides, the aeration process also affects the pH value of the compost. In general, oxygen supply is essential in aerobic reactions to ensure that the decomposition of organic matter can occur more efficiently. Lack of oxygen supply will cause an incomplete oxidation process to occur, which in turn increases the population of anaerobic bacteria in the compost pile. This anaerobic condition will reduce the pH value and will be a limiting factor to the composting process [53].

In the final composting stage, the pH will decrease to a range of 7.0–8.0 in matured compost. This decrease refers to the release of ammonium ions and hydrogen ions in the nitrification process performed by nitrogen-fixing bacteria at the end of the process [50].

7.4.5 Aeration Requirement and Turning Mechanism

In the aerobic composting process, the presence of oxygen is required by microorganisms to decompose organic matter. Oxygen is needed to ensure the growth of the population of microorganisms while controlling the temperature in the compost mass [54]. Turning the pile, placing the pile on a series of pipes, or including bulking agents such as wood chips and shredded newspaper all help aerate the pile. Aerating the pile allows decomposition to occur at a faster rate than anaerobic conditions. Care must be taken, however, not to provide too much oxygen, which can dry out the pile and impede the composting process. The aeration of compost can be done through natural, passive, active, and even forced aeration methods. Natural aeration usually involves the mechanism of diffusion and convection of air from the surface into the compost pile. This method is inexpensive, is convenient, and does not require special aeration equipment. Meanwhile, passive aeration involves the construction of perforated pipes under compost piles to encourage the diffusion of oxygen gas to the composted waste. Active aeration is similar to passive aeration, the difference being that a fan is mounted on a network of pipes to induce air throughout the compost pile. Some researchers argued that in the active composting process, a timer would be used to control the aeration rate. The excessive air supply will cause the compost to cool, and this condition will contribute to the problem of nitrogen loss to the environment. Meanwhile, lack of aeration would impact the temperature stabilization mechanism in compost piles [54]. Compared to the active aeration mode, the composting rate is higher when the passive aeration mode is used. Passive aeration mode does not produce a cooling effect on the compost mass and can reduce nitrogen loss to the environment. This method is cheaper, but its aeration operation is comparable to the active aeration method.

According to Agamuthu [2], the aeration rate also affects the moisture content and the type of organic waste composted. Apart from the four types of aeration that have been mentioned, turning and mixing the compost manually is also one of the techniques used to provide composted organic waste with air requirements. For example, organic waste with high moisture content requires a more frequent turning and over a longer period than organic waste with low moisture content. Tiquia et al. [46] stated that the frequency of turning over the compost mass is among the factors that also affect the composting rate and quality of compost produced. Turning is a preliminary mechanism in controlling aeration and temperature in composting systems [55]. During the composting process using the windrow method, where the pile formed is high, a turning operation should be carried out to prevent anaerobic conditions in the center and bottom of the compost pile. This state arises because the rate of oxygen uptake into the compost pile is too limited for microbial metabolic activity. The turning mechanism must be performed either manually or by mechanical equipment to allow oxygen to enter. Meanwhile, for organic waste with high moisture content (70%), the turning mechanism needs to be conducted more regularly when the value exceeds 70%. This method can help to break down large particles of organic matter while providing maximum surface area for microbial action [5]. The temperature of the compost pile will be affected by the turning frequency. Compost turning operations control aims to ensure that elevated temperatures in the pile can be maintained. Excessive turning will make the compost mass cold and dry, which indirectly destroys fungi and actinomycetes [56]. Also, the frequency of turning the composted organic waste will lead to the loss of nitrogen to the environment in the form of ammonia gas [57]. This condition will cause the resulting compost to be deficient in essential nutrients such as nitrates that are known to be useful for plants.

7.4.6 Particle Size

Grinding, chipping, and shredding materials increase the surface area on which microorganisms can feed. Smaller particles also produce a more homogeneous compost mixture and improve pile insulation to help maintain optimum temperatures (see below). If the particles are too small, however, they might prevent air from flowing freely through the pile. Tchobanoglous et al. [44] stated that small particle sizes are capable of providing microorganisms with the optimum surface area to decompose organic matter. Particle size will influence bulk density, internal friction, flow, and drag forces within the mass of the composted organic waste. However, the particle size that is too small is considered inappropriate because this will cause the structure of the composted organic waste to become too dense. As a result, the air space between the organic waste reduces and, in turn, inhibits the movement of oxygen in the compost pile. Lack of oxygen in the compost pile will cause a slow decomposition reaction and contribute to anaerobic conditions.

Suitable particle size is required in the composting process to ensure better decomposition. Although large-sized organic matter can still be composted, organic matter of less than 5 cm in size is considered appropriate and can produce optimal reactions [44]. Diaz et al. [48] stated that the particle size of the composted material could reach up to 15 cm or larger, but this size depends on the type of organic material to be composted. For example, the appropriate particle size for organic materials such as leaves, grass, and agricultural waste is 5 cm and 5–10 cm for woody and fibrous organic matter, respectively. Meanwhile, for materials consisting of twigs and tree branches, the size reduction process should be done to ensure that the size obtained is 1 cm in diameter, 2 cm wide, and 1–6 cm thick. However, since the content of lignin and cellulose is slow to decompose, and this will lengthen the composting duration, this part is less common to choose as composting material.

The use of equipment such as a hammer mill and shear shredder can help to produce the desired particle size before the materials are composted. Besides, to minimize the size of organic waste, the application of a rotating drum will also aid. For microbial activity, the turning process applied to the organic waste in the reactor allows for blending and breaking down the particle of organic waste to smaller sizes.

7.4.7 Seeding

There are various types of microorganisms found in MSW, sludge, animal droppings (dung), and other compostable materials. However, between each distinct phase transition, there is a delay time needed to multiply the population of the required microorganisms (i.e., the seed) in a batch composting method. Therefore, the starter "seed" needs to be added at the beginning of the process, where it comprises a portion of the compost that is taken up during the active composting process [21]. In general, introducing seed at startup by intimate mixing reduces the time it takes for stabilization and reduces the time it takes to achieve a sufficiently stable operation. Nevertheless, the seed used as a "starter" for the composting process is not always available. Composting plant operators usually obtain this seed from a small pilot plant or another composting facility. Generally, for a fast startup, comparatively high volumes of seed (10–33%) must be added [58]. If the seed is kept at a low temperature and conditions are such that the microbial population is not substantially decreased, optimum benefits may be achieved.

7.5 Classification of Composting Systems

The composting systems can be categorized based on oxygen consumption, temperature, technological approach, operational modes, raw material, and operating methods.

7.5.1 Oxygen Requirement

Microbiological and biochemical transformations occur either aerobically or anaerobically. The difference between biochemical reactions in the aerobic and anaerobic systems in the composting process needs attention to ensure that the operation is in satisfactory condition.

7.5.1.1 Aerobic System

The aerobic composting process has great potential for processing municipal solid waste as well as yard waste [59]. This process involves the decomposition of organic matter in the presence of oxygen. The result of this reaction consists of carbon dioxide, water, and a variety of oxidized end products as well as heat. Figure 7.3 shows the input–output analysis of the aerobic composting process.

Oxygen consumption is high at the beginning of the process and then decreases as the compost has reached maturity [4]. In this process as well, aerobic bacteria will use carbon from organic waste as a source of energy while nitrogen will be recycled [2]. In general, the process of aerobic decomposition can be explained through Eq. (7.1):

$$\begin{aligned}&\mathrm{C}_a\mathrm{H}_b\mathrm{O}_c\mathrm{N}_d\mathrm{S}_e\mathrm{P}_f + 0.5(nw + 2s + r - c + t)\mathrm{O}_2 \longrightarrow \\&n\mathrm{C}_u\mathrm{H}_v\mathrm{O}_w\mathrm{N}_x\mathrm{S}_y\mathrm{P}_z + s\mathrm{CO}_2 + r\mathrm{H}_2\mathrm{O} + (d - nx)\mathrm{NH}_3 \\&+(e - ny)\mathrm{SO}_4^{2-} + (f - nz)\mathrm{PO}_4^{3-}\end{aligned} \tag{7.1}$$

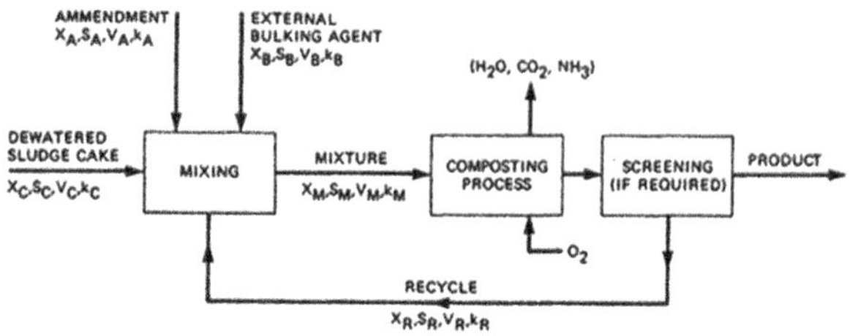

Note: RECYCLE is defined as finished compost for the windrow and mechanical systems and as recycled wood chips for the aerated pile system.

The exact value for these parameters must be determined from samples of the sludge, external bulking agent, amendment, and estimated for the recycle values unless otherwise known.

Process Variables and Range of Average Values (in Parenthesis)

X_C = Total wet weight of sludge cake produced/day.

X_A = Total wet weight of amendment/day.

X_R = Total wet weight of recycle/day.

X_B = Total wet weight of external bulking agent/day.

X_M = Total wet weight of mixture/day.

S_C = Fractional solids content of sludge cake (0.20 to 0.55).

S_A = Fractional solids content of amendment (0.50 to 0.95).

S_R = Fractional solids content of recycle (0.60 to 0.75).

S_B = Fractional solids content of external bulking agent (0.50 to 0.85).

S_M = Fractional solids content of mixture (0.40 to 0.50).

V_C = Volatile solids content of sludge cake, fraction of dry solids (0.40 to 0.60) - Digested; (0.60 to 0.80) - Raw.

V_A = Volatile solids content of amendment, fraction of dry solids (0.80 to 0.95).

V_R = Volatile solids content of recycle, fraction of dry solids (0.00 to 0.90).

V_B = Volatile solids content of external bulking agent, fraction of dry solids (0.55 to 0.90).

V_M = Volatile solids content of mixture, fraction of dry solids (0.40 to 0.80).

k_C = Fraction of sludge cake volatile solids degradable under composting conditions (0.33 to 0.56).

k_A = Fraction of amendment volatile solids degradable under composting conditions (0.40 to 0.60).

k_R = Fraction of recycle volatile solids degradable under composting conditions (0.00 to 0.20).

k_B = Fraction of external bulking agent volatile solids degradable under composting conditions (0.00 to 0.40).

k_M = Fraction of mixture volatile solids degradable under composting conditions (0.20 to 0.60).

Fig. 7.3 Input–output analysis of composting process. (Source: USEPA)

where

$r = 0.5\ [b - nv - 3(d - nx)]$

$s = a - nu$

$t = 4(e + f - ny - nz)$

The terms $C_aH_bO_cN_dS_eP_f$, and $C_uH_vO_wN_xS_yP_z$ reflect the empirical mole composition of the organic solid wastes present at the beginning of the process and at the end (compost), respectively [21]. The following expression applies if complete aerobic oxidation conversion is achieved:

$$\begin{aligned}&4C_aH_bO_cN_dS_eP_f + (4a + b - 2c + 5d + 6e + 5f)O_2 \longrightarrow \\ &4aCO_2 + (2b - 2d - 4e - 6f)H_2O + 4dNO_3^- + 4eSO_4^{2-} \\ &+4fPO_4^{3-} + (4d + 8e + 12f)H^+\end{aligned} \quad (7.2)$$

It is visible from Eq. (7.2) that the entirely oxidized end products are carbon dioxide, water, nitrates, sulfates, phosphates, and other stable end products; i.e., materials in which further biological degradation is no longer possible. Equation (7.2) represents a complete biological oxidation proces. An ideal composting process, however, is an incomplete biological degradation process represented by Eq. (7.1), in which $C_aH_bO_cN_dS_eP_f$, and $C_uH_vO_wN_xS_yP_z$ are the original solid organic waste and the end-product compost, respectively. The compost can be recycled for reuse as a soil conditioner. In general, aerobic biological decomposition is marked by elevated temperatures; thus, shorter detention time is needed compared with anaerobic decomposition, resulting in a less aggressive, less odorous, and more readily stabilized end product. Many current composting methods are essentially aerobic or aim to be because of the benefits of aerobic systems [21].

Detention time, mixing equipment, and air movement are among the other factors that need to be considered in aerobic composting. Oxygen is supplied to the process at a practical level by interaction with oxygen or by air renewal. In most cases, this is done by mixing or forced aeration. Generally, if a mixture of solid waste stays undisturbed, the biological demands are such that oxygen is locally exhausted and anaerobic conditions result, especially in a moist condition (e.g., moisture content above 70%). The method of mixing, rotating, tumbling, agitating, or pushing air through or into the composting material is among the efforts that need to be made to ensure that aerobic composting can be successful [21].

Equation (7.2) can be modified to represent an alternative solid waste disposal process, incineration/combustion, if nitrate, sulfate, and phosphate are further oxidized to NO_x, SO_2 and P_2O_5, respectively. The modified Eq. (7.2) will show that the organic solid waste, $C_aH_bO_cN_dS_e$ can be completely oxidized by incineration/combustion forming CO_2, NO_x, SO_2, P_2O_5, other gaseous and ashes. A comparison between aerobic composting process and incineration/combustion process will show that the aerobic composting process requires less heat energy, emits less carbon dioxide gas, produces no NO_x, SO_2 and P_2O_5, and generates usefill compost.

7.5.1.2 Anaerobic System

Anaerobic composting is the decomposition process of organic matter that takes place without the presence of oxygen. This process takes longer than the aerobic composting process. The resulting final product usually consists of CH_4, CO_2, NH_3, acidic gases and produces a foul odor. Under anaerobic conditions, the decomposing substances tend to be more acidic. In the past, this method was commonly used to process animal manure and human feces. But now, this method has begun to be practiced to process municipal solid waste and yard waste. The overall conversion is as shown in Eq. (7.3):

$$
\begin{aligned}
&C_aH_bO_cN_dS_e \longrightarrow \\
&nC_uH_vO_wN_xS_y + mCH_4 + sCO_2 \\
&+rH_2O + (d - nx)NH_3 + tH_2S \\
&+(-4nu - nv + 2nw + 3nx + 2ny + 4a + b - 2c - 3d - 2e - 8m)H^+
\end{aligned} \tag{7.3}
$$

where

$s = a - nu - m$
$r = c - nw - 2s$, and
$t = e - ny$

Again, the terms $C_aH_bO_cN_dS_e$ and $C_uH_vO_wN_xS_y$ reflect the empirical mole composition of the organic solid wastes at the start and the end product (compost) of the process, respectively.

Low temperatures (unless heat is added from an external source) characterize anaerobic decomposition, which typically occurs at a slower pace than aerobic composting does. Anaerobic composting produces more odorous and unpleasant end products than aerobic composting. The early agricultural activities connected to composting were anaerobic in the 1920s and 1930s and needed extended completion time. Aerobic composting is the most common form of modern municipal composting. However, anaerobic composting provides two main benefits over aerobic composting. These include that (1) the anaerobic process should be conducted with a minimum of attention so that it can be sealed off from the environment, and (2) compared to aerobic composting, more cellulose compounds can be destroyed by anaerobic composting since bacteria in the anaerobic environment have a long time to digest, hydrolyze, and disintegrate the waste content [21].

7.5.2 Temperature

Temperature variations in the composting materials are a natural feature of the composting process. Composting is either mesophilic or thermophilic, classed according to temperature range. Mesophilic composting occurs in the temperature range of 10–40 °C, and most cases occur at ambient temperatures. Meanwhile, thermophilic composting occurs in the range of 40–70 °C. In actual practice, the composting process takes place in both mesophilic and thermophilic temperature ranges. The temperature rise is caused by microbial activities on the composting material as well as the existing physical condition. Self-heating masses are dynamic concerning moisture, oxygen, substrate, and other abiotic factors [21]. A description of the two temperature ranges involved in this composting process has been detailed in Sect. 7.4.1.

7.5.3 *Technological Approaches*

Composting technology is divided into two main categories, namely, open composting and mechanical/enclosed/in-vessel composting [48]. Open composting involves two methods, namely static pile (windrow) and aerated static pile (aerated static pile). While mechanical composting consists of a vertical reactor, a horizontal reactor, and a rotating drum method. Apart from these two main categories, two other methods are increasingly popular in organic waste composting, namely vermicomposting and thermophilic composting.

7.5.3.1 Open On-Site Composting

Parks, camp sites, mobile home sites, golf courses, farms, homeowners with big lots, etc. can all compost small amounts of wasted food, leaves, etc. on-site and reuse the produced compost products. Composting can significantly reduce the amount of wasted foods, leaves, etc. that are thrown away. Yard trimmings and small quantities of food scraps can be composted together on-site. Animal products and excessive quantities of food scraps are not appropriate for on-site composting. The basic concepts are that (1) compost is organic material that can be added to soil to help plants grow; (2) food scraps and yard waste together currently make up more than 30% of what we throw away and could be composted instead; and (3) producing compost may keep these materials out of landfills where they take up space and release methane, a potent greenhouse gas.

A successful open on-site composting requires three basic ingredients:

1. Browns: This includes materials such as dead leaves, branches, and twigs.
2. Greens: This includes materials such as grass clippings, vegetable waste, fruit scraps, and coffee grounds.
3. Water: Having the right amount of water, greens, and browns is important for compost development.

A recommended compost pile should have an equal amount of browns to greens. The composter operator should also alternate layers of organic materials of different-sized particles. The brown materials provide carbon for your compost, the green materials provide nitrogen, and the water provides moisture to help break down the organic matter. The recommended solid wastes to be composted include fruits and vegetables; eggshells; coffee grounds and filters; tea bags; nut shells; shredded newspaper; cardboard; papery; yard trimmings; grass clippings; house plants; hay and straws; leaves; sawdust; wood chips; cotton and wool rags; hair and fur; and fireplace ashes.

The following solid wastes are not recommended to be composted:

1. Black walnut tree leaves or twigs because they release substances that might be harmful to plants
2. Coal or charcoal ash because it might contain substances harmful to plants
3. Dairy products (e.g., butter, milk, sour cream, yogurt) and eggs because they may create odor problems and attract pests such as rodents and flies

4. Diseased or insect-ridden plants because the diseases or insects might survive and be transferred back to other plants
5. Fats, grease, lard, or oils because they may create odor problems and attract pests such as rodents and flies
6. Meat or fish bones and scraps because these solid wastes may create odor problems and attract pests such as rodents and flies
7. Pet wastes (e.g., dog or cat feces, soiled cat litter) because the pet solid wastes may contain parasites, bacteria, germs, pathogens, and viruses harmful to humans
8. Yard trimmings treated with chemical pesticides because these solid wastes might kill beneficial composting organisms

The producer of a large quantity of dairy product wastes, infective solid wastes, oil and grease wastes, meat and fish wastes, and pet solid wastes should check with the local composting or recycling coordinator to see if these organics are accepted by the community curbside or drop-off composting program. The climate and seasons changes will not have a big effect on on-site composting. Small adjustments can be made when changes happen, such as when the rainy season approaches. On-site composting takes very little time or equipment. Education is the key. Local communities might hold composting demonstrations and seminars to encourage homeowners or businesses to compost their own properties. Creating compost can take up to 1 year, but manual turning can speed up the process to between 3 and 6 months. Compost, however, should not be used as potting soil for houseplants because of the presence of weed and grass seeds. There are many benefits of the compost produced on-site because the compost may (1) enrich soil, helping retain moisture and suppress plant diseases and pests; (2) reduce the need for chemical fertilizers; (3) encourage the production of beneficial bacteria and fungi that break down organic matter to create humus, a rich nutrient-filled material; and (4) reduce methane emissions from landfills and lower your carbon footprint.

Open On-Site Composting Process Operation

A successful open on-site composting process operation includes the following operational procedures:

1. Obtaining the helpful tools of pitchforks, square-point shovels or machetes, and water hoses with a spray head connected to a water source
2. Selection of a dry, shady spot near a water source for your compost pile or bin
3. Addition of brown and green materials as they are collected, making sure larger pieces are chopped or shredded
4. Moistening dry materials as they are added
5. Establishing a compost pile
6. Mixing grass clippings and green waste into the pile and burying fruit and vegetable waste under 10 inches of compost material
7. Covering top of compost with a tarp to keep it moist
8. Harvesting the produced compost anywhere between 6 months and 2 years to use when the material at the bottom is dark and rich in color

7.5.3.2 Aerated (Turned) Windrow Composting

Aerated or turned windrow composting is suited for large volumes such as that generated by entire communities and collected by local governments and high volume food-processing businesses (e.g., restaurants, cafeterias, packing plants). It will yield significant amounts of compost, which might require assistance to market the end-product. Local governments may want to make the compost available to residents for a low or no cost.

This type of composting involves forming organic waste into rows of long piles called "windrows" and aerating them periodically by either manually or mechanically turning the piles. The ideal pile height is between 4 and 8 ft with a width of 14–16 ft. This pile size is large enough to generate enough heat and maintain temperatures. It is small enough to allow oxygen flow to the windrow's core.

Large volumes of diverse wastes such as yard trimmings, grease, liquids, and animal byproducts (such as fish and poultry wastes) can be composted through this windrow method. Windrow composting often requires large tracts of land, sturdy equipment, a continual supply of labor to maintain and operate the facility, and patience to experiment with various materials mixtures and turning frequencies.

In a warm, arid climate, windrows are sometimes covered or placed under a shelter to prevent water from evaporating. In rainy seasons, the shapes of the pile can be adjusted so that water runs off the top of the pile rather than being absorbed into the pile. Windrow composting can work in cold climates. Often the outside of the pile might freeze, but in its core, a windrow can reach 140 °F. Leachate is liquid released during the composting process. This can contaminate local ground water and surface-water supplies. It should be collected and treated. Windrow composting is a large-scale operation and might be subject to regulatory enforcement, zoning, and siting requirements. Compost should be tested in a laboratory for bacterial and heavy metal content. Odors also need to be controlled. The public should be informed of the operation and have a method to address any complaints about animals or bad odors.

This windrow method is simple and does not require high technological expertise. It is suitable for processing any type of organic waste, including animal manure, sewage sludge, and garden waste [6], and is capable of converting waste in large quantities into compost. The Indore and Van Maanen methods are among the pilot projects that have featured this type of open composting method since the 1920s.

The windrow composting is a simple solid waste treatment process and can be modified according to the location. Through this method, organic waste will be piled up in the form of large, elongated piles. Windrows piles can be made in the form of a triangle (delta windrow) with a height of 2 m and a width of between 2.5 and 3 m. Meanwhile, a rectangular pile (trapezoidal windrow) can reach a height of up to 3 m, and its width is between 10 and 12 m with the sides tilted slightly, as shown in Fig. 7.4. However, the dimensions of the pile depend on the organic waste used and the weather conditions. The turning mechanism is performed to supply oxygen to the composting pile [46] in addition to breaking down the larger sizes of organic waste into smaller sizes for decomposition by microorganisms or other

Fig. 7.4 Aerated (turned) windrow composting process. (Source: USEPA)

decomposing organisms. The turning mechanism also plays an important role in enhancing the porosity properties of the composted organic materials. This is to prevent deposition and compaction from occurring on the materials while allowing heat, water vapor, and gases trapped in the compost pile to be released. The porosity of the composted organic waste greatly affects the airflow in the pile. For example, denser organic waste such as animal manure requires a smaller pile shape to minimize anaerobic zones, while more porous and lightweight organic waste such as yard waste will be composted into larger pile forms. However, this method is seen to have many shortcomings [60]. These include requiring (1) large space, (2) large equipment or machinery to handle the turning process, and (3) additional costs to pay for labor carrying out turning work.

Figure 7.4 shows that oxygen is supplied to a "static pile" of an aerated (turned) windrow process through a mechanical turning mechanism for periodic mixing along the pile. It is important to note that the entire pile of an aerated (turned) windrow composting process system can be totally turned/mixed by a different mechanical means, so the pile does not have to be a "static pile."

7.5.3.3 Aerated Static Pile Composting

The aerated static pile method was developed by the US Department of Agricultural Research Service Experimental Station in Beltsville, Maryland [44]. In the early stages, the method that is similar to the windrow composting method is used to process organic waste that has a high moisture content, such as sewage sludge.

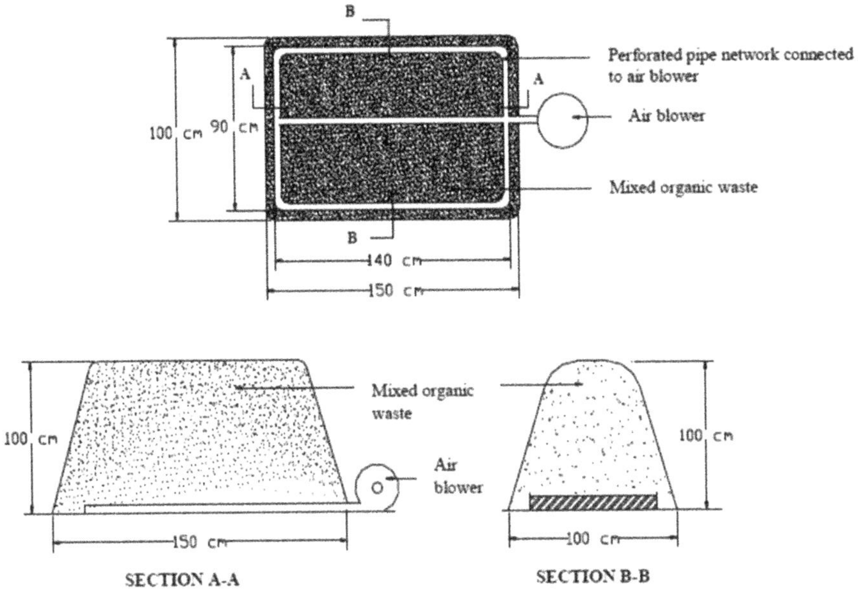

Fig. 7.5 Schematic diagram of the entire composting structure using the aerated static pile (ASP) method [61]

However, this newer method was later extended to treat other organic wastes such as yard waste and municipal solid waste that had been segregated. Aerated static pile composting produces compost relatively quickly (within 3 to 6 months). It is suitable for a relatively homogeneous mix of organic waste and works well for larger quantity generators of yard trimmings and compostable municipal solid waste (e.g., food scraps, paper products), such as local governments, landscapers, or farms. This method, however, does not work well for composting animal byproducts or grease from food processing industries.

The aerated static pile method consists of a perforated pipe network and an air blower to supply air to the composting pile (Fig. 7.5). The incoming air will supply oxygen to assist the decomposition of organic waste while controlling the temperature inside the pile [44]. The aeration inside of the composting pile can be controlled by using a timer that is connected to the blower. Indirectly, the airflow rate can be controlled to produce the desired temperature profile in the composting pile [62]. The piles formed will be covered with screened compost to avoid odor problems. Bulking agents such as wood chips have long been used to increase the cavity in composting piles as well as to absorb excess moisture [60]. In comparison to the windrow method, this technique is seen to provide many advantages in the composting of organic waste. Tiquia and Tam [60] claimed that this approach is very acceptable for use because, opposed to the windrow process, it does not need wide space and requires less labor. In the meantime, Cegarra et al. [63] have accepted that this approach should be used to accommodate the oxygen supply in the pile as well as to eliminate heat, gases, and water vapor. This approach is ideal for use in modern composting on large- and medium-scale farms [62]. The formation of water reservoirs at the base of the pile due to the turning process is not carried out and causes the presence of anaerobic conditions that indirectly create a foul odor, among the shortcomings of this method [63]. The absence of turning mechanisms also causes inconsistencies of compost mass, such as the distribution of microorganisms, nutrients, and water.

In summary, in aerated static pile composting, organic waste is mixed in a large pile. To aerate the pile, layers of loosely piled bulking agents (e.g., wood chips, shredded newspaper) are added so that air can pass from the bottom to the top of the pile. The piles can also be placed over a network of pipes that deliver air into or draw air out of the pile. Air blowers might be activated by a timer or temperature sensors. In a warm, arid climate, it may be necessary to cover the pile or place it under a shelter to prevent water from evaporating. In the cold, the core of the pile will retain its warm temperature. Aeration might be more difficult because passive airflowing is used rather than active turning. Placing the aerated static piles indoors with proper ventilation is also sometimes an option. Since there is no physical turning, this method requires careful monitoring to ensure that the outside of the pile heats up as much as the core. Applying a thick layer of finished compost over the pile may help alleviate any odors. If the air blower draws air out of the pile, filtering the air through a biofilter made from finished compost will also reduce any of the odors. This method may require significant cost and technical assistance to purchase, install, and maintain equipment such as blowers, pipes, sensors, and fans. Having a controlled supply of air allows the construction of large piles, which require less land than the windrow method.

7.5.3.4 Enclosed or Mechanical Composting (In-Vessel Composting)

The mechanical composting takes place in a closed container (in-vessel). In-vessel composting can process large amounts of waste without taking up as much space as the windrow method, and it can accommodate virtually any type of organic waste (e.g., meat, animal manure, biosolids, food scraps). This composting process involves feeding organic materials into a drum, silo, concrete-lined trench, or similar equipment. This allows good control of the environmental conditions such as composting temperature, moisture, and airflow. The material is mechanically turned or mixed to make sure the material is aerated. Air can be injected into the vessel when needed. The size of the vessel can vary in size and capacity. This in-vessel composting method produces compost in just a few weeks. It takes a few more weeks or months until it is ready to use because the microbial activity needs to balance and the pile needs to cool. Some in-vessel composting units are small enough to fit in a school or restaurant kitchen. Large food processing plants often use very large in-vessel processing equipment, similar to the size of a school bus. Careful control, often electronically, of the climate allows year-round use of this method. It is possible to use it either in extremely cold weather or for indoor with insulation. The in-vessel composting process equipment produces very little odor or leachate, uses much less land and manual labor than windrow composting, but is expensive and may require technical expertise to operate it properly.

There are two types of reactors found in the enclosed system, namely silo and agitated bed type, which both rely on aeration and turning mechanisms throughout the process. This method is designed to reduce odor problems and shorten the composting period by controlling airflow, temperature, and oxygen content [44]. Mechanical or enclosed composting is increasingly popular and is the choice of developing countries in processing organic waste because it has the potential to shorten mesophilic and thermophilic phases, more efficient process control, use less space than other methods, and minimize the number of pathogens or harmful bacteria in the resulting compost. However, because of the use of computerized technology and the need for skilled labor, the drawbacks of this approach are the high capital and operating costs. In general, this method can be categorized into three main types, namely the type of vertical reactor, horizontal reactor, and rotating drum.

Vertical reactor composting techniques have been commonly used to compost sludge together with bulking agents such as wood chips [6]. In this method, organic matter is fed through the top of the reactor and distributed in stages to the bottom of the reactor. Forced aeration is performed on organic waste composted from the bottom of the reactor. The height of this reactor is normally more than 4 m. This height makes it difficult to control the process as high airflow rates are required to be uniformly distributed over the surface of the composted organic matter. However, by increasing the uniformity of airflow distribution in the reactor as well as the collection system, this condition can also be resolved. This method involves changing the direction of airflow from the vertical to a horizontal state between the inflow and the pipes attached to the air blower. The use of a bulking agent is also practiced in this procedure, similar to the aerated static pile process, which attempts to disperse the air within the reactor evenly. In regulating oxygen content and temperature, a uniform mix of feeding material with a bulking agent such as wood chips will minimize this problem. The Earp-Thomas, Frazer-Eweson, Jersey (John Thompson), Ebara Multiplex Paddling Fermenter, and Beccari systems are among several examples of vertical reactors used in composting systems a long time ago [6].

Horizontal reactors use a mechanism that is very similar to the vertical reactor, only the direction of the feed of organic matter differs. In this method, the organic material is fed horizontally, and this will shorten the airflow path as well as the composted material until the end of the process, as shown in Fig. 7.6. Indirectly, this short path can avoid the problem of lack of oxygen as occurs in the vertical reactor. This type of reactor is designed using static, stirred, or even forced aeration. The static method requires a mechanism to feed and remove waste into the reactor. Meanwhile, a turning process is normally implemented by the stirring system to continuously move organic matter within the reactor. The aeration system located at the base of this reactor uses temperature and oxygen content as control variables. This method is capable of decomposing organic matter from nonuniform municipal solid waste. The most common composting systems using the world-famous horizontal reactor process include the Dano biostabilizer, Fairfield-Hardy, Compost-A-Matic, Metro-Waste, and Dynatherm systems [6].

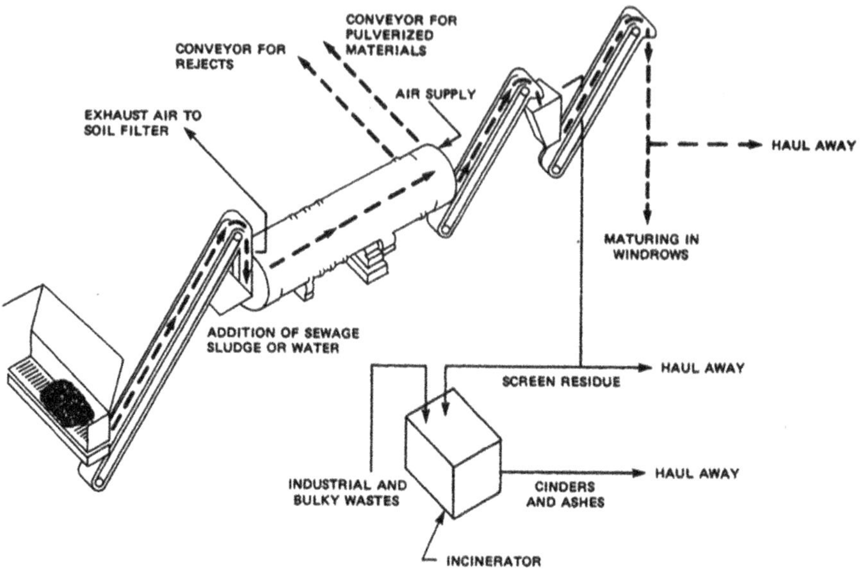

Fig. 7.6 Horizontal in-vessel composting reactor. (Dano biostabilizer; source: USEPA)

Dano, Fermascreen, Eweson, Ruthner, and Voest-Alpine systems are among the municipal solid waste composting systems that use the rotating drum method [6]. In this method, organic matter will be mixed, aerated, and mobilized throughout the system. Composting process occurs faster in the drums. Organic materials that have been turned and crushed in the drum are then removed and allowed to continue to decompose and mature using the windrow or even aerated static piles method. In this method, air will be supplied into the reactor and will mix with organic matter during the turning process. The movement of air is opposite to the direction of movement of organic matter. Figure 7.6 shows a Dano biostabilizer in operation. Figure 7.7 shows a general flow diagram of a mechanical in-vessel composting system. The incoming solid waste can be sewage sludge, agricultural biosolids, ground dead animals, and/or screened refuse. The compost generated at the outlet of the reactor will be cooled by the outside air. Meanwhile, in the central section of the reactor, organic matter will obtain comparatively hot air to allow the aerobic decomposition reaction to proceed. Organic materials applied to the reactor will obtain warmer air to continue the aerobic decomposition process. In a small-scale composting process, the drums can be designed from used materials such as concrete mixing containers, old cement furnaces, or even using a perforated barrel to produce natural ventilation. The Dano system is one of the oldest ways of organic waste composting using the rotating drum process [6]. This system was developed by Dano Ltd. in Denmark in 1933. This reactor is referred to as a biological stabilizer as well. It was built in a slightly tilted position, with a diameter of 2.7–3.6 m in length that reaches 45 m. The volume of the drums is between 1030 and 1800 m^3.

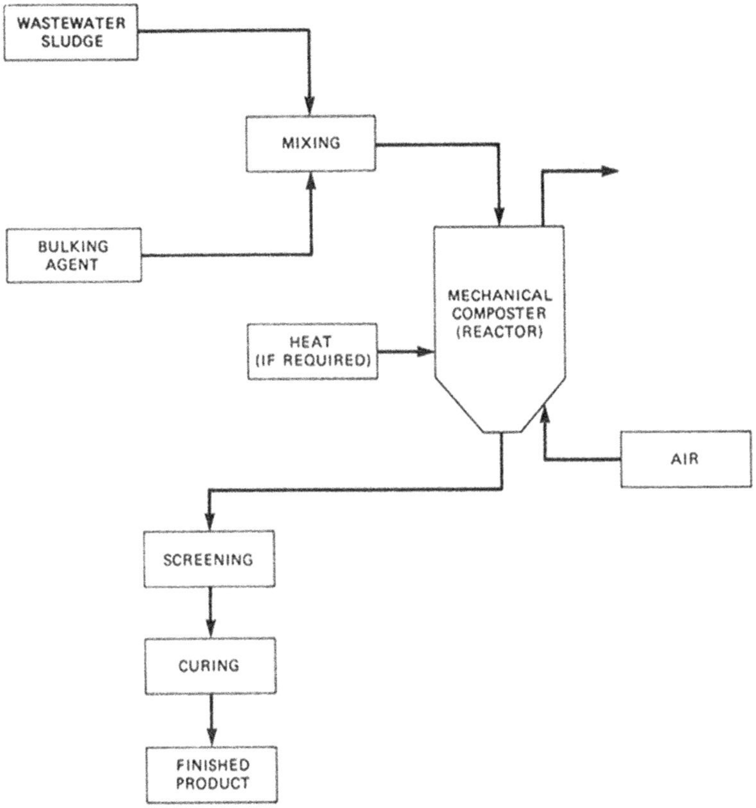

Fig. 7.7 Mechanical in-vessel composting system flow diagram. (Source: USEPA)

The reactor was filled only half full of organic waste, and the rotation rate is 0.1–1.0 rotation per minute (rpm). Usually, the decomposition of organic matter in the drum takes 1–5 days. The addition of water, nutrients, and air is carried out during the composting process. The Dano method has been extended further to countries such as Italy, England, and the United States to treat municipal solid waste to this day.

In general, these three composting methods can be distinguished based on several criteria such as capital cost, operating cost, land requirements, operational control, and so on. Table 7.4 describes the comparison of these three methods.

7.5.3.5 Vermicomposting

Vermi means worm. In a vermicomposting process unit, red worms in bins are used to feed on food scraps, yard trimmings, and other organic matter to create compost. The worms break down this material into high-quality compost called castings. Worm bins are easy to construct and are also available for purchase. One pound of

Table 7.4 The comparison between windrow, aerated static pile, and mechanical method

Criteria	Windrow	Aerated static pile	Mechanical (*in-vessel*)
Capital cost	Low	Low (small scale system) High (large scale system)	High
Operation cost	Low	High (if the organic material to be composted is sludge)	Low (bulking agent is used)
Land requirement	High	High	Low; may increase if the resulting compost needs to be matured in a static pile
Aeration control	Limited; unless forced aeration is performed	Not limited	Not limited
Operation control	Turning frequency, parameter adjustment (moisture content, pH, C:N), the addition of recycled compost	Airflow rate	Airflow rate, dynamic mixing, parameter adjustment (moisture content, pH, C:N), the addition of recycled compost
Sensitivity to climate change	Sensitive; need to install a roof	Can operate in any weather conditions	Can operate in any weather conditions
Odor control	Depending on the organic feed material used; involves a large area	Involves a large area but can still be controlled	Good
Problems in operation	Easily affected by unpredictable weather	Critical air supply control, the potential for short circuit	Potential for the occurrence of short circuits, complex mechanical processes

Source: [4]

mature worms (approximately 800–1000 worms) can eat up to half a pound of organic material per day. The bins can be sized to match the volume of food scraps that will be turned into castings.

It typically takes 3–4 months to produce usable castings. The castings can be used as potting soil. The other byproduct of vermicomposting, known as "worm tea," is used as a high-quality liquid fertilizer for houseplants or gardens.

Vermicomposting or vermiculture using worms to convert organic food wastes to organic fertilizer has long been commercially applied in countries such as India, Indonesia, Korea, Cuba, New Zealand, and the United States for managing their organic solid food wastes. This method is inexpensive and suitable for processing organic waste [64]. The most popular species to produce quality vermicast are tiger worms *Eisenia foetida*, red worms or fecal worms *Lumbricus rubellus*, and

Amynthas gracilis [65]. In this process, organic waste such as vegetable waste, fruit, leaves, grass, sludge, and paper would be decomposed by worms. This waste is constantly turned and mixed with the worms present in the medium. Worms can replicate rapidly; for instance, under ideal conditions, 8 worms can produce 1500 new worms in 6 months. According to Jais [65], 1 kg of worms is capable of decomposing 1 kg of waste in just 24 h and produces approximately 350 g of vermicast, which is one-third of the total organic waste. Vermicast contains nitrates, potassium, calcium, phosphorus, and magnesium in high concentrations [66]. This product is also ideal as a plant fertilizer that is equivalent to the chemical fertilizers currently available on the market. Vermicast also contains high humic acid, which can be used on plants to attain good stability and maturity [66, 67].

In India, the Bhawalkar Earthworm Research Institute (BERI) has developed six composting plants using this method to assist 5000 farmers in their agricultural activities [68]. This method was later used by Indian Aluminum Co. Ltd. to process organic solid waste and sewage from 500 houses. The system developed by BERI is also used by Venkateshwara Hatcheries Ltd. to treat 4 tons of poultry residues (feathers, claws, and other residues) every day. The resulting vermicast is marketed under the brand "Biogold," which is sold at a higher price than the conventionally produced compost price. While in Indonesia, this technology has been developed to help people in low-income areas and have a large population to process organic solid waste. This pilot project involving 60 families around Jakarta uses a 60-litre container filled with 0.5 kg of worms with organic waste generated in their respective homes. After 1–2 months, 20 kg of vermicast would be generated in a container containing 0.5 kg of worms that can be used for agricultural purposes or also for sale to generate family income.

The use of worms in organic waste composting activities has contributed to some benefits. These include [48, 69] (1) reducing labor, (2) effective odor control, (3) reducing particle size, (4) increasing the nitrogen content in compost, (5) increasing nutrient and mineral content, (6) increasing control over pathogen growth, and (7) getting rid of old bacteria found in compost and replacing them with new bacteria. However, some drawbacks in the use of this composting technique are still present: (1) it is a slow and time-consuming process, (2) it is difficult to isolate worms and vermicast after the composting process, (3) the presence of fruit flies and predatory animals such as centipedes and lizards interferes throughout the composting process, (4) problems exist to maintain a temperature between 13 and 25 °C to increase the decomposition activity of organic matter by worms, (5) heavy metals accumulate in the body tissues of the worm if the organic waste is not separated first from the metallic waste, and (6) the presence of pathogens in the vermicast results from processes occurring at inappropriate temperatures.

In a vermicomposting process (vermiculture) unit, red worms in bins are used to feed on food scraps, yard trimmings, and other organic matter to create compost. The worms break down this material into high-quality compost called castings. One pound of mature worms (approximately 800–1000 worms) can eat up to half a pound of organic material per day. Food scraps, paper, and yard trimmings such as grass and plants can be easily and cost-effectively processed by vermicomposting.

The process is ideal for apartment dwellers or small offices. Schools can use vermiculture to teach children conservation and recycling. It is important to keep the worms alive and healthy by providing the proper conditions and sufficient food. Worms are sensitive to changes in climate, so extreme temperatures and direct sunlight are not healthy for the worms. The best temperatures for vermicomposting range from 55 to 77 °F (12.8 to 25 °C). In hot, arid areas, the bin should be placed under the shade. Vermicomposting may be operated indoors to avoid some operational problems.

7.5.3.6 Thermophilic Composting

Thermophilic composting has been commonly practiced for the treatment of solid organic waste. Typically, the temperature of this type of composting exceeds 50 °C and can reach up to 70 °C [70, 71]. However, these temperature changes depend on the diversity of microbial species and the rate of decomposition of organic matter. For public health, the temperature rise is necessary as all pathogens in the compost will be killed during the composting process. In this method, in addition to changes in the microbial community, physical and chemical parameters such as moisture content, oxygen levels, and carbon dioxide also undergo changes throughout the molecular decomposition process of complex organic matter to simpler molecules. According to the study of Mohaibes et al. [72], in large-scale composting plants, all pathogens can be destroyed within 40 days at a temperature of 45 °C, while increasing the temperature to 50 °C can lessen the composting period to 3 days. Meanwhile, it is estimated that at a temperature of 55 °C, the composting period can be shortened to 15 h, 60 °C to 2 h, and 70 °C to 7 min.

Typically, this type of composting process involves bacteria or microorganisms that can live in the thermophilic temperature range of between 50 and 70 °C. At 50 °C, fungi, bacteria, and actinomycetes are the most abundant thermophilic microorganisms present in the composting mixture. A rise in temperature to 65 °C allows the population of fungi to decline, but the bacteria and actinomycetes population continues to increase [5]. *Chaetomium thermophile*, *Humicola lanuginosa*, *Malbranchea pulchella* var. *sulfurea*, and *Talaromyces duponite* are among the group of fungi present in the thermophilic temperature phase. Meanwhile, the most common group of bacteria present is *Bacillus* spp. and *Clostridium* spp. These thermophilic microorganisms are generally not dangerous to humans and can also live under aerobic and anaerobic circumstances. In an exothermic reaction, thermophilic microorganisms will use oxygen for the respiration process and then produce heat. The resulting heat will increase the temperature beyond the mesophilic range, at which point mesophilic microorganisms become inactive, and thermophilic microorganisms will take place to continue the decomposition of organic matter.

Elango et al. [73] have conducted a thermophilic composting process on the MSW collected from the Perungudi yard in Chennai metropolitan city. The composting process is carried out inside a thermophilic bioreactor having dimensions of $1 \times 1 \times 1$ m^3, where all four sides of this bioreactor are covered with wire mesh.

Organic waste incorporated into the bioreactor was shredded to obtain a size between 5 and 10 cm. In this process, thermophilic microorganisms consisting of *Bacillus megaterium* and *Pseudomonas fluorescens* were incorporated into the bioreactor to decompose organic matter during the composting process. They found that thermophilic composting could achieve a 70% volume reduction in MSW organic fraction within 40 days. Also, the physicochemical analysis reveals that further composting on the final compost from the thermophilic bioreactor produces good humus that can enhance the physical properties of the soil, thus supplying plants with essential nutrients. Wadkar et al. [74] have conducted the thermophilic composting process on MSW using a cylindrical wire mesh reactor. The organic waste (including dry vegetable waste) used was obtained from the MSW ramp at Koregaon Park, Pune, India. In their study, three phases were involved in the thermophilic composting process. In the first phase, the process of feeding the organic waste and dried leaves into the reactor is conducted alternately with a thickness of 10 cm, respectively. Bacteria culture (mesophilic bacteria) is then sprayed onto the surface of organic waste. When the temperature rises to 35–40 °C, there is an increase in the population of thermophilic bacteria. This growth will continue until the temperature reaches 55–60 °C. In the second phase, the mixing process will be conducted every 4 days, accompanied by water spraying on the composted organic waste to maintain its moisture content. The mixing process plays a role in spreading thermophilic bacteria throughout the organic waste inside the reactor. In the third phase, the size of waste particles decreases and falls out of the wire mesh, which will be collected in the tub at the bottom of the reactor. Wadkar and his co-workers found that the addition of bacteria culture accelerated the composting process, where the maturation period was 42 days. Also, the resulting compost is suitable for ornamental plants such as azalea, gardenia, camellias, etc.

The use of this approach of treating organic solid waste has many advantages. The following was included [72]: (1) a simple and potentially destructive process of pathogens, (2) exothermic process in which this process is capable of regenerating heat, (3) cheap air supply, (4) the process of converting organic matter to beneficial materials such as compost, (5) processes that can reduce the risk of environmental pollution and are cleaner, and 6) moderate per capita investment. However, there are some disadvantages of using this method [72]: (1) requiring continuous maintenance, (2) requiring high energy, (3) release of ammonia gas and nitrous oxide, and (4) forming aerosols in the aeration system.

7.5.3.7 Two-Stage In-Bin Composting System

A typical two-stage in-bin composting system consists of two stages of aerobic composting treatments in series, mainly for composting treatment of dead animals. The first stage, also called the primary composter, is made up of equally sized bins in which the dead animals and amendments are initially added and allowed to the first aerobic compost. Dead animal composting must reach a temperature in excess of 130 °F (54.4 °C) to destroy pathogens and break down organic matter at the same time. The mixture in the primary composter bins is moved from the first-stage to the second-stage aerobic composter when the compost temperature begins to decline. The second stage can also consist of a number of bins, but it is most often one bin or concrete area or alley that allows compost to be stacked for the second-stage composting with a volume equal to or greater than the sum of the first stage bins. The details of the two-stage in-bin composting systems and their design features are introduced later in Sect. 7.13.7.

7.5.4 Operational Mode

The municipal solid waste composting can also be classified based on the operational mode. Batch and continuous flow are two types of operational modes that are mostly used in the composting process. Batch composting involves the process of mixing all the organic waste to be composted at once and allowing it to decompose without adding any additional organic waste (other than water) so that all the organic waste is converted into compost. Batch composting is ideal for small populations in rural areas with sufficient land area and low cost involved [21]. The main advantage of this mode is that all the materials in the system will be completely composted around the same time, and there will be no contamination from new organic waste. This technique is particularly helpful when working with large amounts of organic waste (either at a particular time or ongoing basis). However, there is a drawback with the use of batch composting that it is first important to store organic waste delivered to the composting site before the amount is adequate to operate the next batch. Besides, the composting operator must pay attention to the C:N ratio and the overall properties of the organic waste mixture. The mixing process needs to be performed more carefully to achieve the optimal effects.

In a continuous flow composting system, composting materials are continuously fed to the reactor, and the resulting compost is continuously discharged as well. In larger municipal areas or where there is abundant and continuous solid waste input or where the process must adhere to the schedule, a continuous flow system is often the main choice compared to the batch method for municipal solid waste composting process. Continuous-flow processes are typically more heavily mechanized, with more sophisticated designs, engineering, and planning, as well as stricter system control and more highly trained personnel [21].

7.5.5 Raw Materials

Based on the raw material used, the composting process may also be classified. This involves municipal solid waste (MSW) composting, sludge composting, manure composting, yard waste composting, combined MSW-sludge composting, and so on.

In sludge composting, sludge is usually mixed with bulking agents such as sawdust, wood chips, shrub clippings, tree branches, paper wastes, agricultural waste, etc. No addition of other organic matter to the composting system is performed. According to Conghos et al. [75], agricultural waste (containing high lignin) is difficult to manage and dispose of because it is produced in bulk quantities and has low commercial values. Therefore, these materials have the potential to be bulking agents that serve to balance the moisture content of the sludge and increase the porosity to facilitate the airflow in the composted sludge. Bulking agents may also be used to balance the C:N ratio of sludge to supply extra carbon to increase microbial growth [76, 77]. Besides, the use of bulking agents in sludge composting can also prevent excessive compaction from occurring between composting materials [78]. Among the sludge types that are commonly used in the composting process are raw, secondary, dewatered, and digested sludge. Ucaroglu and Alkan [79] have investigated the composting of wastewater treatment sludge using different bulking agents resulting from agricultural activities (wheat straw, plane leaf, corncob, and sunflower stalk) with a ratio of 60%:40% (sludge:bulking agent). They found that the mixture of sludge and corncob recorded the highest organic matter degradation (37.6%), loss of dry matter (29.6%), and temperature (64 °C). They also found that a mixture of sludge and sunflower stalk also gave encouraging results. The use of bulking agents in the composting of wastewater sludge can solve problems related to process efficiency, economy, and agricultural waste disposal. The use of bulking agents in the composting process for the management of wastewater treatment sludge is an important concern in terms of process quality, economy, and disposal of agricultural waste. One of the issues that frequently occur when the composting process is carried out on anaerobically digested sludge is the development of unpleasant odors. Wood chips and forced ventilation may be added to the composting pile to solve this issue. According to Goldstein [80] and Komilis et al. [59], the odor produced by the composting process is associated with the release of volatile organic compounds (VOCs), for example, terpenes, alcohols, ketones, sulfur-containing compounds, and amines, as well as ammonia (NH_3). Maulini-Duran et al. [81] studied the emissions of the VOCs and NH_3 during the composting process for sludge produced in wastewater treatment plants, i.e., anaerobically digested sludge (ADS) and raw sludge (RS). The sludge used is a raw sludge obtained from a wastewater treatment plant located in Manresa (Barcelona, Spain) while anaerobically digested sludge is taken from a wastewater treatment plant located in Sabadell (Barcelona, Spain). The ratio of the wood chips (bulking agent) to sludge is 1:3. The composting process is then performed in commercial cylinder reactors with an operating volume of 50 L. This study revealed that NH_3 and VOCs emitted were higher during the RS composting process (19.37 and 0.21 kg Mg^{-1} sludge, respectively) compared to ADS composting (0.16 and 0.04 kg Mg^{-1} sludge). Significant differences were found in the VOC compositions emitted in ADS and RS composting. Rincon et al. [82] identified and analyzed the patterns of odor generation and odorant composition throughout the different operational steps of ADS composting at a pilot scale. The anaerobically digested sewage sludge (ADS) used in this study was collected from a wastewater treatment plant (WWTP) located in Britanny,

France. The wood chips were employed as a bulking agent. From their findings, the odor emission rate (OER) recorded is between 30 and 317 OU_E h^{-1} kg^{-1} ADS in the early stage of the process. Dimethyl disulfide, dimethyl sulfide, and methanethiol are the predominant odors at this early stage of operation. For the middle and later active phase, second turning, and curing phase, odor potential and composition changed, where OER fluctuated from 0.18 to 12.6 OU_E h^{-1} kg^{-1} ADS. Hydrogen sulfide showed the most important odor contribution. The most concentrated air contaminants were ammonia and volatile sulfur compounds (VSCs), comprising 55.5 and 20.6%, respectively, of the total mass released.

According to Kashmanian and Rynk [83], manure composting is one of the methods of animal waste management that is gaining attention in efforts to reduce environmental effects due to manure production. Composting benefits the environment because the nutrients contained in animal manure are converted to a more stable form. Manure composting starts from ancient times until the mid-twentieth century. It is becoming less popular due to increasing mechanization and the introduction of chemical fertilizers. However, the tendency for manure composting increased again in the 1980s when municipal solid waste composting was carried out to overcome the problem of the shortage of space in landfills [84]. The concept of the 3Rs (reduce, reuse, recycle) has been applied where composting is part of the concept in raising awareness of the environment in modern society. In general, composting is a method for converting organic waste (animal manure) into a value-added product that can be recycled to the ground. Manure composting uses two methods, namely windrow and in-vessel or enclosed composting. In the windrow method, wet material (manure) is placed into a long narrow space to allow the manure to be dried by the wind. Windrow is built near the livestock facility on an open-air earthen pad. Meanwhile, in-vessel composting refers to the methods of placing manure inside a building, container, or vessel and relying on mechanical turning and forced aeration.

Composting animal manures is an effective way to kill pathogens, parasites, weed seeds, pesticide residues, and malodors. Larney et al. [85] found that there was a rapid decline in the level of *E. coli* in the first 7 days of composting with the removal of more than 99.95%, although the temperature is between 34 and 42 °C. After 1 month, *E. coli* can no longer be detected by the culturing method. McGinn et al. [86] stated that an issue that is often raised about fresh manure is the production of odors during land application. One of the benefits of compost for land application, when compared to fresh manure, is that it is odorless and smells like soil. Rynk [87] reported that the odor resulting from the composting process is due to the malodorous feedstocks, ammonia volatilization, and the formation of anaerobic conditions in the composting pile. A mixture of composting material (manure and bulking agent) that is too wet and lacks porosity contributes to anaerobic conditions. The turning mechanism can help release odors trapped in the composting pile.

Yard waste or green waste composting is the best alternative to recycle yard waste. This is because the resulting compost is an organic substrate that can be reintroduced into the economic system through compost marketing, minimizing disposal problems while helping to overcome greenhouse gas emissions [88]. Furthermore, yard waste often indicates a low content of micropollutants. This favors the manufacturing of compost with suitable properties that can conform to quality standards and limits the use of compost in organic farming systems [89]. Yard waste consists of tree trunks and bark, pruning of young trees and shrubs, dead and green leaves, grass clippings, and soil, which usually originated from municipal parks, gardens, reserve areas, and domestic residences [88]. According to the US Environmental Protection Agency [90], yard waste contributes the highest fraction of municipal solid waste along with other wastes such as food waste. Zhang and Sun [91] reported that yard waste decomposition rates are low because of the relatively high lignocellulosic content in it. Yard waste contains organic compounds that are recalcitrant to biodegradation, such as cellulose and hemicellulose. The decomposition process becomes slow, requires large areas for treatment, produces malodorous gases, and produces low-quality products unsuitable for commercial use; these are among the effects if composting is not handled and handled properly (e.g., oxygen supply, moisture content, nutrient balance) [88]. Lopez et al. [92] reported that the composition of the yard waste produced is constantly changing. This depends on the source, season, and method of collection. Indirectly, variations in yard waste composition will affect its decomposition rate.

Factors such as capital and maintenance costs, land availability, and operational sophistication are the basis for choosing a particular composting process [93]. Windrow, aerated static pile, and in-vessel techniques are among the yard waste composting techniques that have been documented. Windrows are a commonly used method for the composting of yard waste. In particular, due to its low capital expense, relatively easy operation, and high-quality compost production, windrow composting has many advantages [94]. The composting of yard waste using the windrow method takes between 3 months to 1 year to complete [95]. Since 1988, a large-scale yard waste composting plant has been working on a 40-acre site in the town of Islip, New York [96]. Every year, about 60,000 tons of grass, leaves, and wood debris are gathered from city residents, municipal authorities, and commercial landscapers and transported to the facility by packer trucks. For windrow formation, 25 acres of the facility have been sited. The nature of the feedstock material and the time of year composting takes place determine the size of the windrow formed. The windrow sizes formed for leaves and grass residues are 12 ft high by 26 ft wide and 6 ft high by 14 ft wide, respectively. Windrow composting for leaves requires at least 4 months to complete the decomposition process before being left to mature in curing piles. Meanwhile, the grass composting process takes between 1.5 and 2 months, and another 1–1.5 months is needed to mature the resulting compost. By strictly controlling the decomposition rate and windrow size, this facility can ensure continuous processing of fresh material delivered to the site. The composting process is always kept in aerobic conditions. For this, the turning process is carried out using a rotary drum turning machine for smaller windrows, while the front-end loader is used to turn the composting material piled up in large windrows. The windrow size, feedstock composition, stage of decomposition, and moisture content are among the factors that determine the frequency of the turning mechanism and further maintain the aerobic condition of composting piles. When the compost has matured, the screening process is performed on the final product to get rid of woodchips or other unwanted material such as plastic. The available finished compost is distributed free of charge to Islip residents, while the rest is sold to landscape contractors, turf growers, topsoil suppliers, and nurseries for $6 per yard.

Aerated static piles are a comparatively high-technology solution, often referred to as forced aeration windrows, that can be used to compost yard waste. This technique is successful where space is scarce, and the method of composting must be completed within a year [96]. High-technology approaches in which composting is carried out inside a fully sealed system are in-vessel methods. With this method, all vital environmental conditions are mechanically controlled, and they are also completely automated for most in-vessel systems. Since maintaining this degree of control is costly, this method is rarely used to compost yard waste.

Co-composting of municipal solid waste (MSW) and sludge is practiced to balance moisture content and nutrients (C:N ratio). The optimal ratio of C:N and moisture content for the successful composting process are in the range of 25–30 [47] and 50–60% [42], respectively. According to Kumar et al. [97], the composting process synergistically can be enhanced when the process of mixing two or more organic waste is performed. In this MSW-sludge composting, it was found that MSW contains more biodegradable organic matter, low moisture content, and higher biomass porosity that can provide better conditions for aerobic composting to occur compared to sludge [98]. Meanwhile, sludge has a high nutrient content and microbes that can increase microbial activity and, in turn, shorten the composting period [99]. High water content (80%), as well as low organic content in sludge, is a technical challenge in sludge composting. Sludge containing 95% moisture content was mixed with MSW, which had 25% moisture content to produce a moisture content of 60% for composting mass. Heavy metals are commonly found in sludge, which may impair the composting process. The heavy metal content in sludge limits its direct use to crops and even poses a problem to the environment if this waste is disposed of. The use of sludge to increase the organic matter (OM) content of the soil will be limited because the heavy metal content in it can enter the food chain or be carried to an aquifer [100].

There are several composting studies involving co-composting of MSW-sludge that have been conducted by the researchers. Zhang et al. [101] have studied the composting performance of sewage sludge with MSW at different proportions. To enhance the co-composting process, 15% (of total wet weight) cornstalk has been added to the composting mixture. The composting process is carried out in a 60-L stainless steel composting reactor with a height of 0.6 m and an inner diameter of 0.36 m. They reported that higher sewage sludge proportions could initiate composting quickly, while increasing the percentage of MSW enhanced organic content for biodegradation. However, excessive MSW requires a longer composting period to ensure the desired maturity and quality of compost. An increase between 55% and 85% of MSW will improve the quality of compost in terms of compost salinity and plant toxicity. Lu et al. [102] studied the characteristics of the co-composting of municipal solid waste (MSW) and sewage sludge (SS) involving four main influencing factors (aeration pattern, the proportion of MSW and SS, aeration rate, and mature compost recycling). Lu and his colleagues observed that a constant aeration pattern was preferable to an intermittent aeration pattern during composting because the latter slowed the composting process. The mixing ratio of 3:1 (MSW:SS) is the best mixture because it can maintain the highest temperature for a long time, achieve the fastest degradation of organic matter, and record the highest N content in the final compost. The best aeration rate recorded is 0.5 L/min kg VS, where the initial aeration process occurs rapidly while maintaining a moderate moisture content for microorganisms. The structure and moisture content of composting materials is also enhanced when the mature compost formed is recycled as a bulking agent mixed with new composting materials. Zorpas et al. [100] investigated the physicochemical characteristics of the compost produced from co-composting of dewatered anaerobically stabilized primary sewage sludge (DASPSS), organic fraction of municipal solid waste (OFMSW), and zeolite (clinoptilolite). The results of their investigation found that compost produced through the co-composting of DASPSS (80% w/w), OFMSW (60% and 40%, respectively), and clinoptilolite (20% w/w) produced a high-quality final product than compost produced from DASPSS. The compost produced through the co-composting process also contains total humic acid and organic matter of high concentration compared to DASPSS compost. In addition, final co-composting products recorded low concentrations of heavy metals. The function of zeolite (clinoptilolite) during the co-composting process is to uptake heavy metals, where this material helps in cation exchange.

7.5.6 Operating Methods

The composting process can also be classified based on operating methods, namely conventional composting and nonconventional composting. In recent decades, the conventional process of composting, such as windrow, aerated static pile, and in-vessel composting, has been well practiced worldwide [103]. In general, windrow composting refers to an outdoor composting scheme that depends heavily on mechanical aeration. The organic waste is mixed and then placed in long, narrow piles. The compost turner is used to turn the composting materials. Of these three methods, it is the least sophisticated and often becomes the great candidate when composting is required to handle a high volume of organic waste. However, windrow composting takes a long time to reach maturity as well as a large area of land for windrow construction. Aerated static pile (ASP) composting involves forcing or pulling air from the environment through a composting pile. The method of pile formation is the same as the windrow system, but no mechanical turning is done. This method is found to be suitable for composting municipal sewage sludge. To increase porosity and airflow in the composting pile, bulking agents such as wood chips and sawdust are added. Typically, ASP will be combined with a windrow or other system during the maturation phase, where composting materials will continue to undergo decomposition to become finer and then be removed through a screening process. The in-vessel composting takes place in a closed container. The in-vessel composting is more efficient, does not require large space, and has short composting period and good operating control system, and these are the advantages of this system compared to windrow and ASP. However, the in-vessel method requires a higher cost than the other two methods. The three systems differ in cost, labor, energy, the release of greenhouse gases, and composting duration. Differences in criteria for these three methods related to capital cost, operating cost, land requirements, operational control, and so on have been discussed previously in Sect. 7.5.3.2. From an economic point of view, the windrow system is the best option if the composting process is conducted on a large scale and has sufficient land area. From an environmental point of view, more improvements need to be made on windrow composting as this method contributes to the release of greenhouse gases during the turning process. Among the improvements that can be made are increasing the use of bulking agents [104], changing the frequency of turning according to temperature changes [105], and covering the composting piles with mature compost [106].

Nonconventional composting is a composting method other than the three methods described earlier. The earliest nonconventional method introduced was hydropulping [21, 107], which had successfully composted MSW. This facility is located in Altoona, Pennsylvania. In this process, a large amount of water or wastewater is mixed with solid material and then mechanically diminished with a pulper to a liquid slurry. The composting process is carried out when the slurry has been dewatered; then, the resulting pulp is fed into the digester. The decomposition process usually occurs faster and produces quality compost that is suitable for use as a soil fertilizer.

Recently, a nonconventional method known as two-stage composting began to gain attention among researchers. The two-stage composting system is a technique by which two separate systems are integrated with a single composting process to increase the consistency of the finished product, the reliability of the process, and the environmental effect of the conventional composting process. Combining two composting technologies (e.g., a combination of in-vessel and windrow/ASP; a combination of vermicomposting and other conventional methods) or other methods involving several mechanical treatments such as waste pretreatment, mechanical biological treatment, or anaerobic digestion before the composting process is among several two-stage systems that can be considered.

Kulikowska and Klimiuk [108] have conducted a composting process using a two-stage composting method on sewage sludge. In the first stage, sewage sludge is composted in an aerated bioreactor (1 m^3) and followed by the second method, i.e., turned windrow (0.8 m^3). The bioreactor used can process around 500 kg of composting materials. Rape straw and grass in different proportions are used as amendments in this composting process. From their investigation, Kulikowska and her co-worker found that the temperature and decomposition of organic matter during composting in bioreactors as well as the formation of humic substances such as humic acid (HA) during the maturation of compost in windrow are influenced by feedstock composition. Total HA content shows an increase based on first kinetic order, while labile HA content is constant and does not exceed 12% of total HA. During composting, it was found that temperature was a major factor influencing the polymerization of fulvic acid to HA and only increased when thermophilic conditions were obtained. Zhang et al. [109] conducted a study to determine the physical and chemical properties of yard waste compost produced through two-stage composting with the addition of brown sugar (at 0, 0.5, and 1%) and calcium superphosphate ($Ca(H_2PO_4)_2.H_2O$) (at 0, 3, and 6%) during the second stage. The yard wastes used in this study consisted of fallen leaves and cuttings of branches around Beijing in the spring of 2011 during the gardens and parks maintenance process. During the first stage, yard waste particles, urea, microbial inoculum, and water are added to each of the three digester cells (noncovered containers with a dimension of 48 m long, 4 m wide, and 2.5 m high). This digester has an automatic turning and watering system where it operates daily. Meanwhile, in the second stage, the compost is removed from the digester cells and arranged in a windrow (with a dimension of 2 m long, 1.5 m wide, and 1 m high). The aeration process is done by turning the compost every 3 days, while the moisture content of the compost (60–70%) is controlled by watering the windrow during the turning process. The results obtained showed that all compost showed two peaks of fermentation temperature within 30 days, either with or without the addition of brown sugar and calcium superphosphate. When compared to conventional methods, this method can increase the duration of fermentation of high temperature longer, minimize the maturity period, and reduce costs. High-quality compost concerning C:N ratio, pH, organic matter content, electrical conductivity, particle size distribution, and other characteristics is produced when as much as 0.5% brown sugar and 6% calcium superphosphate are added to the composting pile. In conclusion, nonconventional methods such as two-stage composting can shorten the composting period as well as minimize extensive land use, greenhouse gas emissions, and labor. Besides, this method can also overcome transportation problems for windrow composting and save costs and power consumption [103]. The quality of compost produced through the two-stage composting method is better and safer for use in agriculture, landscaping, and so on.

7.6 Design Approaches

7.6.1 General Approach

In general, raw solid waste, as well as the type of preprocessing or postprocessing applied to waste, affects the quality of the resulting compost. Quality control on solid waste transported to the composting facility must be performed first before the composting process begins. In solid waste management systems, an effective solid waste collection process is when control over the solid waste is implemented starting at the consumer level. The segregation process is done at the waste collection site by separating the compostable solid waste from other noncompostable solid waste such as construction and demolition waste, tree trunks, large metal objects, and others. If large noncompostable solid waste is not segregated first, there would be the issue of damage to equipment such as the grinder and shredder [21].

The composition of solid waste generated in an area depends on several factors. Among them, the solid waste produced in highly urbanized areas is different from the solid waste produced in suburban and rural areas. Besides, the geographical position and weather changes, as well as solid waste input from industrial and commercial areas, also influence the composition of the solid waste generated. Therefore, to assess the solid waste generation rate in an area where the generation rate is different among different areas, data collection needs to be conducted.

According to Tchobanoglous et al. [44] and Cardenas and Wang [21], the compostable materials in the municipal solid waste consist of water-soluble substances (sugar, starch, amino acids, etc.), hemicellulose, cellulose, esters from alcohol and higher fatty acids (fat, oil, and wax), lignin, lignocellulose, and protein. Meanwhile, rubber, textiles, leather components, and plastics are part of noncompostable materials. Also, the working period affects the solid waste produced. For example, a plant that operates 5 days a week needs to do careful planning by increasing solid waste acceptance at a rate of 1.4 times compared to a plant that operates 7 days a week. Many pretreatments on solid waste such as sorting, grinding, shredding need to be adapted to this operating schedule as most composting facilities often operate according to the solid waste collection schedule. The grinder outputs, movement of waste materials, and waste storage need to be well planned to ensure that the operation is on track. For example, for composting plants that operate in a continuous flow, storage facilities are essential to store solid waste throughout the weekend or even during long holidays. Precautions should be taken to ensure that solid waste stored in storage facilities does not undergo premature decomposition or produce odors to the surrounding area. There are five major process operations in municipal solid waste composting, as shown in Fig. 7.8. This process consists of pretreatment of the solid waste received followed by digestion (composting) of solid waste, curing, finishing, or upgrading, and finally storing the resulting compost.

7.6.2 Pretreatment

According to Cardenas and Wang [21], an ideal processing system for composting can remove all glass, aluminum, ferrous and nonferrous metals, plastics, leather, rubber, and a majority of textiles. However, in practice, this condition can never be achieved because all this solid waste cannot decompose biologically and contribute little to the quality of the resulting compost. The Varro process (Fig. 7.9) shows at least one process performed before the composting process in which MSW is shredded and then added to the digester. In this process, ferrous metal is removed from the MSW. However, before the composting process is performed, pretreatment or sample preparation is identified as an important step that must be considered.

Receiving, sorting, size reduction, and ferrous removal (if it is to be performed), as well as storage and modification of waste properties (e.g., the addition of

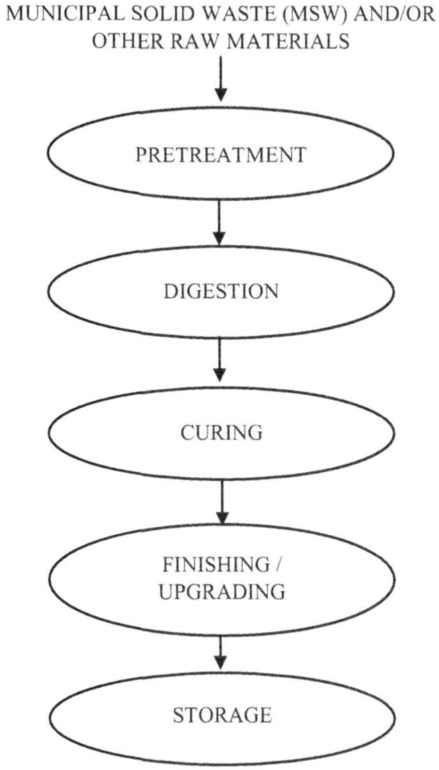

Fig. 7.8 Major process operation in composting process [21]

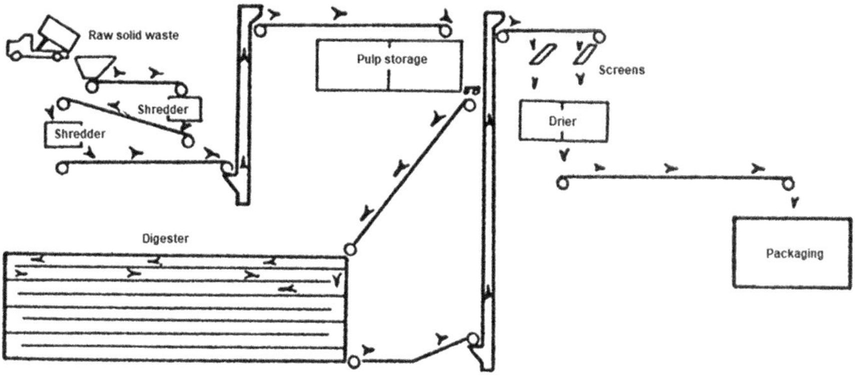

Fig. 7.9 Varro process flow diagram [21]

moisture, C:N ratio, and nutrients), are essential steps in the pretreatment of MSW for composting. Pretreatment strategies are typically devised in conjunction with the digestion process or during the microbiological stabilization stage. Size reduction through grinding, shredding, rasping, or chopping should be performed to produce smaller solid waste particle size and uniform density. Various types of mills such as hammer mills and chain mills are used, including wet pulpers. A hammer mill comprising a high-speed swing-type hammer is the most widely used equipment for chopping or grinding waste, where it pushes waste through grates and subsequently decreases the waste particle size. The use of hammer mills produces noise and vibration. Raspers are preferred in Europe for grinding solid waste because they consume less energy. Wet pulping is done in at least one compost facility in the United States. Water is added to the incoming solid waste to facilitate the grinding process, and the shredding process is done when the solid waste is still wet. Before proceeding with the next processing step, the material needs to be dewatered, and the resulting wastewater needs to be treated [21].

According to Cardenas and Wang [21], the most important pretreatment operation before the composting process is grinding or shredding. The grinding process aims to facilitate the management of solid waste while providing optimal surface area for microbial action during the biological decomposition process. Particle size less than 2.5 cm or smaller is required if rapid composting is carried out. Apart from being able to increase the exposed surface available for microbiological activities, the reduction in particle size also helps in the process of moisture addition and nutrient supply. To protect equipment such as grinders and shredders, the sorting process involving ferrous removal is the best step and should be consistent and in line with the overall solid waste recovery program.

7.6.3 Digestion

Batch and continuous flow are two methods used in the digestion or microbiological stabilization in composting processes [21]. The same nutrient and overall moisture considerations apply to both of these processes. For optimum and continuous production, optimal activity in continuous-flow operations is more reliant on nutrients, moisture, and environmental control.

Digestion in the composting process can be carried out via windrows, pits, trenches, cells, tanks, multi-story or multi-deck towers or buildings, drums, or bins. Some of these processes merge more than one type of digester. Special digester (e.g., batch or continuous flow) is usually merged with storage area for the maturation process either by windrows or bins [21].

7.6.3.1 Batch Operations

The windrow method is the most popular batch technique for MSW composting [21]. The feedstocks are stacked into piles known as windrows in the windrow system after grinding. To speed up the composting process, nutrients, as well as moisture additives, are supplied to the composting pile. The biological activity that takes place in the compost pile will increase the pile temperature between 65 and 80 °C and last for a certain period until the volume of the composting pile gradually decreases. To ensure that composting piles are not anaerobic and uniform aeration can be achieved, windrows will be mechanically turned periodically. Practically, the activity of turning, agitating, or pushing air through or into the composting pile is performed for aeration purposes. Solid waste that goes through a composting process will undergo a change from its appearance, where large and recognizable solid waste is transformed into compost or humus that is rich in nutrients that are good for the soil and darker in color. The drop in temperature in the middle part of the windrow to the atmospheric level will occur when all available organic matter is degraded. The resulting product is considered stable because no more biological degradation occurs.

Minimal sample preparation involving sorting out the huge materials, primary grinding, and moisture adjustment was performed during batch preparations. The quantity of solid waste, bulk density, and process retention time are among the factors considered to estimate land requirements. The amount of nitrogen and phosphorus applied should be dependent on the amount of accessible carbon measured at a ratio of approximately 30–35 to 1 (C:N) when nutrients are to be used; the amount of phosphorus should be about one-fifth of the nitrogen level. Sludge, sewage, or industrial wastes are also added during the composting process. Sludge and sewage serve to provide moisture and nutrients to the composting pile. However, the addition of these materials into the composting system has the potential to contribute to pathogenic microorganisms, heavy metals, and other polluting organic compounds. Accordingly, the compost produced should be evaluated first before use. Generally, the available nitrogen and phosphorus from such sources, along with the addition of moisture, should be measured. Moisture content must be 35–65% for windrow operation. In reality, however, little attempt is made in windrow operations to tightly monitor the process. Batch processes often involve the use of machines in recovery operations as well as occasional turning of the feedstocks. Screens are often used to organize or distinguish the material or finished product [21]. Usually, for "curing," wide spaces are usually needed, which will be addressed in more detail later.

7.6.3.2 Continuous-Flow Operations

Continuous-flow, semi-continuous, and even intermittent composting process operations can be used [21]. In general, ground or shredded MSW is retained and continuously applied into the digester or composting system after going through the pretreatment process. Compared to batch operations, continuous-flow composting processes use shorter detention periods to achieve stability. This operation can be performed in either open pits or enclosed digesters. Enclosed digesters have a number of benefits, including better control of the end product, with shorter detention times and more control over the final product [21].

Open-Pit Digestions

Open-pit digestion is usually performed in an open space, either in a tank or an open pit. In this system, organic solid waste that has been ground or sorted is added to the pit either continuously or semi-continuously. Mechanical equipment is used to turn and aerate the composted solid waste regularly. There is a rise in temperature in the composting pile and a drop in volume after a certain time, as is the windrow method. The use of additives was introduced to speed up the composting process. According to Olds [110], sewage sludge was applied to the composting system at the Metro plant located in Houston but has now stopped operations. This material is added to the digester through conveyers, where the tracked mixer will mix the composting mixture at least once a day. The digester detention time is about 5 days. After the decomposition process in the digester, the composting materials will undergo an extra curing period during which biological activity decreases. The composting materials become more stable and uniform and look like humus or soil.

Enclosed Digesters

Enclosed digesters are normally highly mechanized but provide the benefit of strict control of the environment. Mixing or agitating of the composting materials must be done in an open-pit digester. For closed-digester construction, a broad range of configurations has been used. Figure 7.10 shows some types of configurations that are

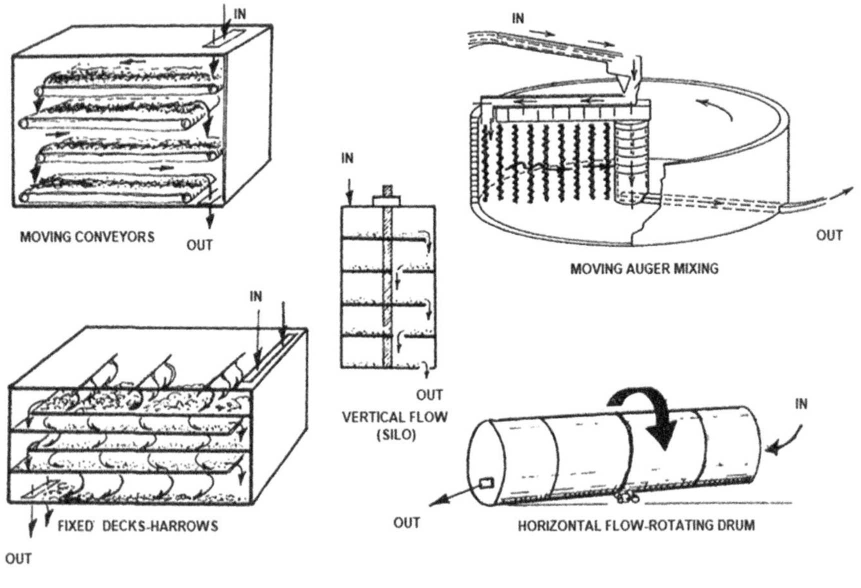

Fig. 7.10 Compost digesters, closed configurations [21]

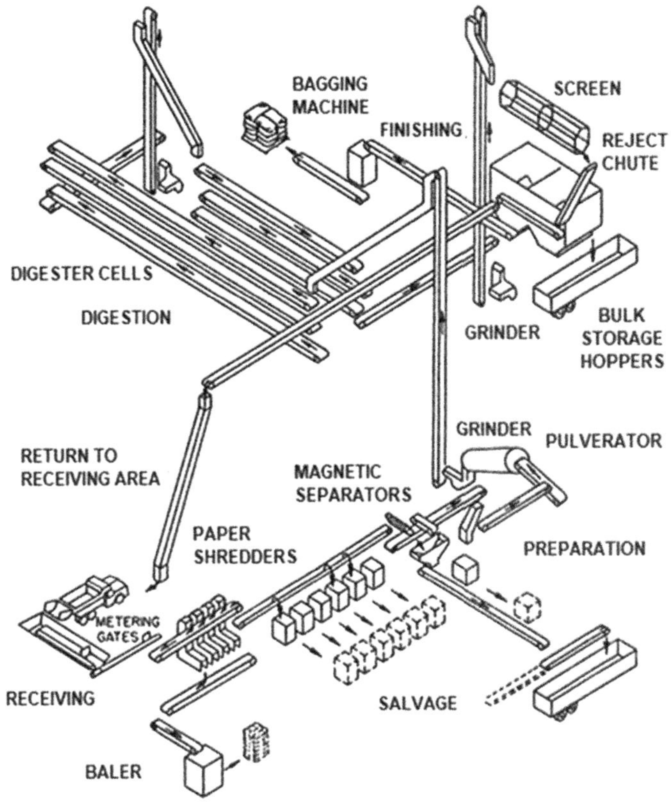

Fig. 7.11 Schematic diagram of the Naturizer system [21]

often used in the composting process. The process of agitating and mixing using mechanical methods, such as screws, plows, rakes, bucket lifts, etc., is done to ensure uniform airflow inside the digester.

According to Drobny et al. [111], as in the Naturizer process, the composting materials can be agitated using conveyors, as can be seen in Fig. 7.11. In this process, the conveyors are arranged in such a way that material is discharged to the conveyor below from the previous conveyor or from the incoming pretreated waste, which in turn releases the material to the conveyor belt underneath it. Other mechanical equipment such as projecting bars, rakes, etc. are used in the process of mixing the digesting materials. The Naturizer process is handled by the International Disposal Corporation at St. Petersburg, Florida (see Fig. 7.11). The organic solid waste is relocated or shifted once a day in this operation, while the fan is used to ventilate the digesting material inside the reactor. In the first grinding, digested sewage sludge, raw sewage sludge, water, or segregated wet garbage are added to control dust and moisture. Results obtained from the Norman Compost Facility in Oklahoma show that the composting process can be finished within 6 days [112].

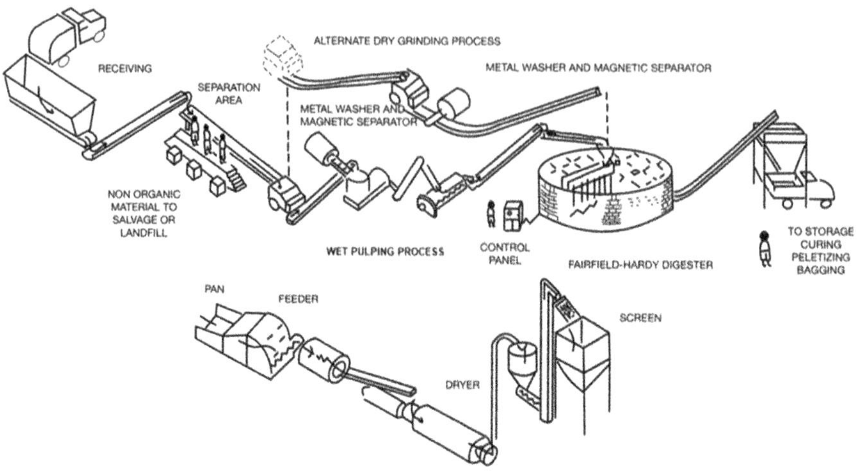

Fig. 7.12 Fairfield-Hardy composting system [21]

The research or experimental wet-pulp Fairfield-Hardy (see Fig. 7.12) located in Altoona, Pennsylvania, is one of the most successful composting plants in the United States. The daily capacity of the plant is 41 metric tons, and the Altoona FAM Incorporation has been operating the plant for around 25 years. Every day, the plant processes between 22.3 and 40.8 metric tons of solid waste. Before collection, trash, paper, and other waste are segregated from glass and cans. Hammer mill is used for initial grinding. While to comminute or reduce waste, wet pulpers are used. The pulper consists of a large bowl containing a round steel plate studded with teeth that can be rotated. Raw solid waste is added after the bowl is half-filled with water and the steel plate is rotated at a speed of about 650 rpm. When solid waste enters the bowl, it will be moved to the teeth found on the rotating steel plate and shredded. Water and sewage sludge are mixed together in the pulper to maintain a slurry of around 5% solids. The slurry is released through a bar screen to filter noncompostable materials such as cans, glass, and so on. The material is dewatered to about 58% before proceeded with the digestion [21]. The resulting wastewater is then discarded.

The digester operating in Altoona comprises an 11.4-m-diameter circular vat in which the pulped and dewatered solid waste is inserted semi-continuously. Material is poured into the tank to a depth of 1.8 m, and mixing is accomplished with augers placed on a movable bridge. Slowly, the augers turn over the pulp, moving the material upward and into the center of the tank, combining the digesting, pulped waste. Temperatures are normally between 60 and 77 °C in the digester. By using blowers at the bottom of the tank, the air is introduced into the digester. Within the process, stabilization is achieved after about 5 days of detention [113]. After the composting process is completed in the digester, the resulting compost is matured for about 3 weeks or more in the windrows before it is marketed.

Figure 7.10 shows several vertical silo variations that have been used. The milled or shredded waste is added to the silo or circular tank, as shown in Fig. 7.10. The material can be rotated or moved within the digester with the aid of rotating arms or platforms. If the platform is used, openings between platform levels help material pass through the digester. Forced ventilation is performed on composting material to maintain aerobic conditions. The resulting compost will be removed from the digester for further use after a certain period in the digester.

According to Gotaas [9], the Dano biostabilizer has been commonly used in Europe. This digester consists of a horizontally positioned, rotating steel drum or cylinder placed at a slope to allow the solid waste inserted at the top end to move inside the drum as it rotates. Control of moisture content and aeration mechanisms are used in most continuous processes. Typically, the rotational speed of the drum is between 1.5 and 5 rpm. Final composting is done in open windrows after the active composting process is finished in the drum. The periodic turning is performed on the 4th, 8th, and 12th day, and the composting process can be completed within 2 weeks.

The Ecology plant is a continuous-flow mechanized plant (closed) used in a major municipality located in Brooklyn, New York [114]. The fixed-deck plant that processes 136 metric tons/day of solid waste uses harrows for the solid waste mixing process (refer to Fig. 7.9). This plant is an excellent example of a new, fully automated municipal composting facility. There are three main processes conducted in the Ecology plant: (a) pretreatment (receiving, sorting, and storage); (b) digestion or bioreactor operations; and (c) finishing, including final drying, screening, upgrading, and bagging. Pretreatment involving receiving, grinding, and sample preparation must have corresponded to the solid waste collection schedule. Although solid waste is not segregated, local regulations have stipulated that materials placed in bins consist only of solid waste and paper along with bottles, cans, plastics, and rags, excluding construction waste, bulky waste, and objectionable materials (dead animals, etc.). Solid waste is added to the receiving hopper manually or even using "cat" or backhoe. The material was introduced to an inspection station by a 1.6-m moving metal apron with a metal leveling system where the material was checked for any items that could damage or impede the equipment (tires, large metal objects, etc.). Two shredding phases were used to minimize the amount of solid waste produced, which essentially decreased the material to around 1 cm or less. An overhead magnetic head pulley was applied to get rid of ferrous metals. The shredded material was then moved into a compartmentalized open storage tank with a capacity of 409 metric tons using a 79-cm conveyor belt and bucket elevators. The storage tank was operated so that the material could be removed continuously to the digester using a traversing rotating screw at the bottom of the storage container. Conveyor belts and bucket-lift elevators were then used to transfer the material to the top of the digester tank. The biological reactors found in the Ecology plant have eight stacked, fully enclosed fixed decks. The deck on the reactor is 49 m long and 3 m wide, with a distance between the decks is 1.2 m. With the use of harrows or plows, the digesting material was transported along with the fixed decks so that one set of harrows on a chain-driven belt could be used to serve coupled, adjoining decks.

Generally, the digesting pulp depth varied from approximately 16–39 cm. The harrows drive the material forward, mixing and aerating the composting material as the harrows push through the material (Fig. 7.9). The composting material is forced at the end of each deck onto the deck below. In total, it took about 40 h from composting to stabilization in the digester, which corresponds to deck 6 where biological activity begins to decline. The final two drying decks have been reserved for drying. Air was constantly circulated throughout the digester. In this process, moisture was applied to the digesting material, as well as nitrogen and phosphorus supplements.

7.6.4 Curing

Curing is a prolonged and undisturbed stabilization period after the digestion process or active composting phase [21]. The role of the technique is to improve the quality of the finished compost that has been processed thermally. During the curing phase, the decomposition rate is slowed down but still occurs, and the stabilization process is still ongoing. Further microbiological stabilization happens during curing, which normally enhances the finished product whether it is intended for use as a soil enhancer or for agricultural activities. It can take a few weeks or months for the composted product to mature, depending on the end-use. If the resulting compost is used for agricultural activities, it will compete for nitrogen and deplete the nitrogen from the soil. Its application will be detrimental without curing it. Curing can be done in the same place used during processing. Usually, the windrow method is often used for curing. Curing offers several benefits, including lowering the C:N ratio, improving the pH, reducing phytotoxicity, and eliminating any phytotoxic materials in low stability composts after active composting. Besides, curing can also avoid the risk of using immature compost such as nitrogen (N) hunger, oxygen deficiency, and the toxic effects of organic acids on crops.

Generally, in the windrow process, curing is estimated to be sufficient with an additional 2 weeks after digestion. Curing time can vary from 1 week to several months when mechanical processes are used. No curing phase is carried out on compost in the Ecology process (Fig. 7.9). In contrast, compost produced at the Altoona plant (Fig. 7.12) takes 3 weeks for curing [114].

7.6.5 Finishing or Upgrading

While more or less uniform in color and texture, close examination of composted MSW reveals the existence of glass, metals, plastics, wood scraps, rags, and dust. Depending on the final moisture content, the finished compost is typically light to fluffy in consistency. Although the finished compost can be used for many applications in this form, it is also preferable to enhance or upgrade its performance to lessen handling issues as well as to increase quality. This can be accomplished by using a few process operations that can be used to impart the required qualities to the finished compost, alone or in combination.

The most widely used methods to accomplish upgrading are compaction and pelletizing [21]. Usually, compaction is accomplished by pressing the object into huge, heavy rollers that compress the finished compost. Additives can be used to achieve optimal compaction and specific moisture content as well as improving the bulk density of finished compost. The process of cutting or milling and screening is performed to get the correct size after the finished compost is compacted. Adding chemical fertilizer to a certain level to produce commercial fertilizer of a certain composition using composted MSW as a carrier is one way of upgrading the compost. Cardenas [114] reported that this method was implemented at the Ecology composting facility in Brooklyn. Meanwhile, the compost mixture is made in the form of granules at the Altoona plant, which uses starch as an adhesive, dried, screened, and packaged.

7.6.6 Storage

According to Cardenas and Wang [21], in early spring and autumn, compost demand is highest for agricultural activities. Therefore, storage is an important component that must be considered in the design of composting systems. Storage and curing can be combined, and products can be stored for further upgrading or in finished form. There is also a method of storing compost by leaving it in windrows outdoors or in containers. If the finished compost is to be upgraded or bagged, it needs to be protected and stored at about the same moisture level.

Good compost storage methods can protect it from weather or season changes and, at the same time, can improve the quality of compost as a whole. This storage period gives the microbial community time to change several times at the end of the curing phase. In addition, the additional curing period allocated either in the windrows (covered with tarp) or in stable storage containers (e.g., plastic garbage bin, composter) can increase the habitat of microbes, which converts the finished compost into looser and spongy humus.

The storage process provides many benefits to the final compost produced. These include that (1) it is able to free up space that has been used for the composting process and provide additional space for new piles; (2) it can provide protection to finished compost from bad weather, including heavy rain and snow—stored compost will increase the activity of microorganisms and, in turn, convert compost into high-quality humus; and (3) it can assist the compost aeration process when turning activities are carried out while moving the compost to a storage location.

7.6.7 Design Considerations of a Complete Aerated (Turned) Windrow Composting System

7.6.7.1 Description of the Composting System Under Consideration

Aerated (turned) windrow composting is the microbial degradation of solid waste, waste sludge, and other putrescible organic solid material by aerobic metabolism in piles or windrows on a surfaced outdoor area. The piles are turned periodically to provide oxygen for the microorganisms to carry out the stabilization and to carry off the excess heat that is generated by the process. When masses of solids are assembled, and conditions of moisture, aeration, and nutrition are favorable for microbial activity and growth, the temperature rises spontaneously. As a result of biological self-heating, composting masses easily reach 60 °C (140 °F) and commonly exceed 70 °C (150 °F). Peak composting temperatures approaching 90 °C (194 °F) have been recorded. Temperatures of 140–160 °F (60–71 °C) serve to kill pathogens, insect larvae, and weed seeds. Nuisances such as odors, insect breeding, and vermin harborage are controlled through rapid destruction of putrescible materials. It is suitable for converting undigested primary and/or secondary sludge, or agricultural biosolids, or refuse to an end product amenable to resource recovery with a minimum capital investment and relatively small operating commitment. The end product is humified and has earthy characteristics; pathogens, weed seeds, and insect larvae are destroyed.

The US Environmental Protection Agency (USEPA) design criteria of an aerated (turned) windrow composting system are briefly introduced for composting biosolids. Approximate land requirement is 1/3 acre/dry ton sludge daily production, which is roughly equivalent to a population of 10,000 with primary and secondary treatment, if the incoming solid waste is waste sludge. Windrows can be 4–8 ft high, 12–25 ft wide at the base, and variable length. Sludge cannot be composted by itself but must be combined with a bulking agent to provide the biomass with the necessary porosity and moisture content. Combined refuse and waste sludge can be successfully composed. Biomass criteria are as follows: moisture content, 45–65%; C/N ratio between 30 to 35:1; C/P, 75 to 150:1; airflow, 10 to 30 ft^3 air/d/lb VS. Detention time is about 6 weeks to 1 year. Sequential steps involved in composting are (1) preparation and mixing, (2) digestion or composting, (3) curing, (4) finishing and screening, and (5) recycle of woodchips and production of compost. The entire composting system steps are discussed below.

7.6.7.2 Preparation

To be compostable by the aerated (turned windrow composting system under consideration, a solid waste mixture must have at least a minimally porous structure and a moisture content of 45–65%. Therefore, the incoming solid waste (refuse), or sludge cake having about 20% solids, cannot be composted by itself but must be combined with bulking agents, such as soil, sawdust, wood chips, refuse, or previously manufactured compost. Sludge and refuse make an ideal process combination. Refuse brings porosity to the mix, while sludge provides needed moisture and nitrogen, and both are converted synergistically to an end mixture product amenable to resource recovery. The waste refuse/sludge is suitably prepared and placed in piles or windrows.

7.6.7.3 Composting or Digestion

The composting or digestion period is characterized by rapid decomposition. Air is supplied by periodic turnings. The reaction is exothermic, and wastes reach temperatures of 140–160 °F (60–71 °C) or higher. Pathogen kill and the inactivation of insect larvae and weed seeds are possible at these temperatures. The period of digestion is normally about 6 weeks.

7.6.7.4 Curing

This is characterized by a slowing of the decomposition rate. The temperature drops back to ambient, and the process is brought to completion. The period takes about two more windrow weeks.

7.6.7.5 Finishing and Screening

If municipal solid waste fractions containing nondigestible debris have been included, or if the bulking agent such as wood chips is to be separated and recycled, some sort of screening or other removal procedure is necessary. The compost may be pulverized with a shredder, if desired.

7.6.7.6 Recycle of Woodchips and Production of Compost

After the finishing and screening step is over, the woodchips and some other potential bulking agents can be recycled to the initial preparation and mixing step, while the compost end product is stored ready for reuse. The produced compost represents an environmental benefit when used as a soil amendment. Other uses include wallboard production, livestock feed, litter for the chicken industry, and adsorbent for oil spill cleanup.

7.6.8 Design Considerations of a Complete Aerated Static Pile Composting System

7.6.8.1 Description of the Composting System Under Consideration

In an aerated static pile composting system, the solid waste stream (refuse and waste sludge, for instance) is converted to compost in a large pile in approximately 3–8 weeks under aerobic biological conditions using forced air. The bulking agents (e.g., wood chips, shredded newspaper) are added to the solid waste forming piles so that the forced air at the pile bottom can pass from the bottom to the top of the pile. The piles may be placed over a network of pipes that deliver air into or draw air out of the pile. Air blowers can be activated by a timer or temperature sensors. In a warm, arid climate, it may be necessary to cover the pile or place it under a shelter to prevent water from evaporating. In the cold, the core of the pile will retain its warm temperature. Aeration might be more difficult because passive airflowing is used rather than active turning. Placing the aerated static piles indoors with proper ventilation is also sometimes an option. Since there is no physical turning, this method requires careful monitoring to ensure that the outside of the pile heats up as much as the core. Applying a thick layer of finished compost over the pile may help alleviate any odors. If the air blower draws air out of the pile, filtering the air through a biofilter made from finished compost will also reduce any of the odors. This method may require significant cost and technical assistance to purchase, install, and maintain equipment such as blowers, pipes, sensors, and fans. Having a controlled supply of air allows construction of large piles, which require less land than the windrow method. At elevated temperatures, biosolids may be stabilized after as short as 21 days. Maximum temperatures of between 60 and 80 °C are produced during the first 3–5 days, during which time odors, pathogens, and weed seeds are destroyed. Temperatures above 55 °C (131 °F) for sufficient periods can effectively destroy most human pathogens. The finished compost is a humus-like material, free of malodors, and useful as a soil conditioner containing low levels of essential plant macronutrients such as nitrogen and phosphorus and often adequate levels of micronutrients such as copper and zinc.

The US Environmental Protection Agency (USEPA) has developed the design criteria for composting biosolids using the aerated static pile composting system, assuming construction of the pile for a 10 dry ton/d (43 wet tons) operation: (1) a 6-in. layer of unscreened compost for base; (2) a 94-ft loop of 4-in. diameter perforated plastic pipe is placed on top (hole diameter 0.25 in.); (3) pipe is covered with 6-in. layer of unscreened compost or wood chips; (4) loop is connected to a 1/3 hp blower by 14 ft of solid pipe fitted with a water trap to collect condensate; (5) timer is set for cycle of 4 min on and 16 min off; (6) blower is connected to conical scrubber pile (2 cubic yard wood chips covered with 10 cubic yard screened compost) by 16 ft of solid pipe; (7) wet sludge versus wood chip mixture in a volumetric ratio of 1:2.5 is placed on the prepared base; (8) a 12-in. layer of screened compost is placed on top for insulation; (9) airflow: 100 cubic feet per hour per ton of sludge; (10) land area requirement for 10 dry tons processed daily: 3.5 acres, including runoff collection pond, bituminous surface for roads, mixing, composting, drying, storage, and administration area; (11) pile dimension: 53 ft × 12 ft × 8 ft high; and (12) population equivalent, 100,000. The composting system converts solid waste to compost in the following five-step process: (1) preparation, (2) composting or digestion, (3) drying and screening, (4) curing, and (5) finishing.

7.6.8.2 Preparation

Raw solid waste (refuse and/or sludge) is mixed with a bulking material such as wood chips or leaves in order to facilitate handling, to provide the necessary structure and porosity for aeration, and to lower the moisture content of the biomass to 60% or less. Following mixing, the aerated pile is constructed and positioned over a porous pipe through which air is drawn. The pile is covered for insulation.

7.6.8.3 Digestion

The aerated pile undergoes decomposition by thermophilic organisms, whose activity generates a concomitant elevation in temperature to 60 °C (140 °F) or more. Aerobic composting conditions are maintained by drawing air through the pile at a predetermined rate. The effluent air stream is conducted into a small pile of screened, cured compost where odorous gases are effectively absorbed. After about 21 days, the composting rates and temperatures decline, and the pile is taken down, the plastic pipe is discarded, and the compost is either dried or cured depending upon weather conditions.

7.6.8.4 Drying and Screening

Drying to 40–45% moisture facilitates clean separation of compost from wood chips. The drying process is weather-dependent and requires at least two rainless days. The unscreened compost is spread out with a front-end loader to a depth of 12 inches. Periodically, a tractor-drawn harrow is employed to facilitate drying. Screening is performed with a rotary screen. The chips are recycled.

7.6.8.5 Curing

The compost is stored in piles for about 30 days to ensure that no offensive odors remain and to complete stabilization. The compost is then ready for utilization as a low-grade fertilizer, a soil amendment, or for land reclamation.

7.6.8.6 Finishing

The produced compost is ready for reuse. The compost is suitable for converting digested and undigested sludge cake to an end product of some economic value. Potential odor problems can occur for a brief period between the time a malodorous sludge arrives at site, is mixed, and is covered by the insulating layer. Human pathogen generation and aerosol distribution potential dictate careful attention to downwind land use.

7.6.9 Design Considerations of a Complete Two-Stage In-Bin Composting System

7.6.9.1 Description of the Composting System Under Consideration

The first-stage primary aerobic composting process reactors (primary composter, or PC) are many equal size open wooden bins under a roof for rain protection. The influent raw solid waste (dead animal, poultry, or fish, or cut livestock) is mixed with bulking agents (wheat straw and manure cake, for instance), forming a mixture or "PC influent," which is then fed to the primary composter's bins for aerobic composting reaction in the presence of sufficient oxygen, microorganisms, nutrient, and moisture. The process temperature in PC will gradually increase from room temperature to about 60–71 °C (140–160 °F) within about 3–6 days. During the first-stage primary composting, the organic matter is broken down and the volume of the mixture (solid waste + bulking agents) is significantly reduced, and the pathogens are killed. The first-stage composting process is over when the process temperature starts to drop from the high-temperature readings (7–9 days). The effluent of PC bins is the primary compost or "PC effluent," which is then transferred to the second-stage open secondary composter (SC) for subsequent processing. The total volume of the second-stage secondary aerobic composting process reactor (secondary composter volume, or SCV) is slightly greater than or equal to the total volume of the first-stage primary composter (PCV). The PC effluent is the SC influent. Due to this reactor transfer from PC to SC, much more oxygen is added to the primary compost (PC effluent or SC influent). In the SC, aerobic biological reactions are restored, the SC process temperature starts to increase again to about 145 °F within 3 days, and it lasts about 1 week. During the secondary composting, the solid waste volume (SC influent) is further reduced by biological oxidation, and pathogens are further killed by high temperature. The SC effluent goes through a curing process step and finishing process step, producing a final product, compost. The complete two-stage in-bin composting process system is broken down into the steps of (1) preparation and mixing, (2) first-stage composting or digestion, (3) second-stage composting or digestion, (4) curing, and (5) finishing.

7.6.9.2 Preparation and Mixing

A simple mixture of litter or manure cake, straw, and dead chickens will satisfy the requirements of certain readily available microorganisms (bacteria, fungi, worms, etc.) to convert these materials to an inoffensive and useful compost. According to DW Murphy of Department of Poultry Science, University of Maryland, each pound of dead chicken will need 1.5 lb chicken manure (cake) and 0.1 lb of wheat straw being the bulking agents.

7.6.9.3 First-Stage Composting or Digestion

Dead chicken disposal operations consist of adding correct volumes of birds, manure, and straw (or like substance) to a primary composter (PC) bin in layers. Depending on bird weight to be disposed of, partial layers, full layers, or full bins of compost mixture may be added. Within 2–4 days of loading, temperatures should increase rapidly and reach peaks of near 60–71 °C (140–160 °F). As mortality accumulates, successive primary composter bins are loaded. A 36-inch probe-type thermometer is needed to monitor the temperature in the pile on a daily basis.

It is important that the elements of the compost mixture be placed in the bin in the order of straw, dead birds, and litter or "cake" (read Sect. 7.13.7). Note that the first layer has straw, litter, straw, birds, and litter. This double layer of straw and a 6-in layer of manure is placed in the bottom of the composter to help prevent compaction and to help soak up extra moisture. Ideally, the composter will be sized so that the average day's mortality will equal one layer of dead chickens in the primary bin. Each successive day, the dead chickens are layered in the bin with the other elements added as described (straw, chickens, manure; straw, chickens, manure).

7.6.9.4 Secondary Stage Composting or Digestion

When the last primary composter (PC) bin is filled, the first one normally will have undergone 7–10 days of composting and volume reduction and will be ready to move to a secondary composter (SC) bin. After a flock is moved to market, continue to turn more primary composter bins into secondary composter bins.

Again a 36-in probe-type thermometer is needed to monitor the temperature in the pile on a daily basis. As mentioned earlier, the pile reaches its high temperature of 140–160 °F rapidly, which pasteurizes the compost. Once the temperature begins to fall from this high reading (7–9 days), it is time to move the material to the secondary treatment bin for aeration and reheating. As additional compost from the primary bins is added to older compost in the secondary treatment bin, the loader both mixes and aerates the compost.

7.6.9.5 Curing

After 7–10 days in the secondary composter (SC) bin, the next step is to place the compost outside, covered with plastic for curing. An average of about 20 min per day is required for loading and managing a composter that is sized for a broiler farm having a 100,000–130,000 bird capacity. This average does not include the time necessary to pick up the dead chickens.

7.6.9.6 Finishing

The cured compost (secondary composter effluent) may be applied to the land as supplemental fertilizer using the same guidelines as applied to poultry manure (read Sect. 7.13.7). Compost will generally average lower in nitrogen and higher in P_2O_5 and K_2O than broiler litter. This is due to the fact that nitrogen is given off as ammonia during the composting process. Also, P_2O_5 and K_2O will be higher in compost since the volume of the mass will reduce 25–30% during the composting process.

7.7 Process Control

Temperature, moisture content, nutrient content, aeration and mixing, pH, and so on are some of the parameters involved in process control to ensure that composting operations occur at optimum conditions. Most process control parameters are related to the microbiological stabilization process and moisture addition [21].

The temperature increases very quickly in the initial stages of the composting process and drops as the material progresses to completion or stabilization [21]. Intensive microbiological activity in the early stages of the composting process contributes to an increase in temperature in biological reactors or windrows up to 60 °C or more. After a certain period, there is a decrease in temperature, and the composting process will change from the decomposition phase to the maturation phase. The optimum temperature for the composting process is between 60 and 65 °C. However, an increase in temperature up to 70–80 °C should be avoided to produce better compost quality [115]. The diversity of microorganisms in the compost material will be limited and will further reduce the decomposition rate in the event of a very high-temperature increase in the composting pile. However, this situation can be controlled by daily temperature monitoring, and the turning process is done periodically to ensure that the optimum temperature can be achieved. Frequent turning mechanisms and aeration are the activities that can maintain the optimum thermophilic temperature (60 °C) in composting piles or biological reactors. The turning mechanism carried out will ensure the uniformity of the compost as well as transfer the cooler outer layer into the core composting pile at a higher temperature. The drop in temperature will occur after turning the composting pile but will immediately increase again. The turning operation will distribute the decomposing organic

Fig. 7.13 Temperature observation by inserting the thermometer probe into the center of the composting pile [61]

matter and optimize the distribution of oxygen in the pile and, in turn, promote biological activities in the pile. The longer thermophilic phase can prevent the extinction of thermophilic microorganisms, provide better control against pathogens and weed seeds, and, in turn, produce high-quality compost [115]. Factors such as the initial mixture containing high carbon, too dry, too wet, or insufficient oxygen in the pile also contribute to the insufficient temperature of the pile. A quality composting process is characterized by a good evolution of temperature in which the temperature rises rapidly at the beginning of the process and then slowly decreases. A sharp drop in temperature in the composting pile indicates that there are problems in the process, such as insufficient or excessive moisture content. Fuchs et al. [115] emphasize the importance of measuring the temperature correctly. The measurement process is done by inserting a thermometer at a depth between 40 and 50 cm at 3–10 places in the composting pile. The location of the temperature determination depends on the dimensions of the pile. During temperature measurements for larger piles, the thermometer will be placed in the center of the pile, at half of the height (Fig. 7.13). In temperature monitoring, the type of monitoring equipment used depends on the level of management that the operator wants to provide. In the composting process, a thermometer is needed to set a normal temperature profile, turning schedule, or microbial activity. A dial thermometer with a 3-ft-long pointed probe is the simplest and cheapest thermometer [116]. However, the disadvantage of this thermometer is that it takes time to take a stable reading, especially when the compost temperature reading needs to be taken several times. A fast-response thermometer can be used to overcome this problem. Analog or digital thermometer readings and probe lengths that can range from 3 to 6 ft are the two main features that distinguish between thermometers available in the market [116]. In general, thermometers consist of dial or analog thermometers (cheapest), intermediate digital rapid response thermometers, and thermocouple coupling thermometers (most expensive).

According to Cardenas and Wang [21], the moisture content is one of the most important parameters that must be controlled in the early stages of the composting process. The evaporation that occurs in large quantities and the addition of water that occurs due to precipitation are among the factors that influence the change of moisture conditions in the composting pile throughout the composting process [116]. Some problems will arise due to this improper moisture, such as the formation of anaerobic zones in the pile and odor production, and then inhibit the composting process. Meanwhile, the dry pile condition will contribute to the formation of dust that transmits odors and even fungal pathogens such as *Aspergillus fumigatus* as well as affects the microbial activity in the pile, which requires moisture content in the range of 40–60%. Climate also affects the efforts to maintain the compost moisture level. In summer or dry weather, the loss of moisture from the compost pile due to evaporation makes it difficult to maintain moisture. The composting process can be stopped prematurely if the pile is too dry. In contrast to areas with humid climates, excessive moisture is a major problem at composting sites. This condition can be overcome by increasing the frequency of turning or active aeration of the pile [115], installing a roofing structure to protect the composting pile as well as adding a large number of dry amendments [116]. In general, the moisture content of the outer layer of the pile is different from other parts of the pile. During the moisture content test, test samples are usually taken from a depth of 40–50 cm [115]. The squeeze test is a simple method in determining the moisture content in compost at the composting site [115, 116]. During this test, the compost will be held as tightly as possible between the palms and fingers. Compost is considered too wet if water flows out through the fingers. When the finger is opened, if the compost ball breaks on its own, then the compost is considered too dry. If the compost ball stays compact and does not break, the compost is considered to have an optimum moisture content. Meanwhile, the traditional method often used is drying compost in a drying oven at a temperature of 103 °C overnight. Although this method is considered satisfactory, in terms of operational requirements, this method takes too long to determine the moisture content in the compost sample. In line with the development of technology, rapid moisture determination has begun to be introduced by several previous researchers. Cardenas [117] has suggested the use of commercially available microwave ovens in determining the moisture content in compost. This inexpensive and quick method only takes 10 min to determine the moisture content of the compost sample. McCartney and Tingley [118] have recommended two methods for rapid moisture determination for compost. They have made improvements to the existing drying method using oven drying at 103 °C by reducing the drying time from 19 to 9 h by minimizing the amount of sample used. Meanwhile, the use of the new infrared (IR) method can shorten the moisture content determination period to 33 min. Indirectly, this new method will attract those who need moisture content value at an immediate rate. The use of Rapid Moisture Meters today provides an advantage in determining the moisture content at the composting site. Through this method, the water content is determined by the gas pressure developed by the reaction of calcium carbide (absorbent) with the free water found in the compost sample [119]. The percentage of water in the total amount of compost sample (wet) was obtained from the calibrated scale of pressure gauge, and the same percentage is then converted into water content or moisture content on the dry mass of soil.

Adequate aeration and mixing are very important for the continuation of the composting process. An adequate aeration process can maintain aerobic conditions in the composting pile and promote the biological activity of microorganisms but does not cause excessive cooling of composting materials [21]. Factors such as the depth of the compost pile and the air temperature in the digester affect the aeration process. Only a few process controls need to be done once the proper aeration and mixing rate are successfully performed. Typically, an increase in temperature will occur when the mixing rate is low and the static duration of the composting pile is longer. Aeration rates have a significant effect on the production and release of ammonia (NH_3) [120, 121]. This is supported by several composting studies conducted by Gu et al. [122] and Jiang et al. [123]. Gu et al. [122] reported that about 16–76% nitrogen (N) was lost during aerobic composting of poultry waste-straw stalks, sludge-straw stalks, and poultry waste-sawdust. Jiang et al. [123] revealed that there was a loss between 9.6% and 46% of N in the form of NH_3 emission during the composting of pig feces from the Chinese Ganqinfen system. The release of NH_3 into the environment will result in a final compost that has a low nitrogen content and, in turn, reduces its value as fertilizer. Nitrogen loss, quality of compost products, and energy consumption are directly influenced by the aeration rate. According to Xiong et al. [124], inadequate aeration can contribute to anaerobic conditions in the composting pile due to lack of oxygen, while excessive aeration will slow down the composting process through the loss of heat, water, ammonia, and nitrogen and even increase operating costs. Xiong et al. [124] have conducted a composting process on cow dung and turfgrass. They found that aeration rates significantly affected oxygen content under different conditions. It is found that the aeration rates affect the water content, nitrate, and nitrogen loss. As the aeration rates increase at high temperatures, there is an increase in NH_3 emissions due to nitrogen loss. According to these findings, the aeration rate had a significant effect on total N and NH_3 emissions ($p < 0.05$). Therefore, they suggest that the optimization of aeration methods be carried out to increase the seed germination rate.

7.8 Pathogen Survival

Bacteria (both pathogens and indicators), spore-forming fungi, actinomycetes, parasitic worms (including cysts and eggs), protozoa or helminths, and viruses, especially in municipal solids waste and sewage sludge, have also been researched [21]. Exposure to heat, time-temperature exposure, nutrient competition, microbial antagonism (including antibiotic production and direct parasitism), organic acid production, and ammonia are among the factors influencing pathogen destruction during the composting period [125, 126]. Wilkinson [126] identified temperature and duration of exposure as the most critical factors in pathogen inactivation. Furthermore, measuring the temperature during composting is relatively easy. When all parts of the composting material are exposed to temperatures of around 60 °C or above, pathogenic bacteria, for example, are quickly destroyed. According to Deportes et al. [127], exposure to an average temperature of around 55–60 °C for at least 3 days while composting is normally adequate to destroy the majority of the enteric pathogen. Pathogens are heat-sensitive, and the heat generated during the composting process kills them, resulting in a pathogen-free finished product [128]. The pathogen content of compost is significant because, if not properly handled, it may be a pathogen vector in the ecosystem, posing a danger to humans and animals.

Quality assurance requirements differ significantly across the world [129]. In the United Kingdom, compost must retain a temperature of 65 °C or above for a total of 7 days (which need not be consecutive) at a moisture content of 50% w/w to comply with UK PAS100 guidelines [130]. Temperature is a critical factor in pathogen survival in compost [131]. In the standard quality of composting, *Salmonella* spp. and *Escherichia coli* are two well-known pathogenic indicators. The European Commission Decisions identified the importance of *Salmonella* spp. and *E. coli* in determining the ecological quality of the final compost. According to Briancesco [132], limits for its density (<103 MPN g^{-1} and absence in 50 g, respectively) are set to provide seal quality in Commission Decision 2001/688/EC; Commission Decision 2005/384/EC. In the meantime, the UK composting standard states that compost sold in the United Kingdom must be free of *Salmonella* spp. and have less than 1000 colony-forming units of *E. coli* per gram of material [130].

Pereira-Neto et al. [125] have conducted a composting study on garbage and sludge using aerated static pile systems. The pile temperature was monitored every 2 days along the pile diameter, and compost samples were collected at the top, middle, and bottom of the pile to test the *Salmonella* content in it. In the first trial, *Salmonella* levels could not be detected at the top and middle of the pile where the temperature exceeded 55 °C while the temperature never reached 55 °C at the bottom of the pile. Meanwhile, in the second trial, the *Salmonella* level at the bottom was reduced below the detection limit within 7 days, while in the other experiment, the *Salmonella* lasted up to 15 days. Nell et al. [133] have carried out a composting process on sewage sludge using a full-scale windrow method. They found *Salmonella* and *E. coli* undetectable in the composting pile after 2 and 5 weeks, respectively. Further studies conducted showed that *Salmonella* could be inactivated at a temperature of at least 55 °C and sustained for 3 days. Wiley and Westerberg [134] reported that no *Salmonella* Newport could be detected after 25 h of sewage sludge composted in the composting drum. The temperature inside the drum was maintained between 60 and 70 °C. These researchers concluded that the finished product was safe and tested for its suitability as a soil conditioner or fertilizer [21].

According to Wichuk and McCartney [135], in fecally contaminated materials, pathogenic and nonpathogenic strains of *E. coli* can be found. The position of *E. coli* as a bacterial indicator of fecal pollution must be considered since these bacteria have a strong association with the existence of other enteric bacteria with similar characteristics [128]. Cekmecelioglu et al. [136] have investigated the survival of *E. coli* during windrow composting of food waste. The study conducted in summer showed an average temperature increase that quickly surpassed 55 °C and lasted for 9 weeks. *E. coli* O157: H7 decreases rapidly to nondetectable levels within a few days to reach a temperature of 55 °C. No pathogen regrowth was observed beyond day 21. In contrast, during the study conducted in the winter, *E.coli* levels showed a fluctuating trend even though the pile temperature recorded 55 °C for at least 24 days. *E. coli* can still be detected at the end of the composting process. Although composting at thermophilic temperatures (>55 °C) is the best method for destroying pathogens found in piles, some studies show good results where pathogens cannot be detected even when composting takes place below 55 °C. For example, Pereira-Neto et al. [125], who composted garbage and sludge using an aerated static pile system, found a reduction in *E. coli* levels from over 10^7 organisms per gram to levels below the detection limit (10^2 organisms per gram) even though the temperature at the bottom of the pile remained lower than 55 °C. Jones and Martin [137] have cited a study in which 10^7 organisms per gram of *E. coli* O157: H7 were added to bench-scale manure composting. They found no *E. coli* strains could be detected after composting lasted for 48 h at 45 °C. The same condition was also observed even when composting was continued at a temperature of 25 °C.

Apart from *E. coli* and *Salmonellae*, feces-contaminated compost feedstock materials can harbor a wide range of pathogenic bacteria, viruses, protozoa, and helminths. While there is little or no research on the survival of most other pathogenic bacteria, the thermal inactivation of a few of them has been investigated. Morgan and Macdonald [138] conducted windrow composting of MSW and sewage sludge. They reported that *Mycobacterium tuberculosis* could be destroyed within 10 days at an average temperature of at least 60 °C. However, there is a windrow that takes 21 days to reduce these bacteria to nondetectable levels. There are several bacteria, such as *Bacillus* and *Clostridium*, that can produce resistant spores that can survive for a long time. Krogstad and Gudding [139] reported *Bacillus cereus* could be detected after 7 days of composting MSW and biosolids at temperatures below 70 °C. However, these bacteria are only undetectable when the temperature rises above 70 °C for 2–3 days. Meanwhile, Jones and Martin [137] revealed that in slightly acidic conditions, clostridial spores could survive for almost 2 h at a temperature of 100 °C. Therefore, it is expected that the spores of *Clostridium botulinum* and *Clostridium perfringens* can survive in conditions where other bacteria are not able to survive in the same composting conditions.

Wiley and Westerberg [134] added a laboratory culture of poliovirus type 1 to composting drums used to compost sewage sludge at temperatures between 60 and 70 °C. Observations made found that the virus could not be detected after 1 h of the process of adding a culture of poliovirus type 1 was done. Senne et al. [140] discovered that temperatures below 55 °C were enough to destroy certain viruses. The survival of highly pathogenic avian influenza (HPAI) and the adenovirus that causes egg-drop syndrome-76 (EDS-76) was investigated using pilot-scale bin composting of chicken carcasses. HPAI virus was not observed in any of the 20 samples after 10 days of composting, during which the upper layer reached 55 °C for 3 consecutive days, and the bottom layer remained below 41 °C, and only one was positive for adenovirus EDS-76. Even though the measured temperature in the bottom layer of the composter did not surpass 43 °C after turning and composting for a second 10-day period, no HPAI or EDS-76 adenovirus was found in any sample.

Pathogenic protozoa are often found in fecally contaminated feedstock materials. Protozoan cysts and oocysts are commonly thought to be easily killed by simple treatment or even environmental factors like drying. As a result, although a few studies have looked into the effects of temperature on protozoan survival during composting, it is widely assumed that these parasites are not a problem in finished composts because they can effectively be reduced to levels below detection [135]. Rimhanen-Finne et al. [141] have assessed the levels of *Cryptosporidium* oocysts and *Giardia* cysts after 10 weeks and 30 weeks of sewage sludge composting. The results obtained showed that approximately 37.5% and 44% of the samples analyzed contained *Cryptosporidium* oocysts and *Giardia* cysts, respectively, after 10 weeks of the composting process took place. After 30 weeks, *Cryptosporidium* oocysts and *Giardia* cysts recorded a decrease to 10% and 35%, respectively. Jones and Martin [137] reported that protozoa have the potential to survive during composting. In one study, they found that *Giardia* cysts could be detected after the composting process at a temperature of 52–53 °C. On the other hand, other protozoa such as *Entamoeba histolytica* and *Endolimax nana* and even a small number of helminths have been destroyed and cannot be detected after a few days through composting under the same conditions.

Helminth ova are of particular concern in compost since they are often found isolated in the fecal feedstock, where they are abundant and highly resistant to a range of chemical and physical agents [135]. Several studies report *Ascaris* ova can be decreased to nondetectable levels if the composting temperature reaches 55 °C or higher for 3 days. If the composting process is performed by windrow, the high temperature should be sustained for 15 days. Deportes et al. [142] have investigated *Ascaris* ova at three locations and four sampling points during the turning events performed at a municipal solid waste composting facility. At each of the three locations, the temperature monitoring process was performed at a depth of 50 cm. Deportes and his co-workers found that *Ascaris* ova was not detectable within 27 days, where the minimum temperature recorded exceeded 55 °C for at least 2 days but less than 21 days. Some studies reported that helminth ova reduction could occur at temperatures lower than 55 °C. For example, Tharaldsen and Helle [143] found that after 2 weeks, at 37 °C, the viability of *Ascaris* eggs could be reduced. The *Ascaris* eggs were destroyed after 31 days at the same temperature. From the findings of this study, they concluded that heat conduction occurred better inside the slurry of liquid manure (feedstock used in this study) than in the drier feedstock.

The temperatures and survival times of several pathogens other than those discussed above have been summarized and depicted in Table 7.5 [21, 128].

The enteric pathogens can regrow after depleting below detectable limits, and this represents a health hazard in the use of certain composts [128]. According to Sidhu et al. [144], moisture content, temperature, and indigenous microorganisms are among the factors influencing pathogen regrowth in compost. Stability is the

Table 7.5 Thermal death points of some common pathogens and parasites

Pathogens and parasites	Thermal death points and contact times
Salmonella typhosa	Death within 30 min at 60 °C; no growth beyond 46 °C
Shigella (Groups A to D)	Death within 1 h at 55 °C
Escherichia coli	Most die within 1 h at 55 °C and within 15–20 min at 60 °C
Entamoeba histolytica	Thermal death point is 68 °C
Vibrio cholerae	Very sensitive to change of environment; outside human body dies in a few hours in the feces at room temperature
Trichinella spiralis	Infectivity reduced as a result of 1 h exposure at 50 °C, thermal death point 67 °C
Necator americanus	Death within 20 days at 45 °C
Ascaris lumbridcoides	Death within 20 days at 45 °C, 2 h at 50 °C, 50 min at 55 °C, 3.5 min at 60 °C
Taenia saginata	Death within 5 min at 71 °C
Micrococcus pyogenes	Death within 10 min at 50 °C
Streptococcus pyogenes	Death within 10 min at 54 °C
Mycobacterium tuberculosis	Death within 15–20 min at 66 °C
Corynebacterium diphtheriae	Death within 45 min at 55 °C
Brucella abortus	Death within 3 min at 61 °C

Source: [21, 128]

most important element in preventing pathogen regrowth. When the compost materials are too dry to support high concentrations of microbiology activity, they tend to be stable. The rewetting of these compost products would create a suitable habitat for bacteria to repopulate [126]. In comparison to viruses, protozoa, and helminths, regrowth is only a problem for such bacterial pathogens, including *Salmonella* spp. and *E. coli*. For example, observations made by Russ and Yanko [145] show that there is regrowth of *Salmonella* from undetectable levels even if the compost resulting from the sludge composting is kept in a dry condition for about a year. Sunar et al. [128] reported that *Salmonella serovar Enteritidis* was able to survive for at least 3 months in stable compost after 8 weeks of the composting process. Turner [146] showed that while large-scale composting was performed at mesophilic temperatures (i.e., 45 °C), *E. coli* concentrations rose from their original amounts. As a result, Turner concluded that caution should be taken in managing conditions to avoid inducing bacterial pathogen formation. Low-temperature regions in compost are therefore a source of concern, not just because pathogens may not be adequately decreased but also because environments may be suitable for bacterial pathogen growth.

In conclusion, the relationship between the composting process and pathogens is crucial. It is critical that the process be properly handled and maintained across all processing stages to eliminate pathogenic microorganisms to an appropriate level.

7.9 Cost Considerations

The accuracy of capital cost estimating for constructing, extending, or improving a composting facility varies depending on the scope of comprehensive design and construction bids. Site construction, processing equipment (both stationary and mobile), and process monitoring equipment are the key categories of capital costs to be calculated [147]. Since some composting methods take up more space, then the selection of an appropriate composting method should be emphasized in determining the cost for site construction. For example, the use of tractor-pulled turners to turn windrows requires specialized space as opposed to the small space required by straddle turners. Meanwhile, aerated static pile (ASP) systems require less space. Therefore, the approach to selecting the composting method should be in line with the on-site assessment. If the composting process is carried out in a wider area, then the use of less capital-intensive equipment such as tractor-pulled windrow turners is seen as more appropriate. Coker [147] noted that there is one composting facility in Washington that uses straddle-turned windrows on a 300 × 350-foot pad (2.4 acres). Approximately 7500 cubic yards of pad capacity can be used to produce compost using this method. On the other hand, when the facility converted the windrow system to ASP, the pad capacity was increased by threefold (22,750 cubic yards). Currently, site development cost estimation can be done using construction estimation software. The use of this program is intended to make it easier for construction contractors to prepare bids that include more detailed drawings and design specifications.

Front-end loaders, grinders, turners, mixers, depackagers, blowers and piping, screens, contaminant removal, and bagging are among the processing equipment used in composting. In general, equipment costs vary according to technological sophistication, size, and capacity. Apart from being purchased, the equipment to be used can also be leased for a certain period. However, if the equipment is leased, the cost is not included under capital cost but rather part of the operating cost. Composting equipment is also available on the market as used equipment at a reduced cost, in addition to being purchased. The wear issue that equipment components are experiencing, on the other hand, would result in higher operating costs, possibly even a significant amount of money. Table 7.6 shows the capital cost of the equipment required for composting [147].

Among the basic equipment required in the monitoring process involve a dial-face temperature probe ($150–$250) and an iPad ($750–$1000). Wireless temperature probes ($2000+) and Supervisory Control and Data Acquisition (SCADA) computer interfaces are more advanced and state-of-the-art composting monitoring equipment. Some composting facilities set up weather stations at their facilities to record wind speed and direction data as well as rainfall data. The estimated price for this weather station is $700–$2000. Small laboratories in composting facilities usually have facilities to conduct simple tests such as bulk density measurement and free air space, moisture content, and pH test.

Composting is primarily a material handling activity. Each activity in the compost manufacturing process involves a specific amount of time and costs a specific

Table 7.6 The range of capital cost of the equipment required for composting

Equipment required	Capital cost ranges ($)
Loaders	150,000–600,000
Grinders	300,000–750,000
Turners (pull behind)	30,000–75,000
Turners (straddle)	250,000–950,000
Mixers	250,000–400,000
Depackagers	300,000–450,000
Blowers/piping	2000–10,000 (per pile)
Screens	50,000–650,000
Contaminant removal	200,000–600,000
Baggers	50,000–900,000

Source: [147]

Table 7.7 Time-and-motion projection of operating cost

Material handling—mixing	
Daily volumes coming into the facility	121 CY/day
Number of loader bucket movements to load mixer	40 buckets/day
Assumed time spent per loading event	2 min/bucket
Assumed time to load yard truck	2 min/bucket
Time spent handling feed stocks Convert to hours	162 min/day 2.7 h/day
Mixer run time	1.5 h/day
Total machine time	4.2 h/day
Total labor time	$18,937
Labor cost/year (at $22.50/h)	$72,029
Machine cost/year	

Source: [148]

amount of money. Operating costs are all costs incurred to produce a cubic yard of finished compost. Generally, these costs involve capital investment in the composting facility and the cost of money earned to finance the capital. In this section, the operating costs that will be emphasized are the cost of fuel, labor, electricity, and maintenance, which are noncapital related costs [148]. Time and motion projection is the method used in estimating costs for new composting facilities. Operating expenditures, as well as detailed cost accounting of equipment costs, should be calculated in the same way for existing facilities. The determination of the loaded labor rate and the machine rate depends on the time and cost to perform each task. Table 7.7 shows the time-and-motion projection of operating costs for the activity of mixing feedstocks in a mechanical mixer to be delivered to ASP composting pads in a yard truck at a proposed 10,000 tons/year food scraps composting facility in Martha's Vineyard, Massachusetts. The advantage of this costing exercise is that it can help in making equipment selection. The calculation steps are repeated for each task in compost production and then summarized to give a projection of the annual operating cost, as shown in Table 7.8.

Table 7.8 Annual operating cost projection

Process	Labor summary hours/day	ASP composting Labor cost ($)	Machine costs ($)	Utilities ($)	Total ($)
Waste receipt	0.7	4734	11,572		16,306
Grinding/shredding	0.7	7020	34,320		41,340
Mixing	4.2	18,937	72,029		90,966
Transport to pad	2.1	14,913	36,453		51,365
Building ASPs	2.4	17,043	41,660		58,703
Electricity for ASPs				78,122	78,122
Moving compost to curing	1.5	10,439	25,517		35,956
Managing curing piles	1.3	9326	22,797		32,123
Screening compost	1.5	12,079	10,410		22,489
Moving screened compost to storage	1.1	7516	18,372		25,888
Move over to a storage	0.3	1879	4,593		6472
Product marketing and sales	0.8	5616	13,728		19,344
Total	16.6	109,501	291,453	78,122	479,075

Source: [148]
Assuming 85% efficiency of site workers, 16.6 h/day actually equates to 19.5 h/day
Total cost of $479,075 divided by 10,623 annual tons equates to $45.10/ton

7.10 Compost Stability and Maturity

Finished compost has a wide range of applications as a soil enhancement. The resulting compost must be of high quality to ensure that this compost is truly beneficial and marketable. Stability and maturity are key requirements to ensure safety in the use of compost for agricultural purposes [3]. The extent of microbial biomass activity affects compost stability. Several methods have been suggested to establish the biological stability of organic matter. Determination of compost stability using physical methods [149] takes only a short time, while chemical methods are also often preferred because they are not expensive [150]. Biological tests such as enzymatic methods have begun to be used to determine the stability of compost [151]. Compost maturity, on the other hand, refers to the degree of humification and denotes the absence of phytotoxic compounds as well as pathogens [152]. Since microorganisms in unstable composts develop phytotoxic compounds, the relative stability of the substance has an impact on maturity [153]. Phytotoxicity tests are used to assess the degree of compost maturity, although a variety of different procedures have often been used before. Immature and unstable compost can trigger a lot of issues. According to Butler et al. [154], immature compost poses a danger to crops as well as provides phytotoxic effects such as inhibiting root growth, reducing seed germination, and minimizing above-ground plant yield. Immaturity-related phytotoxicity is exacerbated by high amounts of intermediate decomposition by-products such as ammonia and short-chain organic acids [155, 156]. The use of unstable compost (which has a high concentration of water-soluble organics) in the field contributes to the problem of groundwater and surface water pollution [157]. Besides, unstable compost use also promotes the regrowth of pathogenic bacteria [135]. Also, the application of unstable compost into the soil can lead to the occurrence of oxygen depletion because the aerobic decomposition of organic matter that occurs continuously involves high oxygen consumption. This will result in an insufficient supply of oxygen to reach the plant roots and reduce the transfer of nutrients from the soil to the plant [158, 159]. The importance of accurately assessing the stability and maturity of compost becomes a priority before the compost is used in agriculture due to the existence of various potential adverse effects of the use of immature and unstable compost. The ability to guarantee the safety and effectiveness of compost would certainly help compost gain consumer and regulatory approval for use in a range of applications, while the inability to do so could obstruct compost use.

7.10.1 Compost Stability and Maturity Testing Methods

Various methods have been proposed in determining the stability and maturity of compost. Sullivan and Miller [155] and Wichuk and McCartney [135] classify this method of determination into three main categories, namely physical, chemical, and biological methods (Table 7.9).

In modern composting, various standards have been established to ensure that the compost produced is of good quality and safe for the environment. Among them are the standards of the California Compost Quality Council, the Compost Council of Canada, the California Fertilizer Association, the US Composting Council, and the Woods End Research Laboratory [155].

7.10.1.1 Physical Methods

Physical methods involve monitoring changes in pile temperature and sensory indicators such as color and odor. The temperature profile of composting systems is usually characterized by a rapid initial increase to thermophilic temperatures, followed by a prolonged high-temperature duration, and finally a drop to near-ambient temperatures. Exothermic reactions associated with decomposition of microbial biodegradable organic matter in feedstock compost cause an initial increase and high temperature in the pile. As the easily biodegradable substrate is utilized and more bio-resistant materials like cellulose and lignin remain, the temperature decreases. According to Lasaridi et al. [160], the compost is considered to be entering a stable or mature state when the temperature drops to near-ambient temperature and no reheating occurs during the turning process. The final drop in pile temperature during yard trimming composting correlated well with a variety of parameters widely used for compost stability and maturity assessments [161]. They concluded

Table 7.9 Compost maturity test category

Test category	Potential tests for maturity
Physical (including sensory)	Pile temperature
	Color
	Odor
Biological (plant and microbiological)	Respiration (including self-heating capacity)
	Phytotoxicity (plant bioassay)
	Enzyme activity
Chemical	C:N ratio
	Organic matter (OM)
	Cation exchange capacity (CEC)
	Ammonia and nitrate
	Electrical conductivity (EC)
	Humification parameters
	pH
	Spectroscopy
	Dissolved organic carbon (DOC)

Source: [135, 155]

that temperature monitoring could be used to evaluate compost maturity in an easy and fast manner. This is also agreed by Boulter-Bitzer et al. [162] that the method of monitoring the temperature along with the oxygen profile during the composting process is accurate, inexpensive, and simple and can provide an assessment of the development of compost until it reaches a stable and mature state. According to Haug [6], pile size, the extreme temperature during the thermophilic phase, oxygen depletion, drying, excessively high moisture content, and ambient conditions (especially in cold climates) are among the factors influencing pile temperature. For example, moisture loss and oxygen depletion hinder microbial activity and lead to premature pile cooling [157]. Pile cooling can also happen when the water content is too high and, in turn, results in conditions of oxygen deficiency, limited gas diffusion, and decreased microbial activity [163]. Ambient conditions (low temperatures and strong winds) also contribute to heat loss and premature pile cooling [157]. However, there are several examples of pile temperatures providing inaccurate evidence regarding the degree of compost stability and maturity that can be found in the literature. For example, Chefetz et al. [164] reported that pile temperature reached ambient temperature after 60 days of composting, but compost products reached stability after 110 days of composting. Inbar et al. [165] revealed that the substrate decomposition process still occurs even though the pile temperature has dropped to ambient levels after 60 days of composting. The results obtained from the plant bioassay, electrical conductivity, and alkalinity test show that the compost is still not stable and mature enough even though it has been 91 days through the composting process.

When compost stabilizes and matures, changes in color and odor can be detected. The odors of compost change from unpleasant and ammonia-like to rich and earthy, and the color darkens. The color of compost darkens during composting and is highly influenced by the feedstocks used. For example, when yard trimming and manure composts reach maturity, they usually turn into a dark black and brownish color, respectively [155]. Compost color and odor can also be determined using standardized test techniques as specified in the Test Methods for the Evaluation of Composting and Compost (Method 9.03A) [166].

7.10.1.2 Biological Methods

Methods for determining the stability and maturity of compost biologically include respiration, phytotoxicity, and even enzyme activity tests. In determining compost stability, respiration indices can be used. These indices include direct or indirect measurements of the amount of biological activity in a sample, which represents the degree of substrate decomposition under ideal conditions. The higher the decomposition level of the substrate, the lower the microbial activity as well as the respiration rate. Respiratory rate is closely related to oxygen consumption and carbon dioxide release during aerobic composting. Oxygen uptake rate, carbon dioxide evolution, Dewar self-heating, and the Solvita® respiration test are the most commonly used respiration indices.

Tiquia [167] defines oxygen uptake rate (OUR) as a measure of the amount of oxygen consumed by microbes in a solid mass for a given period. Theoretically, this parameter is used to evaluate the stability of compost due to the occurrence of microbial activity. Oxygen consumption is decreased when a biodegradable organic matter has been utilized. The specific oxygen uptake rate (SOUR) test conducted on moist compost is based on Test Methods for the Evaluation of Composting and Compost (Method 9.09B) [168]. These tests require special apparatus and are not widely available in commercial laboratories. This test is also influenced by the moisture of the compost and requires pre-preparation of the sample. The duration of the experiment ranged from 60 to 90 min. The SOUR test requires volatile solid (VS) determination. The suggested values for mature compost are <0.5 mg O_2 g VS^{-1} h^{-1} (very stable), 0.5–1.5 mg O_2 g VS^{-1} h^{-1} (stable), 1.5–3.5 mg O_2 g VS^{-1} h^{-1} (moderate unstable), and 3.5–6.0 mg O_2 g VS^{-1} h^{-1} (unstable). Lasaridi and Stentiford [169] evaluated SOUR as stability indices in compost slurry (aqueous compost suspension). This method is not affected by the moisture content of the compost. The method introduced by these two researchers provides data similar to that of TMECC 9.09B with greater precision. Since it is based on a wastewater technique for biological oxygen demand (BOD), the approach is commonly available. This method requires computer assisted-control to control the oxygen input and the measurement of dissolved oxygen. The test duration is 20 h. According to Wichuk and McCartney [135], the oxygen uptake rate is a better indicator of biological activity than the carbon dioxide (CO_2) evolution rate. This is because microbial activity does not always result in the conversion of organic carbon to CO_2. Furthermore, since CO_2 can be released by both aerobic and anaerobic activity, OUR is best used to monitor only aerobic respiration.

Carbon dioxide (CO_2) is formed as a result of microbial activity. The rate of CO_2 evolution is expected to decrease as compost stabilizes and microbial activity decreases; thus, CO_2 evolution can be used as a stability indicator. The simplest quantitative respiration test, according to Sullivan and Miller [155], is the standard alkaline (KOH or NaOH) trap method. Titration is used to assess the volume of carbon dioxide trapped in KOH or NaOH. The CO_2 evolution test is guided by Test Methods for the Evaluation of Composting and Compost (Method 9.09C). The suggested value for mature compost is <2 mg CO_2-C g VS^{-1} d^{-1} (very stable), 2–8 mg CO_2-C g VS^{-1} d^{-1} (stable), 8–15 mg CO_2-C g VS^{-1} d^{-1} (moderate unstable), and 15–40 mg CO_2-C g VS^{-1} d^{-1} (unstable) [168]. According to several authors, CO_2 evolution is a strong indicator of stability. Brewer and Sullivan [170] looked into using a colorimetric CO_2 detection tube for fast, low-cost analysis. They discovered that it was reliable and that it could be used as a stability measure (with a CO_2 evolution rate of less than 2 mg CO_2 g^{-1} C d^{-1}). The CO_2 evolution test was recommended by Switzenbaum et al. [171] as the preferred respiration test for the majority of the composting facilities, and Goyal et al. [172] also agreed that CO_2 evolution is one of the most reliable compost maturity indices.

Microbial respiration and remaining organic matter can be estimated indirectly by a compost self-heating test. The Dewar self-heating test, which is based on Test Methods for the Evaluation of Composting and Compost (Method 9.11), is a standardized technique for measuring compost heat generation [168]. It is an indirect method of determining the rate of respiration. In this experiment, the temperature rise for moist compost found in an insulated vacuum bottle was recorded for 2–9 days [155]. This test is easy to do but takes a long time. When compared to short-term O_2 or CO_2 respirometry, the Dewar test allows for the growth of a natural compost microflora succession similar to that found in a composting pile. When compared to pile temperature tests, temperature monitoring for the Dewar self-heating test was performed on a small amount of compost sample placed in a closed container under controlled conditions rather than on the entire composting pile [135]. The benefit of this approach is that it can replicate what happens naturally in a compost pile under controlled conditions. It is also simple to carry out, recognize, and apply to composts made from a wide variety of feedstocks [155]. Meanwhile, the Dewar self-heating test has the drawback of being insensitive; while it can differentiate very stable from very young composts, judging stability levels between these two extremes is difficult [158]. The results of the Dewar self-heating test are linked to quantitative respiration measurements [155]. The Dewar test shows that "raw" compost has a respiration rate of more than 20 mg CO_2-C g compost-C^{-1} d^{-1}. For "finished" compost, a Dewar test of 0–4 mg CO_2-C g compost-C^{-1} d^{-1} is typically observed. The Dewar test indicates that "active" compost has a respiration rate of 8–20 mg CO_2-C g compost-C^{-1} d^{-1}.

The Solvita® test, which is developed by Woods End Research Laboratory, is a rapid method for evaluating compost stability and maturity. This test provides a semi-quantitative assessment of the CO_2 evolution rate. The test is carried out in a closed vessel (125 mL) with a specific volume of compost for 4 h [173]. Temperature and moisture in the compost are not strictly regulated by the Solvita® procedure.

Before the inspection, the sample is not "pre-conditioned." There are eight categories in the interpretation scale provided, ranging from "raw" compost (category 1–2), "active" compost (category 3–6), and "finished" compost (category 7–8). "Raw" compost is interpreted as poorly decomposed and possibly phytotoxic, while "finished" compost is ready for application on crops. According to Sullivan and Miller [155], in Washington State, Texas, California, Minnesota, Maine, and Illinois in the United States, as well as Germany and Denmark, the Solvita® test is used following agency compost maturity requirements. The Solvita® method is currently being reviewed by 18 states in the United States for use in compost research protocols. Brinton and Evans [158] discovered that the Solvita® test could foresee growth in seedling tests. Changa et al. [174] reported that the Solvita® index would be beneficial for on-site maturity testing with a standardized form of feedstock, even though it did not reliably predict numerical values of CO_2 evolution or NH_3 concentration. The maturity index correlated well with a few indices of stability and (or) maturity. The Solvita® test has the drawback of using a fixed volume of compost rather than mass, which means that it does not account for changes in bulk density, moisture content, and total carbon content of compost as it matures, which could affect results.

A variation of short-chain organic acids is formed during the early stages of organic matter decomposition. Many species can experience growth inhibition if they are planted in a medium containing immature compost because these compounds can be harmful to plants [135]. According to Sullivan and Miller [155], as the compost matures, these acids will decompose further and, in turn, lower the phytotoxicity effects. Chen and Inbar [175] reported that the phytotoxicity test is considered the "ultimate" test for compost maturity. Focus is given especially when compost products will be used in agriculture or horticulture. Seed germination and (or) plant growth in compost mixtures or extracts are evaluated in plant bioassays. The experiments are carried out under strict conditions, and the findings are expressed as a percentage of germination or growth relative to control [155, 157].

A combination of phytotoxicity factors in compost, including NH_3, soluble salts, short-chain organic acids, and pH, is assessed using standardized germination and growth test. According to Keeling et al. [176], highly unstable composts hinder the growth of most plant species and cultivars. Variation in plant species susceptibility to phytotoxic factors becomes more significant as compost becomes more stable. Under specified environmental conditions, plant growth inhibition caused by compost can be estimated directly through germination and growth tests. The phytotoxicity test is based on Method 9.05 as stated in the Test Methods for the Evaluation of Composting and Compost [177]. The duration of this test is between 1 and 14 days and depends on the method used. If compared to direct seeding tests, compost extract tests are typically quicker and reproducible. However, they need more time for extract preparation. To get rid of bacteria and to avoid the rapid degradation of short-chain organic acids, the compost extract must be free from contamination and sterilized; then, millipore filtration is performed. The majority of tests are semi-quantitative, with germination and growth inhibition scores grouped into 2–4 categories: none, mild, strong, and severe. The selection of crop species also plays an important role because the use of compost that has high soluble salt will have a significant impact on germination and growth test results. Seed germination and root growth can be hindered or reduced by short-chain organic acids (acetic, butyric, propionic, and valeric acids) produced during the decomposition of organic matter. These acids are also responsible for the unpleasant odor associated with anaerobic decomposition. They are produced as a natural byproduct of organic matter decomposition in their early stages. The short-chain organic acids in compost decompose and are lost as it matures [155].

All microorganisms utilize enzymes as catalysts in their respiratory chains. For example, dehydrogenase plays a major role in the oxidation or fermentation of glucose, protease assists in protein breakdown, peroxidase promotes the breakdown of lignin and benzyl alcohols, and cellulase is capable of hydrolyzing polysaccharides such as cellulose [135]. The level of microbial activity in a compost sample can be determined using enzyme activity. The pattern in enzyme activity varies depending on the enzyme used. For example, a high level of dehydrogenase activity indicates that there are significant quantities of readily biodegradable content (e.g., glucose) left in the compost, and its activity is supposed to decrease when this material is utilized, and the compost stabilizes. Benito et al. [178], on the other hand, reported that peroxidase activity showed an increase during composting as lignin increased (the less degraded part of the material). Enzyme activity analysis is an easy, quick, and low-cost technique [179]. This analysis was carried out using the method as described in Test Methods for the Examination of Composting and Compost [180]. It is worth noting that different feedstock materials can result in composts with a variety of enzymatic activities.

7.10.1.3 Chemical Methods

Compost maturity has been measured using several chemical indicators. C:N ratio, organic matter (OM) or volatile solid (VS) reduction, cation exchange capacity (CEC), and inorganic nitrogen (such as ammonium and nitrate) are the most commonly used chemical indicators.

As composting progress, a declining trend in the carbon to nitrogen (C:N) ratio can be seen; with eventual stabilization, it leads to the removal of CO_2 as organic substrates decompose, causing the loss of carbon from the system [181]. Chefetz et al. [164] and Goyal et al. [172] stated that there are various values of the C:N ratio for compost produced in the literature ranging from <20:1 to <10:1. According to Sullivan and Miller [155], the limit for this parameter should be dependent on the ratio of stable soil organic matter to total soil organic matter, which is usually between 10:1 and 15:1. The C:N ratio is not a sensitive indicator of maturity in many composting systems [182, 183]. Since C loss as CO_2 and N loss as NH_3 occur simultaneously in compost production systems with pH > 7.5, the C:N ratio can change very little during composting. As compost stability and (or) maturity measure, this criterion has a range of limitations. First, nitrogen fixation by compost microorganisms may result in more nitrogen in the compost, possibly reducing the

C:N ratio until bioavailable carbon is depleted [135]. Second, variations in this ratio are influenced by other compost properties, such as pH. At simple pH (>7.5), carbon loss as CO_2 and nitrogen loss as NH_3 occur simultaneously, resulting in a steady C:N ratio during composting [155]. Third, although this ratio can change dramatically during the thermophilic stage, it can remain relatively constant during curing, even though other maturity indices indicate that biodegradation is still occurring [164]. Finally, the C:N ratio does not always obey the predicted pattern or achieve the desired final value. Hutchinson and Griffin [184] revealed that after the thermophilic stage, the C:N ratio increased to a final value of 35:1 to 45:1. It is possible that NH_3 volatilization caused this rise during the composting period [185]. Finally, the large variety of feedstocks contributes to variation in final C:N ratios in various composts, rendering it impossible to set an absolute C:N ratio that will apply to all feedstocks [172]. This analysis was carried out using the method as described in Test Methods for the Examination of Composting and Compost (Method 9.02A) [155].

Carbohydrates, proteins, lipids, and lignin are among the organic compounds found in compost feedstock resources. About half of the organic material is transformed to CO_2 and released during the composting process, while the rest is gradually transformed into more stable compounds. Therefore, the removal of readily available organic matter (OM) will potentially be used to track the stabilization process. The volatile solids (VS) composition of compost can be used to measure the OM content [155, 186]. The OM analysis was carried out using the method as described in Method 9.10 in Test Methods for the Examination of Composting and Compost [186]. The suggested values for mature compost are 45 to 60+%. According to Benito et al. [187], there is a strong connection between CO_2 evolution and OM reduction. However, it appears that using OM reduction as a metric of compost stability has substantial drawbacks. Organic matter determinations can be influenced by inert plastic content and inorganic carbon (i.e., carbonates), which will increase the actual value [186]. The feedstock characteristics and operating requirements have an effect on the reduction in OM content during composting [188]. This evaluation's usefulness can be harmed due to these factors. Instead of being used as an individual test, TMECC (2002e) [186] suggests combining OM measures with other stability and (or) maturity indices such as C:N ratio, respirometry, pH, ammonium to nitrate ratio, etc.

Humic substances have a high potential for adsorbing positively charged ions, which can then be readily substituted for other cations on the same adsorption sites. This capacity is known as "cation exchange capacity" (CEC). The CEC has been identified as an indicator of compost stability and maturity. As organic materials are humified and carboxyl and phenolic functional groups are produced, CEC usually increases during composting [154]. The CEC analysis was carried out using the method as described in Method 8.03 in Test Methods for the Examination of Composting and Compost [155]. The feedstocks determine the maximum CEC in mature compost. For mature MSW composts, a target CEC of 60 meq/100 g of compost volatile solids (ash-free basis) has been recommended [189]. Iglesias Jimenez and Perez Garcıa [190] explained that CEC is a good and rapid method for assessing compost maturity conditions, and preferably pile temperature measurements are performed at the same time. CEC and other possible maturity indicators, such as the C:N ratio and humic compound levels, were found to have a strong connection [191]. Nevertheless, there are some disagreements regarding the CEC. Since they did not see a clear trend throughout the composting process of pruning waste with leaves and grass clippings, Benito et al. [178] concluded that CEC is not a good method for assessing stability or maturity. Mathur et al. [157] also suggested that referring CEC values to OM content could not be accurate because not all the OM is humus and that complexing ions like copper, iron, or aluminum might obstruct cation exchange sites allowing CEC readings to vary. Furthermore, even though this parameter is found to be suitable for stability determinations, determining a single threshold value can be problematic because CEC values are influenced by both composting methods and feedstock materials [154].

Another basic chemical indicator of maturity is the amount or ratio of NH_4–N to NO_3–N. Ammonium–N levels are typically highest in the early stages of composting, then gradually decrease as the compost stabilizes. The lower respiration rates observed in mature compost are more desirable for NO_3 generation through nitrification and less desirable for NO_3 loss through denitrification [155]. The release of relatively high levels of ammonium ion (NH_4^+) results from the decomposition of proteins during active composting. Ammonium undergoes nitrification to obtain nitrate (NO_3) as composting progress. According to Brinton [129] and Sullivan and Miller [155], as compost matures, the decline in NH_4^+ and rise in NO_3 results in an increase in the NO_3: NH_4^+ ratio. The NO_3: NH_4^+ ratio is very helpful in assessing the quality of compost, with a ratio of less than 2:1 indicating an immature product. Brinton [129] stated that high NO_3 levels are an indication of maturation. Bernal et al. [192] suggested a cutoff between mature and immature composts of 0.16:1 or less (i.e., a NO_3: NH_4^+ ratio of 6.25:1 or higher). Meanwhile, Ko et al. [193] used a 1:1 NH_4^+ :NO_3 ratio (i.e., 1:1 or higher for NO_3: NH_4^+) and found that this ratio is more useful for assessing the state of compost than the C:N ratio. The inorganic N analysis was conducted using the method as described in Method 9.02C in Test Methods for the Examination of Composting and Compost [155]. In general, mature composts contain more NO_3–N than NH_4–N.

7.11 Marketing of Composts

Compost may be used as a soil conditioner in a variety of applications. There are no set limits in marketing good, quality compost. However, factors such as the cost of production, transportation, and use of compost will go beyond the benefits of compost use itself. Therefore, an effective marketing program is the main basis for the success of a composting project. Past history has shown that most failed composting projects are due to ineffective marketing strategies [4]. Assessing the potential of the existing local market is the first step in developing a marketing strategy. Therefore, knowledge of the compost product produced, its uses, limits of use, and the skill of estimating the value of the product to consumers is an important aspect that needs to be considered. In addition, compost marketing strategies should also meet local needs by considering soil characteristics, types of agricultural activities carried out, weather, and transportation costs. The marketability of the resulting compost product is usually influenced by the following factors: (i) soil condition and fertility, (ii) government policy on import permits as well as the provision of financial assistance in the use of chemical fertilizers, (iii) availability and cost of other soil improvement materials such as animal manure and agricultural waste, (iv) transportation costs, (v) agricultural activities of the local community, (vi) quantity of compost produced, (vii) quality of compost produced such as nutrient content, particle size, and maturity, and (viii) the amount of organic waste produced and the composition of the waste.

The use of compost as a soil improvement material can help improve fertility, aeration, texture, nutrient content, and soil water retention capacity. Compost has a wide range of uses and can be used by a variety of market segments due to its advantageous characteristics. The compost markets cover agriculture (small- and large-scale), landscaping, gardening (residential, community), nurseries, top dressing (e.g., golf courses, parks, median strips), and land reclamation or rehabilitation (landfills, surface mines, and others) [194, 195]. Table 7.10 summarizes potential applications, relative market size, and potential barriers to widespread use of finished compost by the market segment [195].

Quality and consistency are the two main elements that influence the use and marketability of compost. According to Eggerth et al. [194], the chemical, biological, and physical characteristics of a particular form of compost determine its quality. Provided that a composting process is properly conducted, the quality finished product is defined by (i) the composition and characteristics of the raw material used in the compost processing and (ii) the form and comprehensiveness of the process used to extract impurities. Color, uniform particle size, earthy odor, absence of contaminants, adequate moisture, nutrient concentration, and amount of organic matter are some of the physical characteristics that are usually desired for final compost.

Much of the market size for compost depends on the quality of the compost and the type of use of the material. Composts made from various substrates (e.g., yard waste, source-separated MSW) have different properties and, as a result, different market potential [194]. Compost derived from the organic fraction of MSW has the largest potential market in the agricultural sector (small or large scale). This condition is due to two factors: (i) the nature and quality of MSW compost and (ii) the comparatively large amounts of substrate that would be available if composting were extensively used as a method of handling the organic fraction of MSW. Meanwhile, home gardeners, landscape contractors and suppliers, sod and sodding services, retailers of soil conditioners, nurseries, and public agencies are among the organizations or individuals who show a deep interest in using yard waste compost. A report released by the US Environmental Protection Agency shows that the highest compost market demand is from the agricultural sector with 684 million m^3/year compared to the market demand for the nursery industry, which

Table 7.10 Compost markets, applications, and potential constraints

Market segment	Applications	Potential market size	Primary constraints
Agriculture	Soil conditioning, fertilizer amendments, and erosion control for vegetable and field crops and forage grass Development of marginal lands Mulching after conservation seeding	Very large, estimated at 895 million cubic yards/year. The demand for compost for agricultural purposes within a 50-mile radius of the 190 largest US cities would exceed the supply of compost	Contaminant concentrations for crop production and cumulative loading limits Cost of transportation to end-user Bulk application equipment requirements and costs
Silviculture	Landspreading as soil conditioner for evergreen establishment Mulching for woodlot soil improvement and maintenance	Very large, estimated at 104 million cubic yards/year. This segment's potential demand could exceed the available supply of compost	Transportation cost and distance Bulk application equipment requirements and costs
Sod production	Blending with topsoil to reduce the amount of fertilizer needed to establish sod	Moderate, estimated at 20 million cubic yards/year. Market potential will be dictated by the rate at which sod producers deplete existing topsoil	Transportation cost Bulk application equipment requirements and costs
Residential retail	Soil amendment to enrich planting areas Top dressing for lawns	Moderate, estimated at 8 million cubic yards/year. Much of topsoil sold in bags is currently made with compost; thus, this market has already been penetrated	Postprocessing requirements (e.g., screening and bagging) and associated costs Consistent quality assurance Contaminant levels must be low enough to meet requirements for unrestricted distribution

(continued)

Table 7.10 (continued)

Market segment	Applications	Potential market size	Primary constraints
Nurseries	Potting mixes Topsoil amendment for areas in which field-grown trees are harvested on a periodic basis	Small, estimated at 0.9 million cubic yards/year	Consistent pH balance, nutrient content, particle size, shrinkage, and water-holding capacity required Complete and continuous testing requirements to ensure high-quality product and associated costs Compost suppliers will need to be sensitive and responsive to specific growing requirements
Delivered topsoil	Blending with marginal topsoil to produce topsoil used for establishing new lawns and planting trees and shrubs	Small, estimated at 3.7 million cubic yards/year	Consistent supplies of compost required to meet seasonal demands
Landscaping	Soil amendment for lawn establishment Top dressing Mulch	Small, estimated at 2 million cubic yards/year	Quality assurance that compost does not contain harmful amounts of contaminants Physical contaminants that might be visible on lawns Consistent supplies of compost required to meet seasonal demands
Landfill cover and surface mine reclamation	Topsoil amendments for lower grade and nonuniform compost products	Small, estimated at 0.6 million cubic yards/year. There are only a limited number of landfills or mines that are undergoing reclamation at any given time	Transportation cost

Source: [195, 196]

is less than 0.8 million m^3/year [197]. It is estimated that about 80–85% of compost produced in the United States is successfully marketed. A study conducted by the Composting Council showed that 31% of compost use was through sectors involved with landscaping activities, food crop production (25%), landfills (14%), and nurseries (11%) [197]. Compost can be purchased in bulk or in bags. The amounts used influence preferences. Landscape contractors, nurseries, producers, and park districts are examples of large-scale consumers who tend to purchase compost in bulk (in large quantities and not bagged). Small-scale consumers, such as home gardeners, prefer the bagged type. Bagged compost usually comes in sizes ranging from 0.03 to 0.06 m^3, but some larger bags are available. Nurseries and garden supply stores also sell bagged compost [194].

In an effort to market the compost produced, compost operators realize the main issue that needs to be considered is the price of the compost. In the United States, pricing typically depends on promotion, packaging, and distribution channels. The market demand and the price that customers are willing to pay for it are both influenced by the raw material used in its production. Table 7.11 shows the market price of compost in the United States produced from various types of organic waste.

Table 7.12 displays a number of composting case studies from Asia, including China, Taiwan, Malaysia, Indonesia, and Sri Lanka. Different composting systems were introduced at various scales in the studies, as well as potential income [201]. Several case studies of composting plants in Asian countries have been published, with a variety of composting systems operating on a small to large scale, utilizing windrow and other composting systems (e.g., in-vessel composting, bin composting). By selecting composting facilities of different capacities, the performance of the operating modes of government-affiliated composting units and private firms is compared.

According to Table 7.12, all private businesses are profitable, but some have a long payback period. To maximize benefit, all private companies sell compost in the medium to high price range. Chen [199] stated that one of the case studies on private firms in Taiwan sells compost at the highest price relative to other private facilities. Due to the facility's use of direct marketing to avoid shipping costs, the selling price hits USD 287/ton. The other two private companies, according to Chen [199], sell their compost via wholesale systems at wholesale prices of USD 94/ton and USD 124/ton, respectively. The high selling price of compost, set at USD 180/ton in a Sri Lankan case study by Pandyaswargo and Premakumara [198], was due to good compost quality approved by the Ministry of Agriculture and an effective marketing

Table 7.11 Compost market price in the United States

Composting raw materials	Market price (per ton)
Yard trimmings	USD 32
Source separated organic wastes	USD 39
Municipal solid waste (MSW)	USD 3
Biosolids	USD 7

Source: [197]

Table 7.12 Different composting systems at various scales and their potential income for Asian countries (China, Taiwan, Malaysia, Indonesia, and Sri Lanka)

Country/state	Operating mode	Incentive/ subsidy	Input capacity* (ton/day)	Scale	Compost output capacity* (ton/day)	Compost selling price** (USD/ton) [price category]	Profitability and payback period	References
Windrow composting system								
Indonesia	Government affiliated	Yes	0.6	Small	0.1	70 [medium]	Yes [6 years]	[198]
Indonesia	Government affiliated		200	Large	30	106 [medium]	Yes (6 years)	[198]
Taiwan, China	Government affiliated	Yes	9	Medium	3.6	93 (TWD 3000/ton) [medium]	No	[199]
Taiwan, China	Government affiliated	No	50	Medium	27	94 (TWD 3040/ton) [medium]	Yes (55 years)	[199]
Other composting systems								
Malaysia	Government affiliated	Yes	0.2	Small	0.05 (18 ton/year)	250 (MYR 1000/ton) [high]	Yes (N/A)	[200]
Taiwan, China	Government affiliated	Yes	1.5	Small	0.2	93 (TWD 3000/ton) [medium]	No	[199]
Sri Lanka	Private firms	Yes	1	Small	0.3	180 [high]	Yes (7 years)	[198]
Taiwan, China	Government affiliated	Yes	8	Medium	2	93 (TWD 3000/ton) [medium]	No	[199]
	Private firms	No	5	Small	2.36	287 (TWD 9270/ton) [high]	Yes (13.6 years)	[199]
Indonesia	Government affiliated	Yes	51	Medium	15.3	53 [medium]	Yes (3 years)	[198]
Taiwan, China	Private firms	No	100	Medium	50	124 (TWD 4000/ton) [medium]	Yes (110 years)	[199]
China, mainland	Government affiliated	Yes	638	Large	465.4	7.8–12.6 [low]	No	[198]

Source: [201]

program. These case studies indicate that, depending on marketing strategies and compost quality, the selling price of compost set by private composting facilities will vary greatly.

In contrast, government-affiliated facilities have reported losses due to lower compost sale costs, resulting in low revenue that is inadequate to keep the composting operation going. Chen [199] presented case studies focused on three government-affiliated composting facilities, where the compost generated is not sold but instead given away to farmers who participate in waste recycling as part of a national development program. The sale prices for compost provided at these facilities were lower than the selling prices set by private Taiwanese firms. There was no financial pressure or profit incentive in these facilities because there was no market-oriented goal. When compared to private composting facilities, the majority of government-affiliated composting facilities showed lower compost prices (ranging from low to medium). Only Zulkepli et al. [200] recorded a relatively high compost price (USD 250/ton) from a government-affiliated facility, though the reason for this is not specified. The majority of government-affiliated composting facilities barely making any profit, depending on external incentives or financial aid from the government or other agencies. In comparison to smaller and larger capacity plants, the findings in Table 7.12 showed that most medium-scale composting plants have the best chance of becoming financially viable. One of the most important aspects to consider when designing composting plants is the size of the facility. Other factors that can influence the composting plant's viability include incentives or subsidies offered, compost marketing, and selling price [201].

7.12 Compost Utilization

Compost has long been used as a valuable soil improvement material. Its use in a more effective way will increase the quality of crop production, reduce the use of chemical fertilizers, save costs, and indirectly preserve natural resources. Its ability to improve the physical, chemical, and biological properties of the soil results in compost with great potential to improve soil quality and crop yield. However, the use of compost today is not only limited as a crop fertilizer but has been further developed in controlling soil erosion, as a landfill cover, landscape improvement, chemical fertilizer substitute, used in road construction projects, etc.

7.12.1 Erosion Control, Turf Remediation, and Landscaping

According to the US Department of Agriculture, as much as 2 million tons of topsoil is eroded each year in the United States due to erosion. Erosion caused by wind and rain causes the topsoil of the open areas and hillsides to be carried out to the rivers, streams, and lakes nearby. As a result, this nutrient-rich layer will

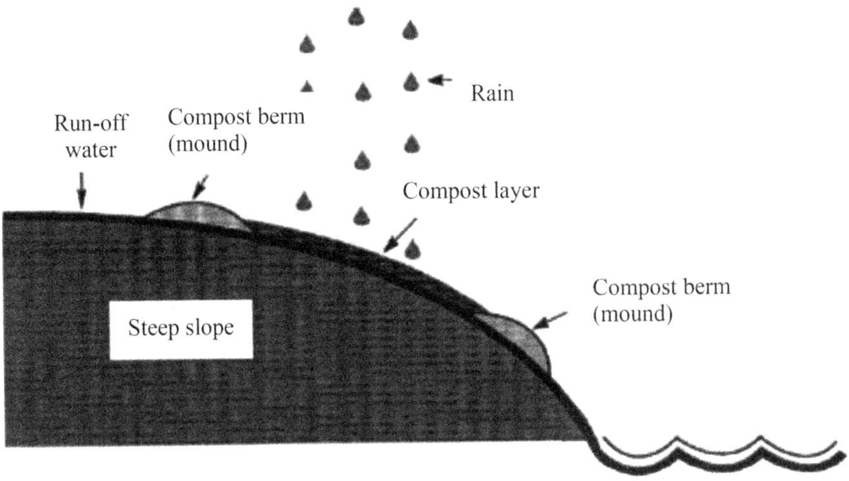

Fig. 7.14 Construction of compost berm (mound) to control erosion on a steep slope [202]

contaminate the river water surface, sediment deposits at the bottom, and can even threaten the health of aquatic organisms (because excess sediment carried during erosion contains chemical fertilizers or toxic substances). Erosion will also affect the commercial, recreational, and esthetic value of water resources. Therefore, avoiding erosion is very important for waterway protection as well as soil quality and productivity [202].

Erosion is a natural occurrence, although it is often exacerbated by activities such as road development and new construction. Some construction projects start with the removal of all vegetation and topsoil, leaving the subsoil susceptible to erosion. Compost can be more effective than conventional hydromulch at reducing erosion and creating turf on steep embankments along roads and highways because compost forms a thicker, more permanent growth due to its ability to enhance soil infrastructure. A 2- to 3-inch layer of mature compost, screened to 1/2 to 3/4 inch and spread directly on top of the soil, has been shown to mitigate erosion by promoting the growth of planted or volunteer vegetation. Compost berms (mounds) at the top and bottom of steep slopes may be used to regulate the velocity of water and provide additional protection for receiving waters (Fig. 7.14). Compost also protects soil from wind erosion and droughts because of its capacity to hold moisture. Erosion control using compost has been applied in road construction in suburban Washington, DC. Compost generated from yard waste is used to cover the cleared surface to prevent erosion on a steep slope.

Intensive turf maintenance is required to provide safe, uniform playing surfaces for recreational activities such as golf, football, soccer, and other field sports. Turfgrasses used for recreation purposes are exposed to a lot of wear and tear, making them difficult to maintain and vulnerable to turf diseases, pests, and soil compaction. Turf managers have previously used a variety of fertilizers, pesticides, fungicides, and aeration methods to solve these issues, which has resulted in high costs and the potential for negative environmental consequences. Compost is being used to replace peat moss in topdressing applications by some turf managers due to its demonstrated ability to suppress plant disease. Unlike peat moss, compost has the characteristics of rich nutrients and microorganisms that stimulate the establishment of turf and increase resistance to common turf diseases like snow mold, brown patches, and dollar spots. The utilization of compost as a topdressing at the Country Club of Rochester, New York, has successfully eliminated the reliance on the need for fungicides to eradicate such diseases after 3 years of using compost [202].

Maintaining a lawn or garden after it has been built can be difficult for both home gardeners and commercial landscape contractors. Even though aeration, topdressing, and chemical fertilizer applications are some of the most popular landscaping methods, compost can be a viable alternative. Compost can promote plant growth, minimize pests and plant infestation, and improve soil structure when used as a topdressing or planted into the soil. Compost may also be used as a mulch in the garden. Compost mulch, when applied over the roots of plants, conserves water and regulates soil temperatures. Compost mulch also keeps plants healthy by preventing soil erosion, managing weeds, and providing a gradual release of nutrients. The United States National Park Service transformed a poorly compacted 40-acre tract of land in Washington, DC, into a landscaped park in 1973, using a compost mixture consisting of digested sewage sludge, wood chips, leaf mold, and a small quantity of topsoil [202].

7.12.2 Bioremediation and Pollution Prevention

The use of a biological system of microorganisms in a mature, cured compost to sequester or break down pollutants in water or soil is known as compost bioremediation. Contaminants in soils, ground and surface waters, and the air are consumed by microorganisms. Contaminants are digested, metabolized, and converted into humus and inert byproducts, including carbon dioxide, water, and salts. Many forms of pollutants, including chlorinated and nonchlorinated hydrocarbons, wood-preserving chemicals, solvents, heavy metals, pesticides, petroleum products, and explosives, have been successfully degraded or altered using compost bioremediation. In bioremediation, compost is referred to as "tailored" or "designed" compost because it is made specifically to handle particular pollutants at specific locations. Any remediation project aims to restore the site to its pre-contamination state, which also requires revegetation to stabilize the treated soil [203]. Compost contributes to this goal by facilitating plant growth in addition to lowering contaminant levels. Compost serves as a soil conditioner as well as a source of nutrients for a wide range of plants. An example of the use of compost involving soil remediation due to heavy metal contamination was conducted by Dr. Rufus Chaney, a senior research agronomist at the US Department of Agriculture. The remediation process was carried out at a denuded site near the Burle Palmerton zinc smelter facility in Palmerton, Pennsylvania. Dr. Chaney and his team launched a remediation project to rehabilitate 4 square miles of barren soil polluted with heavy metals. Researchers used lime fertilizer and compost made from a mixture of municipal wastewater treatment sludge and coal fly ash to plant Merlin Red Fescue, a metal-tolerant grass. The remediation effort was successful, and Merlin Red Fescue and Kentucky Bluegrass are now flourishing in the area [203].

Compost bioremediation technologies have also been introduced to eliminate volatile organic compounds (VOCs) from the air that cause unpleasant or unhealthy odors [203]. The polluted air is removed by moving it through "designed," "tailored" compost. Compost plays a role as an organic medium (containing microorganisms) that digests odor-causing compounds. Every year in the United States, billions of aerosol cans are produced and used in homes, businesses, and industries. Paints, lubricants, solvents, cleaners, and other VOC-containing products are commonly found in these cans. One technology that has commonly been used to handle these cans before disposal is activated carbon. As an alternative technology for handling aerosol cans, vapor-phase biofilters using compost are gaining popularity. Biofilters, unlike traditional VOC protection systems like activated carbon, actually degrade dangerous pollutants into harmless items. They also have low capital, life-cycle, and operational costs, as well as low maintenance and energy requirements. Over 18 months, the Metro Central Household Hazardous Waste recycling facility in Portland, Oregon, saved about $47,000 in hazardous waste disposal costs by transitioning from landfilling to using a vapor-phase biofilter. About 38,000 aerosol cans were remediated with vapor-phase biofilters at the plant. As a result, it reduced its disposal costs from $505 per loose-packed drum to $265 per drum (from $2.35 per can to $1.30) since the cans were no longer dangerous and no longer needed special handling [203].

7.12.3 Disease Control for Plants and Pest Control

Compost technology is a useful method that farmers involved in sustainable agriculture are already using to increase yields. Skilled growers are now discovering that compost-enriched soil can also aid in disease suppression and pest control [204]. These compost applications can help farmers save money, reduce pesticide use, and conserve natural resources.

It is reported that more than 10% of vegetables grown in the United States are lost due to root rot every year [204]. Other soil-borne plant pathogens, including the microbes that cause ashy stem blight and chili pepper wilt, cause additional crop losses. Compost can aid in the control of plant diseases and the reduction of crop losses. Four potential mechanisms for disease control with compost have been proposed: (1) effective competition for nutrients by beneficial microbes; (2) antibiotic development by beneficial microbes; (3) successful predation against pathogens by beneficial microbes; and (4) activation of disease-resistant genes in plants by composts. By enriching compost with various disease-fighting microorganisms or other amendments, scientists have improved its natural capacity to suppress diseases.

This "tailored" compost can then be applied to crops that have been contaminated with identified diseases. "Tailored" compost substantially decreased or eliminated the use of pesticides, fungicides, and nematicides, all of which have the potential to damage water resources, food safety, and worker safety. Chemical soil treatments, such as methyl bromide, can be more costly than using "tailored" compost. Compost-treated soil keeps irrigation water better, which saves money on water.

Dr. Harry Hoitink, with his team from the Ohio State University, has studied the effects of compost on plants suffering from *Pythium* root rot. They found that the use of "tailored" compost had a very good effect on plant growth and was able to prevent the spread of disease. A group of researchers from the New Mexico State University has applied compost from municipal wastewater sludge and yard trimmings on chili trees that suffer from *Phytophthora* root rot or chili wilt. They discovered that the salt content of compost is important in preventing disease and rising crop yields. The 10-ton and 20-ton compost applications suppressed chili wilt the most and produced the highest yields. Salt concentrations in compost should be measured and application rates modified accordingly for better performance. Plant salt sensitivity necessitates a salt-concentration-controlled compost.

In addition to its use in disease control, compost can also be used to kill certain pests, such as parasitic nematode (worm) infections [204]. Chemicals that destroy nematodes or prevent their eggs from hatching can be contained in specially formulated (tailored) compost. Most of the compost assists in the control of parasitic nematodes by supplying nutrients to the soil, which promote the growth of fungi and other species, which compete with or kill parasitic nematodes. Compost also increases the basic health of plants, rendering them less resistant to pests. A study conducted by a group of researchers at the University of Florida found that the utilization of compost can significantly reduce root-knot nematodes, even without fumigants.

7.12.4 Reforestation and Wetland Restoration

Natural phenomena and human activities such as erosion, flooding, and logging are contributing to the loss of large amounts of organic matter in soils in the United States [205]. Compost, on the other hand, may help to restore barren soils. Compost provides plants with the infrastructure, humus, and nutrients they need to re-establish themselves in damaged areas. Over the last 30 decades, organic matter in the soils of wetlands in the United States has gradually decreased. Over 100 million acres of wetlands in the United States have been drained, and watersheds now hold less than half of the organic matter they did in the seventeenth century. This has resulted in worsening annual floods, declining groundwater quality as well as wildlife diversity. Compost can absorb up to four times its weight in water and can be used to substitute essential organic material in wetlands due to its high organic content. Compost will also assist in the restoration of forests and the revitalization of ecosystems. By offering an excellent growing medium for young seedlings, compost will help with reforestation efforts. Similarly, compost will aid in restoring barren habitats, supplying vital nutrition to native wildlife populations. Compost promotes native plant growth, which provides food for native and endangered animal populations, by improving the chemical and mineral properties of soil.

The US Forest Service, Bureau of Indian Affairs, Cherokee Tribal Council, and the US Environmental Protection Agency (USEPA) collaborated on a 3-year study (1995–1998) to compare the efficacy of straw versus three different types of composts in stimulating tree seedling growth and decreasing soil erosion. The three composts were used as a 2-inch mulch on white pine softwood, chestnut oak, and Chinese chestnut hardwood seedlings and were made from yard trimmings, municipal wastewater sludge (biosolids), and municipal solid waste (MSW) [205]. The project was carried out at three separate locations in Cherokee, North Carolina: the Cheoah Ranger District, the Nantahela Forest, and the neighboring Qualla Cherokee Reservation. Since the research sites had compacted, eroded areas or disturbed steep slopes, they were selected. The seedling forms were grown on each plot, and each of the three composts and the straw were tested on two plots each. The height, diameter, and survival rates of seedlings planted in the composted test plots outperformed those planted in the straw test plots after 20 months. Besides, in the composted plots, volunteer revegetation by herbaceous plants was exceptional. Erosion was visible in the straw plots after 30 months but not in the composted plots.

The use of compost is also performed extensively in wetland restoration. The Clean Washington Center funded a 2-year project in Everett, Washington, from 1994 to 1996 to test two forms of compost in the restoration of damaged wetlands [205]. The restoration site is composed of two massive wetlands connected by a 550-foot-long, 18-inch-deep culvert. A sawmill once stood in the sandy field between the wetlands. The area around the mill was left largely empty after it was destroyed, leaving the rail-road tracks and bike path adjacent to the upper wetland susceptible to flooding. A yard debris compost and a mixed compost made of biosolids and yard debris were deposited into 14 different test plots. There was also a control plot generated with no compost. Throughout 1996, a variety of indigenous wetland plant species were introduced into each plot, and their growth was monitored every 6 months. From the observation, the compost-enriched soils closely resembled the natural wetland substrate. Furthermore, plants in both compost test plots grew 20% faster and survived 10% to 15% longer than control plots. The site was also able to deal with the flow of heavy winter rain in 1996 and subsequently prevent floods on railroad tracks and bike paths from occurring.

7.12.5 Soil Amendment and Renewable Energy via Compost Pelletization

Compost is typically used as a fertilizer or soil enhancer, while other methods such as anaerobic digestion are used to generate energy from organic waste. Composting, on the other hand, will provide a new source of energy fuel while also addressing the issues of low-quality compost and compost overproduction [206]. Using superheated steam produced by a heat recovery steam generator, compost can be combusted, and electricity can be generated [207]. Different types of composts have been studied, including spent mushroom compost (SMC) [208], sewage sludge compost [209], and green waste compost [210]. With a heating value of 3300 kcal/kg, food waste compost has recently been used for this purpose and has demonstrated its ability [211]. Food waste and biowaste compost can be generated by using biowaste that contains essential nutrients rather than dumping it in landfills. Grinding and sieving are used to increase the contact surface area of these composts to make them more presentable for distribution. Composts in powder form, on the other hand, are difficult to handle, maintain, and store because they are heavy and bulky [212]. Among the properties of the compost produced is that it has a low density and produces little odor from the type of waste used, as well as dust pollution from fine particles during the grinding process. Handling compost during transportation and storage processes can have harmful effects on human health.

Pelletization of compost, which involves compressing or molding the compost into solid pellets, is one way to solve this [213]. Pelletization may result in a 50–90% reduction in volume and weight of compost, as well as an increase in bulk density [214]. Compost pelletizing has several benefits, including reduced dust production, reduced pollution, improved fertilizer release control, easier transportation, and longer shelf life. Pelletizing of compost enhances process safety by reducing powder formation and removing pathogens and parasites that are present in the microbes responsible for composting [215]. Furthermore, it improves fertilizer application precision while also reducing compost consumption [216]. When opposed to traditional compost powder, which can be easily washed away by rain or surface runoff, compost pellets appear to stay in the soil [217]. Compost pellets are a slow-release fertilizer that improves nitrogen absorption, reduces leaching losses, and improves plant and soil health [218].

Compost pelletization is also carried out to transform the compost into energy [206]. After researching the energetic properties of compost, Zajonc et al. [219] looked into pelletizing it to improve its fuel properties. Zajonc et al. [220] investigated the energetic properties of compost and found that cellulose-rich materials, such as waste cardboard, could reduce carbon loss during the composting process, achieving a gross calorific value (GCV) of 12 MJ/kg. In the dry matter of compost, the total content of elements that pose a danger was also decreased (10% in Ni, Zn, and Cd, and 20% in Hg). To use the compost for energy, it must be prepared in a protected area to prevent overwatering due to precipitation, and it must be dried to a certain amount that is appropriate for treatment and pelletization, with the optimum moisture for pelletization being about 25–30% depending on the compost's composition [219]. Furthermore, composts combined with high calorific value items and pelletized can be used as a heating fuel in cold weather. A suitable mixture of compost and compounds, especially high calorific products like coal or wood chippings, must be combined for the compost to be used as an energy source efficiently [219]. Combustible solid fuels made from processed compost may be used to generate heat.

7.13 Selected Composting Case Studies Around the World

This section highlights several successful composting schemes and practices currently in use around the world. These composting systems, which are based on technologies such as in-vessel reactors, tunnel composting, windrows, and aerated static piles, are drawn from Asian countries, EU Member States, and the United States. It focuses on full-scale, centralized composting facilities that handle MSW and other biodegradable organic waste, such as green or yard waste. This section includes details about the name of the plant and location, as well as its design capacity, the types of waste it handles, and the composting technology it utilizes.

7.13.1 Case Study 1: The Takakura Composting Method (Surabaya, Indonesia)

The Japan International Cooperation Agency (JICA)—Kyushu International Centre, Kitakyushu City, the Kitakyushu International Techno-Cooperative Association (KITA), and the Institute for Global Environmental Strategies (IGES) are collaborating to promote KitaQ System Composting in Asian cities, intending to reduce landfill dependence and develop more sustainable material-cycle society [221]. Surabaya, Indonesia's second-largest city, was the first to try out the KitaQ System Composting. KitaQ System Composting, in comparison to larger-scale, high-tech composting, focuses on decentralized, community-managed, and easy technologies [221]. Residents in Kampong Rungkot Lor (an urban community with about 200 households adjacent to Surabaya's largest industrial area) were encouraged by a local nongovernmental organization (NGO), Pusdakota, to separate waste at the source. The organic waste is then segregated and processed at a nearby composting facility using the Takakura Composting Method (TCM), a simple, low-tech, and low-cost composting method developed by Koji Takakura of J-Power Company, JPec Co., Ltd.

The Takakura Composting Method (TCM), named after its founder, Koji Takakura, was first implemented in Kitakyushu, Japan, and has since extended to many other cities. TCM was then introduced to Surabaya, Indonesia, as part of a city-to-city cooperation program, and successful implementation has been established over the years [222]. The average volume of waste disposed per day at Benowo Landfill in Surabaya has decreased by 30% over 5 years (2004–2009) [223]. Because of its potentials compared to other traditional techniques, TCM has been replicated in several cities in Indonesia, including Semarang, Medan, Makassar, Palembang, Jakarta, and Balikpapan, as well as other cities in Thailand (Bangkok), the Philippines (Bogo, Cebu, Talisay, Puerto Princessa), Malaysia (Sibu), and Nepal (Lalitpur) [224].

TCM's key innovation is the use of fermentative microbes as seed compost, which were originally cultured from local fermented foods including soy sauce, yogurt, fermented beans, which are known in Indonesia as *tempe* and *tape* (fermented rice), fruits and vegetable peels, rice bran, and rice husks [223]. Fermentative microorganisms are introduced to wet waste after it has been obtained from storage containers. When the waste is being delivered to the composting sites, the fermentation process starts. When the waste arrives at the composting facility, it is positioned in well-ventilated containers lined with grain bags and then piled on top of one another to finish the fermentation process. After the waste has been fermented, it is put in piles with controlled temperature and oxygen levels to finish the composting phase. The TCM produces high-quality compost in 10–14 days [225]. The utilization of TCM offers several advantages, including (1) TCM can be completed in 1 or 2 weeks, which is much quicker than the windrow method or other composting processes, which typically take months to complete, (2) because of its versatility and high efficiency, TCM can be maintained over a smaller area of land, requiring less land for composting centers, resulting in lower costs, (3) TCM compost is half-matured, which means that it matures in the fields rather than in composting facilities, (4) TCM is naturally low-cost since it relies on local resources, uses simple technology, and is labor-intensive, and (5) TCM can be used in both homes and composting centers. The main difference is that it employs fermentative microorganisms, which can be used in a small ventilated basket at home or in a larger area (heap or windrow) in composting facilities [223, 224].

A composting facility costs an average of IDR 194,000,000 (USD 20,000) to construct. Residents pay one fee to the community group, which is in charge of garbage collection, and another to the city, which is in charge of disposal and transportation. The fees are usually charged as a lump sum payment following other community fees (local taxes), and the rates are collectively agreed upon and set by the community members [225]. The compost generated at the community plants has been used extensively in public green spaces.

7.13.2 Case Study 2: Botarell Composting Scheme (Tarragona, Spain)

The Botarell composting system is situated in the Baix Camp region of Tarragona province in northern Spain. The Baix Camp region is a Catalonian administrative division that offers centralized services to the local municipalities. Biodegradable waste from 50,000 homes, as well as biodegradable waste from hotels, schools, markets, and factories, is sent to the composting system. The biodegradable fraction of waste is collected using a house-to-house kerbside separate collection method and shipped by lorry to a composting plant near Botarell village. The Botarell composting facility was opened in June 1997. Approximately 7000 tons of kitchen waste and 3000 tons of garden waste were composted in the first 2.5 years of operation, yielding 900 tons of compost. However, as more municipalities have introduced separate collection systems, the volume of biodegradable waste collected at the plant has risen over time [226].

The composting technology used in this plant is an aerated static pile. The plant can manage 30,000 tons of biodegradable kitchen waste and 5000 tons of garden waste per year. The mixture of these materials is composted for 2–3 weeks after the two are mixed. The mixture is then screened through a trommel screen with an 80 mm diameter. The rejected stream is disposed of in a landfill, while the screened fraction is placed in aerated static piles for 12–14 weeks to mature. A mechanical mixer is used to provide aeration. A 25-mm trommel screen and a densimetric table are used to screen the finished compost. The trommel screen separates the finished compost into different fractions based on demand. Buildings are used to contain the aerated static piles. Biofilters are used to handle the air in these buildings.

Other than the legal definition of compost for agricultural purposes and the size required by various clients, there are no clear requirements for compost quality. Private gardens and individual farmers are the current markets for compost, which is mostly used in fruit and olive orchards. It was also sold for public work projects such as landfill closure and road revegetation. The compost that is currently being sold costs about 12€ per ton. For farmers, the compost is too costly (there is an abundance of manure available in the area), but it is relatively inexpensive for private gardeners. As a result, attempts are made to promote the product to retailers [226, 227].

The operating costs are covered by a flat rate of 20€ charged to households by each municipality for the treatment of the compostable, biodegradable fraction, as well as charges for waste from municipalities outside the Baix Camp zone. When the amount of biodegradable waste being processed increases, so does the revenue generated from compost sales. Compost revenues received a total of 10,850€ during the first 2.5 years of operation. Since manure is abundant in the region, the price of compost is low [226].

The Botarell composting scheme is generally recognized as one of Spain's most popular composting facilities. One of the key reasons for success, apart from technological excellence, was the large number of people who participated in the segregation of the biodegradable fraction of waste. This was accomplished by appropriate Catalonian legislation that mandates separate collection for municipalities with populations greater than 5000 residents, as well as a comprehensive public awareness campaign that included the door-to-door distribution of leaflets and brochures, a bus roadshow, as well as radio and press campaigns [227].

7.13.3 Case Study 3: St. Oedenrode Composting Scheme (Netherlands)

The GICOM company in the Netherlands developed the St. Oedenrode composting system. Van Kaathoden B.V. is in charge of the plant's operation. The composting facility was installed in 1991 and has been expanded three times since then in 1992, 1995, and 2002. Both municipal solid waste (MSW) and green waste are processed at the plant. The plant now has a 35,000 ton/year treatment capacity as a result of the expansions. The construction requires 6 months to finish, and the extension required 4 months. Compost tunneling is the technology that was used for the high rate stage of composting. Currently, 14 composting tunnels with dimensions of $30 \times 5 \times 5$ m^3 and $30 \times 4 \times 5$ m^3 (length, width, and height) are in service. The tunnels are located inside the building. Scrubbers and biofilters play a role in purifying the outgoing process air from ammonia and odors [227].

The tunnel's climate control program is initiated after the tunnel has been filled with compost feedstock. The composting tunnels are regulated in terms of temperature, oxygen, and moisture levels to maximize the process, achieve pathogen mitigation criteria, and meet process goals. After the tunnels have been filled with compost, the variable speed blower attached to the tunnel begins to work, pushing air into the tunnel's aeration floor system. The air is recirculated by blowing it into the concrete plenum behind each tunnel, then into the floor piping system, and finally into the compost mix. It passes from the headspace to an external recirculation loop ductwork through an open recirculation damper, then back through the blower and into the plenum. Several parameters are measured as the air reaches the circulation path, including air temperature, air humidity, oxygen and carbon dioxide concentrations, and pressure [227]. Composting tunnels are also used for the maturation process. The high rate and maturation stages both take 3–5 weeks to compost, with the high rate stage taking 5–7 days.

7.13.4 Case Study 4: Castle-Morpeth Composting Scheme (England)

The Environmental and Planning Department of Castle Morpeth Borough Council is in charge of the Castle-Morpeth composting scheme. The scheme serves a 3000-ha area and comprises 25% of the borough's population, including 5000 of the 20,400 households. The program, which started in March 1993, aims to reduce the amount of biodegradable waste sent to landfill in the borough to a minimum. The scheme was first piloted on a 468-house housing estate, with a leaflet outlining the pilot scheme's objectives and suggesting that if people in the region did not want a permanent scheme after the trial, it would be removed. Morpeth Borough Council also arranged an exhibition at a local school, which was attended by approximately 80 people. Following the pilot's success, the scheme was extended to include other householders, with a leaflet acting as a marketing tool. The scheme is now widely acknowledged to be self-publicizing, with nearby residents asking to be included in the program [226].

The scheme aims to help Castle Morpeth achieve its recycling goals set by the government. Biodegradable waste (kitchen and garden waste) was selected as a target because it makes up a significant portion of the waste stream and has the greatest potential for pollution if landfilled. The scheme uses a two-bin system, with each householder getting green and a grey bin. Both bins are 240-L wheel bins, with the green one collecting biodegradable kitchen and garden waste and the grey one collecting all other waste. A standard refuse vehicle is used to collect biodegradable waste from residents, as well as the contents of grey wheel bins. The biodegradable waste is then transported to a central composting site, where larger items, such as plastic bags, are discarded. The waste is then shredded on-site, then placed in covered windrows, and turned on a regular basis. It is ready to be screened and bagged after 7 weeks. The biodegradable waste is turned using a telescopic shovel at the site. There was no UK standard for compost at that time, but Morpeth Borough Council given comparative figures of levels of substances allowed in the EU eco-label standard, such as heavy metals [226]. Newcastle University is conducting trials to see how effective the compost is at growing different types of shrubs and plants, as well as investigating the heavy metal content of these plants.

Compost generated by all scheme participants is collected and sold in 50-L and 80-L bags at retail price, or even in bulk. The compost is sold directly by Morpeth Borough Council to residents, garden centers, or landscape gardeners. The use of compost for golf course top treatment, combined with sand, has been described as a potential growth market. Morpeth Borough Council sells the compost for GBP 1.99 (EUR 2.99) for a 50-L bag, GBP 2.99 (EUR 4.49) for an 80-L bag, and GBP 50 (EUR 75) for 1.5 tons [226]. These prices are also offered to garden centers as a retail price recommendation. The scheme generates an annual income of GBP 76 600 (EUR 115,000) by producing 3000 tons of compost from 5000 tons of collected biodegradable waste. The success of this scheme can be attributed to its simplicity, as collecting biodegradable waste requires no extra effort from the householder.

7.13.5 Case Study 5: Dodge County Transfer and MSW Compost Facility (Minnesota)

In 2004, the Dodge County Transfer and MSW Compost Facility in Mantorville, Minnesota, started composting at its 50-ton/day solid waste transfer facility [228]. Six 40-yard Nature Tech in-vessel compost digesters with a 3000 tons/year capacity are installed at the plant. The county is still fine-tuning its composting method and determining which waste streams to process. The operation's primary goal is to compost MSW residue left behind by trucks transporting waste to a waste-to-energy facility that serves both Dodge and Olmsted Counties. Broken glass and grit are present in the remaining material.

When compared to the county's MSW transfer station tip fee ($70/ton), the county charges a lower tip fee for organic ($45/ton). This will act as an incentive for commercial haulers. At the transfer point, a materials recovery facility (MRF) separates recyclables into two streams: bottles and cans and mixed paper. Dodge County participates in the Southeastern Minnesota Recyclers Exchange, a 10-county marketing cooperative, to sell recyclables.

The organic fraction of the MSW stream is loaded into the digester and held there for 21 days with either positive or negative aeration, as well as being heated to at least 155 °F for 7 days to meet pathogen reduction requirements. Materials are unloaded from the vessels after 21 days and placed in 16-ft wide, 10-ft height of windrows for 3–6 months of curing. A CAT front-end loader is used to turn the piles once a week. A private contractor screens the compost after it has been cured. On-site landfill cover is made from compost that is still contaminated with broken glass, bottle caps, and other small debris. Landscapers and farmers in the area can buy clean compost for $8 to $10 per yard.

7.13.6 Case Study 6: Rapid City Composting Facility (South Dakota, USA)

Rapid City is an integrated solid waste management facility consisting of an MSW-biosolid composting plant, material recovery facility (MRF), an outdoor yard composting operations, and the landfill that have been opened in phases [228]. The MRF has been in operation since 1996. Two rotating Dano drums were constructed in 1997 by the city, which had previously been used at a privately owned and operated mixed waste composting plant in Portland, Oregon. The drums were used for volume reduction prior to landfilling before the composting portion of the facility was built and allowed to operate. In 2002, an IPS-Siemens agitated bay composting system was constructed in a building next to the MRF, and full-scale MSW fraction operations began in 2003. Noncompostable are picked off the line ahead of the drums in a sorting line. Biosolids were first added to the drums in 2004. MSW and biosolids are processed at a rate of over 200 tons/day. Organics that have been preprocessed account for 157 tons/day, while biosolids account for 56 wet tons/day. The drums will hold a maximum of 192 tons/day of MSW and 9 dry tons/day of biosolids.

After the composting process in the drums, materials are screened and transported to the building housing the 9-bay (each 10 ft wide × 8 ft high × 280 ft long) IPS Composting System from Siemens Water Technologies. Compost is unloaded and moved to a curing building with an aerated floor after about 30 days in the bays. The material is processed via a product refinement system that includes a Bivitec screen (3–8 inches screen size) and a Triple S destoner after a minimum of 1 month. The product is stored in a 3-acre area. Rapid City's compost is being used for erosion and sediment control, stormwater management, and landscaping projects, which is a new opportunity for the business [228].

7.13.7 *Case Study 7: Disposal of Dead Animals, Poultry, or Fish by Two-Stage In-Bin Composting System (Illinois, Maryland, Arkansas, and Indiana)*

7.13.7.1 Needs and Legal Requirements for Disposal of Dead Animals, Poultry, or Fish

The Illinois Department of Agriculture (IDOA) administers the Illinois Dead Animal Disposal Act. The Act requires Illinois owners or those caring for livestock to dispose of dead animals, poultry, or fish within 24 h of death. However, if the death is caused by a highly contagious, infectious, or communicable disease, IDOA must be contacted to recommend a safe method of disposal before the carcass is taken away for disposal. Any Illinois facility that composts dead animal, poultry, or fish requires an Illinois Environmental Protection Agency (IEPA) permit unless the compost is composed of solid waste generated by the facility's own activities, which are treated within the site where such solid wastes are generated. A legal composting process for disposal of dead animals, poultry, or fish involves co-treatment of significant quantities of livestock waste. Facilities used for this co-treatment would be legally defined as potential secondary sources pursuant to the Illinois Environment Protection Act. To protect underground drinking water sources, the Act establishes minimum setback zones for the location of new potential sources. No new potential secondary source may be placed within 200 ft (60.96 m) of any existing community water supply well or other potable water supply. No new potential secondary source may be placed within 400 ft (121.9 m) of any existing or permitted community water supply well deriving water from an unconfined shallow fractured or highly permeable bedrock formation or from an unconsolidated and unconfined sand and gravel formation. An exception to the minimum setback zone is provided, which is applicable when the owner of a private potable well is also the owner of the new potential secondary source. In such instances, a prohibition of 75 ft (22.9 m) shall apply and the owner shall notify the IEPA of the intended action. Any legal matter changes from time to time and is location specific. The readers are required to consult with the local government agency for the most recent legal requirements of composting dead animals, poultry, or fish [229–236].

Composting can be an economical and environmentally acceptable method of handling dead animals. It produces little odor and destroys harmful pathogens. Composting of dead poultry is the most common practice. The process may apply equally well to other dead animals weighing as much as 100 pounds by grinding or cutting them into smaller pieces. Accordingly, larger animals require additional equipment, labor, and handling to cut the animals into smaller pieces to facilitate rapid composting.

Composting of dead animals may be considered when (a) a preferred use, such as rendering, is not available; (b) the mortality rate as a result of normal animal production is predictable, and the composting system will work with normal mortality during all seasons of the year; (c) sufficient land is available for nutrient utilization; (d) federal, state or local regulations permit dead animal composting; (e) other disposal methods are not permitted or desired; (f) the composting system can be constructed by the farm owner at a reasonable cost for materials; (g) the composting system fits into the everyday management chores of the farm; (h) there are no offensive odors or danger to people or firm animals; and (i) the end product of compost is safe and valuable as a crop fertilizer, or marketing of finished compost is feasible.

The composting of dead animals is similar in many ways to other methods of composting, so the same composting process siting and planning considerations apply. Composting of dead animals does, however, have unique problems that require special attention. The animal producer is responsible for procuring a construction permit before installation of the facility begins and an operating permit to operate the facility [229–231].

7.13.7.2 Design Considerations of a Two-Stage In-Bin Composting System

Dead animal composting facilities should be roofed to prevent rainfall from interfering with the compost operation. Dead animal composting must reach a temperature in excess of 130 °F (54.4 °C) to destroy pathogens. The addition of rainfall can elevate the moisture content and result in a compost mix that is anaerobic. Anaerobic composting takes much longer and creates odor problems. All good composters have certain common features: (a) a roof ensures year-round operation and controls rain water and percolation problems; (b) an impervious, weight-bearing concrete foundation is critical to all-weather operation, also secures the composter against rodents, dogs, etc., and prevents contamination of the surrounding area; (c) rot-resistant building materials: preservative-treated lumber resists the biological activity of composting. It, or a similar material, should be specified when materials are ordered.

A typical dead animal composting facility consists of two stages: (1) primary composter (PC) and (2) secondary composter (SC) or secondary digester.

The first-stage primary composter (PC) is made up of many equally sized bins (Figs. 7.15 and 7.16) in which the dead animals and amendments are initially added and allowed to compost. The mixture is moved from the first stage to the second stage, or secondary digester, when the compost temperature begins to decline.

The second-stage secondary digester can also consist of a number of bins, but it is most often one bin or concrete area or alley that allows compost to be stacked with a volume equal to or greater than the sum of the first-stage bins.

The design volume for each stage should be based on peak disposal requirements for the animal operation. The peak disposal period normally occurs when the animals are close to their market weight. The volume for each stage is calculated by multiplying the weight of dead animals at maturity times a volume factor. The volume factor (VF) can vary from 1.0 to 2.5 cubic feet per pound of dead animal, depending on the type of composter, local conditions, and experiences.

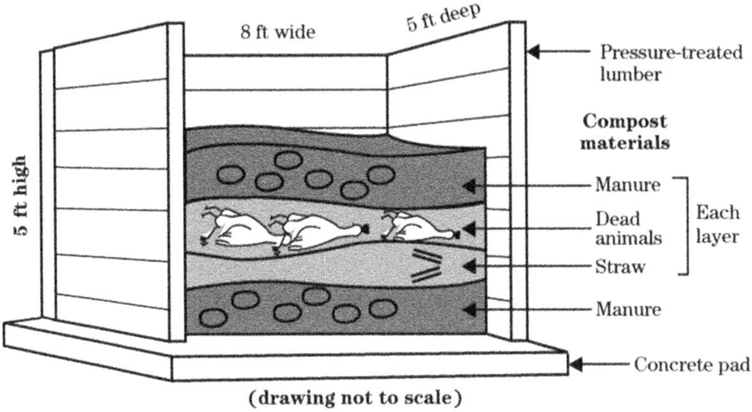

Fig. 7.15 One of the composter bins of the two-stage in-bin composting system [230]

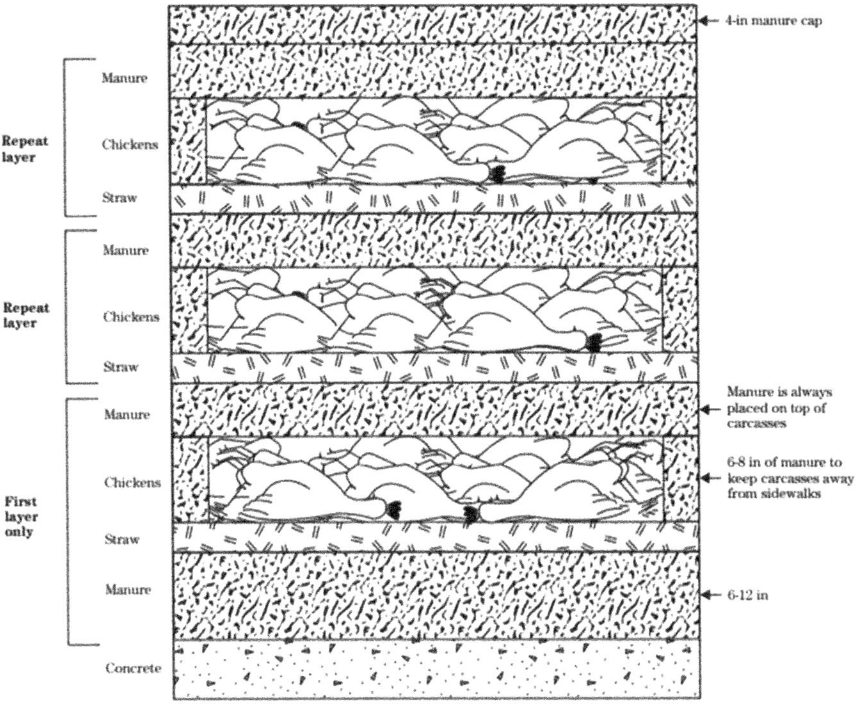

Fig. 7.16 Layers of dead chicken, manure, and straw in a composter bin [230, 236]

It has been introduced that bins for the first-stage primary composters are typically 5 ft high (H), 5 ft deep (L), and 8 ft across the front (W), shown in Fig. 7.15, as a basic bin unit. The width across the front (W) should be sized to accommodate the equipment used to load and unload the facility. Although a typical bin is 5 ft deep (L), the depth should also be sized to accommodate the equipment used. When more composting volume is needed, the operator or farm owner will simply increase the number of the basic bins.

A high volume to surface area ratio of each bin is important to insulate the compost and allow the internal temperature to rise. The bin height and depth should be no less than one-half the width. Shallow bins are easier to unload and load; therefore, the bin depth should be no more than the width.

Technical description and the process steps of a complete composting system have been introduced in Sect. 7.6.9.

Mortality rates of animals vary considerably because of climate and varieties, species, and types of operation. Information provided by the animal producer/operator should be used whenever possible. Table 7.13 gives typical mortality rates, flock life, and market weights for poultry. Composter size is calculated based on animal farm capacity (projected mortality) and certain guidelines for proper sizing. The University of Maryland guidelines [230] recommend the sizing of one pound of dead chicken per cubic foot of primary composter space per day. Any disposal method, including composting, should accommodate the normal mortality of birds of market age. The normal mortality rate of broilers is 0.1% per day, but daily fluctuation in mortality of up to 0.25% is not uncommon. The general rule of volume factor (VF) for the design of dead chicken composter is to build one cubic foot of primary composter capacity and one cubic foot of secondary composter capacity for each pound of (chicken) mortality to be disposed of. Therefore,

$$\mathrm{VF} = \left(1\,\mathrm{ft}^3 \text{ composter capacity}\right) / \left(1\,\mathrm{lb} \text{ of mortality to be disposed of}\right) \tag{7.4}$$

The following illustration demonstrates a method of estimating peak dead bird disposal requirements in birds of market age and weight:

$$\mathrm{PDADR} = \mathrm{TFLW}\left(0.0025\right) \tag{7.5}$$

$$\mathrm{TFLW} = \mathrm{N} \times W_{\mathrm{m}} \tag{7.6}$$

$$\mathrm{PCV} = \left(\mathrm{PDADR}\right)\left(\mathrm{GR}\right) = \left(\mathrm{N} \times W_{\mathrm{m}}\right)\left(0.0025\right)\left(\mathrm{VF}\right) \tag{7.7}$$

$$\mathrm{SCV} = \left(\mathrm{PDADR}\right)\left(\mathrm{VF}\right) \tag{7.8}$$

$$\mathrm{VOL} = L \times W \times H \tag{7.9}$$

Table 7.13 Poultry mortality rates [230]

Animal type	Mortality rate	Growth cycle	Cycles	Market weight
Poultry type	(%)	(d)	(per year)	(lb)
Broiler	4.5–5.0	42–19	5.5–6.0	4.2
Boaster				
Female	3	42	4	4.0
Male	8	70	4	7.5
Laying hen	14	440	0.9	4.5
Breeding hen	10–12	440	0.9	7–8
Breeder male	20–25	300	1.1	10–12
Turkey female	5–6	95	3	14
Turkey male	9	112	3	24
Swine, farrow—prewean	11	20		10
Swine, farrow—nursery to 60 lb	2.6	47		35
Swine, growcr/finisher	6	119	2.5	210
Swine, sow and gilt <250 lb	2.5			
Swine, sow and gilt 250–500 lb	3			
Swine, sow and gilt >500 lb	3.7			
Beef cattle (>500 lb)	1.2			
Beef calf	3.3			
Dairy cattle (>500 lb)	2.8			
Dairy calf	6.4			
Horse <20 years old	1.2			
Horse >20 years old	10.2			
Horse, foal (less than 30 days)	4.9			
Sheep, all causes	6.2			
Sheep, nonpredator	3.9			
Lamb, all causes	10.1			
Lamb, nonpredator	5.5			

where PDADR = peak dead animal disposal requirement (lb); TFLW = theoretical farm live weight (lb); 0.0025 = 0.25% mortality rate; N = head capacity, number of live animal on a farm; W_m = market weight (lb); PCV = primary composter volume (ft^3); SCV = secondary composter volume (ft^3); VOL = volume of a single bin (ft^3); L = length of a single bin (ft); W = width of a single bin (ft); and H = depth of a single bin (ft).

Alternatively, the following equation and Table 7.13 can be used for sizing the volume required for each composting stage:

$$\text{PCV} = (\text{PDADR})(\text{VF}) = (N \times W_m)(M/T)(\text{VF}/100) \tag{7.10}$$

where M = percent normal mortality of animals for the entire life cycle expressed as percentage; T = number of days for animals to reach market weight; M/T = an estimation of the percentage of dead animals to be composted at maturity although other estimators or field experience may be more accurate; and VF = volume factor, between 1.0 and 2.5 cubic feet per pound.

7.13.8 Design Example of a Two-Stage In-Bin Composting System

This design example can be applied to poultry farms of varying sizes and types of birds. The general rule for design is to build one cubic foot of primary composter volume and one cubic foot of secondary composter volume for each pound of mortality to be disposed of. The following technical information is given to determine the size, number, and configuration of primary composting boxes: (1) primary capacity (cubic feet) equals requirement (pounds per day); (2) height of primary and secondary bins is 5 ft; (3) width of primary and secondary bins is determined by the width of manure-handling equipment but should not exceed 8 ft; (4) depth of primary bins should not exceed 6 ft; (5) more, smaller primary bins work more efficiently than few, very large bins; (6) the head capacity of a chicken farm N in any day = 60,000; and (7) market weight W = 4 pounds.

Solution 1 N = head capacity, number of live animal per day on a farm = 60,000

W_m = market weight = 4 pound
0.25% mortality rate = 0.0025
TFLW = theoretical farm live weight = $N \times W_m$ = (60,000) (4 lb) = 240,000 lb
PDADR = peak dead animal disposal requirement, pounds = TFLW × 0.0025
PDADR = (240,000 lb) (0.0025) = 600 lb of mortality to be disposed of

The general rule of volume factor (VF) for design is to build one cubic foot of primary composter volume and one cubic foot of secondary composter volume for each pound of mortality to be disposed of. Therefore,

VF = (1 ft^3 composter volume) / (1 lb of mortality to be disposed of)
PCV = primary composter volume (ft^3) = PDADR (VF)
PCV = primary composter volume = 600 ft^3
SCV = secondary composter volume = 600 ft^3

Selection of a primary composter bin = $L \times W \times H = 5 \times 6 \times 5 = 150$ ft^3 = VOL. A width of 6 ft and a depth of 5 ft (with a height of 5 ft) can easily accommodate equipment on this farm.

Then the number of primary composter bins needed = PCV/($L \times W \times H$) = (600 ft^3)/(150 ft^3) = 4. The four 150 cubic feet primary composter bins can be arranged in any of several configurations to suit the needs of a particular situation (see Figs. 7.17 and 7.18).

To determine the size and shape of secondary composting treatment box(es), the following is given: W = 6 ft, H = 5 ft, and SCV = PCV = 600 ft^3.

The designer has the choice of having (1) a four-bin secondary composter system identical to the calculated four-bin primary composter system (each bin = $L \times W \times H$ = 5 ft × 6 ft × 5 ft) or (2) a giant one-bin secondary composter system with a length L = 20 ft because L = VOL/($W \times H$) = 600 ft^3/(6 ft × 5 ft) = 20 ft. The above design calculations shown can be applied to poultry farms of varying sizes and types of birds if the assumption (general rule of VF) for design is to build one cubic foot of primary capacity and one cubic foot of secondary capacity for each pound of mortality to be disposed of. The properly cured final compost or effluent of secondary composter (SC) can be applied to the land as a fertilizer using the same guidelines as applied to poultry manure (see Table 7.14) in order to have enough nitrogen, phosphorus, and potassium.

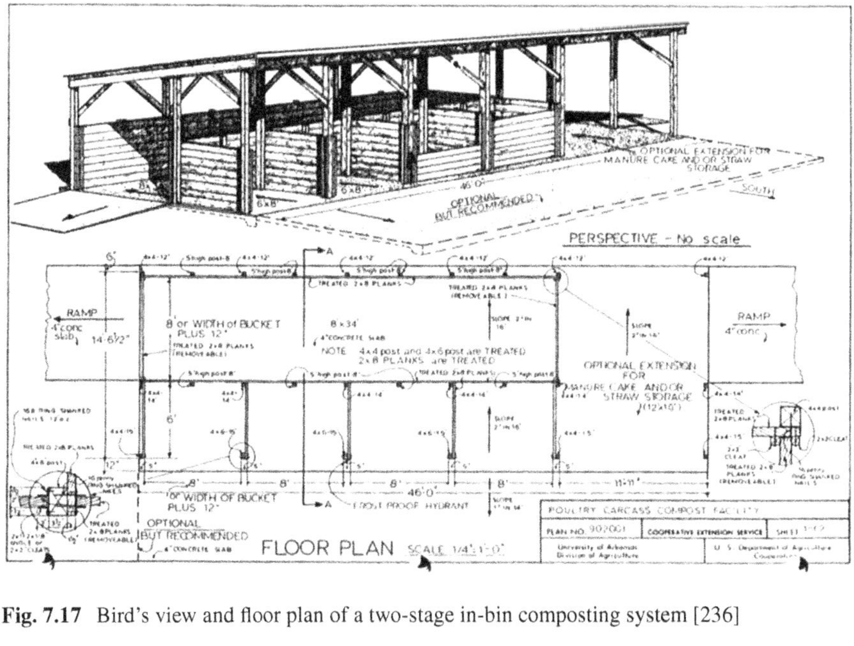

Fig. 7.17 Bird's view and floor plan of a two-stage in-bin composting system [236]

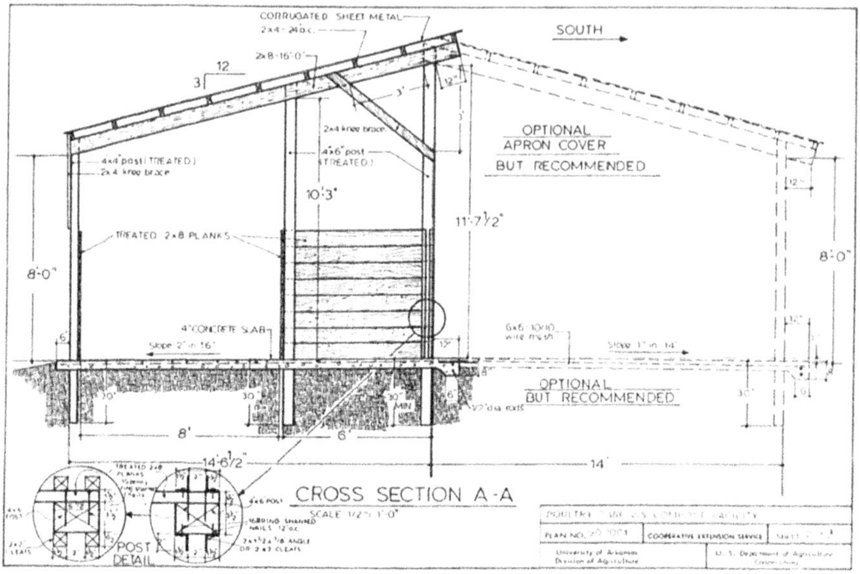

Fig. 7.18 Cross-section view of a two-stage in-bin composting system [236]

Table 7.14 Nutrients in manure and compost [229, 236]

	"Cake"	Compost	12-flock litter
Dry matter %	54.7	53.90 ± 3.37	79.0
NH_4/N %	33.0	27.00 ± 0.02	81.0
N %[a]	3.11	4.08 ± 0.22	5.25
P_2O_5 %[a]	4.94	6.06 ± 0.27	4.81
K_2O %[a]	3.66	4.43 ± 0.15	3.61

Source: D. W. Murphy, Department of Pouitry Science, University of Maryland
[a]Expressed on a dry-matter basis

Rapid composting of dead animals occurs when the C:N ratio of the compost mix is maintained between 10 and 20. This is considerably lower than what is normally recommended for other types of composting. Much of the nitrogen in the dead animal mass is not exposed on the surface initially; therefore, a lower C:N ratio is necessary to ensure rapid composting with elevated temperatures later. Table 7.14 indicates the nutrients in manures (cake) and compost when composting dead animals. The moisture content of the initial compost mixture should be between 45 and 55%. An initial moisture content of higher than 60% would retard the compost process.

Composting of dead animals should remain aerobic at all times throughout the process. Its initial mix should have enough porosity to allow air movement into and out of the compost mix. This can be accomplished by layering dead animals and amendments in the mix. For example, a dead poultry compost mix would be layered with straw, dead birds, and manure or waste cake from the poultry houses. Layers of

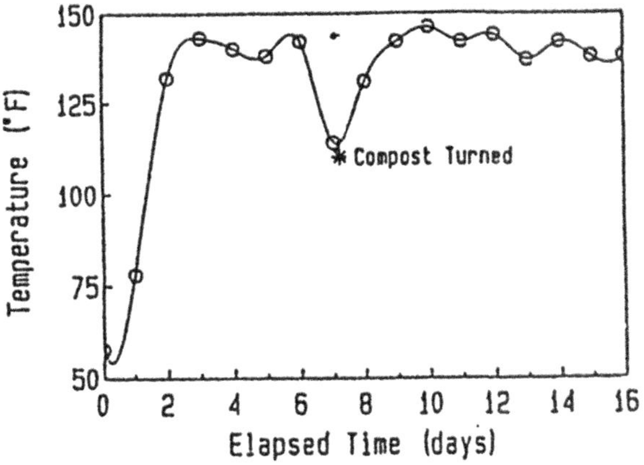

Fig. 7.19 Temperature changes in the typical two-stage in-bin composting system for composting dead birds [229, 236]

such high porosity material as straw, wood chips, peanut hulls, and bark allow lateral movement of air in the compost mix. Figure 7.16 is an example of the commonly recommended layering of manure, straw, and dead poultry. Figure 7.19 shows how the temperature changes in the two-stage in-bin composting system.

7.14 Summary

Solid waste generation is increasing worldwide as a result of population growth, lifestyle, and land scarcity as a landfill and further compounded by the huge expense to manage it, so it is important to find other cheaper alternatives to diversify solid waste disposal methods. The solution to the solid waste problem should not be completely burdened with one or two methods of solution such as landfilling or incineration only. Both of these systems are the final stage of solid waste management, whereas the waste problem starts on the first day it is generated. One of the approaches in solving the generation of solid waste is composting. The composting process is one of the methods or alternatives to conserve or recover resources that have many positive characteristics where the technology can be modified according to the local situation or circumstances. The compost produced is a final product that can be used as soil improvement or as a fertilizer in agriculture. Composting, however, is in line with an integrated solid waste management plan that includes other considerations such as landfilling, incineration, anaerobic digestion, and recycling. Therefore, in the long term, composting will be a suitable and environmentally friendly method to recycle the organic waste produced. The design of the composting process should be refined and combined with the utilization of the final product as well as the marketability of the resulting compost. The readers are referred to the literature for additional technical information concerning composting process description, advancement, and applications [229–253].

Glossary

Actinomycete A community of microorganisms, intermediate between bacteria and true fungi that create a characteristic branched mycelium. Compost has an earthy odor that is caused by these species.

Aerated static pile composting system Process in which decomposing organic material is located in piles over an air supply system that can be used to supply oxygen and regulate temperature for the purpose of generating compost. Piles must be insulated to ensure that all parts of the decomposing material reach and sustain temperatures at or above 55 °C for a minimum of 3 days.

Aeration Bringing about the contact of compost with air through turning or ventilating to enable microbial aerobic metabolism.

Aerobic Taking place in the presence of oxygen. Enough oxygen should be supplied to keep the system aerobic for effective composting. This means that composting takes place quickly and with no odor.

Ambient temperature The temperature of the air in the vicinity of the compost pile.

Anaerobic Occurring in the absence of oxygen. Anaerobic composting takes a long time and produces an unpleasant odor.

Bacteria A community of microorganisms having single-celled or noncellular bodies. Bacteria are usually shaped like a spheroid, rod-like, or curved entity, but they may also take the form of sheets, chains, or branched filaments.

Bin composting Composting method that uses basic structures (bins) rather than freestanding piles to compost mixtures of feedstocks. Bin composting is a form of in-vessel composting that is not normally enclosed. Many composting bins have a forced aeration system.

Biofilter A filter that reduces odors by microbial action. Finished compost is widely used as a biofilter to remove possible odors from active compost systems. It is as simple as layering finished compost on top of a pile of fresh feedstocks. Before being released into the atmosphere, the air in forced aeration systems is usually blown through a biofilter of finished compost.

Bioreactor An enclosed container used for producing compost or performing scientific composting experiments.

Biosolids Solids generated from sanitary wastewater that has been treated by one or more controlled processes that greatly minimize pathogens and volatile solids or chemically stabilize solids to the point where they do not attract vectors.

Bulking agent A composting nutrient content with larger particle sizes than carbon sources, preventing material packing and maintaining sufficient air spaces (around 25–35% porosity) within the compost pile. They should be made up of a three-dimensional matrix of solid particles that can sustain themselves by particle-to-particle interaction.

C:N (carbon-to-nitrogen ratio) The ratio of the weight of organic carbon (C) to that of total nitrogen (N) in organic material.

California Compost Quality Council (CCQC) The council has a unique coalition between compost producers, farmers, landscape contractors, scientists, and recycling advocates dedicated to creating standardized compost quality requirements. The aim is to boost compost production and usage in California. The council administers an independent compost registration that assures users of the highest quality compost.

Cellulose Cellulose is a series of organic compounds containing carbon, hydrogen, and oxygen that are assembled into chains of 1000–10,000 glucose molecules and is the main component of plant cell walls. Cellulose forms the fibrous and woody parts of plants, and it accounts for more than half of all organic carbon in the biosphere.

Compost Council of Canada The council is a national nonprofit, member-driven organization with a mission to promote organic residuals recycling and composting. It acts as a core resource and network for the Canadian compost industry, and its members contribute to the environmental sustainability of the communities in which they operate.

Curing The final stage of composting is when the compost has stabilized, but the rate of decomposition has decreased to the point that turning or forced aeration is no longer needed. Curing usually takes place at lower, mesophilic temperatures.

Dry matter The portion of a material that does not consist of water. The dry matter content (%) is equivalent to 100% minus the moisture content (%).

Fecal coliform Enteric organisms that serve as an indicator of possible presence of pathogens.

Finished compost A stable and hygienic product that has gone through successful composting and curing stages.

Fungus (plural fungi) Molds, mildews, yeasts, and mushrooms are all members of this group. Fungal cells, unlike bacteria, have nuclei. Fungi do not have chlorophyll, but they eat dead organic matter. Fungi are important in compost because they break down tough debris like cellulose and thrive when moisture and nitrogen levels are low during the curing stage.

Hemicellulose A series of organic compounds made up of chains of 50–150 sugar units, including glucose, xylose, and galactose. Hemicellulose covers cellulose in wood and aids in its binding to lignin.

Humus The relatively stable dark or black carbon-rich residue that results from the decomposition of organic matter.

Inoculum (plural inocula) Living organisms or materials containing living organisms (such as bacteria or other microorganisms) that are used to start or speed up a biological process (for example, biological seeding).

Institute for Global Environmental Strategies (IGES) The Institute for Global Environmental Strategies (IGES) was initiated by the Japanese government and supported by Kanagawa Prefecture. The aim of the Institute is to achieve a new paradigm for civilization and conduct innovative policy development and strategic research for environmental initiatives, translating research findings into political decisions for achieving sustainable development in the Asia-Pacific region and around the world.

Japan International Cooperation Agency (JICA) Japanese agency for technical cooperation aimed at the transfer of technology and knowledge around the world.

Kitakyushu International Techno-Cooperative Association (KITA) The organization that promotes environmental conservation on a global scale through international collaboration and technology transfer to developing countries as well as conducting necessary research and survey.

Lignin A series of complex organic polymers that are resistant to microbial degradation. Lignin binds cellulose fibers together in wood and protects them from chemical and microbial degradation.

Mature (or maturation) A chemical condition of the compost. Toxic chemical compounds in immature compost can inhibit plant growth.

Maturity A measure of whether compost has completed both the rapid decomposition process and the longer curing phase, during which gradual chemical changes render the compost more suitable for plant use.

Mesophilic The temperature range most conducive to the maintenance of optimum digestion by mesophilic bacteria, generally accepted as between 10 and 40 °C.

Mulch Any material that is spread on the soil surface to preserve soil moisture, prevent weed growth, moderate temperature changes, or avoid soil erosion, such as compost, bark, wood chips, or straw.

Pathogen Any organism that may cause disease or infection. Many pathogens, which are commonly found in waste, are destroyed by the high temperatures of composting processes.

Phytotoxic A term used to describe a material that is poisonous to plants. Acids or alcohols in immature or anaerobic compost may damage seedlings and sensitive plants.

Phytotoxicity A measure of a substance's ability to inhibit seed germination, damage plant roots, or stunt plant growth.

Protozoa Single-celled, animal-like microorganisms belonging to the kingdom Protista. Many species live in water or aquatic films surrounding soil or compost particles.

Soil amendment Any material used to change the chemical or physical properties of soil in order to improve its productivity. Compost, lime, sulfur, gypsum, and synthetic conditioners are some examples. Chemical fertilizers are usually excluded.

Stability A measure of whether compost has decomposed to the point at which it does not reheat, emit offensive odors, or support high rates of microbial growth when optimal moisture levels are supplied.

Test Methods for the Evaluation of Composting and Compost (TMECC) A laboratory manual modeled after the American Society for Testing and Materials (ASTM). TMECC offers comprehensive guidelines for the composting industry to check the physical, chemical, and biological condition of composting feedstocks, material in process, and compost products at the point of sale.

Thermophilic Heat-loving microorganisms that thrive in and generate temperatures above 40 °C.

US Composting Council The Council focuses on large-scale compost manufacturing and marketing and includes training and education of compost facility operators, certification programs for quality compost, and lobbying and advocacy campaigns at the state and federal level.

Windrow composting This method entails stacking the feedstock in windrows, which are long, narrow, and low piles. The wide exposed surface area of windrows promotes passive aeration and drying. Aeration is accomplished by both convective airflow and turning. The windrow piles work like a chimney, drawing air through the sides as the center heats up.

Windrow A long, relatively narrow, low pile. Windrows have a huge exposed surface area, which encourages passive aeration and drying.

Woods End Research Laboratory The company specializes in the utilization of organic wastes in soil systems and chemicals as well as provides soil testing services.

References

1. Bertran, E., Sort, X., Soliva, M., & Trillas, I. (2004). Composting winery waste: Sludges and grape stalks. *Bioresource Technology, 95*, 203–208.
2. Agamuthu, P. (2001). *Solid waste: Principals and management with Malaysian case studies*. Universiti of Malaya Press.
3. Cesaro, A., Belgiorno, V., & Guida, M. (2015). Compost from organic solid waste: Quality assessment and European regulations for its sustainable use. *Resources, Conservation and Recycling, 94*, 72–79.
4. Hoornweg, D., Thomas, L., & Otten, L. (1999). *Composting and its applicability in developing countries*. The World Bank, Urban Development Division Working Paper Series, No. 8
5. Epstein, E. (1997). *The science of composting*. Technomic Publishing Company Inc..
6. Haug, R. T. (1993). *The practical handbook of compost engineering*. Lewis Publishers.
7. Diaz, L. F., & de Bertoldi, M. (2007). History of composting. In L. F. Diaz, M. de Bertoldi, W. Bidlingmaier, & E. Stentiford (Eds.), *Compost science and technology*. Elsevier.
8. Ravoni, J. L., Heer, J. E., & Hagerty, D. J. (1975). *Handbook of solid waste disposal*. Van Nostrand Reinhold Co.
9. Gotaas, H. B. (1956). *Composting: Sanitary disposal and reclamation of organic wastes* (World Health Organization Monograph Series 31). World Health Organization.
10. Martin-Leake, H. (1949). *How can we use our sewage and our refuse?* (Publication no. 2). Albert Howard Foundation of Organic Husbandry.
11. Stovroff, R. P. (1954). Capitalizing on municipal wastes by composting. *Proceeding American Society of Civil Engineers*, 80 (Separate No. 545), USA.
12. Breidenbach, A. W. (1971). *Composting of municipal solid wastes in the United States* (Publication SW-47r). U.S. Environmental Protection Agency.
13. Epstein, E., Wilson, G. B., Burge, W. D., Mullen, D. C., & Enkiri, N. K. (1976). A forced aeration system for composting wastewater sludge. *Journal of the Water Pollution Control Federation, 48*, 688–694.
14. Finstein, M. S., Miller, F. C., & Strom, P. F. (1986). Waste treatment composting as a controlled system. In W. Schönborn (Ed.), *Microbial degradations* (Biotechnology, Vol. 8) (pp. 363–398). VCH Verlagsgesellschaft (German Chemical Society).
15. Chanyasak, V., Hirai, F. M., & Kubota, H. (1982). Changes of chemical components and nitrogen transformation in water extracts during composting of garbage. *Journal of Fermentation Technology, 60*, 441–448.

16. Nakasaki, K., Sasaki, M., Shoda, M., & Kubota, H. (1983). Change in microbial numbers during thermophilic composting of activated sludge with reference to CO_2 evolution rate. *Proceeding international symposium on biological reclamation and land utilization of urban waste*, Naples, Italy.
17. Hoitink, H. A. J., & Schmitthenner, A. F. (1988). *Effects of composted municipal sludge on soil-borne plant pathogens* (U.S. EPA Project Summary, EPA/600/52-88/024). U.S. Environmental Protection Agency.
18. Browne, C. A. (1933). The spontaneous heating and ignition of hay and other agricultural products. *Science, 77*, 223–229.
19. Waksman, S. A. (1932). *Principles of soil microbiology*. Williams & Wilkins.
20. Galitskaya, P., Biktasheva, L., Saveliev, A., Grigoryeva, T., Boulygina, E., & Selivanovskaya, S. (2017). Fungal and bacterial successions in the process of co-composting of organic wastes as revealed by 454 pyrosequencing. *PLoS ONE, 12*(10), e0186051.
21. Cardenas, R. R., Jr., & Wang, L. K. (1980). Composting process. In L. K. Wang & N. C. Pereira (Eds.), *Handbook of environmental engineering, Volume 2, Solid waste processing and resource recovery*. The Humana Press.
22. Hafeez, M., Gupta, P., & Gupta, Y. P. (2018). Rapid composting of different wastes with Yash activator plus. *International Journal of Life-Sciences Scientific Research, 4*, 1670–1674.
23. Chennaou, M., Salama, Y., Aouinty, B., Mountadar, M., & Assobhei, O. (2018). Evolution of bacterial and fungal flora during in-vessel composting of organic household waste under air pressure. *Journal of Materials and Environmental Science, 9*, 680–688.
24. Gajalakshmi, S., & Abbasi, S. A. (2008). Solid waste management by composting: State of the art. *Critical Reviews in Environmental Science and Technology, 38*(5), 311–400.
25. Golueke, C. G. (1992). Bacteriology of composting. *Biocycle, 33*, 55–57.
26. Nakasaki, K., Sasaki, M., Shoda, M., & Kubota, H. (1985). Change in microbial numbers during thermophilic composting of sewage sludge with reference to CO_2 evolution rate. *Applied Environmental Microbiology, 49*, 37–41.
27. Strom, P. F. (1985). Effect of temperature on bacterial species diversity in thermophilic solid waste composting. *Applied Environmental Microbiology, 50*, 899–905.
28. Tuomela, M., Vikman, M., Hatakka, A., & Itavaara, M. (2000). Biodegradation of lignin in a compost environment: A review. *Bioresource Technology, 72*(2), 169–183.
29. Focht, D. D., & Verstraete, W. (1977). Biochemical ecology of nitrification and denitrification. *Advances in Microbiology and Ecology, 1*, 135–214.
30. Firestone, M. K. (1982). Biological denitrification. In F. J. Stevenson (Ed.), *Nitrogen in agricultural soils* (pp. 289–326). ASA, CSSA and SSSA.
31. Tiquia, S. M., Richard, T. L., & Honeyman, M. S. (2002). Carbon, nutrient, and mass loss during composting. *Nutrient Cycling in Agroecosystem, 62*, 15–24.
32. Beffa, T., Blanc, M., Marilley, L., Fischer, J. L., Lyon, P. F., & Aragno, M. (1996). Taxonomic and metabolic microbial diversity during composting. In M. Bertoldi, P. Sequi, B. Lemmes, & T. Papi (Eds.), *The science of composting, Part I*. Chapman and Hall.
33. Insam, H., & de Bertoldi, M. (2007). Microbiology of the composting process. In L. F. Diaz, M. de Bertoldi, W. Bidlingmaier, & E. Stentiford (Eds.), *Compost science and technology*. Elsevier.
34. Singh, S., & Nain, L. (2014). Microorganisms in the conversion of agricultural wastes to compost. *Proceedings of the National Academy of Sciences, 80*, 473–481.
35. Muthukumar, G., & Mahadevan, A. (1983). Microbial degradation of lignin. *Journal of Science and Industrial Research, 42*, 518–528.
36. Dix, N. J., & Webster, J. (1995). *Fungal ecology*. Chapman and Hall.
37. Limaye, L., Patil, R., Ranadive, P., & Kamath, G. (2017). Application of potent actinomycete strains for biodegradation of domestic agro-waste by composting and treatment of pulp-paper mill. *Advance in Microbiology., 7*, 94–108.
38. Ghosh, S., Kapadnis, B. P., & Singh, N. B. (2000). Composting of cellulosic hospital solid waste: a potentially novel approach. *International Biodeterioration Biodegradation, 45*, 89–92.

39. McGauhey, P. H., & Goleuke, C. G. (1953). *Reclamation of municipal refuse by composting* (Technical bulletin no. 9). University of California.
40. Schroeder, E. D. (1977). *Water and wastewater treatment*. McGraw-Hill.
41. Madejon, E., Diaz, M. J., Lopez, R., & Cabrera, F. (2002). New approaches to establish optimum moisture content for compostable materials. *Bioresource Technology, 85*, 73–78.
42. Liang, C., Das, K. C., & McClendon, R. W. (2003). The influence of temperature and moisture contents regimes on the aerobic microbial activity of biosolids composting blend. *Bioresource Technology, 86*, 131–137.
43. Day, M., & Shaw, K. (2001). Biological, chemical and physical processes of composting. In P. J. Stofella & B. A. Kahn (Eds.), *Compost utilization in horticultural cropping systems*. Lewis Publishers.
44. Tchobanoglous, G., Theisen, H., & Vigil, S. (1993). *Integrated solid waste management: Engineering principles and management issues*. McGraw-Hill.
45. Miyatake, F., & Iwabuchi, K. (2005). Effect of high compost temperature on enzymatic activity and species diversity of culturable bacteria in cattle manure compost. *Bioresource Technology, 96*, 1821–1825.
46. Tiquia, S. M., Tam, N. F. Y., & Hodgkiss, I. J. (1997). Effects of bacterial inoculum and moisture adjustment on composting of pig manure. *Environmental Pollution, 96*, 161–171.
47. Huang, G. F., Wong, J. W. C., Wu, Q. T., & Nagar, B. B. (2004). Effect of C:N on composting of pig manure with sawdust. *Waste Management, 24*, 805–813.
48. Diaz, L. F., Savage, G. M., Eggerth, L. L., & Golueke, C. G. (1993). *Composting and recycling municipal solid waste*. Lewis Publishers.
49. Vourinen, A. H., & Saharinen, M. H. (1997). Evolution of microbiological and chemical parameters during manure and straw co-composting in a drum composting system. *Agriculture Ecosystems & Environment, 66*, 19–29.
50. Wong, J. W. C., Mak, K. F., Chan, N. W., Lam, A., Fang, M., Zhou, L. X., Wu, Q. T., & Liao, X. D. (2001). Co-composting of soybean residues and leaves in Hong Kong, China. *Bioresource Technology, 76*, 99–106.
51. Kumar, S. (2011). Composting of municipal solid waste. *Critical Reviews in Biotechnology, 31*(2), 112–136.
52. Yamada, Y., & Kawase, Y. (2006). Aerobic composting of waste activated sludge: Kinetic analysis for microbiological reaction and oxygen consumption. *Waste Management, 26*, 49–61.
53. Albuquerque, J. A., Gonzalvez, J., Garcia, D., & Cegarra, J. (2006). Effects of bulking agent on the composting of "alperujo", the solid by-product of the two – Phase centrifugation method for olive extraction. *Process Biochemistry, 41*, 127–132.
54. Barrington, S., Choiniere, D., Trigui, M., & Knight, W. (2003). Compost convective airflow under passive aeration. *Bioresource Technology, 86*, 259–266.
55. Tiquia, S. M., Tam, N. F. Y., & Hodgkiss, I. J. (1998). Effects of turning frequency on composting of spent pig-manure sawdust litter. *Bioresource Technology, 62*, 37–42.
56. Ahmad, J. (2001). *Proses pengkomposan bahan-bahan organik dari aliran sisa pepejal domestik dengan menggunakan reaktor aerobik*. MSc thesis. Universiti Sains Malaysia (Malay Version).
57. Parkinson, R., Gibbs, P., Burchett, S., & Misselbrook, T. (2003). Effect of turning regime and seasonal weather conditions on nitrogen and phosphorus losses during aerobic composting of cattle manure. *Bioresource Technology, 91*(2), 171–178.
58. Chang, J. I., Tsai, J. J., & Wu, K. H. (2006). Thermophilic composting of food waste. *Bioresource Technology, 97*, 116–122.
59. Komilis, D. P. (2006). A kinetic analysis of solid waste composting at optimal condition. *Waste Management, 26*, 82–91.
60. Tiquia, S. M., & Tam, N. F. Y. (1998). Composting of spent pig litter in turned and forced-aerated piles. *Environmental Pollution, 99*, 329–337.
61. Sabiani, N. H. M. (2008). *Pengkomposan sisa taman menggunakan kaedah timbunan statik berudara dan deram berputar*. MSc thesis. Universiti Sains Malaysia (Malay version).

62. Zhu, N., Deng, C., Xiong, Y., & Qian, H. (2004). Performance characteristics of three aeration systems in the swine manure composting. *Bioresource Technology, 95*, 319–326.
63. Cegarra, J., Albuquerque, J. A., Gonzalvez, J., Tortosa, G., & Chaw, D. (2006). Effects of the forced ventilation on composting of a solid olive-mill by-product ("alperujo") managed by mechanical turning. *Waste Management, 26*, 1377–1383.
64. Loh, T. C., Lee, Y. C., Liang, J. B., & Tan, D. (2005). Vermicomposting of cattle and goat manures by Eisenia foetida and their growth and reproduction performance. *Bioresource Technology, 96*, 111–114.
65. Jais, H. M. (2003). *Teknologi vermikultur dalam pelupusan sisa dapur dan penghasilan vermikas sebagai baja organik melalui ternakan cacing tanah Eisenia foetida*. Penerbit Universiti Sains Malaysia (Malay Version).
66. Arancon, N. Q., Edwards, C. A., Bierman, P., Welch, C., & Metzger, J. D. (2004). Influences of vermicomposts on field strawberries: Effects on growth and yields. *Bioresource Technology, 93*, 145–153.
67. Kaushik, P., & Garg, V. K. (2004). Dynamics of biological and chemical parameters during vermicomposting of solid textile mill sludge mixed with cow dung and agricultural residues. *Bioresource Technology, 94*, 203–209.
68. Bhawalkar, U. (1993). *Turning garbage into gold – An introduction to vermiculture biotechnology*. Bhawalkar Earthworm Research Institute.
69. Marsh, L., Subler, S., Mishra, S., & Marini, M. (2005). Suitability of aquaculture effluent solids mixed with cardboard as a feedstock for vermicomposting. *Bioresource Technology, 96*, 413–418.
70. Tang, J. C., Kanamori, T., Inoue, Y., Yasuta, T., Yoshida, S., & Katayama, T. (2004). Changes in the microbial community structure during thermophilic composting of manure as detected by quinone profile method. *Process Biochemistry, 39*, 1999–2006.
71. Sundberg, C., Smars, S., & Jonsson, H. (2004). Low pH as an inhibiting factor in the transition from mesophilic to thermophilic phase in composting. *Bioresource Technology, 95*, 145–150.
72. Mohaibes, M., & Tanski, H. H. (2004). Aerobic thermophilic treatment of farm slurry and food wastes. *Bioresource Technology, 95*, 245–254.
73. Elango, D., Thinakaran, N., Panneerselvam, P., & Sivanesan, S. (2009). Thermophilic composting of municipal solid waste. *Applied Energy, 86*, 663–668.
74. Wadkar, D. V., Modak, P. R., & Chavan, V. S. (2013). Aerobic thermophilic composting of municipal solid waste. *International Journal of Engineering Science and Technology, 5*(3), 716–718.
75. Conghos, M. M., Aguirre, M. E., & Santamaria, R. M. (2003). Biodegradation of sunflower hulls with different nitrogen sources under mesophilic and thermophilic incubations. *Biology and Fertility of Soils, 38*(5), 282–287.
76. Mohajer, A., Tremier, A., Barrington, S., Martinez, J., Teglia, C., & Carone, C. (2009). Microbial oxygen uptake as influenced by compost material physical parameters. *Waste Management, 29*, 2257–2264.
77. Huet, J., Druilhe, C., Tremier, A., Benoist, J. C., & Debenest, G. (2012). The impact of compaction, moisture content, particle size and type of bulking agent on initial physical properties of sludge-bulking agent mixtures before composting. *Bioresource Technology, 114*, 428–436.
78. Kulcu, R., & Yaldiz, O. (2007). Composting of goat manure and wheat straw using pine cones as a bulking agent. *Bioresource Technology, 98*, 2700–2704.
79. Ucaroglu, S., & Alkan, U. (2016). Composting of wastewater treatment sludge with different bulking agents. *Journal of the Air & Waste Management Association, 66*(3), 288–295.
80. Goldstein, N. (2002). Getting to know the odour compounds. *Biocycle, 43*, 42–44.
81. Maulini-Duran, C., Artola, A., Font, X., & Sanchez, A. (2013). A systematic study of the gaseous emissions from biosolids composting: raw sludge versus anaerobically digested sludge. *Bioresource Technology, 147*, 43–51.

82. Rincon, C. A., De Guardia, A., Couvert, A., Soutrel, I., Guezel, S., & Le Serrec, C. (2019). Odour generation patterns during different operational composting stages of anaerobically digested sewage sludge. *Waste Management, 95*, 661–673.
83. Kashmanian, R. M., & Rynk, R. F. (1998). Creating positive incentives for farm composting. *American Journal of Alternative Agriculture, 13*, 40–45.
84. Larney, F. J., Sullivan, D. M., Buckley, K. E., & Eghball, B. (2006). The role of composting in recycling manure nutrients. *Canadian Journal of Soil Science, 86*, 597–611.
85. Larney, F. J., Yanke, L. J., Miller, J. J., & McAllister, T. A. (2003). Fate of coliform bacteria in composted beef cattle feedlot manure. *Journal of Environmental Quality, 32*, 1508–1515.
86. McGinn, S. M., Janzen, H. H., & Coates, T. (2003). Atmospheric ammonia, volatile fatty acids, and other odourants near beef feedlots. *Journal of Environmental Quality, 32*, 1173–1182.
87. Rynk, R. (1992). *On-farm composting handbook*. Cooperative Extension, Northeast Regional Agricultural Engineering Service.
88. Reyes-Torres, M., Oviedo-Ocana, E. R., Dominguez, I., Komilis, D., & Sanchez, A. (2018). A systematic review on the composting of green waste: Feedstock quality and optimization strategies. *Waste Management, 77*, 486–499.
89. Bustamante, M. A., Ceglie, F. G., Aly, A., Mihreteab, H. T., Ciaccia, C., & Tittarelli, F. (2016). Phosphorus availability from rock phosphate: Combined effect of green waste composting and sulfur addition. *Journal of Environmental Management, 182*, 557–563.
90. U.S. EPA. (2010). *Municipal solid waste generation, recycling, and disposal in the United States: Facts and figures for 2010* (pp. 1–12). U.S. EPA.
91. Zhang, L., & Sun, X. (2016). Influence of bulking agents on physical, chemical, and microbiological properties during the two-stage composting of green waste. *Waste Management, 48*, 115–126.
92. Lopez, M., Soliva, M., Martínez-Farré, F. X., Bonmatí, A., & Huerta-Pujol, O. (2010). An assessment of the characteristics of yard trimmings and recirculated yard trimmings used in biowaste composting. *Bioresource Technology, 101*(4), 1399–1405.
93. Mohee, R. (2007). *Waste management opportunities for rural communities: Composting as an effective waste management strategy for farm households and others*. Food and Agriculture Organization of the United Nations.
94. Vigneswaran, S., Kandasamy, J., & Johir, M. A. H. (2016). Sustainable operation of composting in solid waste management. *Procedia Environmental Sciences, 35*, 408–415.
95. University of Connecticut Cooperative Extension Service (UConn CES). (1989). *Leaf composting: A guide for municipalities*. State of Connecticut Department of Environmental Protection, Local Assistance and Program Coordination Unit, Recycling Program.
96. U.S. EPA. (1994, May). *Composting of yard trimmings and municipal solid waste* (EPA530-R-94-003). U.S. Environmental Protection Agency.
97. Kumar, M., Ou, Y. L., & Lin, J. G. (2010). Co-composting of green waste and food waste at low C:N ratio. *Waste Management, 30*(4), 602–609.
98. Zhao, L., Gu, W. M., He, P. J., & Shao, L. M. (2010). Effect of air-flow rate and turning frequency on bio-drying of dewatered sludge. *Water Research, 44*(20), 6144–6152.
99. Tweib, S. A. K., Rahman, R. A., & Khalil, M. S. (2013). Physicochemical changes in co-composting process of palm oil mill sludge (POMS) and solid waste (kitchen waste) using bin composter. *Arabian Journal for Science and Engineering, 39*(4), 2455–2462.
100. Zorpas, A. A., Kapetanios, E., Zorpas, G. A., Karlis, P., Vlyssides, A., Haralambous, I., & Loizidou, M. (2000). Compost produced from organic fraction of municipal solid waste, primary stabilized sewage sludge and natural zeolite. *Journal of Hazardous Materials, B77*, 149–159.
101. Zhang, D., Luo, W., Li, Y., Wang, G., & Li, G. (2018). Performance of co-composting sewage sludge and organic fraction of municipal solid waste at different proportions. *Bioresource Technology, 250*, 853–859.

102. Lu, Y., Wu, X., & Guo, J. (2009). Characteristics of municipal solid waste and sewage sludge co-composting. *Waste Management, 29*, 1152–1157.
103. Lim, L. Y., Bong, C. P. C., Lee, C. T., Klemes, J. J., Sarmidi, M. R., & Lim, J. S. (2017). Review on the current composting practices and the potential of improvement using two-stage composting. *Chemical Engineering Transactions, 61*, 1051–1056.
104. Santos, A., Bustamante, M. A., Tortosa, G., Moral, R., & Bernal, M. P. (2015). Gaseous emissions and process development during composting of pig slurry: The influence of the proportion of cotton gin waste. *Journal of Cleaner Production, 112*, 81–90.
105. Ermolaev, E., Pell, M., Smars, S., Sundberg, C., & Jonsson, H. (2012). Greenhouse gas emission from covered windrow composting with controlled ventilation. *Waste Management & Research, 30*, 155–160.
106. Luo, W. H., Yuan, J., Luo, Y. M., Li, G. X., Nghiem, L. D., & Price, W. E. (2014). Effects of mixing and covering with mature compost on gaseous emissions during composting. *Chemosphere, 117*, 14–19.
107. Huang, C. J., & Dalton, C. (1975). *Energy recovery from solid waste*. NASA Contractor Report. NASA CR-2525. National Aeronautics and Space Administration.
108. Kulikowska, D., & Klimiuk, E. (2011). Organic matter transformations and kinetics during sewage sludge composting in a two-stage system. *Bioresource Technology, 102*, 10951–10958.
109. Zhang, L., Sun, X., Tian, Y., & Gong, X. (2013). Effects of brown sugar and calcium superphosphate on the secondary fermentation of green waste. *Bioresource Technology, 131*, 68–75.
110. Olds, J. (1968). Houston compost plant. Second year report. *Compost Science, 9*(1), 18–19.
111. Drobny, H. L., Hull, H. E., & Testin, R. F. (1971). *Recovery and utilization of municipal waste* (Publ. SW-l0c). U.S. Environmental Protection Agency.
112. McGauhey, P. H. (1960). Refuse composting plant at Norman, Oklahoma. *Compost Science, 1*(3), 5–8.
113. Schulze, K. L. (1965). The Fairfield-Hardy composting pilot plant at Altoona, PA. *Compost Science, 5*(3), 5–10.
114. Cardenas, R. R., Jr. (1973). *Proceedings of 1973 solid waste conference* (p. 285). Worcester Polytechnic Institute.
115. Fuchs, J. G., Janmaat, L., & Raviv, M. (2016). What control measures do we need for compost production and use. In A. W. G. van der Wurff, J. G. Fuchs, M. Raviv, & A. Termorshuizen (Eds.), *Handbook for composting and compost use in organic horticulture*. BioGreenhouse.
116. Graves, R. E., Hattemer, G. M., Stettler, D., Krider, J. N., & Chapman, D. (2000). *Composting-Part 637 environmental engineering national engineering handbook natural resources conservation service*. United States Department of Agriculture.
117. Cardenas, R. R., Jr. (1977). Rapid moisture determination in compost for process control. *Compost Science, 18*(1), 14.
118. McCartney, D., & Tingley, J. (1998). Development of a rapid moisture content method for compost materials. *Compost Science & Utilization, 6*(3), 14–25.
119. IS-2720 (Part 2):1973 (Reaffirmed-May 2015) Methods of test for soils: Determination of water content (second revision).
120. Ahn, H. K., Mulbry, W., White, J. W., & Kondrad, S. L. (2011). Pile mixing increases greenhouse gas emissions during composting of dairy manure. *Bioresource Technology, 102*(3), 2904–2909.
121. Shen, Y., Ren, L., Li, G., Chen, T., & Guo, R. (2011). Influence of aeration on CH_4, N_2O and NH_3 emissions during aerobic composting of chicken manure and high C:N waste mixture. *Waste Management, 31*(1), 33–38.
122. Gu, W., Zhang, F., Xu, P., Tang, S., Xie, K., Huang, X., & Huang, Q. (2011). Effects of sulphur and thiobacillus thioparus on cow manure aerobic composting. *Bioresource Technology, 102*(11), 6529–6535.

123. Jiang, T., Schuchardt, F., Li, G., Guo, R., & Zhao, Y. (2011). Effect of C:N ratio, aeration rate and water content on ammonia and greenhouse gas emission during the composting. *Journal of Environmental Sciences, 23*(10), 1754–1760.
124. Xiong, Z. Q., Wang, G. X., Huo, Z. C., Yan, L., Gao, Y. M., Wang, Y. J., Gu, J. D., & Wang, W. D. (2017). Effect of aeration rates on the composting processes and nitrogen loss during composting. *Applied Environmental Biotechnology, 2*(1), 1–8.
125. Pereira-Neto, J. T., Stentiford, E. I., & Smith, D. V. (1986). Survival of faecal indicator microorganisms in refuse/sludge composting using the aerated static pile system. *Waste Management & Research, 4*, 397–406.
126. Wilkinson, K. G. (2008). The biosecurity of on-farm mortality composting. *Journal of Applied Microbiology, 102*, 609–618.
127. Deportes, I., Benoit-Guyod, J. L., & Zmirou, D. (1995). Hazard to man and the environment posed by the use of urban waste compost: a review. *Science of the Total Environment, 172*, 197–222.
128. Sunar, N. M., Stentiford, E. I., Stewart, D. I., & Fletcher, L. A. (2009). The process and pathogen behaviour in composting: A review. *Proceeding UMT-MSD 2009 post graduate seminar 2009*. Universiti Malaysia Terengganu, Malaysian Student Department UK & Institute for Transport Studies University of Leeds, pp. 78-87
129. Brinton, W. F. (2000). Final report on compost quality standards and guidelines prepared for the New York State Association of Recyclers.
130. Avery, L. M., Booth, P., Campbell, C., Tompkins, D., & Hough, R. L. (2012). Prevalence and survival of potential pathogens in source-segregated green waste compost. *Science of the Total Environment, 431*, 128–138.
131. Ceustermans, A., De Clercq, D., Aertsen, A., Michiels, C., Coosemans, J., & Ryckeboer, J. (2007). Inactivation of *Salmonella Senftenberg* strain W 775 during composting of biowastes and garden wastes. *The Journal of Applied Microbiology, 103*, 53–64.
132. Briancesco, R. (2008). Assessment of microbiological and parasitological quality of composted wastes: health implications and hygienic measures. *Waste Management & Research, 26*, 196–202.
133. Nell, J. H., Steer, A. G., & van Rensburg, P. A. J. (1983). Hygienic quality of sewage sludge compost. *Water Science and Technology, 19*, 181–194.
134. Wiley, B. B., & Westerberg, S. C. (1969). Survival of human pathogens in composted sewage. *Journal of Applied Microbiology, 18*, 994–1001.
135. Wichuk, K. M., & McCartney, D. (2007). A review of the effectiveness of current time-temperature regulations on pathogen inactivation during composting. *Journal of Environmental Engineering and Science, 6*(5), 573–586.
136. Cekmecelioglu, D., Demirci, A., Graves, R. E., & Davitt, N. H. (2005). Optimization of windrow food waste composting to inactivate pathogenic microorganisms. *Transactions – American Society of Agricultural Engineers, 48*, 2023–2032.
137. Jones, P., & Martin, M. (2003). *A review of the literature on the occurrence and survival of pathogens of animals and humans in green compost*. Waste and Resource Action Program (WARP)
138. Morgan, M. T., & Macdonald, F. W. (1969). Tests show MB tuberculosis doesn't survive composting. *Journal of Environmental Health, 32*, 101–108.
139. Krogstad, O., & Gudding, R. (1975). The survival of some pathogenic microorganisms during reactor composting. *Acta Agriculturae Scandinavica, 25*, 281–284.
140. Senne, D. A., Panigrahy, B., & Morgan, R. L. (1994). Effect of composting poultry carcasses on survival of exotic avian viruses—Highly Pathogenic Avian Influenza (HPAI) Virus and Adenovirus of Egg Drop Syndrome-76. *Avian Diseases, 38*, 733–737.
141. Rimhanen-Finne, R., Vuorinen, A., Marmo, S., Malmberg, S., & Hänninen, M. L. (2004). Comparative analysis of Cryptosporidium, Giardia and indicator bacteria during sewage sludge hygienization in various composting processes. *Letters in Applied Microbiology, 38*, 301–305.

142. Deportes, B. G., & Zmirou, B. (1998). Microbial disinfection capacity of municipal solid waste (MSW) composting. *Journal of Applied Microbiology, 85*, 238–246.
143. Tharaldsen, J., & Helle, O. (1989). Survival of parasite eggs in livestock slurry utilized for compost heat. *Acta Agriculturae Scandinavica, 39*, 381–387.
144. Sidhu, J., Gibbs, R. A., Ho, G. E., & Unkovich, I. (2001). The role of indigenous microorganisms in suppression of salmonella regrowth in composted biosolids. *Water Research, 35*, 913–920.
145. Russ, C. F., & Yanko, W. A. (1981). Factors affecting salmonellae repopulation in composted sludges. *Applied and Environmental Microbiology, 41*, 597–602.
146. Turner, C. (2002). The thermal inactivation of E-coli in straw and pig manure. *Bioresource Technology, 84*, 57–61.
147. Coker, C. (2020a). *Composting business management: Capital cost of composting facility construction*. Retrieved from https://www.biocycle.net/composting-business-management-capital-cost-composting-facility-construction/. Accessed on 15 Mar 2021.
148. Coker, C. (2020b). *Composting business management: Composting facility operating cost estimates*. Retrieved from https://www.biocycle.net/composting-business-management-composting-facility-operating-cost-estimates/. Accessed on 16 Mar 2021.
149. Fernandez, J. M., Plaza, C., Polo, A., & Plante, A. F. (2012). Use of thermal analysis techniques (TG-DSC) for the characterization of diverse organic municipal waste streams to predict biological stability prior to land application. *Waste Management, 32*, 158–164.
150. Lopez, M., Huerta-Pujol, O., Martinez-Farre, F. X., & Soliva, M. (2010). Approaching compost stability from Klason lignin modified method: chemical stability degree for OM and N quality assessment. *Resources, Conservation & Recycling, 55*, 171–181.
151. Komilis, D., Kontou, I., & Ntougias, S. (2011). A modified static respiration assay and its relationship with an enzymatic test to assess compost stability and maturity. *Bioresource Technology, 102*, 5863–5872.
152. Bernal, M. P., Paredes, C., Sanchez-Monedero, M. A., & Cegarra, J. (1997). Maturity and stability of composts prepared with a wide range of organic wastes. *Bioresource Technology, 63*, 91–99.
153. Bernal, M. P., Alburquerque, J. A., & Moral, R. (2009). Composting of animal manures and chemical criteria for compost maturity. *Bioresource Technology, 100*, 5444–5453.
154. Butler, T. A., Sikora, L. J., Steinhilber, P. M., & Douglass, L. W. (2001). Compost age and sample storage effects on maturity indicators of biosolids compost. *Journal of Environmental Quality, 30*(6), 2141–2148.
155. Sullivan, D. M., & Miller, R. O. (2001). Compost quality attributes, measurements, and variability. In P. J. Stofella & B. A. Kahn (Eds.), *Compost utilization in horticultural cropping systems*. Lewis Publishers.
156. Ge, B., McCartney, D., & Zeb, J. (2006). Compost environmental protection standards in Canada. *Journal of Environmental Engineering and Science, 5*(3), 221–234.
157. Mathur, S. P., Owen, G., Dinel, H., & Schnitzer, M. (1993). Determination of compost biomaturity. I. Literature review. *Biological Agriculture and Horticulture, 10*, 65–85.
158. Brinton, W. F., & Evans, E. (2001). How compost maturity affects container grown plants. *BioCycle, 42*(1), 56–60.
159. Hogg, D., Barth, J., Favoino, E., Centemero, M., Caimi, V., & Amlinger, F. (2002). *Comparison of compost standards within the EU, North America and Australasia*. Main Report. The Waste and Resources Action Programme.
160. Lasaridi, K. E., Stentiford, E. I., & Evans, T. (2000). Windrow composting of wastewater biosolids: process performance and product stability assessment. *Water Science and Technology, 42*(9), 217–226.
161. Tiquia, S. M., Wan, J. H. C., & Tam, N. F. Y. (2002). Dynamics of yard trimmings composting as determined by dehydrogenase activity, ATP content, arginine ammonification, and nitrification potential. *Process Biochemistry, 37*(10), 1057–1065.

162. Boulter-Bitzer, J. I., Trevors, J. T., & Boland, G. J. (2006). A polyphasic approach for assessing maturity and stability in compost intended for suppression of plant pathogens. *Applied Soil Ecology, 34*(1), 65–81.
163. Tiquia, S. M., Tam, N. F. Y., & Hodgkiss, I. J. (1996). Microbial activities during composting of spent pig-manure sawdust litter at different moisture contents. *Bioresource Technology, 55*(3), 201–206.
164. Chefetz, B., Hatcher, P. G., Hadar, Y., & Chen, Y. (1996). Chemical and biological characterization of organic matter during composting of municipal solid waste. *Journal of Environmental Quality, 25*(4), 776–785.
165. Inbar, Y., Hadar, Y., & Chen, Y. (1993). Recycling of cattle manure: The composting process and characterization of maturity. *Journal of Environmental Quality, 22*(4), 857–863.
166. TMECC. (2002a). Organic and biological properties – 05.06 odour. In W. H. Thompson, P. B. Leege, P. D. Millner, & M. E. Wilson (Eds.), *Test methods for the examination of composting and compost*. United States Department of Agriculture, and Composting Council Research and Education Foundation.
167. Tiquia, S. M. (2005). Microbiological parameters as indicators of compost maturity. *Journal of Applied Microbiology, 99*(4), 816–828.
168. TMECC. (2002b). Organic and biological properties – 05.08 respirometry. In W. H. Thompson, P. B. Leege, P. D. Millner, & M. E. Wilson (Eds.), *Test methods for the examination of composting and compost* (pp. 05.08-1–05.07-24). United States Department of Agriculture, and Composting Council Research and Education Foundation.
169. Lasaridi, K. E., & Stentiford, E. I. (1998a). A simple respirometric technique for assessing compost stability. *Water Research, 32*, 3717–3723.
170. Brewer, L. J., & Sullivan, D. M. (2001). *Maturity and stability evaluation of composted yard debris*. MSc thesis, Oregon State University, Corvallis, Ore.
171. Switzenbaum, M. S., Moss, L. H., Epstein, E., Pincince, A. B., & Donovan, J. F. (1997). Defining biosolids stability. *Journal of Environmental Engineering, 123*(12), 1178–1184.
172. Goyal, S., Dhull, S. K., & Kapoor, K. K. (2005). Chemical and biological changes during composting of different organic wastes and assessment of compost maturity. *Bioresource Technology, 96*(14), 1584–1591.
173. Woods End Research Laboratory. (1999). *Guide to Solvita® testing for compost maturity index*. [Online]. Available at https://cdn.sare.org/wp-content/uploads/20171204134211/982665solvita-compost-test-manual.pdf. Accessed on 22 Mar 2021.
174. Changa, C. M., Wang, P., Watson, M. E., Hoitink, H. A. J., & Michel, F. C. J. (2003). Assessment of the reliability of a commercial maturity test kit for composted manures. *Compost Science & Utilization, 11*(2), 125–143.
175. Chen, Y., & Inbar, Y. (1993). Chemical and spectroscopical analyses of organic matter transformations during composting in relation to compost maturity. In H. A. J. Hoitink & H. M. Keener (Eds.), *Science and engineering of composting: Design, environmental, microbiological and utilization aspects*. Renaissance Publications.
176. Keeling, A. A., Paton, I. K., & Mullett, J. A. (1994). Germination and growth of plants in media containing unstable refuse-derived compost. *Soil Biology & Biochemistry, 26*, 767–772.
177. TMECC. (2002c). Organic and biological properties – 05.05 biological assays. In W. H. Thompson, P. B. Leege, P. D. Millner, & M. E. Wilson (Eds.), *Test methods for the examination of composting and compost*. United States Department of Agriculture, and Composting Council Research and Education Foundation.
178. Benito, M., Masaguer, A., Moliner, A., Arrigo, N., Palma, R. M., & Effron, D. (2005). Evaluation of maturity and stability of pruning waste compost. *Soil Science, 170*(5), 360–370.
179. Gomez-Brandon, M., Lazcano, C., & Dominguez, J. (2008). The evaluation of stability and maturity during the composting of cattle manure. *Chemosphere, 70*(3), 436–444.
180. TMECC. (2002d). Organic and biological properties – 05.04 enzyme activity and analysis. In W. H. Thompson, P. B. Leege, P. D. Millner, & M. E. Wilson (Eds.), *Test methods for*

the examination of composting and compost. United States Department of Agriculture, and Composting Council Research and Education Foundation.

181. Khan, M. A. I., Ueno, K., Horimoto, S., Komai, F., Tanaka, K., & Ono, Y. (2009). Physiochemical, including spectroscopic, and biological analyses during composting of green tea waste and rice bran. *Biology and Fertility of Soils, 45*(3), 305–313.
182. Lasaridi, K. E., & Stentiford, E. I. (1998b). Biological parameters for compost stability assessment and process evaluation. *Acta Horticulturae, 469*, 119–128.
183. Forster, J. C., Zech, W., & Wurdinger, E. (1993). Comparison of chemical and microbiological methods for the characterization of the maturity of composts from contrasting sources. *Biology and Fertility of Soils, 16*(2), 93–99.
184. Hutchinson, M., & Griffin, T. S. (2008). Evaluation of fiber content relative to other measures of compost stability. *Compost Science & Utilization, 16*(1), 6–11.
185. Namkoong, W., Hwang, E.-Y., Cheong, J.-G., & Choi, J.-Y. (1999). A comparative evaluation of maturity parameters for food waste composting. *Compost Science & Utilization, 7*(2), 55–62.
186. TMECC. (2002e). Organic and biological properties – 05.07 organic matter. In W. H. Thompson, P. B. Leege, P. D. Millner, & M. E. Wilson (Eds.), *Test methods for the examination of composting and compost*. United States Department of Agriculture, and Composting Council Research and Education Foundation.
187. Benito, M., Masaguer, A., Moliner, A., Hontoria, C., & Almorox, J. (2009). Dynamics of pruning waste and spent horse litter co-composting as determined by chemical parameters. *Bioresource Technology, 100*(1), 497–500.
188. Richard, D., & Zimmerman, R. (1995). Respiration rate-reheating potential: a comparison of measures of compost stability. *Compost Science & Utilization, 3*(2), 74–79.
189. Harada, Y., Inoko, A., Tadaki, M., & Izawa, T. (1981). Maturing process of city refuse compost during piling: Application of composts to agricultural land. *Soil Science and Plant Nutrition, 27*, 357–364.
190. Iglesias Jimenez, E., & Perez Garcıa, V. (1991). Composting of domestic refuse and sewage sludge. I. Evolution of temperature, pH, C:N ratio, and cation exchange capacity. *Resources, Conservation and Recycling, 6*(1), 45–60.
191. Inbar, Y., Chen, Y., Hadar, Y., & Hoitink, H. A. J. (1990). New approaches to compost maturity. *BioCycle, 31*(12), 64–69.
192. Bernal, M. P., Paredes, C., Sanchez-Monedero, M. A., & Cegarra, J. (1998). Maturity and stability parameters of composts prepared with a wide range of organic wastes. *Bioresource Technology, 63*(1), 91–99.
193. Ko, H. J., Kim, K. Y., Kim, H. T., Kim, C. N., & Umeda, M. (2008). Evaluation of maturity parameters and heavy metal contents in composts made from animal manure. *Waste Management, 28*(5), 813–820.
194. Eggerth, L. F., Diaz, L. F., Chang, M. T. F., & Iseppi, L. (2007). Marketing of composts. In L. F. Diaz, M. de Bertoldi, W. Bidlingmaier, & E. Stentiford (Eds.), *Compost science and technology*. Elsevier.
195. USEPA. (1999). *Organic materials management strategies* (EPA 30-R-99-016. Solid waste and emergency response (5306W)). United State Environmental Protection Agency.
196. Buhr, A. R., McClure, T., Slivka, D. C., & Albrecht, R. (1993). Compost supply and demand. *BioCycle*, 54–58. ISBN:0276-5055. JG Press. USA.
197. Criner, G. K., Allen, T. G., & Schatzer, R. J. (2001). Compost economics: Production and utilization in agriculture. In P. J. Stofella & B. A. Kahn (Eds.), *Compost utilization in horticultural cropping systems*. Lewis Publishers.
198. Pandyaswargo, A. H., & Premakumara, D. G. J. (2014). Financial sustainability of modern composting: The economically optimal scale for municipal waste composting plant in developing Asia. *International Journal of Recycling of Organic Waste in Agriculture, 3*(66), 1–14.
199. Chen, Y. T. (2016). A cost analysis of food waste composting in Taiwan. *Sustainability, 8*(11), 1210.

200. Zulkepli, N. E., Muis, Z. A., Mahmood, N. A. N., Hashim, H., & Ho, W. S. (2017). Cost benefit analysis of composting and anaerobic digestion in a community: A review. *Chemical Engineering Transactions, 56*, 1777–1782.
201. Sabki, M. H., Lee, C. T., Bong, C. P. C., & Klemes, J. J. (2018). A review on the economic feasibility of composting for organic waste management in Asian countries. *Chemical Engineering Transactions, 70*, 49–54.
202. USEPA. (1997a). *Innovative uses of compost erosion control, turf remediation, and landscaping* (EPA530-F-97-043. Solid Waste and Emergency Response (5306W)). United State Environmental Protection Agency.
203. USEPA. (1997b). *Innovative uses of compost bioremediation and pollution prevention* (EPA530-F-97-042. Solid Waste and Emergency Response (5306W)). United State Environmental Protection Agency.
204. USEPA. (1997c). *Innovative uses of compost disease control for plants and animals* (EPA530-F-97-044. Solid waste and emergency response (5306W)). United State Environmental Protection Agency.
205. USEPA. (1997d). *Innovative uses of compost reforestation, wetlands restoration, and habitat revitalization* (EPA530-F-97-046. Solid waste and emergency response (5306W)). United State Environmental Protection Agency.
206. Chia, W. Y., Chew, K. W., Le, C. F., Lam, S. S., Chee, S. C. C., Ooi, M. S. L., & Show, P. L. (2020). Sustainable utilization of biowaste compost for renewable energy and soil amendments. *Environmental Pollution, 267*, 115662.
207. Williams, B. C., McMullan, J. T., & McCahey, S. (2001). An initial assessment of spent mushroom compost as a potential energy feedstock. *Bioresource Technology, 79*, 227–230.
208. Mccahey, S., Mcmullan, J. T., & Williams, B. C. (2003). Consideration of spent mushroom compost as a source of energy. *Developments in Chemical Engineering and Mineral Processing, 11*, 43–53.
209. Kliopova, I., & Makarskiene, K. (2012). Generation of solid recovered fuel from sewage sludge compost. *Environmental Research, Engineering and Management, 62*, 39–52.
210. Raclavska, H., Juchelkova, D., Skrobankova, H., Wiltowski, T., & Campen, A. (2011). Conditions for energy generation as an alternative approach to compost utilization. *Environmental Technology, 32*, 407–417.
211. Yeo, J., Oh, J.-I., Cheung, H. H. L., Lee, P. K. H., & An, A. K. (2019). Smart food waste recycling bin (S-FRB) to turn food waste into green energy resources. *Journal of Environmental Management, 234*, 290–296.
212. Obidzinski, S. (2012). Pelletization process of postproduction plant waste. *International Agrophysics, 26*, 279–284.
213. Zafari, A., & Kianmehr, M. H. (2013). Application of densification process in organic waste management. *Waste Management and Research, 31*, 684–691.
214. Hara, M. (2001). *Fertilizer pellets made from composted livestock manure*. Food & Fertilizer Technology Center.
215. Hammed, T. (2013). The effect of locally fabricated pelletizing machine on the chemical and microbial composition of organic fertilizer. *Biotechnology Journal International, 3*(1), 29–38.
216. Zafari, A., & Kianmehr, M. (2014). Factors affecting mechanical properties of biomass pellet from compost. *Environmental Technology, 35*, 478–486.
217. Tim, S., Ngo, B., Tim, S., Yun, H., Mihara, M., & Siriwattananon, L. (2012). Study on suitable moisture for pellet compost making and mechanical mixing technique. *International Journal of Environmental and Rural Development, 3*, 96–101.
218. Villegas-Pangga, G., & Cedillo, N. O. (2018). Fertilizer value of compost and densified fertilizers made from swine and poultry manures on crops and chemical properties of clay soil. *Journal of the International Society for Southeast Asian Agricultural Sciences, 24*, 43–55.
219. Zajonc, O., Frydrych, J., & Jezerska, L. (2014). Pelletization of compost for energy utilization. *IERI Procedia, 8*, 2–10.

220. Zajonc, O., Raclavsk, H., Juchelkov, D., & Volfov, M. (2011). Energetic properties of compost. *Inzynieria Mineralna, 12*, 77–88.
221. Premakumara, D. G. J., Kazuyoshi, H., & Tamura, E. (2011) *KitaQ system composting and its applicability in Asia: Lessons learned and future potential.* 22nd annual conference of Japan Society of Material Cycles and Waste Management, 3–5, November 2011.
222. Kurniawan, T. A., Puppim De Oliveira, J., Premakumara, D. G. J., & Nagaishi, M. (2013). City-to-city level cooperation for generating urban co-benefits: The case of technological cooperation in the waste sector between Surabaya (Indonesia) and Kitakyushu (Japan). *Journal of Cleaner Production, 58*, 43–50.
223. Nuzir, F. A., Hayashi, S., & Takakura, K. (2019). Takakura composting method (TCM) as an appropriate environmental technology for urban waste management: Case study of Hai Phong, Vietnam. *International Journal of Building, Urban, Interior and Landscape Technology (BUILT), 13*(1), 67–82.
224. Maeda, T. (2009). *Reducing waste through the promotion of composting and active involvement of various stakeholders: Replicating Surabaya's solid waste management model.* IGES Policy Brief. Retrieved from https://www.iges.or.jp/en/pub/reducing-waste-through-promotion-composting/en. Accessed on 28 Mar 2021.
225. Altman, V. (2016). *Municipal composting schemes: International case studies.* Swinburne University of Technology. Retrieved from https://apo.org.au/sites/default/files/resource-files/2016-11/apo-nid240311.pdf. Accessed on 28 Mar 2021.
226. European Commission. (2000). *Success stories on composting and separate collection.* European Commission.
227. Loizidou, M., Malamis, S., Hapeshis, K., Kapetanios, E., Tsolomiti, G., & Malamis, D. (2016). *Review of best practices and success stories of composting of sludge and other BOW in the EU and internationally.* Retrieved from http://uest.ntua.gr/archive/morocomp/1st_progress-report_morocomp_all_pdf_eng/deliverable_4/review-of-best-practices-and-success-stories-composting-of-sludge-and-other-BOW-in-the-EU-and-internationally.pdf. Accessed on 29 Mar 2021.
228. Goldstein, N. (2005). *Mixed MSW composting facilities in the U.S.* Retrieved from: https://www.biocycle.net/mixed-msw-composting-facilities-in-the-u-s/ Accessed on 28 Mar 2021.
229. Taylor, A. G. (1992). *Dead animal disposal.* University of Arkansas and the US Department of Agriculture.
230. Editor. (1994). *Environmental seminar.* Growmark, Inc. and Illinois Department of Energy and Natural Resources.
231. Wang, L. K., & Wang, M. H. S. (1988). *Guidelines for disposal of solid wastes and hazardous wastes* (Vol. 1). Department of Commerce, National Technical Information Service. Technical Report No. PB88178066. OCLC No. 18438829. 234 pages. USA.
232. Wang, L. K., & Wang, M. H. S. (1988). *Guidelines for disposal of solid wastes and hazardous wastes* (Vol. 2). Department of Commerce, National Technical Information Service. Technical Report No. PB88178074. OCLC No. 18438829. 349 pages. USA.
233. Wang, L. K., & Wang, M. H. S. (1988). *Guidelines for disposal of solid wastes and hazardous wastes* (Vol. 3). Department of Commerce, National Technical Information Service. Technical Report No. PB88178082. OCLC No. 18438829. 233 pages. USA.
234. Wang, L. K., & Wang, M. H. S. (1988). *Guidelines for disposal of solid wastes and hazardous wastes* (Vol. 4). Department of Commerce, National Technical Information Service. Technical Report No. PB88178090. OCLC No. 18438829. 285 pages. USA.
235. Wang, L. K., & Wang, M. H. S. (1988). *Guidelines for disposal of solid wastes and hazardous wastes* (Vol. 5). Department of Commerce, National Technical Information Service. Technical Report No. PB88178108. OCLC No. 18438829. 259 pages. USA.
236. Editor. (1994). *Composting Poultry Carcasses.* US Department of Agriculture and University of Arkansas, Cooperative Extension Service.

237. Wang, L. K., Kurylko, L., & Wang, M. H. S. (1992). *Gas dissolving and releasing liquid treatment system*. US Patent No. 5167806, US Patent and Trade Marks Office, Washington, DC, USA.
238. Wang, L. K., Kurylko, L., & Wang, M. H. S. (1994). *Sequencing batch liquid treatment*. US Patent No. 5354458, US Patent and Trade Marks Office, Washington, DC, USA.
239. Wang, L. K., Wang, M. H. S., Wang, P., & Clesceri, N. L. (2021). Biological and physicochemical sequencing batch reactors using sedimentation or flotation. In L. K. Wang, M. H. S. Wang, Y. T. Hung, N. K. Shammas, & D. B. Aulenbach (Eds.), *Environmental flotation engineering* (pp. 397–417). Springer Nature. Switzerland.
240. Shin, H. S., Han, S. K., Song, Y. C., & Lee, C. Y. (1996). *A novel composting method of organic waste using multi-step sequencing batch two-phase anaerobic technique*. First IAWQ specialized conference on sequencing batch reactor technology, US Environmental Protection Agency, Washington, DC, USA.
241. Wang, L. K., Pereira, N. C., & Hung, Y. T. (2009). *Biological treatment processes* (pp. 459–729). Humana Press. USA.
242. Wang, L. K., Shammas, N. K., & Hung, Y. T. (2009). *Advanced biological treatment processes* (pp. 157–410). Humana Press. USA.
243. Wang, L. K., Tay, J. H., Tay, S. T. L., & Hung, Y. T. (2010). *Environmental bioengineering* (pp. 123–192). Humana Press. USA.
244. Wang, M. H. S., & Wang, L. K. (2016). Glossary of land and energy resources engineering. In L. K. Wang, M. H. S. Wang, Y. T. Hung, & N. K. Shammas (Eds.), *Natural resources and control processes* (pp. 493–622). Springer Nature. Switzerland.
245. Wang, M. H. S., & Wang, L. K. (2021). Glossary of natural resources and environmental pollution control. In L. K. Wang, M. H. S. Wang, Y. T. Hung, & N. K. Shammas (Eds.), *Environmental and natural resources engineering* (pp. 421–494). Springer Nature. Switzerland.
246. Shammas, N. K., & Wang, L. K. (2007). Biosolids composting. In L. K. Wang, N. K. Shammas, & Y. T. Hung (Eds.), *Biosolids treatment processes* (pp. 645–688). Humana Press. USA.
247. Wang, L. K., Hung, Y. T., & Li, K. H. (2007). Vermicomposting process. In L. K. Wang, N. K. Shammas, & Y. T. Hung (Eds.), *Biosolids treatment processes* (pp. 689–704). Humana Press. USA.
248. Haug, R. T. (1980). *Compost engineering principles and practice* (655 pages). Technomic Publishing Co. USA.
249. Brown, G. (2014). *The compost-powered water heater* (162 pages). The Countryman Press. USA.
250. Vandergholm, D. H., Day, D. L., Muehling, A. J., Hung, Y. T., & Butler, E. (2016). Management of livestock wastes for water resource protection. In L. K. Wang, M. H. S. Wang, Y. T. Hung, & N. K. Shammas (Eds.), *Natural resources and control processes* (pp. 1–72). Springer Nature. Switzerland.
251. Glick, M. (2021). *Now you can compost human bodies too: Washington State burial facilities pioneer the latest eco-burial method*. Sierra.
252. Wang, L. K., & Wang, M. H. S. (2022). Disposal of dead animal, poultry and fish using two-stage in-bin composting system. In *Evolutionary progress in science technology engineering, Arts and mathematics (STEAM)*. Lenox Institute Press. USA.
253. Adams, D., Felgal, C., & Noll, S. (2021). *Composting Poultry Carcasses* (Manual NCR530) (5 pages). Purdue University. USA.

Chapter 8
Sanitary Landfill Operation and Management

Mohamad Anuar Kamaruddin, Faris Aiman Norashiddin, Mohd Suffian Yusoff, Mohamad Haziq Mohd Hanif, Lawrence K. Wang, and Mu-Hao Sung Wang

Abstract As part of the transformation to achieve sustainable resource recovery and waste management, landfills play an important role. The landfill's primary function is to accept solid wastes that cannot be "avoided, reduced, reused, recycled, or recovered." Recognizing that residual waste composition has changed and will continue to evolve over time in response to technological advancements in recovery operations, it is critical that a precautionary approach be taken to properly mitigate the environmental risks of landfill facilities. Landfills must be built to have the least amount of negative environmental effects possible. The landfill's design must take into account the surrounding area, the amount and nature of waste to be disposed of, the host community's concerns, adjacent land use, and economic and social factors. Landfills should be planned and maintained in such a way that pollutants such as landfill gas, leachate, and stormwater are effectively managed. Monitoring is crucial for having a better understanding and trust in the site's controls and risks, which advises management and treatment options. Rather than a monitoring program that validates impacts that have occurred, a monitoring program should be developed to cover all emissions and put a priority on monitoring to verify the efficacy of current controls, such as by monitoring leachate content, leachate levels, and surface water (groundwater monitoring).

Keywords Solid waste · Municipal solid waste (MSW) · Landfill · Waste management · Waste disposal

---

M. A. Kamaruddin (✉) · F. A. Norashiddin · M. H. M. Hanif
School of Industrial Technology, Universiti Sains Malaysia, Pulau Pinang, Malaysia
e-mail: anuarkamaruddin@usm.my

M. S. Yusoff
School of Civil Engineering, Engineering Campus, Universiti Sains Malaysia, Pulau Pinang, Malaysia
e-mail: suffian@usm.my

L. K. Wang · M.-H. S. Wang
Lenox Institute of Water Technology, Latham, NY, USA

L. K. Wang et al. (eds.), *Solid Waste Engineering and Management*, Volume 1, Handbook of Environmental Engineering 23,
https://doi.org/10.1007/978-3-030-84180-5_8

Nomenclature

ADC	Alternate daily cover
BOD	Biochemical oxygen demand
BOD_5	Biochemical oxygen demand 5 days
C_2H_5COOH	Propionic acid
$CH_3COCOOH$	Pyruvic acid
CH_3COOH	Acetic acid
CH_4	Methane
CO	Carbon monoxide
CO_2	Carbon dioxide
COD	Chemical oxygen demand
FID	Flame ionization detection
GHG	Greenhouse gas
GPS	Global positioning system
H_2S	Hydrogen sulfide
ISWA	International Solid Waste Association
LFG	Landfill gas
MSW	Municipal solid waste
N_2	Nitrogen
NH_3	Ammonia
PPE	Personal protective equipment
RDF	Refuse derived fuel
VOAs	Volatile organic acids
WTE	Waste-to-energy

8.1 Introduction

Many cities around the world, especially in developing Asian countries, are currently struggling to introduce a long-term solid waste management system. Final waste disposal technologies such as sanitary landfill and incineration, which are at the bottom of the sustainable waste management ladder, are still relevant, despite the fact that massive amounts of waste can be reduced using the 3Rs. Landfilling is widely regarded as a better waste disposal option than incineration since the latter only decreases waste volume while also producing residuals and gaseous contaminants that must be disposed of eventually, which means returning to landfills. Because of its straightforward operating protocol and cost-effectiveness, it is favored in solving daunting MSW conundrums [1].

Initially, waste management systems were implemented primarily for the purpose of removing food and breeding media for pests, as well as removing waste from residential or living spaces and storing it far from public view or "out of sight" [2]. Waste management systems are introduced with the aim of improving public sanitation and health. However, as industrial society progresses, waste disposal waste becomes an inadequate and unsustainable option in the future because disposal dumps are major sources of environmental pollution due to leachate and gas emissions into groundwater, surface water, and the atmosphere, making dumps unsuitable and unsustainable disposal practices. As a result, the idea of developing comprehensive landfilling systems to handle and control incoming wastes as well as landfill by-products such as leachate formation and gas emissions is implemented.

Waste generation has risen annually in tandem with population growth, and as a result of widespread urbanization, disposal problems have become more difficult, as more land is required for the ultimate disposal of these solid wastes. Current landfill sites have caused problems in many major cities in developing Asian countries. Rapid urbanization and shifting habits have resulted in a shift in waste composition from primarily organic to primarily complex plastics, paper, and packaging materials. As the types and sources of waste generated diversify, as does the availability of disposal sites within collection areas, storage and collection systems are becoming more complex and expensive. The ability to design solid waste disposal methods requires knowledge of the characteristics of solid waste. Data on solid waste compositions are difficult to come by, and even when it is, it is often out of date. Furthermore, to correct data differences between sanitary landfill sites and waste dumps, a better classification system for landfills is needed. We present an overview of waste disposal issues in several Asian cities in this report, with a focus on waste characteristics and disposal patterns.

In the case of Malaysia, landfilling is currently the most effective way of disposing of urban solid waste, with the majority of landfill sites using an open dumping scheme, as shown in Table 8.1. Since existing landfill sites are filling up at a rapid

Table 8.1 Total landfills in Malaysia [4]

State	Landfill in operation	Landfills that have ceased operation
Johor	13	21
Kedah	10	5
Kelantan	13	4
Melaka	2	5
Negeri Sembilan	8	10
Pahang	19	13
Perak	20	9
Perlis	1	1
Pulau Pinang	1	2
Sabah	21	1
Sarawak	51	12
Selangor	6	12
Terengganu	9	12
Wilayah Persekutuan Kuala Lumpur	1	7
Wilayah Persekutuan Labuan	1	0
Total	176	114

pace, the disposal of wastes by landfilling is becoming more difficult as the amount of solid waste produced grows. The current situation of land scarcity, coupled with higher land prices, especially in urban areas, follows. As a result, municipal governments have agreed to provide infrastructure to facilitate the recovery of materials before solid wastes are transported to landfill sites in order to maximize the use of landfill sites, which can only accept organic materials. This effort, however, has yet to be fully implemented because it requires careful planning and coordination among different agencies in order to fully utilize recovery materials facilities. Materials recovery and recycling are typically not handled by local governments or landfill operators in many developing Asian cities. Scavengers or illegal waste pickers at landfill sites, on the other hand, minimize the amount of recyclable products like paper, plastics, glass, and metals in the waste [3].

Since the 1970s, a total of 111 disposal sites have been closed because of their inability to accept MSW due to their unsuitable venue. The regulated dumps are currently being upgraded to Class IV status to comply with the strict regulations imposed by local authorities and the ministry. It also aims to lessen the environmental effects of landfill grounds. More sanitary landfills are currently being proposed to meet the country's growing waste disposal needs. Table 8.2 shows the sanitary landfills in Malaysia, as well as the sanitary landfills in other countries.

Various industrial and domestic solid wastes cause serious environmental problems, especially in urban areas [4, 5]. Solid waste disposal in sanitary landfills is the most widely used and favored form of waste disposal among all available techniques [7]. Due to its cost-effectiveness in terms of upkeep, technology, and awareness, landfilling is the most common and well-practiced form of waste management, which is typically characterized as a waste storage facility where waste is compacted by layers in engineered constructed cells either on the land surface or on excavated inland surfaces [8]. It disposes of a variety of municipal, private, and mixed industrial wastes, as well as solid waste from institutional building and

Table 8.2 Sanitary landfills in Malaysia [4]

Landfill	Status of disposal facilities	In operation
Bukit Tagar sanitary landfill	Operating	2006
Air Hitam sanitary landfill	Closed	1995
Jeram sanitary landfill	Operating	2008
Seelong sanitary landfill	Operating	2004
Pulau Burong sanitary landfill	Operating	2001
Mambong sanitary landfill	Operating	2000
Bintulu sanitary landfill	Operating	2002
Sibu sanitary landfill	Operating	2002
Kota Kinabalu sanitary landfill	Operating	2001
Tanjung Langsat sanitary landfill	Operating	2005
Tanjung 12 sanitary landfill	Operating	2010
Miri sanitary landfill	Operating	2006

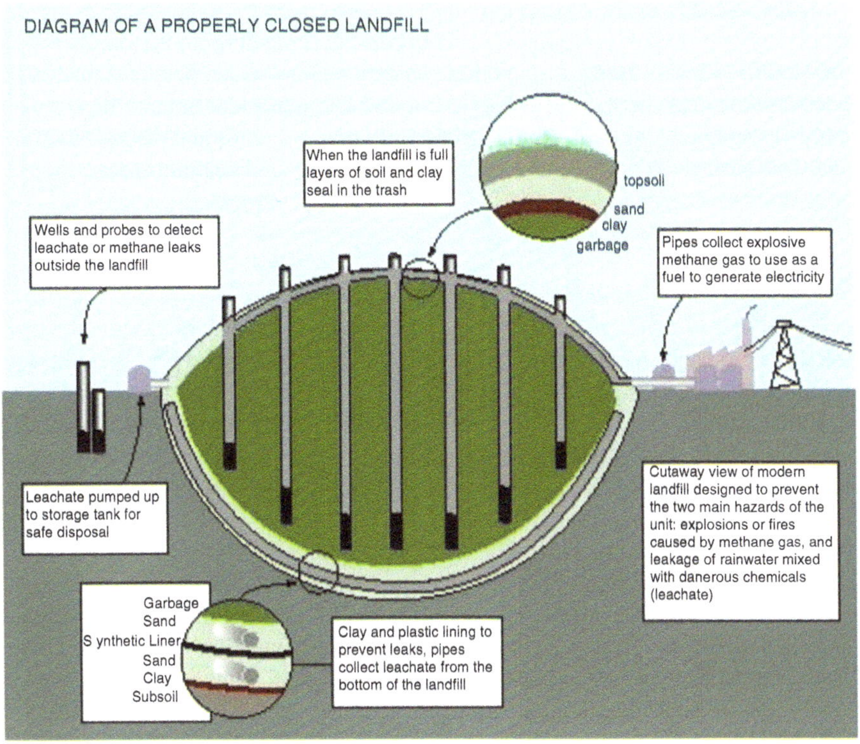

Fig. 8.1 Structure of a landfill. (Source: USEPA)

demolition [9, 10]. Figure 8.1 shows the complex structure of a landfill as depicted by the United States Environmental Protection Agency (USEPA) [11].

Landfill activities may have ecological consequences, such as landscape changes, habitat degradation, and fauna displacement. Inhaled landfill gas from methane extraction and landfill leachate interaction with polluted groundwater pose possible health risks [8]. Landfill pollutants, both gaseous and aqueous, are highly complex mixtures that can be extremely harmful to their environment, with characteristics that differ depending on the landfill's waste composition and age. Furthermore, all landfill sites that accept solid waste for disposal must meet a number of criteria before they can begin operations. Some of the basic criteria included: site pre-siting requirements (1), stability (2), and soil and water safety by leachate and landfill gas management (3), as well as nuisances and hazards management (4). The bottom liner system, which separates garbage and subsequent leachate from ground water, leachate collection and management system, road network, drainage system, and final capping system are the basic components of a landfill. The collection pipes of leachate are a significant component because the conditions in these landfills are prone to pollution, as these sites are often hydrologically connected to surface streams or groundwater sources [12].

Since landfills need a large amount of space, land shortage for housing and declining land market prices have been a problem depending on the distance from the landfill [13, 14]. Bees, strong odors, smoke, and noise can be a nuisance in the housing area near the landfill [15]. Despite the fact that modern landfills are well-designed and fitted with systems to reduce pollution, residents and workers who live and work near them will still be concerned about the sites' health impact.

8.2 Landfills

Landfills are used to dispose of all urban solid waste. In Malaysia, there are four levels of landfill. Regulated dumping (level 1), sanitary landfill with regular cover (level 2), sanitary landfill with leachate circulation (level 3), and sanitary landfill with leachate treatment (level 4) are the different types of sanitary landfills. In general, the majority of landfills in Malaysia operate at level 3. All of the solid waste generated will be collected and transported to landfills for disposal. However, some local municipal councils have established transfer stations to recover valuable materials through the implementation of a segregation mechanism prior to the transfer of unnecessary waste to landfills. The aim of a landfill is to stabilize solid waste and make it sanitary by properly storing waste and using natural metabolic functions. There are many concerns and problems with the landfilling management system as more urban waste is diverted to landfills.

8.2.1 Landfill in Sanitary Conditions

A sanitary landfill is a form of solid waste disposal that is regulated. The location must be appropriate in terms of geology, hydrology, and climate. In general, there are three types of sanitary landfills that are widely used: anaerobic landfills, semi-aerobic landfills, and aerobic landfills, which will be discussed in more detail in the sub-sections below.

8.2.1.1 Anaerobic Landfill

An anaerobic landfill is a decomposition site for solid waste that contributes to negative environmental effects and human health issues by producing hazardous substances with high concentrations of organic material. Aside from that, this mechanism emits a lot of methane and carbon dioxide, contributing to global warming. There are three types of anaerobic landfill systems: anaerobic landfill (1), anaerobic sanitary landfill (2), and enhanced anaerobic sanitary landfill (3). Anaerobic landfills, in general, receive waste and are dug in a plane field or gorge. In anaerobic conditions, the wastes are then filled with water. Biodegradation occurs

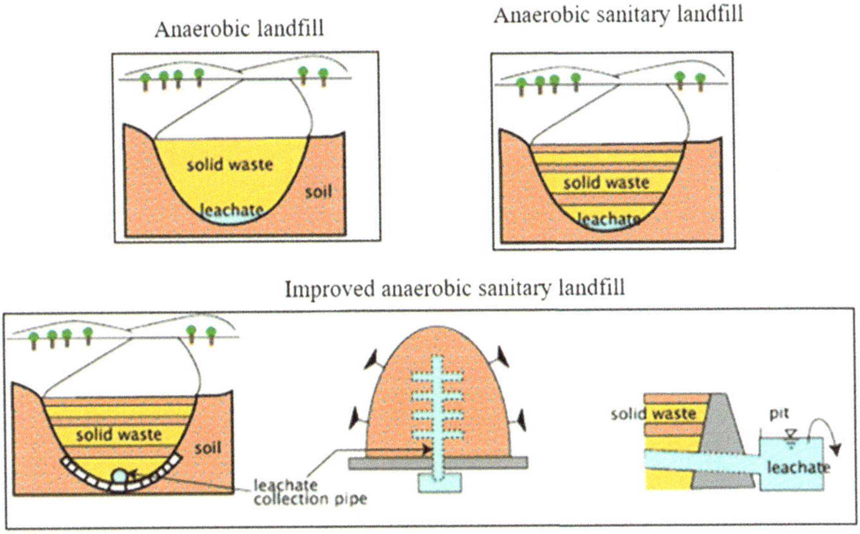

Fig. 8.2 Anaerobic landfill structures [17]

anaerobically (without oxygen) and creates landfill gas [16]. The structures of anaerobic landfills, anaerobic sanitary landfills, and enhanced anaerobic sanitary landfills are shown in Fig. 8.2.

8.2.1.2 Semi-aerobic Sanitary Landfill

A leachate collection pipe (perforated pipe) is provided at the bottom of the site filled with gravel for semi-aerobic landfill systems. Landfill leachate is channeled into the perforated pipe, which prevents the leachate from being absorbed into the soil and keeps it in the layer. This also prevents leachate from seeping into the original dirt, retaining leachate in the solid waste layer, and allowing air to enter the solid waste layer through the collection pipe. It purifies leachate in the solid waste layer prior to collection [18]. It also acts as an air entry route from the outside into the solid waste layer in landfill sites, in addition to the perforated pipe. This also aids in the expansion of aerobic sections and the activation of aerobic bacteria, resulting in a faster rate of waste decomposition. A convection method may be used to create a semi-aerobic system. The above entails the decomposition of organic matter inside the landfill, which would result in a temperature rise. The temperature difference between the inside and outside of the landfill would cause a heat convection current to flow into the landfill through the leachate pipe [19]. The structure of semi-aerobic sanitary landfill systems is depicted in Figs. 8.3, 8.4, and 8.5.

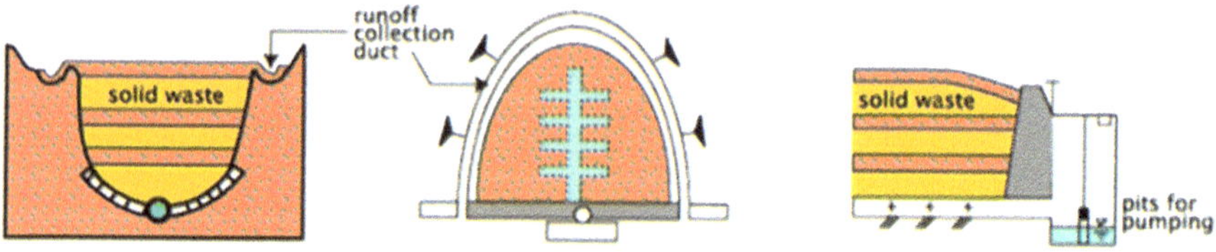

Fig. 8.3 Semi-aerobic sanitary landfill structures [17]

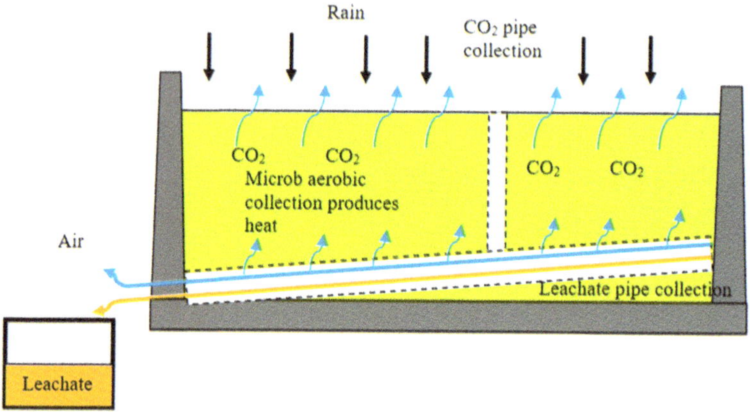

Fig. 8.4 Semi-aerobic sanitary landfill mechanism [17]

8.2.1.3 Landfill with Aerobic Properties

The next form of sanitary landfill is an aerobic landfill, which usually has a perforated pipe for collecting leachate at the bottom of the site. It is set up in such a way that leachate created by refuse can be collected and flowed out of the landfill site in a short period of time. In addition, an aeration pipe is placed underneath the soil bed to provide air to the solid waste layer. Leachate recycling is done to keep moisture in the system and provide nutrients for microorganisms to use in the biodegradation process. Microorganisms will benefit from this situation as they turn organic waste into biodegradable materials and humus. Leachate from aerobic landfills can increase its consistency, emit less methane gas, and improve solid waste stability. As a result, the decomposition process becomes quicker, extending the landfill site's life [18]. The aerobic landfill systems are depicted in Figs. 8.6 and 8.7.

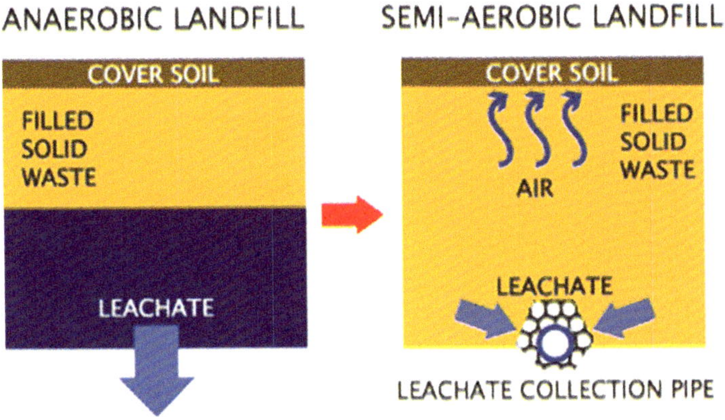

Fig. 8.5 Difference between an anaerobic landfill and semi-aerobic sanitary landfill [17]

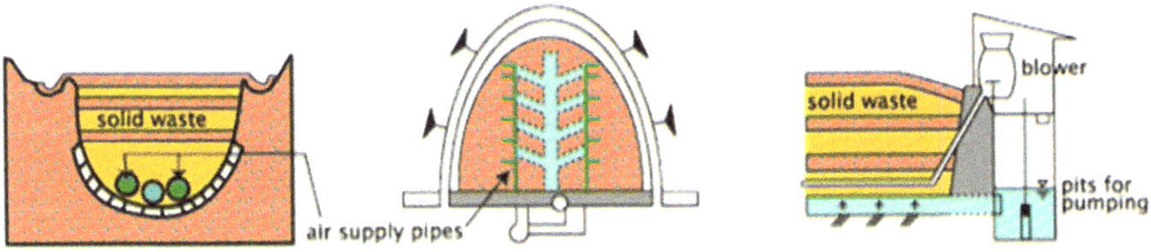

Fig. 8.6 Aerobic landfill structures [17]

8.2.2 Municipal Solid Waste Composition

Here, the focus is on the composition of urban solid waste (MSW) that is consistently collected at landfills on a regular basis. According to the most recent statistics available, solid waste generation in the United States totaled approximately 49,670 tons per day, including household and industrial waste [4]. Several solid waste characteristics have been studied, and the results show that food waste accounts for approximately half of all waste produced, as shown in Fig. 8.8.

8.3 Landfill Decomposition Process

Solid waste decomposition in landfills is a complicated process that varies from one location to the next. This takes into account a variety of variables, including waste composition, landfill operations, changing weather and site hydrology, seasons, landfill age, temperature, waste moisture, and pH [20, 21]. The differences and factors that influence the disintegration process are critical in the design, operation, and management of a leachate treatment plant. When a landfill is completed and closed,

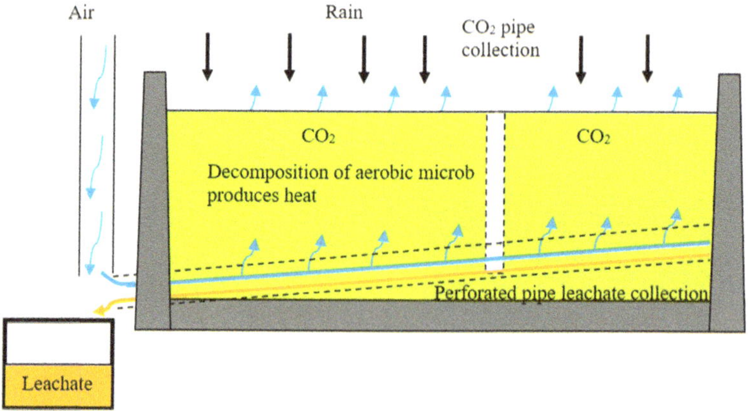

Fig. 8.7 Aerobic landfill mechanism [17]

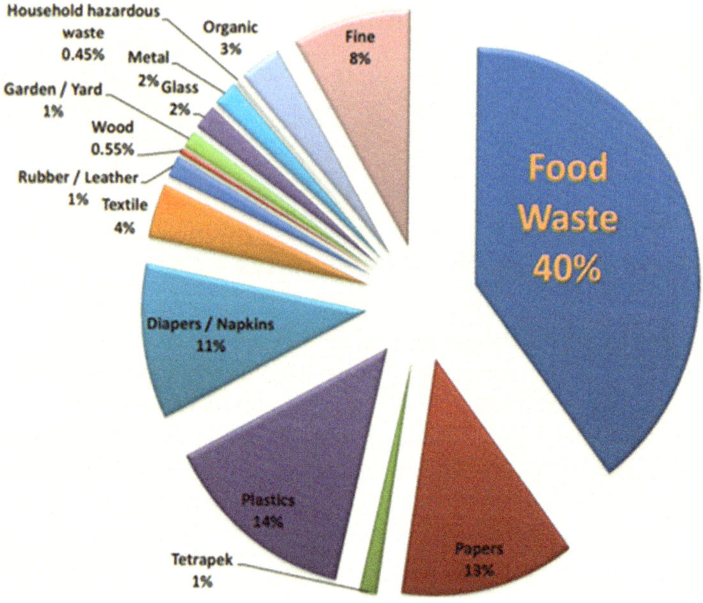

Fig. 8.8 Solid waste composition in Malaysia [4]

the degradation process of waste refuse with the soil matrix causes chemical and physical changes [21].

Decomposition begins when waste is discarded and covered in a landfill, causing a sequence of complex biological and chemical reactions [9]. The decomposition process in landfills can be divided into five phases: aerobic, anaerobic, acidogenic, methanogenic, and methane reduction. Each phase's waste generation rate, characteristics, and gas output differ and are linked to the microbial activities that occur during that phase. Decomposition is physically, chemically, and microbiologically based on landfill activities in each process. When the trash is buried in a landfill, it decomposes in a complex sequence of biological and chemical reactions [9]. The decomposition process in landfills can be broken down into five stages: aerobic, anaerobic, acidogenic, methanogenic, and methane reduction. The characteristics and rates of waste generation, as well as the output of gas from the landfill site, differed by process and were closely related to the microbiological response that occurred at each phase. Physical, chemical, and microbiological factors influence decomposition rates in each process at the landfill site over time.

In general, after the solid waste tipping process is completed, an anaerobic environment develops in which an air pocket in the landfill decomposes the solid waste by aerobic biological processes. The transition step is the second level. When the oxygen in the landfill site is reduced or depleted, the transition from aerobic to anaerobic occurs. The tip became anaerobic when the biodegradation of solid waste used oxygen, and there was no substitution of the free oxygen available [22]. The acidogenic process follows, during which the complex organic materials in the tipped waste biodegrade into simpler organic materials such as acetic acid (CH_3COOH), propionic acid (C_2H_5COOH), pyruvic acid ($CH_3COCOOH$), and other basic organic acids [22].

Methanogenic bacteria use the end products from the first stage of anaerobic decomposition to produce methane and carbon dioxide in the fermentation or methanogenic process. It can happen in an anaerobic setting that is categorically exergonic [23]. The final step of solid waste is the methane reduction phase, during which the rate of methane production peaks and then drops as the pool of soluble substrate (carboxylic acids) diminishes. The rate of CH_4 development in this step is determined by the rate of cellulose and hemicellulose hydrolysis [9]. In the following portion, we will go through the decomposition of solid waste in landfills in more detail. Figure 8.8 depicts the leachate characteristic of solid waste degradation.

8.3.1 Aerobic Phase

The presence of oxygen is rapidly absorbed within the empty spaces present within the freshly covered wastes when this process first starts, releasing CO_2 and causing a rise in temperature within the cell [9]. When exposed to oxygen, biodegradable materials react quickly, producing carbon dioxide, water, and other by-products. With decreasing oxygen levels, anaerobic microbes begin to initiate processes, which may last for a long time [25]. The upper layer is only active in aerobic metabolism, while fresh waste is oxygen-rich, which is introduced through the presence of precipitation. This process is usually brief, with no substantial leachate output [26]. The presence of oxygen, combined with aerobic bacteria, converts organic compounds from solid waste into organic compounds. The development and replication of new cells will be accelerated if energy and nutrients are supplied by a carbon source [27].

8.3.2 Anaerobic Phase

The transition from aerobic to anaerobic phases will take anywhere from 6 to 18 months. This time period begins when waste is deposited on the landfill's top layer. When the oxygen supply is depleted, the organism enters the anaerobic process. The decomposition processes in the anaerobic phase are influenced by a number of factors, including waste characteristics, moisture content, temperature, pH, nutrients available, microbes, and inhibitors such as oxygen, metals, and sulfates [27].

When oxygen becomes scarce, it indicates that aerobic decomposition is coming to an end and that anaerobic decomposition is about to begin. The heat generated in the previous phase, combined with a decrease in the moisture content of the waste, creates an ideal environment for anaerobic microbes to thrive. Furthermore, electron acceptors are transferred from oxygen to nitrate and sulfate, and carbon dioxide is converted from oxygen, resulting in a reduced environment. Assessable concentrations of COD (480–18,000 mg/L) and volatile organic acids (100–3000 mg/L) are detected in the leachate by the end of this step, with ammonia released consistently and not transformed within the anaerobic environment [27].

8.3.3 Phase of Acid Formation

VOAs, ammonia, hydrogen, and CO_2 are generated in substantial amounts during this process as a result of hydrolysis and the biodegradation of organic matter by microorganisms. Strict and facultative microbes work hard to disintegrate waste, lowering the waste's redox potential and facilitating the growth of methanogenic microorganisms [28, 29]. The pH of the leachate will decrease as a result of this operation. During this period, the highest concentrations of BOD (1000 to 57,700 mg/L) and COD (1500 to 71,100 mg/L) are usually reported. The development of acidogenic bacteria and significant substrate and nutrient degradation are the most important characteristics during this process [9].

8.3.4 Methanogenic Phase

During this step, under strict anaerobic conditions, methane is produced, which is known as an exergonic reaction [23]. When methane is emitted in measurable amounts, the methanogenic process begins. This is most likely due to the pH of the waste being sufficiently neutralized, allowing methanogenic bacteria to expand at a minimum. And, since different landfills have different biochemical behaviors, biodegradation in landfills may not be uniform. Lower pH suggests no methane formation and only acid formation, while other areas of the tip may have a healthy population of methanogenic bacteria that use the organic acids to keep the pH neutral [22].

8.3.5 *Phase of Maturation*

This process occurs after 4 to 10 years and lasts for a long time [30]. Methane-forming or methanogenic bacteria ingest complex organic acids and transform intermediate acids to methane and carbon dioxide during this process. Inorganic salts lose their soluble properties and precipitate as a result of this process. COD and BOD concentrations begin to decrease when organic acids are converted to steam. During the anaerobic process, some organic compounds that do not decompose are essential elements for adsorption and complicated reactions that remain as landfill residues [31]. During this step, the pH rises while the bicarbonate buffering mechanism that supports methanogenic bacteria activity is inhibited. Methanogenic bacteria decompose organic matter slowly but efficiently over time, resulting in the incremental removal of heavy metals by precipitation [32].

8.4 Management of Sanitary Landfills

There are a few aspects that must be handled sustainably before, during, and after the construction of a sanitary landfill. These factors are critical in ensuring that sanitary landfills run smoothly for the next 20 to 30 years while causing no pollution to the ecosystem.

8.4.1 *Management of Landfill Leachate*

Before introducing any method of landfill leachate treatment, a thorough understanding of leachate characteristics is needed to comprehend the varying results obtained when treating the leachate using biological, physical, or physicochemical methods. Biological treatment is unquestionably the preferred method due to its numerous benefits, including a wide range of sources and the ease and pace with which microorganisms can be cultured and produced [33]. These systems are classified as aerobic (oxygenated) or anaerobic (oxygen-depleted). Microorganisms or bacteria are used to eliminate leachate pollutants via an assimilating method, in particular. This process aids in the increase of microbial metabolism and living cell building blocks. As a result, leachate parameters can be removed by the metabolic conditions of living cells. Regardless of the treatment method used, choosing the right biological treatment involves careful consideration of how to cultivate and sustain safe biomass, flow rate tolerance, and the organic loads to be handled.

Fig. 8.9 Leachate collection pond prior to treatment

Biological therapies are still one of the acceptable methods for handling leachate because they are of low cost in terms of both capital and operational costs. Furthermore, the use of biological treatment has been shown to completely destroy organic sulfides, organic compounds, and toxicity. Figure 8.9 shows the leachate collection pond before undergoing any further treatment.

8.4.1.1 Composition of Landfill Leachate

The composition of the produced leachate varies greatly as the degradation of solid wastes progresses. Furthermore, the probability of producing concentrated leachate is influenced by a variety of factors that influence its quantity and consistency, including percolating water through wastes, biochemical processes in landfill cells, and the degree of compaction [34–36]. Leachate parameters are usually different depending on the landfill's age.

Table 8.3 shows the leachate parameters of semi-aerobic and anaerobic landfills, which varies depending on the landfill age, waste classification, site hydrology, landfill type, and landfill operations. Landfill networks are divided into four categories. A managed dumping landfill is on the first floor, while a daily cover of a sanitary landfill is on the second. A sanitary landfill with a leachate circulation system is the third level, and a sanitary landfill with a leachate treatment system is the

Table 8.3 A typical landfill leachate characteristics of semi-aerobic and anaerobic landfill in Northern Malaysia [27], DOE Malaysia

	Average values		
Landfill Parameters	Semi-aerobic Pulau Burung (aerated)	Anaerobic Kulim (unaerated)	Discharge limit DOE Malaysia[a]
Phenols (mg/L)	1.2	2.6	–
Ammonia-N	483	300	–
Total nitrogen (mg/L)	542	538	–
Nitrate-N (mg/L)	2200	1283	–
Nitrite-N (mg.L)	91	52	–
Total phosphorus (mg/L)	21	19	–
BOD_5 (mg/L)	83	326	50
COD	935	1892	100
BOD_5/COD	0.09	0.205	0.5
pH	8.2	7.76	5.5–9
Turbidity (TNU)	1546	855	–
Color	3334	1936	–
Total solids (mg/L)	6271	4041	–
Suspended solids (mg/L)	1437	6336	–
Total iron (mg/L)	7.9	707	100
Zinc (mg/L)	0.6	5.3	5
Total coliform	–	0.2	1
E. Coli	–	0.81×10^{-4}	–

[a]DOE Malaysia

Table 8.4 Landfill leachate characteristics [37, 38]

	Recent	Intermediate	Old
Age (years)	<5	5–10	<10
pH	6.5	6.5–7.5	>7.5
COD (mg L^{-1})	>10,000	4000–10,000	<4000
BOD_5/COD	>0.3	0.1–0.3	<0.1
Organic compounds	80% volatile fat acids (VFA)	5–30% VFA + humic and fulvic acids	Humic and fulvic acids
Heavy metals	Low–medium	Low–medium	Low
Biodegradability	Important	Medium	Low

fourth level. The majority of landfills in Malaysia are now using third-level technology.

The composition of leachate varies between sanitary landfills. COD is made up of organic molecules that are not biodegradable, such as humic and fulvic acids [38]. COD values in Table 8.4 range from 100 to 70,900 mg/L, indicating extreme toxicity. The BOD_5/COD ratio is used to calculate the leachate's state, with lower ratios (0.1) indicating the existence of significant amounts of "strong" COD. With the aging of the landfill, the BOD_5/COD ratio (from 0.70 to 0.04) can rapidly decrease, resulting in low bio-treatability. The contents of the leachate are oxidized by the chemicals in the COD test but are badly oxidized in the BOD_5 test if the ratio decreases to 0.1. Leachate material has high organic breakdown products with a high BOD_5/COD ratio in the early stages of decomposition. It has a high pH due to the abundance of volatile fatty acid fermentation products [37, 39].

Furthermore, the composition and characteristics of leachate differ and change depending on the landfill's process. In the aerobic process, oxygen is trapped in fresh waste and supplied by rainwater diffusion in the upper layer of the landfill. The material formed by aerobic bacteria is then decomposed by anaerobic bacteria into acetic acid, lactic acid, formic acid, and alcohols such as methanol and ethanol when the landfill enters the anaerobic phase. The majority of the acid produced during this process lowers the pH of the leachate produced. COD concentrations in the leachate will drop from 18,000 mg/L to 480 mg/L at the end of this process, while volatile organic acids (VOAs) will drop from 3000 mg/L to 100 mg/L.

A wide variety of bacteria and anaerobic bacteria use the acids produced in the next step, the acidic phase, to keep the atmosphere stable by preventing the pH from dropping. Organic acids, alcohols, hydrogen, and CO_2 are all converted to acetate by fermenting bacteria. During this process, BOD and COD concentrations are heavy, with BOD concentrations ranging from 1000 mg/L to 55,700 mg/L, COD concentrations ranging from 1500 mg/L to 71,100 mg/L, and a BOD/COD ratio of 0.4 to 0.7 [26]. Acid phase leachate is chemically reactive due to its acidic pH and can increase the solubility of many compounds (Kjeldsen et al., 2002). The composition of the leachate during the acidic process is shown in Table 8.5.

As the pH rises during the methanogenic process, less inorganic constituents remain in the leachate. This would have an effect on the amount of heavy metals in the leachate [40]. The composition of the leachate during the methanogenic process is shown in Table 8.6.

When compared to the methane fermentation (methanogenic) process, the metal concentration between acid formations (acidogenic) is lower. However, as shown in Table 8.7, there were no noticeable variations between the two phases for the majority of metals. However, as shown in Table 8.8, some other compounds in the leachate were not affected by phase changes. The water balance components that occur inside the landfill are depicted in Fig. 8.10.

Table 8.5 Leachate composition during acetogenic phase [24]

Parameter	Range values
pH	4.5–7.5
COD	6000–60,000
BOD_5	–
BOD/COD	194–3610
SO_4	70–1750
Fe	20–2100
Mg	50–1150
Ca	270–6240

Table 8.6 Leachate composition during methanogenic phase [24]

Parameter	Range values
pH	6.8–8.2
COD	622–8000
BOD_5	97–1770
BOD/COD	–
SO_4	10–420
Fe	3–280
Mg	40–350
Ca	20–600

Table 8.7 Composition of heavy metals during the acidogenic phase and methanogenic phase [41]

Parameter	Acidogenic phase	Methanogenic phase
As	<0.001–0.148	<0.001–0.485
Cd	<0.01–0.10	<0.01–0.08
Cu	0.020–1.10	<0.02–0.62
Ni	<0.03–1.87	<0.03–0.6
Pb	<0.04–0.65	<0.04–1.9
Cr	0.03–0.3	<0.03–0.56
Zn	0.09–140	0.03–6.7
Hg	<0.0001–0.0015	<0.0001–0.0008

8.4.1.2 Leachate Management in Landfills

Physical, biological, and chemical processes have all been used to treat landfill leachate, depending on the cost-effectiveness, complexity of the treatment system, durability, and leachate characteristics. As a result, finding a suitable approach for leachate treatment that does not endanger the ecosystem or cause any significant changes has always been a top priority [21]. A facility to handle leachate and hazardous contaminants is normally present at a fully operational landfill site. Figure 8.11 depicts the various methods for treating landfill leachate that is currently in use.

Because of their low capital and operational costs, biological treatments have attracted interest as a reliable tool. To extract the majority of leachate containing high levels of BOD, systems that are divided into aerobic (with oxygen) and anaerobic (without oxygen) are used. Microorganisms can be cultured to degrade organic compounds into smaller units in the form of sludge and carbon dioxide aerobically and eventually produce biogas (a mixture of CO_2 and methane) anaerobically in biological treatment [33, 37]. Such therapies have been shown to effectively destroy organic compounds [21]. Biological therapies are, without a doubt, the most

Table 8.8 Substance composition during the acetogenic phase and methanogenic phase [41]

Parameter	Acidogenic phase	Methanogenic phase
Fatty acid (C)	963–22,414	<5–146
Conductivity (μS/cm)	5800–52,000	5990–19,300
Alkalinity ($CaCO_3$)	2720–15,870	3000–9130
Phosphate, P (mg/L)	0.6–22.6	0.3–18.4
Sulfate, SO_4 (mg/L)	<5–1560	<5–322

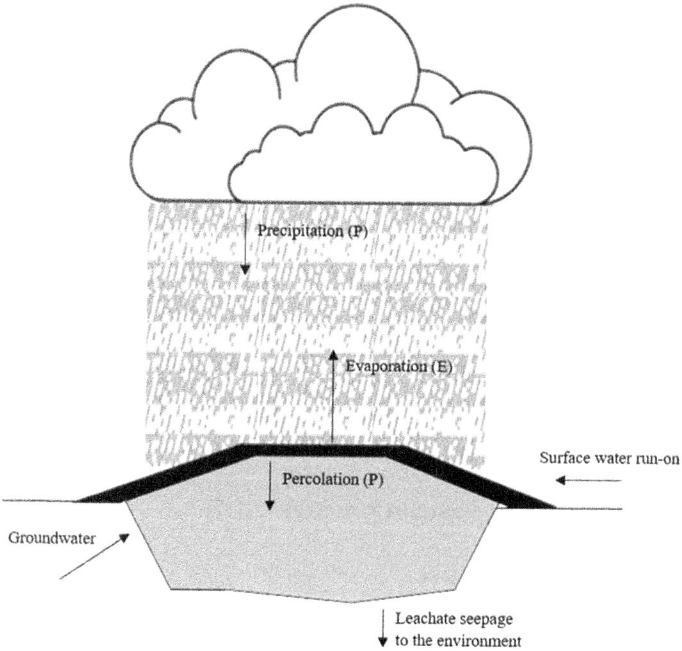

Fig. 8.10 Schematic of water balance components within a landfill

successful method of treating elevated BOD_5 concentrations [37]. However, depending on the nature of the leachate contaminants, sludge bulking can occur in a traditional aerobic system, interfering with leachate treatment. Furthermore, this treatment approach has been shown to be very effective in extracting organic and nitrogenous matter from immature leachate, with a BOD/COD ratio of >0.5. The presence of refractory compounds (primarily humic and fulvic acids) tends to reduce the efficacy of this mechanism over time [44].

Physical treatments, which use biological changes in the leachate to improve leachate consistency, are the next method of treating landfill leachate. Air stripping, adsorption, and membrane filtration are some of the most popular physical treatment methods. Such treatment has been used as a pretreatment or in the final

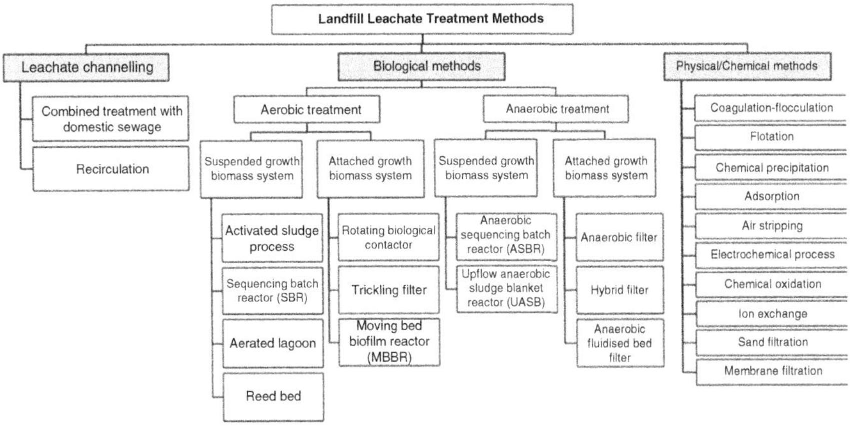

Fig. 8.11 Available techniques of landfill leachate treatments [43]

purification/polishing process to treat particular contaminants, such as ammonia stripping. In the screening process, physical treatments with metal grit traps maintained significant impurities before moving on to the next level of treatment. Physical and chemical treatments are often combined as an additional treatment (pretreatment or purification phase) or to treat a particular pollutant (stripping for ammonia). Furthermore, combining physical and chemical therapies is more successful than treating them separately.

Chemical leachate treatments require the addition of chemicals to increase the consistency of the leachate. During the process, the chemicals neutralize the leachate's condition by treating it with acids or bases [21]. Coagulation–flocculation, chemical precipitation, and chemical electrochemical oxidation are all common chemical treatments. The oldest process, coagulation–flocculation, is commonly used as a pretreatment method prior to any biological treatment and as a polishing phase to extract biodegradable organic matter before final discharge. Due to its ability to remove a wide range of leachate parameters, this allows insoluble end products to shape and through an ionic exchange. It has been used to treat both stabilized and old landfill leachates with effectiveness [45–47]. It is often used as a pretreatment [48–50] step before biological or reverse osmosis or as a final polishing treatment step to eliminate the nonbiodegradable organic matter.

8.4.2 Landfill Gas Management

Landfill gas (LFG) is classified as a gas formed by anaerobic conditions in landfills [51]. As a result, LFG is primarily composed of methane and carbon dioxide. It also includes trace constituents that have been vaporized from solid and liquid wastes into the gas phase. Some biological degradation by-products, such as ammonia

(NH_3) and hydrogen sulfide, are also present in LFG but in low concentrations (H_2S). LFG is typically saturated water vapor, with moisture content varying depending on temperature. LFG is in the vapor process at temperatures of 40 °C and 60 g water/m^3. To avoid odors caused by trace gases, landfill gases, especially those emitted during the early stages of deterioration, must be diluted with a factor of 1:106 until the odor is no longer detectable. Due to the various degradation phases and waste quality disposed at landfills, the quality of LFG can change over time.

Because of the effects of these gases on the environment, especially air quality, LFG extraction and utilization should be a standard procedure at all municipal solid waste landfills. By doing so, greenhouse gas emissions can be reduced dramatically, electricity can be saved, odors can be avoided, and explosion risks can be reduced. The gas wells had been mounted directly on top of the bottom liner, according to Rettenberger's (2004) research. After 11 to 15 months of landfill activities, the s(CH_4) to s(CO_2) ratio increased to 1. As a result, the gas extraction systems must be accessible within the first year of landfill service. Given that the half-life value of LFG output is in the early stages of a landfill, between 3 and 4 years, landfill operations without a gas extraction method result in the loss of nearly half of the gas in this time period [53].

The s(CH_4)/s(CO_2) ratio over time at four gas wells (K1, K2, K4, K5) measured since deposition began [52]. The flame ionization detector (FID) system can be used to detect gas emissions. This monitoring system works by putting a small cap on top of the soil at various points in the landfill and injecting small quantities of gas through an FID detector. In general, the detector can detect concentrations of CH_4 ranging from 0 to 10,000 parts per million. This approach can be used to identify gas-emitting spots. Gas extraction rates may be changed, more wells may be constructed, and/or the areas where the gas is leaking must be sealed to minimize unregulated gas emissions. Controlling diffusive emissions by calculating methane concentrations outside of the landfill and using meteorological data to determine the emission rate using gas distribution models is an alternative to the FID process. The downside of this approach is that it is impossible to pinpoint the hotspots where the majority of the pollution occurs.

Gas collection can be accomplished in either an active or passive device. Gas is collected from the landfill using blowers in an active system, resulting in vacuum pressures in the landfill body ranging from −1 to −30 hPa. Positive pressure in the landfill is used in passive gas collection systems to transport gas out under semi-controlled conditions. Passive systems are only viable if methane gas output is less than 0.5 L/m^2.h and a gas-tight top cover has been mounted. Figure 8.13 depicts LFG and air part gradients across the depth of a landfill body or cover soil [52]. The ratio of s(CH_4) to s(CO_2) shifts as the depth of the landfill increases, as seen in Fig. 8.14, indicating that methane oxidation occurs. Technically, this procedure is used to biologically oxidize residual methane in landfill covers.

LFG also includes a large number of trace constituents [54, 55]. These compounds may be organic, such as halogenated hydrocarbons, or inorganic, such as hydrogen sulfide and ammonia, which come from a variety of sources, including solvents, propellants, and organic silicon gases deposited in landfills. Some of these

Fig. 8.13 Different landfill operation equipment and machinery

gases are generated naturally in the landfill, while others are caused by human activity. Since some of the gases are acutely poisonous, carcinogenic, or genetically dangerous, LFG is known as a danger. The use of LFG has caused problems in gas engines and other equipment, such as clogging and corrosion. This is mostly due to the presence of sulfur, chlorine, fluorine, and siloxane-containing gases. As a result, gas treatment or increased LFG maintenance may be needed. Controlling the combined concentrations of these compounds is therefore very normal and important. Oxygen-containing gases will appear early in the gas generation process, as shown in Table 8.9.

Figure 8.15 shows how the concentration of the main gas compounds in a landfill body varies over time, as depicted by a scheme demonstrating the progression. The composition of LFG varies over time and can be classified into nine phases:

- Phases I–III: LFG development begins to take shape.
- Phase IV: A stable gas output occurs in the landfill, and pores and tiny voids in the landfill fill up with LFG. Gas emissions can be analyzed at the surface.
- Phase V: Gas output falls but remains steady. Biologically, waste components that are easily degradable can be minimized. LFG emissions will be reduced. Since the easily degradable fraction has almost fully degraded and more difficult organics are present, the gas composition changes.
- Phase VI: As gas output declines, air can migrate into the landfill body. From the surface to the deeper layers, this phase occurs. Processes that are aerobic will

Fig. 8.14 An example of a landfill fire occurring

begin. LFG emissions have decreased and can now be observed in certain regions, but only at very low levels.

- Phase VII: Carbon dioxide will be generated as a result of aerobic processes. Microbiological processes can destroy residual levels of methane. As a result, the $s(CH_4):s(CO_2)$ ratio will continue to fall. Only very small amounts of LFG will be released.
- Phase VIII: The landfill body is on the verge of being aerobic. Low amounts of carbon dioxide will be released because some organic material will still be left in the landfill.
- Phase IX: The landfill waste is nearly inert. The pore gas content would be close to that of natural soil (air and some carbon dioxide).

8.4.2.1 Collection of Landfill Gas

In most landfills, each well is linked to a distribution station by a pipe, which connects the transportation pipes of many wells to the main header pipe that transports the gas to the blower station. A valve in the distribution station may change the flow rate. Technical facilities for calculating pipe pressure, flow rate, gas composition, and temperature should be mounted in the pipe just in front of the valve. It may be necessary to differentiate between lean gas and "normal" gas in the gas distribution station, as this may occur in landfills with an older and younger component, for example. The distribution station may be connected to two header pipes in this

Table 8.9 Oxygen-containing constituents in landfill gas [56]

Compound	Concentration range (mg/m^3)
Ethanol	16–1450
Methanol	2.2–210
1-Propanol	4.1–630
2-Propanol	1.2–73
1-Butanol	2.3–73
2-Butanol	18–626
Acetone	0.27–4.1
Butanone	0.078–38
Pentanal	0.8
Hexanal	4.04
Acetic ester	2.4–263
Butyric ester	<0.9–350
Acetic butyl ester	60
Butyric propyl ester	<0.1–100
Acetic propyl ester	<0.5–50
Acetic acid	<0.06–3.4
Butyric acid	<0.02–6.8
Furan	0.01–2.4
Methylfuran	0.06–170
Tetrahydrofuran	<0.5–8.8

situation, resulting in two different gas collection systems. Most of the time, this is not required as long as the gas collection system can be calibrated to achieve high methane concentrations.

Since the LFG is generally water-filled and cools when it exits the landfill, the condensate must be removed from the pipes at their deepest points. These can be found at gas wells, distribution stations, and utilization units. Of course, if more knockout facilities are needed, such as in the header pipes, they must be mounted. When there is more gas produced than used, the excess must be flared. Flaring may be needed if the gas utilization plant experiences a breakdown or is undergoing maintenance. Since gas extraction rates change over time, consumption units cannot be able to use all of the gas extracted. Flaring lean gas that is not used for energy production is also recommended before methane concentrations are low. Standard flares will burn gas until methane concentrations exceed 20% to 25%; newly created special flares can burn gas with methane concentrations as low as 8% to 10% or even lower. If the gas cannot be thermally handled, it will be released into the atmosphere, causing odor problems and adding to the environment gas load. Flares should be mandatory at all landfills for all of these reasons.

8.4.2.2 Landfill Gas Treatment

Either for pollution control or damage prevention, gas treatment is needed. The cost of using LFG is increased by all treatment types. Since oil functions as a sorbent, pretreatment is often replaced by adjusting the oil in gas engines more frequently. In general, LFG leaves the landfill at temperatures ranging from 35 °C to 45 °C, in a nearly vapor-saturated state. If the gas is intended for use in gas engines, the moisture content does not exceed 50%. The emitted gas is cooled to 10 °C and then heated to 20 °C to meet this criterion. Since this temperature shift occurs "naturally" during gas collection, no separate technical moisture reduction is needed in most plants in moderate or cold climates (in the latter case, freezing of the condensate in pipes or other technical devices must be avoided). In tropical countries, for example, where separate cooling devices may be needed, the situation may be different. In either case, the condensate must be removed from the pipes. As a result, passive dewatering systems must be installed. This can be accomplished by mounting a siphon at the lowest point of the pipe system. Figure 8.13 depicts an example of a siphon system installed in the landfill (designed for negative pressure in the pipes of up to 100 hPa). Condensate may also be diverted back into the landfill through the gas wells if the slopes are sufficient.

Flaring, which is the combustion of flammable gases by converting methane in LFG into carbon dioxide through combustion with oxygen, is a widely used method of LFG treatment. Depending on the LFG's composition (H_2S, N_2), very harmful gases (SO_x, NO_x, CO) may be emitted during combustion, causing harm to the atmosphere [57]. Flaring must take place at temperatures of 1200 °C or above, as temperatures below this threaten the formation of toxic compounds such as dioxins [58]. Flaring is common in landfills that do not catch LFG for reuse, which is common in older landfills that did not have this built-in from the outset. LFG that contains trace quantities of CH_4, such as that present in aerobic and semi-aerobic landfills, cannot be flared without first adding methane. As a result, this approach is better suited to anaerobic LFG.

8.4.2.3 Landfill Gas Use

Many people believe that LFG is a significant source of energy that should be extracted for use when it is environmentally, technically, and economically feasible. Over a 15–20-year cycle, approximately 60–80 m^3 of LFG per ton of wet municipal solid waste (MSW) can be used.

Massive quantities of gases are flared in cases where LFG collection systems have not been built, often in old landfills. This is due to the relatively low cost of energy in these countries, making the investment and operation of a gas utilization plant uneconomical. Furthermore, gas extraction and flaring are only carried out because of a carbon credit program's financial support. This is common in economically developing countries, where the paradoxical situation exists where energy is desperately needed on the one hand and is squandered on the other. LFG has a variety of applications, including power generation, heating, and pipeline quality gas. Pretreatment is relatively simple in the case of LFG power generation. Condensate removal and filtration are the only treatments available. If there are corrosive or harmful trace constituents, a more thorough cleaning may be needed [59].

According to Han et al. (2010), the use of LFG for power production in China resulted in a CO_2 reduction of approximately 25,000 tons in 2009, with a maximum reduction of nearly 12,5000 tons in 2019. Although the emissions may not be important, LFG generation continues for decades after the plant closes, and it must be effectively controlled to prevent any unintended consequences. After decades, the cumulative emissions can be high and detrimental to the atmosphere. The best way to use LFG is to use a heat engine to produce both heat and power [61]. When compared to other options, the results showed that it offered the most GHG reduction. Incineration, on the other hand, has been commonly stated to be dangerous. Anaerobic LFG, on the other hand, contains methane at a concentration of 45 to 50% on average and has around half the heating value of natural gas [62].

8.4.2.4 Disposal of Gas

Passive gas venting can be useful in old landfills where very low gas output rates indicate that active gas extraction is no longer possible. When a landfill's surface is lined, a certain amount of positive pressure will build up within the landfill. The use of positive pressure to monitor and treat gas emissions is possible. Thermal treatment is not a choice if the landfill has low pressure and flow rate. Gas can migrate out of the landfill body by passing through the landfill cover at certain collection points, such as former gas wells. Until reaching the atmosphere, methane is biologically oxidized in the soil cover. To accomplish this theoretically, the gas must move into a form of gas distribution layer (sand or gravel-filled layer) where gas quality and flow are more or less equalized. The gas migrates from this layer into the cultivation layer, which is topsoil, where it is oxidized. An example of a full-scale application is shown in Fig. 8.12.

8.5 Operations of Sanitary Landfill

In order to ensure an effective landfill in terms of operations and management, it is essential to have a robust operation and monitoring system in place, as well as proper siting and design. A landfill's technological activity necessitates the integration of a number of components, including machinery, waste filling sequences and positioning methods, waste compaction, and regular cover placement. Other factors that must be addressed and enforced to ensure the smooth operation of operations include temporary road placement, protection, health and security, waste input control, and stormwater management. The technological process of efficiently depositing waste in the landfill is accompanied by environmental practices and standards

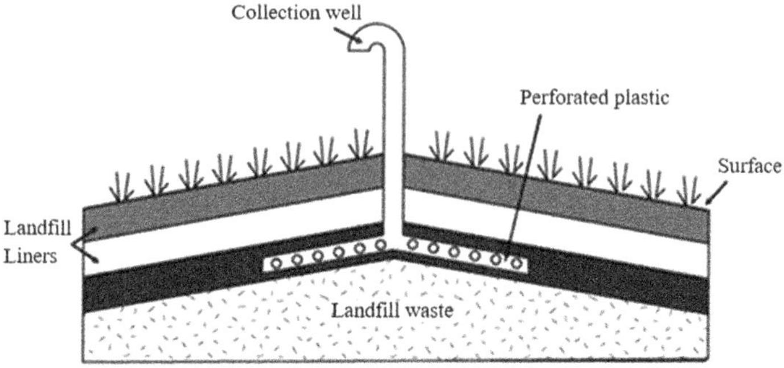

Fig. 8.12 Passive gas venting using methane oxidation processes in the top cover

that must be followed at the site in terms of noise, odors, burning, garbage, dust, and vectors. Landfill output in terms of leachate, air, groundwater, liner, and settlement must all be tracked. The results of these monitoring activities will be sent to local governments and made publicly accessible.

8.5.1 Types of Waste Disposed

A list of wastes must be specified for each landfill, with details on the types of waste accepted for disposal. Sanitary landfills typically accept waste from municipal solid waste, household waste, industrial waste, and nonhazardous waste that has been accepted by the local government and landfill operator. Prior to transporting waste from the source to the landfills, waste collectors must seek permission from municipal authorities. In a standard landfill operation, waste should be inspected both at the entrance and at the dumping site (to detect hidden components). Staff at the landfill must have mirrors with long sticks and ladders (to search open vans and containers) as well as walkie-talkies for the contact between the entrance and the dumping site. The waste must be registered and recorded at the entrance.

In Malaysia, landfills are also divided into waste types. In Malaysia, there are three types of landfills: Class I for hazardous waste, Class II for designated waste, and Class III for municipal solid waste [63]. This is analogous to the European Commission's Waste Landfill Directive (1999), which divides landfills into hazardous, nonhazardous, and inert waste categories.

8.5.2 Equipment

Heavy machinery is on hand at all times to transfer, position, spread, and compact incoming waste, as well as regular and final cover soils. Excavators, service and water trucks, and grinders are among the machinery or equipment that are needed during the operation of a sanitary landfill. Figure 8.13 depicts the various types of landfill equipment and machinery that are used on a regular basis to ensure that sanitary landfill operations operate smoothly. The crawler tractor is the best at spreading both unloaded waste and covering soil from waste transporters. Additionally, it can compress both materials to a degree, with waste densities ranging from 475 to 725 kg/m^3 [40]. Both the crawler tractor and the wheel loader will excavate, which is useful in certain situations where the cover soil source is a borrow pit on-site or must be excavated at a nearby off-site spot.

Crawler tractors and wheel loaders may also be used to clear the landfill and build temporary roads. Compactors are specifically built for landfill operations, and they are widely used and favored [42]. They are big and heavy, with knobbed steel wheels that help shred, scatter, and compact large volumes of waste efficiently. Compactors, on the other hand, are ineffective at excavating and hauling objects. As a result, additional equipment is needed at a site where cover soil must be excavated from on-site borrow pits or hauled long distances. At landfills, pans and scrapers have been used to excavate and position daily cover soils. These machines are built to grind up cohesive soil that can be removed in layers, transport it to the landfill, and then run down the compacted slope, essentially spreading dirt over the waste. Hydraulic excavators and trucks have been shown to carry dirt faster and at a lower cost than scrapers over longer distances [64]. Many US landfills have adopted the articulated (or artic) dump truck as their favorite hauling vehicle [65].

8.5.3 *Waste Disposal*

The landfill is able to accept waste once the design and construction are considered complete. There are several waste disposal options. The excavated cell method, the area method, and the canyon/depression method are the three most popular methods. Excavated cells are packed with waste after being excavated, lined with synthetic and/or low permeability clay liners to prevent subsurface movement of leachate and gases. This landfilling technique is ideally suited to areas where the water table is not close to the surface, and the excavated soil can be used as a cover material.

Next, the area method differs from the excavated method in that no land excavation is required; instead, the waste is put above grade. Where the soil is unsuitable for excavation or the groundwater table is too close to the surface, this approach is preferred and used in most developed countries. A liner and leachate collection system are built above grade to prevent leachate contamination of groundwater. Cover materials may be transported to the landfill from off-site or excavated on-site [40]. The unloaded solid waste would be spread and compacted by any landfill equipment that is suitable. At the end of the day's activities, a compactor is used to haul cover material. In most sanitary landfills, a portable fence is installed to trap any blowing debris. Furthermore, refuse is compacted in a lined pit, and the compacted refuse is covered with an earthen cover on a regular basis. Unloaded refuse is typically compacted with bulldozers or other heavy equipment before being coated with compacted soil. Depending on the soil composition, the daily cover is between 6 and 12 inches thick.

Finally, the canyon/depression process, also referred to as the trench method, entails depositing waste in an existing canyon, ravine, borrow pit, quarry, or cell. The geometry of the depression dictates the filling and compaction procedures for this technique [40]. The waste collection trucks dump their load into the landfill cell, where the compactor spreads and compacts it. This approach necessitates stringent surface and groundwater monitoring, as well as a pumping system capable of pumping leachate out of the landfill for an indefinite period of time. Since a liner may not be entirely impermeable and therefore not be a long-term solution, leachate pumping may become necessary until the landfill is filled with an impermeable liner. While this approach is widely used, it has the potential to trigger issues as landfill operations progress.

8.5.4 Compaction

As waste begins to be disposed of in landfills, compaction is a required process to allow the maximum amount of waste to be disposed of. It also avoids and reduces vermin (no holes/caves, low waste surface), fires (low air intrusion), odors, littering (highly compacted waste surface), and so on. Furthermore, less to no cover soil can be used, saving energy and landfill maintenance. Waste should be disposed of in thin layers at a slight slope after shredding and compaction; tipping over the edge is not appropriate and reflects dumping.

It was needed that a layer of soil (10–25 cm) or alternative cover material be put on the waste after the last load of waste for that day. This cover has several advantages, and it is needed because it regulates the amount of water infiltrated into the waste, lowering the early generation of leachate. Second, it reduces litter, dust, and air pollution while also preventing air intrusion. Finally, it prevents predators from burrowing and emerging. In addition, between 15% and 20% of the landfill is filled with soil and other inert materials. The aforementioned advantages of regular cover can be accomplished to a large extent with detailed high compaction using a heavyweight compactor. In this scenario, only small quantities of daily cover soil, if any, should be used.

Soil is commonly used as a day-to-day cover material. However, not all soil qualities, such as cohesive soils with high clay content, are suitable for use as a regular cover. Using this type of soil may cause serious operational issues if it rains. Alternative cover materials may be used when soil is inaccessible or not cost-effective as a daily cover material [40]. Compost and mulch have been used as frequent cover materials with great success. Alternative cover materials such as tarps, building and demolition waste, and agricultural residues have also been used to save money. A permeable or removable regular cover is needed to ensure sufficient vertical moisture movement.

8.5.5 Fences and Temporary Roads

Temporary access roads are needed to ensure the smooth operation of the landfill, allowing collection vehicles and other landfill machinery to continue filling the landfill as the working face shifts. Road building is normally completed by landfill workers using on-site equipment. The location of the road changes as the working face moves. Sometimes, materials received at the landfill's tipping area, such as concrete rubble, may be used to build the lane. Long-term access roads can also be paved or constructed with temporary concrete plates.

8.6 Monitoring of Sanitary Landfills

It is important to keep meticulous records of the type and volume of waste disposed of daily, as well as the landfill section in which it is disposed of. It is important to keep an eye on things both before and after the closure. Leachate (quantity, quality, leakage), local air quality, groundwater quality both upstream and downstream of the landfill, and landfill performance criteria such as head-on the liner and cap integrity are all subject to regulation. Furthermore, as landfill design shifts to bioreactor technology, in situ parameters, including temperature and moisture content, must be monitored to ensure proper device operation and process control.

8.6.1 Monitoring of Leachate and Groundwater

To assess the negative environmental effects of leachate, it is critical to keep a close eye on its consistency, quantity, leakage through the liner, and outbreaks on side slopes. Landfill operators are responsible for ensuring that leachate is properly handled and that the leachate collection and removal system is in good working order during the landfill's operating time. To ensure that leachate water quality standards are not broken, leachate quality analyses are required. All tests must be performed on leachate collected from a site that is indicative of in situ landfill conditions, such as the leachate collection and removal system, and that has not been modified or altered as a result of sampling storage or transportation. Leachate leakage is also something to keep an eye on. The leak detection system in landfills with a liner system consisting of two different liners with a drain in between is usually made of a highly permeable material. This system collects leachate, and in cases where leachate is monitored, it can be inferred that the primary liner system has failed [68].

In most cases, composite liners are used, which are two liners that are connected directly to each other. Via these means, if the upper liner (usually a geomembrane) leaks locally, the leachate only contacts the second liner (usually a mineral liner) locally; otherwise, the leachate may be spread over the entire surface of the second liner. In addition, direct monitoring (placement of lysimeters under the liner) or indirect monitoring (instruments placed near or under the liner system) that detect changes in moisture content or chemical concentration (salinity in most cases) can be done in the zone between the landfill liner and the groundwater table to detect leachate leakage [69]. These devices are not used very much. As failures of liners installed underneath landfills are discovered, the question is still what to do. It is a huge challenge to repair those who are in this state.

It is also crucial to keep an eye on leachate seeps. Local waste heterogeneities containing impermeable materials, impermeable cover soil placement, or leachate build-up may trigger leachate outbreaks on side slopes. The most popular method of identification is to conduct regular visual inspections of side slopes. If seeps are discovered, drains should be installed, cover material should be sloped toward the inside of the landfill, leachate injection volumes and levels in the landfill should be reduced, or stormwater penetration should be prevented.

8.6.2 Monitoring of Air and Gaseous Emissions

To protect the environment and ensure public health and safety, monitoring both air/gas quality and migration at and around landfill sites is critical. Gaseous pollutants escaping and migrating from landfills may have a number of negative environmental consequences. At landfills, air and gas monitoring typically include ambient air quality, extracted landfill gases, gases in the vadose zone, and any off-gases from a treatment or energy recovery plant.

The concentrations and migration of toxic gases like methane and hydrogen sulfide are tracked in the ambient air. Explosive gas meters, hydrogen sulfide meters, and sample collection devices can be installed at various locations around the landfill to collect samples at various locations. If there are unique issues at a facility, these steps are usually implemented. The ambient air quality in on-site buildings must be monitored on a regular basis. The use of global positioning systems (GPSs) to help in the creation of contour maps depicting plume concentrations of gases measured around the landfill site allows for a more precise assessment of gas migration patterns [70]. Gas samples can be collected in an evacuated canister, syringe, or air collection bag and stored for gas chromatography analysis, or they can be analyzed with a portable infrared gas meter.

8.6.3 Temperature Measurement

Temperature control of landfill conditions is often done to help avoid or track excessive heat generation. In situ temperature profiles may give landfill operators an indication of which areas of the landfill are likely to exceed combustible temperatures, allowing them to take preventative measures. Temperature monitoring may also be used to locate possible and actual subsurface fires. The temperature may be used as an indirect measure of waste degradation since waste degradation requires biochemical reactions that produce large quantities of heat; higher temperatures indicate a region of more rapidly degraded waste. A balance between heat output during biological degradation of organic waste fractions and heat loss to the surrounding soil and atmosphere will decide the temperature inside a landfill [71].

Microbial processes can generate a lot of heat, particularly up to a point where there is a lot of moisture in the air. Temperature is a relatively easy and low-cost parameter to monitor. Thermocouples and thermistors are two popular temperature measurement devices found in landfills. Thermocouples are made up of two metal wires that are connected at one end. A net thermoelectric voltage is produced by a temperature difference between the paired wires. Thermistors are devices that calculate temperature as a function of changes in material electrical resistance as a result of temperature changes. Thermocouples are less costly and more durable than thermistors, but they are significantly less reliable. Unless the instruments are located in existing wells or boreholes, the expense of measuring temperatures is linked to their location in the landfill.

8.6.4 Monitoring the Settlement

Another critical parameter to monitor in order to determine the effect on landfill cap integrity is a settlement. The cap's aim is to maintain the landfill's long-term integrity and to allow for post-closure uses. Waste decomposition and overburden pressures imposed on the landfill as a result of added moisture, waste, or cover soil cause settlement. Since settlement in the landfill is rarely uniform, excessive differential settlement can cause cracks and breaks in the cap. The amount of settlement is determined by compaction, waste characteristics, decomposition, water in the landfill, and the height of the landfill. The majority of landfill settlement occurs during the landfill's intense biodegradation process, but it may last for decades. A variety of methods can be used to determine settlement.

Settlement plates, GPS surveying, and flyover aerial photography are the most popular. Routine surveys of the landfill surface using GPS technology are becoming more popular, and they can be used to estimate the amount of settlement accurately. Both differential and average settlements can be calculated with great precision depending on the number of survey points used. Additionally, as the waste settles, air space can be reclaimed and used to fill the landfill with more waste. The time it takes for settling to occur can be minimized by using in situ aerations in the aftercare process.

8.6.5 Control of Stormwater and Sediment

Landfills are man-made structures that usually result in the formation of a new landform, such as a valley infill or a mound. This almost always happens inside a surface water catchment, so the landfill must be built to handle the rainfall and stormwater runoff during construction, filling, and for the long term after closure. Stormwater from the local catchment usually runs on or toward the footprint of landfills, generating runoff from completed cell areas. All runoff has the potential to generate sediment, particularly in areas where there is no vegetation to stabilize it. The geotechnical components of a landfill, such as batters, toe bunds, and anchor trenches for geosynthetics, can be degraded by poor stormwater management. Poor stormwater management, for example, can obstruct landfill operations by damaging roads. Organic and inorganic materials from waste, as well as leachate reaching surface water drains, have the ability to contaminate runoff from active areas (where waste is being disposed of or in areas where waste is poorly controlled). Contaminated runoff may also occur in inactive areas where there is re-exposed waste or litter. Significant pollution of the site's runoff will eventually contaminate surface water sources and even groundwater.

Primary drainage systems include both natural streams and channels as well as engineered drains that serve as the landfill's permanent external drainage system. Secondary drainage is made up of semi-permanent or permanent subsidiary channels, structures, piped drains, road culverts, and mechanized pumping systems. These features are typically associated with major phases of landfill construction, such as cells, benches, or waste lifts, and are expected to have a 5–20-year service life. Secondary drainage, on the other hand, requires permanent drainage on the final cap. Typically, such systems are designed to strike a balance between construction cost and risk. Such drainage systems are likely to experience drainage and require repair and reinstatement during storm events that are more serious than the selected design life.

8.6.6 Monitoring of Noise

Because of the routine work of collection trucks and noisy engines of landfill equipment, noise levels at landfill sites can be excessive, creating a disturbance to all landfill workers and any residents living near the site. As a result, depending on the situation, noise reduction monitoring should be considered (location, amount and kinds of vehicles, etc.). Employees at landfills should be provided with hearing aids. Several control methods can be used to prevent noise from reaching adjacent residences, including planting trees to create a noise mitigation buffer zone (which also acts as a litter and dust control buffer and reduces the visibility of the landfill), properly maintaining equipment, regulating hours, and ensuring that the working face or tipping area (the loudest area of the laundromat) is protected.

8.6.7 Monitoring of Odor

Odors are a common problem caused by putrescible wastes, disposed sludge, landfill gases, and leachate, but they are particularly bad when organic waste is unloaded in a hot weather collection truck. Controlling odor requires a good overall landfill service, high compaction, regular cover placement, and immediate covering of materials with an offensive odor [72]. Using an efficient gas management system that includes thermal oxidation, wet gas scrubbing, activated carbon filtration, and biofiltration to deodorize the collected gas will also help to reduce odor [73, 74]. In some cases, more regular waste disposal can also help minimize incoming waste odors. Chemicals such as ozone and mixtures containing plant oils and surfactants can be used to reduce odor in serious cases of odor regulation [75]. Odor-neutralizing chemicals are often dispensed in a perimeter misting system where these precautions are only taken in a few instances (Fig. 8.14).

8.6.8 Fire Surveillance

Another aspect that requires the landfill management's attention and intervention in terms of the landfill's protection is this. When methane concentrations in the air range from 5% to 15%, the gas is explosive and can trigger internal fires [40]. In general, this is a unique problem in low compacted landfills, where methane escapes through the surface and air may enter the landfill through diffusion, but primarily through the wind (especially on slopes) and atmospheric pressure changes. Glass fragments (magnifying effect), open fires on the landfill, chemical processes in the landfill, and other potential ignition causes are all possibilities. Many dumps and low-compacted landfills in developed countries are particularly bad examples of this situation. As a result, high compaction is also essential for this purpose. However, oxygen penetration can occur in active landfills where the gas extraction system is not properly maintained, resulting in explosive mixtures. When landfills are aerated, a unique circumstance arises. However, there are no such ignition potentials within the landfill, with the exception of unknown chemical processes. This monitoring is essential to prevent any explosions or large fires that could endanger landfill workers or disrupt operations (Table 8.10).

Fire is one of the most severe hazards that a landfill can encounter over the course of its life. Figure 8.15 shows an example of a landfill fire erupting. While fires are common at dumpsites, serious fires are uncommon at well-managed landfills. Landfill fires can devastate a landfill's infrastructure and pose a significant risk to site workers. Furthermore, landfill fires can cause major problems in the local community (in terms of health, air quality, and social acceptance). Table 8.14 illustrates the dangers of fires at landfills.

Monitoring the internal temperature of a landfill is extremely useful for determining the probability of or duration of a fire, but only if the temperature is

Table 8.10 Hazards of fire at landfills [76]

Hazard	Low severity	High severity
Uncontrolled gas and smoke emission	Additional on-site health and safety precautions required; additional off-site receptor gas risk assessment (chronic effects)	Fire service required; nearby housing evacuated
Rapid settlement	Settlement causes seals around gas infrastructure to fail	Plant falls into underground cavity-causing injury/death
Damage to landfill liner	Reduced lifespan	Immediate loss of integrity
Additional site management	Extra staff required to address subsurface fire issues and liaison with authorities	Emergency response including 24 h supervision and public relations/media management
Uncontrolled chemical reaction	Considerable additional on-site health and safety required; additional off-site receptor gas risk assessment (acute effects)	Explosion

Fig. 8.15 Stray dogs scavenging at landfills. (Source: Rescuers Without Borders)

measured at depth. Drilling a variety of testing wells in and around the suspected fire zone is the safest way to obtain temperature measurements (and gas composition samples). Air rotary rigs should not be used because large amounts of air could speed up the fire and potentially cause a methane explosion. In either case, staff must use protective devices such as respirators and breathing fans when performing such tasks. Temperature control has proved to be an effective method for both preventing landfill fires and confirming that they have been extinguished. The relationship between landfill conditions and temperature is shown in Table 8.11.

Another important monitoring system that landfill managers or workers can implement is gas composition monitoring, which offers valuable insight into fire conditions at depth and the effectiveness of firefighting measures. Methane, oxygen,

Table 8.11 The relation between landfill conditions and temperature [76]

Temperature	Landfill conditions
<55 °C	Normal landfill temperature
55–60 °C	Elevated biological activity
60–70 °C	Abnormally elevated biological activity
>70 °C	Likelihood of landfill fire

carbon monoxide, and hydrogen sulfide are some of the parameters that must be calculated at different times. Carbon monoxide is the most useful indicator of a subsurface fire out of the four gases. An analytical scale is provided in Table 8.12 to aid in the evaluation of fire conditions in demolition landfills. The presence of oxygen at concentrations greater than 1% indicates that established oxygen intrusion barriers (soil or membrane covers) are ineffective and that additional soil cover is needed. However, it is not a real problem for the activation of fire conditions before the oxygen level reaches 5%. A rise in methane levels above 40%, on the other hand, is a good sign that oxygen is being successfully excluded, and the biological regime is reverting to cooler anaerobic conditions.

8.6.9 Control of Litter and Vectors

Litter must also be monitored, as it is a major source of irritation for local residents. Litter is generated by uncovered loads being delivered, wind, and organizational activities. Unloading waste on a small surface area with immediate high compaction can be the most effective process, as well as operating the working face to reduce wind disturbance (creating it in the opposite direction of the wind). Furthermore, placing portable screens near the working face is a popular control system. Other methods of mitigation include covering waste more frequently on windy days, requiring all shipped loads to be covered, and manually collecting litter when required. When using permanent or mobile fences, it is essential to clean them on a regular basis to reduce their susceptibility to wind damage. The landfill buffer areas around the landfills, which are planted with trees, are helpful in preventing long-distance movement of litter outside. It is important to remember that the landfill operator is generally responsible for litter outside of the landfill and is required to collect it.

"Vectors" in landfills can include rats and other rodents, foxes, feral cats and dogs, insects, birds, and other species, all of which can spread disease and pose a public health risk. Birds necessitate unique control techniques, which are discussed in a separate guideline. Figure 8.15 shows an example of vectors found in landfills. Each type of vector has the ability to live and multiply in a landfill, posing a risk to site operators, regulators, public health practitioners, and the general public. Fortunately, vectors are controllable and should only be present on a well-controlled

Table 8.12 The relation between CO concentrations and fire at the landfill [76]

CO concentration (ppm)	Fire indication
0–25	No fire indication
25–100	Possible fire in the area
100–500	Potential moldering nearby
500–1000	Fire or exothermic reaction likely
>1000	Fire in the area

landfill infrequently, if at all. The application of regular cover to landfills is needed to reduce vector problems. All solid waste should be covered except the tipping face as it is being handled. During and after regular operations, a daily cover of at least 150 mm of lightly compacted soil or similar material, or an appropriate layer of alternate daily cover (ADC), should be applied to finished portions of the daily cell. After a careful site-specific assessment, alternative regular cover materials such as tarpaulins, foams, granular waste, and other materials may be effective as vector control.

8.6.10 Bird Management

One of the recommendations in the International Solid Waste Association's (ISWA) landfill activity guidelines was bird management. Birds visiting a landfill site do so primarily for food or whatever is left of it. They are thought to be noisy and messy, and they are always pathogen carriers or the source of local annoyance by fouling roofs, roof-water sources, gardens, and public open space. Furthermore, if landfills are located near commercial airports, birds can pose a threat to aircraft safety. If birds are provided with a consistent food supply and a secure environment (suitable resting or roosting areas), their breeding rate is likely to increase, as seen in Fig. 8.16, attracting more birds from a greater distance around the landfill site.

Landfills have had an impact on the behavior of wild birds, especially white storks. Gilbert et al. (2016) discovered that the year-round availability of food supplies in landfills has encouraged white storks' year-round nest use and is affecting their home ranges and movement behavior. Because of the abundance of landfills around the world, white storks rely on them for scavenging, especially during the nonbreeding season when other food sources are scarcer, and this artificial food supplementation likely aided the establishment of resident populations. This is one of the many negative effects landfills have on the environment, as they alter animal behavior and norms, such as migration patterns.

Many methods exist for reducing the number of birds that flock to the tipping area. When attempting to reduce bird populations, successful management of the working face is the first step., The working area should be kept as limited as possible to minimize the surface area where food may be readily accessible. All waste that may be a food source should be compacted and covered with soil on a daily basis and fully by the end of each working day, preventing access to the food source. Areas of the site that have been restored and areas that are not working need attention as well. There should be no uncovered waste or places where water can collect and allow the birds to stand, drink, and clean themselves. The grass should be allowed to grow to a height of at least 225 mm in the restored area, as this would deprive most birds of resting areas and make it difficult for them to land and take off. Where there is long grass, many bird species are afraid of predators.

Gas weapons, which are used to scare the birds, are another management tool for managing bird counts. They are easy to use and can be very effective for short periods of time. The gas guns must be moved around the site on a regular basis for them to be successful. However, this method of control can be a nuisance to neighbors, particularly if the equipment's operating hours are outside of normal business hours.

8.7 Landfill Safety, Health, and Security

The safety, health, and protection of landfill workers, as well as the general public, are important considerations in landfill management. Communication with dangerous chemicals (pathogens, harmful air contaminants, asbestos), a high risk of accidents, allergies induced by environmental hazards (gas, dust, litter), and unnecessary

Fig. 8.16 Massive rubbish dumps and sprawling landfills have led some birds (storks) to give up on migration

noise and side effects caused by landfill machinery operations are all potential hazards. Employees should be provided with the required personal protective equipment (PPE) as a means of preventing any protection or health-related injuries. Furthermore, since entry to landfills is limited, a health and safety plan must be enforced and clearly displayed in the event of an emergency inside the landfill's compounds. The entire landfill should be surrounded by a fence to prevent unauthorized workers from entering. Table 8.13 shows the dangers associated with landfills as well as the most common risks.

Personnel employed face numerous dangers and health threats. The following are some of the most common hazards and impositions: (1) contact with hazardous substances such as biogas, silicon dust, and infectious substances, (2) accident risk from heavy vehicle collisions, (3) dangerous technical equipment (danger points on pumps, defective hydraulic pipes on scrapers, defective electrical and soldering equipment, (4) vibration caused by heavy vehicles, and (5) accidents.

Apart from that, the treatment or flaring of landfill gas necessitates the implementation of protection and health measures. LFG containing oxygen or coming into contact with air can produce explosive mixtures with serious consequences for humans and the environment. Explosions will reach a pressure of about 7 bar if deflagration occurs, but even higher pressures occur if a detonation has formed. When recognizing possible explosion dangers and hazards at LFG plants, keep in mind that where vacuum pressure exists (gas pipes from the landfill to the blower), air may be drawn into the pipe or plant, potentially resulting in an explosive mixture. Gas can escape the pipe or plant and produce explosive mixtures outside the plant in areas where there is overpressure (most commonly in the area from the blower to the utilization facility or flaring facility). This is particularly likely if a pipe connecting to a building leaks, allowing methane to accumulate and potentially ignite sources such as the blower or an electrical switch. Various situations where and under what circumstances an explosion may occur must be established during the danger and hazard evaluation. The landfill management must enforce safety measures based on the evaluation of the landfill facilities to ensure the safety of their personnel and the environment.

8.8 Landfill Impacts

The location of waste management facilities may be a major problem since all infrastructure projects have the potential to harm the ecosystem of the site where they are built, resulting in vegetation changes, habitat destruction, and fauna displacement. These effects vary widely from one location to the next, necessitating a more detailed analysis on a case-by-case basis [8, 78]. Because of the general removal of topsoil as well as complex process-related modifications, the soils on selected sites appear to suffer from high levels of disruption, and their chemical and physical properties vary from those of the surrounding areas. Soil is a valuable resource that serves a number of ecological, economic, and cultural purposes. The factors that

Table 8.13 Risks of landfilling [76]

Landfill specific risks	Common risks
Environmental risks are those where broader effects are suffered, like leachate and flooding, atmospheric emissions from landfill fire and bio-gas, epidemiological hazard	Common health safety assessment: Many of the activities in the landfill are similar to common activities. Civil works and machinery maintenance are everyday tasks for landfill employees and contractors. Replication of standard rules from transport, construction, and manufacturing industry's procedures can be instructed in the landfill regulations to minimize risk in well-documented activities
Personal risks are caused by individuals and can affect a limited number of people and can come from traffic, biological hazards, scavenging, lack of knowledge, gas leaks, and caves in the landfill, so they are associated with individual's damage	

control soil quality, such as porosity, density, water holding capacity, and aggregate strength, are better formed in the topsoil fraction, while the subsoil is less developed and has a lower ability to sustain plant development. During the building process, this consistency can be harmed. Heavy machinery movements may cause excessive compaction of topsoil and subsoil, which may only be reversible over longer time periods in deeper soil. The destruction of existing vegetation has a significant effect on flora and fauna during the construction process of landfills. However, after the landfills are closed, this damage could be recoverable. During the operating and closing phases of landfills, studies have shown that they can host diverse and rich fauna, including exotic species [79].

The landfilling activities have the potential to have serious negative environmental consequences. Improper landfill tipping causes water to become clogged, providing a breeding ground for insects. This can spread harmful diseases like cholera and dengue fever to workers and residents in the area of the landfill. Furthermore, vectors like rats and vermin are easily drawn to insufficient solid waste disposal. The odors created by landfilling operations make staff and residents living near the landfill site feel uncomfortable. Open burning on the landfill site releases a variety of toxic gases such as methane, sulfide, and carbon monoxide, all of which degrade the air quality in the local area.

Leachate output declines steadily, and some criteria can be of environmental significance for decades or centuries after the landfill closes. Dissolved methane, fatty acids, sulfate, nitrate, phosphates, calcium, sodium, chlorine, magnesium, potassium, and trace metals such as chromium, manganese, iron, nickel, copper, zinc, cadmium, mercury, and lead are the primary constituents of landfill leachate. Due to the lack of a proper liner system or damage to the liners, leachate may migrate through the soil to groundwater or even surface water, posing a serious problem because aquifer recovery takes a long time. Furthermore, soil can hold leachate constituents such as metals and nutrients, which can have negative consequences for the environment. Metals retained in the soil are taken up by plants, providing a crucial route for metals to enter the food chain. Trace metal deposition in plants can affect crop growth and productivity, as well as posing a greater risk to animal health. Plant uptake is influenced by soil pH and salinity, and the uptake of cadmium and lead is aided by the metals' chloride complexation in the leachate.

Despite the fact that methane and carbon dioxide are the two main constituents of landfill gas, there is evidence that it comprises a variety of other constituents in trace quantities high enough to trigger environmental and health concerns. Normal illegal dumping is to blame for the presence of these contaminants in landfill gas. Biodegradation by-products inside the landfill, according to microbial investigations, may also lead to the formation of many of these chemicals. Air pollution and possible health risks are the main issues of trace gas emissions. VOC emissions are thought to increase cancer risks in local communities and lead to the production of atmospheric ozone [80]. Trace gases can also reduce methane production by inhibiting methanogen growth and corroding gas recovery equipment [81]. Table 8.14 lists the pollutants of flare and their health effects.

Gas flaring is one of the most difficult sources of electricity, with major environmental and human consequences. The environmental effects of gas flaring have a significant effect on local communities, often resulting in serious health problems. Gas flaring is usually noticeable and produces both noise and heat.

Ghadyanlou and Vatani (2015) used commercial software to measure the thermal radiation and noise level as a function of distance from the flare, as shown in Table 8.15. This demonstrates that the effect of flaring in terms of noise and heat is extremely dangerous at a distance, necessitating mitigation steps by landfill gas operators to ensure the safety and health of plant personnel.

Table 8.14 Pollutants of flare and their health effects [82]

Chemical name	Health Effects
Ozone in land	In low densities, the eye will stimulate, and in high densities, especially in children and adults, it will cause respiratory problems.
Sulfide hydrogen	In low densities, it affects the eye and nose, which results in insomnia and headache.
Dioxide nitrogen	It will affect deep within the lung and respiratory pipes and aggravates symptoms of asthma. In high densities, it will result in meta-hemoglobin, preventing oxygen absorption by the blood.
Particle matter	There is this belief that it will result in cancer and heart attack.
Dioxide of sulfur	It will stimulate the respiratory system, thus aggravating asthma and bronchitis.
Alkanes: methane, ethane, propane	In low densities, it will result in swelling, itching, and inflammation, and in high densities, it will result in eczema and acute lung swelling.
Alkenes: ethylene, propylene	It will result in weakness, nausea, and vomiting.
Aromatics: benzene, toluene, xylene	It is poisonous and carcinogenic. It influences the nerve system, and in low densities, it will result in blood abnormalities, and it will also stimulate skin and result in depression.

Table 8.15 Thermal and noise emissions from flaring

Distance, m	Thermal radiation, kW/m^2	Noise level, dB
10	5.66	86.30
20	5.87	86.19
30	6.04	86.02
40	6.14	85.78
50	6.17	85.50
60	6.14	85.18
70	6.04	84.83
80	5.88	84.46
90	5.67	84.08
100	5.42	83.68

Methane and carbon dioxide emissions from landfills contribute greatly to global warming, also known as the greenhouse effect. Methane has long been regarded as the primary cause of global warming because it has a molecular effect 20 to 25 times that of carbon dioxide [84, 85], is more efficient at trapping infrared radiation, and appears to survive longer in the atmosphere due to other organisms (carbon monoxide) having a higher affinity for hydroxyl ions, the oxidizing agent for methane [86]. Due to recent rises in atmospheric methane concentrations, detailed characterization studies of global methane sources and sinks have been conducted. Methane concentrations in the atmosphere have been observed to rise at a rate of about 1% to 2% per year [87]. Methane is thought to be responsible for about 18% of total global warming. Solid waste landfills are becoming a major contributor to atmospheric methane unless recovery management mechanisms are introduced due to continuing trends in population growth and urbanization.

8.9 Final Remarks

Landfilling has both advantages and disadvantages. As a result, in order to avoid further contamination or pollution of the ecosystem and its surroundings, environmental management and activity must be carried out. This chapter has gone into great detail about the factors that require effective management techniques as well as effective treatment approaches in order to ensure the environment's protection and purity. Furthermore, long-term planning and maintenance are essential for ensuring the everyday smoothness of landfill operations so that this form of solid waste management can maintain the highest level of public hygiene and cleanliness. Implementing good management practices and technical controls at the landfill facility will help prevent leachate migration and pollution of ground and surface water. Operational practices divert local precipitation, and surface water runoff to the waste mass is a good way to cut down on the amount of leachate generated.

Glossary

Aerobic composting A method of composting organic wastes that involve the use of bacteria that require oxygen. This necessitates exposing the waste to sunlight, either by turning it or pushing air into pipes that pass through it.

Anaerobic digestion A form of composting that does not necessitate the use of oxygen. Methane is generated by this composting process. Anaerobic composting is another name for it.

Ash Solid by-products of incineration or other burning processes that are noncombustible.

Autoclaving A pressurized, high-temperature steam process is used to sterilize the products.

Baghouse An emission control system for a combustion plant that consists of a series of fabric filters that carry flue gases via an incinerator flue. Particles are suspended, preventing them from entering the atmosphere.

Basel Convention The Basel Convention is a treaty that was signed by over 100 countries on the management of transboundary movements of hazardous wastes and their disposal, which was drafted in March 1989 in Basel, Switzerland.

Biodegradable material Any organic material that microorganisms can break down into simpler, more stable compounds. The majority of organic wastes (such as food and paper) are biodegradable.

Bottom ash The incinerator residue that accumulates on the grate of a furnace is relatively coarse, noncombustible, and generally toxic.

Bulky waste Big wastes, such as machinery, furniture, and trees and branches, cannot be processed using standard MSW methods.

Cell The fundamental building block of a landfill. It is where incoming waste is flipped, scattered, compacted, and sealed.

Cleaner production Processes that aim to reduce the amount of waste produced during processing.

Co-disposal Generation of both electricity and steam from the same fuel source in a single plant.

Collection Paper, plastics, wood, and food and garden wastes are all combustible materials in the waste stream.

Combustion Materials are burned in an incinerator.

Commingled After being isolated from mixed MSW, mixed recyclables are collected together.

Communal collection A waste disposal system in which people carry their trash to a central location where it is processed.

Compactor vehicle To minimize the amount of solid waste, a recycling vehicle with high-power mechanical or hydraulic equipment is used.

Composite liner A land-fill liner system made up of an engineered soil layer and a synthetic sheet of material.

Compost The content that results from a composting. Compost, also known as humus, is a soil conditioner that can also be used as a fertilizer in some cases.

Composting Biological decomposition of solid organic materials into a soil-like substance by bacteria, fungi, and other species.

Construction and demolition debris Waste includes bricks, asphalt, drywall, lumber, miscellaneous metal parts and sheets, packaging products, and other materials.

Controlled dump A proposed landfill with some of the characteristics of a sanitary landfill: hydrogeological suitability, grading, compaction in some cases, leachate control, partial gas management, frequent (but not always daily) cover, access control, simple record-keeping, and managed waste picking.

Curbside collection Compostables, recyclables, and garbage are collected at the edge of a sidewalk in front of a home or business.

Disposal Following collection, sorting, or incineration, the final handling of solid waste. The most common method of disposal is to deposit waste in a landfill or a dump.

Emissions Gases that have been emitted into the atmosphere.

Energy recovery The method of extracting useful energy from waste, usually by using the heat produced by incineration or landfill methane gas.

Environmental impact assessment (EIA) An assessment aimed at determining and forecasting the effect of a policy or project on the climate, human health, and well-being. Risk evaluation, as well as economic and land use assessments, are all possible components.

Environmental risk assessment (EnRA) A study of the relationships between agents, humans, and natural resources. Usually assessing the probabilities and magnitudes of harm that may be caused by environmental pollutants, it is made up of human health risk assessment and ecological risk assessment.

European Commission's Waste Landfill Directive Aims to protect both human health and the environment. With the goal to prevent, or reduce as much as possible, any negative impact from landfill on surface water, groundwater, soil, air, and human health by introducing rigorous operational and technical requirements.

Flaring At a landfill, methane released from storage pipes is burned.

Fluidized-bed incinerator The stoker grate is replaced by a bed of limestone or sand that can withstand high temperatures in this form of an incinerator. The word "fluidized" comes from the fact that the bed is heated and high air velocities are used, causing the bed to bubble.

Fly ash The extremely toxic particulate matter captured by an air pollution control device from an incinerator's flue gas.

Geomembrane A low permeability synthetic membrane liner or barrier used with any geotechnical engineering related material to control fluid migration in a human-made project, structure, or system.

Groundwater Water that fills underground pockets (known as aquifers) and supplies wells and springs under the earth's surface.

Hazardous waste Reactive, poisonous, corrosive, or otherwise harmful to living things and/or the atmosphere waste. Hazardous manufacturing by-products abound.

Heavy metals Metals with a high atomic weight and density that are poisonous to living organisms, such as mercury, lead, and cadmium.

Household hazardous waste Items used in homes that are harmful to living organisms and/or the atmosphere, such as paints and certain cleaning compounds.

Incineration The method of burning waste by reducing the weight and volume of solid waste while still producing energy under regulated conditions.

Inorganic waste Sand, dust, glass, and a variety of synthetics are examples of waste made up of materials other than plant or animal matter.

Integrated solid waste management Usage of a coordinated collection of waste management strategies, each of which may play a role in a larger MSVVM strategy.

In-vessel composting Composting in a closed vessel or drum with a balanced internal setting, mechanical mixing, and aeration are all options.

Landfill gas (LFG) Consists of a mixture of different gases produced by microorganisms within a landfill as they decompose organic waste. It is approximately 40–60% methane, with the remainder being mostly carbon dioxide.

Landfill gases Methane, carbon dioxide, and hydrogen sulfide are the main gases produced by the decomposition of organic wastes. Landfills can experience explosions as a result of these gases.

Landfilling The final disposal of solid waste by depositing it in a regulated manner in a long-term location. This concept is used in the Source Book to describe both supervised dumps and sanitary landfills.

Leachate pond A pond or tank built at a landfill to collect leachate from the surrounding area. Typically, the pond is built to handle the leachate in some way, such as allowing solids to settle or allowing for aeration to facilitate biological processes.

Leachate Any liquid, in the course of passing through matter, extracts soluble or suspended solids or any other component of the material through which it has passed. Referring to liquid produced from landfills or dumpsites.

Liner A protective layer made of soil and/or synthetic materials, which is built along the bottom and sides of a landfill to prevent or minimize leachate from entering the atmosphere.

Lysimeter A device used to measure the amount of actual evapotranspiration.

Materials recovery facility (MRF) A facility for manually or mechanically separating commingled recyclables. Some MRFs are planned to distinguish recyclables from mixed municipal solid waste. The recovered materials are then baled and sold by MRFs.

Materials recovery Obtaining goods that are recyclable or can be reused.

Methane Is an odorless, colorless, flammable, and explosive gas formed by landfills anaerobically decomposing MSW.

Mixed waste Materials that have been discarded into the waste stream without being sorted.

MSW The term "municipal solid waste" refers to all solid waste generated in a given region, except industrial and agricultural waste. Construction and demolition debris, as well as other special wastes, can sometimes join the municipal waste stream. Hazardous wastes are generally excluded, except to the degree that they join the industrial waste stream. Occasionally, the term is used to refer to all solid wastes for which a city government takes responsibility in some way.

MSWM Municipal solid waste management.

Municipal solid waste (MSW) Commonly known as trash or garbage, which consists of everyday items we use and then throw away. The sources include homes, schools, hospitals, and businesses.

Open dump An impromptu "landfill" with a few, if any, of the characteristics of a managed landfill. Usually, there is no leachate monitoring, no access control, no cover, no management, and a large number of waste pickers.

Organic waste Is described as carbon-based waste, which includes paper, plastics, wood, food waste, and yard waste. In MSWM practice, the term is often used in a more limited context to refer to material that is derived more directly from plant or animal sources and can be decomposed by microorganisms.

Pathogen Organism that is capable of causing disease.

Processing Using processes such as baling, magnetic isolation, grinding, and shredding, MSW materials are prepared for future use or management. Separation of recyclables from mixed MSW is another word for the same thing.

Putrescible Decomposition or decay is a term used to describe the process of decomposition or decay. Food wastes and other organic wastes that decompose easily are often referred to as "biodegradable."

Pyrolysis In the absence of oxygen, heat causes chemical decomposition of a material, resulting in various hydrocarbon gases and carbon-like residue.

Recyclables Things that can be reprocessed into new product feedstock. Paper, glass, iron, corrugated cardboard, and plastic containers are all common examples.

Recycling The method of converting materials into raw materials for the production of new goods that may or may not be identical to the original.

Refuse A word that is often interchanged with solid waste.

Refuse-derived fuel (RDF) MSW that has been processed and is used to make diesel. Separation of recyclables and noncombustible materials, shredding, size reduction, and pelletizing are all examples of processing.

Resource recovery Utilization of resources and energy from wastes is referred to as resource recovery.

Reuse The use of a commodity in its original form more than once for the same or a different reason.

Rubbish Solid waste is referred to as "waste" in general. Food wastes and ashes are sometimes excluded.

Sanitary landfill An engineered method of disposing of solid waste on land that meets most of the standard requirements, such as proper siting, comprehensive site planning, proper leachate and gas management and tracking, compaction, regular and final cover, full access control, and record-keeping.

Secured landfill A waste management facility that is built to keep wastes out of the atmosphere indefinitely. This involves burying the wastes in a landfill with clay and/or synthetic liners, leachate collection, gas collection (if gas is produced), and an impermeable cover.

Sewage sludge A semi-liquid residue found at the bottom of canals and pipes containing sewage or industrial wastewaters, as well as the bottom of wastewater treatment tanks.

Site remediation Removing hazardous solids or liquids from a contaminated site or handling them on-site.

Source reduction The process of designing, manufacturing, acquiring, and reusing materials in order to reduce the amount and/or toxicity of waste generated.

Source separation To promote reuse, recycling, and composting, compostable and recyclable materials are separated from the waste stream before being collected with other MSW.

Special wastes Wastes that are preferably kept out of the MSW stream but occasionally find their way in and must be dealt with by local governments. Household hazardous waste, medical waste, building and demolition debris, war and earthquake debris, tires, oils, wet batteries, sewage sludge, human excreta, slaughterhouse waste, and industrial waste are all examples of these.

Tipping fee Unloading or dumping waste at a landfill, transfer station, incinerator, or recycling plant is subject to a levy.

Tipping floor A place, usually on the outskirts of a neighborhood, where small collection vehicles move waste to larger vehicles for transport to disposal sites.

Vectors Organisms that bear pathogens that cause disease. The key vectors that disperse pathogens outside the landfill site are mice, flies, and birds.

Virgin materials Any raw material for industrial processes that has never been used before, such as wood pulp trees, iron ore, crude oil, and bauxite.

Waste characterization study The analysis of samples from a waste stream to determine its composition is known as waste characterization.

Waste collector An individual hired by a municipality or a private company to collect trash from homes, businesses, and community bins.

Waste management hierarchy A rating of waste management operations based on the environmental or energy benefits they have. The waste management hierarchy was created with the aim of making waste management activities as environmentally friendly as possible.

Waste picker An individual who separates recyclables from mixed waste wherever it is temporarily accessible or discarded.

Waste reduction All methods of minimizing the amount of waste generated at the outset and collected by solid waste authorities. This includes everything from regulations and product design to community-based initiatives aimed at keeping recyclables and compostables out of the final waste stream.

Waste stream A community's, region's, or facility's complete waste flow.

Waste-to-energy (WTE) plant A plant that generates energy from solid waste materials (processed or unprocessed). Incinerators that produce steam for district heating or industrial use, as well as facilities that convert landfill gas to electricity, are examples of WTE plants.

Water table The depth below which the earth's crust becomes filled with water.

Wetland For at least part of the year, an area that is constantly wet or flooded and has a water level that is at or above the ground surface.

Working face The length and width of the waste-disposal row at a landfill. The tipping face is another name for it.

Yard waste Yard and garden waste includes leaves, grass clippings, prunings, and other natural organic matter.

References

1. Aziz, H. A., Alias, S., Assari, F., & Adlan, M. N. (2007). The use of alum, ferric chloride and ferrous sulphate as coagulants in removing suspended solids, colour and COD from semi-aerobic landfill leachate at controlled pH. *Waste Management & Research, 25*(6), 556–565.
2. Fauziah, S. H., & Agamuthu, P. (2012). Trends in sustainable landfilling in Malaysia, a developing country. *Waste Management & Research, 30*(7), 656–663.
3. Idris, A., Inanc, B., & Hassan, M. N. (2004). Overview of waste disposal and landfills/dumps in Asian countries. *Journal of Material Cycles and Waste Management, 6*(2), 104–110.
4. JPSPN. (2015). Jabatan Pengurusan Sisa Pepejal Negara. Accessed from: http://jpspn.kpkt.gov.my
5. Abd Manaf, L., Samah, M. A. A., & Zukki, N. I. M. (2009). Municipal solid waste management in Malaysia: Practices and challenges. *Waste Management, 29*(11), 2902–2906.
6. Buenrostro, O., & Bocco, G. (2003). Solid waste management in municipalities in Mexico: Goals and perspectives. *Resources, Conservation and Recycling, 39*(3), 251–263.
7. Alrukaibi, D. S., & Alsulaili, A. D. (2017). GIS-based modelling for appropriate selection of landfill sites. *Journal of Engineering Research, 5*(2).
8. Crowley, D. (2003). Health and environmental effects of landfilling and incineration of waste: A literature review. 1014.
9. Kjeldsen, P., Barlaz, M. A., Rooker, A. P., Baun, A., Ledin, A., & Christensen, T. H. (2002). Present and long-term composition of MSW landfill leachate: a review. *Critical Reviews in Environmental Science and Technology, 32*(4), 297–336.
10. Ngoc, U. N., & Schnitzer, H. (2009). Sustainable solutions for solid waste management in Southeast Asian countries. *Waste Management, 29*(6), 1982–1995.
11. United States Envitonmental Protection Agency. Municipal Solid Waste Landfills. United States Envitonmental Protection Agency. Viewed 10 June 2021. https://www.epa.gov/landfills/municipal-solid-waste-landfills#whatis
12. Artiola, J. F. (2019). Industrial waste and municipal solid waste treatment and disposal. In *Environmental and pollution science* (pp. 377–391). Academic.
13. Akinjare, O. A., Oluwatobi, A. O., & Iroham, O. C. (2011). Impact of sanitary landfills on urban residential property value in Lagos State, Nigeria. *Journal of Sustainable Development, 4*(2), 48.
14. Reichert, A., Small, M., & Mohanty, S. (1992). The impact of landfills on residential property values. *Journal of Real Estate Research, 7*(3), 297–314.
15. Danthurebandara, M., Van Passel, S., Nelen, D., Tielemans, Y., & Van Acker, K. (2012). Environmental and socio-economic impacts of landfills. *Linnaeus Eco-Tech, 2012*, 40–52.
16. Timothy, A. (2008). *A guide for Development, implementation and monitoring*. The Florida Development, Implementation and Monitoring.
17. Kamaruddin, M. A., Yusoff, M. S., & Ahmad, M. A. (2012). Optimization of preparation conditions for Durian peel-based activated carbon for the removal of COD and ammoniacal nitrogen (NH_3-N) using response surface methodology. *Kuwait Journal of Science and Engineering, 39*(1B), 37–58.

18. Shimaoka, T., Matsufuji, Y., & Hanashima, M. (2000, December). Characteristic and mechanism of semi-aerobic landfill on stabilization of solid waste. In *Proceedings of the first intercontinental landfill research symposia, Lulea, Sweden*.
19. Aziz, H. A., Yusoff, M. S., Adlan, M. N., Adnan, N. H., & Alias, S. (2004). Physico-chemical removal of iron from semi-aerobic landfill leachate by limestone filter. *Waste Management, 24*(4), 353–358.
20. Hunce, S. Y., Akgul, D., Demir, G., & Mertoglu, B. (2012). Solidification/stabilization of landfill leachate concentrate using different aggregate materials. *Waste Management, 32*(7), 1394–1400.
21. Kamaruddin, M. A., Yusoff, M. S., Aziz, H. A., & Hung, Y. T. (2015). Sustainable treatment of landfill leachate. *Applied Water Science, 5*(2), 113–126.
22. Smith, P.G., & Crawford, J. B. (1985). *Landfill technology*.
23. Deublien, D., & Steinhauser, A. (2008). *Biogas from waste and renewable resources* (pp. 27–83). Wiley-VCH Verlag GmbH & Co. KGaA.
24. Daniel, D. E. (1993). Landfills and impoundments. In *Geotechnical practice for waste disposal* (pp. 97–112). Boston: Springer.
25. Bhambulkar, A. V. (2011). Municipal solid waste collection routes optimized with arc GIS network analyst. *International Journal of Advanced Engineering Sciences and Technologies, 11*, 202–207.
26. Jördening, H. J., & Winter, J. (Eds.). (2005). *Environmental biotechnology: Concepts and applications*. Wiley.
27. Aziz, S. Q., Aziz, H. A., Yusoff, M. S., Bashir, M. J., & Umar, M. (2010). Leachate characterization in semi-aerobic and anaerobic sanitary landfills: A comparative study. *Journal of Environmental Management, 91*(12), 2608–2614.
28. Adhikari, B., Dahal, K. R., & Khanal, S. N. (2014). A review of factors affecting the composition of municipal solid waste landfill leachate. *International Journal of Innovative Science Engineering and Technology, 3*(5), 272–281.
29. Dos Santos, R. E., Dos Santos, I. F. S., Barros, R. M., Bernal, A. P., Tiago Filho, G. L., & da Silva, F. D. G. B. (2019). Generating electrical energy through urban solid waste in Brazil: An economic and energy comparative analysis. *Journal of Environmental Management, 231*, 198–206.
30. Kale, S. S., Kadam, A. K., Kumar, S., & Pawar, N. J. (2010). Evaluating pollution potential of leachate from landfill site, from the Pune metropolitan city and its impact on shallow basaltic aquifers. *Environmental Monitoring and Assessment, 162*(1), 327–346.
31. Mayakaduwa, S. S., Siriwardana, A. R., Wijesekara SSRMDHR, B. B., Vithanage, M. (2012, October). Characterization of landfill leachate draining from Gohagoda municipal solid waste open dump site for dissolved organic carbon, nutrients and heavy metals. In *Proceedings of the 7th Asian Pacific Landfill Symposium, Bali, Indonesia*.
32. Welikannage, K., & Liyanage, B. C. (2009). Organic waste composing by low cost semi aerobic converted trench method at Central Province. *Proceedings of Annual Academic Sessions, Open University of Sri Lanka*, 52–55.
33. Zhao, X., Qu, J., Liu, H., Wang, C., Xiao, S., Liu, R., Liu, P., Lan, H., & Hu, C. (2010). Photoelectrochemical treatment of landfill leachate in a continuous flow reactor. *Bioresource Technology, 101*(3), 865–869.
34. Abbas, A. A., Jingsong, G., Ping, L. Z., Ya, P. Y., & Al-Rekabi, W. S. (2009). Review on landfill leachate treatments. *Journal of Applied Sciences Research, 5*(5), 534–545.
35. Li, W., Hua, T., Zhou, Q., Zhang, S., & Li, F. (2010). Treatment of stabilized landfill leachate by the combined process of coagulation/flocculation and powder activated carbon adsorption. *Desalination, 264*(1–2), 56–62.
36. Xu, Z. Y., Zeng, G. M., Yang, Z. H., Xiao, Y., Cao, M., Sun, H. S., Ji, L. L., & Chen, Y. (2010). Biological treatment of landfill leachate with the integration of partial nitrification, anaerobic ammonium oxidation and heterotrophic denitrification. *Bioresource Technology, 101*(1), 79–86.

37. Renou, S., Givaudan, J. G., Poulain, S., Dirassouyan, F., & Moulin, P. (2008). Landfill leachate treatment: Review and opportunity. *Journal of Hazardous Materials, 150*(3), 468–493.
38. Shouliang, H. U. O., Beidou, X. I., Haichan, Y. U., Liansheng, H. E., Shilei, F. A. N., & Hongliang, L. I. U. (2008). Characteristics of dissolved organic matter (DOM) in leachate with different landfill ages. *Journal of Environmental Sciences, 20*(4), 492–498.
39. Castrillón, L., Fernández-Nava, Y., Ulmanu, M., Anger, I., & Marañón, E. (2010). Physico-chemical and biological treatment of MSW landfill leachate. *Waste Management, 30*(2), 228–235.
40. Tchobanoglous, G., Theisen, H., & Vigil, S. (1993). *Integrated solid waste management.* New York: McGraw-Hill.
41. Williams, P. T. (2005). *Waste treatment and disposal.* Wiley.
42. Worrell, W., & Vesilind, P. (2011). *Solid waste engineering, SI Version.* Nelson Education.
43. Kamaruddin, M. A., Yusoff, M. S., Rui, L. M., Isa, A. M., Zawawi, M. H., & Alrozi, R. (2017). An overview of municipal solid waste management and landfill leachate treatment: Malaysia and Asian perspectives. *Environmental Science and Pollution Research, 24*(35), 26988–27020.
44. Vilar, V. J., Rocha, E. M., Mota, F. S., Fonseca, A., Saraiva, I., & Boaventura, R. A. (2011). Treatment of a sanitary landfill leachate using combined solar photo-Fenton and biological immobilized biomass reactor at a pilot scale. *Water Research, 45*(8), 2647–2658.
45. Silva, A. C., Dezotti, M., & Sant'Anna, G. L., Jr. (2004). Treatment and detoxification of a sanitary landfill leachate. *Chemosphere, 55*(2), 207–214.
46. Kang, Y. W., & Hwang, K. Y. (2000). Effects of reaction conditions on the oxidation efficiency in the Fenton process. *Water Research, 34*(10), 2786–2790.
47. Monje-Ramirez, I., & De Velasquez, M. O. (2004). Removal and transformation of recalcitrant organic matter from stabilized saline landfill leachates by coagulation–ozonation coupling processes. *Water Research, 38*(9), 2359–2367.
48. Tatsi, A. A., Zouboulis, A. I., Matis, K. A., & Samaras, P. (2003). Coagulation–flocculation pretreatment of sanitary landfill leachates. *Chemosphere, 53*(7), 737–744.
49. Amokrane, A., Comel, C., & Veron, J. (1997). Landfill leachates pretreatment by coagulation-flocculation. *Water Research, 31*(11), 2775–2782.
50. Zamora, R., de Velasquez, M. O., Moreno, A., & Ramirez, I. (2000). Treatment of landfill leachates by comparing advanced oxidation and coagulation–flocculation processes coupled with active carbon adsorption. *Water Science Technology, 41,* 231–235.
51. Rettenberger, G. (2018a). Quality of landfill gas. Solid waste landfilling: Concepts, processes, technology. In *Solid waste landfilling* (p. 439). Amsterdam: Elsevier.
52. Rettenberger, G. (2018). *Collection and disposal of landfill gas. Solid waste landfilling* (pp. 449–462). Amsterdam: Elsevier.
53. A. for T.S. and D.R. ATSDR, Chapter 5: Landfill Gas Control Measures. In: Landfill gas primer – An overview environment health professional, 2001.
54. Janson, O. (1991). Analytik, Bewertung und Bilanzierung gasförmiger Emissionen aus anaeroben Abbauprozessen unter besonderer Berücksichtigung der Schwefelverbindungen. In *Stuttgarter Berichte zur Abfallwirtschaft* (Vol. 32). Berlin: Erich Schmidt Verlag GmbH & Co.
55. Janson, O., & Bardtke, D. (2005). Analytik der Spurenstoffe im Deponiegas in Rettenberger, G. (publisher): Neue Vrfahren und Methoden zur Sicherung und Sanierung von Altlasten am Beispiel der Deponie Gerolsheim e Kurzfassungen des Verbundvorhabens, Trierer Berichte zur Abfallwirtschaft, vol. 15. Verlag Abfall aktuell, Stuttgar.
56. Rettenbeger, G., & Stegmann, R. (1996). Landfill gas components. In T. H. Christensen, R. Cossu, & R. Stegmann (Eds.), *Landfilling of waste: Biogas.* E&FN Spon. ISBN 0419194002.
57. Solov'yanov, A. A. (2011). Associated petroleum gas flaring: Environmental issues. *Russian Journal of General Chemistry, 81*(12), 2531–2541.
58. Ménard, C., Ramirez, A. A., Nikiema, J., & Heitz, M. (2012). Biofiltration of methane and trace gases from landfills: A review. *Environmental Reviews, 20*(1), 40–53.
59. Brown, K. A., & Maunder, D. H. (1994). Exploitation of landfill gas: A UK perspective. *Water Science and Technology, 30*(12), 143.

60. Han, H., Long, J., Li, S., & Qian, G. (2010). Comparison of green-house gas emission reductions and landfill gas utilization between a landfill system and an incineration system. *Waste Management & Research, 28*(4), 315–321.
61. Niskanen, A., Värri, H., Havukainen, J., Uusitalo, V., & Horttanainen, M. (2013). Enhancing landfill gas recovery. *Journal of Cleaner Production, 55*, 67–71.
62. Jewaskiewitz, B. (2010). Landfill gas recovery, green energy, and the clean development mechanism. *Civil Engineering Management South African Institution of Civil Engineering, 18*(7), 19–23.
63. Ujang, Z., 2004. Part 1—conventional landfill design & operation from design to problem solving. In *Workshop on New Technologies for Cost-effective Landfill Management. UTM, Kuala Lumpur (Vol. 7)*.
64. Taylor, M. A. (2000). Landfill final cover storm drainage. *MSW Management*, p. 7. Retrieved from: https://foresternetwork.com/weekly/mswmanagement-weekly/landfill-management/landfill-final-cover-stormwater-drainage
65. Bader, C. D. (1998). The articulated dump truck: workhorse of the landfill. *MSW Management*, 56–61.
66. Ramke, H. G. (2001). Appropriate design and operation of sanitary landfills. In University of Applied Sciences Ostwestfalen-Lippe, Germany. *Prepared for the international conference on sustainable economic development and sound resource management in Central Asia* (p. 30).
67. Salvato, J. A., Nemerow, N. L., & Agardy, F. J. (2003). *Solid waste disposal. Environmental engineering*. Wiley.
68. Koerner, R. (2003). Linear system design and materials. In *USEPA Workshop on bioreactor landfills*. Arlington: (CDROM).
69. Bagchi, A., & Bagchi, A. (1994). *Design, construction, and monitoring of landfills* (2nd ed.). Wiley.
70. Curro, J. (2000). Using the global positioning system in landfill gas system monitoring. *MSW Management*.
71. Rees, J. F. (1980). Optimisation of methane production and refuse decomposition in landfills by temperature control. *Journal of Chemical Technology and Biotechnology, 30*(1), 458–465.
72. Farquhar, G. J., Rovers, F. A., & McBean, E. A. (1995). *Solid waste landfill engineering and design*. PTR Prentice Hall.
73. Frechen, F. B. (1989). 6.4 Odour emissions and controls. Sanitary landfilling: Process, technology and environmental impact, 425.
74. Feinbaum, R. 2000. Odor control: Going the extra mile. *Journal for Municipal Solid Waste Professional*. Santa Barbara.
75. O'Malley, P. G. (2003). Controlling odor at transfer stations and MRFs: There's more than one way to reduce complaints. *MSW Management, 13*(2), 44–52.
76. International Solid Waste Association. (2019). ISWA – Landfill operational guidelines, 3rd edition. A report from ISWA's working group on landfill 2019.
77. Gilbert, N. I., Correia, R. A., Silva, J. P., Pacheco, C., Catry, I., Atkinson, P. W., Gill, J. A., & Franco, A. M. (2016). Are white storks addicted to junk food? Impacts of landfill use on the movement and behaviour of resident white storks (Ciconia ciconia) from a partially migratory population. *Movement Ecology, 4*(1), 1–13.
78. Treweek, J. (1999). *Ecological impact assessment*. Oxford.
79. Mellanby, K. (1992). *Waste and pollution: The problem for Britain*. Somerset: Harper Collins.
80. Shen, T. T., Nelson, T. P., & Schmidt, C. E. (1990). Assessment and control of VOC emissions from waste disposal facilities. *Critical Reviews in Environmental Science and Technology, 20*(1), 43–76.
81. Dent, C. G., Scott, P., & Baldwin, G. (1986). *Study of landfill gas composition at three UK domestic waste disposal sites. US and UK Departments of Energy, Chilton*. Oxfordshire: DoE, Harwell Research Lab.
82. Emam, E. A. (2015). Gas flaring in industry: An overview. *Petroleum & Coa, 57*(5).

83. Ghadyanlou, F., & Vatani, A. (2015). Flare-gas recovery methods for olefin plants. *Chemical Engineering, 122*(5), 66.
84. Lagerkvist, A. (1987). Sanitary landfills. A Swedish point of view-In: Process, technology, and environmental impact on sanitary landfill. *Proceedings of the International Symposium*, 1–7.
85. Augenstein, D. (1991). Greenhouse effect contributions of US landfill methane. In: *Landfill gas: Energy and environment.*
86. Gardner, N., Manley, B. J. W., & Pearson, J. M. (1993). Gas emissions from landfills and their contributions to global warming. *Applied Energy, 44*, 165–174.
87. Thornoloe, S., & Peers, R. L. (1990). Landfill gas and the greenhouse effect. In G. E. Richards & Y. R. Alston (Eds.), *Landfill gas: Energy and environment* (pp. 361–368). Harwell Laboratories.
88. Shammas, N. K., Wang, L. K., Wang, M. H. S., & Chen, S. L. (2020). Ecological impact and anagement of solid waste landfill gas. In Y. T. Hung, L. K. Wang, & N. K. Shammas (Eds.), *Handbook of environment and waste management: Acid rain and greenhouse gas pollution control* (pp. 455–482). Singapore: World Scientific.
89. Wang, L. K., Shammas, N. K., & Hung, Y. T. (2008). *Biosolids engineering and management.* Totowa, NJ, USA: Humana Press. 800 pages.
90. Hung, Y. T., Wang, L. K., & Shammas, N. K. (2014). *Handbook of environment and waste management: Land and groundwater pollution control.* Singapore: World Scientific. 1091 pages.

Chapter 9
Solid Waste Systems Planning

Sharifah Akmam Syed Zakaria, Hamidi Abdul Aziz, Jarir S. Dajani, Dennis Warner, and Yung-Tse Hung

Abstract This chapter focuses on the relevance and importance of proper and sustainable solid waste system (SWS) planning as a part of the solid waste management (SWM) field. Complexities and uncertainties in the construction industry require SWS planning to be highly considered and performed due to rapid population growth and urbanization trends. These trends influence the generation of solid waste in terms of its quantity, types, treatment, disposal, etc. that require appropriate solutions in the collection, transport or treatment, besides other variations in time to be considered when designing a SWS. As a key function of SWM, SWS planning is crucial for economic development, the protection of human and environmental health for sustainable development. Improvements in SWS can be achieved on different levels based on a large scope in optimization and improving cost efficiency based on robust planning with a systematic process of SWS planning, integrated SWM plan enrichment, SWS implementation plan, strategic technology plan of SWS and their influencing factors. A case study on the cost-benefit analysis of SWS technology adoption is presented with insights on an assessment plan in a solid waste technology implementation for sustainable development.

S. A. S. Zakaria (✉)
School of Civil Engineering, Engineering Campus, Universiti Sains Malaysia, Nibong Tebal, Pulau Pinang, Malaysia
e-mail: akmam@usm.my

H. A. Aziz
School of Civil Engineering and Solid Waste Management Cluster, Engineering Campus, Universiti Sains Malaysia, Nibong Tebal, Pulau Pinang, Malaysia
e-mail: cehamidi@usm.my

J. S. Dajani
Department of Civil Engineering, Stanford University, Stanford, CA, USA

D. Warner
Gannett, Fleming, Corddry, and Carpenter, Inc., Harrisburg, PA, USA

Y.-T. Hung
Department of Civil and Environmental Engineering, Cleveland State University, Cleveland, OH, USA

L. K. Wang et al. (eds.), *Solid Waste Engineering and Management*, Volume 1, Handbook of Environmental Engineering 23,
https://doi.org/10.1007/978-3-030-84180-5_9

Keywords Solid waste systems planning · Solid waste management · Integrated solid waste management plan

Acronym and Nomenclature

CBA	Cost-benefit analysis
EIA	Environmental impact assessment
GIS	Geographic information system
GPRS	General Packet Radio Service
GSM	International system for mobile
IoT	Internet of things
ISWMF	Integrated solid waste management framework
MSW	Municipal solid waste
NGO	Non-governmental organizations
N-P-K	Nitrogen, phosphorus and potassium
PFI	Private financing investment
PPP	Public-private partnership
RFID	Radio-frequency identification
SDG	Sustainable development goals
SWM	Solid waste management
SWS	Solid waste systems
USEPA	US Environmental Protection Agency

9.1 Introduction to Solid Waste Systems

The overincreasing volume of solid waste produced by human due to the rapid growth of population, urbanization and development is one of the most pressing issues the world faces today. Waste leads to public health at risk and pollutes the environment, and its disposal costs societies, cities and countries a fortune they cannot afford. There is an increase in the global population that already struggles to deal with tons of municipal solid waste produced every year, while the waste collection is not available throughout every location in the world [1]. Countries that have waste collection often cannot keep up with the challenges presented by a growing number of people living in cities and the new streams of hazardous waste resulting from rapid industrialization [2]. This requires more efficient and effective management of solid waste systems (SWS) to handle the impacts of climate change, improvements of air quality, the quality of human life, water quality and the ecosystem or diversity of plant and animal life.

Solid waste management (SWM) includes all activities that are incorporated as a systematic structure that intends to handle and organize solid waste from its collection until its disposal [3]. This systematic structure is important to reduce and minimize the environmental, economic and health impacts of solid wastes. Solid waste, which is also known as rubbish, trash, garbage and refuse, is categorized as a material in a non-liquid form that literally has no value [4]. Due to the stages of industrial revolutions, with rapid production and mass consumption, these generate wastes that are not reused, recycled and fixed. Thus, landfills are loaded with waste that may impact the natural environment as waste will lay in the landfills for years into the future.

In order to comprehend a solid waste system, firstly, it is vital to determine the challenges in solid waste management that require a robust plan of solid waste systems to be developed. One of the major challenges in the management of the solid waste system is terms of the practical and systematic implementation of solid waste services [5]. Specifically, the operational inefficiencies of services and inadequate service coverage are two major aspects that require attention and actions. Secondly, solid waste systems are not well supported with reducing and reuse practices, thus limiting the utilization of recycling activities. Lastly, although there are regulations in place, there are inadequate and inefficient management of hazardous waste, healthcare waste and landfill disposal.

The rapid growth of the world's population and urbanization have resulted in increasing waste generation. Therefore, most states, cities and areas realize that it is difficult not only to handle or cope with the existing situation but also to anticipate the future conditions and trends of SWS and strategic plans for SWS. In the process of SWM planning, it should focus on the resources, commitments and unity of purpose. Officials need to support SWS planning to ensure the efficiency and effectiveness of coordination mechanisms. This requires various aspects of waste management that are often the responsibility of various government agencies and stakeholders [6]. Thus, it is important to have good coordination among all of them. This will ensure all relevant parties are aligned with and actively supporting the implementation of this plan.

The major challenge of SWS planning is to develop a reliable waste collection mechanism that ensures and improve public health [7]. This is due to the fact that deficient treatment or disposal will definitely pollute the environment and contributes to global warming. Thus, the key issue of SWS starts with waste generation, collection, transport, treatment or recycling and landfill disposal [3], which requires the involvement and active role of stakeholders as well as the efficient and effective management of resources, especially financial aspects. In this case, SWS coordination does involve not only the services and infrastructure of solid waste but also people in terms of their attitudes and culture, SWS technology adoption and interactions with external environmental factors [8]. SWS plan is the utmost important aspect in a SWM as it provides a well-structured, comprehensive and holistic overview of all technical and non-technical aspects that are involved in SWM and their interdependencies.

9.2 Solid Waste Systems Planning Process

In the field of the solid waste system (SWS), planning is the fundamental task of the overall management of solid waste that involves designing, decision-making, scheduling, resource allocation, alternative identification and the development of coordination, monitoring and control mechanisms to achieve specific goals in fulfilling the needs of the community regarding solid waste management. SWS planning incorporates important factors such as economic, environmental, social and technical aspects that are interrelated. In meeting the future requirements of economic prosperity, people's well-being, environmental protection and partnering implementation, SWS planning serves as a guideline for intended actions with a specified time frame or duration based on the priorities of SWS requirements [9].

SWS planning process involves a systematic flow of work implementation in:

(i) The collection and analysis of data on the SWS in terms of its current performance.
(ii) The recognition of areas in the present SWS that need to be developed, changed and improved.
(iii) The generation of alternatives to improve the current situation or to solve or overcome the current problem of SWS.
(iv) The development of suitable SWS strategies to be implemented in a specified duration.
(v) The implementation of SWS strategies.
(vi) Monitoring and evaluation of actions for SWS improvements based on SWS objectives.

In the planning process of SWM, the first dimension is on the physical components of SWM service from the whole chain, starting from solid waste generation to solid waste collection, treatment and disposal. The second dimension focuses on the governance aspects, and these include the role and responsibilities of stakeholders, financing mechanisms, legislation and policies, as well as social-cultural aspects. In SWS planning, there are various issues and challenges that need to be considered, such as sustainability aspects in terms of environment, economy and society, like creating value from solid waste and treatment options from biowaste [10]. Not only does SWS planning involve solid waste services and infrastructures only, but also it involves people, their attitudes, behaviours and interactions, financial aspects, legislation and other aspects. Therefore, having a comprehensive and well-structured overview of all aspects of SWS planning contributes to a more robust management of solid waste and its interdependencies.

Generally, it is important to understand the two major categories of solid waste management [11]. The first category is municipal solid waste (MSW), which comprises wastes from private and public places such as houses, offices, shops, hospitals and streets, which includes (i) general waste such as paper, plastics, metal, glasses, etc.; (ii) e-waste such as electrical devices, televisions, phones, etc. that are thrown away; and (iii) hazardous waste that degrade the environment and impact on human illness and well-being. The second category is related to the disposal activities of solid waste. There are ways to clean up huge areas of solid waste. Solid waste is eventually disposed of either in a landfill or incinerated, which is normally the responsibility and tasks of the local municipality or other governmental agencies or authorities.

Municipal solid waste management involves institutional and administrative aspects, which consists of planning, organizing, monitoring, leading and controlling various resources such as financial, manpower, material and machinery in the systematic disposal of solid waste. It is important to look at areas where there are huge contaminations due to hazardous waste. Thus, the major focus is on how to minimize or reduce the amount of solid waste. People are familiar with the idea of the 3Rs: reduce, reuse and recycle. These three activities sometimes can be energy-intensive that should be well-considered in the process of SWS planning. All too often, an option to minimize solid waste amount is based on a condition where the option might fit in with respect to a SWS or at least what requirements must be met to reach this goal. Figure 9.1 illustrates the hierarchy of solid waste which reflects the options of minimizing solid waste amount.

This hierarchy illustrates a prioritization of action or activities for solid waste management, giving top priority to preventing solid waste from being generated in the first place, followed by reduction through reuse and recycling, recovering energy through solid waste processing and finally disposal.

Generally, the solid waste hierarchy is about the disposal of solid waste. Technically, energy can also be generated from solid waste. However, there is a more economical way to recycle solid waste and reduce it in the first place. As solid waste normally goes to the landfill, it is important to break through the 3Rs. In order to recycle solid waste, stuff or solid waste that can be used again should be sorted and removed before it is actually transported to the landfill. In this case, closed-loop

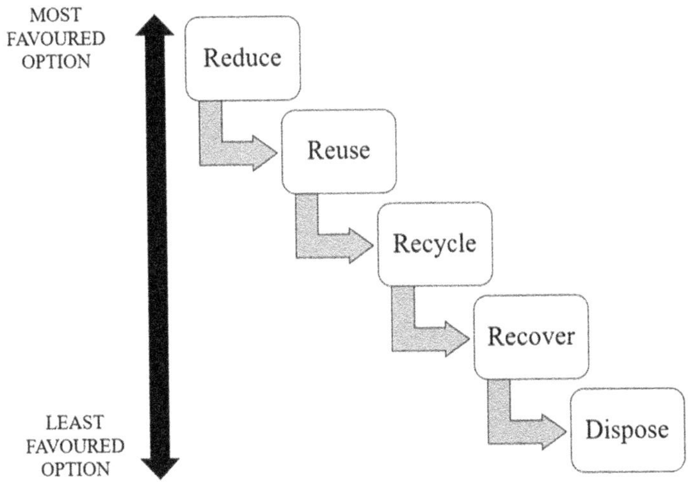

Fig. 9.1 Solid waste hierarchy

recycling can be performed where stuff like aluminium cans can be recycled to produce more aluminium cans. Alternatively, open-loop recycling is performed where plastic bottles, for instance, can be either recycled to make more plastic bottles or also transformed into another form like plastic sheets [12]. Kitchen wastes like vegetable scraps and eggshells can be recycled as plant nutrients through composting processes in the backyard as a way to reduce the amount of solid waste that goes to the landfill.

There are companies that implement a zero-landfill policy as they are not putting anything in landfills through recycling, reuse and reduce activities; besides, a number of them are incinerating solid waste to produce energy where garbage is burnt to produce energy instead of not using it at all [13]. If solid waste has to be disposed of, it is vital to handle and manage it properly. Therefore, solid waste management decisions and actions should be made with sufficient planning which considers various aspects of a situation, households, organization, community, location, district, state and country.

Strategic planning of SWM serves to improve the efficiency and effectiveness of SWM services by taking a broader view and addressing the problem based on their priority. The planning process has a lot to do with the management of relationships and building consensus among stakeholders. In addition, it is an interactive process, including regular revision and updates as a systematic improvement system.

9.2.1 Initial Planning Process: Background of Planning Area

This is the starting point that focuses on the way to prepare a SWM plan. There are two major activities required to start the planning process, namely, administrative support and work organization. In order to ensure a SWM plan materializes, it is important to obtain administrative and managerial support on behalf of the government authorities and decision-makers. In this case, a steering committee is needed to lead and direct the process of SWS planning and provide the required administrative support. In parallel, a working group is also needed to translate and clarify this steering committee into practical measures. By forming and having such committees, the present situation of SWM in a city can be assessed. Furthermore, the scope of the planning framework should be determined. This includes the planning of the geographical area, the duration of the SWS plan, the consideration of waste fracture, the level of service quality and environmental and public health service.

In the planning process of SWS, it is important to obtain information on the capacity of and the number of collection vehicles that are necessary and the types of solid waste generated in households that need to be removed and collected by vehicles. In relation to this, it is also vital to determine the feasibility and the scale of solid waste treatment options, besides what needs to be processed. The focus is on identifying the opportunities of recycling activities based on the composition and value of waste generated. In addition, SWS planning includes the estimation of landfill lifespan, besides the measurement of solid waste amounts that are reaching the landfill. It is also important to determine the lifestyle of people or society in a particular area in relation to the degree of urbanization and income level. Thus, in the initial planning process of SWS, the necessary information on solid waste is needed in a particular area. For instance, if an area needs to be divided into subareas, it is vital to know how the different income groups and their backgrounds affect solid waste generation in terms of its quantity and types.

9.2.2 Situation Analysis: Existing Solid Waste Management Conditions

In developing a SWS plan, the starting point is to assess the present situation of municipal SWM at a city or district level. This step can be divided into a number of tasks, namely, assessing the generated amount and characterization of the waste in the study area, reviewing the current SWM in practice, predicting future solid waste quantities and capacity needs and understanding the current constraints and opportunities for improvements.

Determining the current situation and how well solid waste is being managed is central to developing a SWS plan. The following steps are important for understanding the current SWM conditions and identify the gaps that need to be addressed. Therefore, it is vital to compile waste-related data available, including information on waste generation collection, transportation, disposal, waste classification and composition. If related data is not available, samples, studies and surveys may be needed. In addition, it is also vital to review waste management policies that are currently in place in order to understand the current SWM system and how the new SWS planning is going to be interlinked with the existing SWM system, mechanism and policies. Developing an understanding on the technical capacity of SWS infrastructure and resources available in a city, state or country is also important. This technical capacity can be related to solid waste collection, separation, collection, treatment and disposal. In particular, financial and human resources will be needed to implement improved SWM practices based on the capacity and availability of resources. There are several serious situations or questions that need to be considered, such as how is SWM currently funded? Are the resources adequate? Are there cost recovery schemes in operation through taxes or fees? What are related technical skills available? What is the condition or level of community awareness and interests besides the roles of various bodies or agencies in SWM, public participation in the planning and implementation of SWM initiatives? Publics or society may be requested to separate their solid waste or use special waste bins. In order to do this, it is important to understand the awareness level of solid waste management among members of the public and private sectors and the community as a whole. From this early assessment, situation and gap analysis can be performed in order to make an informed decision towards the development of the SWM plan. Once the vision and scope have been defined, key objectives and targets for the SWM plan can be established.

9.2.3 Issues Identification: Problem Definition and Future Conditions

Currently, it is estimated that by 2050, the generation of worldwide municipal solid waste is expected to increase by roughly 70% to 3.4 billion metric tons, with at least 33% of that are not managed in an environmentally safe manner [14]. Thus, general predictions do not forecast a better future. Moreover, the global population is predicted to rise by 9.7 billion people in 2050 [15], which will be concentrated in Asia and Africa. This will have serious implications on resource consumptions and solid waste generations. It is also imperative that the generation of solid waste increases according to the development of urbanization as the size of cities is also rapidly growing. Therefore, it is important to be well prepared in managing solid waste based on the complexity of current development with the anticipations of the future's conditions. In this case, many waste-related decisions should be taken with much thought on the current situations and the predictions of future changes.

Therefore, the major areas that require superior attention in the identification of SWS issues are:

(i) Institutional – organizational framework, organization administration, networking and the involvement of public and private sectors.
(ii) Waste collection and recycling – service performance, service coverage, infectious and hazardous wastes and recycling supports.
(iii) Solid waste treatment and disposal – technology of waste treatment and proposal, existing site conditions, technical and environmental aspects.
(iv) Cost and financial aspects – financial planning, affordability and payment willingness, investment needs, cost recovery and accounting aspects.
(v) Public awareness – participation and commitment levels.

In this case, it is vital to determine areas that solid waste is not collected from these homes and neighbourhoods, besides the number of people who lack access to more controlled and efficient disposal facilities of solid waste. Thus, people and their ecosystems are the direct victims of this situation's economic, environmental and health impacts. In the case of disease spreads caused by major flooding, as an impact of uncollected solid waste blocking the drains, this situation resulted in various problems, including social and economic losses [16]. Therefore, by identifying problems and issues related to SWM, people can learn from their mistakes and make positive transformations to create a cleaner and sustainable environment. By having this evaluation, it intends to bring some light into various issues related to SWS and to start looking for solutions to the previous issues.

9.2.4 *Involvement of Implementation Entities*

Once the major issues or weaknesses of the SWM system have been identified, the first action is to create an initial agreement with relevant agencies showing the approval and commitment of local administration and leadership with the SWM plan. It is important to engage people who are motivated and dedicated to solving solid waste problems with good knowledge and understanding on the complexity of SWS, and have the suitable tools to improve it. At the city level, this would be preferably the mayor or a district officer, while in the state case, it should be the state leaders and relevant ministers at the country's level. Their approvals are required in the implementation of SWM plans. In this way, the development of the plan came to be truly owned by the local government agencies. Secondly, besides the involvement of administrative bodies, engaging stakeholders in the SWM plan are also needed to assist the development of any SWM planning, supported by agencies with sectoral responsibilities like environmental or health organizations that will play key roles in the SWM system. Apart from government agencies within each country, a great variety of organizations and groups play vital roles in SWM activities, and these relevant stakeholders must be identified, such as waste generators, service providers and supporting entities [17]. Members of these categories can be found in the community, private sectors, non-governmental organizations (NGOs), academia and civil society. The third action is to identify the roles of stakeholders. Identifying the entities or groups who are actively involved in SWM activities, getting them involved in the process of SWS planning from the beginning and determining their roles are very important in the development of a sound and optimistic SWS plan. Figure 9.2 presents the coordination of entities and resources in SWM planning.

Different stakeholders can play their roles and support in unique ways. NGOs, for instance, have experience working with communities at the grassroots level and

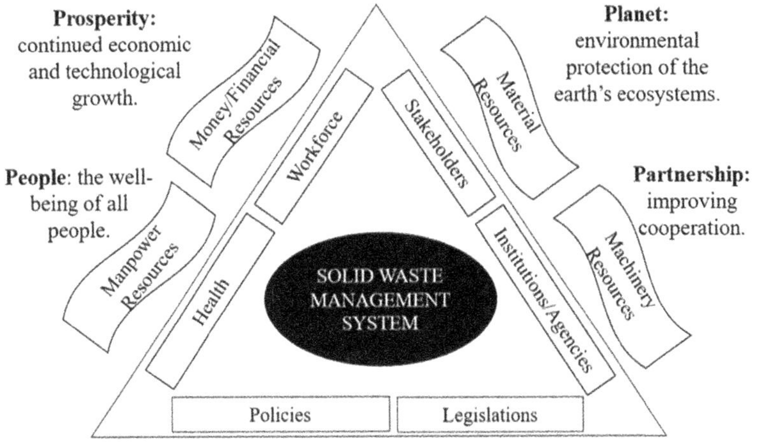

Fig. 9.2 Coordination of entities and resources in SWM planning

can play an active role in complementing the SWM service provided by the government. In some areas, there may be a lot of informal solid waste recyclers collecting materials from households, shops or organizations and selling them for profit [18]. They can be considered as very important stakeholders as they could support recycling activities in the future. Thus, their involvement in the development of the SWM planning should be coordinated with other stakeholders. The fourth action is to establish a coordinating committee to organize all these stakeholders in relation to the SWM activities. The committee should be formed by representatives from the government (federal and state), local municipals, private sectors, academia, NGOs, societies and the community. Based on this formation, their involvements, roles and responsibilities are outlined throughout SWS plans. All key stakeholders should be brought together to create a strategic vision with all the inputs of the members. This vision development focuses on several conditions:

(i) Reasons for the improvement of SWM services.
(ii) The goals of SWM services in the short, medium or long term.
(iii) The improvement mechanism of SWM services.
(iv) The duration of SWM goals achievement.

9.2.5 System Design: Approaches and Alternative Actions

In the aspect of system design, the major focus is on the set of SWM parameters. Important aspects such as the general dimension of SWM, its focus and the scope of SWS plans should be determined at an early stage. The plan should focus on preventing the generation of solid waste while phasing out the use of hazardous substances in products, manufacturing process and other industrial activities, rather than simply on managing what is generated. While some categories of solid waste may be categorized as a higher priority, any type of solid waste, even if it is of minor or major conditions, will need to be managed accordingly [10]. All parties involved in the SWS plan will need to achieve a common agreement or consensus on the overall goals and objectives of SWM. Therefore, it is important to determine the major goals of SWM that should be accompanied by clear objectives and targets to complement the timeline of a SWS plan for the benchmarking and measurement of SWM progress and performance. Alternatives for all the physical components of SWM in terms of solid waste generation, collection, treatment, triple Rs (reduce, reuse and recycle) activities and disposal should be supported by the governance of stakeholder inclusivity, financial sustainability and SWM policies. After coming up with the list of possibilities, the stakeholders should agree on a shortlist of preferred options for each component.

Establishments of targets and indicators are important in the system design of SWS plans [19]. In designing the SWS plan, the target of action plans should be set as a priority on solid waste streams or other issues. Target refers to the quantitative translation of SWM objectives based on a realistic timeframe. For example, the reduction of solid waste generation to be landfilled can be achieved with an increased recycling rate or targets for moving the solid waste from uncontrolled dumpsites to properly engineered landfills. These targets can be used to drive SWM actions, determine the frequency of solid waste collection, create momentum, monitor progress, control actions and alert those who are involved in the implementation of SWS plans to solve waste-related problems. Targets should be SMART, namely, specific, measurable, achievable, relevant and time-bound. For example, an objective could be "to increase the coverage of solid waste collection", whereas the target would be "the expansion of collection services to 95% of the central district area by 2030″.

9.2.6 Action Plan Development: Recommendations for the Solutions

Based on the approaches and alternatives of SWM, its action plan can be developed to turn or implement the strategies of SWS into practical realities. The development of an action plan for SWM is based on its strategic visions, targets and objectives that should be divided into actionable steps. The prior steps have combined information and guidance on the content of each step from the perspective of what needs to be done and the way of delivering it. The focus is on the content of SWM as its purpose is to determine and evaluate the practical options available for addressing each of the components of the SWM framework. Based on the considerations and focus on the planning framework, it is combined with the preferred options of SWM physical components to come out with the overall strategic plan and the most appropriate SWS strategies for the SWM action plan within an appropriate time span. This action plan should focus on the high level issues, especially the detailed aspects of various approaches and alternatives in relation to SWM, which normally has a time span of five (5) years. The action plan of SWM should consist of these major aspects:

(i) The detailed and specific actions that need to be taken to implement the individual components of the overall strategy of SWS.
(ii) Responsibility or person in charge of the process of SWM strategy implementation.
(iii) Time of strategy execution and the duration of its accomplishment.
(iv) An investment plan based on a full review of financial planning for SWM.

9.2.7 Plan Revision and Evaluation

Once the action plans of SWS have been drafted and developed, the SWS planning process is practically complete. It is important that the SWS plan need to be officially endorsed by the government officials and the stakeholders. The plan should be submitted to a full public verification and consultation process, in which the public and the stakeholders are invited to provide feedback and inputs for improvements. In this case, the government officials present the SWS plan to various stakeholder groups to determine their concerns, achieve consensus and obtain commitments for the implementations of the SWS plans. The plan document should be distributed among employees and the workforce dealing with solid waste management. In addition, news on this plan presentation should be published in mass media.

9.2.8 Endorsement

After the process of the SWS plan revision and evaluation, the endorsement of the SWS plan is crucial for implementation to secure the budget and other resources allocated to improving SWM. The SWM plan needs to get an approval from the regional government to make it legal documentations for the next duration of SWM activities. With the final approval, the SWS plan is completed, and the plan is then distributed and circulated among all stakeholders as a printed and digital publication.

The outcome of the process is a SWS plan consisting of a number of elements and an ambitious overall framework and goals for sound SWM in a city, state or country based on specific rules and regulations on SWM. A list of priority solid waste streams and issues are stated in the SWS action plan for each solid waste stream or issue comprising one or more targets with SWM policy actions, cost estimations for each action plan, clearly allocated roles and responsibilities for implementing the identified actions, SWS plans for review, including indicators to measure progress and collection of appropriate data and revision of the plan on a regular basis according to the latest requirements of SWM.

9.3 Integrated Solid Waste Management Plan

According to The US Environmental Protection Agency (US EPA), integrated solid waste management is a comprehensive waste prevention, recycling, composting and disposal program [20]. An effective ISWM consists of ways to prevent, recycle and manage solid waste in the most efficient and effective manner to protect human health and the environment. The solid waste management (SWM) plan is based on a planning hierarchy. This hierarchy presents the conceptual level of different components in the development and implementation of SWM activities and practices. The basis of a SWM plan is an operational plan which consists of a detailed implementation of the SWM plan. An operational plan is required for the detailed implementation plan of the overall SWM strategy. This is followed by the strategic level, which consists of the action plan and strategy of SWM. The strategic plan should operate within the framework of the government policy of SWM. The strategy provides the strategic vision over the next ten (10) years, for instance, to determine the achievement of the SWM process and activities. Meanwhile, the action plan sets out the detailed or specific actions of the SWM plan that are necessary over a certain period to realize that vision. Finally, the highest hierarchy is the policy of the SWM plan as a country's strategy which is established by the central or national government. Therefore, a strategic SWM plan can be defined as a guideline document or a blueprint that determines the activities, needs, priorities and necessary actions that need to be taken to develop and implement SWM practices.

Integrated Solid Waste Management Framework (ISWMF), as illustrated in Fig. 9.3, represents integration and interrelations between the important aspects of SWM in terms of the hardware and software of SWS. The hardware of SWS consists of the physical components, and the software of the system consists of the governance aspects.

Physical elements that need to be considered in an ISWM to ensure the workability of ISWM are:

(i) Public health – the maintenance of health conditions in urban and rural areas through good and proper collection service systems of solid waste to ensure the quality of public health amongst local people.
(ii) Environment – the proper treatment and disposal of solid waste in order to ensure the protection of the physical environment.

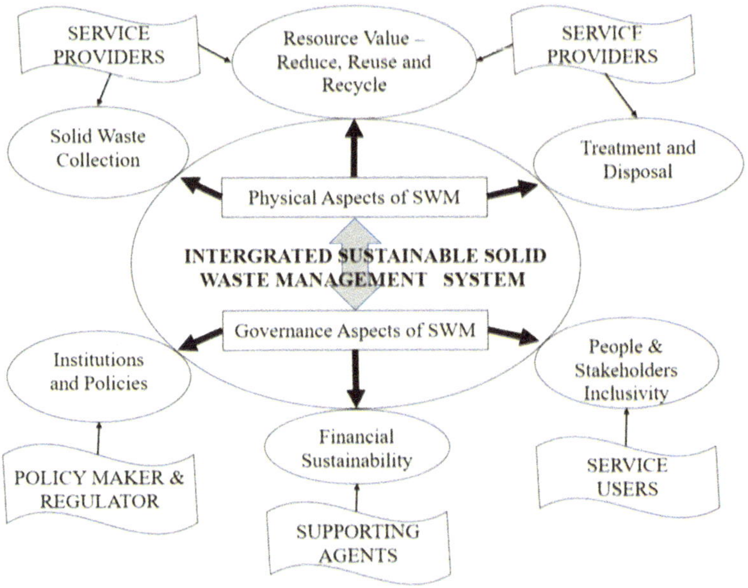

Fig. 9.3 Integrated Solid Waste Management Framework (ISWMF)

(iii) Resources management – the practice of 3Rs and the use of organic materials to reduce and prevent solid waste growth while ensuring a high rate of organic and natural recovery.

ISWM serves as a blueprint on comprehensive solid waste collection, treatment, recovery and disposal method that aims to develop environmental sustainability, economic affordability and social cooperation for any specific regions. It consists of three major aspects, namely, (i) waste prevention, (ii) recycling and composting and (iii) disposal (landfilling and combustion). ISWM strategy includes the entire issues about the SWM system while ISWM plan serves as a basis of determining:

(i) The ideal capacity of the on-site storage containers and the type of collection vehicle.
(ii) The size of a composting plant to treat the generated organic wastes.
(iii) The potential income from recovering plastics or other recyclable items.
(iv) The viability of building an incineration plant.

These are the areas that need to be considered when planning for solid waste infrastructure. Therefore, this requires information on the generated amount of solid waste and its composition.

9.3.1 Source of Waste

In SWS planning, it is important to understand the source of solid waste generation to determine the characteristics of each solid waste in terms of chemical or biological properties for the purpose of determining the types of solid waste treatment or disposal and the design of primary or secondary collection facilities. It is important to have the information on waste generation amounts as this will have various impacts on the planning process of SWS. Further, it is also viable to design the type of required or suitable vehicle and solid waste treatment facilities based on different sources or categories of solid waste such as [21]:

(i) Residential waste – consists of waste generated from household activities and consists of food leftovers, plastic, paper, glasses, cans, clothes, etc.
(ii) Commercial waste – consists of wastes generated in offices, business premises, restaurants, hotels, markets, warehouses and other commercial activities or organizations.
(iii) Municipal waste – consists of wastes resulting from municipal functions such as road sweeping, street wastes, abandoned vehicles, dead animals, etc.
(iv) Industrial waste – consists of wastes from industrial operations, discarded solid materials of manufacturing process, etc.
(v) Hospital – consists of healthcare waste from clinics, wards, doctor offices, etc. Hospital waste includes hazardous materials such as sharps, infectious, chemical, radioactive, etc. These types of solid waste need to be segregated as a priority task, besides its other municipal solid waste.

(vi) Institutional - consists of waste generated from educational institutions, administrative and public buildings including schools, offices, colleges, universities, etc., which includes paper, plastics, glasses, and aluminium.

(vii) Construction and demolition – consists of waste of heavy materials and high-density materials generated by the construction, demolition, refurbishment and repair of houses, commercial buildings and other structures, which include concrete, wood, steel, and dirt.

(viii) Agricultural – consists of wastes generated from farms, fields, orchards, and vineyards, which include spoiled foods, grains and vegetables, agricultural remains, and litter.

Specifically, based on solid waste categories and their respective risks and by determining the targets of solid waste types and quantity from these sources where the greatest amount of solid waste can be estimated, treatment facilities design could be possibly made [22]. Moreover, with the different nature of solid waste treatment between wet and dry solid waste conditions, if they are not properly degraded or treated, the contamination level might be higher. Thus, planning on special or specific segregation, storage and treatment can be performed more efficiently based on each particular source of solid waste for wet and dry solid waste.

In addition, it is also important to estimate the rates of solid waste generation based on per capita or other factors for each of these sources [23]. For the purpose of SWS planning to be undertaken, it is important to calculate the quantity of solid waste generation as these data become inputs to possibly forecast future waste generation. Hence, from the sources of solid waste, the calculation of waste generation rates within the council or any particular area can be determined [24].

9.3.2 *Composition of Waste*

Determination of solid waste composition serves as a basis in SWS planning and the direction of SWM, besides quantifying how much solid waste has been generated by a community in a particular area. This is important to determine the measures that can be used for solid waste prevention and reduction. The current estimation of solid waste composition is needed to match with the capacity of existing recycling rates, composting activities and landfilling system, besides estimating the generation of future solid waste. In the establishment of the SWS plan, the determination of solid waste properties is very important in identifying the capacity needs of SWS in terms of its quantities as well as the processing types of solid waste which will be used in order to determine the selection of an appropriate technology for SWM. The composition of solid waste varies from place to place and country to country, depending on various factors such as the standard of living, location, culture, season, climate, and storage time. The various compositions of solid waste are based on its [25]:

(i) Physical characteristics such as metal, glass, ceramics, paper, plastic, rubber, food and kitchen waste, organic materials, combustibles and non-combustibles. Data on the physical of solid waste includes determination on the content of various ingredients of the solid waste. Specifically, the major physical properties of solid waste are based on its specific weight (per unit volume, kg/m^3) or density, particle size, moisture content percentage, shapes and colour. In terms of SWS planning, these are important aspects and considerations to determine the selection of solid waste treatment facilities, their operation process and the design of solid waste disposal facilities. For example, it is essential to determine solid waste density in order to design an efficient SWS, which requires the compaction of solid waste to an optimum density. Thus, waste density affects SWS planning in terms of landfill designs, solid waste storage, and the types of vehicle for solid waste collection and transportation. For example, a higher level of moisture content in the solid waste will lead to a higher level of collection and transport costs. As the thermal treatment of such waste consumes energy for water evaporation, this also affects the economic feasibility of solid waste treatment.

(ii) Chemical characteristics such as pH, total carbon, carbon/nitrogen ratio, nitrogen, phosphorus and potassium (N-P-K). It is essential to obtain information on the behaviour of solid waste and its chemical properties, including its biodegradable factor, to determine and plan the economical use of solid waste as fuel or for any other purposes. For example, the quantification of lipid content in solid waste is used to determine the process of energy recovery. As for paper and plastic, for instance, the chemical content in terms of the percentage of carbon, oxygen, nitrogen, sulfur and ash is quantified to plan for the biological conversion processes of solid waste. Meanwhile, for plastics in particular, since they are highly resistant to biodegradation with high heating value, they are more suitable for incineration. Therefore, the planning tasks of SWS can be performed in a more efficient manner based on the chemical characteristics of solid waste from the perspectives of conversion types and processes. This is important to plan for the allocation of budgets, technical skills, technology adoption types, manpower availability, etc. to execute the overall process of SWS.

(iii) Biological characteristics of solid waste components that are biodegradable in terms of their decomposition into gases by the action of microbes. This is an important input for the purpose of SWS planning to comply with related regulations and specific laws. Moreover, these characteristics can be used to regulate policies and management decisions on SWS in terms of what can be decomposed using specific methods and tools.

(iv) Solid waste quantity and new decomposition opportunities. Specifically, depending on locations, decomposition activities should be carefully planned to address the requirements, standards or norms of SWS to be followed. Based on this information, the transformation plan of solid waste can be performed in terms of composting or digestion. Biowaste, for instance, is categorized as the largest fraction of municipal solid waste and may impact negatively on the environment while degrading. It pollutes groundwater and surface water, and also promotes disease-carrying vectors. Therefore, the treatment of biowaste needs to be well-planned in a systematic manner using the most efficient and effective treatment technologies as it becomes public health threat.

In terms of solid waste composition, it comes from four different income groups, namely, low-, lower-middle-, upper-middle- and high-income countries [26]. In low-income countries, for instance, the organic fraction is significantly higher than the organic fraction in higher-income countries. Meanwhile, packaging waste, paper and plastic waste in high-income countries are higher than in low-income countries. Thus, the composition of solid waste may vary significantly depending on the location or area. This condition shows how solid waste composition affects its collection system. In SWS planning, based on solid waste composition, the waste collection vehicles are adapted according to the properties of solid waste, such as the percentage of moisture content and the density of solid waste (kg/m^3), which is due to the moisture [27]. This information is important to determine the type of vehicle for solid waste collection. If the selected compactor truck is unsuitable for a specific type of solid waste, it might not fulfil any purpose. This truck, for instance, is very useful in high-income country conditions where the solid waste is bulky and needs to be compacted to efficiently transport the waste.

9.3.3 Solid Waste Storage Facilities

In terms of the storage of solid waste, it is stored in a variety of containers. In planning for SWS, it is vital to determine the type of container for solid waste storage based on the types of solid waste stored, the amount and density of solid waste, time limits for storage, collection methods, transport types, the frequency of collection and local regulations or ordinances [26]. There are a variety of containers to store solid wastes, which depends on the types of containers according to the type of solid waste. For example, the wet solid waste and the dry waste need to be collected and stored separately. At a household level, for a small quantity or generators of solid waste, trash bins or wheeled carts can be used for its storage like garbage cans or recycling bins. Thus, the SWS plan should quantify that different types of waste materials will be stored in different types of containers, and that's how solid wastes are segregated. As for biomedical waste, it needs to be managed separately based on the rule of biomedical waste management which specified that the waste needs to be incinerated [28].

For a large quantity or large generators of solid waste, the storage of solid waste involves carts, dumpsters, big wire bins, open-top roll-off boxes, closed top compactor boxes and hauled containers that are normally used in shopping malls factories, warehouses etc. [29]. Solid waste will be taken to the recycling facility, where it will be separated. In this case, the right selection of these storage containers is becoming high-tech. Nowadays, due to rapid urbanization, it is essential to adopt information technology in SWM, such as having sensors incorporated into smart bins. Once the solid wastes achieve a certain level or percentage when it is almost full, the sensor will send a message to the collector that it is time to empty the bin.

9.3.4 Solid Waste Collection: Primary and Secondary

Solid waste collection is one of the basic and crucial services in urban areas. Once the waste is produced, it has to be collected. The objective is to collect the generated waste and then transport it to a treatment or recycling facility or a place for safe disposal. It contributes to a hygienic environment and quality of life and supports public health. Specifically, the services of municipal solid waste management must be provided to all. Thus, the services need to be carefully planned as health and hazard protection in order to avoid exposure to waste and vectors. In many cities of the developing world, collection coverage often remains below 50% [30]. It is expected that with a proper plan of SWS, good service of solid waste collection from waste generators can be provided for all the urban residents, which is regular, reliable and efficient. The service of solid waste collection includes street sweeping, cleaning of drains and cleansing of other public places. Collaboration of the residents is needed in the collection process of solid waste, and it should be as cost-effective as possible, besides ensuring impacts that are as little as possible in the environment.

Thus, with a variety of settlement situations and patterns, SWS planning needs to adapt the solution of solid waste collection and transportation to the local situation. For example, in areas with a variety of neighbourhoods, it is necessary to plan for a split of the solid waste collection into two stages [29]:

(i) Primary collection – provides service in the neighbourhood with smaller, simpler and more appropriate vehicles. Simple collection vehicles can be human-powered, animal-powered or even motorized. The collected waste is then disposed of at a transfer or collection point where it is stored in larger containers. The solid waste is then transported with a larger vehicle to a recycling, treatment or disposal facility. In planning for SWS, it is vital to determine the specific type of primary collection system to be used at a particular point of solid waste collection. The first type is a hauled system that is removed and replaced with an empty container, and the second type is a stationary system that is emptied and left at the same spot.

(ii) Secondary collection – this type of SWM service involves solid waste collection from a number of major collection areas and is taken from a transfer station to the final disposable site. It involves the transfer of solid waste into a larger transport vehicle prior to dumping at the disposal site for safe disposal. The collection process needs to be well-planned as it should be as cost-effective as possible and impact as little as possible on the environment. Therefore, in order to design and develop an ideal transfer station, it must best fit the intended area, location, station capacity, vehicle types, sources of collection, mode of unloading, segregation facility, etc.

Coordination and cooperation between both collection systems are very important in the planning of SWM. In order to plan and design a robust SWM that is cost-efficient, sustainable and effective, it is also important to forecast and determine a solid waste collection system based on the present and future growth of population, population density, coverage area, generation of solid waste per capita/day, generation of solid waste quantity per day and the chemical and physical characteristics of solid waste. Accordingly, in order to develop an efficient and cost-effective solid waste collection system, it is vital to synchronize the primary and secondary collection system by planning the route schedule of solid waste collection, machinery capacities, manpower requirements and the monitoring systems of machinery and vehicle performance [4].

9.3.5 Solid Waste Transportation

Solid waste transportation involves logistics in terms of moving the solid waste from one place or point to another and how to do it in a more efficient manner. Providing the efficient service of solid waste collection requires enough space to provide access for waste collection and for a truck to circulate. Costs of the solid waste collection are strongly influenced by choices in equipment and by time inefficiencies, like the idle time of collection trucks waiting to be loaded with solid waste.

Therefore, in order to handle the factors of inefficiencies in solid waste collection, the planning of its transportation system with the focus on how to improve on cost efficiency is based on [31]:

(i) Number of vehicles in terms of their type, size and capacities for the transportation of solid waste.
(ii) Number of vehicles used in a particular trip.
(iii) Number of trips made by a vehicle in a particular shift.
(iv) Quantity of solid waste transported by each vehicle.
(v) Quantity of solid waste transported by each shift.
(vi) Total quantity of solid waste transported each day.

In the SWS plan, it is vital to determine areas that can be serviced by trucks but might face the challenge of traffic congestion. Where space and road infrastructure permit, a direct collection service with a larger vehicle is possible. One of the important points in SWS planning for solid waste transportation is how to route collection vehicles because vehicle routing is one of the very critical components in terms of solid waste collection [32]. If the routing can be planned and performed properly, the number of vehicles required for collection can be optimized, besides the optimization of other resources, particularly in terms of financial aspects and budgets.

In terms of vehicle routing, its planning tasks can be performed by either manual technique using heuristic routing or computer-assisted routing using related software. A number of factors should be considered in designing the pickup routes of solid waste, such as loading and unloading time, volume per truck or compaction rating, travel time to transfer station and routing time from one point to another. Thus, there are a lot of decision variables in the design of solid waste transportation to meet various objectives. Minimizing the routing time, for instance, is imperative because time is money. If routing time can be minimized, this will also minimize the costs of solid waste operations. This is important to improve the performance of SWM from the productivity and quality perspectives.

In SWM services, it is important to consider various factors when developing the plan for solid waste transportation. Another factor is transport distance. This is the distance from when a truck is full after collection to the point of solid waste delivery as a disposal site or treatment facility. When the transport distance increases beyond 20 kilometres, it is advisable to consider a transfer station [33]. At this point, the solid waste is transferred from a smaller vehicle to a larger one. In this case, transporting a larger vehicle is more cost-efficient. Understanding this situation can assist in planning the loading quantity of solid waste that is suitable for a truck. In addition, once the solid waste arrives at the disposal point, the efficiency of the vehicle will also depend on how fast the truck can be uploaded. If the solid waste is manually unloaded, the productivity level of SWS is low, unlike unloading the truck body with hydraulic tipping. Therefore, in the planning of SWS, the selection of the vehicle used for solid waste collection is very important. There is a wide variety of vehicles available on the market. However, the choice of solid waste vehicle should be based on its suitability with the conditions of local context so that it can be operationalized, serviced and maintained easily. Additionally, other selection criteria that influence the choice of solid waste vehicle depends on its compatibility with street width and conditions, body volume (m^3) and payload, as these will depend on solid waste amounts and the length of vehicle routes and the number of trips per day that the vehicle should make [34]. The number of crew that is needed to operate the vehicle needs to be planned and estimated accordingly as it impacts the operational cost of the solid waste vehicle through the wages and salary allocations of the vehicle crews.

Lastly, there are also the costs of the vehicle and its operation, such as fuel consumption, maintenance, repairing, etc. All these aspects will reflect on the costs per ton of solid waste collected. Thus, it is important to compare other factors in terms of performance based on the cost per day of the vehicle and compare this with the solid waste per day that a vehicle can collect, which gives the cost per kilogram or ton of waste collected. In SWS planning, another aspect of increasing work productivity is on the plan to reduce the downtime of vehicles, in terms of the time that the vehicle is out of service, broke down and waiting to be repaired. In this condition, it is vital to rely on well-represented brands that ensure quality aspects in terms of the supply of vehicle spare parts, and the availability of skilled mechanics. Also, it is important to include and consider the procurement process, rules and regulations of vehicle acquisition which are clear and organized to assist in spare parts purchasing.

9.3.6 Waste Processing, Disposal, and Recycling

The activities of solid waste processing should be well-planned, and it requires a proper selection of processing techniques and equipment for each type of solid waste. This is important as it contributes towards achieving the best possible benefit and returns on technology investment in SWM. The processing of solid waste involves a procedure by which the physical properties of solid waste are transformed to make it best suited for technology adopted for its treatment. Various techniques that are used in solid waste processing are compaction, shredding, metal segregation, drying, etc. In order to derive a maximum economic value of solid waste processing, it is vital to determine solid waste that is suitable for further use that can be processed. Thus, the planning of solid waste processing as part of SWS needs to be incorporated in a SWM because solid waste processing improves the efficiency of SWM through more effective solid waste storage and transportation. Indeed, solid waste processing helps in achieving resource recovery, and it also improves the performance of solid waste treatment.

From the perspective of SWS planning, in order to plan an effective technique of solid waste processing, various aspects need to be considered and evaluated to ensure their applicability and suitability in the local situation of socio-economy, such as solid waste processing that serves as fuel or energy recovery or biogas production or compost production, etc. Before waste papers can be reused, for instance, they are usually baled to reduce the requirements of storage volume and transportation. This process intends to reduce haul costs at the disposal site where solid wastes are compacted to suit the capacity of available landfills. In order to plan and design an efficient operation of landfills that are well-engineered, sustainable and least costly with scientifically and environmentally sound, it can be based on various factors such as the type of land use, sufficient land area, land ownership, accessibility, infrastructure and surroundings.

As for the development of SWS planning, there is a good understanding of solid waste disposal. After performing solid waste treatment through various methods, there are some solid waste residuals that cannot be processed and have to be disposed of in landfills. There are a lot of problems that occur if solid waste throughout the country is improperly managed. Thus, it is important to ensure that anything that cannot be treated or if the proper treatment system of solid waste is not in place, at least the remaining solid waste can be placed in a good engineered landfill or dumpsite [29]. Disposal activity should be carefully planned or designed, which can be performed through controlling disease spread, odours and fires by sanitary landfill techniques and modern landfill techniques to control groundwater contamination and gas emissions. Therefore, landfill planning should also consider geotechnical aspects in a landfill design and be able to propose measures that can upgrade and improve the landfill. Developing a solid waste placement plan requires a system development where solid waste is unloaded in certain specific areas with roads inside the site that can be easily accessible. Development of green buffer with trees and bushes around the landfill site which acts as a vegetation filter reduces the visual impacts for residents and to reduce smell emissions. In planning for a better dumpsite or improving a better site, several measures can be taken, such as controlling access, improving waste placement and developing a green buffer.

Planning for recycling activities is another important component of the resource recovery aspect or triple R (reduce, reuse and recycle). Reduce and reuse are straightforward. Reduction is mainly achieved by policy interventions, and reuse of solid waste is normally practised as it is easy to understand and implement. Recycling paper and cardboard, glass, plastic, and metal is important as it preserves resources, prevents extraction of raw materials and protects natural habitats. These solid wastes are classified as dry recyclables, while wet recyclables consist of food waste and garden waste. Therefore, recycling activities should be carefully planned as they lead to energy savings as less energy is needed to produce from recycled materials than is needed for new products. With the introduction of the zero-waste concept as a solid waste strategy agenda, it means that zero waste is sent to landfills as recycling is practised at its maximum level. Obviously, the amount of energy saved by recycling varies greatly by material, with some recycled materials saving a huge amount of electricity. By understanding the properties of dry recyclables, it is viable to decide and plan the most beneficial recycling from an energy perspective, which is aluminium such as tins and cans [35].

9.3.7 Financial Aspects

In developing countries, due to their economic conditions, financial resources are scarce, and households have more needs to be fulfilled. This part focuses on the financial aspect of solid waste management, particularly on costs aspects (i.e. operation and maintenance) and SWM services, besides various methods to finance the investment costs of SWS. In SWM, in terms of the financial aspect, the main focus of SWM services in a city is the total costs of providing efficient and effective service, considering all the physical elements. In planning a SWS, it is important to assess the way to improve the cost-efficiency of solid waste services. There are various factors that contribute towards this goal, but the main consideration is the budgeting of SWS. In this case, the SWS budget refers to an estimation of revenue from different sources and expenses to finance the operations of SWS over a specified future period.

Financial planning for SWS involves the determination of capital costs (investment costs) and recurrence costs (operational and maintenance costs). For instance, operational costs for each type of solid waste need to be quantified accordingly, such as cost for collection/ton or volume, cost of transportation/ton or volume and cost of disposal/ton or volume. In addition, SWS planning should also consider externalities or hidden costs of no or poor solid waste management, like water contamination, public health impacts and impacts on tourism [36]. Therefore, expanding the coverage of waste collection or upgrading disposal facilities has costs implications in the future that need to be considered in budgeting activities. It is cheaper and more cost-effective to manage solid waste in an environmentally sound manner rather than to leaving the environment to be polluted and trying to clean it later. In terms of SWM service expenditures by the municipality or the organization providing the service of SWM, various sources are needed to cover the investment costs, such as from the central government, local government, private sector, donors, sponsors and financial institutions or a combination of this. The financing received can either be in terms of a grant or loan. Normally, central government funding is a very important source of investment financing for a SWS. However, local government or municipality can obtain funding from local sources of income such as property tax fees, parking fees, reserves, etc. Reserves, for instance, are based on current revenues, which are saved for financing future investment. Ideally, most costs for investment in SWS should be financed through the use of reserves. However, this is not always suitable for all types of capital financing for SWS, as it requires good and smart planning. Capital investment in SWM involves huge financial allocation for the acquisition and procurement of heavy machinery, equipment, technology, skills or know-how, etc. In this case, SWM planning should include the capital budgeting of these kinds of assets with the consideration of their useful life, depreciation, interest rates, cash in-flow and out-flow. Accordingly, local, state or central governments have other priorities than providing solid waste services with limited budgets. Thus, local municipalities often need borrowings or loans which are lent by local or international financial institutions. Some of the most active international institutions in the investment of SWS are the World Bank and the Asian Development Bank. In terms of the investment costs of SWM and the population of the countries receiving solid waste services, this investment is regarded as a tiny portion as not all countries are creditworthy. However, donor financing often comes with requirements such as achieving full costs recovery or using public-private partnership (PPP) or other private financing investment (PFI) financing [37]. In relation to the financial planning of SWM, the element of investment risks need to be considered as compared to the operation and maintenance costs of SWS, or other investments made into SWS technologies that are not suitable with the local conditions or solid waste characteristics. For example, investments in developing an incineration plant where SWM is predominantly wet or obtaining compactor trucks for an area with narrow unpaved streets with a high density of solid waste. In terms of assets utilization for SWS, the operational and maintenance costs of old solid waste vehicles are often several times higher than operating a more appropriate but maybe not high-tech and sophisticated vehicle.

9.3.8 *Tools and Manpower*

In the planning of SWS technology adoption, the supply and availability of resources, especially tools and manpower aspects, are crucial considerations in ensuring the monetary and non-monetary returns of technology implementation. Thus, it is important to ensure the adequate supply of tools and manpower in relation to SWS technology at both the national and local levels with the technical expertise necessary for SWM operations. This is to ensure that with adequate and well-trained staff, SWS technology adopted by external consultants, for instance, can be implemented efficiently and effectively. Besides manpower development in the technical aspects of SWS technology, it is also important to provide them with non-technical knowledge, particularly in the area of financial planning and management.

When the best option of SWS technology has been decided, it is necessary to plan it in detail, including the specification of equipment, location and manpower requirements. This includes the detailed plan of SWS technology implementation, which consists of the profile, types and size of the workforce, the location of solid waste for treatment and the types and specifications of SWS technology tools. In relation to this, work allocation can be performed by identifying the activities of each group of manpower. This is also related to the determination of manpower costs, vehicle operating costs, maintenance costs of machines etc. that are involved in the implementation of SWS technology.

From another perspective, technology adoption may lead to workforce deployment. Therefore, it is important to gain control of the number of employees. With the adoption of automation systems in solid waste collection, for instance, it may be necessary to freeze manpower recruitment as a part of manpower planning for SWS technology adoption. Freezing recruitment of permanent staff due to technology substitution can be an effective method of reducing workforce size, as labour costs can be shifted to cover the costs of technology adoptions [38]. The increase in solid waste generation can be covered by engaging private sector operators or by employing temporary staff. In the case of SWS technology adoption, planning the optimum size of manpower requires the determination of productivity level for each category of workforce who are involved in a SWM.

9.4 Managing and Sustaining SWM Plan Implementation

The implementation phase of the SWS plan is worthless if there is no implementation phase. Thus, it is important to move from planning to implementation, to regularly revise and update the strategic plan of SWM and to monitor the performance of SWM systems. Policies and strategies of SWM and their translation into legislation are fundamentals to solid waste governance and management in terms of their plan implementation [39]. Not only the legislation of SWM must be clear, but also it must be enforceable. All policies and strategies start with a clear definition of goals that are based on certain principles related to SWM. Instruments of regulation and enforcement, social mobilization and economic leverage are needed in managing and sustaining the implementation of SWM plan. This must be supported with a clear definition of the role and responsibilities of all stakeholders and agencies, for instance, the different levels in government agencies, based on their roles and responsibilities related to SWM.

9.4.1 Solid Waste Policy Guideline

The guideline of implementing solid waste policy consists of important guidelines on SWM administration, enforcement, solid waste processing concessions, hospital waste and handling of legal matters essential for day-to-day management. Managing and sustaining SWM plan implementation involves the governance aspects of SWM in terms of the formulation of policies and their translation into legislation and regulations overarching the actions and activities of SWM. These become the fundamentals or backbone of SWM. The policies of SWS are based on the system's goals and guiding principles in ensuring public health, environmental protection and recovery of resource value from discarded products and waste materials [40]. The guiding principles in SWS policies may include solid waste prevention, precautionary principle, service coverage and cost recovery. It is not compulsory for them to be compatible with each other as they are based on priorities settings based on the concepts of the solid waste hierarchy. The hierarchy, for instance, highlights the most preferred action at the top of avoiding waste, then down to the least preferred action of solid waste disposal in landfills. Figure 9.4 presents the flow of SWM plan implementation.

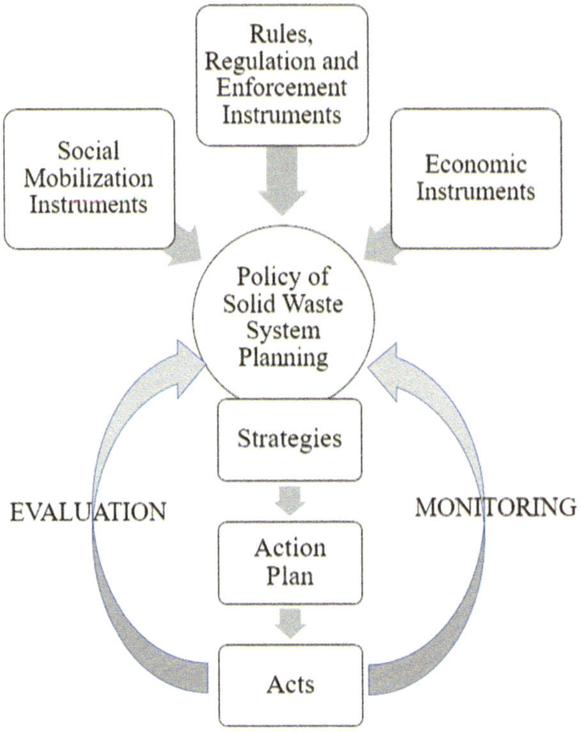

Fig. 9.4 Plan implementation of solid waste management

The policies and strategies of SWS are based on their translation into legislation. Laws and regulations and their enforcement are just one way to support the implementation of SWS policies and strategies through instructions and commands [41]. Other policy instruments that can help support SWS plan implementation are the instruments of social mobilization such as education, awareness, communication, training, empowerment and engagement to change perception and behaviour. In addition, economic instruments can be used to achieve certain SWM policy goals through incentives, rewards and disincentives. Depending on the goal of SWS, a mix of these instruments can be used in implementing the SWS plan. All these SWS policy instruments are prepared or implemented by various government agencies.

Legislation and regulations on SWS can be adopted to give effect to any aspects of SWM policy, as the bottom line is set to protect human health and the environment. There is also an alternative regulation either to regulate the specification that can only define any accepted technologies on solid waste treatment or define specific ways on how a solid waste treatment facility must be constructed or operated. The other approach is to regulate by a function that prescribes standards to which a solid waste technology has to perform. In the case of landfill design and operation, for instance, a classification system was developed where landfills can be classified into different categories [42]. Another important role of legislation and regulation is a clear allocation of authority related to SWS. The legislation established duties and authorities of appropriate government agencies at all levels, starting from the various national ministries down to provincial institutions and municipal authorities. For example, there is a provision specifying that certain areas need to be serviced with solid waste collection. There is also a provision on the way of managing a particular waste stream, as a certain landfill, for instance, is restricted to the disposal of non-biodegradable solid waste while biodegradable solid waste will be processed in specified treatment facilities [43]. The most important focus in SWS plan implementation is that regulation must also be enforced, which is considered very challenging in certain countries. Another common example of the role of legislation enforcement is to ban or prohibit the use of certain substances in materials, like a plastic bag ban. There are also countries that have committed to some international aspirations like sustainable development goals (SDGs) for greenhouse gas emission reduction and thus need to implement a strategy to comply with this international obligation.

9.4.2 Strategy of Solid Waste System

As society becomes a more resource-efficient society, it is important to improve waste management beyond collection and disposal activities in a SWS. Literally, reducing the amount of waste that needs to be disposed of by reuse and recycling and improving treatment and disposal using environmental technologies are the best solutions in a SWM. However, the best solution to handle this complex issue is to develop a SWS strategy in order to bring positive changes in the SWM systems of areas, cities and countries.

Generally, SWS strategy consists of five (5) goals:

(i) Accelerate waste collection and 3Rs activities.
(ii) Proper management of household, industrial and other waste.
(iii) Stop open burning and disposal.
(iv) Awareness, capacity building and empowerment.
(v) Regular monitoring and controlling.

Once the SWS plan has included the parameters of its implementation, the next task is to generate and develop the actual SWM strategy. It is important to consider each issue identified and related to the SWM and what actions can be taken to address it based on the goals set. There are several questions that need to be considered, such as:

(i) How is the issues of solid waste are determined and managed at present?
(ii) What are the opportunities for solid waste prevention, recovery, and recycling?
(iii) What policy options that are available to address these issues?

SWS strategy should highlight the benefit of better SWM and the implementation process of the SWS plan. These should include waste reduction, lowering operating costs, resources efficiency, poverty reduction, new economic opportunities, reduced environmental impacts and improved health. Adequate finances and other resources for the strategy development of SWS are important to ensure the implementation of SWM plans is not interrupted [44]. The capacities necessary to implement the SWS system need to be considered as well. A country or state with limited capacities available may choose to limit the scope coverage of solid waste collection. Additionally, it is important to have a clear timeline for SWS strategy adoption, implementation and completion with the inclusions of milestones over the duration of 1, 3, 5, 8 and 10 years.

In regards to SWS planning, the aspect of SWM policy coherence is important. The potential interaction of SWM policy and other policies need to be considered and developed. Positive relations and interactions are to be encouraged, and issues or conflicts will need to be handled and resolved, respectively. It is important to consider whether the SWM strategy would be linked to existing plans of municipal waste management, public health, environment and urban development. SWS plans should determine priorities as to how and when to deal with a different kind of solid waste. A municipality, for instance, may have to urgently address or handle waste types or streams that are especially problematic. Most countries or areas will have several SWM priorities which need urgent attention and action, while other solid waste streams will be addressed later. Municipal solid waste management is considered a priority for any city, state or country because the provision of these services is important to ensure their smooth operations [45]. This is due to the condition that the municipal waste issue is within the mandate of local municipalities as the leading agency in the formulation of SWM strategy. Meanwhile, dealing with hazardous and industrial waste requires the corporation of more governmental agencies, making SWS implementation more complicated.

In the implementation of the SWS plan, strategic actions must be identified to determine choices about the management of each solid waste issue that should be carefully analyzed. In this situation, the aim is to ensure compliance with regulations and to optimize the use of limited resources, besides managing solid waste in better ways. This intends to avoid the generation of solid waste by considering it as a resource waiting to be recovered and reuse. This strategic action should be guided by a clear solid waste hierarchy which indicates an order of reference for actions to reduce and manage solid waste.

Once the action has been identified, an action plan should be prepared for each solid waste stream or issue with accurate budgets and responsibilities for implementation. Necessary policy instruments should be identified as well in these action plans that will need to be integrated into existing SWM related rules, regulations and laws through appropriate amendments or new regulations and laws accordingly [46]. Therefore, it is important to identify initial actions that can be easily implemented so that the partners can start working together and achieve early results.

SWS planning requires a robust strategy as a key component in SWM to reduce environmental impact and improve socio-economic conditions. The strategy is needed as a holistic system of SWM that acts as an organizational structure and procedures for the continuous improvement of environmental performance. Based on Fig. 9.4, a basic strategy in SWS that focuses on solid waste reduction can be generated based on SWM policy, for example, to reduce solid waste by 10–15%. In the planning of SWS, it is important to determine the target duration of solid waste reduction, for example, on a yearly basis and the measurement method of solid waste reduction. This is followed by the stages of solid waste implementation and operation, checking, monitoring, follow-up and review. Having the hierarchy or sequence of SWS gives an idea and guidance in SWM. The basic level in solid waste minimization or reduction is through prevention measures, followed by reuse, recycle, energy recovery by burning or incineration and disposal.

9.4.3 Cost-Benefit Analysis

Cost-benefit analysis is often a methodology used to determine and analyze the costs and benefits of managing and sustaining SWM plan implementation. The analysis consists of the estimation of capital costs and operating costs in a SWM. Operation and maintenance costs occur at every step of a SWM and can represent up to 60–70% of the total costs, which comprises labour, fuel, energy, customer care, revenue collection, etc. [34]. For solid waste collection, for instance, the cost of performing this task needs to be quantified, which includes operational and maintenance costs for solid waste collection, sanitary landfill, composting, etc. The money required to cover this cost is through revenue generations which mainly comes from users or public sources. As for public sources, the revenue comes from the central or national government (through subsidies or grants), local government (property taxes, municipal fees, parking fees) and users, which can be the residents themselves or private companies or businesses. Users can be charged directly for the service of SWM, and this is called direct charging on a weekly, monthly or yearly basis. Direct charging is advantageous as it raises public awareness about the real cost of solid waste service, and they tend to make the service accountable [47]. User charges can either be based on a flat rate where everybody pays the same or based on the quantity of waste discarded, which may serve as incentives for weights prevention. This is also called "pay as you throw". Generally, an increase in fees needs to be gradual and connected to the real improvements in the SWM service as well that are actually perceived as a good service by users. In reality, the collection of the fee of SWM services is comparatively low as not everyone is willing to pay or able to pay higher or significant amount of fees. In this case, the aspects of affordability, cost recovery and willingness to pay are considered as important factors in the cost/benefit analysis of SWM in managing and sustaining SWM plan implementation.

It is also important to determine how much of the operational and maintenance costs should be covered by user charges in terms of a certain percentage of the SWM collection service or disposal or treatment. These decisions should be made based on the availability of another revenue source to cover the costs of SWM operations and maintenance based on the affordability of the citizens. Public financing can be performed in SWM, such as profit-sharing in public, private partnership (PPP) and reselling recyclables.

9.4.4 Implementation Schedule

The implementation schedule of the SWS plan includes its related monitoring and evaluation activities. SWS plan should be implemented as planned by the responsible parties. Monitoring the targets and indicators is crucial. Based on the results of this monitoring, adjustments in terms of improvements and suitability need to be performed regularly to ensure the plan is effectively leading and achieving the SWM goals.

Unplanned SWM leads to public health deficiencies, low coverage of collection service, low managerial and organizational performance of the solid waste service, no financial margin to allow progressive improvements in service quality and environmental protection deficiencies, and the waste collection department has only capacity to solve daily problems rather than monthly and yearly problems.

9.5 Strategic Planning of Solid Waste Technology

SWM plays an important role in our daily life, and it requires strategic planning in order to provide innovative solid waste management services to residents, businesses and industries, which is efficient and effective that meets users' expectations. This requires technology adoption and implementation based on a holistic planning to create environmental sustainability, socio-economic developments, promoting diversion and maintaining a clean area and country.

Generally, SWM services are also responsible for transporting, processing, composting and disposing of solid waste using various technologies. The adoption of SWS technologies supports SWM activities by planning and implementing SWM strategies to provide the necessary tools to reduce solid waste such as equipment, vehicles, facilities, etc. Therefore, the strategic planning of solid waste technology is a mechanism to keep an eye on the industry trends. This is important to determine the way SWM programs can be enhanced, what new SWS technology is available and when is a new policy needed or bylaw required. Obsolete, abandoned, non-functional and mismanaged technologies are common in many developing and underdeveloped countries. Thus, careful planning on SWM strategy prepares for the future needs of a city, state or country. There are also solid-waste management technologies that are failed in their implementation and maintenance due to a number of reasons. Many of them could be avoided by having proper planning in terms of the assessment on the suitability of SWS technology with the local characteristics and needs of SWS of each location or area.

Monitoring solid waste production is a primary step in any solid waste management strategy, which involves high-end waste monitoring technologies that have been developed to enhance the performance of SWM [48]. These technologies include the adoption of automatic waste collection systems based on pneumatic conveying technology, radio-frequency identification (RFID), ultrasonic sensors, geographic information system (GIS) and international system for mobile/general packet radio service (GSM/GPRS) [49]. However, these technologies of SWS involve high costs and risks. In this case, it is important to generate enough revenue from users, fees, the sale of recyclables and industry funding to cover operating costs and capital investment costs.

The implementation plan of SWS technology should consist of:

(i) Planning SWS technology initiative, which consists of action plans for SWS technology to reduce solid waste generation and to handle the process of solid waste disposal. It starts from having an overall vision of implementing SWS technology in terms of its methods and strategies in effective ways for better environmental performance.
(ii) Determining SWS technology needs.
(iii) Selecting SWS technology solutions.
(iv) Implementing SWS technology needs assessment.
(v) Safeguarding SWS technology adoptions.
(vi) Maintaining SWS technology implementations.
(vii) Training for SWS technology advancements.
(viii) Integrating SWS technology innovations and developments.

As SWS focuses on the 3Rs—reduce, reuse and recycle—another activity, namely, recovery, is also important in SWS. For example, recovering additional benefits from solid waste can be obtained by producing energy from solid waste. SWS needs a long-term waste management strategy to handle waste diversion processing and the need for solid waste disposal for the next 20 years. Thus, all options and best practices for new and emerging waste diversion and disposal methods using related technology will be evaluated and considered. The goal of strategic planning of SWM technology is to find solutions that are cost-effective, environmentally sustainable and socially acceptable.

The strategy development of SWM technology consists of five aspects that involve the determination of environmental requirements, socio-economic values, technical implementation, technology procurement and partnerships.

9.5.1 Environmental Requirement

In the planning of the SWM strategy, it is important to evaluate the environmental requirements of SWS. As the generation of solid waste is increasing at a rapid rate, especially in urban settlements, this situation requires more focus to be given to technology adoption based on the environmental requirements of SWS. Evaluation of alternate SWS technologies serves as guidance in SWM strategic plan. In this case, the choice of SWS technology must best fit the environmental policy [50]. This is to ensure that the development of a specific plan on SWS technology adoption complies with the environmental conditions and requirements in a SWS and that environmental risks are properly managed.

In SWS technology adoption, it is vital to plan for environmental impact assessment and consent conditions or pollution control approvals. Authorities and organizations who are responsible for SWM must be aware of any statutory obligations related to environmental impact assessment. In determining any environmental aspects and impacts, the adoption of any SWM technologies needs to consider if the technologies will have any effect on air, water or land [51]. The technology adoption plan should also consider:

(i) Type of emission to air from the use of SWM technology such as noise, exhaust fumes or smoke.
(ii) Quantity of water use in the use of SWM technology and its discharge such as to rivers, ponds, streams, and dams.
(iii) Impacts on soil, biodiversity, wildlife and plants.
(iv) Level of energy consumption in the use of SWM technology.

These issues should be anticipated, and action plans need to be developed. In addition, all of these must be documented and communicated in terms of determining and preparing:

(i) Environmental risks are involved in the use of SWM technology.
(ii) The plan to manage these environmental risks.
(iii) Leaders and team members are responsible for ensuring the SWM technology plans are followed.
(iv) Environmental management plan addressing the requirements of environmental impacts.
(v) The specific training plan that is related to SWM technology implementation.
(vi) If subcontractors are involved, notify them of environmental requirements based on an environmental management plan.

Conclusively, in the planning of SWS technology, it is important to evaluate how this technology performs for every attribute in the environmental context. This represents the environmental performance of SWS technologies based on their high contribution to SWM.

9.5.2 Socio-Economic Value

The acquisition and adoption of SWS technology involve the consideration of economic factors, scientific necessities, technical practicalities and human and social considerations, among all other factors. In the planning of SWS technology, it is imperative to choose the best alternative that optimizes the total socio-economic value of SWS technology adoption with the considerations of sustainability measures [52]. These values are illustrated in Fig. 9.5.

Another important aspect is the determination of how compatible are the SWS technologies are with the goals and objectives of SWM. Therefore, in planning SWS technology, the socio-economic values can be determined based on the performance measures of SWS technology adoptions either quantitatively or qualitatively. For example, with the adoption of SWS technology, high technical reliability or performance can be achieved by reducing the days of downtime per year. Meanwhile,

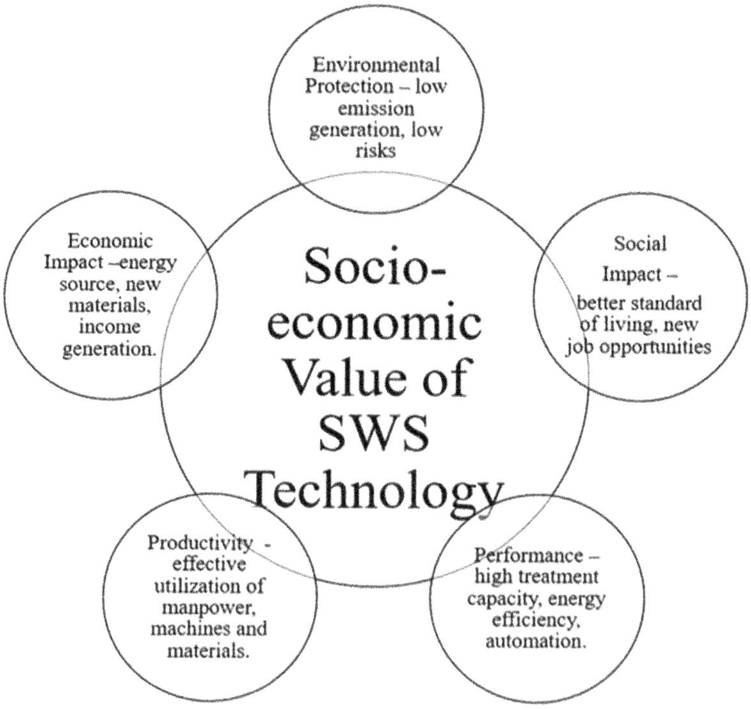

Fig. 9.5 Socio-economic value of SWS technology

high job creation can be achieved through the number of workers that are required for each ton of solid waste per day treated using a particular SWS technology. High-capacity treatment, for instance, could be achieved based on a higher percentage of solid waste collected that the technology can treat. In terms of economic impacts, it can be evaluated from a higher income-expenditure ratio [53]. Therefore, from the perspective of socio-economic, SWS planning requires the comparisons of available SWS technologies in terms of its return on investment, maintenance, space, time and labour requirements to obtain an insight into which technology could suit the needs and constraints of an area.

Generally, SWM is limited to waste collection and disposal with the aim of collecting as much waste as possible to keep the neighbourhood, streets and cities clean. To many countries, especially developing and underdeveloped countries, there never seem to have enough capacity, resources and infrastructure to deal with the growing amount of solid waste. Although solid waste collection keeps up with the growing amount of waste, the practice of SWS is just transferring solid waste collection from the streets to landfill sites. In the strategic planning of solid waste technology, solid waste can be considered a valuable resource [54]. For instance, if recyclables are separated at home, they can be sold as income generation while keeping our environment clean. In a wider perspective, to best consider socio-economic value, recycling can be a strategic action with technology adoption as a solid waste solution. In SWM, a fundamental step in relation to developing socio-economic value is the generation of a recycling strategy. The strategy is a road map that will guide the process of creating appropriate solutions based on stakeholders' involvements in the local context for sustainable developments.

9.5.3 Technical Implementation

In order to ensure SWS technology can be implemented efficiently and effectively, it is important to understand the overall technology know-how, its technical and non-technical requirements [55]. Figure 9.6 illustrates sectoral involvement in the technical implementation of SWM technology.

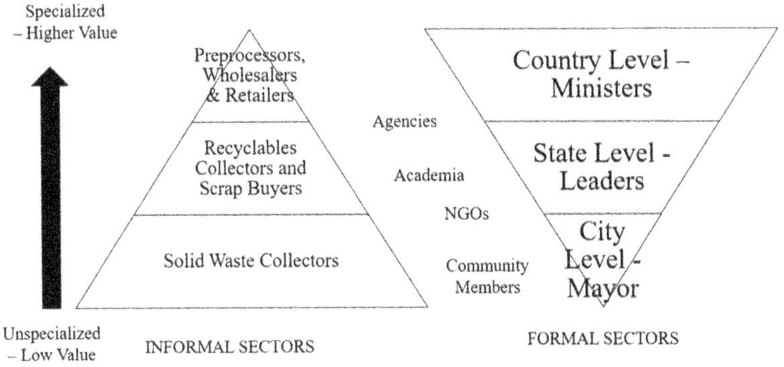

Fig. 9.6 Sectoral involvement: technical implementation of SWM technology

Before any adoptions of SWM technology can be implemented, it is important to determine and prepare the following:

(i) Identification of the person in charge of the SWS technology plan or the technology adoption project, the contractor's profiles or identity and ensuring that it is followed. In this case, if there is more than one contractor, the details of the principal contractor should be included.
(ii) The SWS technology plan accessibility so that it can be referred to by the site staff, implemented and updated, especially when there are site circumstances changes.
(iii) Identification of different types of solid waste and its quantities that will be produced in an area throughout a particular duration transported for disposal.
(iv) Identification of solid waste systems changes in terms of their types and materials specifications.
(v) Regular or periodic environmental inspections or assessment plans pertaining to the implementation of SWM technology.
(vi) Training plan for internal and external staff to ensure that everyone understands the requirements and process of SWS technology adoption.
(vii) Environment incident form if there is an environmental incident.
(viii) Internal audit plan to ensure that all environmental and other statutory requirements are met satisfactorily on a regular basis to ensure the work has progressed in accordance with the plan.
(ix) Monitoring plan on SWS technology implementation to ensure that it is being followed and updated.

For example, it is vital to verify that biowaste treatment technologies are suitable according to the circumstances of an area as they consist of various default biowaste treatment technologies such as windrow composting, in-vessel composting, vermicomposting, anaerobic digestion and slow pyrolysis, as well as instructions on how to incorporate new technologies or how to discard some of them.

9.5.4 Technology Procurement

In the technology planning of SWS, it is important to point out the need for clear objectives of SWS technology adoption to determine the criteria of tender documents for technology procurements. This is a process that includes the science of rational decision-making. In the adoption of SWS technology, the first step is to identify the needs of a particular SWS technology [56]. This is related to the decision-making of the most appropriate solid waste technology for a given context which involves information gathering on the type of solid waste that requires a particular SWS technology, its amount, its collection frequency, its quantity and staff involved in managing this type of solid waste. In this case, it is important to consider sustainability elements such as environmental impacts, economic conditions and society at the beginning of the technology procurement process [57]. In addition, there are two major considerations, namely, the boundary of SWS (solid waste amount) and the profile of solid waste (solid waste characterization). Based on these considerations, a suitable SWS technology can be identified based on the availability of resources to ensure that the technology procurement is really necessary and what the key requirements and performance are. From here, justifications can be made if there are also other alternatives available and whether SWS technology procurement can take place. For example, there are questions that need to be clarified, such as, can a machine be leased or refurbished or shared rather than bought? What are the problems that the SWS technology should remediate?

Secondly, it is to define the sourcing strategy of SWS technology. In the procurement of SWS technology, by assuming that the procurement is really required, it is important to explore the marketplace to obtain information on the sustainability of SWS technology throughout a life cycle [58]. In this case, it is not only about the use and needs of a particular SWS technology but also in terms of its specifications, performance, endurance, maintenance and flexibility from its beginning up until its disposal. This is to ensure that common and relevant objective that a particular SWS technology should fulfil.

Thirdly, it is to identify the suppliers of SWS technology so that the supplier can meet the SWS technology needs. Sustainability needs, for instance, should be built throughout the procurement process of SWS needs in terms of specific competencies and outcomes in tender scorings, costings and to secure in an agreement, for example, requirements on a solution of more energy-efficient technology or with high recycling content or better waste performance.

Fourth is in terms of the award stage of technology procurement. At this stage, it is important to ensure that the supplier is really receptive. This provides an opportunity to gain support for any requirements or specific commitments that could not be delivered through the tender. Evaluation on suppliers is required for any improvement targets so that what the supplier actually commits to deliver as a part of the tender will be realized in practice.

Fifth is the practice of performance and relationships throughout the procurement process of SWS technology. In this case, all specifications and requirements of SWS technology with its sustainability indicators are measured besides other business performance indicators such as cost, quality, time, technicalities and service requirements.

The final stage is the review of SWS technology procurement. This is a part of continuous improvement, which is the cornerstone of good practice in the process of technology procurement. As SWM and its related technologies are considered as a rapidly evolving field, it is important to capture a good practice of ongoing or continuous improvement.

9.5.5 Partnership

In the strategic planning of solid waste technology, the management team of a SWS should identify key leaders to develop this strategic plan. This is important as they will engage with the key government and non-government agencies dealing with solid waste to form a partnership [59]. The key leaders of SWS will be given responsibilities for driving the SWM strategy process with external parties who are involved in the adoption of SWS technology. This partnership should have the administrative power, financial resources or budgets and capable team members or workforce to provide focus and follow through with a high level of commitments and unity of purpose in the planning and implementation of SWS strategies [60]. This is important in ensuring the coordination of parties involved in the partnership of solid waste technology planning. A group discussion with all involved in the partnership is desired in order to arrive at a maximum level of consensus on which and how SWS technology should be implemented. Normally, the aspects of SWM are the responsibility of various government agencies, so it is important to have internal coordination among all of them in order to collaborate and trade with technology developers in a partnership. This will ensure all relevant parties are aligned with and actively supporting the development of this strategy.

The major focus is to identify all those stakeholders that could participate and contribute towards the implementation of new SWS technology based on what skills and resources, such as financial and technical aspects that the stakeholders should contribute to the planning and implementation of new SWS technology. In the adoption of an organic waste treatment technology, for instance, some possible stakeholders that may support the technical implementation of SWS technology are organic waste generators, community-level authorities, NGOs, SWM service providers etc., as they will create a stakeholder cluster [61]. A wide representation of stakeholders in a SWM helps in the implementation of a particular solid waste technology based on their role, supports, expertise and networking. Therefore, in SWS planning, the activities and mandate of each stakeholder should be identified, including strategies to be implemented to ensure their appropriate involvement and active participation.

9.6 Factors Influencing Solid Waste Systems Planning

One of the most important aspects in the development of a solid waste system plan is that the planning function must fully understand and comprehend the main technical and non-technical issues of solid waste systems. The planning function must not only focus on the advantages and benefits of the SWS but also deal with the system's limitations relative to current systems and other new system developments.

First is to determine the service coverage of solid waste collection. In some areas, this service only serves a limited part or area of the urban population due to a lack of financial resources. When there are insufficient funds from the central government or state government due to limited budgets and inadequate fees charged, these situations lead to inefficient solid waste organizations and poor waste management capacity. The reality of inadequate SWM disposal is also due to resources constraints, particularly on the financial aspects. Therefore, SWS planning serves as a foundation to synergize the availability of financial resources with other resources such as manpower with well-trained personnel, physical resources like machineries and materials to ensure the efficiency of solid waste disposal activities.

Second, there are some economic factors that affect the potential of resources recovery for recycling due to the high costs of waste or materials separation, their purity, quantity and location. In this case, the reuse of organic waste materials is gaining popularity because of its benefits in reducing the operational costs of disposal facilities, reducing the environmental impact of disposal sites, methane problems and leachate pollutions, besides the increase of sites life span.

Third, there is a need to develop a well-controlled MSW collection and disposal activity that makes more economical use of the available space, avoiding unpleasant and hazardous smoke from slow-burning fires. This is crucial when designing a waste collection for the protection of human and environmental health. Improvements in solid waste collection can be achieved at different levels with optimization to achieve cost efficiency.

Fourth, as there is an increasing amount of generated waste produced by the rapid growing cities, solid waste management becomes more challenging when the low-income population remains without having proper waste collection services and without the adoption of appropriate solid waste technologies. In addition, rapid urbanization developments with suburban settlements and housing demands require more organized logistic systems to be implemented in these areas.

Fifth, the location of landfills is another factor that influenced SWS planning. The landfill's location, which is significantly far away from the central areas of solid waste collection, has greater cost implications in terms of higher transfer costs, extra wages for workers and additional maintenance costs for vehicles. Therefore, besides location, landfills need to be planned in terms of their design and operation systems if they require upgrading to increase SWS efficiency and effectiveness.

Sixth, careful planning for the disposal of hazardous solid waste and healthcare wastes is required to ensure their safe handling as a part of SWS. Management of these kinds of solid waste involves legislation enforcement and technology adoptions. Special attention is needed in this case because of their physical, chemical and biological nature that require special incinerators or other solid waste treatment technologies which incur higher operating costs of the plants, thus involving funding and financial instruments.

9.7 A Case Study: Cost-Benefit Analysis for the Adoption of Solid Waste System (SWS) Technology for Sustainable Development in Malaysia

Sustainability should not be seen as a separate element in solid waste system (SWS), and sustainability consideration should be formed as a part of the overall decision-making and performance monitoring process in solid waste management (SWM). Sustainability is an agenda where collaborations in various fields are of value.

Owing to the depletion of natural resources and concerns about climate change, solid waste has become a major environmental problem. These situations have paved the way for the solid waste management field to meet the aspirations of the United Nation's sustainable development goals (SDGs), allowing for resource optimizations and recoveries through a more dynamic and holistic way of managing solid waste. For instance, many countries throughout the world have been deriving benefits through power generation from wastes using technologies such as anaerobic digestion and designed landfills [62–64]. As a result, after the classification and quantification of wastes from different sources, the implementation of a sustainable waste management strategy has the potential to yield economic benefits as well as resource conservation.

Through the use of technologies such as designed landfills for methane gas processing, energy recovery through incineration, pyrolysis, gasification, and anaerobic digestion, manure production through organic waste composting and material recovery through recycling, solid waste management has the potential to significantly contribute to a country's economic growth. There have been many debates about how waste management practices have progressed in order to protect the environment and achieve the goal of "sustainability". Starting from solid waste disposal and the end-of-pipe treatment, these problems progressed to waste management, waste minimization, cleaner production and, finally, zero-emission systems. This path is also used to develop sustainable waste management strategies. These solutions begin with environmentally sound waste management, followed by the implementation of zero-emission industrial ecosystems and other automated procedures of solid waste or trash sorting or disposal by using artificial intelligence, robotics, automated solid waste sorter, RFID tags to identify solid waste, Internet of Things (IoT) sensors, computer vision programs in solid waste collections, etc. for smart and sustainable solid waste management.

Thus, in the decision-making of SWM technology adoption, cost-benefit analysis (CBA) is a management tool that can be used to explore the potential implementation of any solid waste system technology from a sustainability perspective. For example, in Malaysia, the Klang River has been categorized as the 50th most polluted river. It is considerably polluted because of deep siltation caused by human waste from informal settlers of the riverbank and even some businesses without septic tanks or sewage treatment plants, as well as soil brought by mudflows from the mountains [65]. Starting from October 2019, the active Interceptor is currently removing debris from the Klang River, a highly polluted Malaysian river that runs through the capital Kuala Lumpur and its surrounding area [66]. Dutch non-profit organization The Ocean Cleanup has launched the Interceptor, an autonomous system for collecting plastic pollution from rivers before it reaches the sea. According to The Ocean Cleanup, each Interceptor can extract 50,000 kilograms of trash from a river each day, going up to 100,000 kilograms "under optimized conditions". Interceptors are secured to the riverbed and are made up of floating barriers connected to processing plants that look like barges. The barriers direct plastic waste into the plant's mouth, which is powered by solar panels and runs without human operators. A conveyor belt removes the waste from the water and transports it to a shuttle, which dumps it into containers on a separate barge docked below. The waste then was sent to local waste management facilities.

The example of a CBA presented here incorporates various insights from the academic and industry literature on the adoption and the performance of SWS technologies, with the considerations of known gaps, emerging opportunities and possible future developments. This case study briefly presents the result of CBA that was performed qualitatively based on a series of literature reviews for adopting an available existing sustainable SWS technology in Malaysia, rather than inventing or developing our own or local SWS technology. CBA is an assessment tool that can be used to justify either to adapt and adopt an existing SWS technology or to develop a totally new SWS technology. In this case, a CBA was conducted by investigating relevant factors based on these two options from economic, environmental, social and technology perspectives. It is important to highlight and assume that the quantitative data of the actual cost of licencing or developing a SWS technology is unknown. However, many effects, in terms of fees, charges, budgets etc., besides time savings, resources utilization, returns, etc. associated with a SWS technology can be based on the market price or conditions, besides can be valued by individuals or society as a whole and, as such, should be included in a CBA.

Based on various analyses on SWS technologies, there is a possibility to determine and select the most economical and available SWS technology that can be adapted and adopted for sustainable development in the Malaysian construction industry. This requires the quantification and monetization of positive effects, and this can be performed by assessing the willingness to pay for or invest in SWS technology based on the monetary and non-monetary benefits of adapting and adopting such technology. Moreover, this is important if the adoption of a SWS technology needs to be implemented within a short period of time. Table 9.1 summarizes some of the possible effects of adapting and adopting a SWS technology based on the potential sources of monetization values and setbacks for them.

Based on Table 9.1, from the economic perspective, it shows that the potential costs of adopting an existing SWS technology are higher than the potential benefits. This is valid and reliable due to the acquisition of a SWS technology which is already established in the developed countries that needs to be procured and transferred to our country. In this case, the consideration of higher cost is significant as risk elements have to be well quantified and justified. Among the highest costs are technology licencing fees, exchange rates for an imported SWS technology and operational costs. Although these costs outweigh the potential benefits of SWS technology adoptions, the savings of time and effort will be significant. In addition, by adopting an established SWS technology, the management of the overall SWS can be performed in a more efficient and effective way.

From the perspective of the environmental aspect, it is discovered that the potential costs and benefits of adopting an existing SWS technology are equal. It can be concluded that benefits or returns from investing in an existing SWS technology will contribute towards the environmental well-being of the country satisfactorily due to an established mechanism on the assessment of environmental aspects like the Environmental Impact Assessment (EIA). This is substantial as an established

Table 9.1 Key items for a cost-benefit analysis of adapting and adopting a SWS technology

Sustainability aspects:	Cost-benefit aspects: Potential cost (C)	Potential benefit (B)
Economic (E)	EC1: Feasibility studies	EB1: Higher business or organizational values and image
	EC2: Consultation fees	EB2: Enhanced marketability of the SWS technology
	EC3: Cost of project negotiations	EB3: Provide more cost-effective SWS solutions
	EC4: The price of goods or equipment associated with the SWS technology	EB4: Save time and effort
	EC5: Increase in operating costs	EB5: Higher productivity in SWS operations
	EC6: Interest and exchange rates	EB6: Higher return on investment in a long run
	EC7: International trade/monetary issues on importation	
	EC8: Taxation specific to product/services	
	EC9: Higher prices for imported SWS technology	
Environmental (E)	EC1: Assessment on climate issues	EB1: Protection of the environment
	EC2: Specific environmental disputes	EB2: Reducing climate change and emissions
	EC3: Cost of the waste management system	EB3: Effective solid waste reduction
	EC4: Charge on environment assessment	EB4: Better control on the impact of environmental changes
	EC5: Pollution control costs	EB5: Management of environmental risk
	EC6: Consultation on environmental legislation	EB6: Complementary with other mainstream standards
	EC7: Maintenance for the environmental protection measure	EB7: Support from experienced consultants
Social (S)	SC1: Staffing costs (wages, training, etc.)	SB1: Greater health and Well-being
	SC2: The cost of the resources spent on a SWS technology acquisition	SB2: Improved learning and healing environments
	SC3: The cost of adjusting an established routine	SB3: Moral imperative of being green
	SC4: Law changes affecting social factors	SB4: Higher quality of life and standard of living
		SB5: Better lifestyle and trends
		SB6: Empowerment process for staff

(continued)

Table 9.1 (continued)

	Cost-benefit aspects:	
Sustainability aspects:	Potential cost (C)	Potential benefit (B)
Technology (T)	TC1: Cost for technology transfer and legislation	TB1: Use of information and advanced technologies
	TC2: Technology access and licensing	TB2: Knowledge from technology adoption and know-how
	TC3: Intellectual property issues	TB3: Keep on track with technology trends
	TC4: Certification mechanisms/ technology	TB4: Source of technology know-how for further innovations
	TC5: Technology maintenance and development costs	

technology that has been implemented on a worldwide basis with a good reputation based on its performance in various countries and locations.

As sustainability aspiration implies, the direct benefits are meant for social development as a whole. In this case, adopting an existing SWS technology might create remarkable benefits for the society as compared to the cost aspects. These benefits will be tangible in the long run as social development is a longitudinal process that requires investments in sustainable development as well. There is also a major challenge in this condition when the adoption of an existing SWS technology has to deal with the societal aspects such as acceptance, training, culture and legislation that need to be best fit with the level of civilization amongst the society.

Acquiring an existing SWS technology has to involve a certain level of technology adoption. In this case, the costs of technology transfer or licensing that are associated with the available SWS technology are substantial, but the actual benefits are focusing on the performance and contributions of an existing SWS technology. This is due to the matter of fact that technology transfer from an existing SWS technology might take a long time, and there are also some barriers to technology transfer with issues such as licencing, royalties, training and other related charges.

From this analysis, it can be concluded that adapting and adopting an existing SWS technology is reliable for a short-term period, but in the long run, the benefits of developing a new SWS technology that is more affordable and suitable with the local condition will create more tangible benefits for the sustainable development of the Malaysian construction industry.

9.8 Conclusion

This chapter highlights on SWS plan, which is integrated into the whole aspects of SWM. SWS planning consists of several steps and aspects such as framing the problems of SWS, deciding objectives and eliciting their weights, assessing the performance of the selected alternatives and scoring them. The primary goal of SWS planning is to prepare a community and a country to effectively manage solid waste with the consideration of the environmental, economic and social aspects. The chapter focuses on the importance of developing a strategic SWM plan with its relevant steps. It is imperative that negative impacts can be reduced if the right management options are chosen and implemented. Several economic mechanisms have been established to manage pollutions issues related to solid waste. Our society needs knowledge, skills and empowerment in SWS planning as sustainable solutions are needed in the coming decades. This requires a holistic SWS planning in order to come up with some informed and robust decisions on how to manage solid waste in a more efficient and effective manner based on these actions: deciding on the objectives of SWS, assessing the different types of solid waste and validating the performance of the current SWS versus the objectives.

Glossary

Aerobic decomposition Is a type of decomposition that requires oxygen.

Anaerobic digestion Is a type of decomposition that does not use oxygen. Anaerobic decomposition creates odour problems; aerobic decomposition does not.

Combustion Oxidation of combustible materials at elevated temperatures.

Collection Is obtaining materials from the curbside or drop-off centres and bringing that material to an unloading point.

Cost-benefit analysis (CBA) Is a systematic approach to estimating the strengths and weaknesses of alternatives used to determine options that provide the best approach to achieving benefits while preserving savings.

Demolition The removal of existing structures and utilities as required to clear the construction site. The removal of the facilities proposed for destruction in the justification for the new construction.

Environmental Impact Assessment (EIA) Is the assessment of the environmental consequences of a plan, policy, program or actual projects prior to the decision to move forward with the proposed action.

US Environmental Protection Agency (USEPA) Is an independent executive agency of the United States federal government tasked with environmental protection matters.

General Packet Radio Service (GPRS) Is a packet-oriented mobile data standard on the 2G and 3G cellular communication network's global system for mobile communications.

Geographic information system (GIS) Is a conceptualized framework that provides the ability to capture and analyse spatial and geographic data.

Global system for mobile (GSM) Is a standard developed by the European Telecommunications Standards Institute to describe the protocols for second-generation digital cellular networks used by mobile devices such as mobile phones and tablets.

Global warming Is the long-term heating of Earth's climate system observed since the pre-industrial period due to human activities, primarily fossil fuel burning, which increases heat-trapping greenhouse gas levels in Earth's atmosphere.

Greenhouse gas (GHG) Is gases in the atmosphere that absorb and reemit infrared radiation; they cause the GHG effect that results in the heating up of the atmosphere.

Incineration A combustion of waste, in Europe preferentially in grate furnaces.

Industrial waste Is a solid waste that is generated during the manufacture of products. Material spills, dusts, sludges, defect products, etc. belong to this group of wastes.

Internet of things (IoT) Is the network interconnection that is embedded with sensors, software, and other technologies for the purpose of connecting and exchanging data with other devices and systems over the Internet.

Integrated sustainable waste management (ISWM) Is a systems approach to waste management that recognizes three important dimensions of waste management included stakeholders, waste system elements and sustainability aspects.

Leachate A liquid that has percolated through solid waste or another medium and has extracted, dissolved or suspended materials from it. Because leachate may include potentially harmful materials, leachate collection and treatment are crucial at municipal waste landfills.

Material recovery facility Is a special type of transfer station where recyclables are processed before transport.

Municipal authority Is a form of a special-purpose governmental unit. The municipal authority is an alternate vehicle for accomplishing public purposes without the direct action of counties, municipalities and school districts.

Municipal solid waste management (MSWM) The collection, segregation, storage, transportation, processing and disposal of municipal solid waste, including reduction, re-use, recovery, recycling in a scientific and hygienic manner.

Municipal solid waste (MSW) Commonly known as trash or garbage in the United States and rubbish in Britain, it is a waste type consisting of everyday items that are discarded by the public. "Garbage' can also refer specifically to food waste, as in a garbage disposal; the two are sometimes collected separately.

Non-governmental organizations (NGO) Organizations which are independent of government involvement are known as non-governmental organizations or non-government organizations, with NGO as an acronym.

Public-private partnership (PPP) A public–private partnership is a cooperative arrangement between two or more public and private sectors, typically of a long-term nature. In other words, it involves government and businesses that work together to complete a project and/or to provide services to the population.

Private financing investment (PFI) A private finance initiative is a method of providing funds for major capital investments, where private firms complete and manages public projects.

Public health Is defined as "the science and art of preventing disease", prolonging life and improving quality of life through organized efforts and informed choices of society, organizations, communities and individuals.

Radio-frequency identification (RFID) Radio-frequency identification uses electromagnetic fields to automatically identify and track tags attached to objects. An RFID system consists of a tiny radio transponder, a radio receiver and a transmitter.

Reuse is the actual reuse of a material in its present form. Some examples are printing draft copies on the backside of previously used paper, using incoming pallets as an outgoing pallets, or using incoming boxes as collection containers for recyclables.

Recycling Is the process by which materials otherwise destined for disposal are collected, reprocessed or remanufactured and are reused.

Residential waste Waste generated in single- and multiple-family homes.

Solid waste system (SWS) The system of solid waste is to reduce the amount of natural resources consumed and confirm that any materials that are taken from nature are reused as many times as possible and that the waste created is kept to a minimum.

Sorting Is waste particle separation usually carried out adopting optical-electronic recognition devices and logics.

Sustainable development goals (SDG) Known as global goals are a collection of 17 interlinked global goals designed to be a "blueprint for achieving a better and more sustainable future for all".

Storage This is keeping waste material at the generation point until it can be collected.

Transport Moving wastes long distances for treatment, disposal or recycling.

Waste is that which cannot be sold but the owner wants to or is required to get rid of.

Waste reduction This is a broad term encompassing all waste management methods like source reduction, recycling and composting, which results in a reduction of waste going to a combustion facility or landfill.

Waste stream A term describing the total flow of solid waste from homes, businesses, institutions and manufacturing plants that must be recycled, burned or disposed of in landfills or any segment thereof, such as the "residential waste stream" or the "recyclable waste stream".

Waste characterization This is the intrinsic properties of waste materials and how the materials will influence and be influenced by different environments.

Waste generation rate This is the amount of waste that is produced over a given amount of time. For example, a district may have a generation rate of 100 tons per day.

References

1. Joshi, R., & Ahmed, S. (2016). Status and challenges of municipal solid waste management in India: A review. *Cogent Environmental Science, 2*(1), 1139434.
2. Ribić, B., Voća, N., & Ilakovac, B. (2017). Concept of sustainable waste management in the city of Zagreb: Towards the implementation of circular economy approach. *Journal of the Air & Waste Management Association, 67*(2), 241–259.

3. Rada, E. C. (Ed.). (2016). *Solid waste management: Policy and planning for a sustainable society*. USA: CRC Press.
4. Kumar, S. (2016). *Municipal solid waste management in developing countries*. USA: CRC press.
5. Das, S., Lee, S. H., Kumar, P., Kim, K. H., Lee, S. S., & Bhattacharya, S. S. (2019). Solid waste management: Scope and the challenge of sustainability. *Journal of Cleaner Production, 228*, 658–678.
6. Kim, S. Y., & Nguyen, M. V. (2020). *A performance evaluation framework for construction and demolition waste management: Stakeholder perspectives*. Engineering, Construction and Architectural Management.
7. Ferronato, N., Ragazzi, M., Portillo, M. A. G., Lizarazu, E. G. G., Viotti, P., & Torretta, V. (2019). How to improve recycling rate in developing big cities: An integrated approach for assessing municipal solid waste collection and treatment scenarios. *Environmental Development, 29*, 94–110.
8. Mintz, K. K., Henn, L., Park, J., & Kurman, J. (2019). What predicts household waste management behaviors? Culture and type of behavior as moderators. *Resources, Conservation and Recycling, 145*, 11–18.
9. De Gisi, S., Casella, P., Sabia, G., Farina, R., Landolfo, P., Notarnicola, M., & De Feo, G. (2017). Assessing the public perception of islanders regarding the implementation of new technologies to optimize the municipal solid waste management system: A Mediterranean case study. *Journal of Cleaner Production, 164*, 1586–1601.
10. Tsai, F. M., Bui, T. D., Tseng, M. L., Wu, K. J., & Chiu, A. S. (2020). A performance assessment approach for integrated solid waste management using a sustainable balanced scorecard approach. *Journal of Cleaner Production, 251*, 119740.
11. Rhyner, C. R., Schwartz, L. J., Wenger, R. B., & Kohrell, M. G. (2017). *Waste management and resource recovery*. USA: CRC Press.
12. Huysveld, S., Hubo, S., Ragaert, K., & Dewulf, J. (2019). Advancing circular economy benefit indicators and application on open-loop recycling of mixed and contaminated plastic waste fractions. *Journal of Cleaner Production, 211*, 1–13.
13. Miller, Z. D., Lawhon, B., Taff, B. D., Schwartz, F., & Newman, P. (2020). Identifying strategies to reduce visitor-generated waste in national parks of the United States: The zero landfill initiative. *Applied Environmental Education & Communication, 19*(3), 303–316.
14. https://www.statista.com/topics/4983/waste-generation-worldwide/
15. https://www.un.org/development/desa/en/news/population/world-population-prospects-2019.html
16. Martínez-Gomariz, E., Russo, B., Gómez, M., & Plumed, A. (2020). An approach to the modelling of stability of waste containers during urban flooding. *Journal of Flood Risk Management, 13*, e12558.
17. Diaz-Barriga-Fernandez, A. D., Santibañez-Aguilar, J. E., Radwan, N., Nápoles-Rivera, F., El-Halwagi, M. M., & Ponce-Ortega, J. M. (2017). Strategic planning for managing municipal solid wastes with consideration of multiple stakeholders. *ACS Sustainable Chemistry & Engineering, 5*(11), 10744–10762.
18. Aparcana, S. (2017). Approaches to formalization of the informal waste sector into municipal solid waste management systems in low-and middle-income countries: Review of barriers and success factors. *Waste Management, 61*, 593–607.
19. Rodrigues, A. P., Fernandes, M. L., Rodrigues, M. F. F., Bortoluzzi, S. C., da Costa, S. G., & de Lima, E. P. (2018). Developing criteria for performance assessment in municipal solid waste management. *Journal of Cleaner Production, 186*, 748–757.
20. USEPA, 2020. *What is integrated solid waste management?* Retrieved from http://www.epa.gov/climatechange/wycd/waste/downloads/overview.pdf
21. Kaza, S., Yao, L., Bhada-Tata, P., & Van Woerden, F. (2018). *What a waste 2.0: A global snapshot of solid waste management to 2050* World Bank Publications.

22. Sadef, Y., Nizami, A. S., Batool, S. A., Chaudary, M. N., Ouda, O. K. M., Asam, Z. U. Z., … Demirbas, A. (2016). Waste-to-energy and recycling value for developing integrated solid waste management plan in Lahore. *Energy Sources, Part B: Economics, Planning, and Policy, 11*(7), 569–579.
23. Kawai, K., & Tasaki, T. (2016). Revisiting estimates of municipal solid waste generation per capita and their reliability. *Journal of Material Cycles and Waste Management, 18*(1), 1–13.
24. Kannangara, M., Dua, R., Ahmadi, L., & Bensebaa, F. (2018). Modeling and prediction of regional municipal solid waste generation and diversion in Canada using machine learning approaches. *Waste Management, 74*, 3–15.
25. Pichtel, J. (2005). *Waste management practices: Municipal, hazardous, and industrial*. CRC press.
26. Ozcan, H. K., Guvenc, S. Y., Guvenc, L., & Demir, G. (2016). Municipal solid waste characterization according to different income levels: A case study. *Sustainability, 8*(10), 1044.
27. Tumuluru, J. S., Yancey, N. A., & Kane, J. J. (2021). Pilot-scale grinding and briquetting studies on variable moisture content municipal solid waste bales–impact on physical properties, chemical composition, and calorific value. *Waste Management, 125*, 316–327.
28. Shammi, M., Behal, A., & Tareq, S. M. (2021). The escalating biomedical waste management to control the environmental transmission of COVID-19 pandemic: A perspective from two south Asian countries. *Environmental Science & Technology, 55*(7), 4087–4093.
29. Diaz, L. F., Golueke, C. G., Savage, G. M., & Eggerth, L. L. (2020). *Composting and recycling municipal solid waste*. USA: CRC Press.
30. Aleluia, J., & Ferrão, P. (2016). Characterization of urban waste management practices in developing Asian countries: A new analytical framework based on waste characteristics and urban dimension. *Waste Management, 58*, 415–429.
31. Sulemana, A., Donkor, E. A., Forkuo, E. K., & Oduro-Kwarteng, S. (2018). Optimal routing of solid waste collection trucks: A review of methods. *Journal of Engineering, 2018*.
32. Louati, A. (2016). Modeling municipal solid waste collection: A generalized vehicle routing model with multiple transfer stations, gather sites and inhomogeneous vehicles in time windows. *Waste Management, 52*, 34–49.
33. Ibáñez-Forés, V., Bovea, M. D., Coutinho-Nóbrega, C., & de Medeiros, H. R. (2019). Assessing the social performance of municipal solid waste management systems in developing countries: Proposal of indicators and a case study. *Ecological Indicators, 98*, 164–178.
34. Pires, A., Martinho, G., Rodrigues, S., & Gomes, M. I. (2019). *Sustainable solid waste collection and management*. Springer.
35. Kumar, A., & Samadder, S. R. (2017). A review on technological options of waste to energy for effective management of municipal solid waste. *Waste Management, 69*, 407–422.
36. Pérez-López, G., Prior, D., Zafra-Gómez, J. L., & Plata-Díaz, A. M. (2016). Cost efficiency in municipal solid waste service delivery. Alternative management forms in relation to local population size. *European Journal of Operational Research, 255*(2), 583–592.
37. Saadeh, D., Al-Khatib, I. A., & Kontogianni, S. (2019). Public–private partnership in solid waste management sector in the West Bank of Palestine. *Environmental Monitoring and Assessment, 191*(4), 1–19.
38. Hannan, M. A., Al Mamun, M. A., Hussain, A., Basri, H., & Begum, R. A. (2015). A review on technologies and their usage in solid waste monitoring and management systems: Issues and challenges. *Waste Management, 43*, 509–523.
39. Haregu, T. N., Ziraba, A. K., Aboderin, I., Amugsi, D., Muindi, K., & Mberu, B. (2017). An assessment of the evolution of Kenya's solid waste management policies and their implementation in Nairobi and Mombasa: Analysis of policies and practices. *Environment and Urbanization, 29*(2), 515–532.
40. Malinauskaite, J., Jouhara, H., Czajczyńska, D., Stanchev, P., Katsou, E., Rostkowski, P., … Spencer, N. (2017). Municipal solid waste management and waste-to-energy in the context of a circular economy and energy recycling in Europe. *Energy, 141*, 2013–2044.

41. Cetrulo, T. B., Marques, R. C., Cetrulo, N. M., Pinto, F. S., Moreira, R. M., Mendizábal-Cortés, A. D., & Malheiros, T. F. (2018). Effectiveness of solid waste policies in developing countries: A case study in Brazil. *Journal of Cleaner Production, 205*, 179–187.
42. Fatimah, Y. A., Govindan, K., Murniningsih, R., & Setiawan, A. (2020). Industry 4.0 based sustainable circular economy approach for smart waste management system to achieve sustainable development goals: A case study of Indonesia. *Journal of Cleaner Production, 269*, 122263.
43. Velvizhi, G., Shanthakumar, S., Das, B., Pugazhendhi, A., Priya, T. S., Ashok, B., ... Karthick, C. (2020). Biodegradable and non-biodegradable fraction of municipal solid waste for multifaceted applications through a closed loop integrated refinery platform: Paving a path towards circular economy. *Science of the Total Environment, 731*, 138049.
44. Harir, A. I., Kasim, R., & Ishiyaku, B. (2015). Exploring the resource recovery potentials of municipal solid waste: A review of solid wastes composting in developing countries. *International Journal of Scientific and Research Publications, 5*(4), 2250–3153.
45. Rodić, L., & Wilson, D. C. (2017). Resolving governance issues to achieve priority sustainable development goals related to solid waste management in developing countries. *Sustainability, 9*(3), 404.
46. Mavrotas, G., Gakis, N., Skoulaxinou, S., Katsouros, V., & Georgopoulou, E. (2015). Municipal solid waste management and energy production: Consideration of external cost through multi-objective optimization and its effect on waste-to-energy solutions. *Renewable and Sustainable Energy Reviews, 51*, 1205–1222.
47. Welivita, I., Wattage, P., & Gunawardena, P. (2015). Review of household solid waste charges for developing countries–a focus on quantity-based charge methods. *Waste Management, 46*, 637–645.
48. Gallardo, A., Carlos, M., Peris, M., & Colomer, F. J. (2015). Methodology to design a municipal solid waste pre-collection system. A case study. *Waste Management, 36*, 1–11.
49. Lella, J., Mandla, V. R., & Zhu, X. (2017). Solid waste collection/transport optimization and vegetation land cover estimation using Geographic Information System (GIS): A case study of a proposed smart-city. *Sustainable Cities and Society, 35*, 336–349.
50. Antonioli, D., Caratù, F., & Nicolli, F. (2018). Waste performance, waste technology and policy effects. *Journal of Environmental Planning and Management, 61*(11), 1883–1904.
51. Wang, Y., Zhang, X., Liao, W., Wu, J., Yang, X., Shui, W., ... Peng, H. (2018). Investigating impact of waste reuse on the sustainability of municipal solid waste (MSW) incineration industry using emergy approach: A case study from Sichuan province, China. *Waste Management, 77*, 252–267.
52. Al-Khateeb, A. J., Al-Sari, M. I., Al-Khatib, I. A., & Anayah, F. (2017). Factors affecting the sustainability of solid waste management system—The case of Palestine. *Environmental Monitoring and Assessment, 189*(2), 93.
53. Massarutto, A. (2015). Economic aspects of thermal treatment of solid waste in a sustainable WM system. *Waste Management, 37*, 45–57.
54. Kinnaman, T. C. (Ed.). (2017). *The economics of residential solid waste management.* Routledge.
55. Dlamini, S., Simatele, M. D., & Serge Kubanza, N. (2019). Municipal solid waste management in South Africa: From waste to energy recovery through waste-to-energy technologies in Johannesburg. *Local Environment, 24*(3), 249–257.
56. Ajayi, S. O., Oyedele, L. O., Akinade, O. O., Bilal, M., Alaka, H. A., & Owolabi, H. A. (2017). Optimising material procurement for construction waste minimization: An exploration of success factors. *Sustainable Materials and Technologies, 11*, 38–46.
57. Alidoosti, Z., Govindan, K., Pishvaee, M. S., Mostafaeipour, A., & Hossain, A. K. (2021). Social sustainability of treatment technologies for bioenergy generation from the municipal solid waste using best worst method. *Journal of Cleaner Production, 288*, 125592.
58. Margallo, M., Taddei, M. B. M., Hernández-Pellón, A., Aldaco, R., & Irabien, A. (2015). Environmental sustainability assessment of the management of municipal solid waste incin-

eration residues: A review of the current situation. *Clean Technologies and Environmental Policy, 17*(5), 1333–1353.
59. Mohan, G., Sinha, U. K., & Lal, M. (2016). Managing of solid waste through public private partnership model. *Procedia Environmental Sciences, 35*, 158–168.
60. Christensen, D., & Bach, L. T. (2015). A Danish–Vietnamese partnership for business and technology development in solid waste management. *Resources, Conservation and Recycling, 105*, 123–133.
61. Bommareddy, S. S. R., & Asadi, S. S. (2017). Solid waste management through public-private partnership model. *Technology, 8*(1), 263–272.
62. Kiyasudeen, K., Ibrahim, M. H., Quaik, S., & Ismail, S. A. (2016). An introduction to anaerobic digestion of organic wastes. In *Prospects of organic waste management and the significance of earthworms* (pp. 23–44). Cham: Springer.
63. Rasapoor, M., Young, B., Brar, R., Sarmah, A., Zhuang, W. Q., & Baroutian, S. (2020). Recognizing the challenges of anaerobic digestion: Critical steps toward improving biogas generation. *Fuel, 261*, 116497.
64. Cremiato, R., Mastellone, M. L., Tagliaferri, C., Zaccariello, L., & Lettieri, P. (2018). Environmental impact of municipal solid waste management using Life Cycle Assessment: The effect of anaerobic digestion, materials recovery and secondary fuels production. *Renewable Energy, 124*, 180–188.
65. Mohamed, I., Othman, F., Ibrahim, A. I., Alaa-Eldin, M. E., & Yunus, R. M. (2015). Assessment of water quality parameters using multivariate analysis for Klang River basin, Malaysia. *Environmental Monitoring and Assessment, 187*(1), 1–12.
66. https://www.thestar.com.my/news/nation/2019/11/05/giant-vacuum-cleaning-up-klang-river

Chapter 10
Practices of Solid Waste Processing and Disposal

Harris Ramli, Hamidi Abdul Aziz, and Yung-Tse Hung

Abstract Around the world, the amount of waste generated has been steadily increasing. The composition of municipal solid waste (MSW) varies significantly between municipalities and countries. This variation is influenced by various factors like lifestyle, economics, waste management laws, and industrial structure. Waste control is usually handled by municipalities. They must have a system that is both reliable and usable for the residents. Nonetheless, they are often confronted with various issues beyond the municipal authority's capacity to manage MSW. Additionally, a lack of understanding, particularly regarding the quantity, characteristics, calorific value, and generation rate of MSW, can further cause impractical solid waste treatment and final disposal strategies. As a result, most developed countries would have much better solid waste management efficiency than developing countries due to the lack of these data. Apart from the complexity of MSW, the composition of e-waste is well known to be complex and varies by product type. It is made up of over a thousand different compounds classified as hazardous or nonhazardous that would increase treatment and disposal challenges when mixed with MSW. Solid waste processing and treatment before disposals, such as bioconversion (composting, vermicomposting, anaerobic digestion, and fermentation) and thermal conversion (incineration either with or without energy recovery, pyrolysis, and gasification), have demonstrated an environmental and economic benefit from waste material. An incinerator can be used for both concepts in MSW management, either processing or disposal of solid waste. Regardless of the incinerator's construction or intended use, the by-product of the incinerator must eventually be disposed of in a landfill. Landfilling is regarded as an effective method of waste management. Despite its benefits in terms of resolving solid waste prob-

H. Ramli · H. A. Aziz (✉)
School of Civil Engineering and Solid Waste Management Cluster, Engineering Campus, Universiti Sains Malaysia, Nibong Tebal, Pulau Pinang, Malaysia
e-mail: cemhr@usm.my; cehamidi@usm.my

Y.-T. Hung
Department of Civil and Environmental Engineering, Cleveland State University, Cleveland, OH, USA

L. K. Wang et al. (eds.), *Solid Waste Engineering and Management*, Volume 1, Handbook of Environmental Engineering 23,
https://doi.org/10.1007/978-3-030-84180-5_10

lems, there are several environmental concerns. Furthermore, landfills can only accommodate for a certain amount of time, and landfill reclamation can take many decades. Each method of solid waste processing, treatment, and disposal has several advantages and disadvantages.

Keywords Municipal solid waste · Waste bioconversion · Waste thermal conversion · Incineration · Landfill management

Acronyms

CR	Compaction ratio
EOL	End of life
EU	The European Union's
GHE	Greenhouse gases
HHW	Household hazardous wastes
LFG	Landfill gas
LFM	Landfill mining
MSW	Municipal solid waste
NIMBY	"Not in my backyard"
OECD	Organization for Economic Cooperation and Development
USEPA	US Environmental Protection Agency
WEEE	Waste electrical and electronic equipment
WTE	Waste-to-energy

10.1 Introduction

As of this moment, solid waste management is a global problem. The issues are aggravated further by the disproportional increase in municipal solid waste generation, particularly in rapid urbanization, population growth, and economic globalization. Every metropolitan or growing city in the world, from Asia to Europe, has a similar issue; countries such as China, India, South Africa, Brazil, and Russia are among the countries on this list. Over the last three decades, several nations, particularly Brazil, Russia, India, China, and South Africa, have experienced rapid urbanization. This urbanization has resulted in a massive increase in trash (Fig. 10.1). Public habits, industrialization levels, local environment, and economic development all influence the rate of municipal solid waste generation.

The municipal solid waste (MSW) consists of residuals and residues of household operations, street sweeping, and residential and public cleaning. Increased municipal solid waste generation is a developing concern in cities globally, owing to the increased necessity of special municipal management solutions [1, 2].

MSW management is a multidisciplinary endeavor encompassing the production, separation, storage, collection, transportation, processing, and disposal of waste materials. As a result, MSW management has become a global concern, particularly in the urban districts of growing economies. For authorities in both small and large cities in developed countries, the issue of solid waste management is the

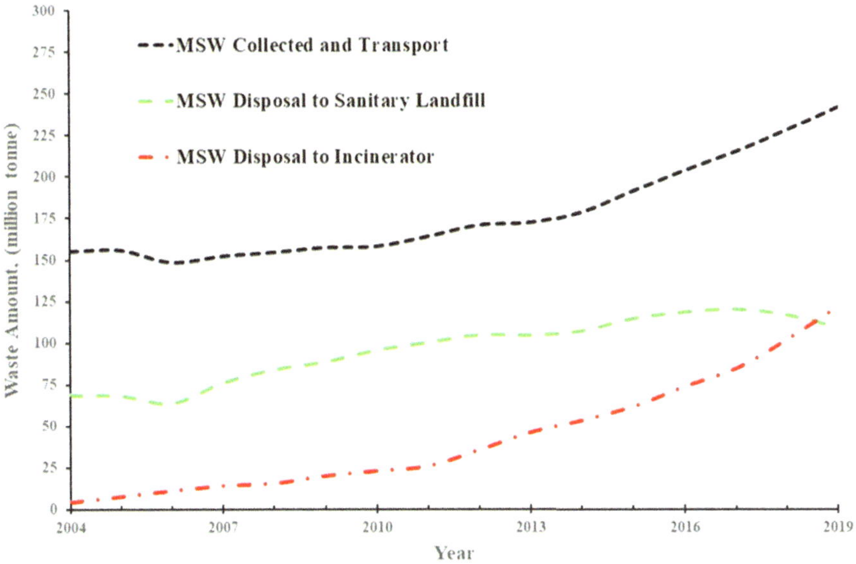

Fig. 10.1 MSW collected, transport, and disposal for China from 2004 to 2019 [3]

most complicated. This is primarily due to increased solid waste generation and the resultant financial strain on municipalities. Apart from the high costs, solid waste management is harmed by a lack of knowledge of the numerous variables that influence the entire handling scheme. Population development, rapid urbanization, a booming economy, and living standards have all accelerated the pace, volume, and quality of MSW generation in developing countries [4].

MSW management encompasses six fundamental services: storage, collection, transfer and transportation, processing or treatment, and, lastly, disposal of products that cannot be economically recovered for recycling or reuse. Therefore, in order to adopt an effective municipal solid waste management policy, it is necessary to know the quantity and composition of MSW.

Additionally, MSW management may be defined as the discipline concerned with regulating the generation, storage, collection, transfer and transport, processing/treatment, and disposal of solid waste in a manner consistent with the best principles of health, economics, engineering, conservation, aesthetics, and other environmental considerations, while also being accountable. Thus, MSW management encompasses all planning, administrative, legal, financial, and engineering activities associated with resolving municipal solid waste-related problems in metropolitan settings.

Technically, solid waste is an undesirable material or by-product of any economic or social activity. Solid waste can generally be group based on its source or characteristic. The common groups of solid waste are municipal solid waste, construction and demolition waste, hazardous waste, and medical waste. This chapter will focus on MSW and waste associated with the MSW disposal and process.

10.2 Solid Waste Composition, Characterization, and Generation

One of the most pressing environmental concerns is municipal solid waste. Waste control is usually handled by municipalities. They must have a system that is both reliable and usable for the residents. Nonetheless, they are often confronted with various issues that go beyond the municipal authority's capacity to manage MSW [5]. This is mostly due to a lack of financial capital, coordination, and complexity.

The composition of MSW varies significantly between municipalities, as well as from one country to the next. This variation is influenced by a variety of factors like lifestyle, economics, waste management laws, and industrial structure. MSW generation is any solid, nonhazardous substance or object generated inside an urban region, except wastewater sludge. While the primary constituents of MSW generated globally are similar, the density, the quantity generated, and the proportion of streams vary significantly between countries, according mainly to culture and tradition, income and lifestyle, predominant weather conditions, and geographic location.

10.2.1 Municipal Solid Waste (MSW)

Various researchers and institutions define MSW as a word that is typically used to refer to a diverse collection of wastes generated in metropolitan areas, the nature of which varies by region. The differences in wastes between regions or within the same region are due to the fact that the quantity, characteristics, and quality of solid waste generated are a function not only of the region's inhabitants' living standards and lifestyles but also the region's natural resources' type and abundance.

It is clear that the term MSW refers to either the source of garbage or its composition or both. Thus, MSW is defined as waste generated by streets, residential, and economic activities in an urban area that enters and/or exits the municipal waste stream. It is critical to understand the volume and composition of MSW generated in order to plan and design an appropriate solid waste management policy in a particular area. Additionally, a lack of understanding, particularly regarding the quantity and characteristics of MSW, can render specific steps impractical, such as treatment and final disposal. In 2013, the Environmental Protection Agency (EPA) stated that the United States generated 254 million tons of MSW. MSW composition and classification are depicted in Fig. 10.2.

In terms of MSW generation, it is influenced by economic and behavioral variables and population considerations associated with population expansion and concentration in metropolitan regions. According to Sharholy et al. [6], MSW generation per capita in India ranges between 0.2 and 0.5 kg/inhabitant/day, with an annual growth rate of 1–1.33%. Additionally, in some cities, the generation may be considerably higher due to the high standards of urbanization. Both population and economic expansion have had a considerable impact on the quantity and kinds of trash

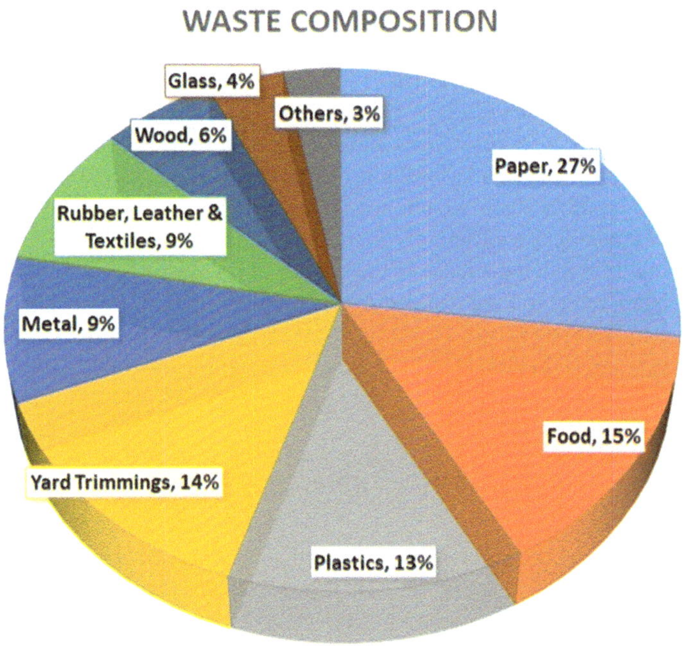

Fig. 10.2 MSW generated by the United States in 2013: composition and classification (by material) (USEPA)

generated, transforming waste management into an issue that must be managed by developing and implementing appropriate strategies for each type of garbage generated. Zhang et al. [7] also note that China has had difficulties managing MSW as a result of the growth in per capita generation, which is currently at 1.134 kg/inhabitant/day. Russia generates 0.63 kg per capita per day, while Brazil generates 1.062 kg per capita per day. In terms of composition, the majority of MSW created in China, India, Brazil, South Africa, and Russia is organic, reflecting these populations' consumption patterns. This scenario is predictable, given that Sharholy et al. [6] and Singh et al. [8] found that in countries classified as undeveloped or developing, the rate of MSW generation is increasing, with organic material accounting for the majority of the increase.

10.2.2 Solid Waste from Food

Large quantities of waste created worldwide can be used to provide a sustainable and significant source for many industrial chemicals. Food residues and waste, such as kitchen compost, trash, and swill, are often referred to as food by-products and solid waste. These wastes are produced during food processing, cooking, and distribution, as well as during consumption. Food waste, on the other hand, and its meaning differ greatly between cities and countries. In the European Union, food waste is described as "raw or cooked food substances that are discarded or that are intended or required to be discarded."

On the other hand, the US EPA defines food wastes as "Uneaten foods and food preparation wastes from residences and commercial establishments including grocery stores, restaurants, produce stands, institutional cafeterias and kitchens, as well as industrial sources such as employee lunchrooms." Additionally, the United Nations acknowledges a variety of "food loss" and "food waste" outlets. Food loss is a term that refers to a decrease in the quality or quantity of food. Food waste, on the other hand, refers to food losses incurred by the producer or by consumer behavior [9]. Uncooked raw products, wasted foodstuffs, and edible materials from grocery stores or the wet market are all examples of food waste.

10.2.3 Solid Waste Calorific Value

The moisture content of solid waste is defined as the mass of water or dry materials per unit amount of moisture [10–12]. Therefore, it is a critical aspect in determining MSW collection and transportation strategies. Moisture is transferred between waste bins and collector trucks during the storage and transit of MSW, and as a result, the moisture content of individual components varies over time.

Moisture content is also critical for MSW breakdown and treatment. For instance, moisture content has an effect on the amount of heat generated during composting, which might impact the compost's quality [13]. Leachate is created in a landfill when the moisture content of the garbage exceeds the field capacity. Additionally, numerous researchers have noted that excessive moisture content is a significant impediment in waste-to-energy (WTE) thermal conversion, as moisture content affects the calorific value of the solid waste to be burnt. The energy value of solid waste is determined by its calorific value, which is affected by the waste's hydrogen content and moisture [14–16]. Accordingly, the ability of waste to sustain a combustion process without the addition of additional fuel is dependent on several chemical and physical parameters, the most important of which is the lower (inferior) calorific value [17, 18]. However, the minimum calorific value required is depending on the configuration of the furnace for controlled incineration. Therefore, when considering MSW incineration and other WTE methods, the calorific value of MSW is crucial for the recovery of energy from MSW. For instance, the high proportion of organic waste in Ghana's MSW stream results in increased moisture content (over 50% on average) of the MSW, which is consistent with the waste stream in other developing nations. As a result, the application of MSW incinerators may be less efficient than in nations where MSW has a lower moisture content.

10.2.4 E-Waste

Waste electrical and electronic equipment) WEEE or end-of-life (EOL) electronics are examples of e-waste. ICT serves as a hub for information exchange, networking, and access to remote resources with the aid of newer technology such as cell phones, the internet, iPads, laptops, and other devices. Any appliance that uses an electric power supply that has reached its end-of-life, according to the Organization for Economic Cooperation and Development (OECD), falls under WEEE.

E-waste is a catchall term for a variety of electrical and electronic equipment (EEE) that has reached the end of its useful life and is no longer useful to its owners. Growing demand for newer electronic appliances in developed and developing countries is also contributing to large e-waste mountains [19]. The composition of e-waste is well known to be complex and varies by product type. It is made up of over a thousand different compounds that are classified as hazardous or nonhazardous.

E-waste production is estimated to be about 40 million tons per year. In developing countries, e-waste accounts for 1–2% of overall solid waste production, with that percentage projected to rise to 2% by 2010 [19]. E-waste consists of both highly useful and toxic nonrenewable substances that can be recovered and recycled, resulting in lucrative business opportunities, and most developed countries take the simple route by sending WEEE to developing countries for processing. By removing/recycling e-waste in a formal, systematic, and environmentally friendly manner, there are endless market opportunities and scope for academicians in developing sustainable models [20].

The high toxicity of WEEE component materials causes a slew of socioeconomic issues, particularly when burned or recycled in an unregulated manner. E-waste disposal in a haphazard manner will degrade ecological and human systems. Actually, the composition of e-waste is very diverse and complex. There are over 1000 chemicals in e-waste that can be categorized as hazardous or nonhazardous materials.

The following types of electrical and computer equipment can be classified:

(i) Large appliances for the home (refrigerator, freezer, washing machine, cooking appliances, etc.)
(ii) Small appliances for the home (vacuum cleaners, watches, grinders, etc.)
(iii) Equipment for information technology and telecommunications (PCs, printers, telephones, telephones, etc.)
(iv) Consumer electronics (TV, radio, video camera, amplifiers, etc.)
(v) Equipment for lighting (CFL, high-intensity sodium lamp, etc.)
(vi) Tools for electrical and electronic work (drills, saws, sewing machine, etc.)
(vii) Toys, recreational, and sporting goods (computer/video games, electric trains, and so on).
(viii) Medical facilities (with the exception of all implanted and infected products, radiotherapy equipment, cardiology, dialysis, nuclear medicine, etc.)

(ix) Instruments for monitoring and control (smoke detector, heating regulators, thermostat, etc.)
(x) Dispensers that work on their own (for hot drinks, money, hot and cold bottles, etc.)

Ferrous material (38%) is the most common substance contained in electric and electronic waste, followed by nonferrous material (28%), plastic (19%), glass (4%), and other materials such as wood, rubber, ceramics, and so on (11%). E-waste is more varied and complex in nature due to the various elements present in it.

There are two types of substances present in WEEE: hazardous and nonhazardous products. The substances within the mentioned components are heavy metals like mercury, lead, chromium (VI), cadmium, polychlorinated biphenyls, brominated flame retardants are used on circuit boards (BFRs), halogenated substances (e.g., CFCs), and plastics. BFR can generate dioxins and furans during incineration. Other elements and contaminants that may be present include arsenic, asbestos, nickel, and copper. These substances can act as a catalyst, causing more dioxins to form during incineration.

10.3 Solid Waste Generating Issues Affect MSW Disposal

According to the report by World Bank in 2012 [21], worldwide MSW generation now stands at approximately 1.3 billion tonnes per year and is predicted to increase to nearly 2.2 billion tons per year by 2025 (Table 10.1). This would imply a significant increase in per capita trash creation rates over the following eight years, from 1.2 kg to 1.42 kg per person per day (Table 10.2). Thus, waste management problems in the majority of developing countries are anticipated to deteriorate unless proper measures are put in place to deal with this accelerating generation rate.

According to Eiselt and Marianov [22], numerous developing countries' per capita garbage generation rates have surpassed one kilogram per day. Unfortunately,

Table 10.1 Regional waste generation per capita (Report by The World Bank in 2012) [21]

Region	Waste generation per capita (kg/capita/day)		
	Lower boundary	Upper boundary	Average
Organization for Economic co-operation and Development (OECD)	1.10	3.70	2.20
Europe and Central Asia region (ECA)	0.29	2.10	1.10
Latin America and the Caribbean region (LCR)	0.11	5.50	1.10
The Middle East and North Africa region NA)	0.16	5.70	1.10
East Asia and Pacific region (EAP)	0.44	4.30	0.95
Africa region (AFR)	0.09	3.00	0.65
South Asia region (SAR)	0.12	5.10	0.45

the majority of municipal governments lack the capacity to manage this waste appropriately. For example, sub-Saharan Africa generates roughly 62 million tons of waste each year; while per capita waste creation is generally modest in the region, it varies widely, ranging from 0.09 to 3.00 kg per person per day on average [23, 24]. Furthermore, in OECD countries, the average amount of waste generated per capita per day is 2.2 kg, and the rate of solid waste generation is expected to be between 0.5% to 0.7% annually [21]. Similarly, solid waste generation in the Middle East and North Africa (MENA) is comparable to the rest of the world, at 63 million tons per year, with per capita waste generation spanning from 0.16 kg to 5.70 kg per person per day, with an average of 1.1 kg/capita/day [21].

MSW should include the only waste that does not require special treatment because it is managed or regulated by municipal or local government. Clinical waste, construction, and demolition waste will worsen many developing countries' MSW management challenges when incorporated into the MSW system. MSW is defined broadly as waste that municipalities or local governments collect. MSW is often comprised of residential waste, nonhazardous commercial waste, yard, and park garbage as well as street sweepings.

MSW is mainly composed of rubbish generated by households about 60% to 90%, including waste generated by trade or public institutions. Construction and demolition wastes are typically excluded from municipal solid waste due to their bulky nature. Additionally, the treatment of this material before disposal requires specialized machinery. Medical waste should be processed and dispose of separately because it requires particular care and management. However, it also

Table 10.2 Waste generation forecasts by region for 2025 (Report by The World Bank in 2012) [21]

Region	Current Available data			Projection for 2025			
	Total Urban Population (millions)	Municipal Solid Waste Generation		Projected Population		Projected Municipal Solid Waste	
		Per Capita (kg/capita/day)	Total (tons/day)	Total Population (millions)	Urban Population (millions)	Per Capita (kg/capita/day)	Total (tons/day)
East Asia and Pacific region (EAP)	777	0.95	738,958	2,124	1,229	1.50	1,865,379
Organization for Economic Co-operation and Development (OECD)	729	2.20	1,566,286	1,031	842	2.10	1,742,417
Latin America and the Caribbean region (LCR)	399	1.10	437,545	681	466	1.60	728,392
South Asia region (SAR)	426	0.45	192,410	1,938	734	0.77	567,545
Africa Region (AFR)	260	0.65	169,119	1,152	518	0.85	441,840
The Middle East and North Africa region (MENA)	162	1.10	173,545	379	257	1.43	369,320
Europe and Central Asia region (ECA)	227	1.10	254,389	339	239	1.50	354,810
Total	2,980	1.20	3,532,252	7,644	4,285	1.40	6,069,703

sometimes is classified as municipal solid waste in some regions, especially in developing countries. Medical waste usually processes and dispose of similar to hazardous waste.

Household or municipal wastes are typically created from a variety of sources as a result of various human activities. According to several studies, households in developed countries produce the majority of municipal solid waste (55–80%), followed by markets or industrial areas (10–30%). The latter is made up of a diverse set of quantities generated by factories, highways, businesses, and other sources [25]. The solid waste produced by these sources is frequently very diverse.

As a result, depending on their origin, they exhibit a wide variety of physical and chemical characteristics. Yard waste, food waste, plastics, wood, metals, papers, rubbers, leather, batteries, inert materials, textiles, paint containers, demolition and construction materials, and a host of other difficult-to-identify materials are all mixed in. The heterogeneity of such treated solid waste is a significant drawback in terms of sorting and material usage. As a result, proper fractionation and sorting of these wastes are necessary prior to any meaningful treatment process. Sorting and separating such wastes is a crucial and proven method for establishing the integrity of the separated fractions for potential solid waste management applications.

However, the effectiveness of any solid waste segregation strategy is heavily dependent on public awareness and active involvement on the part of waste producers in various communities (i.e., how they conform to basic waste sorting and separation principles) [26]. The generation of solid waste is a problem and a source of concern throughout the world, especially in cities. This method of solid waste generation is widely regarded as one of the most challenging issues confronting the majority of developing countries, which suffer from severe environmental contamination as a result of large amounts of solid waste generation [27].

Storage is a critical technique for effective MSW management. As a step prior to collection, how solid waste is stored has an effect on how it is carried. It is recommended that trash be placed in waste containers selected for their properties at the point of generation. For example, the waste and the container must be chemically and mechanically compatible.

In India, storage is insufficient, with trash being dumped in open spaces and streets without separation at the source; in some regions, community waste bins are employed [28]. Whereas in China, the utilization of trash containers, collection sites, and transfer stations all stand out as a means of temporary storage to assist in the movement of MSW from small collection vehicles to bigger ones [29]. In Russia, this is accomplished through the use of home dumpsters or trash deposits with common trash cans at residencies around 0.75 m^2 container [30].

Increased solid waste generation in cities had a big effect on sanitary issues and essential services like transportation infrastructure, water supply, sanitation, and waste management [31]. According to several reports, solid waste collection, transportation, storage, and final disposal are major problems in metropolitan cities and areas. Most developing countries, as well as towns in East and North Africa, face similar severe solid waste generation problems. The fundamental cause of these issues is these countries' weak economy, which accounts for their poor solid waste management performance. Due to scarce resources and conflicting priorities for those resources, the majority of these developing countries have issues with solid waste management. Thus, solid waste generation is indeed one of the grave and major issues many cities face globally.

On the other hand, other socioeconomic variables such as employment status, monthly income, average family size, and the number of rooms all have an impact on the generation and composition of solid waste. It has also been asserted that the composition of solid waste and community social activities are inextricably linked [32]. On the other hand, other socioeconomic variables such as average family size, number of rooms, monthly income, and employment status all have an impact on the generation and composition of solid waste.

Sociocultural, economic, legal, political, and environmental factors, as well as available resources, all have an effect on MSW management in all countries. As a result, any new technology for MSW management and solid waste generation should take into account the effects and consequences on the sociocultural and economic well-being of the community.

Owing to changes in people's consumption habits and rapid technological advances, the volume, and composition of MSW have also changed. From 2001 to 2010, the European Environmental Agency looked at the per capita annual MSW generated by 32 European countries; they discovered that this waste increased in 21 countries and decreased in 11 countries. The study also looked at the amount of waste produced in 26 countries between 2001 and 2008, finding that in six of them, the amount of waste produced decreased. As a consequence, depending on the factors mentioned above, such as people's consumption habits, the quantities, and characteristics of waste varied from country to country and region to region, even within the same city [33].

The characterization of solid wastes has been extensively studied [34–37]. There is also research on solid waste's socioeconomic use to determine the potential revenue from these wastes [38]. A study where the waste composition was classified into three classes based on residents' socioeconomic status in Lahore, Pakistan, discovered significant variations in the composition of collected solid waste based on socioeconomic factors and income level [39]. They are divided into various income groups. They classified the people into three income groups: low, middle, and strong. As a result, they measured solid waste components and proportions using the income of each group. As a result, solid waste composting can be classified according to income levels: low, middle, and large. Seasonal differences were used by Gomez et al. [40] to characterize the characteristics of solid wastes. In their research, they looked at three different socioeconomic classes.

Composition and characterization investigations are critical in waste management because they aid in the selection and preparation of the best solid waste transportation, storage, and disposal methods. Meanwhile, characterizing any potential environmental effects, including those on nature and culture, is critical. The content of plant nutrients in the majority of MSW is between 0.5% and 0.7% nitrogen, 0.5% and 0.8% phosphorus, and 0.5% and 0.8% potassium.

10.4 Solid Waste Processing and Treatment Prior Disposal

Inadequate management of MSW leads to economic losses and puts public health and natural resources at risk. Any activity involving the treatment of MSW reduces not just the overall amount of solid waste created but also the associated costs of disposal. China, India, and South Africa, for example, affirm the presence of composting. However, several issues are discovered, such as inadequate product quality and a lack of market for composted material. Composting should be explored in these places due to the high organic content of MSW. Numerous academics argue that composting should be enhanced, as it is critical for the separation of waste material from its source, with recyclable elements being routed to recycling processes and organic material being composted.

On the other side, incineration is a prevalent practice in China's large cities. In contrast, MSW has a limited heat-generating capacity in small towns. However, incineration has increased in recent years as a result of government subsidies and commercial investments targeted at lowering the volume of MSW and generating electricity. Narayana [41] asserts that this activity has not been as prevalent in India as it should be due to limited quantities, poor heat-generating content, and high humidity levels that do not fulfill the requirements of central incinerator plants. In Brazil, incineration is used as a secondary method of treatment. Within the country, incinerators are mostly used to process garbage classified as requiring special treatment, such as medical waste.

In terms of recycling, the informal sector plays a significant role. According to Wang et al. [42], recyclable material collectors are critical to the sustainability of the overall recycling system in developing nations. As a result, there is a need to properly integrate these critical components of the system. These workers frequently suffer from authorities' neglect and require legalized employment and professional training. Brazil stands out among these countries as a model of social inclusion for recyclers via cooperatives and groups.

In another MSW processing technique, new MSW processing and treatment possibilities are based on bioconversion or thermal conversion, as detailed in Table 10.3. The bioconversion process is primarily applicable to organic waste, where it can be used to create compost or biogas such as methane. The generation of refuse-derived fuels (RDF), plasma pyrolysis and palletization, pyrolysis and gasification, and incineration with or without heat recovery are examples of thermal conversion technologies.

Thermal conversion technologies are often incompatible with MSW containing a significant proportion of organic matter, as its moisture content determines the calorific value of the solid waste. Opinions on the efficacy of various technologies for the processing and treatment of waste vary considerably. This is due to the fact that waste infrastructure has a long lifespan, and special consideration must be given during the design phase to guarantee that systems can adapt to long-term changes. Thus, the adaptability of technology to future development is frequently the deciding factor in its selection. Nonetheless, the optimal choice of technology is

Table 10.3 The available technology for processing and treating MSW

Bioconversion technologies	Thermal conversion Technologies
Fermentation	Pyrolysis
Anaerobic digestion	Gasification
Composting	Incineration

contingent on various considerations, including social and environmental acceptability, technological efficiency, and economic gain.

10.4.1 Bioconversion Technology for MSW

Biochemical conversion of MSW utilizes biological agents such as microorganisms and enzymes to degrade organic matter in order to generate biogas and gather value-added goods. MSW and other biomass wastes such as plastic, agricultural leftovers, sewage sludge, and tires can be converted to usable products such as ethanol, hydrogen, and acetic acid via these processes. In addition, any bioconversion method produces either clean energy in the form of biogas that can be turned into heat and electricity via a gas engine or compost that may be utilized as a soil conditioner.

Fermentation is a technique that is used to control waste and generate energy. It is mainly utilized in companies that manufacture food and beverages in a number of nations. In the presence of yeast and bacteria, it is a metabolic process that transforms sugar into alcohol, acids, and gases. As with anaerobic digestion, the MSW fermentation process uses yeast and bacteria to act on the waste in the absence of oxygen to produce acids, ethanol, and trace gases that are environmentally friendly fuels. Despite this, in the majority of developing countries, fermentation as a waste treatment technology still seems to be limited to breweries. There is no evidence of its utility for MSW in general, even in developed countries. Nonetheless, this is a growing technique that developing countries should investigate for MSW management.

Anaerobic digestion is a naturally occurring biological process that converts organic waste into biofertilizers and biogas without oxygen. Anaerobic digestion is a wet process for waste containing more than 85% moisture or a dry procedure for waste containing less than 80% moisture. Anaerobic digestion is fast gaining traction as the primary technology for treating manures, slurries, and wet household organics. It is particularly well suited to organic waste, such as food waste, which typically contains a high moisture content. Anaerobic digestion enables the biochemical degradation of organic waste from various sources under carefully controlled, oxygen-free conditions, resulting in the creation of biogas that can be utilized to generate electricity and heat. Compared to aerobic composting, anaerobic digestion techniques consume far less energy and produce significantly less biologically generated heat, although additional heat may be necessary to maintain appropriate temperatures in an anaerobic digestion process. For many years, anaerobic digestion technology has been widely deployed throughout the world. While some deployments of anaerobic digestion have been effective, others have been abject failures, particularly in some developing countries. According to Mudhoo et al. [43], Tanzania has successfully established an anaerobic digestion project called "TAKA" (waste). They explain that this project addresses the growing problem of municipal solid waste by producing biogas for electricity generation.

Composting is the aerobic decomposition of biodegradable organic matter by yeasts, bacteria, fungus, and other organisms in a warm, moist atmosphere. To achieve the desired decomposition rate and completeness, factors affecting decomposition rate and completeness are regulated in line with local needs and constraints. Trash selection or exclusion, particle size reduction, mixing, seeding, moisture addition, and aeration are only a few of these variables.

Composting creates a biologically stable product free of viable pathogens and plant seeds, which can be used to boost soil nutrients in agricultural areas. More expensive facilities (often in affluent nations) prepare waste mechanically and promote decomposition, whereas less expensive facilities (mostly in developing countries) prioritize natural processes, minimizing mechanical requirements. Composting is often divided into three stages: preprocessing, including size reduction and nutrient addition; breakdown and stabilization of organic matter; and post-processing, which includes grinding and screening. These methods significantly reduce up to half of the weight and volume of the waste and, at the same time, still provide a stable material to be used in agriculture. There are numerous composting technologies available, but the most prevalent are aerobic, anaerobic, and vermicomposting.

Vermicomposting is a relatively new technique for managing municipal solid waste and sludge [1, 44]. It is essentially the decomposition of organic materials by certain earthworm species [44–46]. Due to the worms' droppings and the broken organic waste, vermicomposting is more nutrient-dense than other types of compost and can thus be used as a soil conditioner and natural fertilizer. Numerous scholars assert that composting is the bedrock of garbage sector sustainability and, as such, argue that it should be a more prevalent practice in developing nations due to its ability to be performed on both local and big stages. However, large-scale and centralized composting plants are frequently uneconomical in underdeveloped countries because of high operational, maintenance, and transportation costs. Commercial composting is typically only viable if there is a ready market for the final compost product. Furthermore, subsistence farming is still widely practised in the majority of underdeveloped countries, with farmers relying on manure from their own animals. Therefore, compost demand may not be sufficient to cover the cost of production in the majority of developing countries.

For example, Taiwo [47] suggests that MSW composting has failed in several places in Nigeria due to a lack of funding for upkeep and a lack of markets for the compost generated. On the other side, Ghana, which is located in the same West African subregion as Nigeria, is lowering its reliance on imported fertilizers through composting. Farmers in Ghana used to receive subsidies worth more than US$ 63 million per year, but the usage of ACARP's compost fertilizer, which is now widely used throughout the country, has allowed the government to lessen its reliance on imported fertilizers [48]. While not a cure for today's waste management challenges, composting should be a critical component of MSW management systems in developing countries.

Despite its long history and extensive study and development, the US EPA notes that little is known about the amount of decomposition when it comes to composting's environmental implications. As a result, the quantity of gases produced during decomposition is unknown, and only broad data and theoretical gas composition estimates are frequently employed. Ammonia (NH_3) and Carbon dioxide (CO_2) gases are the composting process's principal metabolic by-products. Although CO_2 is a well-known greenhouse gas, little is known about the exact yields and CO_2 and NH_3 production rates during composting. Additionally, composting facilities produce negligible amounts of leachate when the compost is covered, and the moisture content is kept around optimal levels. Given that the amount of leachate created within a composting facility is typically assumed to be modest, it follows that the leachate produced in composting facilities is generally ignored.

10.4.2 Thermal Conversion Technology for MSW

Thermal conversion is accomplished through three primary methods: gasification in reduced air, pyrolysis in the absence of air, and combustion in excess air. Incineration is a prevalent method for generating both heat and electricity from garbage. Thermal conversion solutions are widely utilized in industrialized countries but are rarely used in developing countries due to the high costs associated with their development, operation, and maintenance.

Pyrolysis and gasification are advanced thermal treatment procedures that serve as an alternative to incineration. They are defined by the transformation of waste into product gas that serves as an energy carrier for subsequent combustion in a boiler or a gas engine. These technologies have a number of potential advantages over conventional cremation. For example, pyrolysis or gasification-based WTE techniques provide a reduction in dioxin and NOx emissions. Pyrolysis is a form of thermal breakdown in which biomass is burnt to temperatures between 400 and 550 degrees Celsius without oxygen to produce aerosols, vapors and incondensable gases, and char. Maintaining this temperature usually necessitates the use of an external source of heat. Pyrolysis of raw municipal waste often requires mechanical preparation and separation of inert materials, metals, and glass before the remaining waste is processed. By and large, the primary environmental hazards associated with solid waste disposal are the release of gas and leachate by decomposing trash. These factors contribute to all forms of pollution from water to air, soil, landscape, and climate (Table 10.4).

Incineration is primarily the process of destroying garbage in a furnace by managing combustion at high temperatures to generate steam, which is then used to generate electricity via steam turbines. By incinerating garbage, roughly 70% of the total waste mass and 90% of the total volume can be eliminated, leaving only a tiny amount of waste disposed of in a landfill, alleviating strain on and demand for landfills. Incineration is particularly well-suited for treating particular hazardous wastes (medical waste), as the high temperature destroys pathogens and poisons that cause disease. In nations such as Japan, where landfill space is scarce, waste incineration is popular, while Sweden and Denmark have been utilizing the energy created by waste incineration for decades. Nonetheless, incineration is widely employed in some developed countries and is limited to the burning of medical waste in some developing countries, such as Ghana, due to the high failure rate of incineration deployment in poor African countries. For example, a WTE incinerator built and put into service in Tanzania recently with the assistance of international experts has been found to be inoperable [43]. The incinerator's operation costs and high maintenance contributed to the project's demise.

Additionally, linked environmental issues such as air pollution are a significant impediment to global incineration. Generally, considerable opposition to the construction of incineration plants near human settlements exists because of the potential for adverse air pollution effects on residents living near the plants. This would be particularly perilous in the majority of poorer countries, which have weak legal systems and ineffective environmental regulation.

The amount and nature of MSW are important factors determining how these wastes should be handled and managed. Such knowledge is necessary and beneficial for the municipality's establishment of municipal solid waste to energy conversion plants. The calorific value, as well as elemental composition of MSW, will be used by engineers and scientists to assess its utility as a fuel.

Table 10.4 The most significant environmental consequences of MSW processing and disposal [49]

Activity	Climate	Landscape	Air	Soil	Water
Waste Transportation	Significant contribution of CO_2		CO_2, SO_2, NO_x, Dust, Odour, Noise	Spills	Spills
Recycling	Recycling	Minor emissions	Dust, Noise	Landfilling of residues	Wastewater
Composting	Small emissions of GHG	Some visual effect	CO_2, CH_4, VOCs, Dust, Odour, Bioaerosols	Minor impact	Leachate
Incineration	GHG	Visual effect	SO_2, NO_x, N_2O, HCl, HF, CO, CO_2, Dioxins, Furans, PAHs, VOCs, Odour, Noise	Fly ash, Slags	Fall-out of atmospheric pollutants
Landfilling	Worst option for GHG Emission	Visual effect, vermin	CO2, CH4, Odour, Noise, VOCs	Heavy metals, Synthetic organic compounds	Leachate (Heavy Metals, synthetic organic compounds)
Open dumping	Small emissions of GHG	Vermin, insects	Bioaerosols, Dust, Odour	Bacteria, Viruses, Heavy Metals, PAHs, PCBs	Bacteria, Viruses, Heavy Metals

CO_2 = carbon dioxide; CH_4 = methane; VOCs = volatile organic compounds; SO_2 = sulphur dioxide; NOx = nitrogen oxides; N_2O = nitrous oxide; HCl = hydrochloric acid; HF = hydrofluoric acid, CO = carbon monoxide; PAHs = polycyclic aromatic hydrocarbons

In the meantime, such knowledge can help in the prediction of the composition of gaseous emissions. After that, the MSW is treated with energy conversion technologies such as gasification and incineration. The presence of potentially hazardous compounds in the ash, on the other hand, should be carefully considered. In this scenario, the waste composition may help determine if the material is suitable for composting or biological conversion to produce biogas as a fuel. Meanwhile, the structure of MSW is greatly influenced by the passage of time. The rate of biodegradation of such MSW over time defines the rate of recyclable material, especially the organic contents.

Effective waste management and resource recovery require reliable data on waste characteristics and generation rates. Accurate forecasting of MSW generation and knowledge of waste characteristics provide the foundational data for designing, planning, and operating a waste management system. However, reliable data on MSW characteristics and generation that may be used to prepare for optimal waste management are frequently unavailable in the majority of developing nations. This is mainly because MSW generation trends vary according to regional consumption patterns. Numerous other elements, such as urban population, consumption rate, economic development, administrative systems, and geographic location, contribute significantly to MSW formation. Among these elements, economic conditions and urban population appear to be the two most significant determinants of MSW volume.

10.5 Incinerator

An incinerator can be used for both concepts in MSW management, either processing or disposal of solid waste. In solid waste processing, an incinerator may significantly reduce MSW volume, alleviating the problem of limited landfill space. When used in conjunction with waste-to-energy conversion, an incinerator can help produce energy. However, regardless of the incinerator's construction or intended use, the bottom ash, a by-product of the incinerator, must eventually be disposed of in a landfill.

Incineration is a method of handling MSW that reduces the need for landfilling and recovers the energy contained in the materials being burned. With the implementation of new device designs over the last few decades, incineration technology has advanced significantly. Each change to these processes has the potential to change the physical and chemical characteristics of the residue streams.

Incineration of urban waste is rapidly becoming the preferred approach for removing more than 90% of its volume in a hygienic and cost-effective manner, leaving an inert, solid residue or slag. Furthermore, the heat produced in modern refuse incinerators is typically recovered and used. The incineration process produces significant amounts of particulate and gaseous emissions, which must be efficiently separated from the combustion gases to meet increasingly stringent emission requirements. Renewable energy sources, in the conventional sense, are those that are replenished by nature, such as hydropower, wind power, solar power, and biomass. Municipal solid waste (MSW) is a term used to describe the products discarded in urban areas, mostly household waste with some industrial waste thrown in for good measure. MSW is collected and disposed of by municipalities. MSW is a biomass source and includes a considerable amount of paper, food waste, wood and yard trimmings, cotton, and leather. Waste management strategy is often articulated in the following order of diminishing priorities:

(i) Substitution and cleaner technology initiatives are used to reduce waste production, potential hazards, and energy usage.
(ii) Utilization or recycling.
(iii) Energy recovery from incineration.
(iv) Garbage disposal.

Incineration and landfilling, despite being ranked third and fourth in this priority list, play significant roles in waste management in many parts of the world and will continue to do so in the future. The incineration process is clearly not a final waste management stage, and the numerous incineration residues must be used or disposed of. According to the priority order, using residues rather than landfilling is preferred in practice, as long as it does not have undesirable environmental or health consequences. In reality, there are a number of reasons that prevent incinerator residues from being used:

(i) Regulations in place.
(ii) A scarcity of financial incentives.
(iii) Concerns over liability.
(iv) Practices for separating residues.
(v) Uncertainties over the residues' functional properties.
(vi) Uncertainties about the scope and acceptability of environmental effects and health risks.

As a result, in some countries, landfilling or storage are the most popular MSW incinerator residue management solutions. In the United States, for example, the vast majority of incinerator residues are actually landfilled as mixed ash. In comparison, in some European countries (e.g., Denmark, France, Germany, and the Netherlands), large amounts of bottom ash from incinerators are used for road building and other purposes (40–60% or more). In most countries, air pollution control system residues are landfilled, but in the Netherlands, about half of the fly ash produced is used as a filler in asphalt. To minimize the risk of undesirable environmental effects, incinerator residues need to be handled prior to or during disposal in certain situations.

The formation and release of leachate, as well as fugitive dust pollution, are the two most significant possible environmental impacts associated with the disposal of incineration residues. Fugitive dust issues are most common during the landfill's relatively short duration of actual deposition. The use of covered or closed transport containers and preserving sufficient moisture content in residues are widely regarded as effective methods for reducing fugitive dust. The formation of leachate, on the other hand, may be a short- and long-term issue that can be mitigated by implementing a proper disposal plan as well as proper landfill design and service.

MSW also contains materials derived from fossil fuels, such as plastics, rubber, and fabrics. MSW is classified as a renewable energy resource by the US Environmental Protection Agency because it would otherwise be disposed of in landfills [50]. MSW is only counted as green energy by the US Department of Energy if the energy content of the MSW source stream is biogenic. The nonrenewable component must be isolated or approved as part of the fuel (80), and almost all wastes in MSW are considered renewable after resource recovery and recycling.

Waste-to-energy (WTE) technologies use direct combustion (e.g., incineration, pyrolysis, and gasification) or the processing of combustible fuels such as methane, hydrogen, and other synthetic fuels to extract energy from waste (e.g., anaerobic digestion, mechanical biological treatment, and refuse-derived fuel). The two main WTE technologies that have been used widely around the world are incineration and gasification. Around 130 million tonnes of MSW are expected to be combusted annually in over 600 WTE facilities around the world, generating electricity and steam for district heating, as well as recovered metals for recycling [51].

WTE incineration has long been recognized as a viable alternative to landfilling and composting for solid waste management [51, 52]. Table 10.5, 10.6 and 10.7 compare the benefits and drawbacks of the three main MSW disposal technologies: landfilling, composting, and incineration. MSW incineration in WTE facilities avoids the aqueous and gaseous waste that comes with landfilling and provides a reliable, clean energy source.

WTE has been widely used in Europe and developed Asian countries such as Japan and Singapore as an established, environmentally sound technology. In China, the demand for long-term urban development is unprecedented: 300 million people will migrate from the countryside to cities (18–20 million people each year), necessitating the construction of over 400 new cities over the next two decades [53, 54]. By 2050, it is projected that 70% of China's population, or approximately 1.0 billion people, will live in cities.

Dealing with the volume of MSW produced due to urbanization and people's improving lifestyles is a daunting task. Simultaneously, China, the world's second-largest energy user and third-largest oil importer face huge energy demand to fuel its economic growth. In a carbon-constrained environment, discarded MSW is a viable energy source for electricity generation; thus, an MSW management technology that recovers energy from waste is a promising alternative MSW disposal issue in China. WTE is gaining traction in China, owing to its ability to minimize the amount of MSW that must be disposed of in landfills, as well as the country's reliance on fossil fuels and greenhouse gas (GHG) emissions.

China faces several environmental challenges, including air pollution, water and soil pollution, waste disposal, water scarcity, and massive energy demand, due to its rapidly growing population, rapidly developing economic and social systems, accelerated urbanization, and need for improvements in both living standards and surrounding ecosystems [55]. MSW management is one of the big issues affecting China's environmental quality and the long-term growth of its cities.

Table 10.5 Advantages and disadvantages of landfilling [38]

Advantages of landfill	Disadvantages of landfill
• A one-size-fits-all solution for waste disposal.	• The cost of a liner, a leachate collection and removal method, and tighter regulations increases dramatically.
• Cost-effective and easy to implement.	• Requires a sizable amount of land.
• Complements other technology solutions for residual waste management.	• It does not accomplish the goals of minimizing MSW volume and transforming it into reusable resources.
• Possibility of obtaining landfill gas as a by-product for domestic and. • Industrial use,	• This could result in secondary pollution issues, such as groundwater contamination, air pollution, and soil contamination.
• Costs associated with landfill expansion on a gradual basis.	• Possibility of providing a breeding ground for pests and diseases.
	• There are lengthy postclosure care obligations and unknowns, and the project imposes long-term limits on the site's land use.
	• The position of the site may be constrained by the geology of the area and the natural stability of the underground soil.
	• Due to public acceptance and space constraints, landfills are often located far from the sources of waste, necessitating long-distance transport.

Since 2005, MSW generation has increased at an annual rate of 8–10%, with over 150 million tonnes of MSW generated each year [56]. The percentage of MSW handled by MSW management facilities has risen from around 5% in the 1980s to approximately 55% today.

In China, MSW is handled by a combination of landfilling, composting, and incineration. In China, landfilling is the most common method of waste disposal, accounting for more than 80% of all handled MSW. However, due to the lack of leachate collection and treatment facilities in over half of the current landfills, significant surface and groundwater pollution has occurred. Land availability, on the other hand, restricts the development of new lined landfills in many cities.

Composting has fallen out of favor as an MSW management method as it falls from 17% to 4% between the years 2001 to 2006, respectively. Due to a lack of waste sorting and materials separation, compost products with low nutrient content and high heavy metal levels. Meanwhile, the amount of MSW incinerated has steadily increased, and incineration has gradually surpassed landfilling as the second most effective MSW management method.

The vast majority of WTE plants in China are focused on incineration, which is a more mature and simple technology than others. Incineration converts heterogeneous wastes into more homogeneous residues (flue gas, fly ash, and bottom ash), with the primary advantage of significant weight (up to 75%) and volume reduction in the waste (up to 90%).

Table 10.6 Advantages and disadvantages of an incinerator [38]

Advantages of incinerator	Disadvantages of incinerator
• Provides a significant reduction (by 90%) in the overall waste needing landfill disposal.	• High capital, operating, and repair costs in comparison to alternatives that do not include incineration.
• Requires minimum waste preprocessing.	• Operator expertise is needed.
• Bottom ash from incineration is biologically safe and stable, making it suitable for road construction and the construction industry.	• To treat the flue gas, air pollution control equipment is needed, and the fly ash must be disposed of in hazardous waste landfills.
• Burning is a very stable method that can be used to dispose of almost any waste, and the burning process can be properly managed.	• Additional raw materials must be used to replace those that have been incinerated, and this method does not result in long-term energy savings because resources are not recycled.
• The heat generated by combustion can be used to generate steam and. • Electricity.	• May sometimes discourage recycling and waste reduction. • Public opinion can be detrimental, especially when dioxins are emitted.
• Incineration facilities may be situated close to residential areas, minimizing the expense of shipping MSW to waste disposal sites.	
• Air emissions are highly controllable.	
• More effective land use and resource alignment than landfilling.	

Table 10.7 Advantages and disadvantages of composting [38]

Advantages of composting	Disadvantages of composting
• Converts biodegradable organic waste into organic fertilizer.	• It consumes more space than any other method of waste management.
• Reduces the amount of waste that must be landfilled and works well in conjunction with landfilling and materials recovery.	• It is also prohibitively expensive to introduce and sustain and offers no environmental or economic benefits over incineration.
	• Requires waste to be reduced in size and some waste to be separated.
	• There are perception problems, such as odor and bioaerosol pollution during the composting process, as well as disease-producing plants, weeds, and insects control.
	• The quality of the fertilizer produced is substandard, and the quantity produced is excessive, resulting in an inadequate market demand.
	• Compost materials contain heavy metals and pathogens that can contaminate soil.

MSW is combusted at high temperatures in a specially constructed chamber with a constant air supply to ensure turbulence and total combustion of the components to their stable and normal molecular forms. The solid contaminants may be disposed of in landfills or washed and reused off-site for specific building projects [57]. Particulate matter, heavy metals, dioxins, sulfur dioxide, and hydrochloric acid may be present in large quantities in flue gases.

The most critical environmental issue associated with the incineration of MSW used to be dioxins. Incinerators can now operate with virtually no dioxins thanks to major improvements in incinerator design and pollution control prompted by stricter legislation in developed countries. Other contaminants in the air can be efficiently managed and eliminated by the flue gas cleaning system during the combustion phase.

The heat produced by MSW combustion can be collected and used for power generation or heating. The selling of electricity/steam produced as a by-product of the incinerator process helps to offset the cost of incineration. Except for a few small-scale (up to 200 tonnes/day) furnaces used as a support treatment in integrated waste management plants, all MSW incineration facilities in China will generate electricity [56].

Incineration produces carbon dioxide and nitrous oxide, while the anaerobic decomposition of MSW in landfills produces methane (which is 21 times more potent than carbon dioxide over 100 years). Approximately up to 4% of global anthropogenic GHG emissions come from methane emitted at solid waste disposal sites. In 2004, China's MSW management emitted 1.87–3.37 Mt. of methane [58].

By preventing the release of methane from landfills and offsetting pollution from fossil fuel power plants, WTE will reduce MSW's contribution to GHG emissions compared to landfilling. WTE will mitigate up to 1.3 tonnes of carbon equivalent per ton of MSW by avoiding methane release from landfills and offsetting pollution from fossil fuel power plants, according to comparative studies of WTE and landfilling. According to the US EPA study in 2006, for every ton of MSW handled by WTE rather than being landfilled in 2003, a net emission reduction of 0.15 tons of carbon equivalent reduction was achieved.

China has implemented policies in recent years to slow the rise of its greenhouse gas emissions. Even though MSW management contributes a small percentage of GHG emissions (1%) [58], WTE will help China reduce its overall GHG emissions. WTE can also reduce the pollution and fuel consumption associated with transporting MSW to distant landfills.

The first modernized WTE plant in China (Shenzhen, Guangdong) is an excellent example of technological advancement in incinerating unsorted MSW with high moisture levels and low heat contents. In the late 1980s, two 150 t/d incinerators were imported from Japan. To properly incinerate the local waste, they were run with long drying and incineration periods. They had issues such as grate blockage and large temperature variations in the combustion chamber. To help combustion, additional fuel was needed, resulting in a significant increase in operating costs. The incinerator's performance was generally unreliable, and its power output was limited (500 kW). In 1996, a third incinerator was installed, with over 80% of parts made in the United States, as well as modifications to the imported incinerators and the addition of a 3 MW generator unit. Unsorted MSW with calorific values of >3300 kJ/kg and moisture contents of 55% is adequately incinerated at the facility following this significant upgrade, producing approximately 200 kWh electricity per ton of MSW.

WTE has been increasingly adopted as an alternative to landfilling in China, especially in the relatively more developed cities, as urbanization. The main MSW incineration technologies currently in use in China are the Stoke grate (more than half) and fluidized bed. The majority of them use imported equipment and have modest 500 tonnes/day or high 1000 tonnes/day incineration capacities.

In China, over a hundred businesses, research institutes, and universities are researching and developing WTE incineration technologies and related equipment. China has made considerable progress in the creation of novel incineration technologies in recent years. On the basis of domestic incineration technologies, more than 20 WTE facilities have been installed. China has gradually accepted incinerators based on domestically developed technologies, especially for circulating fluidized bed incinerators [56, 57]. The improvement in co-firing performance is expected to make domestic WTE technologies much more efficient.

10.6 Disposal of Solid Waste

The characteristics of solid wastes have been greatly impacted by inefficient bin collection processes, transfer, and transport systems. In addition, insufficient route planning, an absence of information about the timetable for collection [59], the amount of solid waste collection vehicles and deplorable road conditions [60], and inadequate facilities [61] may all affect the characteristics of solid wastes.

Sharholy et al. [6] investigated and published on reliable and affordable waste collection services. One of the most relevant factors influencing solid waste management is authorities' knowledge of treatment [62]. Their findings revealed that the availability of waste disposal facilities has a significant impact on waste disposal choices. Due to a shortage of waste containers and the longer distance needed to transport them, waste is more likely to be discarded in open areas and along the roadside along the journey. Inadequate financial resources, a lack of legislation, and poorly equipped and designed landfills all contribute to the limitation of safe solid waste disposal, according to Pokhrel and Viraraghavan [63].

10.6.1 Disposal of Municipal Solid Waste

MSW generation, management, and disposal are major environmental concerns in urban areas. The absence of MSW management and disposal is causing grave environmental problems, including contamination of the soil, air, water, and aesthetics.

As a result of increased greenhouse gas emissions, such environmental concerns are linked to human health problems [64].

Hazardous substances in household waste vary from waste streams derived from industrial sources. Hazardous waste laws such as the European Hazardous Waste Directive 91/689/EEC and the United States Resource Conservation and Recovery Act 1976 (RCRA) do not apply to them. Household hazardous wastes (HHW) are disposed of alongside municipal solid waste in landfills. The amount, type, and significance of such disposal are all unknown. The volume of household hazardous waste is usually believed to be negligible, so the disposal risks are minimal. However, the separation of municipal solid waste, industrial, and other wastes stresses the toxic and hazardous elements found in these wastes. The presence of many chemicals in household goods has sparked widespread concern.

About 71% of municipal solid waste is disposed of in landfills around the world [65]. Items like mercury-containing batteries, pharmaceuticals, oils, automobile repair materials, paints, and many other items are covered under MSW. Magazines, hardboard paper, food, and yard waste, on the other hand, make up more than 53% of landfilled wastes and are biodegradable by anaerobic bacteria. As a result, in Europe and the United States, landfilling is the primary waste disposal method.

The New Source Performance Standards of the Clean Air Act, Subtitle D of the Resource Conservation and Recovery Act, as well as other state regulations, govern the design and operation of landfills in the United States. As a result, landfills have progressed from straightforward, open pits to sophisticated waste-containment facilities and sites. They are fully isolated from the rest of the planet, collect contaminated water from waste contact (leachate), and regulate gas migration. A typical landfill site is dug and lined with a structure that comprises layers for groundwater protection by preventing leachate migration to the groundwater level, as well as layers for collecting and treating leachate collected from the landfill site itself. Fig. 10.3 depicts a cross-section of a typical landfill configuration.

10.6.2 Disposal of Plastics Waste

Plastic waste management is a significant global environmental problem. Each year, Europe, the United States, and Japan produced 50 million tons of post-consumer plastic waste. Dumping these plastic wastes in landfills has been determined to be unsustainable for the environment. Furthermore, landfill sites and resources are increasingly diminishing.

Plastics, which are basically hydrocarbons, have calorific values ranging from 30 to 40 MJ/kg. As a result, they may be burned or incinerated to produce power and heat in municipal or other dedicated wastes. In certain production methods, such as blast furnaces and cement kilns, they can also be used as an alternative to fossil fuels. These thermal applications are capable of completely destroying these plastic wastes. This method of burning plastic waste will eventually take the place of fossil fuels. However, this necessitates the implementation of additional advanced

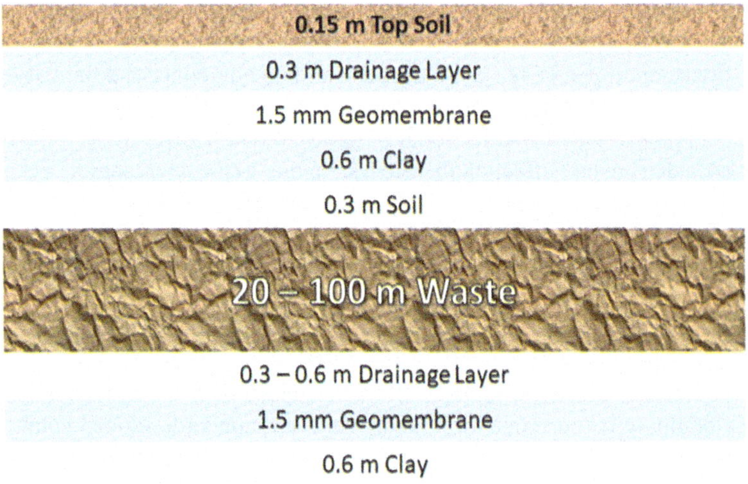

Fig. 10.3 Typical cross-section of a landfill

emission control steps [66]. Nonetheless, effective waste management will help to reduce greenhouse gas emissions [67]. Plastics and other nonbiodegradable products can be disposed of in landfills. Biological solids (biosolids) can also be used as a source of energy by anaerobically converting them to landfill biogas. As a result, burning plastics and other nonbiodegradable products are more harmful than landfilling because it emits more greenhouse gases.

10.6.3 Solid Waste Management Issues in Rural Communities in Developing Countries

Historically, waste disposal was not a serious issue due to a small population and a vast quantity of land available for waste assimilation. However, as populations migrated from scattering geographical areas to gather in settlements, man's requirement for efficient waste treatment and disposal became apparent. Thus, while safe waste disposal has become a global standard, MSW treatment and disposal remains a neglected topic in many developing nations.

In underdeveloped countries, improper waste disposal manifests itself through the dumping of waste into bodies of water and wetlands, as well as the burning of rubbish to reduce its volume. These practices are known to have negative environmental consequences ranging from contaminating natural resources and ecosystems to generating health problems that may result in creating a public nuisance, long-term public health complications, and degrading the environment and aesthetics.

Municipal solid waste is a severe and pervasive problem in many developing countries' cities and rural areas. As sources of organic and inorganic household waste, many canals and drains are widely used to discharge different forms of solid waste as open spaces. Since there are no permanent garbage disposal systems in place, drains and open canals are clogged with large amounts of solid waste. As a result, they are no longer functional. These solid wastes are primarily composed of plastic and paper, with a small amount of toxic content. However, such toxic products have a detrimental effect on the ecosystem due to the degradation of their biodegradable constituents, which results in large BOD loads being added to the local ecosystem.

Nonetheless, improper trash disposal is not unique to underdeveloped countries; it has occurred in every country at some point. Thus, every country has faced the difficulty of incorrect waste disposal at some point in history. For example, cholera epidemics were reported in the 1950s and 1960s in the United Kingdom as a result of inadequate sanitation, particularly MSW management [68]. At the moment, open rubbish dumping is the standard in Ghana and other underdeveloped countries [6, 69, 70]. Open dumping is an illegal practice in which any form of waste, such as domestic waste, rubbish, tires, metal, construction and demolition waste, or any other material, is put anywhere other than a designated landfill or facility.

Open dumps are non-engineered sites that lack leachate and landfill gas management. They have a detrimental influence on the environment, resulting in long-term contamination of the groundwater, air, and soil. As a result, in developing countries, landfilling is the most frequently applied method of MSW treatment and disposal. It is the simplest and typically least expensive way of waste disposal. The primary considerations in the design, construction, operation, and decommissioning of landfills, which are landfill gas emissions control and groundwater contamination prevention, are frequently overlooked, owing to the high capital costs and lack of technical expertise required for landfilling in some developing countries. As a result, most developing countries employ non-engineered landfilling, which is a veiled form of open dumping.

Despite extensive efforts over the last few decades in a number of developing nations, with technical and financial assistance from certain international organizations and rich countries, significant reforms in the disposal of MSW have yet to be achieved. This failure can be ascribed to the lack of an enabling environment for MSW management, such as waste management governance in terms of legal, policy, institutional and financial frameworks, as well as enough technical capability, which is a necessary component of sustainable waste management.

10.7 Disposal of Solid Waste to Landfill

For decades, the amount of waste generated around the world and in EU countries has been steadily increasing. Due to technological, economic, and legal reasons, landfilling remains the most realistic waste management solution in most EU countries (Table 10.8) [20, 71].

Landfilling is considered an efficient waste management system [72, 73]; however, there are many explanations why landfilling tends to be the least rational waste management method. Landfills can only accommodate for a certain amount of time, and landfill reclamation can take many decades [74]. Biogas, as well as leachates, may have a significant negative effect on the climate [74, 75]. Furthermore, MSW is dumped in landfills without being sorted in a number of countries. Furthermore, for the vast majority of landfills around the world, foul odors and air pollution cause severe sanitary issues. They are the primary contributors to the "not in my backyard" (NIMBY) phenomenon in neighboring neighborhoods [76]. Waste management has progressed significantly, and legislation-imposed regulation (e.g., setting recycling targets and limiting the amount of biologically degradable waste that can be landfilled) has slowed the pace of landfill expansion. Evaluation of the environmental effects of landfills (Fig. 10.4) is a critical issue to related parties that has recently gotten more attention due to rising environmental concerns.

The majority of MSW was disposed of in open dumps or tip sites around the world after WWII. In 1959, sanitary landfilling was described as the controlled operation by the American Society of Civil Engineers (ASCE) in which the MSW was dumped in certain levels, each of which was compacted and covered in the soil before being deposited in the next layer. A landfill differs from a dump in that there is no effort to remove the waste from the underlying soil or rock strata, and waste is deposited directly into the groundwater where the hole reaches below the groundwater surface.

A sanitary landfill, on the other hand, is a well-designed structure with final covers, leachate collection and disposal systems, and bottom liners. Landfills are intended to hold on to ensure as well as process waste. The migration of leachate and landfill gas causes much of the possible danger from MSW landfills; thus, the environmental effects of the many landfills that occur around the world cannot be overlooked. Biological processes in major emissions (leachates and biogas) have a significant impact on them. If MSW is disposed of in a landfill without being pretreated, pollutants form during the landfill's operation and continue to exist even after the landfill is closed [77, 78].

The technique of sanitary landfilling for final waste disposal is still widely recognized and practised. Despite this, the scientific evidence on the environmental and health effects of waste remains inconclusive. The European Union's (EU) Waste Landfill Directive defined specific goals for waste volume reduction and placed stringent conditions on landfilling and landfill sites. Environmental impact assessment of landfills is a crucial area of research that has recently received increased attention as a result of growing environmental concerns.

While sanitary sites could be regarded as the best sites to date, most developing countries still rely on the conventional landfill as waste disposal because of the expensive construction of the sanitary landfill. Open dumping remains the practice for solid waste disposal in some poor developing countries.

Table 10.8 Municipal solid waste generated in 1995 and 2016 (kg/capita) and share of landfill disposal (%) [20, 70]

Country	1995	2016	2017	Waste Treatment- Landfill, 2016 (share of landfill disposal)
Malta	387	584	604	92%
Greece	331	498	-	82%
Cyprus	595	592	637	81%
Romania	254	228	272	80%
Latvia	184	367	438	72%
Slovakia	294	344	378	66%
Bulgaria	531	404	416	64%
Spain	365	443	462	57%
Hungary	377	380	385	51%
Czech Republic	312	339	344	50%
Portugal*	351	483	487	49%
Poland	284	307	315	37%
Lithuania	542	422	455	31%
Italy	468	436	489	28%
United Kingdom*	501	476	-	28%
Slovenia**	469	434	471	24%
France	476	510	513	22%
Ireland*	430	615	-	22%
Luxemburg	587	614	607	17%
Estonia	370	327	390	12%
Austria	480	552	570	3%
Finland	437	504	510	3%
Belgium	446	414	409	1%
Denmark	521	777	781	1%
Germany	623	625	633	1%
Netherlands	509	518	513	1%
Sweden	386	442	452	1%

*** Data from 2014; ** Data from 2015.**

10.7.1 Landfill Operation

A good landfill needs not only proper siting and construction but also proper operation and monitoring. A landfill's technological activity necessitates the integration of a number of components, including machinery, solid waste filling sequences,

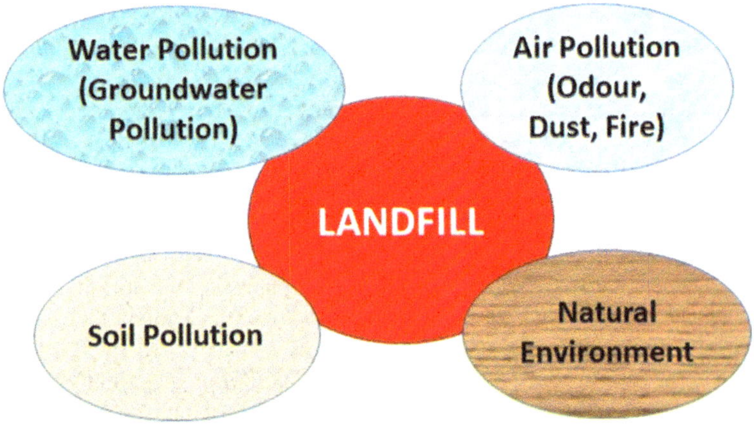

Fig. 10.4 Potential impact of landfills on the environment

methods of positioning and compaction, and daily cover placement. Aspects including protection and safety, waste input control, placement of temporary roads, and stormwater management must also be discussed and enforced. Environmental management processes make sure the noise is reduced; prevent fires and garbage; manage smells, mosquitos, and vermin; and use barriers or barriers established around the landfill; all help run the landfill operation smoothly.

Leachate and air quality must be tested, as well as landfill's settlement rate must be examined since it can have an effect on landfill stability. Results must also be made accessible to the appropriate parties, public agencies and also made available to the general public. The following issues must be addressed in the organization of a landfill:

(i) The term "waste admitted to disposal" is described as "waste that has been accepted for disposal".
(ii) Regulation of landfill entrances for municipal solid waste.
(iii) Techniques of waste management.
(iv) Biological pretreatment of municipal solid waste if required.
(v) Toxic, hazardous waste, or radioactive waste.
(vi) Documentation and recording of landfill operations, as well as general improvement of landfill conditions.

In general, waste from home, industrial wastes such as domestic waste, and nonhazardous waste that the local government and management permit to be disposed of on municipal solid waste landfills can be disposed of. Nonmunicipal waste must be identified and listed before it can be disposed of. In addition, there should be a list of industrial and hazardous wastes that includes wastes that are not allowed to be disposed of at a municipal landfill. The waste forms mentioned should be classified according to national or international waste classification systems.

The type of waste delivered must match the list of wastes that can be disposed of. This list must be posted at the landfill and included in the operation manual. Delivered waste that has been deemed questionable by the yardmen should be unloaded near the entrance area for further inspection (sampling, etc.). The waste should be inspected both at the point of entry and at the point of disposal (to detect hidden components). The waste is registered and recorded at the entrance. According to the EC-Directive on the Landfill of Waste, the operator must follow the following reception procedures in EU countries:

(i) In the case of municipal waste, the transport papers (company and vehicle) are checked, as well as the waste documents in the case of nonmunicipal waste.
(ii) Calculation of the waste's weight or volume.
(iii) Checking the waste for compliance with the definition given in the waste generator's documentation (odor, color, quality, materials; in some instances, it is also for radioactive substances) at the entrance and the deposit location.
(iv) Keeping a record of the amounts and types of waste deposited, as well as displaying a bill.

For the disposal of special waste in smaller municipal solid waste landfills, the following guidelines are made:

Recyclable Materials
Suppose there are any recycling options in the region. In that case, materials that are recyclable should be refused for disposal. There should be a separate, secure area for the temporary storage of recyclable materials and containers at the front of the facility, as well as a wide-open space for other items. The word "recyclable waste" refers to paper, green waste, plastic, tires as well as other products that are already being separated.

Bulky Waste
Unique forms of bulky wastes (such as refrigerators or iron materials) should be disposed of in such a way that landfill dumping is avoided.

Debris or Construction Waste
Debris or building waste should be disposed of as little as possible in municipal landfills. Debris can only be approved for service road maintenance or to cover dumping areas. In the event that debris cannot be recycled, special landfills for inert materials would be established.

Sludge
Sewage sludge disposal can be problematic. The toxicity of the sludge sometimes results in the failure to meet the acceptance requirements. Layers of spread sludge can also pose a threat to the landfill's stability (risk of sliding, etc.). A good rule of thumb is that sludge does not account for more than 10% of the total weight of waste disposed of.

Industrial Waste and Other Nonhazardous Waste of Nonmunicipal Origin
Nonhazardous waste from outside the municipality (mining waste, slag, etc.) must be refused for disposal in municipal landfills. The generator with a local government allowance and the user, who must accept waste properties and landfill capacities, are the only exceptions.

Hazardous Waste
Hazardous waste disposal is not permitted on municipal landfills, according to the EC-Directive definitions on Waste Landfilling.

10.7.2 Importance of Waste Placement and Daily Cover

It is important to use machinery to lift, position, spread, and compact the waste, as well as regular and final cover soils. Excavators, service and water trucks, and grinders are examples of service or support equipment needed to keep a landfill running. Both waste and cover soil are spread easily with the crawler tractor. Additionally, it can compress both materials to a degree, with waste densities ranging from 475 to 725 kg/m^3. As compared to compactors, densities and shredding effects are both minimal. Another feature of a wheel loader and crawler tractor is the ability to excavate; depending on the position of the cover soil source, this soil source can be whether it is drawn from on-site or excavated from an off-site pit. Both machines may also construct the landfill site and temporary roads for future activities such as digging, preparing, and deposit.

Temporary access roads are needed in most cases to enable collection trucks to work. Road building is normally completed by landfill workers using on-site equipment. The location of the road changes as the working face moves. Materials from the landfill, such as concrete rubble, are often used to build the lane. Long-term access roads can be paved or constructed.

The compactor is wide and heavy, with knobbed steel wheels that shred, scatter, and large compact volumes of waste effectively. Densities of 725 to 950 kg/m^3 are frequently achieved. Compactors, on the other hand, are ineffective at excavating and hauling objects. As a result, additional equipment is needed at a site where cover soil must be excavated or transported long distances from on-site borrow pits.

Pans and scrapers were used in landfill sites for excavating and positioning of soil cover daily. They are designed for machining the cohesive soil that can be removed in layers, transported to the site, and then prepared for the soil cover for deposit waste at a slope.

In contrast to the United States, the soil is not excavated to generate landfill volume and daily cover material in Germany and other European countries, so machinery is not needed. Unlike many other countries, Germany uses no or just a limited amount of regular cover content. Owing to this, the compactor is mostly used as a universal piece of equipment. Additional tasks can necessitate the use of crawler tractors. The exact form, amount, and size of equipment required are highly dependent on the specific needs of each landfill.

The landfill is able to accept waste once the site has been prepared (i.e., liner placement). There are several waste placement options. The excavated cell method, the area method, and the canyon/depression method are the three most popular methods.

When using the excavated cell technique, first, the soil is excavated in a cell arrangement. The bottom of the landfill is lined with clay liners that have a very low soil permeability value. The function of this clay liner is to prevent leachate and landfill gases from contaminating the underlying subsurface. The selection of this method is usually when the landfill area has a low groundwater level. Furthermore, the excavated soil during cell preparation can be used as a soil daily cover material. This is considered common practice in the United States and a large number of other countries.

The area method is distinguished from the cell technique by the absence of excavation; waste is literally piled on top of the earth. This method is chosen when the groundwater table is very high, close to the ground surface, and when the excavation work is very difficult to be conducted safely. In order to comply with the leachate management issue, a liner and leachate collection system is installed above the ground. The daily soil cover may be either imported in from another location or excavated on-site. In Europe, for example, the area method is used to maintain that leachate continues to drain naturally into culverts even after the landfill is closed.

The final method is the canyon method, where the waste is dumped into an existing canyon, quarry, or borrow pit. The geometry of the depression dictates the placement and compaction procedures for this technique. This approach necessitates stringent surface and groundwater management, as well as the continuous pumping of leachate from the landfill. Leachate pumping may be needed later when the landfill is fully filled and covered with an impermeable liner, as the liner may not be completely sealed and may not last indefinitely. This approach should be reconsidered because it has the potential to trigger serious long-term issues. Thin waste layers are often used to store waste. For the work surface to accommodate many types of machinery, it must be wide enough to ensure they can be unloaded simultaneously. A minimum of 4–6 meters is needed per vehicle.

The first layer, which is immediately above the leachate collection system, is made up of preselected waste that has had sharp and heavy items removed. It must be mounted and compacted in such a way that no equipment can damage the leachate collection system. This layer, also known as the organizational layer, serves to secure the leachate collection system. The filling starts in small, highly compacted lifts after that. Construction can begin as part of an active gas extraction plan during landfilling if a horizontal gas collection system is required after the placement of several lifts.

One of the main important aspects of landfills is volume. The volume would determine the life span of a landfill. Therefore, when waste is deposited in a landfill, it is best to reduce the total increment of volume due to newly deposit waste. The only way to achieve this at landfill sites is by compacting the waste as compact as possible. A compacted waste would reduce its own volume, thus making more space for another waste at the same location of the landfill. The following results are just as significant as using less volume: vermin avoidance/reduction, fires, odors, and littering. Furthermore, little to no cover soil may be used.

When the waste is compacted, the possibility of holes or caves created by vermin could be avoided. Furthermore, the compacted waste reduces air intrusion that may reduce the possibility of landfill fire and reduce odor problems due to low waste surface.

Factors affecting waste placement performance, such as cell geometry and depth, must be considered in addition to compaction. Compaction raises the density of waste and can be measured in terms of a compaction ratio (CR). The CR is defined as the volume of waste before compaction (V_i) divided by the volume of waste after compaction (V_f).

Waste moisture content and composition formed slope angle, waste placement height, the waste compaction effort of the compactor machine, and the type of compaction machine all affect the CR. In the area where the slope is flatter, the CR value would be much higher than the steeper slope because it is a lot easier to compact waste in a flatter slope area. With the tremendous amount of compaction energy transfer to the waste, the CR value would increase. Up to a certain percentage, higher moisture contents result in higher CRs than drier waste. Furthermore, as the uncompacted waste layer increases, the amount of waste that can be compacted would also decrease. This would reduce the CR value. This is due to the fact that compaction energy is not being able to penetrate into a thick layer. In order to solve this, a heavier compactor is required. Thus, it can be said that the heavier the compactor, the higher the CR value. Similarly, the more pass of compaction work is conducted on the waste, the higher the CR value would be achieved. This is also due to the fact that the shredding effect is growing.

The density of typical real weights of landfilled waste with no daily soil cover should be around 800 kg/m^3. Mineral waste that has been compacted as bottom ash from waste incineration can have densities of over 1000 kg/m^3. Landfill tipping reports and surveys are often used to assess the specific weight of waste placed in the landfill. Cover soil makes up 15% to 20% of the number of certain landfills. The cover soil has a density of about 2600 kg/m^3. The waste and soil daily cover density should be taken into consideration when measuring the in-place density of a whole landfill. In Europe, for example, waste is usually distributed in 300–500 (700) mm lifts and shredded with multiple compactors passes. The waste should be put into a maximum of half a meter lift and be well compacted with about up to around 6 to 10 passes of compaction by the compactor machine.

Many places demand that a soil layer (100–250 mm) or alternate cover material be applied to the waste after the final load of waste is deposited in the landfill for the day. The main function of the daily cover is to reduce the amount of water infiltrating into the waste. Such water would increase the amount of leachate generate because the infiltrating water will mix with the original leachate underneath the waste mass. In a tropical country where precipitation is very high with annual rainfall over 3000 mm in a year, the additional water would increase the amount of leachate generated from the landfill if there is no daily cover.

The daily cover would also reduce the dust and landfill gas emissions into the air. Thus, the environmental impact from landfills can be reduced. Dust created from a landfill site is a severe problem as the landfill usually has a vast open space. Any contaminant particle or dust from the landfill will easily be blown from one point to another if the waste is not covered underneath the daily cover. The well-compacted daily cover will also avoid landfill gas escape into the atmosphere. As the gas may not be able to escape, the daily cover also prevents air from entering the waste mass. This would reduce the possibility of landfill fire start to initiate when the amount of oxygen entering the waste mass would be blocked. The final benefit of daily cover is to keep small animals and insects from burrowing and emerging, such as mice and flies.

On the other side, dirt or other inert content makes up 15% to 20% of the landfill. In many countries, the use of regular cover is needed. In countries where it is not needed, the use of large quantities of regular coverage may be reconsidered. Since the abovementioned advantages of regular cover are important, it is accomplished with thorough high compaction using a heavy-duty compactor. In this scenario, only small quantities of daily cover soil, if any, should be used. In Germany, this type of operation is common. Intermediate cover may be installed if landfill parts are not in operation for an extended period of time.

Soil is commonly used as a day-to-day cover material. However, not all soil qualities, such as cohesive soils with high clay content, are suitable for use as a regular cover. Using this type of soil may cause serious operational issues if it rains. In areas where obtaining soil for daily cover is difficult or where the expense of transporting the soil makes daily cover uneconomical, an alternative material can be used as cover. Cover materials have been employed frequently with success in compost or mulching. Alternative cover materials also included tarps, building and demolition waste, and agricultural residues. Removable geosynthetic tarpaulins do not take up precious airspace that could be used to put more waste. The use of a permeable or reusable regular cover is essential to ensure sufficient vertical moisture movement.

10.7.3 Sanitary Landfill Operation

Landfilling is particularly difficult in tropical countries with high precipitation rates or in countries where waste with high moisture content, mostly organics, and must be landfilled. The procedure must be adapted to the circumstances. As in China, there are primarily two types of landfills: mounds constructed on flatlands and valleys built below level. The topography of the region determines the form of landfills. Landfills are constructed above ground in most northern cities, such as Beijing and Zhengzhou, and can reach heights of up to 50 meters. Landfills are installed directly in valleys in southern cities like Suzhou and Shenzhen.

Different types of garbage vehicles transport MSW to landfills. The total weight of the waste truck is registered as it reaches the weighbridge at the landfill entrance. The weighbridge system is normally linked to the Smart Environmental and Sanitation Management Network, which is used by the local City Environmental and Sanitation Bureau; as part of the platform's simple data collection, the garbage truck's related data is automatically sent.

The working units or cells of a sanitary landfill are often divided. Some landfills divide their working units into distinct units for different forms of waste, while others divide their working units strictly for administrative purposes. Once the waste is deposited, a bulldozer distributes and flattens the waste in layers less than two feet thick, and compactors continuously compact the waste until it reaches a final density of at least 600 kg/m^3. Several of the compacted layers lie on top of the other on top of the original pile of waste to a height of 2 to 4 meters, but no higher than 6 meters. A landfill's various covers include daily cover, intermediate cover, and final cover. The compacted waste on the working open-cell must be covered with 200–250 mm soil or HDPE/LDPE membranes no less than 0.5 mm thick on a regular basis. When a unit is packed with waste, it is protected by a minimum of 300 mm of clay or 0.5 mm thick HDPE/LDPE membranes, known as intermediate cover. A final cover must be installed after the landfill has reached its final height.

Many landfills in China have long struggled with the control and disposal of leachate. Often due to the fact that waste composition is highly on organic waste, this contributes to the high moisture content of the waste. Large quantities of leachate are produced, and the amount varies greatly depending on the season and environment. A pump collects leachate at the bottom of the landfill and transports it to the surface. Leachate is typically stored in a leachate-conditioning tank for a limited time before flowing to the leachate treatment facility. The majority of leachate conditioning pools are exposed to the air, but some landfills cover them to avoid odor emissions.

Leachate is recirculated in some landfills to eliminate leachate and improve water quality. A surface spray device is often used. This technology is very cost-effective and can be scaled easily to landfills with restricted leachate treatment space; hence, it is widely utilized in landfills. A solution to the problem may lead to various issues, including the emission of odors, destabilization of the soil, and soil runoff.

10.7.4 Safety and Security

When running a landfill, the health and safety of both landfill workers and the general public are important considerations. Communication with dangerous chemicals (pathogens, harmful air contaminants, asbestos), a high risk of accidents, allergies induced by environmental hazards such as landfill gas and fine particle dust, and unnecessary noise and side effects caused by landfill machinery activity are all potential hazards [79]. Employees should be provided with the required protective equipment. In addition, landfill access is usually restricted. A health and safety strategy must be created and shown prominently. Each landfill should be surrounded by a fence to prevent prohibited people from entering. This is also important in terms of protection and responsibility in the event of an accident.

Since all rainfall and stormwater runoff contribute to the total amount of leachate produced and the amount that must be handled, landfills need a comprehensive stormwater management plan. In addition, poor stormwater management will lead to erosion of the cover material. Using a correctly daily cover with an acceptable gradient, usually 3–5%, is the most efficient control measure. Runoff can be easily redirected away from the landfill's tip using concrete-lined trapezoidal ditches. Plan slopes to absorb and hold stormwater, allowing it to penetrate into the waste and serve as a moisture source, whether the landfill is being used as a bioreactor. Naturally, the necessity of these measures is highly dependent on the trend and annual total rainfall. Runoff should be intercepted in ditches constructed around the landfill; the water quality should be tested and handled if it does not meet the target values.

Because of the regular collection trucks' movement and the rumbling engines of landfill equipment, noise levels at landfill sites can be unacceptable, creating a nuisance to both landfill workers and nearby residents. As a result, depending on the case, a noise reduction program should be considered based on the surrounding area and the number of machinery at the landfill. Employees at landfills should be provided with hearing aids. Numerous measures may be used to prevent noise from reaching neighboring homes, including tree planting to establish a buffer zone for noise mitigation (which should also occur for other purposes, such as litter and dust control, and reducing the landfill's visibility), maintaining equipment properly, controlling hours, and locating the working face (the landfill's loudest area). According to Chander et al. [80], 1.9 km setbacks are needed to reduce noise disruptions.

There are odors from leachate seeps, disposed sludge, putrescible wastes, and landfill gases well as when organic-rich waste is unloaded by the collection truck, particularly in hot climates. Controlling odor requires a good overall landfill service, high compaction, regular cover placement, and even the immediate covering of materials with an offensive odor. The use of a good gas control device would also help to reduce odor. Activated carbon filtration, thermal oxidation, biofiltration, and wet gas scrubbing can all be used to deodorize collected gas. More regular waste disposal can help to reduce odors in the incoming waste in some cases. In extreme situations, chemicals such as ozone and mixtures containing plant oils and surfactants may be used to reduce odor [81]. A perimeter misting machine is often used to dispense odor-neutralizing chemicals. Only in a few instances are these steps taken. In most situations, bad odors at landfills in countries with mild climate zones are the product of inefficient activity.

In most cases, landfill conditions are anaerobic. However, although anaerobic processes produce less energy than aerobic processes, as biological activity rises, internal temperatures rise as well. Heat tends to collect in waste because of its high insulating properties, as high as 65 °C. This must be considered when selecting suitable landfill construction materials, such as tubing, liners, and testing equipment.

When methane concentrations in the air are between 5% and 15%, it is explosive and can cause internal fires. In general, this is a special issue on low compacted landfills, where methane emits through the landfill and air-surface may reach into the landfill through diffusion, but primarily through the wind (especially on slopes) and atmospheric pressure changes. Glass fragments (magnifying effect), open fires on the landfill, chemical processes in the landfill, and other potential ignition causes are all possibilities. Many dumps and low-compacted landfills in developed countries are particularly bad examples of this situation. As a result, high compaction is also essential for this purpose. However, oxygen penetration can occur in inactive landfills where the gas extraction system is not properly maintained, resulting in explosive mixtures. When landfills are aerated, a unique circumstance arises. However, there are no ignition potentials within the landfill, with the exception of unknown chemical processes. As the temperature in the landfill body rises above 60–70 °C, there could be a low ignition potential, according to past experience. As a result, attempts should be made to maintain temperatures below 50 degrees Celsius.

Dust problems can be caused by unloading waste, earthmoving and compaction operations, wind, and traffic on roadways. Excessive dust can cause health issues for landfill workers (allergies and lung problems), increased equipment maintenance costs and frequency, as well as annoyance for local residents. Spraying access roads with water is the most common control measure. Other ways to reduce dust include using dust-free roads such as asphalt, using speed bumps to enforce speed limits, using vegetation to reduce wind speed at ground level, and moving soils at the right time to reduce dust.

10.7.5 Environmental Issue Due to Landfill

The formation of leachates is influenced significantly by rainfall. Precipitation percolates through the accumulated waste and, through a series of physical and chemical reactions, binds to dissolved and non-dissolved waste constituents. Leachates are often formed by groundwater tributaries, surface runoff, and biological decomposition. Leachates are formed when liquid fractions in waste interact with the moisture content of the soil cover.

Moisture may be extracted from the landfill by consuming water in the LFG formation or by removing leachate through the drainage system. As a result, leachate discharge is closely linked to rainfall, surface runoff, and groundwater penetration into the landfill. The method of landfilling (waterproof covers, specifications for insulation layers such as clay (cohesive mineral soil), geotextiles (GCL), or plastic materials) is critical for ensuring that water does not penetrate the landfill's upper layers and therefore for contamination risk mitigation. Climate change has a big impact on leachate production because it affects the amount of precipitation that goes into the landfill and how much evaporates. Furthermore, the processing of leachates is influenced by the nature of the disposed waste, specifically the water content and the degree of compaction of the upper landfill layers [82].

A leachate treatment system based on the "dilute and disperse" system was used as a standard on landfills developed in Europe in the twentieth century. Landfills were often not properly sealed, allowing leachates to spill into the natural environment, where they mixed with groundwater and dispersed. Such dilution and dispersion facilities are no longer built; however, existing ones have left old environmental burdens in the form of landfills across Europe, which may pose a risk of pollution, especially in areas with a high groundwater table [83–85].

Due to the leaching of toxic chemicals, landfill leachate has been shown to be a major source of contaminants in several studies [71, 82, 86]. Nutrients (specifically nitrogen), volatile organic compounds (VOCs), heavy metals (HM), and toxic organic compounds (TOCs) are the four main components of leachates [87, 88]. One of the priority substances to be removed in order to reduce leachate toxicity is nitrogen in the form of NH3 [89–91].

The quality of leachates is influenced by a number of factors, including landfill age, seasonal weather variations, total precipitation amount, waste type, and composition [92]. The composition of leachates varies greatly depending on the age of the landfill. According to the age of the landfill, there are three forms of leachates (Table 10.9). The concentration of organic compounds in landfill leachates decreases as the landfill ages, while the concentration of NH_3 rises.

Leachates are complex mixtures of substances that include dissolved organic matter, inorganic macro-components, heavy metals (HM), and a variety of xenobiotic organic compounds. Many of the chemicals used in landfill leachates are dangerous to human health and the environment. Furthermore, chemicals will build up in species and then be transferred down the food chain, ultimately reaching humans [87].

Traditionally, risk assessment from landfill leachates has been focused on the chemical analysis of individual chemical substances. Although the strategy is critical, it does have some drawbacks. For starters, certain chemical pollutants can be present in the leachates at concentrations below the detection limits of chemical analysis. Owing to the limitations of analytical methods, detecting them can be difficult. Changes caused by frequent refluxes may be required for the continuous chemical sampling regime; however, this method is very costly and labor-intensive.

Table 10.9 Classification of landfill leachates [91]

Parameter	Old landfill	Mid-age landfill	Low-age landfill
pH	>5	1–5	<1
BOD_5/COD	<0.1	0.1–0.5	0.3–1
COD (g O_2.Dm^{-3})	<3	3–15	>15
TOC/COD	>0.5	0.3–0.5	<0.3
NH_3-N (mg.Dm^{-3})	>400	400	<400
Heavy metals (mg.Dm^{-3})	<2	<2	>2

Furthermore, these chemical methods do not predict how contaminants may affect recipients (ecosystems). Ecotoxicological studies of the responses of model biological organisms to toxic substances may provide valuable knowledge in addition to traditional chemical analyses [85]. Unlike a chemical study, a biological toxicity test may take into account all present compounds' biological effects as well as their biological availability [85, 86].

Chemical analysis alone offers only a small amount of knowledge regarding the environmental fate of complex leachates refluxes [85]. Ecotoxicological tests can thus serve as a critical link between conventional chemical analyses and large-scale field studies involving biological organisms in regulated (defined) environments. In 1980, one of the first studies on leachate toxicity highlighted the limitations of conventional chemical testing, leading to the creation of a test with rainbow trout salmon (*Oncorhynchus mykiss*) and the establishment of a standard protocol for assessing leachates' biological toxicity. On the model fish organisms, the key constituents of leachate toxicity (non-ionized ammonia, tannins, and copper) are described. The toxic effects of NH3 on the environment are unknown. NH3 has been shown to be extremely toxic to marine species, both acutely and chronically.

A few studies have also established that NH3 is one of the main causes of toxicity in MSW landfill leachates. Furthermore, NH3 is released from leachates by volatilization, and higher NH_3 concentrations in the atmosphere can have negative effects on vegetation. Comparative toxicological tests showed that these leachates are mutagenic. Helma et al. [93] found that the genotoxic capacity of leachates from MSW landfills is greater than that of cellulose-making wastewater, industrial wastewater, polluted surface and ground waters, or even drinking and bathing water samples. The abovementioned properties of landfill leachates highlight the importance of implementing a high-quality strategy for landfill leachate management (collection, recirculation, and final treatment prior to discharge into the environment) as a final strategy for pollutant management. There is a lot of scientific literature on how to extract, store, and properly handle landfill leachates. To treat landfill leachate, a variety of technologies are available, all of which seek to meet the legal requirements.

A landfill is made up of waste and covering material, which is normally dirt. While it is almost impossible for soil to start a fire, it is possible for a landfill to catch fire. Landfill fires are, in fact, surprisingly common [73, 94]. However, many environmental aspects of landfill fires have received little attention in the scientific community, or studies are difficult to perform due to the large number of variables involved. There were 840 fire events registered in the United States between 2004 and 2010, resulting in material, equipment, and human health losses. In Poland, however, more than 60 landfill fires have recently been recorded. According to the authorities, many of them were most likely set up on purpose to consume illicit waste smuggled from other nations. There are few published studies on the environmental effects of landfill fires due to hazardous material emissions (Table 10.10).

Landfill fires can cause major environmental damage by releasing toxins into the atmosphere, soil, and water. The type of burning waste, the location of the landfill, and the type of fire all influence the risk factors. These fires usually occur at low temperatures and in anoxic environments. Under such settings, hydrocarbons, chlorinated products, and pesticides emit a wide range of toxic gases, including dioxins/furans, polynuclear aromatic hydrocarbons, respirable particulates [96], and HM [97], as well as other hazardous compounds [98]. The smoke emitted by the landfill fire may contain toxic gases like CO, H_2S, and CH_4 as well as carcinogenic substances like dioxins. The emitted foul odors and smoke annoy the neighbors and can endanger human health, especially among vulnerable populations such as the elderly, infants, pregnant women, and people with chronic respiratory conditions.

Biogas, also known as landfill gas (LFG), is one of the components of biological processes in MSW landfills [78]. Biogas is a by-product of biologically decomposable organic matter decomposition [99]. Over the landfill's lifespan, the rate of biogas production and its composition change. During the anaerobic decomposition of organic matter, LFG is generated from MSW. MSW contains around (150–250) kg of organic carbon per ton of waste, which is converted into landfill gas by microorganisms during anaerobic processes. The average heating value of the produced LFG with (40–60%) methane is nearly 18,000 kJ/Nm3. The emitted electric energy is approximately 2.5 kW h/Nm3 with a 34% energy conversion efficiency [100]. The following are the key factors that influence the amount of LFG: waste composition, moisture content, temperature, landfill age, and so on. LFG production begins one to two years after waste is deposited in a landfill and lasts for fifteen to twenty-five years [101, 102]. "LFG shall be obtained from all landfills receiving biodegradable waste, and the LFG must be stored and used," according to Directive 31/1999/CE. The collected LFG must be incinerated if it cannot be used to generate energy.

Table 10.10 The environmental effects of a landfill fire [95]

Type of landfill	Landfill location	Year	Environment
MSW, industrial and construction waste	Western Norway	2003	Landfill leachates
MSW	Tagarades, Greece	2006	Landfill surrounding area/soil and vegetation samples
Landfill's shredded tire drainage layer	Low City, United States	2012	Air
MSW	Niger Delta, southern Nigeria	2013	Air
MSW	Iqaluit, northern Canada	2014	Air
MSW, electronic waste, and bulky waste	Araraquara City, Brazil	2015	Soil, dust, leachate, and well water
Tire landfill	Sesena, Toledo, Spain	2016	Air/soil
MSW	Talagante, Chile	2016	Air

10.7.6 The Benefits of Landfill Mining

Europe is moving away from landfilling and toward recycling and reuse in the waste management hierarchy [103, 104]. The directive, which has been in effect since 1999, includes provisions for reducing the volume of waste disposed of in landfills. Since 2016, member countries have been prohibited from dumping more than 35% of biodegradable MSW that was landfilled in 1995. Some countries were able to meet these objectives four years later. The primary goal of the waste management strategy is to avoid the formation of waste. Waste should be reused or recycled if mitigation is not feasible. If this isn't possible, waste should be converted into electricity (thermo-valorization). Waste can only be disposed of in landfills if there are no other options for its management.

Landfills can still play a part in the waste management system, regardless of what prevention steps, reuse, or recycling society can achieve. Furthermore, landfill rates are still high globally, even in EU countries, while waste prevention and recycling rates are too poor.

Waste reduction, reuse, and recycling (3R) behaviors have been generally recognized as waste management techniques, according to several experts [105]. In all circumstances, however, adequate capacity for recycling and reuse of all forms of waste would not be economically viable. Investing in recycling and reusing wastes that would vanish in the future will be economically disadvantageous in the search for greater prevention. Furthermore, the amount of waste generated varies from year to year.

The amount of waste that can be recycled, reused, or incinerated often exceeds the power. Some wastes cannot be recycled or incinerated. A landfill is their only choice for some of them. The waste should not be left in residential areas if the recycling plant or incinerator is out of service due to maintenance, repair, or breakdown. This demonstrates that even though recycling and reuse are in place, some waste must be disposed of in landfills [106]. These landfills function as "security networks" in a sound waste management system. Landfills should be built using environmentally friendly approaches to ensure that future generations are not burdened by them.

The use of impermeable membranes to insulate landfills becomes a European standard. All processes in the landfill are halted by insulation. These membranes can last up to 500 years and have a lifespan of up to 50 years. They will, however, fail at some point in the future, and if they are interrupted, the emission-generating processes will restart. As a result, future generations will be spared from potential pollution.

Aftercare landfill maintenance is controlled by law in many countries, and it must be ensured for at least 30–60 years after the landfill is closed. Some countries demand that aftercare be provided as long as the appropriate state administration authority deems it essential. Aftercare is expected for a period of time that exceeds one generation. A more long-term solution would be a safe one. As a result, a community aiming for long-term sustainability needs long-term sustainable landfills.

There are no internationally agreed-upon definitions of what constitutes a sustainable landfill. When it comes to landfills, words like stability, completion, end technologies, and environmental hazard are widely used in discussions regarding sustainability [107].

According to the Solid Waste Association of North America's (SWANA) subcommittee for stability, a landfill is "functionally stable" as long as the waste does not pose a danger to human health or the environment after it is closed. The quality and quantity of leachates, gas output and composition, cover, slope gradient, slope and nature of insulation, site geology and hydrogeology, environment, potential recipients, ecosystem and human exposure to landfill impacts, and other factors related to particular localities must all be evaluated. According to the researcher, aftercare should be discontinued when the potential for pollution production is so poor that real emissions do not affect the ecosystem.

While different meanings have somewhat different interpretations, it appears that there is a general consensus that a sustainable landfill, or one in which aftercare termination is deemed secure, is one that will achieve a state within a limited time span in which its undisturbed material will no longer pose a threat to human health or the environment. This is the point at which the landfill's aftercare can be ended (often referred to as completion). It is important to recognize that this requirement is in line with the spirit of EU waste legislation. In landfills of inert waste, Annex II of the Landfill Directive does not require insulation or aftercare, and inert waste is described similarly to not posing a threat to human health or the environment [108]. Since research in this area is limited, it remains to be seen if inert waste poses a risk to human health and the environment.

The most frequent threats associated with landfill activity have been identified previously. Now is the time to consider claims in favor of minimal waste disposal in landfills. Landfilling, particularly over the long term, has the potential to convert a significant portion of waste flow into short-term gas generation and long-term carbon supply. Additionally, landfill aftercare will encourage the renovation of brownfields, which will eventually result in new possibilities for improving the landscape's condition and the ecosystem.

Waste management options are not interchangeable. Diverse solid waste management methods are commonly considered the hierarchy of recycling opportunities, with waste reduction at the source being the best alternative and landfilling being the worst. However, activists, landfill managers, and waste providers are currently critical of alternative waste disposal methods.

Despite the fact that the possibility of zero waste is widely debated [109, 110], Song et al. [111] claim that converting current over-consumptive practices to zero waste is still difficult. It is difficult to effectively implement recycling at a fair cost since it requires careful separation of individual waste fractions both at the source and after processing.

Organic matter is thought to be a significant source of pathogens as it is separated from waste, whether biodegradable or combustible. Composting, as the most efficient method of recycling biologically degradable waste, eventually creates bioaerosols that are harmful to the health of workers and residents who live near large composting facilities. Incineration is the most popular alternative to landfills, but it has a range of drawbacks. It is a very expensive technology that is needed to address all potential threats, such as air pollution. It has also been criticized for possible toxic emissions, the failure to remove pathogenic chemicals, the lack of heavy metal immobilization, and other waste management economy flaws. Furthermore, the public's understanding of threats vastly outnumbers real risks. Municipal waste incinerator ash may be disposed of in landfills with or without pretreatment, but the possibility of inhaling ash directly from the source in landfills has been determined to be negligible.

Landfilling has the cheapest operating costs. Its secondary effects (measured in terms of hygienic, social, and environmental impact) are historically referred to as important. However, their numerical expression is debatable, and any considerations should not take into account the emerging developments that are now available [74]. It's worth noting that the existing massive amounts of waste dumped in landfills may be viewed as potential resource reserves for metals, high-quality recycled aggregates, and waste-derived fuels if landfill mining is used. There are 150,000–500,000 active landfills in the EU, with a total amount of waste estimated to be 30–50 Gm^3. Landfills can be seen as urban stocks, with resource reservoirs for future regeneration, as well as a bank account for future generations.

Landfill mining (LFM) is a hot subject these days. LFM has been suggested as a revolutionary method for reducing environmental risks associated with landfills, recovering secondary raw materials and energy from stored waste, and enabling high-valued land uses on the site.

Landfills are mined for raw materials such as metals or energy resources such as plastics. Additionally, the LFM concept is intended to close material loops toward a circular economy (CE) by landfill waste recycling. However, according to the most recent research, LFM has a number of drawbacks. Furthermore, the results of 80% of the produced LFM scenarios are negative. As a result, one of the technical challenges for LFM's future growth is the development of a treatment plan that allows for full resource recovery while still being environmentally and economically reliable. Strategic policy decisions and customized support mechanisms, including combined incentives for material recycling, energy use, and nature conservation, are needed for LFM to achieve its full potential.

10.8 Concluding Remarks

Increased solid waste generation in cities had a big effect on sanitary issues and essential services like transportation infrastructure, water supply, sanitation, and waste management. Around the world, the amount of waste generated has been steadily increasing. Although landfilling is considered an efficient waste disposal option management system, there is many environmental impact concern. Thus, a few other solid waste disposal techniques are developed, such as composting and incinerator. All the disposal methods have their own advantage and disadvantage.

Even that, landfills still consider the main disposal technique for solid waste management. Therefore, it is important to ensure the management of landfill operation is at its very best, as well as the constant environmental monitoring around the landfill area.

Glossary

The European Union's (EU): is a political and economic union of 27 member states that are located primarily in Europe.

Environmental Protection Agency (EPA): is an independent executive agency of the United States federal government tasked with environmental protection matters.

References

1. Alshehrei, F., & Ameen, F. (2021). Vermicomposting: A management tool to mitigate solid waste. *Saudi Journal of Biological Sciences, 28*, 3284–3293.
2. Bahçelioğlu, E., Buğdaycı, E. S., Doğan, N. B., Şimşek, N., Kaya, S., & Alp, E. (2020). Integrated solid waste management strategy of a large campus: A comprehensive study on METU campus, Turkey. *Journal of Cleaner Production, 265*.
3. Guo, W., Xi, B., Huang, C., Li, J., Tang, Z., & Li, W. (2021). Resources, conservation & recycling solid waste management in China: Policy and driving factors in 2004 – 2019. *Resources, Conservation & Recycling, 173*, 105727.
4. Minghua, Z., Xiumin, F., Rovetta, A., Qichang, H., Vicentini, F., & Bingkai, L. (2009). Municipal solid waste management in Pudong new area. *China, Journal Waste Management, 29*, 1227–1233.
5. Sujauddin, M., Huda, S., & Hoque, A. R. (2008). Household solid waste characteristics and management in Chittagong. *Bangladesh, Journal Waste Management, 28*, 1688–1695.
6. Sharholy, M., Ahmad, K., Mahmood, G., & Trivedi, R. (2008). Municipal solid waste management in Indian cities–a review. *Journal Waste Management, 28*, 459–467.
7. Zhang, D. Q., Tan, S. K., & Gersberg, R. M. (2010). Municipal solid waste management in China: Status, problems and challenges. *Journal of Environmental Management, 91*, 1623–1633.
8. Singh, J., Laurenti, R., Sinha, R., & Frostell, B. (2014). Progress and challenges to the global waste management system. *Waste Management Research, 32*, 800–812.
9. Zhang, R., El-Mashad, H. M., Hartman, K., Wang, F., Liu, G., & Choate, C. (2007). Characterization of food waste as feedstock for anaerobic digestion. *Journal Bioresource technology, 98*, 929–935.
10. Cai, L., Gao, D., Chen, T.-B., Liu, H.-T., Zheng, G.-D., & Yang, Q.-W. (2012). Moisture variation associated with water input and evaporation during sewage sludge bio-drying. *Bioresource Technology, 117*, 13–19.
11. Beneroso, D., Bermúdez, J., Arenillas, A., & Menéndez, J. (2014). Influence of the microwave absorbent and moisture content on the microwave pyrolysis of an organic municipal solid waste. *Journal of Analytical Applied Pyrolysis, 105*, 234–240.
12. Yermán, L., Wall, H., & Torero, J. L. (2017). Experimental investigation on the destruction rates of organic waste with high moisture content by means of self-sustained smoldering combustion. *Proceedings of the Combustion Institute, 36*, 4419–4426.
13. Rada, E. C., Ragazzi, M., Villotti, S., & Torretta, V. (2014). Sewage sludge drying by energy recovery from OFMSW composting: Preliminary feasibility evaluation. *Waste Management, 34*, 859–866.

14. Tyagi, V. K., & Lo, S.-L. (2013). Sludge: A waste or renewable source for energy and resources recovery? *Renewable Sustainable Energy Reviews, 25*, 708–728.
15. Roberts, D. (2015). Characterisation of chemical composition and energy content of green waste and municipal solid waste from greater Brisbane. *Australia, Waste management, 41*, 12–19.
16. Shi, H., Mahinpey, N., Aqsha, A., & Silbermann, R. (2016). Characterization, thermochemical conversion studies, and heating value modeling of municipal solid waste. *Waste Management, 48*, 34–47.
17. Shahir, S., Masjuki, H., Kalam, M., Imran, A., Fattah, I. R., & Sanjid, A. (2014). Feasibility of diesel–biodiesel–ethanol/bioethanol blend as existing CI engine fuel: An assessment of properties, material compatibility, safety and combustion. *Renewable Sustainable Energy Reviews, 32*, 379–395.
18. Brunner, P. H., & Rechberger, H. (2015). Waste to energy–key element for sustainable waste management. *Waste Management, 37*, 3–12.
19. Bhat, V., & Patil, Y. (2014). E-waste consciousness and disposal practices among residents of Pune City. *Procedia - Social and Behavioral Sciences, 133*, 491–498.
20. Karak, T., Bhagat, R., & Bhattacharyya, P. (2012). Municipal solid waste generation, composition, and management: The world scenario. *Journal Critical Reviews in Environmental Science Technology, 42*, 1509–1630.
21. Hoornweg, D., P. Bhada-Tata (2012). *What A Waste: A Global Review of Solid Waste Management*, Washington, DC.
22. Eiselt, H. A., & Marianov, V. (2015). Location modeling for municipal solid waste facilities. *Computers Operations Research, 62*, 305–315.
23. Hoornweg, D., Bhada-Tata, P., & Kennedy, C. (2013). Environment: Waste production must peak this century. *Nature News, 502*, 615.
24. Miezah, K., Obiri-Danso, K., Kádár, Z., Heiske, S., Fei-Baffoe, B., & Mensah, M. (2017). Municipal solid waste management in a low income economy through biogas and bioethanol production. *Waste and Biomass Valorization, 8*, 115–127.
25. Nabegu, A. B. (2010). An analysis of municipal solid waste in Kano Metropolis. *Nigeria, Journal of Human Ecology, 31*, 111–119.
26. Valkenburt, C., C.W. Walton, B.L. Thompson, M.A. Gerber, S.B. Jones, D.J. Stevens Municipal solid waste (MSW) to liquid fuels synthesis, volume 1: Availability of feedstock and technology.
27. Al-Khatib, I. A., Monou, M., Zahra, A. S. F. A., Shaheen, H. Q., & Kassinos, D. (2010). Solid waste characterization, quantification and management practices in developing countries. *A case study: Nablus district–Palestine, Journal of environmental management, 91*, 1131–1138.
28. Zia, H., & Devadas, V. (2008). Urban solid waste management in Kanpur: Opportunities and perspectives. *Habitat International, 32*, 58–73.
29. Jiang, J., Lou, Z., Ng, S., Luobu, C., & Ji, D. (2009). The current municipal solid waste management situation in Tibet. *Waste Management, 29*, 1186–1191.
30. Starostina, V., Damgaard, A., Rechberger, H., & Christensen, T. H. (2014). Waste management in the Irkutsk region. *Siberia, Russia: Environmental assessment of current practice focusing on landfilling, Waste Management, 32*, 389–396.
31. Liyala, C. M. (2011). *Modernising solid waste management at municipal level*. Wageningen Academic Publishers.
32. Gidarakos, E., Havas, G., & Ntzamilis, P. (2006). Municipal solid waste composition determination supporting the integrated solid waste management system in the island of Crete. *Journal Waste Management, 26*, 668–679.
33. Khan, D., Kumar, A., & Samadder, S. (2016). Impact of socioeconomic status on municipal solid waste generation rate. *Journal Waste Management, 49*, 15–25.
34. Nas, S. S., & Bayram, A. (2008). Municipal solid waste characteristics and management in Gümüşhane. *Turkey, Journal Waste Management, 28*, 2435–2442.

35. Rigamonti, L., Ferreira, S., Grosso, M., & Marques, R. C. (2015). Economic-financial analysis of the Italian packaging waste management system from a local authority's perspective. *Journal of Cleaner Production, 87*, 533–541.
36. Thanh, N. P., Matsui, Y., & Fujiwara, T. (2010). Household solid waste generation and characteristic in a Mekong Delta city, Vietnam. *Journal of Environmental Management, 91*, 2307–2321.
37. Ferreira, S., Cabral, M., Cruz, N. F., Simões, P., & Marques, R. C. (2014). Life cycle assessment of a packaging waste recycling system in Portugal. *Journal Waste Management, 34*, 1725–1735.
38. Cheng, H., & Hu, Y. (2010). Municipal solid waste (MSW) as a renewable source of energy: Current and future practices in China. *Bioresource Technology, 101*, 3816–3824.
39. Jadoon, A., Batool, S. A., & Chaudhry, M. N. (2014). Assessment of factors affecting household solid waste generation and its composition in Gulberg town. *Lahore, Pakistan, Journal of material cycles waste management, 16*, 73–81.
40. Gómez, G., Meneses, M., Ballinas, L., & Castells, F. (2009). Seasonal characterization of municipal solid waste (MSW) in the city of Chihuahua, Mexico. *J Waste management, 29*, 2018–2024.
41. Narayana, T. (2009). Municipal solid waste management in India: From waste disposal to recovery of resources? *Waste Management, 29*, 1163–1166.
42. Wang, J., Han, L., & Li, S. (2008). The collection system for residential recyclables in communities in Haidian District. *Beijing: A possible approach for China recycling, Waste Management, 28*, 1672–1680.
43. Mudhoo, A., Somaroo, G. D., & Mohee, R. (2015). *Treatment of solid waste: Principles, key technologies and selected case studies from Africa*. Oxford, UK: African Books Collective South Africa.
44. Mohee, R., & Soobhany, N. (2014). Comparison of heavy metals content in compost against vermicompost of organic solid waste: Past and present. *Resources, Conservation Recycling, 92*, 206–213.
45. Huang, K., Li, F., Wei, Y., Fu, X., & Chen, X. (2014). Effects of earthworms on physicochemical properties and microbial profiles during vermicomposting of fresh fruit and vegetable wastes. *Bioresource Technology, 170*, 45–52.
46. Sequeira, V., & Chandrashekar, J. (2015). Vermicomposting of biodegradable municipal solid waste using indigenous Eudrilus sp. earthworms. *Int J Curr Microbiol Appl Sci, 4*, 356–365.
47. Taiwo, A. M. (2011). Composting as a sustainable waste management technique on developing countries. *Journal of Environmental Science and Technology, 4*, 93–102.
48. Banful, A.B. (2009). Operational details of the 2008 fertilizer subsidy in Ghana—Preliminary report, .
49. Giusti, L. (2009). A review of waste management practices and their impact on human health. *Waste Management, 29*, 2227–2239.
50. Usepa, U. (2006). *Edition of the drinking water standards and health advisories*, Washington, DC.
51. Themelis, N. J. (2003). An overview of the global waste-to-energy industry. *Waste Management World, 3*, 40–47.
52. Denison, R. A. (1996). Environmental life-cycle comparison of recycling, landfilling, and incineration: A review of recent studies. *Annual Review Energy and the Environment, 21*, 191–237.
53. Brookins, C. L. (2007). Major forces and factors shaping global markets: The emerging agricultural world. *Journal of Agribusiness, 25*, 1–16.
54. Cheng, H., Hu, Y., & Zhao, J. (2009). Meeting China's water shortage crisis: Current practices and challenges. *Environmental Science & Technology, 43*, 240–244.
55. Liu, J., & Diamond, J. (2005). China's environment in a globalizing world. *Nature, 435*, 1179–1186.

56. Nie, Y. (2008). Development and prospects of municipal solid waste (MSW) incineration in China. *Front, Frontiers of Environmental Science & Engineering in China, 2*, 1–7.
57. Cheng, H., Zhang, Y., Meng, A., & Li, Q. (2007). Municipal solid waste fueled power generation in China: A case study of waste-to-energy in Changchun city. *Environmental Science & Technology, 41*, 7509–7515.
58. Gao, Q., Du, W., Lu, S., Zhang, Z., Zhang, E., Wu, J., & Ren, Z. (2007). Methane emission from municipal solid waste treatments in China. *Advances in Climate Change Research, 3*, 70–74.
59. Hazra, T., & Goel, S. (2009). Solid waste management in Kolkata. *India: Practices and challenges, Waste Management, 29*, 470–478.
60. Henry, R. K., Yongsheng, Z., & Jun, D. (2006). Municipal solid waste management challenges in developing countries - Kenyan case study. *Waste Management, 26*, 92–100.
61. Moghadam, M. A., Mokhtarani, N., & Mokhtarani, B. (2009). Municipal solid waste management in Rasht City. *Iran Journal Waste Management, 29*, 485–489.
62. Chung, S. S., & Lo, C. W. (2008). Local waste management constraints and waste administrators in China. *Journal Waste Management, 28*, 272–281.
63. Pokhrel, D., & Viraraghavan, T. (2005). Municipal solid waste management in Nepal: Practices and challenges. *Waste Management, 25*, 555–562.
64. Weigand, H., J. Fripan, I. Przybilla, C. Marb (2003). Composition and contaminant loads of household waste in Bavaria, Germany: Investigating effects of settlement structure and waste management practice, In: *Proceedings of the 9th International Waste Management and Landfill Symposium*. Cagliari, CD-ROM.
65. Zacarias-Farah, A., & Geyer-Allély, E. (2003). Household consumption patterns in OECD countries: Trends and figures. *Journal of Cleaner Production, 8*, 819–827.
66. Everaert, K., & Baeyens, J. (2002). The formation and emission of dioxins in large scale thermal processes. *Journal Chemosphere, 46*, 439–448.
67. Abdel-Shafy, H. I., & Mansour, M. S. M. (2018). Solid waste issue: Sources, composition, disposal, recycling, and valorization. *Egyptian Journal of Petroleum, 27*, 1275–1290.
68. Griffith, D. C., Kelly-Hope, L. A., & Miller, M. A. (2006). Review of reported cholera outbreaks worldwide, 1995–2005. *The American journal of tropical medicine hygiene, 75*, 973–977.
69. Ogwueleka, T. (2009). Municipal solid waste characteristics and management in Nigeria. *Journal of Environmental Health Science Engineering, 6*, 173–180.
70. Papargyropoulou, E., Colenbrander, S., Sudmant, A. H., Gouldson, A., & Tin, L. C. (2015). The economic case for low carbon waste management in rapidly growing cities in the developing world: The case of Palembang, Indonesia. *Journal of Environmental Management, 163*, 11–19.
71. Brennan, R., Healy, M., Morrison, L., Hynes, S., Norton, D., & Clifford, E. (2016). Management of landfill leachate: The legacy of European Union directives. *Journal Waste Management, 55*, 355–363.
72. Chakravarty, P., M. Kumar (2019). Floral species in pollution remediation and augmentation of micrometeorological conditions and microclimate: An integrated approach. In: *Phytomanagement of Polluted Sites*. Elsevier. pp. 203–219.
73. Chavan, D., Lakshmikanthan, P., Mondal, P., Kumar, S., & Kumar, R. (2019). Determination of ignition temperature of municipal solid waste for understanding surface and sub-surface landfill fire. *Journal Waste Management, 97*, 123–130.
74. Białowiec, A. (2011). Hazardous emissions from municipal solid waste landfills. *Journal of Contemporary problems of management environmental protection, 9*, 7–28.
75. Koda, E., Pachuta, K., & Osinski, P. (2013). Potential of plant applications in the initial stage of the landfill reclamation process. *Journal Polish Journal of Environmental Studies, 22*.
76. Weng, Y.-C., Fujiwara, T., Houng, H. J., Sun, C.-H., Li, W.-Y., & Kuo, Y.-W. (2015). Management of landfill reclamation with regard to biodiversity preservation, global warming

mitigation and landfill mining: Experiences from the Asia–Pacific region. *Journal of Cleaner Production, 104*, 364–373.
77. Ziyang, L., Luochun, W., Nanwen, Z., & Youcai, Z. (2015). Martial recycling from renewable landfill and associated risks: A review. *Journal Chemosphere, 131*, 91–103.
78. Shen, S., Chen, Y., Zhan, L., Xie, H., Bouazza, A., & He, F. (2018). Methane hotspot localization and visualization at a large-scale Xi'an landfill in China: Effective tool for landfill gas management. *Journal of Environmental Management, 225*, 232–241.
79. Otieno, F., & Magagula, C. (2001). Management strategies of odour problems at landfill sites. *Journal African Journal of Science Technology, 2*, 108–112.
80. Chander, R., D. Reinhart, M. Rogoff, R. Wayson, A. Martin (2007). *Urban infilling impacts on solid waste facilities*. In: World environmental and water resources congress 2007: Restoring Our Natural Habitat. . pp. 1–7.
81. O'Malley, P. (2003). Controlling odor at transfer stations and MRFs: There's more than one way to reduce complaints. *Journal African Journal of Science Technology Environment international MSW Management, 13*, 44–52.
82. Costa, F. M., Daflon, S. D. A., Bila, D. M., Fonseca, F. V., & Campos, J. C. (2018). Evaluation of the biodegradability and toxicity of landfill leachates after pretreatment using advanced oxidative processes. *Journal Waste Management, 76*, 606–613.
83. Regadío, M., Ruiz, A., Soto, I. S., Rastrero, M. R., Sánchez, N., & Gismera, M. (2012). Pollution profiles and physicochemical parameters in old uncontrolled landfills. *Journal Waste Management, 32*, 482–497.
84. O'Shea, F. T., Cundy, A. B., & Spencer, K. L. (2018). The contaminant legacy from historic coastal landfills and their potential as sources of diffuse pollution. *Journal Marine pollution bulletin, 128*, 446–455.
85. Rezapour, S., Samadi, A., Kalavrouziotis, I. K., & Ghaemian, N. (2018). Impact of the uncontrolled leakage of leachate from a municipal solid waste landfill on soil in a cultivated-calcareous environment. *Journal Waste Management, 82*, 51–61.
86. (2014). Chemical pollution and toxicity of water samples from stream receiving leachate from controlled municipal solid waste (MSW) landfill. Journal African Journal of Science Technology Environment international Environmental Research 135: 253–261.
87. Budi, S., Suliasih, B. A., Othman, M. S., Heng, L. Y., & Surif, S. (2016). Toxicity identification evaluation of landfill leachate using fish, prawn and seed plant. *Journal Waste Management, 55*, 231–237.
88. Moody, C. M., & Townsend, T. G. (2017). A comparison of landfill leachates based on waste composition. *Journal Waste Management, 63*, 267–274.
89. Paskuliakova, A., McGowan, T., Tonry, S., & Touzet, N. (2018). Microalgal bioremediation of nitrogenous compounds in landfill leachate–the importance of micronutrient balance in the treatment of leachates of variable composition. *Journal Algal research, 32*, 162–171.
90. Singh, S. K., & Tang, W. Z. (2013). Statistical analysis of optimum Fenton oxidation conditions for landfill leachate treatment. *Journal Waste Management, 33*, 81–88.
91. Xiong, J., Zheng, Z., Yang, X., Dai, X., Zhou, T., & He, J. (2018). Recovery of NH3-N from mature leachate via negative pressure steam-stripping pretreatment and its benefits on MBR systems: A pilot scale study. *Journal of Cleaner Production, 203*, 918–925.
92. Koda, E., S. Żakowicz (1998). *Physical and hydraulics properties of the MSW for water balance of the landfill*. In: Proceedings of the 3rd International Congress on Environmental Geotechnics. Lisboa. pp. 217–222.
93. Helma, C., Mersch-Sundermann, V., Houk, V. S., Glasbrenner, U., Klein, C., & Wenquing, L. (1996). Comparative evaluation of four bacterial assays for the detection of genotoxic effects in the dissolved water phases of aqueous matrices. *Journal Environmental science technology, 30*, 897–907.
94. Powell, J. T., Pons, J. C., & Chertow, M. (2016). Waste informatics: Establishing characteristics of contemporary US landfill quantities and practices. *Journal African Journal of Science Technology Environment international Environmental science technology, 50*, 10877–10884.

95. Vaverková, M. D. (2019). Landfill impacts on the environment. *Journal of Geosciences, 9*, 431.
96. Purser, D.A., R.L. Maynard, J.C. Wakefield (2015). *Toxicology, survival and health hazards of combustion products, Royal Society of Chemistry.*
97. Rovira, J., Domínguez-Morueco, N., Nadal, M., Schuhmacher, M., & Domingo, J. L. (2018). Temporal trend in the levels of polycyclic aromatic hydrocarbons emitted in a big tire landfill fire in Spain: Risk assessment for human health. *Journal of Environmental Science Health, Part A, 53*, 222–229.
98. Nadal, M., Rovira, J., Díaz-Ferrero, J., Schuhmacher, M., & Domingo, J. L. (2016). Human exposure to environmental pollutants after a tire landfill fire in Spain: Health risks. *Journal African Journal of Science Technology Environment international, 97*, 37–44.
99. Rada, E. C., Ragazzi, M., Stefani, P., Schiavon, M., & Torretta, V. (2015). Modelling the potential biogas productivity range from a MSW landfill for its sustainable exploitation. *Journal Sustainability, 7*, 482–495.
100. Zappini, G., Cocca, P., & Rossi, D. (2010). Performance analysis of energy recovery in an Italian municipal solid waste landfill. *Journal Energy, 35*, 5063–5069.
101. Scheutz, C., & Kjeldsen, P. (2019). Guidelines for landfill gas emission monitoring using the tracer gas dispersion method. *Journal Waste Management, 85*, 351–360.
102. Thomasen, T. B., Scheutz, C., & Kjeldsen, P. (2019). Treatment of landfill gas with low methane content by biocover systems. *Journal Waste Management, 84*, 29–37.
103. Dhabbah, A. M. (2015). Ways of analysis of fire effluents and assessment of toxic hazards. *Journal of Analytical Sciences, Methods Instrumentation, 5*, 1.
104. Stoeva, K., & Alriksson, S. (2017). Influence of recycling programmes on waste separation behaviour. *Journal Waste Management, 68*, 732–741.
105. Pandey, R. U., Surjan, A., & Kapshe, M. (2018). Exploring linkages between sustainable consumption and prevailing green practices in reuse and recycling of household waste: Case of Bhopal city in India. *Journal of Cleaner Production, 173*, 49–59.
106. Westlake, K. (1997). Sustainable landfill—Possibility or pipe-dream? *Journal Waste Management Research, 15*, 453–461.
107. Meegoda, J. N., Soliman, A., Hettiaratchi, P. A., & Agbakpe, M. (2019). Resource mining for a bioreactor landfill. *Journal Current Environmental Engineering, 6*, 17–34.
108. Scharff, H., Kok, B., & Krom, A. (2007). The role of sustainable landfill in future waste management systems. *Proceedings Sardinia, 2007*, 1–8.
109. Pietzsch, N., Ribeiro, J. L. D., & Medeiros, J. F. (2017). Benefits, challenges and critical factors of success for zero waste: A systematic literature review. *Journal Waste Management, 67*, 324–353.
110. Huber-Humer, M., & Lechner, P. (2011). Sustainable landfilling or sustainable society without landfilling? *Journal Waste Management, 7*, 1427–1428.
111. Song, Q., Li, J., & Zeng, X. (2015). Minimizing the increasing solid waste through zero waste strategy. *Journal of Cleaner Production, 104*, 199–210.

Chapter 11
Landfilling and Its Environmental Impacts

Omar Fawzi Suleiman Khasawneh, Puganeshwary Palaniandy, Mohd Suffian Yusoff, Hamidi Abdul Aziz, and Yung-Tse Hung

Abstract The rapid urbanization and industrial production led to increasing the amount of municipal and industrial solid waste generation. Among the different waste management practices, disposal of solid waste in landfills and open dumpsite 'landfilling' is considered the most common method, especially in developing countries, for the disposal of waste. Although the method of landfilling is considered simple and cost-effective, the process of landfilling poses significant potential impacts on the environment. Researchers all over the world have performed evaluation studies to investigate the impact of landfilling on the environment (e.g. water quality, air quality, soil and plant contamination, and toxic effects on the different organisms). Therefore, it is a crucial step to summarize and assess the findings reported by researchers in this field. In this chapter, the impacts of landfilling on different environmental compartments are assessed and summarized. Further, the toxicological assessment of landfill leachate on different organisms (invertebrates, algal, aquatic organisms, etc.) is summarized. Finally, the methodologies used to assess the environmental impacts of landfilling on the environment are discussed and summarized.

Keywords Landfill · Solid waste · Leachate · Toxicological risk assessment · Water quality index · Environmental impact

Acronyms

ADI	Average daily intake
BDL	Below limit of detection

O. F. S. Khasawneh · P. Palaniandy (✉) · M. S. Yusoff · H. A. Aziz
School of Civil Engineering, Engineering Campus, Universiti Sains Malaysia, Nibong Tebal, Pulau Pinang, Malaysia
e-mail: omar.fa@student.usm.my; cepuganeshwary@usm.my; suffian@usm.my; cehamidi@usm.my

Y.-T. Hung
Civil Engineering Department, Cleveland State University, Cleveland, OH, USA
e-mail: y.hung@gbgconsulting.com; y.hung@csuohio.edu

L. K. Wang et al. (eds.), *Solid Waste Engineering and Management*, Volume 1, Handbook of Environmental Engineering 23,
https://doi.org/10.1007/978-3-030-84180-5_11

BOD	Biochemical oxygen demand
CDI	Chronic daily intake
COD	Chemical oxygen demand
CR	Cancer risk
EC	Electrical conductivity
EC_{50}	Median effective concentration
H_2S	Hydrogen sulfide
HCC	Highest hazardous chemicals
HI	Hazard index
HMEI	Heavy metals evaluation index
HMPI	Heavy metals pollution index
HQ	Hazard quotient
LC_{50}	Median lethal concentration
LFGs	Landfill gases
LOD	LOD of detection
MSW	Municipal solid waste
NH_3	Ammonia
NMVOCs	Non-methane volatile organic compounds
NOx	Nitrogen oxides
NVOCs/SVOCs	Non-semi-volatile organic compounds
ORP	Oxidation-reduction potential
PCA	Principal component analysis
q_i	Quality rating scale
SED-TOC	Sediment toxicity
SOAs	Secondary organic aerosols
TDS	Total dissolved solids
TOC	Total organic carbon
TS	Total solids
TU	Total unit
VOCs	Volatile organic compounds
WPI	Water pollution index
XRF	X-ray fluorescence

Nomenclature

BAC	Biological accumulation coefficient
K_k	Chemical element concentration
K_o	Soil contamination factor
P	Comprehensive pollution index
WF	Phytoaccumulation coefficient
Z_d	Summed concentration coefficient

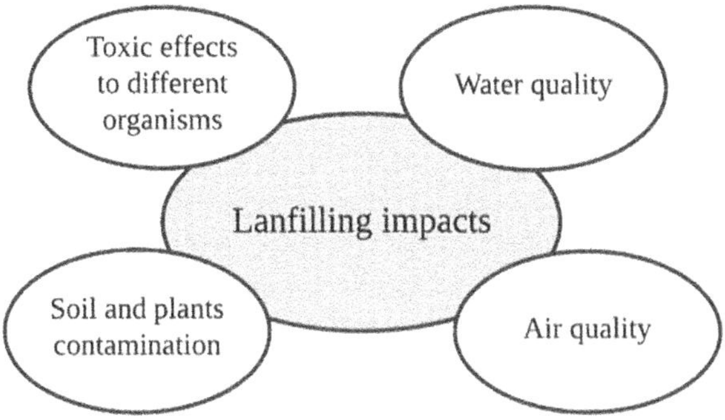

Fig. 11.1 Impact of landfilling on different environmental compartments

11.1 Introduction

The fast urbanization and continuous growth in industrial production in the past decades have resulted in the fast rise of the generation of industrial and municipal waste worldwide. As a result, the appropriate management of solid waste is becoming a public health concern due to its impact on the environment. In the 1990s, it was reported that the yearly amount of produced waste in developed countries ranged from 300 to 800 kg per person, while it was less than 200 kg per person for developing countries [6]. It is estimated that by the year 2025, the annual production of waste could increase by double to around 2.2 billion tonnes, making this a struggle for all the cities around the world [7].

Among the management and disposal practices of solid wastes, landfilling is still the dominant disposal method due to its economic advantages, effectiveness, and simplicity [8]. Landfilling can be defined as 'the dumping, compression, and embankment fill of waste at appropriate disposal sites'. Although landfilling has decreased over the decades, however, landfills remain a considerable element of the solid waste management systems across the globe [9]. Despite the advantages of landfilling, the process of landfilling can pose potential risks to different environmental compartments due to the incidence of hazardous organic and inorganic mixtures in the waste. Also, the poor management strategies of landfills can even cause potential negative risks to the humans residing near the landfill and the surrounding environment. Therefore, the assessment of the impact of landfilling practice on the environment (Fig. 11.1) is an essential step towards reducing its effects on the environment by developing strategies and regulations to manage the issue of landfilling and its risks to the environment.

11.2 Impact of Landfilling on Soil and Plants

11.2.1 Impact of Landfill Leachate on Soil and Plants

Leachate is considered one of the products generated by landfills that can cause contamination to the environment. The composition of leachate differs widely, and it depends on many aspects involving the composition of waste, depth of waste, degree of moisture and oxygen, as well as the design, operation, and age of landfills [10]. The major components of the landfill leachate can include different elements/substances such as dissolved methane (CH_4), sulfate (SO_4^{2-}), nitrate (NO_3^-), fatty acids, phosphates (PO_4^{3-}), nitrite (NO_2^-), calcium (Ca), chloride (Cl), sodium (Na), magnesium (Mg), potassium (K), and other trace metals including chromium (Cr), iron (Fe), manganese (Mn), nickel (Ni), zinc (Zn), copper (Cu), cadmium (Cd), mercury (Hg), arsenic (As), and lead (Pb). The contamination of soil caused by leachate generation can have a major impression on the soil quality. According to several studies [11], in nearly all conditions, the soil surrounding the landfills is deemed as the most part of the ecosystem that is being exposed to contamination. This is because the chemical elements from the landfill are migrated and dispersed when the water percolates through it. Numerous contaminants may accumulate in the soil, including heavy metals, pharmaceutical compounds, and polyaromatic hydrocarbons [12], causing contamination to the soil. In some studies, a variety of these contaminants may be adsorbed onto the soil during their transmission in the soil [13]. The consequence associated with such pollutants, particularly heavy metal contamination, is of great concern in the agricultural production system [8]. The absorbed metal elements by the soil can also be migrated and retained to the plants and vegetation, providing a major route for their entry into the food chain. Moreover, retained metals can affect both the growth and productivity of plants and pose negative effects on animal health. The degree of uptake of metals by plants is mainly influenced by several factors, such as the soil pH and salinity. The uptake of certain metals (i.e., cadmium and lead) could be developed by the complexation of chloride metals present in the leachate [14].

Numerous studies have shown that the environmental impacts of leachate on water sources (i.e. surface water and groundwater) have been observed regardless of a modern landfill site and/or monitoring networks [15]. Other studies have also suggested that landfills that contain hazardous pollutants are regularly monitored by analyzing the groundwater and soil, which has been polluted with landfill leachate [16]. There are a number of studies that have reported the possible impacts of landfill leachate on surface water and groundwater resources. For instance, Aderemi et al. (2011) revealed that the uncontrolled accumulation of leachate over time could be a result of the absence of leachate collectors. This would pose a significant negative effect on the quality of groundwater [10].

A study by Kanmani and Gandhimathi was conducted to evaluate the heavy metal contamination in soil caused by leachate migration in open dumping landfills. The study showed that heavy metals existed in the analyzed soil samples. The order of detected heavy metal was in the following order: manganese > lead > copper > cadmium. It is worth mentioning that these pollutants will continuously exist and migrate through the soil to further contaminant the nearby groundwater sources [17]. Generally, disposal of solid wastes, whether in sanitary landfills, landfills, or open dumpsites, presents a substantial source of heavy metal contamination. In particular, open dumpsites are considered the major source of such contamination. Heavy metal contamination is deemed as a serious problem in both urban and industrial areas due to their toxicity and potential risks to human health, and the environment and soil compartments are considered the ultimate sink for the discharge of heavy metals into the environment due to the fact that many heavy metals are bounded to the soil. In the study of Kanmani and Gandhimathi, the detected concentrations of heavy metals are reported in Table 11.1. A total of 12 samples were analyzed for heavy metals. Manganese was detected in all of the 12 samples, with concentrations ranging from 420.7 to 1711.6 mg/kg. Due to the fact that lead has high chalcophile properties, therefore, it can be present in its natural state in the soil as galena (PbS). However, concentrations ranging from 44.09 to 178.84 mg/kg were observed in most of the analyzed soil samples. Nevertheless, some soil samples from certain locations had very low concentrations of lead (BDL). While out of the 12 collected samples, only 5 samples were observed with copper concentrations (4.53–75.52 mg/kg). Since copper is rather an immobile element in soils. For cadmium, lower concentrations were detected (5.19–47.72 mg/kg). This is because the concentrations of cadmium were found to be low in both the solid waste and the generated leachate [17].

In a recent study, researchers have investigated the level of contamination on the surrounding environment (soil, sediments, and water basins) of a closed landfill (aftercare). The investigated landfill in the study was closed without any attention to the aftercare management. The study took place for a period of 2 years in the Northern part of Lithuania [1]. In this section, the impact of the closed landfill on the soil will be discussed. For the analysis of soil samples, researchers introduced

Table 11.1 Heavy metal concentrations for the collected soil samples

Sample details	Manganese (mg/kg)	Lead (mg/kg)	Copper (mg/kg)	Cadmium (mg/kg)
A: 1 m from GL	151.70	44.09	BDL	43.63
A: 2 m from GL	156.14	116.14	BDL	BDL
A: 3 m from GL	140.52	173.62	BDL	BDL
B: 1 m from GL	140.45	233.32	75.52	5.19
B: 2 m from GL	117.48	291.29	34.31	10.01
B: 3 m from GL	139.51	BDL	39.27	19.37
C: 1 m from GL	171.16	BDL	28.99	30.58
C: 2 m from GL	42.07	BDL	4.53	BDL
C: 3 m from GL	56.52	BDL	BDL	BDL
D: 1 m from GL	151.45	73.27	BDL	38.41
D: 2 m from GL	145.33	123.60	BDL	47.72
D: 3 m from GL	151.32	178.84	BDL	BDL

the X-ray absorption fine structure spectroscopy method (EXAFS) in order to investigate the occurrence of heavy metals in the examined soils. Baziene et al. extracted the EXAFS spectra from the raw X-ray absorption bands using a handheld XRF analyzer. The LOD for this type of method can be as low as 100 mg/kg for the majority of the heavy metal elements, and it often fits the minimal sample preparation procedures [18, 19].

The 'allowed assessment of heavy metals and metalloid soil' can be performed using the ratio between the current content of metalloids and the content of heavy metal elements in the soil samples and the concentration of heavy metals and metalloids of a specific geochemical background. In the study of Baziene et al., the chemical analysis was conducted on three different analytical specimens from 300 g of collected soil samples. A standard deviation was computed using the obtained data from the three replicate samples. Furthermore, the 'soil contamination factor' (K_0) was employed as an analytic analysis of soil data. The soil contamination factor can be defined as 'the total concentration of the metals for the evaluation of soil pollution' [1]. The highest concentration of hazardous chemicals in the soil (*HCC*) is considered the main indicator for the chemical contamination of soil. To simply put this, the greater the content of compounds (mg/kg) in the soil, the greater the threat of soil pollution. The K_0 is determined using these compounds, and it can be calculated in terms of the following formula:

$$K_0 = \frac{C}{HCC} \tag{11.1}$$

where C is the determined chemical substance concentration in soil (mg/kg) and *HCC* represents the highest concentration for a chemical substance in the soil (mg/kg), whereas the factor of the 'chemical element concentration' (K_k) can be calculated using the following equation:

$$K_k = \frac{C}{C_f} \tag{11.2}$$

where C_f presents the 'background content of the chemical element in the collected soil samples' (mg/kg).

In the case where the soil is polluted with additional chemical compound(s)/substance(s)/element(s), then the level of soil pollution can be assessed corresponding to the 'summed contamination coefficient Z_d', as presented in the following equation:

$$Z_d = \sum K_K - (n-1) \tag{11.3}$$

where n represents the number of chemical substances/elements in the soil.

According to the standards in Lithuania '(Lithuania Hygiene Norm HN 60:2004)', the maximum levels for hazardous elements/substances in the soil can be evaluated based on the following standards, as shown in Table 11.2 [1]:

Table 11.3 shows the results of Baziene et al.'s study on the impact of a closed landfill on the quality of the soil. The soil samples were taken at different depths ranging from 0 to 60 cm in a two-year period (2017–2018). Due to the fact that metalloids and heavy metals have the ability to transfer in liquid media, thus, the maximum concentration levels were observed in the deep soil layers (Table 11.3). However, it is worth mentioning that the concentrations of the detected heavy metals and metalloids in the collected soil samples did not exceed the maximum allowable concertation reported in the Lithuania standards (As-10; Cr-100; Cu-100; Pb-100; Ni-75, mg/kg) [20].

The revealed results indicate that the concentration levels exceeded the maximum permissible level of 70%; however, no concentrations surpassing the maximum levels have been reported. According to the reported data, the highest concentrations of heavy metals and metalloids were observed in the deeper layers of the soil at 45–60 cm (Table 11.3). The higher concentration levels of heavy metals and metalloids detected in the year 2017 imply that contaminants migration depends on several factors such as soil type, rainfall rate and intensity, the porosity of the soil, and the type of grass cover. When the individual substances/elements (Pb, As, Cu, etc.) were analyzed in the soil samples, it was noticed that the highest concentrations were detected for copper (Cu) and lead (Pb), insinuating that no waste was identified in the investigated closed landfill and that these concentrations and the level of contamination with heavy metals and metalloids may increase with time, due to the slow and long-lasting degradation processes of the heavy metals and metalloids. The study also suggested that in some parts of the closed landfill (i.e. south-western), higher concentrations of heavy metals and metalloids were observed. This implies that even the location of the sampling point might affect the results due to the fact that at these locations, the dumped and buried wastes contain higher concentrations of heavy metals and metalloids. Another suggestion is that the uneven cover and its distribution in the landfill can also affect the concentrations of these contaminants in certain locations.

Determining the contamination factor of each element is also necessary when analyzing the concentration levels of each substance. The determination of the K_0 allows a more thorough evaluation of the obtained results consistently because the permissible concentration of certain heavy metal elements and metalloids (i.e. Cu,

Table 11.2 Evaluation of soil contamination on chemical elements/substances

Contamination category of soil	Z_d value
Safe	$Z_d < 1$
Average hazard	$1 < Z_d < 3$
Hazard	$3 < Z_d < 10$
Very hazard	$Z_d > 10$

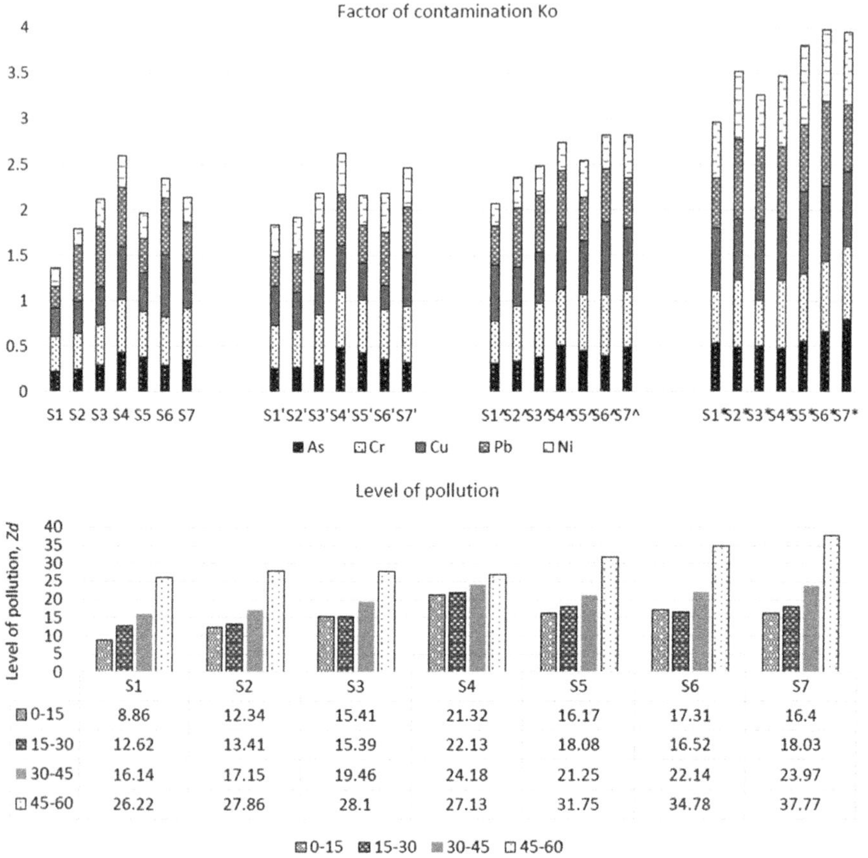

Fig. 11.2 Variations in the K_0 for different elements [1]

Pb) is larger than that for some other elements (i.e. As, Cr). As discussed previously, the contamination factor of the soil can be calculated using Eq. 11.1. In the study conducted by Baziene et al., the K_0 factor was calculated for the investigated elements and depicted in Fig. 11.2.

It can be noticed from Fig. 11.2 that the propensity for the measured concentration of elements is proportional to the sampling depth. Further, it can be observed that the contamination factor for the topsoil layer was less than 3. While the total contamination factor for all the heavy metals and metalloids is greater than 3 at the largest depth of 45–60 cm.

The approximate soil contamination levels are illustrated in Fig. 11.2. It can be observed that both the highest K_0 and the degree of contamination were detected in the deepest layers of the soil (45–60 cm). The conducted assessment of the achieved findings with the levels discovered in the 'hygiene standard (Lithuania HN 60:2004)' indicates that only the point located at the topsoil layer (S1) holds a lethal intensity. While in the other tests, an extremely dangerous level is determined ($Z_d > 10$). Since

Table 11.3 Percentage of heavy metal elements and metalloid from the highest level of concentration in the collected soil specimens

		2017					2018				
Soil sample	Parts of MLC	As	Cr	Cu	Pb	Ni	As	Cr	Cu	Pb	Ni
Samples were taken at 45–60 cm depth											
S1	%	64	45	59	40	89	45	69	79	68	36
S2	%	44	83	48	88	87	55	65	86	87	61
S3	%	54	42	78	92	65	47	59	98	67	51
S4	%	48	82	67	98	92	48	69	68	59	64
S5	%	64	79	98	78	97	49	70	83	68	76
S6	%	79	88	77	98	92	54	69	87	87	66
S7	%	97	89	87	78	77	65	71	78	68	83
Samples taken at 30–45 cm depth											
S1	%	30	35	55	40	16	35	56	68	47	32
S2	%	32	55	42	76	32	36	67	43	53	36
S3	%	46	56	64	77	27	31	62	49	47	40
S4	%	56	63	79	67	31	47	60	58	57	31
S5	%	52	54	56	46	45	39	69	64	48	36
S6	%	45	67	87	78	32	35	68	73	38	42
S7	%	52	67	69	57	60	46	59	69	52	35
Samples taken at 15–30 cm depth											
S1	%	20	39	38	28	32	33	56	48	36	37
S2	%	29	38	32	40	45	25	46	49	45	35
S3	%	28	44	44	58	43	32	67	46	37	40
S4	%	56	60	47	65	57	42	67	52	45	34
S5	%	46	46	39	45	38	41	69	43	38	28
S6	%	37	48	50	67	47	37	59	55	49	38
S7	%	35	56	59	67	60	32	67	58	33	27
Samples taken at 0–15 cm depth											
S1	%	20	38	26	20	16	25	38	37	28	24
S2	%	24	38	24	88	15	26	40	46	36	21
S3	%	31	42	46	92	27	29	45	37	37	37
S4	%	48	60	59	89	43	39	55	59	41	25
S5	%	39	55	41	48	28	39	45	44	26	28
S6	%	25	48	80	87	21	35	58	56	37	23
S7	%	32	58	67	59	27	38	56	39	23	29

the major amounts of heavy metal elements and metalloid compounds were discovered in the deepest layers of the soil, the K_0 was greatest in the layers located at depths ranging from 45–60 cm. While relating to the points of distribution, the pollution is scattered equally across the landfill body, and only the south-western part of the landfill has a greater degree of pollution [1]. This diffusion of heavy metals and metalloids could be attributed to several reasons, but it is mainly caused by the composition of the waste being deposited as well as the properties and characteristics of the lining material in the closed landfill.

Heavy metal elements and the metalloid substances found in soil samples identified during the previous studies [1] imply that the re-developed landfill body shall be monitored regularly. Obtaining such information from the monitoring shows just how far this polluted area can pose risks to the environment. This type of investigation demonstrates that even closed landfills can affect the environment even in the long term, which necessitates regular valuation and monitoring to check if there are new bases of pollution in the vicinity of the landfill body during the long-term biological degradation processes.

11.2.2 Presence of Heavy Metals in Plants

Table 11.4 represents the detected metal concentrations in the leaves and stems of European goldenrod and in grasses near a landfill located in Poland [2]. However, it can be noted that the detected concentrations of metals are relatively low. This is because the investigated plants growing in Poland are located outside the landfill and the direct influence of pollution emitters on the plants is low.

The index of 'biological accumulation coefficient (BAC)', which expresses the ratio of metal concentration in plants to its concentration in soil (0–0.25 m), can be estimated using Eq. 11.4. The BAC index is used to reflect the capability of the plants to absorb metals from the soil [21].

Table 11.4 Heavy metal concentrations (mg/kg) in the European goldenrod (*Solidago virgaurea L.*) and grasses (Poaceae) in Łubna landfill, Poland [2]

Zone no.	Plant	Cd	Cu	Cr	Ni	Pb	Zn
1	Grass	1.0	5.0	1.1	0.9	1.5	73
2	Grass	0.1	2.6	< 0.2	< 0.5	< 0.6	31
	European goldenrod/leaves	0.4	3.7	< 0.2	0.9	0.8	53
	European goldenrod/stem	0.4	2.6	< 0.2	< 0.5	< 0.6	66
3	Grass	0.3	9.1	< 0.2	< 0.5	< 0.6	78
	European goldenrod/leaves	0.2	5.9	< 0.2	< 0.5	< 0.6	42
	European goldenrod/stem	0.3	5.9	< 0.2	0.7	< 0.6	36
4	Grass	0.3	10.9	< 0.2	< 0.5	4.6	30
	European goldenrod/leaves	0.4	10.7	0.2	1.6	2.3	84
	European goldenrod/stem	0.5	9.2	< 0.2	0.7	1.9	92
5	Grass	0.4	6.2	< 0.2	< 0.5	2.6	60
	European goldenrod/leaves	0.3	9.7	0.3	0.6	2.4	90
	European goldenrod/stem	0.3	5.0	< 0.2	< 0.5	<0.6	85
6	Grass	0.1	7.3	< 0.2	< 0.5	< 0.6	34
	European goldenrod/leaves	0.2	8.8	< 0.2	0.7	0.9	55
	European goldenrod/stem	0.3	4.0	< 0.2	0.5	< 0.6	72

Table 11.5 Phytoaccumulation coefficient values: BAC 'ratio of metal concentration plants to metal concentration in soil (0–25 cm)' in Łubna landfill, Poland [2]

Zone no.	Plant	Cd	Cu	Cr	Ni	Pb	Zn
1	Grass	3.58	1.87	0.24	0.43	0.07	6.7
2	Grass	0.41	1.13	0.07	0.38	0.05	4.9
	European goldenrod/leaves	1.88	1.61	0.07	0.69	0.07	8.4
	European goldenrod/stem	2.13	1.13	0.07	0.38	0.05	10.5
3	Grass	0.31	1.60	0.02	0.05	0.02	2.4
	European goldenrod/leaves	0.29	1.04	0.02	0.05	0.02	1.3
	European goldenrod/stem	0.34	1.04	0.02	0.08	0.02	1.1
4	Grass	0.65	1.22	0.02	0.06	0.26	1.4
	European goldenrod/leaves	0.91	1.19	0.02	0.20	0.13	4.0
	European goldenrod/stem	0.97	1.32	0.02	0.09	0.11	4.4
5	Grass	4.75	2.09	0.12	0.42	0.59	27.1
	European goldenrod/leaves	3.75	3.27	0.18	0.50	0.55	17.3
	European goldenrod/stem	4.25	1.68	0.12	0.42	0.14	16.3
6	Grass	2.77	3.04	0.08	0.45	0.07	5.1
	European goldenrod/leaves	4.68	3.67	0.08	0.64	0.10	8.2
	European goldenrod/stem	5.96	1.67	0.08	0.45	0.07	10.8

Table 11.6 Mobility ratio values: 'ratio of metal concentration in soil (0–25 cm)' to its concentration in groundwater in Łubna landfill, Poland [2]

Zone no.	Cd	Cu	Cr	Ni	Pb	Zn
1	456	178	447	39	4202	238
2	493	230	269	90	1983	126
3	418	154	114	1150	2113	310
4	1583	774	855	1337	4425	195
5	160	174	171	590	1105	31
6	116	184	262	1145	2276	108

$$BAC = \frac{Me_{plant}}{Me_{soil(0-25cm)}} \tag{11.4}$$

Similarly, the 'mobility ratio' index can also be used to reflect the infiltration of metals from the soil to the groundwater. The 'mobility ratio' expresses the ratio of metal concentration in soil (0–0.25 m) to its concentration in groundwater, as shown in Eq. 11.5. It is worth mentioning that the literature lacks the data available on the assessment of the mobility ratio.

$$MR = \frac{Me_{\text{soil}(0-25\,\text{cm})}}{Me_{\text{groundwater}}} \tag{11.5}$$

An example of the BAC and mobility ratio was calculated using Eqs. 11.4 and 11.5 [2], and their indexes are shown in Tables 11.5 and 11.6. Mainly, when the phytoaccumulation coefficient (WF) of plants is larger than 1, then they are considered to be accumulators of a given element due to the ability of plants to absorb and accumulate a metal in the plant tissues. However, when the WF ~ 1, in this case, the plants are considered as indicators of a given pollutant. In the case that the WF value is below 1, then it characterizes the plants that limit the intake of a given element/substance. In the study of Gworek et al., the analyses have demonstrated that the intake of some metal elements (i.e. Cr, Ni, and Pb) by European goldenrod and grasses appeared in controlled quantities, and in these cases, the values of the BAC were below 1. In the cases of other metal elements (i.e. Cd, Cu, and Zn), the values of BAC exceeded 1, although the concentrations of these metals in the plants and soil suggest that, in the landfill area, the contamination is low and at a background level. On the other hand, the values of the mobility ratio were high, implying an insignificant impact of the metal concentration in the soil on water contamination [2].

11.2.3 Impact of Landfill Gas on Soil and Plants

The generated gas from landfills can have significant impacts on the surrounding environmental compartments. These impacts are of regional or global significance, and they can mainly contribute to what is called the 'greenhouse effect'. In regard to the effect of gaseous emissions on the soil and vegetation, gaseous pollutants can pose potentially significant risks to the plants and the entire ecosystems. The horizontal migration of gaseous pollutants through the soil and beyond the boundaries of the landfill can trigger oxygen displacement from the soil, resulting in a deterioration of the soil faunal populations as well as burrowing animals. This phenomenon can also cause the dieback of plants [14]. The vegetation surrounding the landfills as well as the newly planted vegetations nearby landfills can be heavily affected and dented due to the repression of air surrounding the roots by the migrated landfill gas [9]. It is worth mentioning that the phenomenon of acidic rain near landfills is mainly caused by the generation of acidic gaseous constituents. Consequently, causing what is known as 'acidification of soils and the ecosystems'. One of the major elements of the formed acid rain is ammonia, which can be often found in landfill gas. Ammonia is considered a 'secondary acidifying agent' that is transformed into nitric acid following its oxidation in the atmosphere. This can have negative impacts on the plants, causing several damages such as reduction in photosynthesis, loss of stomatal control, enzyme inhabitation, depressed growth and yield, and changes in the synthetic pathways [14].

The toxic gaseous pollutant, hydrogen sulfide (H_2S), is also a product of the generated landfill gases; it can also have significant negative impacts on the environment. H_2S is a serious bio-toxic gas that can be effective at even small parts per millions in mammals. Although plants are less sensitive to the direct exposure to

H_2S, however, a threshold of 1 μg/g has been reported [22, 23]. One of the most critical impacts of H_2S on the plants is the ability to inhibit and destruct the growth of roots and vegetation cover, which is mainly due to the anaerobic soil conditions produced by the high concentration levels of sulfides, which could horizontally outflow from landfill sites. Another gaseous pollutant known as 'methane' can laterally migrate through the soil and cause serious hazardous explosions when it is mixed up with an adequate amount of air [14].

11.3 The Impact of Landfill Leachate on Water Sources

The generated leachate from landfills that are constructed without engineered liners and leachate collection systems can have significant negative effects on the quality of surface water and groundwater with severe outcomes on human and ecosystem health [13]. The waste in landfills is exposed to either infiltration from precipitation or groundwater underflow. As the water migrates through the wastes, it collects a mixture of inorganic and organic compounds, which eventually flows out of the waste and gathers at the bottom of the landfill. This results in generating contaminated water 'leachate'. The leachate that accrues at the bottom side of a landfill can percolate through the soil and often could reach groundwater resources. The contamination of groundwater by landfill leachate and landfills' waste can influence the overall water quality, resulting in the water becoming unsuitable for use [8]. Table 11.7 shows the obtained data from the groundwater analyzed near a landfill

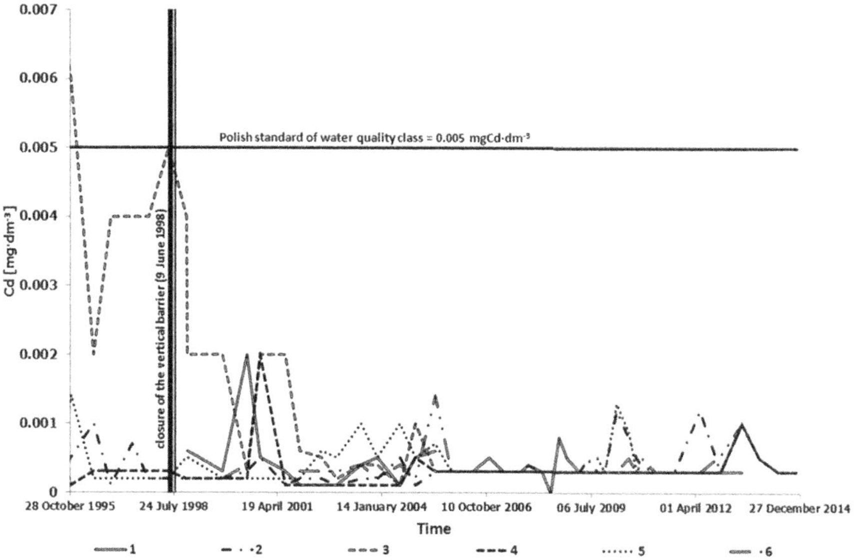

Fig. 11.3 Variations of cadmium content in groundwater in the surroundings of Łubna landfill [2]

located in Poland [2]. It can be seen from the table that the monitoring took place at different sampling points (six points) around the landfill. The researchers analyzed the surrounding groundwater for five different elements, including cadmium (Cd), lead (Pb), chromium (Cr), copper (Cu), and zinc (Zn), as well as the pH and the electrical conductivity (EC) of the groundwater. The obtained results were compared with the admissible limit issued by the Polish institutions.

The reported values of the pH and electrical conductivity can provide a proof that the collected samples were not affected by precipitation or runoff. It is worth mentioning that the top layer of the soil in the investigated Polish landfill is comprised only of thin impermeable soil that can restrict and limit the infiltration process. Nevertheless, it is feasible that both the pH and EC of the groundwater can still be altered and affected by any external pollution sources, such as the existence of any industrial facilitates near the landfill. It can be noted from Table 11.7 that none of the identified heavy metals in the year 2014 have exceeded the maximum allowable concentration reported by the Polish Regulators in 2015 for the quality of water intended for human consumption [2].

Table 11.7 Average values for the concentrations of heavy metals, pH, and EC in groundwater, 1994–2014 [2]

	The average concentration in groundwater during the monitoring phase						
	Sampling period 1994–1998						
Sample Standard	Cd (mg/L)	Pb (mg/L)	Cr (mg/L)	Cu (mg/L)	Zn (mg/L)	pH	EC (μS. cm-1)
Polish standard	*0.005*	*0.025/0.010*	*0.050*	*2.0*	–	*6.5–9.5*	*2500*
S1	–	–	–	–	–	–	–
S2	0.0008	0.011	0.024	0.110	0.204	6.91	1045
S3	0.007	0.080	0.531	0.226	0.586	7.36	14,403
S4	0.0005	0.015	0.023	0.104	0.227	6.83	1110
S5	0.0008	0.035	0.047	0.127	0.615	7.45	1638
S6	–	–	–	–	–	–	–
All piezometers	0.0023	0.035	0.156	0.141	0.408	7.06	4549
	Sampling period 1999–2014						
Polish standard	*0.005*	*0.025*	*0.050*	*2.0*	–	*6.5–9.5*	*2500*
S1	0.0006	0.005	<0.010	0.015	0.086	6.92	7394
S2	0.0004	0.006	<0.010	0.010	0.059	6.07	898
S3	0.0008	0.019	0.216	0.037	0.109	7.03	6349
S4	<0.0003	<0.004	<0.010	0.009	0.190	7.37	957
S5	0.0005	<0.004	<0.010	0.017	0.167	6.91	1064
S6	0.0004	<0.004	<0.010	0.013	0.068	6.62	198
All piezometers	0.0005	0.008	0.044	0.017	0.113	6.61	2810

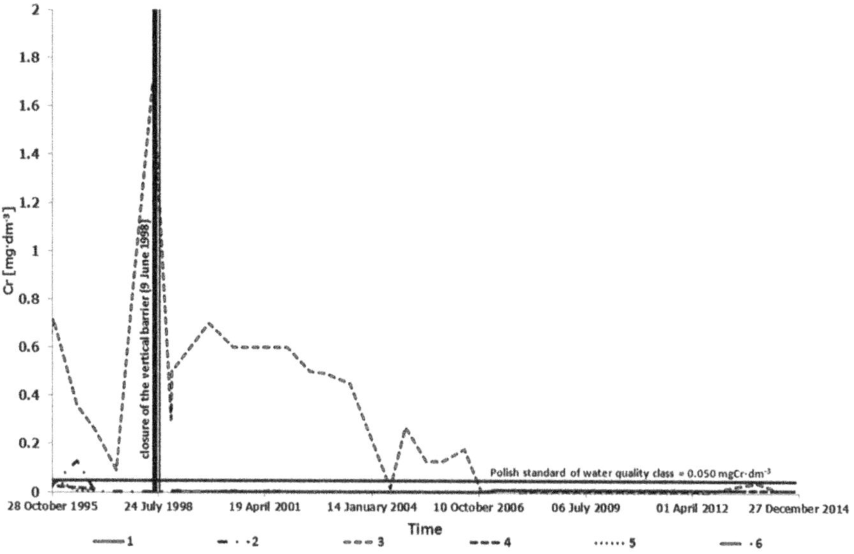

Fig. 11.4 Variations of chromium content in groundwater in the surroundings of Łubna landfill [2]

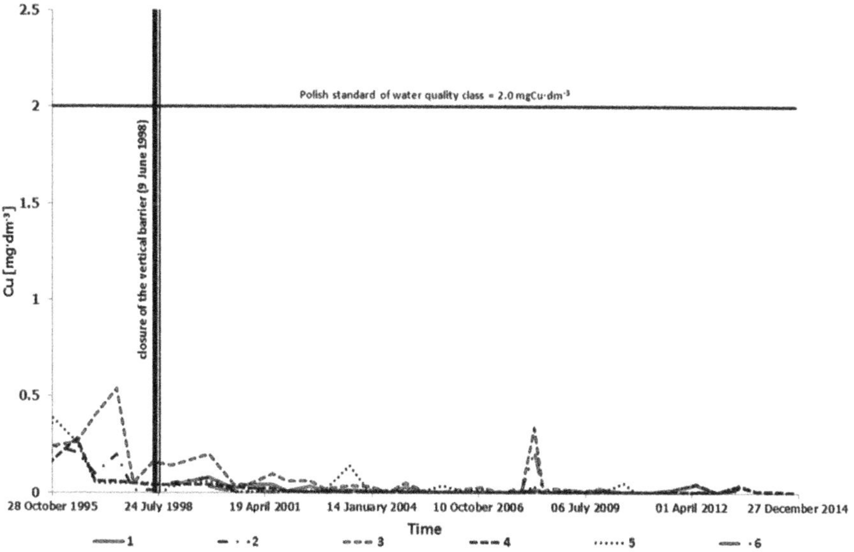

Fig. 11.5 Variations of copper content in groundwater in the surroundings of Łubna landfill [2]

The charts depicted in Figs. 11.3, 11.4, 11.5, 11.6 and 11.7 demonstrate examples of the changes in the concentration of certain heavy metals in the groundwater from the period 1995 to 2014 in a Polish landfill surrounding. It can be observed that after the construction of a leachate drainage system (LDS) and a vertical

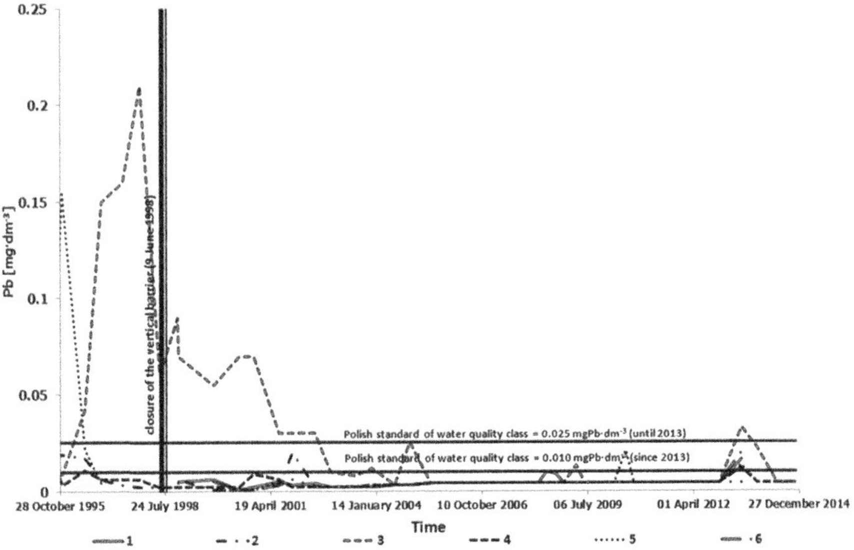

Fig. 11.6 Variations of lead content in groundwater in the surroundings of Łubna landfill [2]

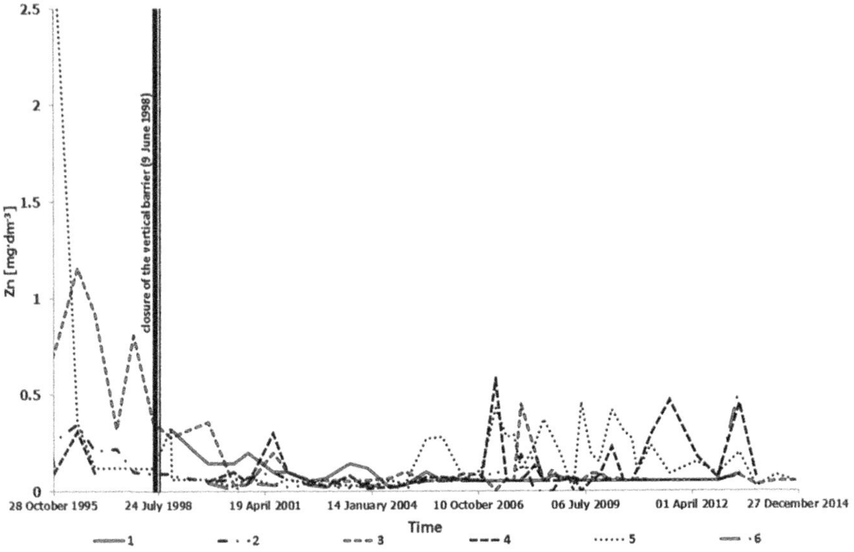

Fig. 11.7 Variations of zinc content in groundwater in the surroundings of Łubna landfill [2]

Table 11.8 Effect of depth, age, and distance on groundwater contamination

Analyzed parameters (mg/L)	1–30 (ft)	30–60 (ft)	60–90 (ft)	90–120 (ft)	> 200 (ft)
	Depth of sampling				
Cl^-	115.9	97.8	81.8	43.9	21.8
NH_4^{2+}	9.8	5.7	4.8	2.6	0.5
COD	128	96.0	40.0	16.0	12.0
Na^+	98.0	81.7	60.0	37.0	35.0
K^+	42.2	31.2	19.0	21.1	16.9
Analyzed parameters (mg/L)	–	1–10 (year)	10–20 (year)	–	20–30 (year)
	Age				
Cl^-		115.9	81.9		51.9
NH_4^{2+}		5.7	9.8		9.1
COD		16.4	64.0		40.0
Na^+		74.9	81.7		70.9
K^+		20.0	42.8		31.0
Analyzed parameters (mg/L)	–	0–1 (km)	1–2 (km)	–	2–3 (km)
	Distance from the landfill				
Cl^-		115.9	97.0		93.0
NH_4^{2+}		9.8	5.7		6.04
COD		128 0.0	96.0		12.0
Na^+		81.7	74.56		98.0
K^+		18.19	42.27		38.76

Table 11.9 Classification for water quality index

P	Level of water quality
≤ 0.2	Cleanness (I)
0.21–0.4	Sub-cleanness (II)
0.41–1	Sligh pollution (III)
1.01–2	Moderate pollution (IV)
≥ 2.01	Severe pollution (V)

barrier, the concentrations of the identified heavy metals decreased significantly. Moreover, a slight decline in the concentration of heavy metals was also observed due to the leachate pumping and transfer to the treatment plant facility. Further, the reported study of Gworek et al. showed water quality improvement in all piezometers due to the installation of a groundwater protection system in 1998 after only a few years. A prove of the water quality improvement can be seen as shown in Table 11.7. As in 1998, the concentrations of the identified heavy metals have exceeded the admissible limit of the Polish standards; however, after the installation of the vertical barrier and the LDS, the migration of heavy metals to the groundwater was reduced noticeably [2].

The movement of chemical pollutants in groundwater is rather a complex process, and it can extend to distances of one kilometre or more. The plume movement can be accidentally accelerated by increasing the flow rate of groundwater due to the rigorous operation of well pumps downstream. Pollution of such wells can often be assuaged by establishing properly positioned new wells that are counteract pumped in order to turn the plume away from the affected areas. Foreseeing the passage of landfill plumes entails studies not only regarding the general region but also

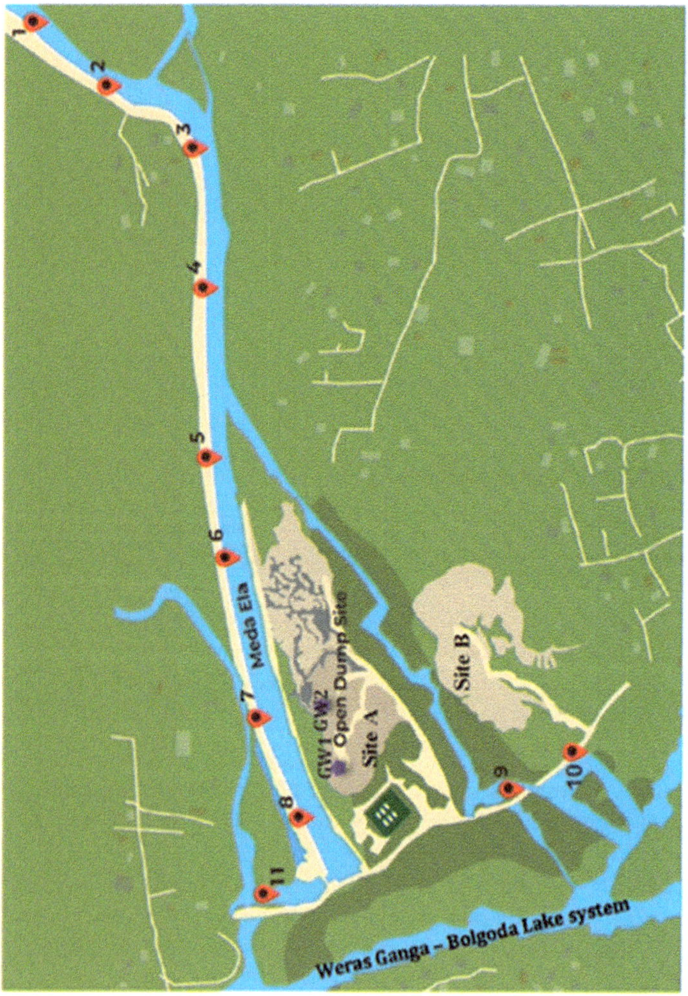

Fig. 11.8 Spots of water sampling locations around the site of Meda Ela [27]. (Reprinted from Koliyabandara S.M.P.A., Cooraya, A.T., Liyanagea, S., Siriwardana, C. Assessment of the impact of an open dumpsite on the surface water quality deterioration in Karadiyana, Sri Lanka, Environmental Nanotechnology, Monitoring & Management 14 (2020) 100371, with permission from Elsevier)

on the local geohydrology of the landfill surrounding areas. In some cases, local geohydrology can help in controlling the geometry of the plume geometry as well as the movement of contaminants. Several factors, including the variations in the aquifers' permeability, intrusion from impeding strata, or corrugation in the underlying bedrock, can be the determinants. Preventing the contamination of groundwater is critical. This is because the aquifer of the groundwater requires decades to cleanse itself. And to artificially remove the pollutants from the groundwater reservoir requires extremely expensive methods and techniques [24].

In another recent study conducted in India, researchers investigated the impact of landfill leachate on the quality of groundwater at a different depth, age, and distance from the landfill [25]. The groundwater samples were collected in two different seasons: pre-monsoon and post-monsoon. The physiochemical characteristics of the groundwater were evaluated, including EC, pH, total dissolved solids (TDS), total alkalinity, total hardness, biochemical oxygen demand (BOD), chemical oxygen demand (COD), calcium (Ca), and magnesium (Mg). The study of Negi et al. showed that the pH of the groundwater varied between 6.8–7.7 and 7.1–7.4 for the pre-monsoon and post-monsoon seasons, respectively, with the majority of samples

Table 11.10 Analysis of water quality parameters and comparison with local and international regulatory authorities

Parameter (variable)	Water quality guidelines for inland fish (FAO)	Guideline for the protection of aquatic life (CCME)	Guideline for water quality and aquatic life (EPA)	WQP[a]	
				Average	Range
Conductivity (mS/cm)	N.A	N.A	N.A	0.29 ± 0.03	0.13–0.31
TDS (mg/L)	N.A	N.A	N.A	156.3 ± 9.8	79.0–160.0
Nitrate (mg/L)	Vary	13	N.A	200.5 ± 16.0	91.2–251.6
DO (mg/L)	Vary	N.A	N.A	4.23 ± 0.45	3.91–7.63
Ammonia (mg/L)	0.5–1.5	0.019	N.A	2.66 ± 0.20	1.84–2.95
pH	6.5–8.5	6.5–9.0	6.5–9.0	7.90 ± 0.19	7.72–8.33
ORP (mV)	N.A	N.A	N.A	118.3 ± 67.9	70.1–402.2
Fe (mg/L)	0.1–0.2	0.3	1.0	1.31 ± 1.52	0.02–4.20
Cu (mg/L)	0.001–0.01	N.A	N.A	0.19 ± 0.16	0.01–0.46
Mn (mg/L)	N.A	0.2	N.A	2.56 ± 0.68	1.10–3.86
Zn (mg/L)	0.01–1.0	0.003	0.12	0.53 ± 0.35	0.05–1.70

[a]WQP: water quality parameter

not exceeding the permissible allowable limit of (6.5–8.5). The acidic nature of the groundwater could be related to the dissolution of carbon dioxide (CO_2) and other organic compounds in the water. The electrical conductivity (EC) of the groundwater samples ranged from 220 to 1550 and from 136 to 1039 μmhos/cm for the pre-monsoon and post-monsoon seasons, respectively.

The high EC values in some samples could be due to the leakage of landfill leachate into the groundwater reservoirs. TDS values ranged from 180 to 679 and from 161 to 664 mg/L for the pre-monsoon and post-monsoon seasons, respectively. The majority of the analyzed groundwater samples (around 50%) exceeded the desirable limit of 500 mg/L. This high concentration in TDS is also related to leachate contamination and seepage into the groundwater. The total alkalinity of the groundwater samples was less affected by the landfilling activities, with concentrations ranging from 24 to 148 mg/L for both seasons. However, the total hardness of the samples was higher than the alkalinity, indicating non-carbonate hardness in the groundwater. High COD concentrations were detected in a number of samples that were collected near the landfill area. While the BOD values were very low in some samples, however, in the area near the landfill, high concentrations of BOD were detected due to leachate contamination [25].

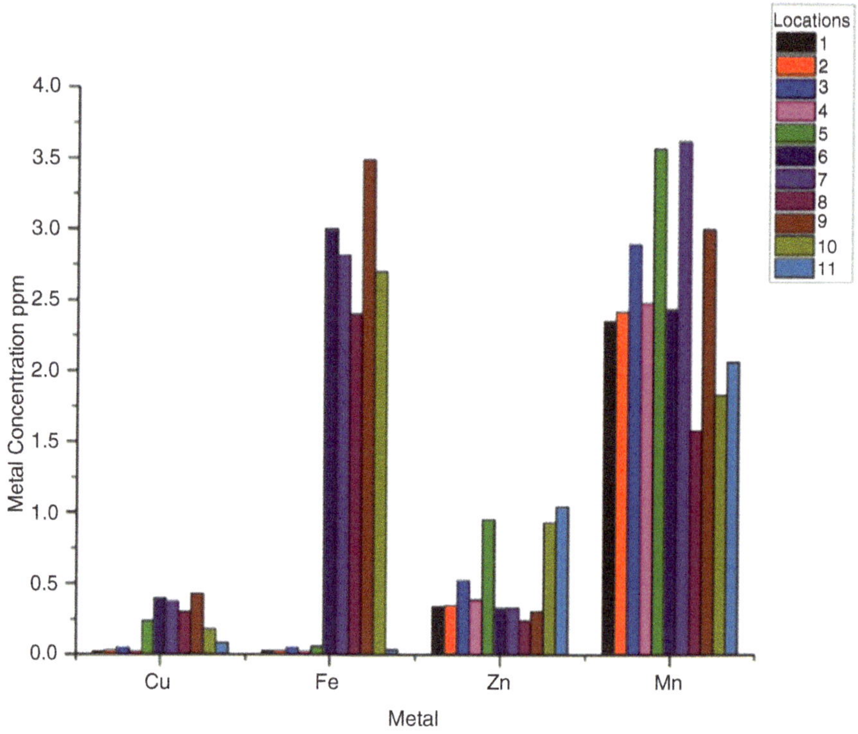

Fig. 11.9 Variations in the concentration of different heavy metals with respect to their sampling locations [27]

Table 11.11 Correlation matrix for different variable parameters

Mn	Zn	Fe	Cu	NH_3	NO^{3-}	DO	TDS	Cond,	ORP	pH	
										1	pH
									1	−0.52	ORP
								1	−0.9	0.55	Cond.
							1	0.87	−0.66	0.25	TDS
						1	−0.87	−0.53	0.3	0.19	DO
					1	−0.84	0.75	0.41	−0.03	−0.07	NO^{3-}
				1	0.03	−0.28	0.65	0.88	−0.96	0.6	NH_3
			1	0.41	−0.08	0.03	0.24	0.5	−0.51	0.13	Cu
		1	0.84	0.41	−0.06	0.08	0.2	0.46	−0.45	0.22	Fe
	1	−0.26	−0.17	0.25	−0.05	0.05	0.13	0.23	−0.24	0.47	Zn
1	−0.01	−0.03	0.24	−0.02	−0.01	−0.17	0.05	−0.01	−0.04	−0.38	Mn

Table 11.12 Experimental loadings of variable parameters on two different principal components

Parameter (variable)	PC1	PC2	PC3	PC4
ORP	0.470	0.025	0.042	0.143
pH	- 0.305	- 0.266	- 0.405	0.161
DO	0.169	- 0.643	- 0.014	- 0.019
EC	- 0.482	0.152	- 0.054	0.024
Nitrate	- 0.462	- 0.034	- 0.109	- 0.093
Ammonia	- 0.097	0.616	- 0.032	0.262
Fe	- 0.303	0.223	0.440	0.251
Zn	- 0.113	- 0.075	- 0.537	- 0.419
Mn	0.007	0.156	0.322	- 0.797
Cu	- 0.312	- 0.180	0.482	- 0.043
Eigenvalue	4.0051	2.0610	1.8162	1.0983
Cumulative	40.1	60.7	78.8	89.8

The occurrence of heavy metals (i.e. Pb, Zn, Cr), cations (i.e. ammoniacal nitrogen, sodium, potassium), and anions (i.e. sulfate chloride) in the groundwater sources near landfilling activities has also been reported [25]. The occurrence of sulfate is mainly due to the presence of 'sulfate-reducing bacteria' in the groundwater or soil that enters with 'methane-producing microbes' for the organic carbon that is available, which results in generating H_2S by reducing sulfates to sulfides. On the other hand, the occurrence of heavy metals in groundwater is mainly attributed to several reasons, including the adsorption capability of the soil, movement of the groundwater, and reaction rate. In some cases, low concentrations of heavy metals can be detected in the groundwater. This is mainly due to the high adsorption ability of the soil, where the metals are adsorbed on the soil or by the organic matter that exists in the soil [25, 26].

11.3.1 Factors Affecting Groundwater Contamination

The contamination of groundwater sources due to the leakage and migration of leachate from open dump landfills is influenced by several factors such as the depth, age, and distance of the water sources as well as the chemical composition of the landfill leachate. Table 11.8 below shows the effect of different factors (depth, distance, and age) on the contamination of a groundwater resource located near an Indian open dumpsite [25]. The results obtained from the table clearly indicate that as the depth of sampling increases, the concentrations of certain parameters (i.e. Cl^-, NH_4^{2+}, and COD) decrease. Moreover, the degree of concentration for these parameters was found to be higher near the dumpsite (0–1 km), compared to the samples taken at longer distances (> 1 km). However, the concertation of some parameters, including potassium and sodium, was observed to be higher at longer distances (> 1 km) and aged 20–30 years. This could be attributed to the fertilizer seepage from nearby agricultural areas. Generally, the risk of groundwater contamination significantly depends on two factors which are the (i) nature of the geological strata and (ii) unsaturated zone thickness [25].

11.3.2 The Impact of Open Landfill (Dumpsite) on the Quality of Surface Water in Sri Lanka (Case Study)

The environmental impact of an open dumpsite (landfill) on the quality of receiving waters in Sri Lanka has been investigated [27]. The investigated site receives around 500 tons of municipal solid waste every day from the surrounding areas. The type of municipal solid waste disposed of in the site is categorized as 80% animal, food, and garden waste (organic waste), while around 10% consists of PVC, polyethylene, and rubber waste. Basic quality parameters including BOD, COD, nitrate, ammonia, and TDS were also analyzed for the dumpsite's leachate. Results from the

Table 11.13 Comprehensive pollution index for the surface water surrounding the landfill

Ci	*Si*	*Ci*/*Si*	Comprehensive pollution index (P)[a]
Conductivity (mS/cm)	0.29	0.225	1.288
TDS (mg/L)	156.34	2100	0.074
Nitrate (mg/L)	200.54	50.0	4.0108
Ammonia (mg/L)	2.66	60.8	0.043
pH	7.9	8.5	0.929
Fe (mg/L)	1.31	3	0.436
Cu (mg/L)	0.19	3	0.063
Mn (mg/L)	2.56	0.4	6.40013.59
Zn (mg/L)	0.53	2	0.265
Cr (mg/L)	0.04	0.5	0.08

[a] $P = \frac{1}{n}\sum_{i=1}^{n}\frac{Ci}{Si}$

Table 11.14 Groundwater water quality monitoring wells

Variable parameter	Concentration in groundwater	Range of concentration
Conductivity (mS/cm)	1.66 ± 0.055	1.55–1.71
Total phosphate (mg/L)	62 ± 11	48–55
Nitrate (mg/L)	151 ± 48	120–283
Ammonia (mg/L)	346 ± 39	288–387
pH	6.75 ± 0.08	6.60–6.88
TDS (mg/L)	129 ± 22	110–172
Fe (mg/L)	7.81 ± 0.68	6.80–8.80
Cr (mg/L)	0.02 ± 0.01	0.0–0.04
Cu (mg/L)	0.16 ± 0.03	0.11–0.21
Mn (mg/L)	0.23 ± 0.04	0.18–0.31
Cd (mg/L)	0.02 ± 0.01	0.01–0.04
Zn (mg/L)	0.05 ± 0.01	0.03–0.07

analysis showed that the concentration values for the analyzed leachate exceeded the allowable limit set by the international and national guidelines for leachate quality [28, 29].

Since the employment of water quality indexes is recognized as a useful and helpful tool in understanding the quality of water. This study used what is known as the 'comprehensive pollution index (P)' as a method to evaluate the quality of water (non-drinking purposes). The classification criteria for water quality based on the comprehensive pollution index method is summarized in Table 11.9. Meanwhile, the comprehensive quality index can be calculated using the equation shown below:

$$P = \frac{1}{n}\sum_{i=1}^{n}\frac{Ci}{Si} \tag{11.6}$$

where Ci is the measured concentration of the targeted contaminant (mg/L), Si is the concentration limit (mg/L) that is permissible by the regulatory authorities, and n represents the number of evaluated parameters.

The lake of Bolgoda is the major receiving lake surrounding the dumpsite, and it is considered one of the main problems with the open landfill (Fig. 11.8). The Bolgoda lake is a brackish water source that is also located close to the dumpsite. The lake consists of two major basins (south and north), with a middle stream connecting both basins. A small stream that crosses the dumpsite known as 'Meda Ela' receives direct leachate from the landfill, and as a result, it can be considered as the major point source of pollution.

Table 11.10 summarizes the obtained measurements of some water quality parameters (e.g. pH, Cond., TDS) and compares the obtained results with the permissible limit provided by different authorities, such as the EPA and FAO.

It can be seen from Table 11.10 that the pH of the samples experienced a small increase of 0.3 units when changing the sampling location from 1 to 8 (Fig. 11.8).

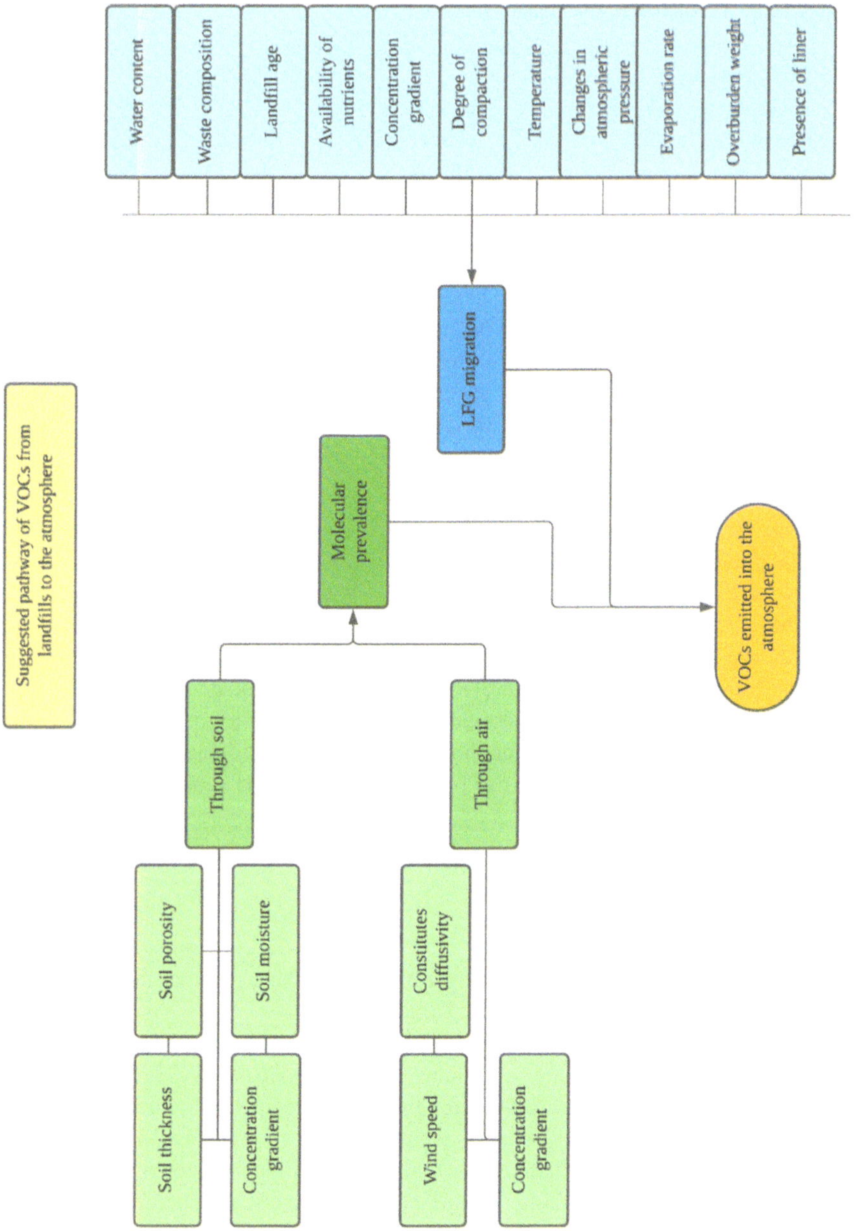

Fig. 11.10 Possible pathway for VOCs. (Adapted and modified from [3])

Further, it can be observed that the pH of water samples collected in locations 9–11 is more basic than the pH of water at locations from 1 to 8. It is a known fact that the conductivity of fresh water should not exceed the limit of 1000 (due to

Table 11.15 Possible VOCs emitted from different types of waste material [34]

No.	Type of disposed waste	Possible emitted VOCs from the landfill site
1	Household cleaning products (solvents)	Limonene, isopentane, isopropanol, 1-propanol, butoxyethanol, limonene, toluene, decane, ethanol, phenol, o-xylene, chlorobenzene, 1,4 dichlorobenzenes
2	Household products (sprays)	Propane, acetaldehyde, isobutyl alcohol, p-xylene, propionaldehyde, butyraldehyde, methyl formate, methyl ethyl ketone, styrene, isovaleraldehyde, benzene, ethyl acetoacetate ethylene acetal, valeraldehyde, ethyl alcohol, methyl isobutyl ketone, methyl butanoate, toluene, butyl acetate, n-octanal, m-xylene,, methyl methacrylate, o-xylene, dl-limonene, benzyl-tert-butanol, and linalool tetrahydride
3	Personal care products (PCPs)	Alcohol, terpene, organic acid, hexane, m, p-xylene, styrene, ethylbenzene,α-pinene, benzene, toluene, camphene, β- pinene, β-myrcene and n-decane, 3-carene, aldehyde
4	Food waste	Hydrogen sulfide, methyl mercaptan, acetaldehyde, propionaldehyde, butyraldehyde, methyl isobutyl ketone, methyl ethyl ketone, isovaleraldehyde, toluene, propionic acid, styrene, dimethyl sulfide and disulfide, Para-xylene, butyl acetate, isobutyl alcohol, butyric acid, isovaleric acid, and valeric acid
5	Textiles	Formaldehyde, tetradecane, acetaldehyde, ethylbenzene, decane, toluene, o-xylene, 1,2,4-trimethyl-benzene, m,p-xylene, acrylonitrile
6	Paints	Propylene glycol, ethylene glycol, o-xylene, methyl propanol, toluene, methyl cyclohexane, benzene, m, p- xylenes, ethylbenzene, 2 methyl-hexane, 2,4 dimethyl pentane, methyl cyclopentane, 3 methyl hexane, o-ethyl toluene, isopropyl benzene, 3 methyl pentane, n-heptane, 2,3,4 trimethyl pentane, 2 methyl heptane, methyl heptane, n-octane, n-nonane, styrene, n-propyl benzene, butyl propionate, cyclohexane, 1,3,5 trimethyl benzene, 1,2,3 trimethylbenzene, m- ethyl toluene, p-ethyl toluene, 1,2,4 trimethyl benzene, m-diethyl benzene, p-diethyl benzene
7	Charcoal (barbecue)	Benzene, toluene, ethyl benzene, formaldehyde, acrolein, acetone, benzaldehyde, acetaldehyde, propionaldehyde, styrene, meta para xylene, crotonaldehyde, butyraldehyde, and valeraldehyde
8	Furniture	Formaldehyde, acetaldehyde, tetrachloroethylene, butoxy ethanol, benzene, toluene, ethyl benzene, xylene, trimethyl benzene, cyclohexanone, dichloro-benzene, benzaldehyde, styrene, butyl-acetate, hexanal, etc.

associated risks to the aquatics); however, in the study of Koliyabandara et al. [27], the measured conductivity ranged from 60 to 310 μS/cm, which is lower than the maximum allowable limit. Levels of ammonia and nitrate were well observed in the water samples, with concentrations exceeding the FAO and CCME guidelines. High concentrations of ammonia (molecular ammonia) can have potential risks towards the fish and cause toxic effects through penetrating the tissues in fish or brains.

The variations in the concentration levels of various heavy metal compounds in the Meda Ela stream at different locations have also been investigated and reported in Fig. 11.9. It can be seen from Fig. 11.9 that Fe and Mn recorded the highest concentrations among other heavy metals at different sampling locations. Nevertheless, it can be seen that the concentration of Fe is reported to be 10 times more in locations from 6 to 10, compared to the other sites. However, the levels of Mn were reported to be five times larger in sites 5 to 11, compared to other sampling locations.

The statistical relationship and analysis between the investigated water quality factors are identified using what is known as 'correlation analysis' as presented in Table 11.11. It can be observed from the table that some parameters experienced significant variation. Generally, a correlation coefficient of less or more than 0.5 is deemed to have a major correlation effect (positive or negative) between the parameters. Nevertheless, some of the studied parameters, such as conductivity and ORP, experienced negative correlation (−0.9); still, the biochemical processes occurring and resulting in such correlation and relationship are truly difficult to explain. These results can be an outcome of complex processes that are not easy to explain.

The studied parameters were extracted in this study using what is known as 'principal component analysis (PCA)'. In PCA, the observed variable factors are created using liner combinations. In other words, the variable parameters are derived from the PCA; then the linear combination of the variable is identified in PCA, which can be defined using underline data structure through the 'number of principal components (PCs)' [30–33]. The results from parameter loading for each component are represented in Table 11.12 [27].

Among the studied components of the PCA, only four of them showed 89.9% of the variance in the whole set of data. The sum total of the variance for each variable component is defined as the 'total variance', and it is explained by the eigenvalue [27]. Only PC components with eigenvalues larger than 1 were selected. Results from the table show that PC1 is negatively loaded with salinity, cond., pH, ammonia, nitrate, and TDS. Meanwhile, PC2 accounted only for 60.7% of the total

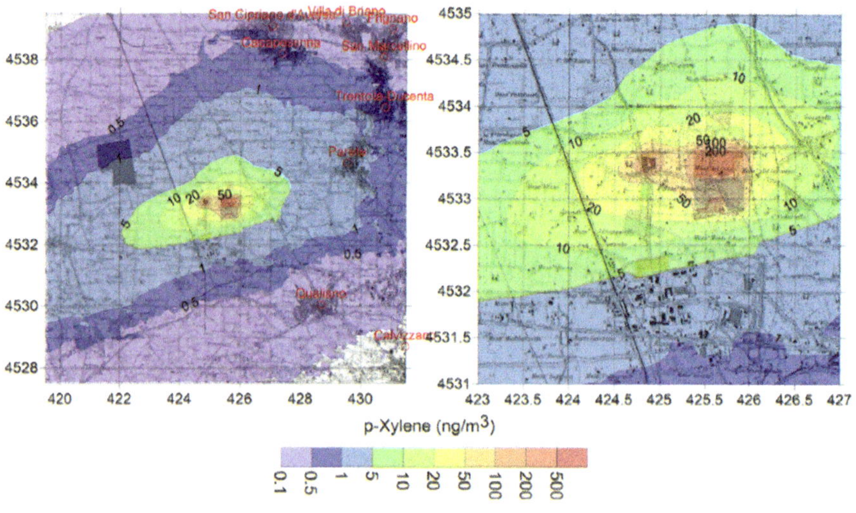

Fig. 11.11 Dispersion of the VOC, p-Xylene, over the landfill of Giugliano (left), and zoomed over a 4 × 4 km^2 area surrounding the landfills (right) [4]

difference while positively loaded with ammonia, ORP, pH, Mn, and DO and negatively charged with nitrate. For PC3, 78.8% of the total variance was also associated with resilient negative loadings on DO and pH, while positive (moderate) loadings were observed with nitrate. The high loading in factors of the biochemical processes and species that are present in the acquired data imply that there is organic pollution in the site. Nevertheless, PC4 exemplifies pollution sources with anthropogenic

Table 11.16 Hazardous pollutants emitted from landfills as reported by EPA [3]

No.	VOCs	Chemical formula	CAS number	Molecular mass (g/mol)
1	Acetone	C_3H_6O	67–64-1	58.08
2	Acrylonitrile	C_3H_3N	107–13-1	53.06
3	Benzene	C_6H_6	71–43-2	78.11
4	Bromodichloromethane	$CHBrCl_2$	75–27-4	163.8
5	Carbon disulfide	CS_2	75–15-0	76.13
6	Carbon tetrachloride	CCl_4	56–23-5	153.81
7	Chlorobenzene	C_6H_5Cl	108–90-7	112.56
8	Chlorofluorocarbons	CCl_2F_2	–	–
9	Chloromethane	CH_3Cl	74–87-3	50.49
10	Chloroethane	C_2H_5Cl	75–00-3	64.51
11	Chloroform	$CHCl_3$	67–66-3	119.37
12	Dichloromethane (methylene chloride)	CH_2Cl_2	75–09-2	84.927
13	Dichlorobenzene	$C_6H_4Cl_2$	106–46-7	147
14	Hydrogen sulfide	H_2S	7783-06-4	34.08
15	Hexane	C_6H_{14}	110–54-3	86.18
16	Methyl isobutyl ketone	$C_6H_{12}O$	108–10-1	100.16
17	Methyl ethyl ketone	C_4H_8O	78–93-3	72.11
18	Methyl mercaptans	CH_3SH	74–93-1	48.11
19	Tetrachloroethylene (perchloroethylene)	C_2Cl_4	127–18-4	165.82
20	Toluene	C_7H_8	108–88-3	92.14
21	Trichloroethylene	C_2HCl_3	79–01-6	131.4
22	Vinyl chloride	C_2H_3Cl	75–01-4	62.50
23	Xylene	C_8H_{10}	1330-20-7	106.16
24	1,1,1-Trichloroethane (methyl chloroform)	$C_2H_3CCl_3$	71–55-6	133.40
25	1,1-Dichloroethane (ethylidene dichloride)	CH_3CHC_{12}	107–06-2	98.954
26	1,2-Dichloroethane (ethylene dichloride)	$C_2H_4C_{12}$	75–34-3	98.95
27	1,1,2,2- Tetrachloroethene	$C_2H_2C_{14}$	79–34-5	167.848
28	1,1-Dichloroethene (vinylidene chloride)	$C_2H_2C_{12}$	75–35-4	96.94
29	1,2-Dichloropropane (propylene dichloride)	$C_3H_6C_{12}$	78–87-5	112.98

activities. Further, it can be clarified that the high presence of organics contributes to consuming huge quantities of oxygen, which leads to the development of ammonia and other organic acids. The occurrence of the hydrolysis process in these acidic materials can cause a decline in the pH of water.

The comprehensive pollution index of the Meda Ela stream water was analyzed. The results from the samples collected from the Meda Ela waters were summarized and categorized according to the used water quality index (P), as shown in Table 11.13.

Since the investigated dumpsite (open landfill) is located in a shallow groundwater table, therefore, the quality monitoring of the groundwater wells was also investigated for a period of ten months in order to realize the effect of the waste depositing into the dumpsite on the quality of groundwater source. Some basic parameters such as cond., total phosphate, TDS, pH, nitrate, and other heavy metals (e.g. Fe, Mn, Cd) were analyzed, and the findings are summarized in Table 11.14. The pH values of the collected groundwater samples were neutral. The high value of TDS of 129 mg/L indicates leakage and leaching of different contaminants into the groundwater. The high values of conductivity are also a result of high amounts of dissolved ions in the water. Such polluted water source with organics, ammonia, nitrate, and heavy metals is considered risky to the consumers and other receiving water bodies and the aquatics.

11.4 The Impact of Landfill Gas on Air Quality

11.4.1 Landfill Gas Generation and Characteristics

The volatile portions of the disposed waste in landfill sites are more likely to be transferred out of the landfill through landfill gasses (LFGs). Mainly, LFGs are formed due to the degradation processes of the disposed waste in the landfill. Several compounds constitute the LFGs, which include methane (CH_4), carbon dioxide (CO_2), and other trace gases, as main products of the degradation processes occurring at the landfill. CH_4 and CO_2 comprise around 50–60% and 30–40% of the total emitted gases [34–36]. There are other formed gases that make only around 1% of the LFGs. These gases are known as 'non-methane volatile organic compounds (NMVOCs)', such as hydrogen sulfide, benzene, trichloroethylene, vinyl chloride, and other hazardous air contaminants [34, 37]. The release of these gases can significantly influence the environment's air quality [38]. In the United States alone, it is estimated that the emission amount of volatile organic compounds (VOCs) generated from landfill sites comprise over 10% of the total VOCs emissions in the entire country [39]. Generally, the generated VOCs have an emission factor that varies between 0.2 and 7.3 kg VOC/Mg of the treated municipal solid waste [35]. The possible pathway for the migration of VOCs from landfills into the atmosphere is illustrated in Fig. 11.10.

VOCs play a significant role in the development of what is known as 'ground-level ozone'. High concentrations of ground-level ozone tend to impede the photosynthesis process, reduce growth, and depress agricultural yields [40]. Moreover, the emissions of VOCs from landfill sites can play a major role in influencing the tropospheric chemistry, as they can contribute to producing photochemical oxidants through a series of photochemical reactions once they are released into the atmosphere [41]. The release of some VOCs into the environment can also contribute to the formation of photochemical smog and ozone (ref). Other types of VOCs, such as the unsaturated VOCs (e.g. alkenes and alkynes) can generate 'secondary organic aerosols (SOAs)', which can lead to particulate contamination [4, 34, 38].

The generation rates of LFGs are generally influenced by several factors, including but not limited to the type of deposited waste, age of the landfill, humidity and ambient temperature, pH, bacterial composition, and available substrates [42]. Most importantly, the concentration and type of VOCs are also dependent on the type of activities and processes used to eliminate and remove the deposited wastes. For instance, these VOCs can be released landfill sites according to the following [43]:

- VOCs released from degrading landfill sites.
- VOCs released from composting sites.
- VOCs released as a result of landfill fires and incineration.

As a result of these factors, the emission rates of LFGs will always vary and demonstrate spatial and temporal irregularity and magnitude. Table 11.15 shows an example of the possible emitted VOCs from different types of wastes at landfill sites. For example, organic waste (e.g. food, wood, paper, etc.) is degraded in several phases by microorganisms. Therefore, the composition and magnitude of emitted VOCs in landfills with organic waste will differ and change during each phase. Some factors, such as high ambient temperature, humidity, and high-water content, are considered favourable for the growth of microorganisms, meaning higher decomposition rate, and hence releasing higher amounts of VOCs [4, 34].

Although LFG's emissions can continue for decades, however, the highest amounts of emitted gases are reported to be in the first 20 years of waste deposition, where almost all of the LFGs are formed and released into the environment. After 20 years, LFGs can still be released, however, only in minor quantities. An example of a VOC diffusion over a landfill site located in Giugliano, Italy, is exemplified in Fig. 11.11. The average annual concentration measured for p-xylene reached 200 ng/m^{-3}, with an average hourly concentration exceeding 2 ng/m^{-3}, whereas the measured concentrations of p-xylene in the nearby residential area were affected by an annual average concentration of 2 ng/m^{-3} for the VOC, p-xylene [4].

Although waste composting can be a friendly management option and considered a good alternative to landfilling and waste incineration, however, the public health risk and possible occupational of composting cannot be ignored. Due to the fact that most composting is performed in open windrows (especially in developing countries), the organic waste will be continuously exposed to air which can generate and emits considerable amounts of VOCs into the atmosphere [44, 45]. The total amounts of VOCs emitted from composting sites can vary from <10 to >150 mg/m^3

[46]. It has been reported that there are more than 100 different types of VOCs being emitted from composting sites, including aromatic hydrocarbons, reduced sulfides, acids, ketonnes, biogenic organics, alcohols, ethers, aliphatic alkanes, esters, aldehydes, halogenated hydrocarbons, and furans.

In the case of insufficient aeration in composting sites, some sulfur compounds such as hydrogen sulfide, dimethyl disulfide and sulfide, carbon disulfide, and methyl mercaptan, which exhibit rigorous odour are emitted and released into the atmosphere. However, when incomplete aeration decomposition occurs in composting site, this can release other compounds into the atmosphere, such as alcohols, organic acids, ketones, and esters [47–49]. Other studies have reported that in the initial phase of composting, certain VOCs are formed and released (e.g. aromatic and aliphatic hydrocarbons, ketones, and terpenes) with percentages of 8%, 41%, 11%, and 31%, respectively. However, in the final phases of composting processes, aromatic compounds constitute around 68%, while trace amounts were observed for other VOCs (e.g. ketones and aliphatic hydrocarbons) [50].

Other potential toxic products that are released from composting facilities are biological aerosols (bioaerosols). Bioaerosols can penetrate the lungs and embed in the alveoli, depending on their particle size. The risks of bioaerosols on human health can include but are not limited to [45]:

- Non-allergic toxic asthma, chronic bronchitis, chronic obstructive pulmonary disease (COPD), mucous membrane irritations (MMI), organic dust toxic syndrome (ODTS), and toxic pneumonitis.
- Rhinitis, allergic asthma, hypersensitivity pneumonitis (HP), allergic bronchopulmonary aspergillosis (ABPA), as well as skin and irritations.
- Infectious aspergillosis and mucormycosis.

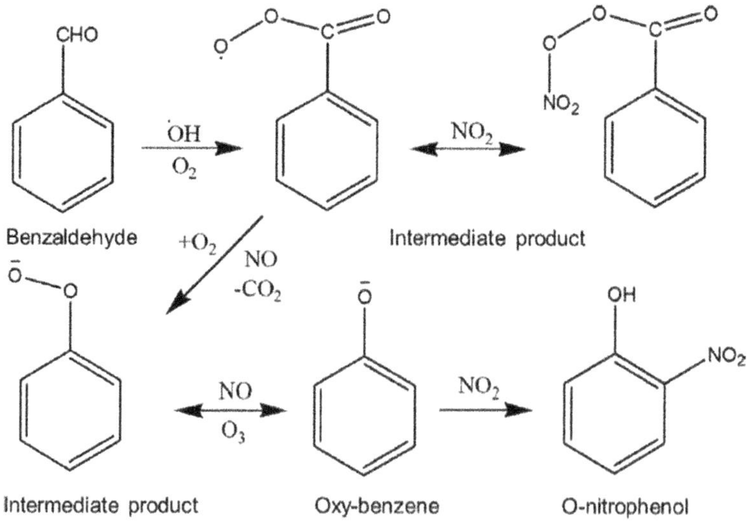

Fig. 11.12 SOA formation due to oxidation of benzaldehyde in troposphere [5]

Table 11.17 OTVs and emission rates of VOCs of concern in Odayeri's MSWL

	OTVs[a]		
Odorous VOCs	ppm	μg/m[3b]	Emission rate (10^{-3} g/s)
Acetone	42.0000	99,826	14.511
Acrylonitrile	8.80000	19,108	13.286
Butane	1200.00	2,854,149	11.565
Carbon tetrachloride	4.60000	28,959	0.000
Carbon disulfide	0.21000	654	0.100
Carbonyl sulfide	0.05500	135	0.068
Chloroform	3.80000	18,566	0.008
Dichloro methanol	160.000	556,163	2.771
Dimethyl sulfide	0.00300	7.628	1.109
Ethyl mercaptan	0.00001	0.022	0.323
Ethyl benzene	0.17000	738	1.115
Ethanol	0.52000	980	2.859
Hexane	1.50000	5290	1.292
Hydrogen sulfide	0.00800	11.1	21.729
Isopropanol	0.09400	231	6.870
Methyl ethyl ketone	0.44000	1298	1.166
Methyl isobuthyl ketone	0.17000	696	30.288
m-xylene	0.04100	178	2.931
Propane	1200.00	2,165,166	1.115
Pentane	1.40000	4133	0.542
Trichloro ethylene	3.90000	20,971	0.844
Tetrachloro ethylene	0.77000	5225	1.411

OTVs: odour threshold values

In 2005, a report was issued by Environmental Protection Agency (EPA) to assess the emission of landfill gases from closed and abandoned landfills. The report 'Guidance for Evaluating Landfill Gas Emissions from Closed or Abandoned Facilities' stated a list of 29 VOCs which are considered hazardous pollutants and are mentioned in Table 11.16 [3].

11.4.2 Impact of Landfill Gas on Human Health

Residential areas near the vicinity of the landfill tend to suffer odour problems as a result of the landfill gas emissions and release into the atmosphere. The main elements that constitute the odour nuisance in landfill sites are limonene, organosulfur compounds (e.g. hydrogen sulfide, carbon disulfide, methyl mercaptan, dimethyl disulfide, and dimethyl sulfide), certain esters, and alkylbenzenes. However, the organosulfur compounds are considered to generate the most rigorous odour in the landfill. The concentration of some elements such as sulfur compounds can

Table 11.18 The ambient concentration levels of odour causing compounds

Category of wind speed (m/s)	Wind speed (cumulative)	Concentrations (hourly) μg/m^3		
		Methyl mercaptan	Ethyl mercaptan	Hydrogen sulfide
5.00	0.302	0.02120	0.02507	1.68738
3.50	0.433	0.03022	0.03573	2.40531
2.75	0.498	0.03840	0.04541	3.05639
2.25	0.562	0.04687	0.05542	3.73054
1.75	0.653	0.06014	0.07111	4.78675
1.25	0.748	0.08395	0.09926	6.68187
0.75	0.878	0.11809	0.13963	9.39919
0.35	0.970	0.12535	0.14821	9.97703
0.10	1.000	0.12435	0.14645	9.89744
Average	–	0.03583	0.04400	2.51834
Standard deviation	–	1.02500	0.74300	0.79100

diminish by the time of disposing of the waste into the landfill. Generally, odour-causing compounds are diminished and dispersed after their release into the atmosphere by the dominant wind force in both vertical and horizontal directions [34, 51, 52].

Besides the odour nuisance phenomenon near landfill adjoining areas, the people living near the landfill area can also be exposed to the inhalation of VOCs. The inhalation of such hazardous compounds can cause potentially severe health effects on human health, including asthma, irritation, allergy, and respiratory diseases [53, 54]. Residents living the landfill downwind can also suffer from severe odours, causing annoyance and decreasing the quality of life and the health of human well-being. Studies have reported that people living nearby landfills can also suffer from emotional stress, headaches, fatigue, stuffy nose, eye infections, dry throat, asthma, coughs, nausea, spontaneous abortions, muscular and joints pain, toothache, shortness of breath, fever, smell disorder, sarcoidosis [54–56].

The exposure and inhalation of VOCs can also increase the risk of cancer in humans. In 2012, a study was conducted to assess the carcinogenic risk of human exposure to volatile organic compounds among landfill workers [57]. Results of the study showed that a total of 275 persons in one million in the area of Deonar and 139 persons out of one million in the area of Malad were at risk of getting cancer. In another study, researchers have performed a cancer risk assessment among landfill workers in South Africa. The results of the cancer risk revealed that each worker in the landfill was exposed to hazardous emissions (e.g. benzene) with values higher than that recommended standards of 0.0004 issued by the Environmental Protection Agency. Two of the landfill workers who worked for more than 25 years had a significant high cancer risk value of 1.028795 and 0.813946, respectively [58].

11.4.3 Impact of Landfill Gas on Air Quality

Generally, the contamination of areas near the vicinity of landfills caused by the emissions of VOCs from landfill sites is rather comparable to moderated contaminated urban areas. The release of VOCs from landfill sites can highly contribute to the formation of secondary organic aerosols (SOAs), ozone (O_3), and photochemical smog [34]. In the presence of sunlight and suitable climatic conditions, the formation of high concentrations of ozone as a result of complex reactions between nitrogen oxides (NOx) and VOCs can be highly promoted in the troposphere layer. The formation and generation levels of O_3 in the atmosphere vary depending on many factors, such as the photochemical reactivity of present VOCs, relative concentrations of NOx and VOCs, the constitutes that comprise the reactive VOCs present in the atmosphere, and reaction rates and pathways of VOCs [59, 60]. Examples of the most reactive VOCs that can contribute to the formation of ozone are phenolic compounds, alkenes, and aromatic hydrocarbons, while ketones and alkanes are considered the least reactive VOCs. One of the main issues with ozone formation is that ozone can react with other chemical compounds in the layer of the troposphere, producing toxic compounds. The formation mechanism of ozone occurs via a series of reactions, as demonstrated in Eqs. (11.7–11.11.):

$$CH_3CHO_{(g)} + OH^{\cdot}_{(g)} \rightarrow CH_3CO^{\cdot}_{(g)} + H_2O \tag{11.7}$$

$$CH_3CO^{\cdot}_{(g)} + O_2 \rightarrow CH_3(CO)OO^{\cdot}_{(g)} \tag{11.8}$$

$$CH_3(CO)OO^{\cdot}_{(g)} + NO_{(g)} \rightarrow CH^{\cdot}_3 + CO_2 + NO_2 \tag{11.9}$$

$$NO_2 + hv + O_2 \rightarrow O^{\cdot} + NO_{(g)} \tag{11.10}$$

$$O^{\cdot} + O_2 \rightarrow O_3 \tag{11.11}$$

In addition, VOCs released from landfill sites can also contribute to the formation of secondary organic aerosols (SOAs). These atmospheric organic aerosols are produced due to the photochemical oxidization of VOCs by the NO_3 and OH radicals, as well as O_3 or Cl atoms in the troposphere layer, generating non- or semi-VOCs (NVOCs/SVOCs) through the exchange of gas-particle products, such as condensation, nucleation, and heterogeneous chemical reactions [61]. The main constitutes of SOAs precursors are carbonyls, aromatic hydrocarbons, oxygenated compounds, alkenes, and alkanes. While the major degradation products are alkyl, alkoxy radicals, and organic peroxyl compounds. The end products of these intermediates are controlled and affected by their chemical and physical properties and their tendency to form SOAs [41, 59]. An example of the atmospheric transformation of benzaldehyde is depicted in Fig. 11.12. It is known that benzaldehyde can form benzoyl through reactions with O_2 and OH, resulting in the generation of what

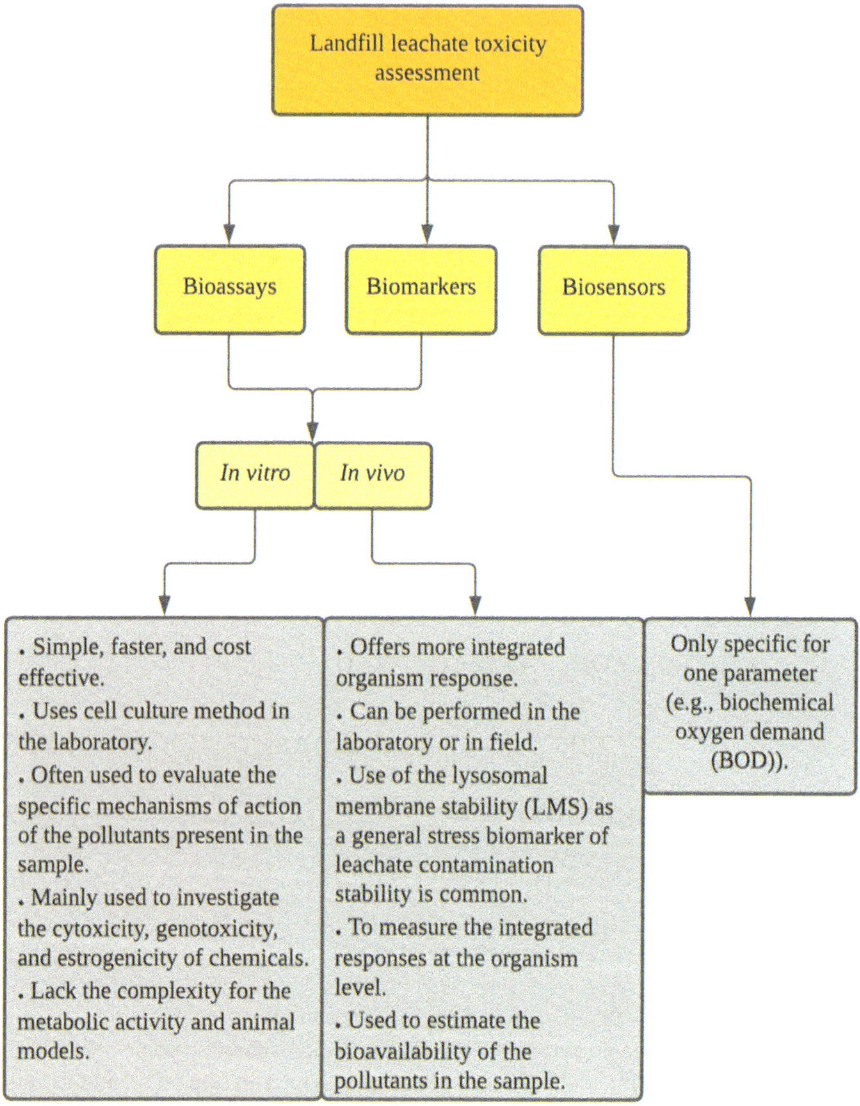

Fig. 11.13 Tools for the assessment of landfill leachate toxicity

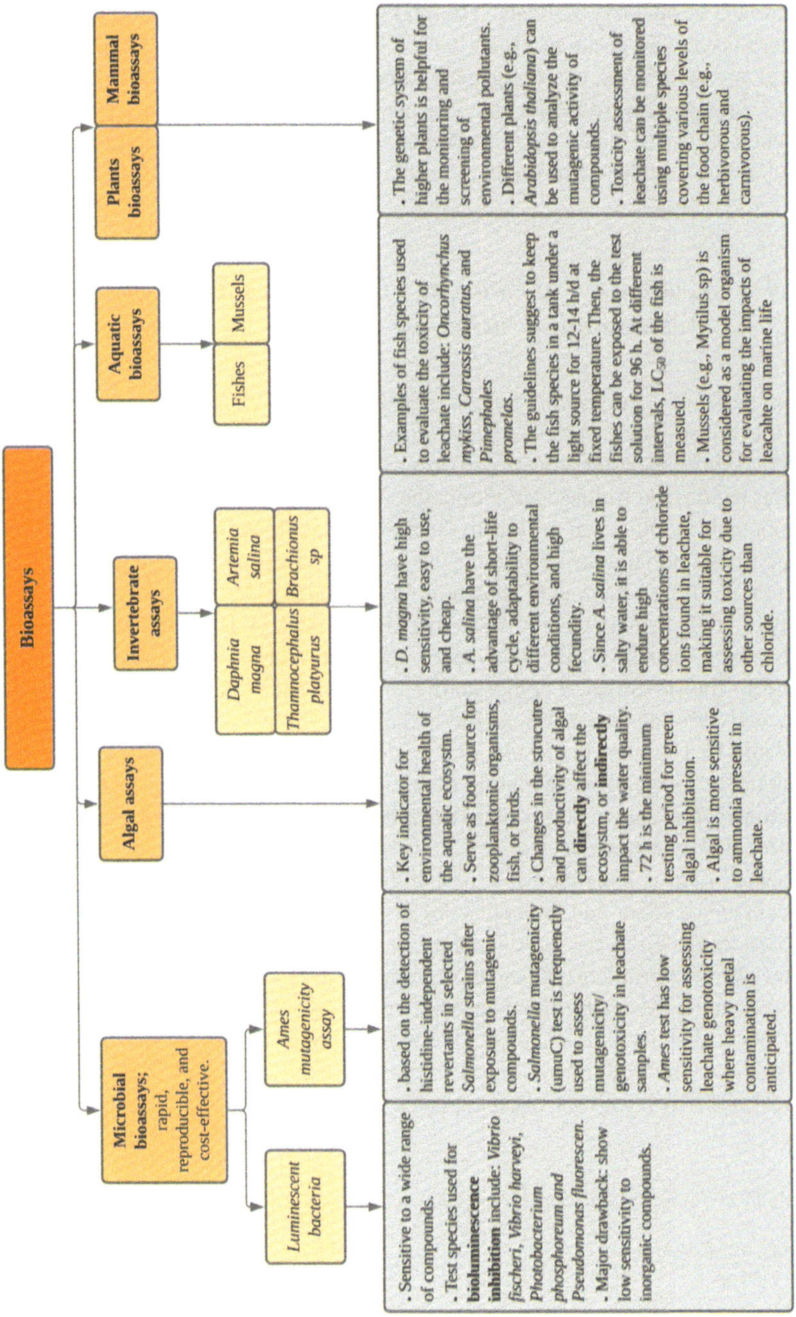

Fig. 11.14 Classifications, types, and major considerations of bioassays

is known as phenoxy radicals. The presence of ozone in the atmosphere (troposphere layer) can promote the formation of oxy-benzene through its reaction with the phenoxy radicals, and the generated product (oxy-benzene) can also react with NO_2, forming O-nitrophenol [5]. The main negative effects of SOAs development are their ability to scatter solar irradiations, forming a phenomenon known as 'cloud condensation nuclei', which can upset the irradiation amount of the entire earth.

11.4.4 Evaluation of Odorous VOCs from a Municipal Solid Waste Landfill Site (Case Study)

The release of odorous compounds into the atmosphere is significantly influenced by the transportation effect of the 'dominant wind' in the horizontal direction, whereas the dispersion process can occur in both directions (horizontal and vertical) of the wind. Due to these activities, the concentrations of certain volatile organic compounds (VOCs) will decline in the transported plume, which as a result will also decrease the odour concentration levels, to beyond a limit where the odour concentration unit falls under odour threshold limit (OTL).

In a study conducted on a landfill site located in Istanbul, Turkey [52], the researchers focused on the analysis of different odorous VOCs released from that landfill. The investigated compounds along with their odour threshold values and total emission rates are summarized in Table 11.17.

Table 11.18 shows the results as a 'matrix peak ambient concentrations' of these compounds located in the Gokturk landfill site. Generally, for this table, each of the column vectors shown in this matrix displays the fluctuations of 'peak ambient concentrations' which are related to certain compounds with odorous properties. The presented data is beneficial for 'probability distribution analysis' to calculate the percentage of odour and time episodes in the Gokturk landfill site.

The results from the conducted work revealed that ethyl mercaptan caused a series of odour problems; however, the odour problems caused by this compound is only accounted for not more than 8.84% of the whole time period. In other words, this means that the 'ambient peak' of ethyl mercaptan causes odour problems 8.84% of the time. On the contrary, other compounds (methyl mercaptan) do not cause any odour problems, with an odour episode percentage of only 0.98% of the time.

Table 11.19 Toxicity classification according to the value of EC

Range of EC_{50} value	Classification
$EC_{50} > 100$	Non-toxic
$100 > EC_{50} > 40$	Potentially toxic
$40 > EC_{50} > 10$	Confirmed toxicity
$EC_{50} < 9$	Severe toxicity

Table 11.20 Toxicity of landfill leachate to different organisms

Location	Species	Test time	ErC50 (%)	TU	Toxicity	Ref.
Toxicity of landfill leachate to invertebrates						
Mustankorkea landfill, Finland	*D. magna*	24	16–28	3.6–6.3	Confirmed toxicity	[72]
Nurmijarvi landfill, Finland	*D. magna*	24	16–29	3.4–6.3	Confirmed toxicity	[72]
Metropolitan landfill, Brazil	*D. similis*	48	2.0–2.3	43.5–50	Severe toxicity	[73]
Metropolitan landfill, Brazil	*D. salina*	48	11.9–25.6	3.9–8.4	Confirmed toxicity	[73]
Kairiai landfill, Lithuania	*D. magna*	48	10.4–25	4.0–9.6	Confirmed/severe toxicity	[74]
Kairiai landfill, Lithuania	*D. magna*	48	37–65	1.5–2.7	Potential/ confirmed toxicity	[74]
Zabrze landfill (old), Poland	*Thamnocephalus platyurus*	24	98	~ 1.0	Potential toxicity, non-toxic	[75]
Zabrze landfill, (new), Poland	*Thamnocephalus platyurus*	24	1.4	71.4	Severe toxicity	[75]
Toxicity of landfill leachate to green algal						
Sandholt-Lyndelse, Denmark	*R. subcapitata*	72	3.8	26.3	Severe toxicity	[76]
Højer, Denmark	*R. subcapitata*	72	2.5	40	Severe toxicity	[76]
Esbjerg, Denmark	*R. subcapitata*	72	6.5	15.4	Severe toxicity	[76]
Junk bay, Hong Kong	*C. vulgaris*	96	7	14.3	Severe toxicity	[77]
Gin drinkers' bay, Hong Kong	*Scenedesmus sp.*	96	> 50	> 2	Potential toxicity, non-toxic	[77]
Nurmijarvi landfill, Finland	*R. subcapitata*	72	10–15	6.7–10	Confirmed toxicity	[72]
Toxicity of landfill leachate to aquatic organisms						
Delhi, India	*Poecilia reticulate*	96	< 3	< 33.3	Severe toxicity	[78]
Sungai Sedu, Malaysia	*Cyprinus carpio*	–	3.8	26.32	Severe toxicity	[79]
B1, Japan (raw leachate)	*Carassis auratus*	96	≥ 100	≥ 1	Non-toxic	[80]
A1, Japan (raw leachate)	*Carassis auratus*	96	≤ 10	≤ 10	Severe toxicity	[80]
A2, Japan (treated leachate)	*Carassis auratus*	96	≥ 100	≥ 1	Non-toxic	[80]

Lastly, similar to methyl mercaptan, hydrogen sulfide also does not cause any odour issues, with an odour episode of 0.34% of the whole time.

11.5 Toxicological Risk Assessment of Landfill Leachate

The physicochemical analysis methods of landfill leachate include parameters such as BOD, DO, COD, pH, EC, heavy metals, and total organic carbon (TOC). Although these types of analyses approaches are necessary and beneficial, especially to monitor the quality of leachate being disposed to adhere with the national and international standards and guidelines, they are insufficient in providing information regarding the effects on biological organisms. Therefore, the use of bioassays as a toxicological risk assessment tool has the advantage over physiochemical analyses in such a way by integrating the biological effects of the present compounds in the sample, while taking into consideration the other factors such as synergism, bioavailability, and antagonism [62]. Therefore, the utilization of bioassays as a monitoring and screening tool for the characterization of pollutants in different environmental matrices (e.g. landfill leachate) has become a prevalent and potent tool in the environmental toxicology field [63].

The use of multiple bioassays is more efficient than using single bioassays for evaluating the toxicity of complex environmental matrices such as landfill leachate. This is because multiple bioassays include organisms that represent different trophic levels, which is more appropriate for these complex matrices. The toxicological assessment of treated and/or untreated landfill leachate is substantial to investigate the impact of landfill leachate disposal on the environment. Further, it can be a good indicator of the effectiveness of the remediation process employed at the landfill leachate treatment plant [64, 65]. Different methods are available and used to perform toxicity assessment, including in vivo and in vitro. Different types of trophic levels have been reported in the literature, including the following [66, 67]:

- Invertebrates: *Daphnia magna* and *Artemia salina.*
- Bacteria: *Vibrio* and *fischeri.*
- Fish: *Carassius auratus.*
- Microalgae: *Pseudokirchneriella subcapitata.*
- Plants: *Vicia faba* and *Hordeum vulgare.*
- Mammals.

11.5.1 Approaches for the Evaluation of Landfill Leachate Toxicity

The toxicity and biological assessment of landfill leachate can be performed by employing the appropriate tools illustrated in Fig. 11.13. As shown in the figure, three main approaches can be used to assess the toxicity of landfill leachatem including (i) bioassays, (ii) biomarkers, and (iii) biosensors. Bioassays describe procedures that include living organisms to measure the toxicity of pollutants. Bioassays involve the exposure of living organisms to various concentrations of toxic pollutants as well as observing the effects of these pollutants on the living organisms' survival and behavior. On the other hand, biomarkers provide an early warning for the biological effects that are caused by chemical exposure to pollutants. The use of commercially available biosensors has emerged as a good technology in the environmental toxicology field. However, as shown in Fig. 11.13, biosensors suffer from a major drawback which limits their use to only specific parameters (e.g. BOD). In contrast, to biosensors, the use of bioassays can offer a measure of the total effluent toxic impact [66–68]. The major types of bioassays are classified as shown in Fig. 11.14.

In order to evaluate the toxicity of the environmental sample, indicators for the toxicity of samples have been developed. The most common are 'the median effective concentration (EC_{50})' and 'the median lethal concentration (LC_{50})'. The EC_{50} represents the concentration levels that influence 50% of the total test population. In the case when the influence is the mortality of the targeted organisms, it is then described as LC_{50}. The term 'EC_{50}/LC_{50}' can be calculated by constructing a 'dose-response' curve. The above-mentioned three terms are frequently used as an eco-toxicological tool for indicating the toxicity of the compound(s) to the environment. The classification of these terms is illustrated in Table 11.19. There are different toxicity classifications that have been developed, such as the sediment toxicity (SED-TOX), potential toxic effects of probe (PEEP), and potential toxicologiae (pT) [69, 70]. However, these classification approaches suffer from a major limitation which is the necessity for continuous maintenance and/or culturing of the live test organisms, making the whole process more expensive [71].

To overcome this limitation, in 2003, researchers developed a new low-cost classification and scoring process to assess the toxicity of landfill leachate percolating from dumpsites and soil [71]. The developed method consists of two main steps of determination and quantification. The first step includes determining the toxicity using non-diluted test samples. Whereas in the second step, a series of dilution test samples are prepared with micro-biotests that can provide for more than 50% effect in the non-diluted samples, then the toxicity tests are conducted on the prepared samples. The obtained results from each micro-biotest are then converted into what is known as 'toxicity unit (TU)'. The TU of a sample is equal to the inverse of its EC_{50} or LC_{50} as illustrated in Eq. 11.12:

Table 11.21 Example of the parameters relative weight for calculating WQI

Parameter	GSA/WHO (mg/L)	Weight (w_i)	Relative weight (Wi)
EC	1000.0	4	0.111
TDS	1000.0	5	0.139
pH	8.5	4	0.111
Dissolved oxygen (DO)	8.0	1	0.028
Alkalinity	300.0	4	0.111
TH	500.0	2	0.056
Turbidity	5.0	5	0.139
Mg^{2+}	50.0	2	0.056
Ca^{2+}	75.0	2	0.056
Cl^-	250.0	3	0.083
Fe	0.5	4	0.111
		$\Sigma\, w_i = 36$	$\Sigma\, W_i = 1$

Table 11.22 Water quality rating

Range of WQI	Water quality rate
< 50	Excellent water quality
50–100	Good water quality
100–200	Poor water quality
200–300	Very poor water quality
> 300	The water is unsuitable for drinking purposes

$$TU = \frac{100}{\%EC_{50} \vee LC_{50}} \tag{11.12}$$

The higher the value of TU, the greater the toxicity of the tested sample to the environment.

The toxicity of landfill leachate on different test species has been reported by researchers. Table 11.20 summarizes some of the studies that have assessed the toxicity of leachate, while including the toxicity unit (TU) as the main classification for whether the toxic effects of leachate on the organisms are severe, confirmed, potential, or non-toxic.

11.6 Methodologies for the Assessment of the Environmental Impact of Landfills

11.6.1 Human Health Risk Assessment

Human health risk assessment can be defined as the process of evaluating and estimating the health consequences and risks of human exposure to different environmental hazards. The EPA method is commonly used to estimate the health risk assessment [81]. The EPA approach includes three main steps. These steps are described in detail in the EPA Framework Report [81]. However, in this section, a discussion on estimating the human health risk to the exposure of landfilling activities will be given as illustrated by other researchers [25].

The three main steps in the process of risk assessment include:

- Hazard identification.
- Exposure assessment.
- Dose-response.

11.6.2 Hazard Identification

The identification of hazards is considered the initial step in exposure assessment. Therefore, in this step, the possible outcomes from the landfilling activities shall be identified. For example, for a population living nearby a landfill, the population is at risk of exposure to different hazards, such as the direct intake of polluted dust and foul odour. Further, the population could indirectly be exposed to drinking contaminated groundwater, since most of the population highly dependent on groundwater as the main source for drinking water. Hence, when assessing the health risk linked with the contaminated groundwater, different parameters should be investigated and determined to examine the quality of the groundwater. These parameters may include heavy metals (e.g. Zn, Pb, Cd, Cu) or other possible pollutants. Moreover, other information that could be related to the 'socioeconomic characteristics' and 'population structure' should be collected and studied.

11.6.2.1 Exposure Assessment

After identifying the possible hazard to the population, the exposure assessment step is then conducted. This step mainly estimates the extent to which the population is exposed to certain contaminants (e.g. Cu, Pb, Zn). Several factors can influence the exposure assessment process, which are:

- Exposure time.

Table 11.23 Example of the HMPI calculation for some elements

Element	W_i	S_i (ppb)	I_i (ppb)
Cu	0.001	2000.0	50.0
Cd	0.100	10.0	3.0
Pb	0.067	15.0	10.0
Fe	0003	300.0	100.0
Mn	0.003	400.0	100.0

The exposure time for any pollutant in the area surrounding the landfill can be determined by considering the variation in time from establishing the landfill to the present while considering a fixed population number.

- Bodyweight.
  The average body weight can be determined based on the available statistics for the investigated country/state near the landfill.
- Water intake rate.
  Based on the Environmental Protection Agency Handbook of Exposure Factor, the water intake is 3 L/d for temperate climates, while it ranges from 6 to 11 L/d for tropical climate areas.

11.6.2.2 Dose-Response

The risk associated with the usage and drinking of groundwater contaminated by landfilling consequences (e.g. leachate generation) is often given by the hazard quotient for non-carcinogenic hazards as well as the cancer risk for carcinogenic hazards. The assessment of human exposure to target chemicals that are present in the contaminated groundwater can be calculated by the 'coupled model' to determine the chronic daily intake (CDI) or what is also known as the average daily intake (ADI), as shown in Eq. 11.13 [81]:

$$CD = \frac{C_{\exp} \times I_R \times E_D \times E_F}{AT \times BW} \tag{11.13}$$

where C_{exp} is the concentration of target chemical(s) (mg/L), I_R represents the water ingestion rate (L/d), E_D is the duration exposure time (year), E_F is the exposure frequency (d/year), AT is the average time span (year), and BW is the average body weight (kg).

The human health risk associated with the consumption of leachate-contaminated groundwater can be calculated using the hazard quotient (HQ) equation [81, 82]:

$$HQ = CDI - R_{Fd} \tag{11.14}$$

where R_{Fd} represents the toxicity reference dose (TRD), which can be obtained from the 'Health Effects Assessment Summary Tables (HEAS)'. For example, the HEAS for the following elements, Pb, Zn, and Cu, is 3.5×10^{-3}, 0.3, and 4×10^{-2} mg/kg, respectively [25]. If the value of the achieved HQ is <1, it implies acceptable risk. However, if the HQ is >1, then it signifies a potentially high risk of adverse non-carcinogen health effects [83].

The 'hazard index (HI)' can be computed using the following equation:

$$HI = \sum HQ \tag{11.15}$$

According to the Environmental Protection Agency, the 'cancer risk (CR)' can be calculated using Eq. 11.16, as illustrated below [81, 82]:

$$CR = CDI \times SF, Risk \leq 0.01; CR = 1 - \exp - CDI \times SF; Risk > 0.01 \quad (11.16)$$

where *SF* represents the cancer slope factor for the target chemical pollutant (mg/kg-d^{-1}).

11.6.3 Water Quality Index

The water quality index (*WQI*) method is applied to assess the overall drinking water quality for each water resource. This quality index is considered very beneficial for ranking the effect of combined water quality parameters on the total quality of water [84, 85]. In order to determine the *WQI*, each collected water sample is assigned with a weight (w_i); then, a relative weight (W_i) and a 'quality ranking scale (q_i)' are estimated. To further illustrate this, Table 11.21. shows the water quality parameters, and for each parameter, w_i values are designated according to the relative relevance of each parameter for the purpose of drinking water quality assessment [86, 87]. This assessment is based on the guidelines and standards issued by the World Health Organization (WHO) and the Ghana Standard Authority (GSA) [88, 89]. As shown in Table 11.21., a value of 5 was assigned for the major parameters which can significantly influence the water quality, while the lowest value of 1 is assigned to parameters with minor effects on the water quality.

The value of the relative weight (*Wi*) for different water resources can be calculated using the formula expressed below [85, 90]:

$$W_i = \frac{w_i}{\sum_{i=1}^{n} w_i} \quad (11.17)$$

where n represents the number of considered parameters in the study. An example of the calculated *Wi* parameters is illustrated in Table 11.21 [90].

The 'quality rating scale (q_i)' which is based on the quality parameters (Table 11.21) can be calculated by dividing the concentration of each quality parameter (C_i) by the corresponding standard values (C_i) issued by GSA/WHO and then multiplied by 100, as shown in Eq. 11.18 below [85]:

$$q_i = \frac{C_i}{S_i} \times 100 \quad (11.18)$$

The final step in this methodology include the estimation of the *WQI*, as illustrated in the following formula [85, 90]:

$$WQI = \sum_{i=1}^{n} q_i \times W_i \tag{11.19}$$

After obtaining the *WQI* values, these values then can be compared with the 'water quality rating', as reported in Table 11.22 [91]:

11.6.4 Heavy Metal Pollution Index

The 'heavy metal pollution index (HMPI)' is employed to assess the degree of contamination of different water sources. The HMPI provides an indication for the overall water quality in regard to heavy metals presence in the water source. The HMPI approach is established by designating a weightage (W_i) for each parameter (heavy metal element). Each designated weightage reflects the relative importance of each quality parameter under consideration, which is inversely proportional to the S_i value of each parameter (obtained from the GSA/WHO standards). To calculate the HMPI, the following Equation is employed [92]:

$$HMPI = \frac{\sum_{i=1}^{n} W_i \times Q_i}{\sum_{i=1}^{n} W_i} \tag{11.20}$$

where Q_i represent sub-index of the *i*th parameter. The value of Q_i can be computed using Eq. 11.21, as shown below:

$$Q_i = \sum_{i=1}^{n} \frac{\{H_i - I_i\}}{S_i - I_i} \times 100 \tag{11.21}$$

where H_i presents the determined concentration of heavy metal in the *i*th parameter; I_i represents the ideal value for drinking water quality; and S_i is the standard value of the drinking water quality. The values of I_i and S_i are obtained from the GSA and WHO standards [88, 89]. An example of the HMPI calculations for some heavy metal elements is demonstrated in Table 11.23 [90]:

11.6.5 Heavy Metal Evaluation Index

Similar to HMPI method, the 'heavy metal evaluation index (HMEI)' method can also be used to provide the overall water quality in relation to heavy metals present in water sources. The HMEI can be estimated using Eq. 11.22 [93]:

$$HMEI = \sum_{i=1}^{n} \frac{M_c}{M_{mac}} \tag{11.22}$$

where M_c describes the measured value of the ith parameter and M_{mac} presents the highest permissible or desirable value described by GSA or WHO for drinking water of the ith parameter.

11.7 Conclusions

This chapter has discussed the impacts of landfilling on different environmental compartments, including soil and plants, surface and ground water resources, and air quality. Further, the potential impacts of the landfilling activities towards human health have been discussed. According to the conducted literature, although employing landfilling as a management and control technology for solid waste can be very effective, however, to achieve a sustainable landfilling, many aspects shall be taken into consideration. This is because the practice of landfilling can potentially affect the environment, especially considerable effects on the water sources and air quality. The generated leachate from landfills can have significant impact on the quality of water sources near the landfills and can contaminate the groundwater wells and nearby surface water. Nevertheless, generated landfill gas has been reported to cause air pollution problems and can potentially affect the life of humans who are residing near the landfill. Therefore, in order to mitigate the effect of landfilling, this chapter also discussed the methods used to assess the risk of toxicity of landfill leachate and the methods used to evaluate the environmental impact of landfilling acitivites on the environment.

Glossary

Biological accumulation factor is used to expresses the ratio of metal concentration in plants to its concentration in soil.

Correlation analysis represents the statistical relationship and analysis between the studied water quality factors.

Landfill refers to the landfill site that is used for the disposal and dumping of waste materials. It is also known as dump, garbage dump, and rubbish dumping ground.

Landfilling is the processes of dumping/discarding, compression, and embankment fill of waste materials at proper disposal sites.

Leachate is one of the products generated by the processes of landfilling, and it can cause pollution to the surrounding environment.

Soil contamination factor refers to the total concentration of the metals for the assessment of soil contamination.

References

1. Baziene K, Tetsman I, Albrektiene R. Level of pollution on surrounding environment from landfill aftercare. International Journal of Environmental Research and Public Health [Internet]. 2020 Mar 18 [cited 2021 Jan 15];17(6):2007. Available from: https://www.mdpi.com/1660-4601/17/6/2007
2. Gworek, B., Dmuchowski, W., Koda, E., Marecka, M., Baczewska, A. H., Brągoszewska, P., et al. (2016). Impact of the municipal solid waste Łubna landfill on environmental pollution by heavy metals. *Water [Internet], 8*(10) Available from: https://www.mdpi.com/2073-4441/8/10/470
3. EPA. (2005). *Guidance for evaluating landfill gas emissions from closed or abandoned facilities*. US Environment Protection Agency.
4. Carriero, G., Neri, L., Famulari, D., Di Lonardo, S., Piscitelli, D., Manco, A., et al. (2018). Composition and emission of VOC from biogas produced by illegally managed waste landfills in Giugliano (Campania, Italy) and potential impact on the local population. *Science of the Total Environment [Internet], 640–641*, 377–386. Available from: http://www.sciencedirect.com/science/article/pii/S0048969718319703
5. Calvert, J., Mellouki, A., & Orlando, J. (2011). *Mechanisms of atmospheric oxidation of the oxygenates*. Oxford University Press (OUP).
6. Ercolano, S., Lucio Gaeta, G. L., Ghinoi, S., & Silvestri, F. (2018). Kuznets curve in municipal solid waste production: An empirical analysis based on municipal-level panel data from the Lombardy region (Italy). *Ecological Indicators [Internet], 93*, 397–403. Available from: http://www.sciencedirect.com/science/article/pii/S1470160X18303637
7. Chan, J. K. H. (2016). The ethics of working with wicked urban waste problems: The case of Singapore's Semakau Landfill. *Landscape and Urban Planning [Internet], 154*, 123–131. Available from: http://www.sciencedirect.com/science/article/pii/S0169204616300275
8. Ololade OO, Mavimbela S, Oke SA, Makhadi R. Impact of leachate from Northern landfill site in bloemfontein on water and soil quality: Implications for water and food security. sustainability [Internet]. 2019. Aug 6 [cited 2021 Jan 15];11(15):4238. Available from: https://www.mdpi.com/2071-1050/11/15/4238
9. Vaverková, M. D. (2019). Landfill impacts on the environment – Review. *Geosciences [Internet]*. Oct 3 [cited 2021 Jan 15];9(10):431. Available from: https://www.mdpi.com/2076-3263/9/10/431
10. Aderemi, A. O., Oriaku, A. V., Adewumi, G. A., & Otitoloju, A. A. (2011). Assessment of groundwater contamination by leachate near a municipal solid waste landfill. *African Journal of Environmental Science and Technology, 5*(11), 933–940.
11. Magaji, J. Y. (2012). Effects of waste dump on the quality of plants cultivated around Mpape dumpsite FCT Abuja, Nigeria. *Ethiopian Journal of Environmental Studies and Management, 5*(4), 567–573.
12. Ghosh, P., Das, M. T., & Thakur, I. S. (2014). Others. In vitro toxicity evaluation of organic extract of landfill soil and its detoxification by indigenous pyrene-degrading bacillus sp. ISTPY1. *International Biodeterioration & Biodegradation, 90*, 145–151.
13. Shaikh, P. R., Bhosle, A. B., & Yannawar, V. B. (2012). The impact of landfill on soil and groundwater quality of the Nanded City, Maharashtra. *Research [Internet], 4*(7), 56–63.
14. Danthurebandara, M., Van Passel, S., Nelen, D., Tielemans, Y., & Van Acker, K. (2012). Environmental and socio-economic impacts of landfills. *Linnaeus Eco-Tech, 2012*, 40–52.
15. Vaverková, M. D., & Adamcová, D. (2015). Evaluation of landfill leachate pollution: findings from a monitoring study at municipal waste landfill. *Journal of Ecological Engineering, 16*(2).
16. Adamcová, D., Vaverková, M. D., Bartoň, S., Havlíček, Z., & Břoušková, E. (2016). Soil contamination in landfills: A case study of a landfill in Czech Republic. *Solid Earth, 7*(1), 239–247.

17. Kanmani, S., & Gandhimathi, R. (2013). Assessment of heavy metal contamination in soil due to leachate migration from an open dumping site. *Applied Water Science [Internet], 3*(1), 193–205. Available from: https://doi.org/10.1007/s13201-012-0072-z.
18. Manceau, A., Marcus, M. A., Tamura, N., Proux, O., Geoffroy, N., & Lanson, B. (2004). Natural speciation of Zn at the micrometer scale in a clayey soil using X-ray fluorescence, absorption, and diffraction. *Geochimica et Cosmochimica Acta [Internet]., 68*(11), 2467–2483. Available from: http://www.sciencedirect.com/science/article/pii/S0016703703008834
19. Sahai N, Carroll SA, Roberts S, O'Day PA. X-ray absorption spectroscopy of strontium(II) coordination: II. Sorption and precipitation at kaolinite, amorphous silica, and goethite surfaces. Journal of Colloid and Interface Science [Internet]. 2000;222(2):198–212. Available from: http://www.sciencedirect.com/science/article/pii/S0021979799965621
20. Manceau, A., Tommaseo, C., Rihs, S., Geoffroy, N., Chateigner, D., Schlegel, M., et al. (2005). Natural speciation of Mn, Ni, and Zn at the micrometer scale in a clayey paddy soil using X-ray fluorescence, absorption, and diffraction. *Geochimica et Cosmochimica Acta [Internet], 69*(16), 4007–4034. Available from: http://www.sciencedirect.com/science/article/pii/S0016703705002528
21. Alagić, S. Č., Šerbula, S. S., Tošić, S. B., Pavlović, A. N., & Petrović, J. V. (2013). Bioaccumulation of arsenic and cadmium in birch and lime from the Bor Region. *Archives of Environmental Contamination and Toxicology [Internet], 65*(4), 671–682. Available from: https://doi.org/10.1007/s00244-013-9948-7
22. Wellburn AR. Air pollution and climate change. Biol Imapce [Internet]. 1994.; Available from: https://ci.nii.ac.jp/naid/10005338874/en/
23. Hester, R. E. (1986). *Understanding our environment [Internet]* (pp. 172–225). Royal Society of Chemistry. Available from: https://www.osti.gov/biblio/5419556
24. Lisk, D. J. (1991). Environmental effects of landfills. *Science of the Total Environment [Internet], 100*, 415–468. Available from: http://www.sciencedirect.com/science/article/pii/004896979190387T
25. Negi, P., Mor, S., & Ravindra, K. (2018). Impact of landfill leachate on the groundwater quality in three cities of North India and health risk assessment. *Environment, Development and Sustainability [Internet], 22*(2), 1455–1474. Available from: https://doi.org/10.1007/s10668-018-0257-1
26. Nagarajan, R., Thirumalaisamy, S., & Lakshumanan, E. (2012). Impact of leachate on groundwater pollution due to non-engineered municipal solid waste landfill sites of erode city, Tamil Nadu, India. *Journal of Environmental Health Science and Engineering [Internet], 9*(1), 35. Available from: https://doi.org/10.1186/1735-2746-9-35
27. Koliyabandara, S. M. P. A., Cooraya, A. T., Liyanagea, S., & Siriwardana, C. (2020). Assessment of the impact of an open dumpsite on the surface water quality deterioration in Karadiyana, Sri Lanka, environmental nanotechnology. *Monitoring & Management, 14*, 100371.
28. Koliyabandara, P. A., Cooray, P. L. A. T., Liyanage, S., & Siriwardana, C. (2017). Spatiotemporal dynamics of biogeochemical species around Karadiyana solid waste landfill, Sri Lanka. In *AGU fall meeting abstracts* (p. B43B-2126).
29. Mandakini LLU, Mannapperuma N, Bandara N, Silva K, Perera MTC. Leachate characterization and assessment of surface water quality near Karadiyana solid waste dump site. 2016.
30. Singh, K. P., Malik, A., Mohan, D., & Sinha, S. (2004). Multivariate statistical techniques for the evaluation of spatial and temporal variations in water quality of Gomti River (India)—A case study. *Water Research [Internet], 38*(18), 3980–3992. Available from: https://www.sciencedirect.com/science/article/pii/S0043135404003367
31. Yidana, S. M., & Yidana, A. (2010). Assessing water quality using water quality index and multivariate analysis. *Environmental Earth Sciences [Internet], 59*(7), 1461–1473. Available from: https://doi.org/10.1007/s12665-009-0132-3
32. Bengraine, K., & Marhaba, T. F. (2003). Using principal component analysis to monitor spatial and temporal changes in water quality. *Journal of Hazardous Materials [Internet],*

100(1), 179–195. Available from: https://www.sciencedirect.com/science/article/pii/S0304389403001043
33. Ouyang, Y. (2005). Evaluation of river water quality monitoring stations by principal component analysis. *Water Research [Internet], 39*(12), 2621–2635. Available from: https://www.sciencedirect.com/science/article/pii/S0043135405001788
34. Nair, A. T., Senthilnathan, J., & Nagendra, S. M. S. (2019). Emerging perspectives on VOC emissions from landfill sites: Impact on tropospheric chemistry and local air quality. *Process Safety and Environmental Protection [Internet], 121*, 143–154. Available from: http://www.sciencedirect.com/science/article/pii/S0957582018307572
35. Cadena, E., Colón, J., Artola, A., Sánchez, A., & Font, X. (2009). Environmental impact of two aerobic composting technologies using life cycle assessment. *The International Journal of Life Cycle Assessment, 14*(5), 401–410. Available from: https://doi.org/10.1007/s11367-009-0107-3
36. Jafari, N. H., Stark, T. D., & Thalhamer, T. (2017). Progression of elevated temperatures in municipal solid waste landfills. *Journal of Geotechnical & Geoenvironmental Engineering, 143*(8), 5017004.
37. AlAhmad, M., Dimashki, M., Nassour, A., & Nelles, M. (2012). Characterization, concentrations and emission rates of volatile organic compounds from two major landfill sites in Kuwait. *American Journal of Environmental Sciences, 8*(1), 56.
38. Ying, D., Chuanyu, C., Bin, H., Yueen, X., Xuejuan, Z., Yingxu, C., et al. (2012). Characterization and control of odorous gases at a landfill site: A case study in Hangzhou, China. *Waste Management [Internet], 32*(2), 317–326. Available from: http://www.sciencedirect.com/science/article/pii/S0956053X11003230
39. Kim, K.-H., Shon, Z.-H., Kim, M.-Y., Sunwoo, Y., Jeon, E.-C., & Hong, J.-H. (2008). Major aromatic VOC in the ambient air in the proximity of an urban landfill facility. *Journal of Hazardous Materials [Internet], 150*(3), 754–764. Available from: http://www.sciencedirect.com/science/article/pii/S0304389407007479
40. Agrawal, S. B., & Agrawal, M. (1999). *Environmental pollution and plant responses*. CRC Press.
41. Ziemann, P. J., & Atkinson, R. (2012). Kinetics, products, and mechanisms of secondary organic aerosol formation. *Chemical Society Reviews, 41*(19), 6582–6605.
42. Majdinasab, A., & Yuan, Q. (2017). Performance of the biotic systems for reducing methane emissions from landfill sites: A review. *Ecological Engineering [Internet], 104*, 116–130. Available from: http://www.sciencedirect.com/science/article/pii/S0925857417301787
43. Nair, A., Sartaj, M., Kennedy, K., & Coelho, N. M. G. (2014). Enhancing biogas production from anaerobic biodegradation of the organic fraction of municipal solid waste through leachate blending and recirculation. *Waste Management & Research [Internet], 32*(10), 939–946. Available from: https://www.scopus.com/inward/record.uri?eid=2-s2.0-84909998571&doi=10.1177%2F0734242X14546036&partnerID=40&md5=5641c489114ed1c932d7de6d67b9a5c1
44. Wei, Y., Li, J., Shi, D., Liu, G., Zhao, Y., & Shimaoka, T. (2017). Environmental challenges impeding the composting of biodegradable municipal solid waste: A critical review. *Resources, Conservation & Recycling [Internet], 122*, 51–65. Available from: http://www.sciencedirect.com/science/article/pii/S0921344917300332
45. Pearson, C., Littlewood, E., Douglas, P., Robertson, S., Gant, T. W., & Hansell, A. L. (2015). Exposures and health outcomes in relation to bioaerosol emissions from composting facilities: A systematic review of occupational and community studies. *Journal of Toxicology and Environmental Health, Part B: Critical Reviews [Internet], 18*(1), 43–69. Available from: https://doi.org/10.1080/10937404.2015.1009961
46. Mustafa, M. F., Liu, Y., Duan, Z., Guo, H., Xu, S., Wang, H., et al. (2017). Volatile compounds emission and health risk assessment during composting of organic fraction of municipal solid waste. *Journal of Hazardous Materials [Internet], 327*, 35–43. Available from: http://www.sciencedirect.com/science/article/pii/S0304389416310767

47. Agapiou, A., Vamvakari, J. P., Andrianopoulos, A., & Pappa, A. (2016). Volatile emissions during storing of green food waste under different aeration conditions. *Environmental Science and Pollution Research [Internet], 23*(9), 8890–8901. Available from: https://doi.org/10.1007/s11356-016-6131-5
48. Hurst, C., Longhurst, P., Pollard, S., Smith, R., Jefferson, B., & Gronow, J. (2005). Assessment of municipal waste compost as a daily cover material for odour control at landfill sites. *Environmental Pollution [Internet], 135*(1), 171–177. Available from: http://www.sciencedirect.com/science/article/pii/S0269749104003768
49. Zhang, H., Schuchardt, F., Li, G., Yang, J., & Yang, Q. (2013). Emission of volatile sulfur compounds during composting of municipal solid waste (MSW). *Waste Management [Internet]., 33*(4), 957–963. Available from: http://www.sciencedirect.com/science/article/pii/S0956053X12005314
50. Scaglia, B., Orzi, V., Artola, A., Font, X., Davoli, E., Sanchez, A., et al. (2011). Odours and volatile organic compounds emitted from municipal solid waste at different stage of decomposition and relationship with biological stability. *Bioresources Technology [Internet], 102*(7), 4638–4645. Available from: http://www.sciencedirect.com/science/article/pii/S0960852411000629
51. Moreno, A. I., Arnáiz, N., Font, R., & Carratalá, A. (2014). Chemical characterization of emissions from a municipal solid waste treatment plant. *Waste Management [Internet]., 34*(11), 2393–2399. Available from: http://www.sciencedirect.com/science/article/pii/S0956053X14003109
52. Saral, A., Demir, S., & Yıldız, Ş. (2009). Assessment of odorous VOCs released from a main MSW landfill site in Istanbul-Turkey via a modelling approach. *Journal of Hazardous Materials [Internet], 168*(1), 338–345. Available from: http://www.sciencedirect.com/science/article/pii/S0304389409002350
53. Su, F.-C., Jia, C., & Batterman, S. (2012). Extreme value analyses of VOC exposures and risks: A comparison of RIOPA and NHANES datasets. *Atmospheric Environment [Internet], 62*, 97–106. Available from: http://www.sciencedirect.com/science/article/pii/S1352231012006012
54. Fielder, H. M. P., Dolk, H., Poon-King, C. M., Palmer, S. R., Moss, N., & Coleman, G. (2000). Assessment of impact on health of residents living near the Nant-y-Gwyddon landfill site: Retrospective analysis commentary: Impact on health needs assessing from different angles. *The BMJ [Internet], 320*(7226), 19–23. Available from: https://www.bmj.com/content/320/7226/19
55. Aatamila M, Verkasalo PK, Korhonen MJ, Suominen AL, Hirvonen M-R, Viluksela MK, et al. Odour annoyance and physical symptoms among residents living near waste treatment centres. Environmental Research [Internet] 2011;111(1):164–170. Available from: http://www.sciencedirect.com/science/article/pii/S001393511000191X
56. Dzaman, K., Wojdas, A., Rapiejko, P., & Jurkiewicz, D. (2009). Taste and smell perception among sewage treatment and landfill workers. *International Journal of Occupational Medicine and Environmental Health, 22*(3), 227.
57. Majumdar, D., & Srivastava, A. (2012). Volatile organic compound emissions from municipal solid waste disposal sites: A case study of Mumbai, India. *Journal of the Air & Waste Management Association [Internet], 62*(4), 398–407. Available from: https://doi.org/10.1080/10473289.2012.655405
58. Moolla, R., Valsamakis, S. K., Curtis, C. J., & Piketh, S. J. (2013). Occupational health risk assessment of benzene and toluene at a landfill site in Johannesburg, South Africa. *WIT Transactions on the Built Environment, 134*, 701–712.
59. Guo, H., Duan, Z., Zhao, Y., Liu, Y., Mustafa, M. F., Lu, W., et al. (2017). Characteristics of volatile compound emission and odour pollution from municipal solid waste treating/disposal facilities of a city in Eastern China. *Environmental Science and Pollution Research [Internet], 24*(22), 18383–18391. Available from: https://doi.org/10.1007/s11356-017-9376-8

60. Li, L., Xie, S., Zeng, L., Wu, R., & Li, J. (2015). Characteristics of volatile organic compounds and their role in ground-level ozone formation in the Beijing-Tianjin-Hebei region, China. *Atmospheric Environment [Internet], 113*, 247–254. Available from: http://www.sciencedirect.com/science/article/pii/S1352231015300947
61. Kourtchev, I., Giorio, C., Manninen, A., Wilson, E., Mahon, B., Aalto, J., et al. (2016). Enhanced volatile organic compounds emissions and organic aerosol mass increase the oligomer content of atmospheric aerosols. *Scientific Reports [Internet], 6*(1), 35038. Available from: https://doi.org/10.1038/srep35038
62. Farré, M., & Barceló, D. (2003). Toxicity testing of wastewater and sewage sludge by biosensors, bioassays and chemical analysis. *TrAC Trends in Analytical Chemistry [Internet], 22*(5), 299–310. Available from: http://www.sciencedirect.com/science/article/pii/S0165993603005041
63. Wilke, B.-M., Riepert, F., Koch, C., & Kühne, T. (2008). Ecotoxicological characterization of hazardous wastes. *Ecotoxicology and Environmental Safety [Internet], 70*(2), 283–293. Available from: http://www.sciencedirect.com/science/article/pii/S0147651307002382
64. Coombe, V. T., Moore, K. W., & Hutchings, M. J. (1999). TIE and TRE: An abbreviated guide to dealing with toxicity. *Water Science and Technology [Internet], 39*(10), 91–97. Available from: http://www.sciencedirect.com/science/article/pii/S0273122399002589
65. Ganey, P. E., & Boyd, S. A. (2005). An approach to evaluation of the effect of bioremediation on biological activity of environmental contaminants: Dechlorination of polychlorinated biphenyls. *Environmental Health Perspectives, 113*(2), 180–185.
66. Bortolotto, T., Bertoldo, J. B., da Silveira, F. Z., Defaveri, T. M., Silvano, J., & Pich, C. T. (2009). Evaluation of the toxic and genotoxic potential of landfill leachates using bioassays. *Environmental Toxicology and Pharmacology [Internet], 28*(2), 288–293. Available from: http://www.sciencedirect.com/science/article/pii/S1382668909001033
67. Ghosh, P., Thakur, I. S., & Kaushik, A. (2017). Bioassays for toxicological risk assessment of landfill leachate: A review. *Ecotoxicology and Environmental Safety [Internet], 141*, 259–270. Available from: http://www.sciencedirect.com/science/article/pii/S0147651317301562
68. Claxton, L. D., Houk, V. S., & Hughes, T. J. (1998). Genotoxicity of industrial wastes and effluents. *Mutation Research [Internet], 410*(3), 237–243. Available from: http://www.sciencedirect.com/science/article/pii/S1383574298000088
69. Marsalek, J., Rochfort, Q., Brownlee, B., Mayer, T., & Servos, M. (1999). An exploratory study of urban runoff toxicity. *Water Science and Technology [Internet], 39*(12), 33–39. Available from: http://www.sciencedirect.com/science/article/pii/S0273122399003157
70. Mankiewicz-Boczek, J., Nałęcz-Jawecki, G., Drobniewska, A., Kaza, M., Sumorok, B., Izydorczyk, K., et al. (2008). Application of a microbiotests battery for complete toxicity assessment of rivers. *Ecotoxicology and Environmental Safety [Internet], 71*(3), 830–836. Available from: http://www.sciencedirect.com/science/article/pii/S0147651308000729
71. Persoone, G., Marsalek, B., Blinova, I., Törökne, A., Zarina, D., Manusadzianas, L., et al. (2003). A practical and user-friendly toxicity classification system with microbiotests for natural waters and wastewaters. *Environmental Toxicology [Internet], 18*(6), 395–402. Available from: https://www.scopus.com/inward/record.uri?eid=2-s2.0-10744226837&doi=10.1002%2Ftox.10141&partnerID=40&md5=156ddfbbf25fd4e74c6526d21529acef
72. Marttinen, S. K., Kettunen, R. H., Sormunen, K. M., Soimasuo, R. M., & Rintala, J. A. (2002). Screening of physical–chemical methods for removal of organic material, nitrogen and toxicity from low strength landfill leachates. *Chemosphere [Internet], 46*(6), 851–858. Available from: http://www.sciencedirect.com/science/article/pii/S0045653501001503
73. Silva, A. C., Dezotti, M., & Sant'Anna, G. L. (2004). Treatment and detoxification of a sanitary landfill leachate. *Chemosphere [Internet], 55*(2), 207–214. Available from: http://www.sciencedirect.com/science/article/pii/S0045653503010014
74. Jurkoniene, S., Maksimov, G., Darginavičiene, J., Sadauskas, K., Vitkus, R., & Manusadžianas, L. (2004). Leachate toxicity assessment by responses of algae Nitellopsis obtusa membrane ATPase and cell resting potential, and with Daphtoxkit F™ magna test. *Environmental*

Toxicology [Internet], 19(4), 403–408. Available from: https://www.scopus.com/inward/record.uri?eid=2-s2.0-4444226153&doi=10.1002%2Ftox.20049&partnerID=40&md5=318b029d91d40c78c3cc8a2514853414
75. Kalka, J. (2012). Landfill leachate toxicity removal in combined treatment with municipal wastewater. Chang KF, editor. *The Scientific World Journal [Internet], 2012*, 202897. Available from: https://doi.org/10.1100/2012/202897
76. Baun, A., Ledin, A., Reitzel, L. A., Bjerg, P. L., & Christensen, T. H. (2004). Xenobiotic organic compounds in leachates from ten Danish MSW landfills—Chemical analysis and toxicity tests. *Water Research [Internet], 38*(18), 3845–3858. Available from: http://www.sciencedirect.com/science/article/pii/S0043135404003574
77. Cheung, K. C., Chu, L. M., & Wong, M. H. (1993). Toxic effect of landfill leachate on microalgae. *Water, Air, & Soil Pollution [Internet], 69*(3–4), 337–349. Available from: https://www.scopus.com/inward/record.uri?eid=2-s2.0-0027643246&doi=10.1007%2FBF00478169&partnerID=40&md5=f85dc1d43da09af82e692f8e8617e6d1
78. Singh, V., & Mittal, A. K. (2009). Toxicity analysis and public health aspects of municipal landfill leachate: A case study of Okhla Landfill, Delhi. In D. Thevenot (Ed.), *8th World Wide Workshop for Young Environmental Scientists WWW-YES 2009: Urban waters: resource or risks? [Internet]*. (WWW-YES; vol. WWW-YES-20). Available from: https://hal.archives-ouvertes.fr/hal-00593106
79. Alkassasbeh, J. Y. M., Heng, L. Y., & Surif, S. (2009). Toxicity testing and the effect of landfill leachate in Malaysia on behavior of common carp (Cyprinus carpio L., 1758; Pisces, Cyprinidae). Am J. *Environmental Science [Internet], 5*(3), 209–217. Available from: https://www.scopus.com/inward/record.uri?eid=2-s2.0-64649095601&doi=10.3844%2Fajessp.2009.209.217&partnerID=40&md5=497ab394514527b6546ae42f273e3825
80. Deguchi, Y., Toyoizumi, T., Masuda, S., Yasuhara, A., Mohri, S., Yamada, M., et al. (2007). Evaluation of mutagenic activities of leachates in landfill sites by micronucleus test and comet assay using goldfish. *Mutation Research – Genetic Toxicology and Environmental Mutagenesis [Internet], 627*(2), 178–185. Available from: http://www.sciencedirect.com/science/article/pii/S1383571806004013
81. USEPA. (2014). *Framework for human health risk assessment to inform decision making*. US Environmental Protection Agency.
82. Xu, Y., Xue, X., Dong, L., Nai, C., Liu, Y., & Huang, Q. (2018). Long-term dynamics of leachate production, leakage from hazardous waste landfill sites and the impact on groundwater quality and human health. *Waste Management [Internet], 82*, 156–166. Available from: http://www.sciencedirect.com/science/article/pii/S0956053X18306196
83. Sun, C., Zhao, W., Zhang, Q., Yu, X., Zheng, X., Zhao, J., et al. (2016). Spatial distribution, sources apportionment and health risk of metals in topsoil in Beijing, China. *International Journal of Environmental Research and Public Health [Internet], 13*(7) Available from: https://www.mdpi.com/1660-4601/13/7/727
84. Kawo, N. S., & Karuppannan, S. (2018). Groundwater quality assessment using water quality index and GIS technique in Modjo River Basin, central Ethiopia. *Journal of African Earth Sciences [Internet], 147*, 300–311. Available from: http://www.sciencedirect.com/science/article/pii/S1464343X1830195X
85. Rana, R., Ganguly, R., & Gupta, A. K. (2017). Indexing method for assessment of pollution potential of leachate from non-engineered landfill sites and its effect on ground water quality. *Environmental Monitoring and Assessment [Internet], 190*(1), 46. Available from: https://doi.org/10.1007/s10661-017-6417-1
86. Srinivasamoorthy, K., Chidambaram, S., Prasanna, M. V., Vasanthavihar, M., Peter, J., & Anandhan, P. (2008). Identification of major sources controlling groundwater chemistry from a hard rock terrain — A case study from Mettur taluk, Salem district, Tamil Nadu, India. *Journal of Earth System Science [Internet], 117*(1), 49. Available from: https://doi.org/10.1007/s12040-008-0012-3

87. Singh, S., Raju, N. J., Gossel, W., & Wycisk, P. (2016). Assessment of pollution potential of leachate from the municipal solid waste disposal site and its impact on groundwater quality, Varanasi environs, India. *Arabian Journal of Geosciences [Internet], 9*(2), 131. Available from: https://doi.org/10.1007/s12517-015-2131-x
88. GSA. (2013). *Specification for drinking water*. Ghana.
89. WHO. (2011). *Guidelines for drinking-water quality*. Vol. 38. Geneva, Switzerland.
90. Amano, K. O. A., Danso-Boateng, E., Adom, E., Kwame Nkansah, D., Amoamah, E. S., & Appiah-Danquah, E. (2020). Effect of waste landfill site on surface and ground water drinking quality. *Water and Environment Journal [Internet], 1*, –15. Available from: https://onlinelibrary.wiley.com/doi/abs/10.1111/wej.12664
91. Sahu, P., & Sikdar, P. K. (2008). Hydrochemical framework of the aquifer in and around East Kolkata Wetlands, West Bengal, India. *Environmental Geology [Internet], 55*(4), 823–835. Available from: https://doi.org/10.1007/s00254-007-1034-x
92. Mohan, S. V., Nithila, P., & Reddy, S. J. (1996). Estimation of heavy metals in drinking water and development of heavy metal pollution index. *Journal of Environmental Science and Health, Part A: Toxic/Hazardous Substances and Environmental Engineering [Internet], 31*(2), 283–289. Available from: https://doi.org/10.1080/10934529609376357
93. Edet, A. E., & Offiong, O. E. (2002). Evaluation of water quality pollution indices for heavy metal contamination monitoring. A study case from Akpabuyo-Odukpani area, Lower Cross River Basin (southeastern Nigeria). *GeoJournal [Internet], 57*(4), 295–304. Available from: https://doi.org/10.1023/B:GEJO.0000007250.92458.de

Index

L. K. Wang et al. (eds.), *Solid Waste Engineering and Management*, Volume 1,
Handbook of Environmental Engineering 23,
https://doi.org/10.1007/978-3-030-84180-5

M

X

The manufacturer's authorised representative in the EU is Springer Nature Customer Service Centre GmbH, Europaplatz 3, 69115 Heidelberg, Germany. If you have any concerns regarding our products, please contact ProductSafety@springernature.com

Printed and bound by CPI Group (UK) Ltd, Croydon, CR0 4YY
30/09/2026
02225445-0017